GW00686432

THE HANDBOOK OF INNOVATION AND SERVICES

The Handbook of Innovation and Services

A Multi-disciplinary Perspective

Edited by

Faïz Gallouj

Professor of Economics, University of Lille 1, France

Faridah Djellal

Professor of Economics, University of Tours, France

Edward Elgar

Cheltenham, UK • Northampton, MA, USA

Published by
Edward Elgar Publishing Limited
The Lypiatts
15 Lansdown Road
Cheltenham
Glos GL50 2JA
UK

Edward Elgar Publishing, Inc.
William Pratt House
9 Dewey Court
Northampton
Massachusetts 01060
USA

A catalogue record for this book
is available from the British Library

Library of Congress Control Number: 2009937896

ISBN 978 1 84720 504 9 (cased)

Printed and bound by MPG Books Group, UK

Contents

Figures

Tables

Contributors

Cristiano Antonelli, Professor of Economics, Laboratorio di Economia dell'Innovazione 'Franco Momigliano', Department of Economics, University of Turin, Italy, and BRICK (Bureau of Research on Innovation, Complexity and Knowledge), Collegio Carlo Alberto, Moncalieri, Italy.

André Barcet, Associate Professor of Economics, University of Lyon, GATE-CNRS, France.

William J. Baumol, Harold Price Professor of Entrepreneurship and Academic Director, Berkley Center for Entrepreneurship and Innovation, New York University, USA: Senior Economist and Professor Emeritus, Princeton University, USA.

Knut Blind, Professor, Berlin University of Technology, Chair of Innovation Economics and Fraunhofer Institute for Systems and Innovation Research, Karlsruhe, Germany, Competence Center 'Regulation and Innovation', Rotterdam School of Management, Erasmus University, the Netherlands, Chair of Standardisation.

Jan Bröchner, Professor, Department of Technology Management and Economics, Division of Service Management, Chalmers University of Technology, Göteborg, Sweden.

John R. Bryson, Professor of Enterprise and Economic Geography, School of Geography, Earth and Environmental Sciences, University of Birmingham, UK.

José A. Camacho, Professor, Department of Applied Economics, University of Granada, Spain.

Peter Daniels, Professor, Deputy Pro-Vice-Chancellor, Co-Director, Service and Enterprise Research Unit (GEES), University of Birmingham, UK.

Faridah Djellal, Professor of Economics, Faculty of Law and Economics, University of Tours, Clersé-CNRS (Centre Lillois d'Etudes et de Recherches Sociologiques et Economiques), France.

Bo Edvardsson, Professor of Business Administration and Director of the Service Research Center, Karlstad University, Sweden.

Rinaldo Evangelista, Associate Professor, Faculty of Law and Political Science, University of Camerino, Italy.

Eileen Fairhurst, Professor of Health and Ageing Policy, Research Institute for Health and Social Change (RIHSC), Manchester Metropolitan University, UK; and Chairman, NHS Salford, UK.

Olivier Furrer, Associate Professor of Strategy, Nijmegen School of Management, Radboud University of Nijmegen, Nijmegen, The Netherlands.

Jean Gadrey, Emeritus Professor of Economics, University of Lille 1, Clersé-CNRS (Centre Lillois d'Etudes et de Recherches Sociologiques et Economiques), France.

Faïz Gallouj, Professor of Economics, Faculty of Economics and Sociology, University of Lille 1, Clersé-CNRS (Centre Lillois d'Etudes et de Recherches Sociologiques et Economiques), France.

Manuel García-Goñi, Assistant Professor of Economics, Department of Applied Economics II, Universidad Complutense de Madrid, Spain.

Lawrence Green, Senior Research Fellow, Manchester Metropolitan University Business School, UK.

Anders Gustafsson, Professor of Business Economics, Service Research Center, Karlstad University, Sweden.

Denis Harrisson, Professor, Organisation and HRM, School of Management, University of Quebec at Montreal (UQAM), CRISES (Centre de recherche sur les innovations sociales), Canada.

Pim den Hertog, Research Coordinator, Amsterdam Centre for Service Innovation (AMSI), Amsterdam Business School, The Netherlands and founding partner Dialogic Innovation and Interaction, Utrecht, The Netherlands.

Christiane Hipp, Professor, Department of Economics and Business Sciences, Brandenburg University of Technology Cottbus, Chair of Organization, Human Resource Management and General Management.

Jeremy Howells, Professor, Executive Director, Manchester Institute of Innovation Research, Manchester Business School, University of Manchester, UK.

Frank M. Hull, Visiting Professor, SPRU (Science and Technology Policy Research), University of Sussex, Brighton, UK.

Juan-Luis Klein, Professor, Department of Geography, University of Quebec at Montreal (UQAM), CRISES (Centre de recherche sur les innovations sociales), Canada.

Per Kristensson, Assistant Professor, Service Research Center, Karlstad University, Sweden.

Paul Leduc Browne, Professor, Department of Social Work and Social Sciences, University of Quebec at Outaouais (UQO) CRISES (Centre de recherche sur les innovation sociales), Canada.

Sylvain Lenfle, University of Cergy-Pontoise and Management Research Centre, Ecole Polytechnique, Paris, France.

Pierre-Yves Léo, Research engineer, CERGAM-GREFI, Paul Cézanne University, Aix-Marseille III, France.

Silvia Massini, Senior Lecturer in Economics and Technology Management, Manchester Institute of Innovation Research (MIoIR), Manchester Business School, University of Manchester, UK.

Christophe Midler, Management Research Centre, Ecole Polytechnique, Paris, France.

Ian Miles, Professor, Technological Innovation and Social Change, Manchester Institute of Innovation Research, Manchester Business School, University of Manchester, UK.

Marcela Miozzo, Professor of Economics and Management of Innovation, Manchester Business School, University of Manchester, UK.

Marie-Christine Monnoyer-Longé, Professor of Management Science, IAE University of Toulouse I, Toulouse, France, CRG and the European Association for Services Research (RESER).

Pier Paolo Patrucco, Assistant Professor of Economics, Laboratorio di Economia dell'Innovazione 'Franco Momigliano', Department of Economics, University of Turin, Italy, and BRICK (Bureau of Research on Innovation, Complexity and Knowledge), Collegio Carlo Alberto, Moncalieri, Italy.

Pascal Petit, Research Director, CNRS, CEPN (Centre d'Economie de Paris Nord), University of Paris XIII and CEPREMAP, France.

Jean Philippe, Professor of Management Science, CERGAM-GREFI, Paul Cézanne University, Aix-Marseille III, France.

Mercedes Rodriguez, Assistant Professor, Department of Applied Economics, University of Granada, Spain.

Federica Rossi, Researcher, Laboratorio di Economia dell' Innovazione 'Franco Momigliano', Department of Economics, University of Turin, Italy; Centre for Innovation Management Research, Birkbeck College, University of London, UK.

Luis Rubalcaba, Professor of Economic Policy, Department of Applied Economics, and Director of the Research Unit on Services, Innovation and Competitiveness, University of Alcalá, Madrid, Spain.

Maria Savona, Associate Professor of Economics, University of Lille 1, Clersé-CNRS (Centre Lillois d'Etudes et de Recherches Sociologiques et Economiques) and SPRU (Science and Technology Policy Research), University of Sussex, Brighton, UK.

Jon Sundbo, Professor of Business Administration and Innovation, Department of Communication, Business and Information Technologies, Roskilde University, Denmark.

Joe Tidd, Professor of Technology and Innovation Management, SPRU (Science and Technology Policy Research), University of Sussex, Brighton, UK.

Marja Toivonen, Research Director, Docent, Helsinki University of Technology Innovation Management Institute (IMI), Finland.

Alexandre Trigo, Researcher of Applied Economics, University of Santiago de Compostela, Spain.

Xavier Vence, Professor of Applied Economics, University of Santiago de Compostela, Spain.

Paul Windrum, Associate Professor in Strategy, University of Nottingham Business School, UK and Visiting Professor, Max Planck Institute for Economics, Jena, Germany.

Lars Witell, Associate Professor, Service Research Center, Karlstad University, Sweden.

Dariusz Wójcik, Lecturer at the School of Geography and the Environment and Fellow, St Peter's College, University of Oxford, UK.

Peter Wood, Emeritus Professor, Department of Geography, University College London, UK.

Foreword

William J. Baumol

Microeconomics is plagued by two major gaps: the absence of appropriately full treatment of either the services or of entrepreneurship. In each case, the evident importance of the subject is inversely proportionate to the space it is assigned in the literature. The entrepreneur, as a vital component of the process of innovation and growth, is often listed as a fourth 'factor of production' and, having once been noticed by name, is often not even mentioned again in the discussion that ensues. In the case of the service sector, the textbooks still prefer to focus on manufactures or even on agriculture, though these now are apt to include, respectively, only some 10 percent and 3 percent of an economically advanced nation's labor force, with the bulk of the remainder devolving upon the services. And here, too, the relation to innovation and growth is overlooked, apparently because services still are typified by what goes on in the barber shop rather than what emerges from the software sector of the economy, that is arguably the services' prime hallmark for the future.

One can understand this bias of the textbook authors. They have at their disposal relatively sophisticated models of the roles of land, labor and capital, but none of the entrepreneur. They have inherited discussions of production functions steeped in their agricultural origins, with some game-theoretic depictions of manufacturer oligopolists, but they do not feel comfortably possessed of anything comparable that is explicitly directed to the services. So the source of the omission is indeed easy to identify, but its identification is no excuse.

This book represents a significant step toward dealing with the lacuna constituted by the inadequacy of the literature on the services. And, as such, it approaches its task from a variety of directions. These differences are readily suggested by the titles in the table of contents. I will now single out a few of the chapters for some passing remarks.

The valuable initial chapter, Chapter 1, of which one of the editors of the volume (Faïz Gallouj) is a co-author, seeks to provide a structure for the field, an organizational approach into which the variety of illuminating contributions in this book and elsewhere can be fitted comfortably. As such, it can contribute substantially to the utility of the literature, and facilitate its further expansion.

I will avoid comment on my own chapter (Chapter 4), which I trust will not succeed just in eliciting tedium, though I flatter myself that it does offer some insights that I have never provided before.[1] For the opposite reason, I will postpone my brief comments on the Gadrey chapter (Chapter 5), because I want to end this Foreword on an upbeat note, in admiration of its important observations.

Predictably, and very appropriately, there are chapters in this book that focus in various ways on innovation as itself a service activity, indeed, for the purpose at hand, as the prime service activity – the one that underlies the growth process and that often leads to persistent intersectoral disparities in productivity growth rates and their now well-recognized differences in rates of real cost increases. The point that is overlooked here is the presence of a critical two-way relationship. The first goes from innovation to services, and amounts to the fact that services are subject to innovation, and are not condemned to languish unchanged forever in their ancient ways.

But the other side of the matter is the fact that the innovation process is itself primarily a service activity. There are a number of chapters here in which this relationship is cited, explored or at least hinted at: Chapter 9 by Harrisson, Klein and Leduc Browne, Chapter 3 by Howells, and Chapter 18 by Camacho and Rodriguez, who are led into this arena by their focus on 'knowledge-intensive services'. Several other chapters follow up on the issue, for example by looking into the relationship in the innovation process between universities and private business (e.g., Antonelli, Patrucco and Rossi, Chapter 7). This last relationship should be of special interest to the Europeans, who apparently are still behind the US in the solidification of a university–business partnership, with all of its promise for the innovation and growth processes.

I emphasize the role of innovation as perhaps the king of the services, among the great variety of topics that are covered in the book, because it is at the heart of the story that must ultimately become a prime focus of the literature on the services. The state of the world's economies in the future is dependent on continuation of the outburst of innovation and productivity growth that emerged after the eighteenth century. But that, in turn, must rest on education, research and utilization (marketing), all of which, evidently, are more or less pure services. This gives us a very different view of the central role of the services in our economic future. They are not merely the remaining repository of jobs, as the productivity explosion in agriculture and manufacturing closes down these traditional avenues for gainful occupation and its paths out of poverty. The dramatic shift of employment to the services is, of course, important in itself. But without understanding the role of the services as the heart of the mechanism that determines productivity growth in the economy as a whole, and its distribution among the activities of the economy, the story loses much of its rhyme and reason, and becomes instead a curious concatenation of tales of haphazard and fortuitous developments.

In short, the fate of the services and the implications of their trajectory is a subject far less narrow than it may appear to be. Indeed, the developments in this sector of the economy have other surprising consequences for our future overall. This is brought out even more sharply than what has just been said in the very illuminating chapter contributed by my friend, Jean Gadrey. His focus is the threat to the environment and even to the future of humanity that is posed by the economic growth process. I need hardly remind the reader of the urgency of the subject, but the connection of the fate of the services with these looming problems, or even the possibility that there is such a connection, is hardly common knowledge.

As Professor Gadrey reminds us, the easiest way to associate the two arenas is to go back to an ancient definition of a service as an output that is not embodied in a material substance. A newly produced hat is made up of cloth, leather or other materials, whereas the cleaning of a table by a waiter in a restaurant or the discovery of the notion of the double helix entails no physical embodiment. It would appear to follow that the restructuring of gross domestic product (GDP) to include a declining share of manufactures and a rising proportion of services offers some hope as part of a promising recipe for containment of the environmental threats to the future, for without use of physical raw materials the problems of waste, emissions and their disposal would appear to be minimized.

Professor Gadrey shows, however, that matters are not so simple. But, first, he notes an opportunity that the preceding observations would seem to reveal. For, taken at their face value, they would appear to say that, by reducing the depletion of resources, the switch from manufacturing to the services helps to control one of the most damaging externality threats that looms over the future of humanity. If so, as we all know, the proper response is taxation of the more damaging activities (the manufactures) and the grant of subsidies to those (the services) that yield beneficial externalities, a step that may do much to offset the cost disease of the productivity-stagnant services. Pigouvian taxes that raise the prices of new cars and fuels, and subsidies that reduce the cost of education and medical research, can serve as an effective counter to the cost disease of the stagnant services.

But as Professor Gadrey makes clear, all this is an oversimplification. A moment's consideration will show how misleading the story can be. Services are not automatically resource-conserving or environment-protecting. Thus, we note that airplane and bus transportation are services and yet are among the leading threats to the environment contributed by technical progress. In addition, there is no reason to believe that the share of service outputs in GDP is really rising materially. It is only their relative money value and their share of the economy's employment that is rising. But there is no reason to expect this to be accompanied by a material decline in the production of steel for building construction or in the use of petroleum products in the production of plastics. And even if there were a continuing and substantial rise in output of services, somehow measured in 'real' terms, rather than mere money values, the rise in their quantity might well offset any alleviation of the environmental threats via the lower resource content of a unit of service output, however defined or measured.

Two things follow from this relatively brief discussion. First, the role of the services and their place in the composition of the output of the economy will yet prove to be of enormous importance. We already understand enough about the matter to be reasonably sure of that. Second, we are only beginning to understand both the nature of the threat and that of the promise. Without investigating them, we may permit the escape of an invaluable opportunity to make for improved prospects for mankind.

Notes

1. I should comment, in connection with my own work, that in the materials with which I have been provided about this book I have found no reference to Oulton's important, albeit paradoxical, theorem. Oulton shows that business services, with their growing importance in the economy, are apt to speed up productivity growth in the economy, even though they themselves suffer from the handicaps to labor-saving productivity growth that beset so many services.

   To hint roughly at the reason, let us use m to symbolize the rate of labor saving in manufacturing, s the rate in services, and b the rate of such productivity growth in the use of services by business firms. If $s < m$, $b < m$, but $s + b > m$, then labor involved in the supply and use of business services will benefit cumulatively from two stages of productivity growth, under the assumed magnitudes, yielding a net overall productivity gain of $s + b$. See Oulton (2001); see also Sesaki (forthcoming).

References

Oulton, Nicholas (2001), 'Must the growth rate decline? Baumol's unbalanced growth revisited', *Oxford Economic Papers*, **53** (4), 605–627.

Introduction: filling the innovation gap in the service economy – a multidisciplinary perspective

Faïz Gallouj and Faridah Djellal

Modern economies are inescapably service economies. For several decades now, services have been our main source of wealth and jobs. The process of deindustrialisation began a long time ago in all the developed countries (1955 in the US, 1950 in the UK, 1973 in France and 1980 in Japan, for example). While it is hardly surprising that the profound economic and social upheavals linked to deindustrialisation have given rise to anxieties, both legitimate and fantastical, the persistence of these anxieties in certain cases is, on the other hand, quite difficult to understand. After all, the service society is still quite frequently associated with negative images of servitude, state bureaucracy and industrial decline. Thus, despite some changes of attitude, it is still regarded with a certain degree of suspicion, in both academic studies and political discourse. This political discourse is based on a number of particularly hardy myths about the service economy and its performance, the quality of its jobs and its capacity for innovation. The purpose of this book is to re-examine these myths and to try to allay certain fears by championing the notion that the service economy is an economy characterised not by decline but rather by high performance and innovation.

0.1 Services and performance

Classical economics, with its focus on manufacturing industry, helped to construct an image of services as deficient in terms of economic performance. Thus Adam Smith (1960 [1776]) contrasted the productive labour of manufacturing industry with the unproductive labour of service activities that fade away as soon as they have been performed. For Smith, services were intangible and therefore incapable of creating value in the same way as manufacturing industries. This analysis, which is based on a notion of services reduced to the activities of domestic servants, of servants of the state and of artists, continues to sustain much contemporary thinking. In modern times, this characterisation of services as deficient is expressed in different forms:

1. Services are said to be non-capital-intensive. They do not require high levels of investment or heavy machinery. The world of services, it is argued, is that of the office, in contrast to the industrial world and its factories and heavy plant and equipment.
2. Productivity growth in services is said to be low. Economists delight in citing the example of hairdressers, whose productivity, they argue, has increased little over the centuries. This characteristic was for a long time (and still is today) regarded as intrinsic to services. Indeed, Jean Fourastié (1949) made it the main criterion in the first non-residual definition of the tertiary sector. It also lies at the heart of Baumol's well-known models, in which it characterized the so-called stagnant sectors (Baumol, 1967; see also Chapter 4 in the present volume).
3. Services are said to be induced activities, and therefore passive or subordinate. They are not economic driving forces, since their output can only be sold locally. In contrast to manufacturing activities, they cannot, given the intangible and interactive nature of their output, export their 'product' to other areas and thereby generate income from outside the immediate locality. Their growth, it is argued, is restricted by 'local purchasing power'.

To these negative myths can be added one positive myth, namely the idea that services (with some exceptions) do not have an adverse effect on the environment, that they are by their very nature 'environment-friendly'. In reality, the future of services is being played out in the environmental sphere as well. The service relation, which is one of the specific characteristics of services, is after all liable to exacerbate environmental problems, since it involves consumers travelling to providers, or vice versa. One important challenge here is that of measuring the environmental impacts of services (see Jean Gadrey's contribution to the present volume in Chapter 5).

Nobody any longer disputes the ability of services to create value. They may even add value to goods themselves (and indeed in some cases may actually constitute the main source of value added). This applies to services that supplement goods or to product-related services, that is, services provided when a manufactured good is produced or sold, whether before, after or during a sale, or even independently of a sale (Mathieu, 2001; Davies, 2004 and, in the present volume, Chapter 29 by Olivier Furrer and Chapter 30 by Sylvain Lenfle and Christophe Midler). However, the other, negative assessments are still current. Nevertheless, the statistics (provided the effort is made to collect and construct them appropriately, which is not self-evident) and the proliferation of empirical investigations can be called on to demythologise the service economy.

With regard to both their intrinsic nature and the ways in which they are delivered, services are too heterogeneous to be captured adequately by a general analysis. However that may be, it can no longer be claimed that services are non-capital-intensive. They have for a very long time been the main users of information and communications technologies. If a broader view of capital investment is taken, then it becomes clear that many services belong to the most capital-intensive group of activities. Examples include transport in its various forms, postal services, electricity, gas and water supply services, some leisure services, etc.

The main criticism made of the service economy is its low productivity. This criticism manifests itself in the contemporary discourse through the diagnosis of a new pathology, namely Solow's paradox, which states that information technologies are to be found everywhere except in the productivity statistics. This criticism can be challenged in various ways.

Firstly, it has to be noted that service activities have for several decades been experimenting successfully with rationalisation strategies that seem to increase both productivity and performance (Gadrey, 1996; Djellal and Gallouj, 2008). These strategies are at work everywhere, whether in knowledge-intensive services (engineering, consultancy) or in more operational services. In the first case, the rationalisation takes what Gadrey calls a 'professional' form (development of methods and 'toolboxes', standardisation of activities and service offer, etc.). In the second case, the rationalisation is 'industrial' in nature (and indeed is sometimes described as 'industrialisation'). In particular, this 'industrialisation' may denote a process leading to the production of tangible goods to the detriment of the provision of intangible services and the implementation in service firms and organisations of a certain mode of production (the type of work organisation and technologies that predominated in the heavy industry of the post-Second World War period).

Secondly, performance in service activities cannot be captured solely through the notion of productivity. Attempts to measure this industrial and technical indicator come up against the difficulty of identifying the output or product of service activities. Thus the level of performance in services is undoubtedly less problematic than that of our methods for measuring that performance. It is preferable, therefore, to adopt a multi-criteria method of evaluation that takes account of the many and various facets of performance, which include technical performance, of course, but also commercial performance (measured in monetary and financial terms), civic performance (measured in terms of fairness, equality of treatment, social cohesion and respect for the environment) and relational performance (measured in terms of quality of interpersonal relations, empathy and relations of trust).

Services are playing an increasingly active role in local and regional development (see Chapters 24 and 25 in the present volume). Increasing numbers of services form part, directly or indirectly, of the 'economic base' of their region, that is, they 'export' their output, selling it outside their own territory (sometimes internationally) and bringing the corresponding revenue back to their local area (Daniels, 1983; Beyers, 2002; Illeris and Philippe, 1993; Gallouj, 1996; see also Chapters 19 to 23 in the present volume). Thus they are acting as an engine for the rest of the economy. This increasing involvement of services in the 'economic base' can be explained in particular by a tendency towards a relaxation (in various forms) of the constraint of proximity, as a result of reductions in transport costs, a considerable increase in the speed and frequency of transport services, and the widespread diffusion of information and communications technologies.

0.2 Services and employment

The relationship between services and employment began to be seen as a fundamental issue and became the subject of considerable polemic from the 1980s onwards. The number of articles denouncing the destruction of manufacturing jobs and their replacement by service jobs proliferated, particularly in the United States. While nobody can deny that services now constitute the main reservoir of jobs, many people still regard service jobs as being largely of mediocre quality: they have been variously described as 'bad jobs', 'hamburger jobs' and McJobs (Bluestone and Harrison, 1986; Cohen and Zysman, 1987; Thurow, 1989). And according to the French philosopher André Gorz (1988), the service society is a society of 'servants'.

These prejudices persist today and the question of services and employment is viewed ambiguously, sometimes with a certain feeling of guilt. Public policies have sought to encourage job growth in the service sector, but as temporary support measures in times of crisis while waiting for 'real' jobs' to be created. Nevertheless, the problem for contemporary developed economies is less their degree of deindustrialisation than their delay in developing as service economies. Such a delay can be observed when employment in Europe and the USA is compared (Gregory et al., 2007).

Here too, these negative judgements do not stand up to statistical analysis. While it is true that the service society creates low-skill jobs, it is equally true that it is now the main source of opportunities for high-level managers and professionals. Thus two employment models or systems can be identified in service activities (including the more operational services, such as cleaning): a neo-Taylorist model characterised by quantitative flexibility, and an (emerging) model based on 'organisational adaptability', which reflects an increase in the level of professionalisation in services. We should also note here the significant development of certain human resource management services whose objective is to improve the skill levels of employees in firms and organisations.

Closely linked to the question of employment is that of entrepreneurship. Here too, the statistics indicate that services are the main source of company start-ups. Most of these entrepreneurs are engaged in traditional service activities (commerce, small-scale service firms). However, there are also new and dynamic forms of entrepreneurship (which bring us back to the question of innovation). 'Cognitive' entrepreneurship, firstly, denotes the activities of experts able to invest in new fields of knowledge, such as a new area of expertise in consultancy or a PhD student seeking to exploit the results of his or her research (whether in the natural sciences or the humanities and social sciences) by setting up a company. 'Social' entrepreneurship, secondly, has as its sphere of action the social and solidarity economy (in which new associations are set up to provide new forms of care for young children, elderly people or people with various physical, psychological or social handicaps). 'Environmental' entrepreneurship, finally, focuses on protection of the environment and sustainable development.

The question of employment can also be considered in terms of its relation to innovation more generally (Djellal and Gallouj, 2007; see Chapters 16, 17, 18 in the present volume). Recent statistical analyses drawing on Community Innovation Surveys show that innovation has an overall positive effect on employment in services (Evangelista, 2000; Evangelista and Savona, 2003; Nählinder, 2005). This positive effect is particularly strong in the case of very high-skill jobs, while the 'job-destroying' effects of technical change mainly affect the least-skilled segments of the employment system. These surveys also show that the impact of innovation on employment varies with size of firm. In large firms, innovation has a negative effect on employment (mainly on low-skill jobs), while in small firms it has a positive effect.

0.3 Services and innovation

The prevailing notion has been that (authentic and spectacular) innovation is the business of manufacturing industry. In this area services are said once again to be in a subordinate position, confining themselves, like Third World countries, to adopting turnkey solutions (vehicles, computers, etc.). In other words, manufacturing industry, the 'driving force' of the economy, produces technical systems from which 'passive' service industries can benefit. From this perspective, services are dominated by manufacturing: they adopt technological innovations but produce few of their own.

This assessment, which seemed to be confirmed by statistical surveys conducted on the basis of Organisation for Economic Co-operation and Development (OECD) directives, is consistent with the prejudices already referred to above (services, allegedly, have low performance levels and service-sector jobs are low-skill). However, it is questionable, on several counts. Firstly, it contradicts the Schumpeterian notion of 'waves of creative destruction', according to which innovative organisations and industries develop at the expense of the others. Thus in today's economies, so it is argued, it is the firms and activities that innovate the least that develop the most. The conventional view is also called into question by in-depth empirical investigations that have identified intensive innovation activities in services, including in the non-profit and voluntary sector and in social enterprises (see Chapter 9 by Denis Harrisson, Juan-Luis Klein and Paul Leduc Browne). True, these innovations may take unusual forms, far removed from the traditional image of the (tangible and spectacular) technical system developed within research and development (R&D) departments.

Innovation in services cannot be reduced to technological innovation, as the following examples, among others, demonstrate: a new insurance contract, new financial instruments, a new area of legal expertise, new formats for restaurants, retail outlets or hotels, a new leisure concept, etc. This does not mean that these innovations are not based or cannot be based on a tangible technology (information or telecommunications systems, for example), but that the innovations and the associated technologies are not one and the same thing and that, in some cases, the technology can be dispensed with. In other words, it cannot justifiably be argued that innovation occurs only when the novelty is embodied in a technical system. Not to accept that this is so is seriously to underestimate the capacity for innovation in services. It is just this error that is to blame for the myopia of national and international indicators of R&D and innovation, which still persists, although it is diminishing thanks to the revisions of the OECD manuals.

It is not that services are unsuited to R&D and innovation, but rather that these highly technologist indicators are unable to capture the innovation that does take place. They are the cause of what might be called an 'innovation gap', that is, the difference between the reality of innovation in services and what can be measured by means of the traditional indicators (R&D, patents) (NESTA, 2006). This innovation gap gives rise in turn to a 'policy gap', in the sense that public policies intended to support innovation in services tend themselves to be technologically biased and to apply to services policies designed for manufacturing industry (Rubalcaba, 2006; den Hertog, forthcoming; see also Chapters 26 and 27 in the present volume).

In services, process innovation, just like product innovation, may be intangible. It may consist of methods, that is, a script defining the words, actions and movements of each individual involved (methods used by consultants as well as in restaurants, cleaning or care protocols, etc.), as in a theatre play or film scenario. Some of these methods could be based on technical systems (computerisation of recruitment methods), while others might be embodied in tools (legal expert systems), but this is not a necessary condition for innovation.

This intangibility (and this non-technological dimension), as well as the importance of the service relationship, are not unconnected with the difficulties associated with appropriating and protecting innovation in services. In our view, however, they do confer an advantage. Detached, to a certain extent, from material and technical contingencies, services are perhaps the last bastion of a certain 'romantic improvisation' when it comes to innovation, to quote Michel Callon (1994). The simplest ideas can still lead to business empires. Examples abound, from the home delivery of pizzas to tourist travel services, via domiciliary care services for the elderly.

Economic theory long championed a linear approach to innovation, in which researchers, producers and sellers were specialists operating in different worlds, each hermetically sealed off from one another. A theoretical concept of this kind is essentially incompatible with the true nature of service activities. Services are by definition interactive, and they have tended almost naturally to organise their innovation activity around an interactive model, that is, one in which actors from different departments interact (Kline and Rosenberg, 1986). This seems to be the rule even for cumbersome bureaucracies such as insurance companies. The development of a new mass-market insurance policy, for example, involves lawyers, actuaries, information technology (IT) personnel, claims specialists, sales staff and customers. To take consultants as an example, it is clear that those who produce innovations are also the ones who sell the services (that is, those who are in contact with clients). It cannot be otherwise in activities in which part of the innovation is produced at the service provider–client interface.

The (quasi-natural) interactivity does not, of course, preclude the existence in some cases of specialist innovation departments, particularly in the largest companies. However, such departments, even when they exist, are seldom the only actors in the innovation process. They are almost always supplemented by (and are in competition with) formalised but non-permanent innovation structures (project groups made up of people from different departments) and, particularly in knowledge-intensive activities, by a high level of informal activity on the part of individuals (Sundbo, 1998; Fuglsang, 2008).

The frequent absence of R&D departments makes it difficult to highlight independent R&D activity. It certainly exists, however. It can be found, of course, in R&D departments when they exist. But it is also to be found in the activities of less permanent structures (project groups, for example). It is usually one of the aspects of innovation projects that may, after all, involve analytical and conceptual work, sometimes accompanied by tests. The humanities and social sciences play a not insignificant role here.

Many service activities today have reversed their subordinate position vis-à-vis manufacturing industry in the sphere of technological innovation. In other words, they are able to produce their own technical systems, either independently or within a power relationship in which they have the upper hand. Examples include bank automated teller machines (ATMs), cleaning robots, cooking and refrigeration systems in fast-food restaurants, and automatic mail processing systems in postal services. Others include certain large retail chains that put pressure on their suppliers and impose on them such tightly drawn specifications that it is no exaggeration to speak of technological suppliers dominated by users (see Gallouj, 2007).

However, there is another phenomenon that demonstrates even more clearly how services can turn the tables when it comes to innovation. The reference here is to the active role played by 'knowledge-intensive services' (engineering, consultancy) in their clients' innovation (particularly in the manufacturing sector). Whether the innovations be organisational, strategic, product-related etc., these service providers assist their clients in various ways, to varying extents and at various points in the innovation process (Muller and Zenker, 2001; Gallouj, 2002b, 2002c; Czarnitzki and Spielkamp, 2003; Toivonen, 2004; Wood, 2005; Miozzo and Grimshaw, 2006; OECD, 2007).

Thus while the association of the terms 'service' and 'innovation' seemed for a long time to be incongruous, since services were associated much more with negative images of servitude and public services, this is no longer the case today. Services and innovation are not two major contemporary phenomena, concomitant with each other certainly, but essentially unconnected. On the contrary, service firms and organisations are the locus of considerable R&D and innovation efforts, which are commensurate with their contribution to national wealth. Thus the growth of the service sector does not pose a challenge to Schumpeter's notion of 'waves of creative destruction', according to which non-innovative firms and industries disappear in favour of firms and industries that introduce new 'productive combinations'. It may quite simply be a new illustration of Schumpeter's thesis.

However, if the full extent of innovation efforts in services is to be properly captured, those engaged in the analysis, whether they be economists, sociologists or management specialists, must themselves make efforts to adapt or even innovate, both conceptually and methodologically. The traditional analytical tools are not always able to account fully for the new service and knowledge economy. This is one of the reasons why analyses are 'locked' in what in organisational sciences is known, metaphorically, as a 'competence trap'. Of course, this need for conceptual innovation is reflected in a need to adapt the actual means of measurement. Thus the efforts of researchers converge and intertwine with those of national and international statistical institutions. Most of the OECD manuals, for example, which set out the guidelines for defining and measuring R&D and innovation, have recently been revised in the light of the new service and knowledge economy, or are in the process of being so revised.

It is important, nevertheless, to recognise that the dynamic of service industries and those of the other sectors of the economy are characterised by a similar dialectic of convergence and divergence. This dialectic is reflected, for example, in the fact that the 'service dimension' or the service relationship or even 'servuction' have a certain degree of universality that extends beyond sectoral boundaries to worm its way into the heart of manufacturing production and agriculture, thereby opening the way for new forms of innovation (intangible products and processes, customised innovation, ad hoc innovation) as well as for new modes of organisation and incentives to innovate. This dialectic is also reflected in the universality of new information and communications technologies (NCTs), which have introduced tangible anchor points into services that facilitate, if not a certain industrialisation of service provision, then at least a form of 'industrial rationalisation' that links them to the traditional forms of innovation without for all that abolishing natural specificities. Thus researchers must concentrate their efforts not only on identifying possible specificities but also on developing integrated analyses of innovation, that is, constructing general theoretical models that are independent of sectoral contexts.

0.4 The contents of the book

Thus contemporary economies have two fundamental characteristics. Firstly, they are service economies. After all, services account for more than 70 per cent of wealth and jobs in most developed countries. Secondly, they are innovation economies. Recent decades have seen an unprecedented development of innovation in all its forms (scientific and technological, organisational, social, etc.).

The aim of this *Handbook* is to link these two major characteristics in order to investigate a question that has hitherto been inadequately explored, one that poses many theoretical and operational challenges, namely innovation in services. A question of such importance cannot be monopolised by a single discipline or by a single theoretical approach.[1] That is why this book contains the views of specialists from a range of disciplines (economics, management, sociology and geography) and draws on a number of different analytical (micro, meso and macro) and methodological perspectives. Thus it brings together some 50 of the best international specialists who are currently engaged in investigating the question of innovation as it relates to services.

The *Handbook* is divided into seven parts. Its first objective is to investigate, on the one hand, the nature of innovation in services from various points of view (concepts, definitions, measurement, sectoral analyses) and, on the other, models of innovation organisation and strategy within service firms and organisations. It then attempts to link the question of innovation in services with a number of fundamental issues, such as employment and skills, national and international spaces, and public policies in support of innovation. The final section contains a number of chapters devoted to the question of innovation in services beyond the service sector itself.

0.4.1 Services and innovation: conceptual and analytical frameworks

Part I is given over to a general theoretical discussion of services and innovation. The aim is not only to compile a state-of-the-art report on the question of innovation in services, but also to sketch out some avenues for research and to provide prospective analyses of the future of the service society as it relates to innovation.

In Chapter 1, Faïz Gallouj and Maria Savona survey the literature on innovation in services. A number of surveys already exist; most of them are based on an analytical framework, developed by Gallouj (1994) and taken up in various studies by the same author and others (see notably Coombs and Miles, 2000; Gallouj and Savona, 2009; Gallouj, forthcoming). In this framework, innovation in services is divided into three categories: assimilation, differentiation and integration. In the assimilative (or technologist) approach, innovation in services is reduced to the adoption of technical systems originating in manufacturing industry. The differentiation or service-oriented approach seeks to identify the specific characteristics of innovation in services. The integrative approach, finally, draws on synthetic analytical frameworks that capture innovation, regardless of the form it takes, in both goods and services. In order to carry out this new updated survey, Gallouj and Savona link the following perspectives:

(1) assimilation vs differentiation approaches; (2) theoretical vs empirical approaches; (3) typological vs analytical approaches; and (4) demand vs supply-oriented approaches.

In Chapter 2, André Barcet also engages in a critique of assimilative (or industrialist) approaches to innovation in services. His chapter begins by showing in what respects the 'so-called industrialist (approach) to innovation' now poses problems, particularly because of its inability to capture uses and dynamics on the clients' or users' side. He then outlines a four-layer model of innovation: Layer 1: service-based innovation (why and for who?); layer 2: the concept of service (what?); layer 3: organisational innovation (how?); layer 4: the methods and resources implemented (with what and with whom?). This model can be used to identify some of the conditions at the level both of 'macrosystems' and of the individual firm that have to be fulfilled if innovation in services is to be effective.

To some extent, Jeremy Howells' chapter (Chapter 3) carries on from the surveys in the previous two chapters by suggesting new avenues that might be explored in order to update current research problematics in the area of innovation in services. Among these promising research avenues, Howells identifies in particular: the contribution of service innovation studies to an integrated and holistic theory of innovation (that is, a further evolution of the integrative approach); the further investigation of interaction in the innovation process and of the individual components of the innovation process as well as the flows linking them together; a greater focus on innovation in public services.

The last two chapters in this first part are prospective in nature. They carry on from Chapter 3 by warning against certain possible dangers posed by the service economy and some of the scenarios for its development.

Thus William Baumol's chapter (Chapter 4) begins by recalling his celebrated cost disease theory. According to this theory, an economy can be broken down into two groups: those activities whose relative price declines as a result of productivity gains and those whose relative prices are constantly increasing. This group of so-called stagnant activities tend to be artisanal or craft-based, which reduces the opportunities for automation. Baumol uses a number of particularly illuminating examples to outline the main strategies deployed against this cost disease in stagnant sectors. They include the introduction of self-service, the development of the 'throw-away society' and constant attempts to automate production processes. However that may be, cost disease does not necessarily lead, according to Baumol, to any reduction in supply and demand, nor incidentally to any decline in the quality of the activities that suffer from it. What happens rather is that the way they are provided changes. This applies particularly to activities for which demand is price-inelastic (for example, education and health). In the progressive sectors, on the other hand, productivity gains and cost reductions have been exceptional since the Industrial Revolution, as a result of mechanisation. The danger of cost disease comes not from the stagnant sectors (services) but from the arms industry, where productivity is constantly increasing and costs declining, putting sophisticated arms within reach of the most ill-intentioned, imperilling in some cases humanity's very future.

In Chapter 5, Jean Gadrey takes the view that the economics of services has two shortcomings, namely that it takes little account of environmental or social considerations. With just a few exceptions, after all, the question of the links between services and the environment and that of inequalities, whether in access to services or those caused by services, have not been priorities on research agendas. In Gadrey's view, the environmental and social crisis demands a revolution in the economics of services. It is going to give rise to a powerful wave of innovations in services, manufacturing and agriculture. Gadrey provides an interesting prospective study in which he shows that certain service activities will inevitably decline (road, air and sea transport, for example) while others will prosper (for example, services for young children and the elderly, local government, hire services). More generally, his analysis leads him 'to call into question the dominant position that economics acquired during the second half of the twentieth century in the definition of progress and the development of the corresponding analytical tools'.

0.4.2 The nature of innovation in services: sectoral analyses and case studies

Part II of the *Handbook* is devoted to a number of (sectoral) case studies of innovation in various types of service activities, focusing on underinvestigated ones: public services, cultural services and the social enterprise or social economy sector.

Thus the chapter by Paul Windrum, Manuel García-Goñi and Eileen Fairhurst (Chapter 6) examines innovation in health services. The authors put forward an amended version of the characteristics-based approach to innovation, which they apply empirically to the provision of new educational services for patients suffering from Type 2 diabetes. The revised model (Windrum and García-Goñi, 2008) is based on studies by Saviotti and Metcalfe (1984) on industrial innovation and Gallouj and Weinstein (1997) and Gallouj (2002a) on innovation in services, which the authors seek to extend by incorporating into the model public service organisations and policy-makers, on the one hand, and the influence of users' preferences on innovation processes, on the other.

The chapter by Cristiano Antonelli, Pier Paolo Patrucco and Federica Rossi (Chapter 7) also focuses on innovation in public services. It is concerned with the strategic importance of cognitive interactions in developed economies and with the new role that universities play in them. Universities are going through major institutional and organisational changes, which are causing the traditional knowledge mode (the 'open science' model) to be challenged by a new knowledge mode (the 'entrepreneurial science' and 'triple helix' models), which is characterised by closer relations between universities and business. In this type of collaboration, universities are playing an increasingly important and strategic role as knowledge-based service providers for firms. The chapter draws on Italian regional data on university–industry collaborations in order to demonstrate the strong and positive link between universities' scientific output and firms' R&D.

Ian Miles and Lawrence Green's chapter (Chapter 8) focuses on the creative industries, such as advertising, architecture, broadcasting, design, entertainment software and many 'cultural' activities. These are essentially service activities that play an important role not only in wealth creation but also in improving quality of life and stimulating an environment favourable to innovation. Despite their closeness to the themes of creation and innovation, these activities, which are important in cultural and media studies, occupy only a marginal place in innovation studies. The aim of this chapter is to help to remedy this paradox by examining the main characteristics of these activities (and particularly the ways in which they differ from other services activities) and their innovation processes.

In Chapter 9, Denis Harrisson, Juan-Luis Klein and Paul Leduc Browne examine the question of social innovation. They note that this notion is not yet solidly grounded theoretically and that it is fairly eclectic and heterogeneous in content. The aim of their chapter is not to remedy these shortcomings but rather to examine one of the facets of social innovation in the service sector, namely the contribution of the social economy, which is also known as the third sector or social enterprise sector. The social innovation that takes place in the social economy involves the adoption of innovative solutions to complex economic and social problems. Part of the chapter is given over to an investigation of the links between social innovation and services. The authors show that the innovation lies not in the service per se but rather in the ways in which the service is created and the nature of its recipients.

0.4.3 Organisational and strategic patterns for service innovation

Part III of the book is given over to analysis of the organisational and strategic models of innovation in services. The various chapters in this third part are informed by the 'assimilation, differentiation, integration' framework outlined in Chapter 1.

In Chapter 10, Marja Toivonen examines the various modes of organising innovation processes in services, whatever the perspective adopted. She provides a general survey and introduction to this part. She identifies in the literature three main innovation process models in services: (1) the stage-gate models, which describe innovation processes as a succession of sequences; (2) the *a posteriori* recognition model; and (3) the rapid application model. Toivonen illustrates these various models with examples drawn from her own empirical investigations. Above all, however, she examines the organisational and managerial implications of each of the models. Thus the main problem with the stage-gate models is that they require (human and financial) resources and time. Slowness seems to be an even more acute problem with the *a posteriori* recognition model, even though firms can put in place systems for the early recognition of novelty. In the rapid application model, one of the difficulties is organising work in such a way as to prevent the subordination of development to immediate market needs.

In Chapter 11, Joe Tidd and Frank Hull adopt an assimilative approach, on the ground that the industrial models might, with certain changes, be applied to services. The sequential model proposed for the development and provision of services is a refinement of the stage-gate models, which themselves are based on industrial models. The underlying hypothesis is that the fundamental stages in the development of a new service are broadly the same as those in the process of developing a new good, with just a few minor differences in the importance attached to certain stages or the modes of implementation. The question of delivery occupies a central position in the model, since delivery is a fundamental element of any service. The proposed model was tested by empirical investigations carried out in 108 service firms in the USA and the UK.

Jon Sundbo's chapter (Chapter 12), evocatively titled 'The toilsome path of service innovation', focuses on the specificities of innovation in services and on their consequences for modes of organisation and innovation strategies in services. Sundbo can be said, therefore, to have adopted a differentiation approach. Like service provision itself, innovation in services is a complex process that brings into play a multiplicity of actors and may follow a number of different trajectories. Thus innovation in services is seen as a labile process that may take a number of different directions; at the same time, however, it may also be easily interrupted. This means that innovation in services requires particular efforts by management, which are examined in this chapter.

Similarly, the chapter by Bo Edvardsson, Anders Gustafsson, Per

Kristensson and Lars Witell (Chapter 13) adopts a differentiation approach, since it investigates one of the specific characteristics of services, namely the role played by clients in the production process, or more precisely the integration of clients into innovation processes. The authors identify the various possible forms of client involvement in these processes and suggest a number of management principles for managing that involvement. This chapter is based on in-depth investigations in many Swedish service firms. However, the analysis can be applied not only to any private or public service provider but also to manufacturing companies effecting the transition from product to service focus.

The last two chapters in this part fall within the scope of economics rather than management. Nevertheless, they investigate the problems inherent in organising innovation processes. They also adopt a differentiation approach, since they emphasise, respectively, the fundamental role played by collaboration in innovation in services and the specific difficulties of protecting innovation in services.

Thus in Chapter 14, Christiane Hipp examines the question of collaboration in service innovation. Collaboration is recognised as an important factor in the success of innovation in services. The forms such collaboration takes can vary, as can its scale and its motivations; it may involve different partners, draw on different sources of knowledge and differ from one service sector to another. The aim of this chapter, which draws on data from the German Community Innovation Survey, is to explore these various questions empirically in an attempt to identify the specific characteristics of collaboration in services.

In Chapter 15, Knut Blind, Rinaldo Evangelista and Jeremy Howells focus on an important element in the innovation process, namely innovation protection. The question of innovation protection regimes in services is an important theoretical and strategic problem, which is far from being resolved. This chapter offers an empirically tested analytical framework that links the types of cognitive regimes that characterise service firms with the specific strategies such firms deploy in order to protect their innovations. The two criteria used to distinguish the various cognitive regimes from each other are, on the one hand, the level of codification and tacitness of 'knowledge' and, on the other, the 'level of tangibility' of the knowledge assets and innovative output of service firms.

0.4.4 Innovation in services and through services: impact analyses (growth, performance, employment and skills)

Part IV of the *Handbook* contains a number of impact analyses of innovation in services. It comprises three chapters. The chapter by Rinaldo

Evangelista and Maria Savona ('Innovation and employment in services', Chapter 16) addresses the difficult question of the relationship between innovation and employment. Its main objective is to re-examine the literature on innovation in services in the light of the employment issue. In particular, it attempts to assess to what extent and in what way this question is implicitly or explicitly addressed in the literature. In pursuit of these objectives, the national and international literature is reviewed and a research agenda proposed.

In Chapter 17, Pascal Petit asks whether, in their relationship with innovation, services are not affected by some specific skill bias, related to the fact that large network services tend to adapt relatively rapidly to general structural changes, leaving little time to consumers to adjust their needs to the new environment. The starting point for his analysis is the hypothesis of a generalised productivity paradox whereby one sees major structural changes with little visible impact on productivity growth. It turns out that it is associated in manufacturing industries with what has been formulated in the literature as a skill bias nature of technological change (SBTC), according to which ICTs are associated with high-skill labour. In the case of some large network services rapid adjustments to structural changes seems to lead to the opposite, favouring low-skill workers. But conversely organisational adjustments are set leaving little room for any upgrading of the clients' skills. A logic of cost reduction prevails while the possibilities to redefine needs and provisions of services are locked in. To overtake this kind of consumers' SBTC effect new norms and regulations of service products have to be designed, debated and implemented. The formulation of such projects would paradoxically clearly enhance the potential of innovation of services.

The chapter by José Camacho and Mercedes Rodriguez (Chapter 18) is devoted to knowledge-intensive services (KIS). The authors seek, firstly, to measure the growing importance of KIS as intermediate consumption in production processes. They then attempt to measure the impact of the use of KIS on productivity and then on the diffusion of knowledge and innovation. In pursuit of these objectives, the authors use the recently available Input–Output Database 2006. The chapter provides both an analysis of developments over time and an international comparison of these questions.

0.4.5 Innovation in services and national and international spaces

Part V is given over to innovation in services as it relates to various dimensions of local, national and international spaces. Peter Daniels's chapter (Chapter 19) investigates innovation in services in a globalised economy. Economic globalisation is both a cause and a consequence of innovation in service enterprises. In order to deal with geographically dispersed markets, even the smallest service firms have to engage in intensive innovation efforts. Much of the innovation in these firms is devoted to overcoming the difficulties caused by geographic distance, even though ICTs have mitigated the tyranny of geographic separation. Peter Daniels notes that, given the interactive nature of service production, it is difficult to consider the innovation process in general terms. The chapter also contains a number of case studies of globally oriented services, with an emphasis on the way in which the firms in question have implemented their innovations.

The chapter by Silvia Massini and Marcela Miozzo (Chapter 20) investigates outsourcing and offshoring in knowledge-intensive services. In contrast to the numerous studies of internationalisation and offshoring in manufacturing industry, there are still relatively few studies of outsourcing and offshoring in services and their links with innovation. The recent outsourcing and, particularly, offshoring of high-level services (R&D services, design services, new software development etc.) that play an active role in innovation processes lead the authors to examine these links closely. The savings in labour costs and other resources are only a part of the issues at stake here. After all, the outsourcing and offshoring of knowledge-intensive business services have fundamental implications not only for the theory (and boundaries) of the firm but also for the globalisation of innovation.

In Chapter 21, Jean Philippe and Pierre-Yves Léo set out to study the links between innovation in services and the internationalisation of service firms. These two strands of research have hitherto tended to be investigated separately. The chapter's aim is to bring them together in order to identify, on both the conceptual and managerial levels, the consequences of internationalisation on innovations in services, particularly business services. The chapter examines how the coupling of innovation to internationalisation operates as regards: (1) R&D activities linked to export; (2) service offering through rationalisation innovation; and (3) international service delivery through relational innovations.

Knud Blind's chapter (Chapter 22) focuses on the role of standards in trade in services. He analyses the complex influence standards exert, not only as facilitators of but also as obstacles to trade in services. The chapter begins by re-examining a number of standard theoretical principles (developed with reference to tangible goods) on the role of standards in trade in services. It then presents a number of examples of the use of standards by a sample of European service firms.

In Chapter 23, Xavier Vence and Alexandre Trigo investigate the role of cooperation in innovation in services and the local and global learning models that result from it. The role of cooperation is established by using empirical data drawn from the third and fourth Community Innovation Surveys. This data shows that service firms in general (and this is even more true of business services) have a higher propensity to cooperate than manufacturing firms. This chapter emphasises the geography of the interactions involved in innovation in services. The authors identify various sectoral models of local or global cooperation in the sphere of innovation processes. It emerges that global interactivity is playing an increasingly important role in innovation processes in certain service industries whose role in the diffusion of knowledge across the world is becoming ever more prominent. This global learning process is closely linked to the globalisation of markets in these service industries.

The last two chapters in this part are concerned with the role of services in the development of large urban centres. One is concerned with one of the major French cities, the other with the British capital. In Chapter 24, Marie-Christine Monnoyer-Longé examines the nature and dynamic of innovation in services in the Toulouse metropolitan area. She is interested in the way in which this innovation contributes to economic development in the metropolitan area. The chapter also includes an evaluation of a local support scheme for new start-up companies providing innovative services and, more generally, an examination of the way in which public policies at local level can encourage innovation and its positive effects on local economies.

Chapter 25 by Peter Wood and Dariusz Wójcik is concerned with London's financial, professional and consultancy services and the major role they play in innovation. The authors take the view that the main factor in regional and urban inequality in the UK since the early 1990s has been the economic success of the London region. They note that London's innovation dynamic does not originate mainly in technological developments but rather in the labour and knowledge-intensive activities of the city's internationally networked service functions (financial services and other professional and business services). The chapter compares the recent innovation dynamics of financial services and those of other types of knowledge intensive business services (KIBS). While there are many similarities, notably in terms of the need for market responsiveness, high quality labour and flexible institutional arrangements, there are also differences. These relate in particular to the role of ICTs, modes of regulation and the autonomy allowed to KIBS firms by clients. Overall, it seems that the distinctive national and regional importance of non-financial KIBS innovation requires greater attention, even though it poses considerable methodological problems.

0.4.6 Innovation in services and public policy

Part VI is given over to public policies in support of innovation in services. It contains only two chapters, but the question of public policies has already been mentioned as a secondary theme in some of the preceding chapters (particularly Chapters 2, 24 and 25). As will become clear, these public policies have to face up not only to an innovation gap, which raises the question of whether specific policies should be introduced for services, but also to a performance gap, which raises questions about the nature of the performance being sought.

The chapter by Pim den Hertog and Luis Rubalcaba (Chapter 26) has two main aims. The first aim is theoretical in nature and has two aspects. Firstly, it uses the assimilation, differentiation and integration analytical framework to examine the policy framework for service R&D and innovation policies in order to establish whether or not specific policies for services should be adopted. The authors then examine the theoretical rationality used to justify the various programmes, and asks whether the analysis in terms of market (and systemic) failures can also be applied to services and R&D and to innovation in services. The second aim is empirical in nature. Drawing in particular on data from the third and fourth Community Innovation Survey (CIS3 and CIS4) and the Innovation Policy Project in Services (IPPS) the authors offer some empirical illustrations at European level of the place of services in innovation support programmes. Although the results vary from one sector to another and from one country to another, these empirical data show that innovation in service industries receives less public support.

In Chapter 27, Faridah Djellal and Faïz Gallouj analyse the innovation–performance relationship in contemporary developed economies. Their analysis reveals a double 'gap' relating to innovation and performance. The 'innovation gap' reflects the difference between the reality of innovation produced in an economy and what traditional innovation indicators (R&D, patents) capture. As for the 'performance gap', this measures the difference between the reality of performance in an economy and the performance assessed by traditional economic tools (mainly productivity and growth). It reflects a hidden performance, invisible to these tools. These two 'gaps' blur the innovation–performance relationship. They are behind a certain number of paradoxes, which this analysis seeks to explain, and they raise questions about the legitimacy of some public policies intended to support innovation.

0.4.7 Service innovation: beyond service sectors

Goods and services are being sold and consumed independently of each other less and less frequently; rather, they are increasingly being offered as solutions, systems and functions. More generally, service or information is now the main component of many goods. Thus the boundaries between sectors are becoming increasingly blurred and there is increasing uncertainty as to the precise nature of 'products'. Thus service innovation can occur in manufacturing industry or in agriculture. Part VII of the *Handbook* is given over, therefore, to analysis and illustration of these phenomena.

John Bryson's chapter (Chapter 28), which is entitled 'Service innovation and manufacturing innovation: bundling and blending services and products in hybrid production systems to produce hybrid products', is a general discussion chapter that serves as an introduction to this part. It highlights in theoretical terms the increasing blurring of the boundaries between the manufacturing and service sectors. This is altering the ways in which goods and services are produced, consumed and innovated, and the ways by which our theoretical models and our managerial and policy strategies should cope with them.

Olivier Furrer devotes his chapter (Chapter 29) to the services provided by manufacturing firms in support of the tangible goods they produce. These 'services around the product' are not confined to after-sale services, such as technical problem-solving, equipment installation, training or maintenance. They also include programmes that help customers design their products or reduce their costs, as well as rebates or bonuses that influence how customers conduct business with a supplier. The aim of this chapter is to provide an improved definition of the notion of 'product services' by incorporating it into a relationship marketing framework, which suggests a consistent and managerially relevant typology of product services strategies. Thus Furrer puts forward a typology of four product service strategies – discount strategy, relational strategy, individual strategy and outsourcing strategy – illustrated by best-practice examples that highlight the value-creation mechanisms. He also analyses the conditions required for successful implementation of 'product service' strategies.

The chapter by Sylvain Lenfle and Christophe Midler (Chapter 30) focuses on the service innovation process at a large European carmaker, which has been developing telematic services since 1998. On the theoretical level, this chapter draws on recent developments in design theory (Hatchuel and Weil, 2002) in order to develop an integrated model of NSD (new service development). The authors illustrate the way in which identification of the design parameters of a new service – that is: (1) use(s) and user(s); (2) associated product; (3) contract signed between the firm and the client; (4) front-office processes; (5) back-office processes; and (6) economic model of the service – helps to improve understanding of the functioning of NSD processes. The model they develop can be used (1) to characterise the types of innovation, to support strategic decision-making and to analyse the innovation process and, consequently, to adapt the organisation.

The final chapter, by Jan Bröchner (Chapter 31), is devoted to innovation in the construction industry, which is increasingly dominated by service activities. Thus even in an activity that closely resembles a manufacturing activity, namely speculative house-building, customisation plays an important role. Construction projects bring together a number of different service providers: architects, structural engineers and other KIBS providers and equipment rental services; moreover, once the building is finished, facilities management services are often bought in. Bröchner outlines some of the results of earlier studies of innovation in the construction industry: the project nature of construction, the lack of knowledge transfer between projects and the insignificant role of intellectual property rights. The openness of construction innovation can be linked to a high degree of customer co-production. ICT progress now appears to create stronger links between technological and non-technological innovation in construction. To conclude, Bröchner draws on the data from a survey carried out in 2006 among Swedish construction contractors in order to analyse the types and levels of innovation.

Note

1. This is acknowledged, once again, by recent important initiatives: the IBM 'Services Science' initiative (Maglio et al., 2006; IBM Research, 2004; Maglio and Spohrer, 2008; Kieliszewski et al., forthcoming) and the MARS (International Monitoring on Activities and Research in Services) inititative (Spath and Ganz, 2008).

References

Baumol, W. (1967), 'Macroeconomics of unbalanced growth: the anatomy of urban crisis', *American Economic Review*, **3** (June), 415–26.
Beyers, W.B. (2002), 'Services and the new economy: elements of a research agenda', *Journal of Economic Geography*, **2**, 1–29.
Bluestone, B. and B. Harrison (1986), 'The Great American Job Machine', Report for the Joint Economic Committee, December.
Callon, M. (1994), 'L'innovation technologique et ses mythes', *Gérer et Comprendre*, March, 5–17.
Cohen, S. and J. Zysman (1987), *Manufacturing Matters*, New York: Basic Books.
Coombs, R. and I. Miles (2000), 'Innovation, measurement and services: the new problematique', in S. Metcalfe and I. Miles (eds), *Innovation Systems in the Service Economy*, Boston, MA: Kluwer, pp. 85–103.
Czarnitzki, D. and A. Spielkamp (2003), 'Business services in Germany: bridges for innovation', *Service Industries Journal*, **23** (2), 1–30.
Daniels, P. (1983), 'Service industries: supporting role or centre stage?' *Area*, **15**, 301–309.
Davies, A. (2004), 'Moving base into high-value integrated solutions: a value stream approach', *Industrial and Corporate Change*, 13, 727–756.
den Hertog, P. (forthcoming), 'Managing the soft side of innovation. Do practitioners, researchers and policy-makers interact and learn how to deal with service innovation?'

in S. Kuhlmann, R. Smits and Ph. Shapira (eds), *The Theory and Practice of Innovation Policy: An International Research Handbook*, Cheltenham, UK and Northampton, MA, USA: Edward Elgar.

Djellal, F. and F. Gallouj (2007), 'Innovation and employment effects in services: a review of the literature and an agenda for research', *Service Industries Journal*, **27** (3–4), 193–213.

Djellal F. and F. Gallouj (2008), *Measuring and Improving Productivity in Services: Issues, Strategies and Challenges*, Cheltenham, UK and Northampton, MA, USA: Edward Elgar.

Evangelista, R. (2000), 'Innovation and employment in services: results from the Italian Innovation Survey', in M. Vivarelli and M. Pianta (eds), *The Employment Impact of Innovation: Evidence and Policy*, London, UK and New York, USA: Routledge, pp. 121–148.

Evangelista, R. and M. Savona (2003), 'Innovation, employment and skills in services: firm and sectoral evidence', *Structural Change and Economic Dynamics*, **14**, 449–474.

Fourastié, J. (1949), *Le Grand Espoir du XXe siècle*, Paris: PUF.

Fuglsang, L. (ed.) (2008), *Innovation and the Creative Process: Towards Innovation with Care*, Cheltenham, UK and Northampton, MA, USA: Edward Elgar.

Gadrey, J. (1996), *Services: la productivité en question*, Paris: Desclée de Brouwer.

Gallouj, C. (1996), 'Le commerce interregional des services aux enterprises: une revue de la literature', *Revue d'économie régionale et urbaine*, **3**, 567–596.

Gallouj, C. (2007), *Innover dans la grande distribution*, Bruxelles: De Boeck.

Gallouj, F. (1994), *Economie de l'innovation dans les services,* Paris: L'Harmattan.

Gallouj, F. (2002a), *Innovation in the Service Economy: The New Wealth of Nations*, Cheltenham, UK and Northampton, MA, USA: Edward Elgar.

Gallouj, F. (2002b), 'Interactional innovation: a neoSchumpeterian model', in J. Sundbo and L. Fuglsang (eds), *Innovation as Strategic Reflexivity*, London, UK and New York, USA: Routledge, pp. 29–56.

Gallouj, F. (2002c), 'Knowledge intensive business services: processing knowledge and producing innovation', in J. Gadrey and F. Gallouj (eds), *Productivity, Innovation and Knowledge in Services*, Cheltenham, UK and Northampton, MA, USA: Edward Elgar, pp. 256–284.

Gallouj, F. (forthcoming), 'Services innovation: assimilation, differentiation, inversion and integration', in H. Bidgoli (ed.), *The Handbook of Technology Management*, New York: John Wiley & Sons.

Gallouj, F. and M. Savona (2009), 'Innovation in services: a review of the debate and perspectives for a research agenda', *Journal of Evolutionary Economics*, **19** (2), 149–172.

Gallouj, F. and O. Weinstein (1997), 'Innovation in services', *Research Policy*, **26** (4–5), 537–556.

Gorz, A. (1988), *Métamorphoses du travail et quête de sens*, Paris: Galilée.

Gregory, M., W. Salverda and R. Schettkat (eds) (2007), *Services and Employment: Explaining the US–European Gap*, Princeton, NJ, USA and Oxford, UK: Princeton University Press.

Hatchuel, A. and B. Weil (2002), 'La théorie C-K: fondements et usages d'une théorie unifiée de la conception', conference on 'Sciences de la conception', Lyon, France.

IBM Research (2004), 'Service science: a new academic discipline?' IBM.

Illeris, S. and J. Philippe (1993), 'Introduction: the role of services in regional economic growth', in P. Daniels (ed.), *The Geography of Services*, London: Frank Cass, pp. 3–11.

Kieliszewski, C., P. Maglio and J. Spohrer (eds) (forthcoming), *The Handbook of Service Science*, New York: Springer.

Kline, S. and N. Rosenberg (1986), 'An Overview of Innovation', in R. Landau and N. Rosenberg (eds), *The Positive Sum Strategy: Harnessing Technology for Economic Growth*, Washington, DC: National Academy Press, 275–305.

Maglio, P.P. and J. Spohrer (2008), 'Fundamentals of service science', *Journal of the Academy of Marketing Science*, **36** (1), 18–20.

Maglio, P., S. Srinivasan, J. Kreulen and J. Spohrer (2006), 'Service systems, service scientists, SSME and innovation', *Communications of the ACM*, **49** (7), 81–85.
Mathieu, V. (2001), 'Service strategies within the manufacturing sector: benefits, costs and partnerships', *International Journal of Service Industry Management*, **12** (5), 451–475.
Miozzo, M. and D. Grimshaw (eds) (2006), *Knowledge Intensive Business Services: Organizational Forms and National Institutions*, Cheltenham, UK and Northampton, MA, USA: Edward Elgar.
Muller, E. and A. Zenker (2001), 'Business services as actors of knowledge transformation: the role of KIBS in regional and national innovation systems', *Research Policy*, **30** (9), 1501–1516.
Nählinder, J. (2005), 'Innovation and employment in services: the case of knowledge intensive business services in Sweden', PhD thesis, Linköping University, Sweden.
NESTA (2006), 'The innovation gap: why policy needs to reflect the reality of innovation in the UK', National Endowment for Science, Technology and the Arts, Research Report, October.
OECD (2007), 'Summary report of the study of globalisation and innovation in the business services sector', edited by G. a. S. Adjustment. Paris.
Rubalcaba, L. (2006), 'Which policy for innovation in services?' *Science and Public Policy*, **33** (10), 745–756.
Saviotti, P. and S. Metcalfe (1984), 'A theoretical approach to the construction of technological output indicators', *Research Policy*, **13**, 141–151.
Smith A. (1960), *The Wealth of Nations*, 1 edn 1776, New York: The Modern Library, Random House.
Spath, D. and W. Ganz (2008) (eds), *The Future of Services: Trends and Perspectives*, München: Hanser.
Sundbo, J. (1998), *The Organisation of Innovation in Services*, Copenhagen: Roskilde University Press.
Thurow, L. (1989), 'Towards a high-wage, high-productivity service sector', Economic Policy Institute, Washington, DC.
Toivonen, M. (2004), 'Expertise as business: long-term development and future prospects of knowledge-intensive business services', PhD, Helsinki University of Technology, Finland.
Windrum, P. and M. García-Goñi (2008), 'A neo-Schumpeterian model of health services innovation', *Research Policy*, **37** (4), 649–672.
Wood, P. (2005), 'A service-informed approach to regional innovation – or adaptation?' *Service Industries Journal*, **25** (4), 429–445.

PART I

SERVICES AND INNOVATION: CONCEPTUAL AND ANALYTICAL FRAMEWORKS

1 Towards a theory of innovation in services: a state of the art

Faïz Gallouj and Maria Savona

1.1 Introduction

This chapter reviews the theoretical and empirical literature on technological change and innovation in services. The survey covers the nature and role of technological change and innovation in service firms and sectors by reviewing the attempts to conceptualise them, the methodological issues related to their quantification, and the existing empirical evidence in this domain.

Though the number of contributions has increased in recent years, the literature on innovation in services has always been more fragmented and less empirically grounded than the literature on innovation in the manufacturing sector. Thus, the task of tracing and systematising its evolution is not straightforward. This chapter outlines the main axes along which this literature has developed so far and provides some suggestions for its future development (see also Chapter 3 in this volume).

Contributions to the service innovation literature can be divided into several groups depending on the approach adopted: (1) assimilation vs demarcation; (2) theoretical vs empirical; (3) typological vs analytical; and (4) demand-vs supply-oriented.

1.1.1 Assimilation vs demarcation

In previous work (Gallouj, 1994, 1998; Gallouj and Gallouj, 1996; see also Gallouj and Weinstein, 1997), one of the authors of this chapter proposed a framework[1] to systematise the literature on innovation in services. This literature has often been classified according to the extent to which the conceptualisation and quantification of innovation in services has borrowed from manufacturing-centred theoretical and methodological frameworks (assimilation or technologist approaches), which mainly consider innovation in services as resulting from technology adoption and use. This view tends to overlook the non-technological aspects of the process of creating novelty. In contrast to the assimilation perspective, several other contributions try to account for the specificity of service products and sectors compared to those related to manufacturing (demarcation or service-oriented approaches).

The very large Schumpeterian literature on manufacturing focuses primarily on the distinction between product and process innovation, in part due to the fact that other types of innovation are less easily identified and measured. Yet, the well-known definition of innovation stemming from the seminal Schumpeterian typology (Schumpeter, 1934 [1912]) is quite broad, and potentially able to encompass the multidimensionality of service innovation beyond the traditional product–process distinction. It should thus be possible to introduce a new and potentially rich stream of investigation in the service domain (Gallouj and Weinstein, 1997; Howells, 2000; Drejer, 2004).

Both the assimilation and the demarcation perspectives on innovation in services are useful, but are somewhat biased. On the one hand, the assimilation perspective undermines much of the specificity of services. On the other hand, the demarcation perspective is overly focused on sectoral case studies and typologies and, therefore, lacks consistency with, or adds very little to, existing innovation theories.

The specificity of service innovation, therefore, should be accounted for in a manufacturing-compatible framework and should also be able to contribute to current innovation theory, through a 'synthetic' or 'integrative' framework that is able to account for the 'non-technological' nature of the innovation activities that occur in services. Syntheses of assimilation and demarcation approaches are still at the embryonic stage, although there have been a few recent attempts to operationalise them (see below).

1.1.2 Theoretical vs empirical

Another way to stylise the different approaches is to distinguish between purely conceptual, and empirical-based studies. Such differentiation obviously would cut across all the categories referred to here.

Nelson and Winter (1982) argue that economic theory has a dual role in economic analysis, depending on how it is used. They distinguish between formal and appreciative theorising, on the basis of the 'guiding aim' of the analyst, the 'use' of theory and the 'style' of theorising. More specifically in the field of the economics of technical change, a large number of studies based on the Schumpeterian heritage and starting with the pioneering works of Nelson and Winter (1982) and Dosi et al. (1988) have encompassed technological change within economic analysis. Also, a flourishing literature has developed around empirical assessment of this complex issue, trying to test the explanatory power of neo-Schumpeterian and evolutionist thought.

Nevertheless, this literature is still strongly biased towards the manufacturing sector, research and development (R&D)-based innovation inputs and patents as indicators of innovation output, which, as we show below, tend to overlook specific aspects of innovation in services (see also Chapter 27 in this volume).

On the one (theoretical, conceptual) side, among the neo-Schumpeterian literature, analysis of the role of technical change and its effect on structural changes in the sectoral composition of economies towards services and aggregate growth has yet to be developed.[2] On the other (empirical, appreciative) side, the fragmentation and lack of comparability of the empirical evidence on innovation in services has often made it difficult to generalise the contributions to this domain to the entire service sector, or has produced rather suggestive and evocative results.[3]

As a result, many authors have proposed new typologies in order to try to make sense of the highly heterogeneous activities and innovation patterns within the service sector. Whether assimilated or demarcated, theoretical or empirical, the contributions on innovation in services can also be classified according to whether they provide typologies of innovation or new analytical tools to advance our understanding of innovation in services.

1.1.3 Typological vs analytical

In the economics of innovation we can interpret the dichotomy between formal and appreciative theorising both in terms of conceptual vs empirical-based studies and in terms of analytical vs typological approaches.

Analytical approaches generally aim to disassemble the evidence into components, making it possible to identify both the nature of the interactions and a hierarchy of causal relationships when the components are reassembled. Analytical approaches aim either to 'demarcate' innovation in services from traditional innovation studies in order to account for the specificity of service products by trying to provide new analytical tools, or to provide 'synthesis' tools able to advance the conceptualisation of innovation beyond the product–process and technological–non-technological dichotomies (Gallouj and Weinstein, 1997).

Compared to analytical tools, typologies aim to reduce the variety of information by systematising and labelling it. Some of the typologies of innovation in service sectors borrow heavily from manufacturing-centred perspectives, thereby 'assimilating' services into the mainstream innovation studies (e.g., Soete and Miozzo, 1989). Others (Van der Aa and Elfring, 1993; Gadrey et al., 1995, Sundbo, 1994, 1997; Gallouj, 1998; Fuglsang, 2002; Raupp de Vargas and Zawislak, 2007) provide 'demarcation' typologies and case studies, by focusing on a limited number of specific service sectors. A third category of typologies builds upon large-scale surveys to come up with taxonomies of sectoral patterns of innovation in services (Evangelista, 2000, among many others). Ideally, scholars should provide a conceptual base for the typologies of innovation and sound empirical evidence to test them in order to bridge and mutually enrich the analytical and typological approaches.

1.1.4 Demand- vs supply-oriented approaches

The opposition between demand- and supply-oriented approaches cuts across all the categories described above. On the one hand, typologies relying mostly on an assimilation perspective and originally ascribed to manufacturing have been simply extended to services, thereby overemphasising supply and overlooking the role of demand-oriented categories. They focus on technological rather than non-technological sources, on processes rather than products, on the scale of activities rather than their position along the vertical chain. The category of 'type of user' often adds a new ingredient to these typologies, recognising the importance of the users' role in the definition of a service product, even before considering their involvement in the innovation process. However, these typologies tend to take no account of the type of user. In demarcated typological approaches, however, demand-side characteristics are afforded more importance, and although often not explored in depth, they are evoked in terms of 'client involvement'.

At the same time, analytical approaches have developed that emphasise the adoption of technology and its impact on productivity. Much of the Schumpeterian heritage has been reduced to the issue of product vs process innovation, and ignores the potential roles of market and organisational innovations. The role of the consumer is also not properly investigated in terms of a full consideration of demand in the analysis.

The empirical literature also tends to be confined to analysing the 'impact' of innovation in services on productivity, rather ignoring output and employment growth.[4] This is certainly the influence of early and enduring concerns about productivity slowdown and Baumol's heritage.[5]

1.1.5 Avenues for a synthetic approach

What could be achieved by a synthetic approach to innovation? To span all the categories stylised above, we could perhaps produce a synthesis of the innovation studies on services: by drawing appropriate analytical implications from conceptually rigorous typologies; by providing empirical grounds for both typologies and analyses; and by drawing aggregate implications from micro- and meso-level evidence.

For the purposes of this chapter, we argue that the crucial issue in the innovation literature on services is to provide a synthetic approach along the following lines:

- By assessing whether the original Schumpeterian taxonomy of innovation holds for services. A number of 'demarcating' contributions, produced in an effort to highlight the alleged specificity of services, risk creating further and redundant varieties of the original Schumpeterian taxonomy of innovation, which add little to the understanding of whether, how and to what extent services are able to exploit technological and non-technological opportunities to create novelty.
- By rebalancing the dominance of productivity-centred impact analysis with a more detailed account of the consequences of technical change in terms of output and employment growth in services.
- By providing a sound empirical grounding for both typologies and analyses, in order to overcome the methodological impasse between assimilation and demarcation, which runs the risk of becoming sterile with respect to a fuller account of the nature and impact of technical change in services.

The review in this chapter organises the various contributions on the basis of the categories discussed above. We consider two pairs of approaches in particular, to systematise the literature – the theoretical vs the empirical and the typological vs the analytical, which we see as particularly relevant for highlighting the methodological and empirical gap that the research agenda on services should attempt to fill.

The chapter is structured as follows. Section 1.2 reviews the theoretical literature, according to the methodology described, that is, analytical as opposed to typological approaches. Section 1.3 reviews the empirical literature on qualitative and quantitative assessments of innovation in services. Section 1.4 offers a summary and some conclusions.

1.2 The theoretical literature

There are fewer theoretical than empirical contributions to the innovation in services literature. The theoretical literature includes a number of typological exercises conducted in line with the taxonomy of technical change proposed by Pavitt (1984), which has been extended to services from a 'top-down' perspective, that is, generally without the support of either qualitative or quantitative empirical evidence. These studies are complemented by 'bottom-up' typologies of innovation in services, which are constructed on the basis of a wide range of empirical evidence, both qualitative and quantitative (reviewed in the next section). At the same time, very few analytical models have been proposed, which perhaps is down to the assimilation, demarcation and integration approaches adopted.

1.2.1 Top-down typologies

This section reviews the attempts to extend to services taxonomic exercises on the nature and impact of technical change on the innovative and economic performance of firms and sectors. The benchmark for such typological exercises in this domain consists of Freeman's (1982; Freeman and Soete, 1987, 1997) typology of technical change, and Pavitt's (1984) widely quoted sectoral taxonomy of technical change. Although the aims of these classificatory exercises are different, both provide precise tools to identify the main sources and characteristics of technical change and its impact on the cycles of capitalistic development and the sectoral variety of innovative performance respectively.

Freeman (1982) proposes a categorisation of types of innovation according to their impact on the stages of capitalistic developments. He distinguishes between incremental innovation, radical innovation, new technological systems and technological revolutions or changes in the techno-economic paradigm. Freeman's typology contributes to the Schumpeterian debate on the discontinuous nature of technical change, focusing on the nature of the technology itself and the characteristics of its use, rather than on the characteristics of the firms and sectors developing and adopting the technology.

Pavitt's seminal taxonomy classifies the subjects (firms and sectors) developing and using technology, rather than the technology itself. Drawing on a dataset of more than 2000 observations of 'significant innovations and innovating firms' in the UK, across the period 1945–80, Pavitt codifies the variety of innovations across firms (and sectors). He identifies five categories of firms: (1) supplier-dominated; (2) production-intensive, including (i) scale-intensive and (ii) specialised suppliers; (3) science-based; and (4) information-intensive, a category added in a later contribution (Pavitt et al., 1989).

The first 'top-down' taxonomic exercise drawing on Pavitt's work and applied to services is by Soete and Miozzo (1989; Miozzo and Soete, 2001). The authors attempt to decompose Pavitt's category of supplier-dominated firms, which in his initial 1984 article included the whole of the services sector. In Pavitt and colleagues' 1989 contribution, parts of this sector were moved to the information-intensive category. Soete and Miozzo reposition the various service industries on the basis of an enlarged version of the set of 'ingredients' employed by Pavitt.

A large chunk of the service sector continues to be classified as supplier-dominated; this includes public and social services such as health, education, public administration, personal services (hotels and restaurants, domestic, repair) and distributive services (which includes the macro retail sector). Soete and Miozzo reprise Pavitt et al.'s 1989 version of the taxonomy and introduce an additional category, the scale/network-intensive sector, which includes those areas depending on large physical networks and information networks, namely transport and financial services. These branches (and especially the latter) are not simple adopters of technology developed elsewhere (embodied in capital equipment or in a large information and communication technology – ICT – infrastructure); rather, they are characterised by an autonomous innovation effort and contribute to defining and specifying the types of innovation introduced by using the network infrastructures adopted.

Overall, Soete and Miozzo's main categories, for instance, 'type of user', are typical of a top-down exercise rather than the bottom-up exercise originally carried out by Pavitt. Nevertheless, their contribution is a first attempt to break down Pavitt's 'supplier-dominated' category, although Pavitt himself acknowledged the increasing role of information-based sectors, among which he included services. In this respect, their taxonomy belongs to the assimilation perspective.

Lakshmanan (1987) provides another typical top-down taxonomy, that draws on evolutionary theory and a pre-existing typology of services developed by Mills (1986). In turn, Mills's typology, as adapted and applied by Lakshmanan (1987), comprises three groups of service activities, which differ according to degree of interaction with customers and degree of information asymmetry between service provider and client. Lakshmanan (1987) distinguishes between service-dispensing activities, task-interactive services and personal-interactive services.

The technological trajectory in service-dispensing activities (distribution, telecommunications, fast-food industry, etc.) is characterised by a tendency towards increased mechanisation and exploitation of economies of scale (Nelson and Winter's, 1982, 'natural' technological trajectory). The technologies trajectories at work in task-interactive services (accountancy, legal and financial services) and personal-interactive services (health, social security) are based mainly on the adoption and use of ICTs and are intended to reduce communication costs and information asymmetries.

Overall, if we assume that there is a certain degree of specificity in service compared to manufacturing innovation strategies, it seems that the underestimation of the innovative dimension of services, due to the specific characteristics of their production and delivery processes and to the consequent misuse of innovative indicators devised for the manufacturing sector, might constitute a crucial issue, that needs to be addressed empirically.

However, if we believe that the specificity of service innovation is a type of conjuring trick used to justify poorer innovative performance in services, then the importance of typological exercises by and large, particularly those carried out within the assimilation (manufacturing-centred) innovation literature, resides in the fact that they allow us to compare services and manufacturing innovation and its impact on economic performance.

Thus, the proliferation of new 'top-down' typologies[6] does not per se translate into a coherent extension of the Schumpeterian heritage to service activities, unless a clear premise in terms of the conceptualisation of service output and production processes, and the formulation of hypotheses regarding the main ingredients of the taxonomy, are provided.

1.2.2 The analytical approaches

Analytical approaches include attempts to contribute to the theory of innovation by providing new conceptual tools aimed at encompassing the specificity of service innovation. Some focus on technology-related processes, which we could perhaps label 'assimilation' approaches. Others start from a 'demarcation' perspective and aim towards a 'synthetic' one.

The reverse product life cycle model Barras's (1986, 1990) reverse product life cycle model describes the cycle of innovation in services as the reverse of the traditional innovation cycle – conceived for manufacturing sectors – formalised by Abernathy and Utterback (1975, 1978). The reverse product life cycle results from subsequent phases of ICT application, which progresses through incremental process innovation, radical process innovation and, finally, product innovation.

Incremental process innovations emerge in the back offices of organisations. They respond to a logic of cost reduction; for example, the computerised recording of insurance policies, computerisation of staff records and payrolls. Radical process innovations mainly affect front offices and are aimed at improving the quality of delivery. Examples include online insurance policy quotations, installation of automated teller machines (ATMs) by banks. Product innovations result in radically new products, for instance online services or home banking.

Barras's reverse product life cycle model acknowledges that the innovation dynamic in services follows a specific pattern. However, this model can be considered to belong to an assimilation perspective to the extent that it accounts only for innovation stemming from the adoption of ICTs. While Abernathy and Utterback's model identifies a product and process innovation life cycle, the reverse cycle, although claiming to be specific to services, relies on the same categories (product and process) which are not necessarily relevant to – or at least are more ambiguous in how they account for – services.[7] Moreover, the ambition of providing a general theory of innovation in services is somewhat compromised by acknowledging only the adoption of ICTs, while other technologies (which deal with the processing of material or biological media, e.g. transport, refrigeration, cooking and cleaning technologies, medical instrumentation, genetics, biotechnology, etc.) may play an essential role in innovation in services. In this respect, Barras's contribution is an important theoretical advance in innovation in services via its identification of a trajectory of innovation stemming from the adoption of ICT in financial services.

Barras's first attempt to provide a theory of innovation in services teaches us that the specific nature of innovation in services can be identified as the ability to adapt a generic technology to specific applications. Therefore, besides the rate of adoption of the enabling technology, what matters is the specific nature of user needs, which in turn affects the nature and type of innovations in service firms. The specificity of service production and delivery processes can be considered in terms of different degrees of standardisation vs specialisation, according to the types of user needs. We return to these issues below.

Sector-specific innovation analyses Within the analytical approaches to innovation in services, it is worth mentioning some of the attempts to provide tools to enrich the conceptualisation of innovation in specific service sectors. These might be considered 'demarcation' attempts to the extent that they do not claim to provide a theory of innovation in services, nor are they able to add to the theory of innovation in general.

Large-scale retailing is a sector which, particularly in the area of management science, has proven fertile for the 'creation' of theories of innovation. Consider, for example, the accordion theory (Hollander, 1966) or the 'wheel of retailing' (McNair, 1958), which considers innovation in shop formats in terms of the evolution of simplified systems (hard discount) towards more complex systems (with higher levels of pre- and after-sales services). Gallouj (2007) examines these theories and highlights their inability to take account of the diversity of forms of innovation in retailing, pointing out that innovative behaviour in retailing is more complex than the mere introduction of technological systems. To take account of this diversity, and also to integrate retailing into the more general analytical model of innovation in services, Gallouj proposes use of the service-based characteristics approach formulated by Djellal and Gallouj (2005, 2008a).

Financial services have also been the object of theoretical models based on the characteristics approach. There are a few models that represent financial products as a vector of characteristics, and innovation as the addition of new characteristics or improvements to existing characteristics (Hardouin, 1973; Niehans, 1983; Desai and Low, 1987). Thus, a financial product can be defined as a vector of n characteristics: $T_i =$

(t_{i1}, t_{i2}. . .t_{ij}. . .t_{in}) in which t_{ij} indicates the extent to which property *j* (e.g. liquidity, return, etc) is incorporated in product *i*. Innovation appears in the following two cases: a variation in t_{ij}, that is, in the extent to which the existing characteristic *j* is incorporated in the product (e.g. the product is more liquid), and the activation of a characteristic that did not previously exist (shift from $t_{ij} = 0$ to $t_{ij} \pm 0$). These models can be considered precursors to the characteristics-based approach to innovation.

The characteristics-based definition of products as a theoretical device to model innovation It has been observed that the boundary between goods and services is becoming more blurred. Goods and services are increasingly less likely to be sold and consumed separately, and are more likely to be sold as integrated solutions, systems or functions. Also, the service or the information provided alongside the supply of many goods has increased substantially the service content of many manufactured goods. This has led some scholars to amend Lancaster's (1966) definition of a product. Saviotti and Metcalfe (1984) adopt a Lancasterian approach to tackle technological innovation in manufactured goods as a result of changes in the vectors of the technical and final characteristics of a product. Gallouj and Weinstein (1997) (see also Gallouj, 2002a) have extended this characteristics-based conception to services. According to them, the product (whether a manufactured good or a service) is defined as a set of vectors of characteristics and competences: service characteristics [Y], internal technical characteristics [T] and external technical characteristics [T'][8] and internal [C] and external competences [C'] (see Figure 1.1).

The general representation in Figure 1.1 makes it possible to include in the analysis both tangible artefacts and intangible products. It can be used to define pure services ([C']—[C]—[Y]), as well as manufactured goods ([T]—[Y]) or integrated goods/services ([C']—[C]—[T]—[Y]) – or even self-service outcomes ([C']—[T]—[Y]).

Innovation emerges as a change in the (technical, service or competence) characteristics brought about by one of a number of 'operations': addition, subtraction, association, dissociation or formatting. These operations lead to different 'models' of innovation: radical, ameliorative, incremental, recombinative and formalised innovation (see Gallouj and Weinstein, 1997; Gallouj, 2002a). Radical innovation denotes the creation of a new set of characteristics and competences. Ameliorative innovation reflects an increase in the prominence (or quality) of certain characteristics, but without any change in the structure of the system of competences and characteristics. Incremental innovation denotes the addition (and possibly also the elimination or replacement) of characteristics. Recombinative innovation is a form of innovation that relies on the basic principles of

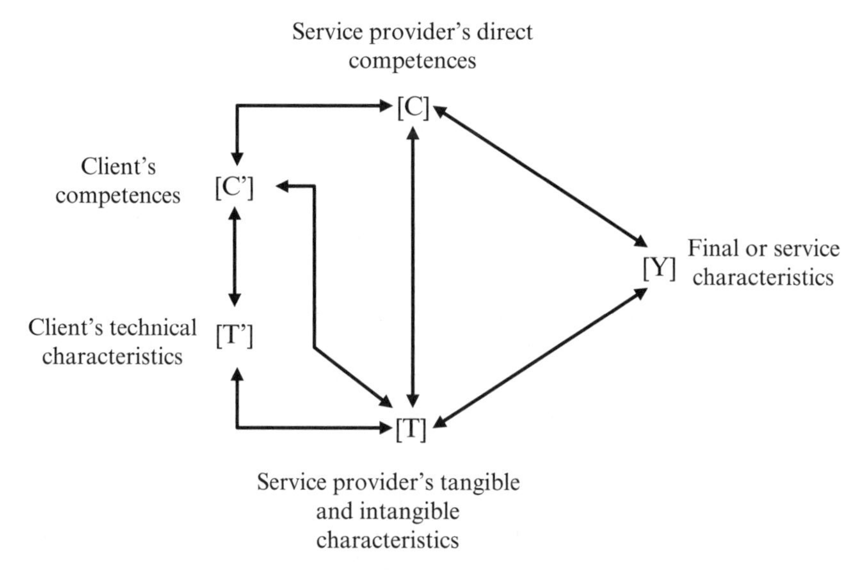

Source: adapted from Gallouj and Weinstein (1997).

Figure 1.1 The product as the conjunction of characteristics and competences

dissociation and association (i.e. the splitting or combining) of final and technical characteristics. Formalised innovation is based on the formatting and standardisation of characteristics.

The use of the characteristics-based definition of products as a theoretical device to identify different models of innovation should be ascribed to the 'synthesis' perspective of innovation, as it is able to encompass technological and non-technological, consumer- and producer-sourced innovation in components of the product, whether it be a good or a service.

The characteristics-based approach has been extended to model different cases of innovation in complex services, such as network services (De Vries, 2006) or public services where government is involved in the innovation (Windrum and Garçia-Goñi, 2008). Djellal and Gallouj (2005, 2008a) also use this model to explain, in a more operational way, innovation in 'assembled' services, that is, services resulting from the combination of a variable number of other basic services (such as hotels, retailing, and hospitals). Finally, Gallouj and Toivonen (forthcoming) add four new aspects: separate categories for process characteristics, the front office–back office division, the customer's technology and process, and the distinction between direct and indirect service characteristics.

1.3 The empirical literature

This section reviews empirical attempts to ground typological and analytical approaches to innovation in services. The first attempts to quantify innovation in services were mainly based on data sources and indicators tailored to manufacturing, including most of the contributions based on Community Innovation Survey (CIS) (first round) data. They are often seen, by default, as 'assimilation' analyses of innovation in services. Qualitative approaches to innovation in services explore the specificities of services or focus on specific case sectors and, therefore, are classified as 'demarcation approaches'. Here, we focus on the empirical literature on innovation in services, the way it tries to identify and quantify appropriate indicators of innovation in services, and how the resulting evidence has contributed to the literature on innovation.

1.3.1 Qualitative typologies and sector-specific case studies

One of the pioneering qualitative typological approaches in the domain of services is by Miles (1993, 1994), and can be considered one of the first 'demarcation' typologies. Miles tries to relate the key features of service activities – service specificity and service heterogeneity – to the traditional statistical classifications and attempts to cover the wide heterogeneity of service products and production processes by taking a broader perspective on the main ingredients of the typology, respectively sets of product-related features.

One of the typological categories proposed by Miles (Miles et al., 1995, Miles, 2002), has become the motivation for in-depth study of the knowledge base of some service activities. Miles points to the importance of what he terms knowledge-intensive business services (KIBS) and the nature of innovation carried out within this branch of the service subsectors. Den Hertog (2000, 2002) develops this category by analysing the modes of innovation specific to KIBS. Drawing on Miles et al. (1995), den Hertog (2002) defines KIBS as private service firms providing highly innovative intermediate products and services, relying on professional knowledge, and specific to a technical discipline or functional domain. In particular, KIBS innovate by co-producing technological and non-technological knowledge in close interaction with their service clients and by boosting knowledge flows across the value chain.

Gallouj (1991, 2002b) and Gadrey and Gallouj (1998) analyse the case of consultancy within KIBS and propose an innovation typology that takes account of the cognitive nature of these activities, which are defined as a machine for handling knowledge to produce knowledge.[9] They distinguish three forms of innovation: ad hoc innovation (co-construction, with the customer, of a new solution to a problem); new expertise field innovation (detecting an emergent field of knowledge and providing consultancy in this field); and formalisation innovation (implementing methods to make a service better defined).

These case studies have sparked attempts to conduct similar qualitative empirical investigation on less knowledge-intensive services, such as transport, cleaning, care of the elderly, hospitals and social services (Djellal, 2000, 2002; Djellal and Gallouj, 2006; Sundbo, 1996; Callon, 1995; Fuglsang, 2007). All these case studies, which use qualitative interviews, demonstrate the importance of non-technological forms of innovation.

Other examples supported by qualitative evidence, include complex services that combine a number of elementary services (Djellal and Gallouj, 2005, 2008a), for instance tourism. Caccomo and Solonandrasana (2001) (see also Sundbo et al., 2007) show that tourism is a mixed good; it links complex goods, defined as temporal sequences of market goods and services (transport, accommodation, catering, attractions, visits), with public goods and services (heritage and cultural sites, nature reserves, transport and signalling infrastructure, tourist offices).

These qualitative empirical studies were extended by the implementation of quantitative surveys once the particular forms of innovation were identified. These surveys, concerned with the identification and measuring of specific forms of innovation in services, have been described as autonomous surveys (Djellal and Gallouj, 1999), in contrast to the subordinated surveys used in assimilationist approaches.

1.3.2 Quantitative typologies and empirical analyses of innovation based on large-scale surveys

Use of quantitative empirical evidence on services, based on large-scale datasets, has historically encountered a variety of problems, including methodological difficulties related to output definition, measurement and data coverage (Griliches, 1992; Djellal and Gallouj, 2008b; Hipp and Grupp, 2005). Data on innovation are based mostly on research and development (R&D) and patent indicators provided by the CIS. The CIS follows the methodological guidelines provided by the Oslo Manual (OECD–Eurostat, 1997; OECD, 2005).

The service sectors were included in the CIS in the early 1990s. A detailed description of the characteristics of this survey (its structure, list of questions and response rate) is beyond the scope of this chapter. However, it constitutes a powerful tool for assessing the presence and extent of innovation activities across sectors and has enabled empirical challenges to the traditional view of innovation as linear (Kline and Rosenberg, 1986), by accounting for a wider range of innovation activities (expenditures) beyond R&D. This approach is particularly useful in the context of the service industries where R&D expenses do not represent the largest share of innovation input.

The data derived from different rounds of the CIS have produced[10] consistent evidence that R&D plays only a marginal role in services, and that patents are used infrequently by service firms to protect their innovative output from imitation (Licht and Moch, 1997; Evangelista and Savona, 1998; Evangelista, 2000; Hollenstein, 2003; Hipp et al., 2000; Tether et al., 2001). This suggests that different and more comprehensive measures need to be applied to firms' innovation activities in order to study the nature of innovation in services and its effects on economic performance in services.

The first quantitative typology, which was based on the Italian CIS (Evangelista and Savona, 1998; Evangelista, 2000), shows sectoral patterns of innovation in services. This was confirmed by cross-country studies based on succeeding rounds of the CIS (Hollenstein, 2003; Hipp et al., 2000; Tether et al., 2001; Tether, 2005; Abreu et al., 2008). These authors identified the following recurrent features:

- 'Science-based sectors', which account for a relatively small share of the whole service sector; that is, overall, few service sectors rely on R&D as the main innovation input. This evidence holds for all the countries studied in these contributions.
- A large group of sectors that rely heavily on ICT expenditures, which after the science-based sectors, are the most innovative. These sectors actively cooperate with client industries and firms positioned downstream along the value chain (retail and financial services).
- A set of poor innovators, that aim at introducing cost-cutting and rationalised hardware technologies, which involve ICTs to a small extent. These sectors include the most traditional service sectors (public services).

Hipp et al. (2000) and Tether et al. (2001) conducted interesting typological exercises based on CIS data for Germany, that focus on the nature of innovation in services according to the degree of customisation vis-à-vis standardisation of service products. Hipp et al. (2000) found that the characteristics of service products and production and delivery processes (as well as firm size), in terms of degree of standardisation, were the main variables affecting the propensity to innovate and the type of innovation introduced. Their findings confirm that taxonomic exercises aimed at mapping service sectors according to the nature of the innovation activities carried out should clearly identify which dimension of service specificity they draw on.

The latest CIS-based analyses confirm the richness and variety of innovation patterns in services and the important role played by non-R&D innovation expenditures and innovation output indicators other than patents (e.g. copyright, licences, lead-time advantages), as well as the significance of cooperation in service innovation (Abreu et al., 2008, 2009; Tether and Tajar, 2008). Service firms tend to cooperate and establish 'open modes' of innovation, and to rely on high-level skills and a particular type of human capital, that is, graduates from the humanities and soft disciplines. All these contributions highlight the role of organisational innovation as an important side-effect of the introduction of innovative services on the market.

Empirical, CIS-based analyses have proliferated since the beginning of the 1990s and, despite the limitations of large-scale innovation surveys and 'assimilation' innovation indicators highlighted by Djellal and Gallouj (1999, 2008b), provide reliable evidence of the intensity, propensity and nature of innovation in services and also its economic impact (Van Der Wiel, 2001; Cainelli et al., 2006; Crespi et al., 2006; Abreu et al., 2008, 2009, among others). Most of these contributions focus on the productivity gains from investment in ICTs by service firms (Cainelli et al., 2006; Abreu et al., 2008). They show empirically that, despite high sectoral and firm heterogeneity, in the service sector innovation plays a significant role in affecting productivity gains at firm level.

With a very few exceptions, empirical assessments of the economic impact of innovation in services are based on firm-level analyses of the returns on ICT investment in terms of productivity. Most of these attempts emanate from the new ideas generated by the concept of the 'New Economy' and the general optimism about the potential productivity gains for services from the adoption and use of ICTs. However, this means that other issues have been relatively unexplored, due mainly, in our view, to the fact that most studies apply proxies for ICT investments in services in place of R&D expenditure in manufacturing.

Most of these studies highlight that innovation might have job-displacing effects in service firms, an important issue for which there is no empirical evidence despite its significance for services and their employment growth potential. The exceptions that try to assess the impact of innovation on employment (e.g., Evangelista and Savona, 2003) are reviewed in Chapter 16 in this volume by Evangelista and Savona, and by Djellal and Gallouj (2007).

It should be noted that if empirical contributions, and the exercise of 'appreciative theorising' more generally, are to be reasonably rigorous, a compromise will be needed between the use of conceptual categories and the availability of empirical evidence. This compromise, as Griliches (1995) points out, is particularly relevant for services.

1.4 Conclusion

This chapter has presented a review of the literature on innovation in services structured around the following groupings:

1. Theoretical vs empirical contributions.
2. Typological vs analytical (whether theoretical or empirical) contributions.
3. Contributions crossing both these strands, and based on the degree of conceptual 'proximity' to the existing innovation theory, which is very much tailored to the manufacturing sector.
4. Contributions focusing on technology-productivity issues or based on a non-technological-user centred perspective.

Typological approaches are frequent in service innovation literature. Their aim is often classificatory rather than purely taxonomic. Some use 'top-down' approaches, extending existing conceptual categories to services, but not based on empirical evidence. Others are based on qualitative investigation, but suffer from fragmentation and lack of generalisability. A few studies are empirically grounded on large-scale surveys in the typologies proposed (i.e. 'bottom-up' exercises). These are mainly based on CIS data, which to date are the most exhaustive source of information on innovation activities covering services, and enable a satisfactory degree of (European) cross-country comparability. The most recent contributions based on the latest round of the CIS shed light on the many specificities of innovation in services (i.e. non-R&D innovation expenditures, importance of specialist human capital, relatively marginal use of patents and traditional intellectual property rights protection, cooperative modes of innovation, organisational innovation) and undoubtedly contribute to the development of more appropriate indicators of innovation for the whole economy, and to innovation theory more generally.

In contrast to typological studies, analytical approaches have attempted to develop and use new conceptual tools, with the aim of accounting for the specificity of service products, processes and modes of innovation. The recurring theme in these attempts relates to the importance of the customer's role in defining service output. Also, the latest developments in the Lancasterian definition of products in terms of characteristics would seem promising in terms of an encompassing theoretical framework for innovation in services and in goods. In the theory of innovation in general and in the domain of services in particular, this should lead to models that go beyond the traditional product or process innovation and take account of the interactive and cooperative process that occurs between client and service provider.

The conceptual and empirical issues that remain to be considered within the domain of services can be summarised as follows:

- Despite repeated attempts to draw attention to the role of the customers in the definition of service products and the measurement of outputs and innovation, it is still not clear, from either a conceptual or an empirical point of view, how and to what extent their role should be taken into account. The imbalance towards production and away from consumption in what shapes innovation performance applies also to analyses of the economic impact of innovation in services.
- An almost exclusive focus on productivity is the main outcome of most empirical evidence. Although the problems of productivity mismeasurements in services have been repeatedly underlined (see Djellal and Gallouj, 2008b for an updated survey), the empirical literature has not proposed a satisfactory solution, nor has it broadened its approach in order to assess the economic impact of innovation in services in terms of output and employment growth.
- Some increasingly crucial areas of analysis have been ignored or their investigation is at an embryonic stage, for example, analysis of innovation in public, social and environmental services (see Chapters 6 to 9 in this volume). These areas might require a completely new approach (yet to be developed) to assess innovation performance and its impact.

We believe that greater effort is required to systematise the complex nature of innovation in services and its economic impact (Gago and Rubalcaba, 2007). The huge significance of these issues is highlighted by the fact that much of the burgeoning literature in this domain tends to emphasise the role of ICTs as the panacea in terms of being the engine of growth and productivity in services, particularly in the so-called KIBS. Considerable empirical efforts are needed before this view can be accepted as valid, and generalised to all the service sectors and all the (advanced) countries, where three-quarters of total employment is concentrated in these activities. Some of the contributions in this volume are aimed at disentangling whether and to what extent service firms and sectors do innovate and what the main economic consequences are.

Notes

1. This was later widely adopted: see, e.g., Coombs and Miles (2000), Miles (2002, 2005), Tether (2005), Howells (2006, 2007).
2. One of the authors of this chapter proposed a theoretical or formal framework in

which the presence of innovation affects sectoral, structural changes in production and consumption and changes in the sectoral composition of the economy (see Lorentz and Savona, 2008; Ciarli et al., 2009).
3. Use of standardised European Community Innovation Surveys (CIS), the first large-scale innovation data set covering the service sectors, has in part filled this gap.
4. We do not explore this issue in depth in this chapter; other parts of this volume are devoted to impact analyses of innovation. Here, it is sufficient to note only that very few scholars have accounted for the impact of innovation on employment; see Djellal and Gallouj (2007); and Evangelista and Savona, Chapter 16 in this volume, for a reassessment.
5. For a recent reassessment of Baumol's heritage see Gallouj and Savona (2009) and Evangelista and Savona, Chapter 16 in this volume.
6. In section 1.3 we describe the attempts made to provide an empirical base for the typological exercises (i.e. 'bottom-up').
7. For a reassessment of the ill-definition and mismeasurement of service output and the ambiguity this brought to the conceptual identification of innovation in services, see Gallouj and Savona (2009).
8. The inclusion of clients' technical characteristics is suggested by De Vries (2006) to take into account the new channels of access to consumption, such as – once again – information and communication technologies.
9. See also Toivonen (2004), Bettencourt et al. (2002).
10. The production of CIS had reached its fifth round in 2006.

References

Abernathy, W.J. and J.M. Utterback (1975), 'A dynamic model of process and product innovation', *Omega*, **3** (6), 639–656.

Abernathy, W.J. and J.M. Utterback (1978), 'Patterns of industrial innovation', *Technology Review*, **80**, 41–47.

Abreu, M., V. Grinevich, M. Kitson and M. Savona (2008), 'Understanding hidden innovation: services in the UK. Indicators, empirical evidence and policy implications', NESTA, National Endowment of Science, Technology and Arts, London, May.

Abreu, M., V. Grinevich, M. Kitson and M. Savona (2009), 'Policies to enhance the "Hidden" innovation in services: evidences and lessons from the UK', *Service Industries Journal*, iFirst article, pp. 1–23 (advanced access online: DOI: 10.1080/02642060802236160).

Barras, R. (1986), 'Towards a theory of innovation in services', *Research Policy*, **15**, 161–173.

Barras, R. (1990), 'Interactive innovation in financial and business services: the vanguard of the service revolution', *Research Policy*, **19**, 215–237.

Bettencourt, L.A., A.L. Ostrom, S.W. Brown and R.J. Roundtree (2002), 'Client co-production in knowledge-intensive business services', *California Management Review*, **2**, 1–29.

Caccomo, J.-L. and B. Solonandrasana (2001), *L'innovation dans l'industrie touristique*, Paris: L'Harmattan, Collection Tourisme et Sociétés.

Cainelli, G., R. Evangelista and M. Savona (2006), 'Innovation and economic performance in services: a firm level analysis', *Cambridge Journal of Economics*, **30**, 435–458.

Callon, M. (1995), 'Pratiques et enjeux de la demarche qualité dans deux filiales de la Sodexho (Bâteaux parisiens, Chèques restaurants)', Research Report CSI, Ecole Nationale Supérieure des Mines de Paris, December.

Ciarli, T., A. Lorentz, M. Savona and M. Valente (2009), 'The effects of consumption and production structure on growth and income distribution: a micro to macro model', *Metroeconomica*, 1–39 (advanced access online: DOI: 10.1111/j.1467-999X.2009.04069.x).

Coombs, R. and I. Miles (2000), 'Innovation, measurement and services: the new problematique', in J.S. Metcalfe and I. Miles (eds), *Innovation Systems in the Service Economy: Measurement and Case Study Analysis*, Dordrecht: Kluwer Academic Publishers, pp. 85–103.

Crespi, G., C. Criscuolo, J. Haskel and D. Hawkes (2006), 'Measuring and understanding productivity in UK market services', *Oxford Review of Economic Policy*, **22** (4), 560–572.

De Vries, E. (2006), 'Innovation in services in networks of organizations and in the distribution of services', *Research Policy*, **35** (7), 1037–1051.

den Hertog, P. (2000), 'Knowledge intensive business services as co-producers of innovation', *International Journal of Innovation Management*, **4**, 91–528.

den Hertog, P. (2002), 'Co-producers of innovation: on the role of knowledge-intensive business services in innovation', in J. Gadrey and F. Gallouj (eds), *Productivity, Innovation and Knowledge in Services*, Cheltenham, UK and Northampton, MA, USA: Edward Elgar, pp. 223–255.

Desai, M. and W. Low (1987), 'Measuring the opportunity for product innovation', in Marcello De Cecco (ed.), *Changing Money: Financial Innovation in Developed Countries*, Oxford: Basil Blackwell, pp. 112–140.

Djellal, F. (2000), 'The rise of information technologies in non informational services', *Vierteljahrshefte zur wirtschaftsforschung*, **69** (4), 646–656.

Djellal, F. (2002), 'Innovation trajectories in the cleaning industry', *New Technology, Work and Employment*, **17** (2), 119–131.

Djellal, F. and F. Gallouj (1999), 'Services and the search for relevant innovation indicators: a review of national and international survey', *Science and Public Policy*, **26** (4), 218–232.

Djellal, F. and F. Gallouj (2005), 'Mapping innovation dynamics in hospitals', *Research Policy*, **34**, 817–835.

Djellal, F. and F. Gallouj (2006), 'Innovation in care services for the elderly', *Service Industries Journal*, **26** (3), 303–327.

Djellal, F. and F. Gallouj (2007), 'Innovation and employment effects in services: a review of the literature and an agenda for research', *Service Industries Journal*, **27** (3–4), 193–213.

Djellal, F. and F. Gallouj (2008a), 'A model for analysing the innovation dynamic in services: the case of assembled services', *International Journal of Services Technology and Management*, **9** (3/4), 285–304.

Djellal, F. and F. Gallouj (2008b), *Measuring and Improving Productivity in Services: Issues, Strategies and Challenges*, Cheltenham, UK and Northampton, MA, USA: Edward Elgar.

Dosi, G., C. Freeman, R. Nelson, G. Silverberg and L. Soete (1988), *Technical Change and Economic Theory*, London: Pinter.

Drejer, I. (2004), 'Identifying innovation in surveys of services: a Schumpeterian perspective', *Research Policy*, **33**, 551–562.

Evangelista, R. (2000), 'Sectoral patterns of innovation in services', *Economics of Innovation and New Technology*, **9**, 183–221.

Evangelista, R. and M. Savona (1998), 'Patterns of innovation in services: the results of the Italian Innovation Survey', VIII Annual RESER Conference, Berlin.

Evangelista, R. and M. Savona (2003), 'Innovation, employment and skills in services: firm and sectoral evidence', *Structural Change and Economic Dynamics*, **14** (4), 449–474.

Freeman, C. (1982), *The Economics of Industrial Innovation*, London: Pinter.

Freeman, C. and L. Soete (eds) (1987), *Technical Change and Full Employment*, Oxford: Basil Blackwell.

Freeman, C. and L. Soete (1997), *The Economics of Industrial Innovation*, London: Pinter.

Fuglsang, L. (2002), 'Systems of innovation in social services', in J. Sundbo and L. Fuglsang (eds), *Innovation as Strategic Reflexivity*, London, UK and New York, USA: Routledge, pp. 164–180.

Fuglsang, L. (2007), *Innovation and the Creation Process: Towards Innovation with Care*, Cheltenham, UK and Northampton, MA, USA: Edward Elgar.

Gadrey, J. and F. Gallouj (1998), 'The provider–customer interface in business and professional services', *Service Industries Journal*, **18** (2), 1–15.

Gadrey, J., F. Gallouj and O. Weinstein (1995), 'New modes of innovation: how services benefit industry', *International Journal of Service Industry Management*, **6** (3), 4–16.

Gago, D. and L. Rubalcaba (2007), 'Innovation and ICT in service firms: towards a

multidimensional approach for impact assessement', *Journal of Evolutionary Economics*, **17** (1), 25–44.
Gallouj, C. (2007), *Innover dans la grande distribution*, Bruxelles: De Boeck.
Gallouj, C. and F. Gallouj (1996), *L'innovation dans les services*, Paris: Economica.
Gallouj, F. (1991), 'Les formes de l'innovation dans le conseil', *Revue d'Economie Industrielle*, **57**, 25–45.
Gallouj, F. (1994), *Economie de l'innovation dans les services*, Paris: Editions L'Harmattan, Logiques économiques.
Gallouj, F. (1998), 'Innovating in reverse: services and the reverse product cycle', *European Journal of Innovation Management*, **1** (3), 123–138.
Gallouj, F. (2002a), *Innovation in Services: The New Wealth of Nations*, Cheltenham, UK and Northampton, MA, USA: Edward Elgar.
Gallouj, F. (2002b), 'Knowledge intensive business services: processing knowledge and producing innovation', in J. Gadrey and F. Gallouj (eds), *Productivity, Innovation and Knowledge in Services*, Cheltenham, UK and Northampton, MA, USA: Edward Elgar, pp. 256–284.
Gallouj, F. and M. Savona (2009), 'Innovation in services. a review of the debate and a research agenda', *Journal of Evolutionary Economics*, 19, 149–172.
Gallouj, F. and M. Toivonen (forthcoming), 'Elaborating the characteristics-based approach to service innovation: making the service process visible'.
Gallouj, F. and O. Weinstein (1997), 'Innovation in services', *Research Policy*, **26**, 537–556.
Griliches, Z. (1992), *Output Measurement in the Service Sector*, Chicago, IL: University of Chicago Press.
Griliches, Z. (1995), 'R&D and productivity: econometric results and measurement issues', in P. Stoneman (ed.), *Handbook of Innovation and Technological Change*, Oxford: Blackwell, pp. 52–89.
Hardouin, J.C. (1973), 'L'apparition de l'innovation financière. Contribution à l'étude de ses elements explicatifs', Complementary PhD. thesis, University of Rennes, France.
Hipp, C. and H. Grupp (2005), 'Innovation in the service sector: the demand for service-specific innovation measurement concepts and typologies', *Research Policy*, **34** (4), 517–535.
Hipp, C., B. Tether and I. Miles (2000), 'The incidence and effects of innovation in services: evidence from Germany', *International Journal of Innovation Management*, **4** (4), 417–454.
Hollander, S. (1966), 'Notes on the retail accordion', *Journal of Retailing*, **42** (2), 24–34.
Hollenstein, H. (2003), 'Innovation modes in the Swiss service sector: a cluster analysis based on firm-level data', *Research Policy*, **32**, 845–863.
Howells, J. (2000), 'Services and systems of innovation', in B. Andersen, J. Howells, R. Hull, I. Miles and J. Roberts (eds), *Knowledge and Innovation in the New Service Economy*, Cheltenham, UK and Northampton, MA, USA: Edward Elgar, pp. 215–228.
Howells, J. (2006), 'Where to from here for services innovation', paper presented at the knowledge intensive services activities (KISA) conference, Sydney, 22 March.
Howells, J. (2007), 'Services and innovation: conceptual and theoretical perspectives', in J.R. Bryson and P.W. Daniels (eds), *The Handbook of Service Industries*, Cheltenham, UK and Northampton, MA, USA: Edward Elgar, pp. 34–44.
Kline, S.J. and N. Rosenberg (1986), 'An overview on innovation', in R. Landau and N. Rosenberg (eds), *The Positive Sum Strategy: Harnessing Technology for Economic Growth*, Washington, DC: National Academy Press, pp. 275–305.
Lakshmanan, T.R. (1987), 'Technological and institutional innovation in the service sector', conference on Research and Development, Industrial Change and Economic Policy, University of Karlstad, Karlstad, Sweden, June.
Lancaster, K.J. (1966), 'A new approach to consumer theory', *Journal of Political Economy*, **14**, 133–156.
Licht, G. and D. Moch (1997), 'Innovation and information technology in services', Discussion Paper, no. 97–20, Mannheim, ZEW, Centre for European Economic Research.

Lorentz, A. and M. Savona (2008), 'Evolutionary micro-dynamics and changes in the economic structure', *Journal of Evolutionary Economics*, **18** (3–4), 389–412.
McNair, M. (1958), 'Significant trends and developments in the post war period', in D. Albert Smith (ed.) *Competitive Distribution in a Free High Level Economy and its implication for the University*, Pittsburgh, PA: University of Pittsburgh Press, pp. 1–25.
Miles, I. (1993), 'Services in the new industrial economy', *Futures*, **25**, 653–672.
Miles, I. (1994), 'Innovation in services', in M. Dodgson and R. Rothwell (eds), *Handbook of Industrial Innovation*, Aldershot, UK and Brookfield, VT, USA: Edward Elgar, pp. 243–256.
Miles, I. (2002), 'Service innovation: towards a tertiarisation of innovation studies', in J. Gadrey and F. Gallouj (eds), *Productivity, Innovation and Knowledge in Services*, Cheltenham, UK and Northampton, MA, USA: Edward Elgar, pp. 164–196.
Miles, I. (2005), 'Innovation in services', in J. Fagerberg, D. Mowery and R. Nelson (eds), *The Oxford Handbook of Innovation*, Oxford: Oxford University Press.
Miles, I., N. Kastrinos, P. Bilderbeek and P. den Hertog (1995), 'Knowledge intensive business services: their roles as users, carriers and sources of innovation', report to DG13 SPRINT-EIMS, PREST, Manchester, March.
Mills, P.K. (1986), *Managing Service Industries: Organizational Practices in a Postindustrial Economy*, Cambridge, MA: Ballinger Publishing Company.
Miozzo, M. and L. Soete (2001), 'Internationalisation of services: a technological perspective', *Technological Forcating and Social Change*, **67** (2), 159–185.
Nelson, R.R. and S.G. Winter (1982), *An Evolutionary Theory of Economic Change*, Cambridge, MA, USA and London, UK: Belknap Harvard.
Niehans, J. (1983), 'Financial innovation, multinational banking, and monetary policy', *Journal of Banking and Finance*, **7**, 537–51.
OECD (2005), *Oslo Manual: Proposed Guidelines for Collecting and Interpreting Technological Innovation Data*, 3rd edn, Paris: OECD.
OECD–Eurostat (1997), *Proposed Guidelines for Collecting and Interpreting Technological Innovation Data – Oslo Manual*, OECD: Paris.
Pavitt, K. (1984), 'Sectoral patterns of technical change: towards a taxonomy and a theory', *Research Policy*, **13**, 343–374.
Pavitt, K., M. Robson and J. Townsend (1989), 'Technological accumulation, diversification and organisation in UK companies', *Management Science*, **35**, 81–99.
Raupp de Vargas, E. and P.A. Zawislak (2007), 'A Dinâmica da Inovação em serviçios hospitalares', in R.C. Bernardes and T. Andreassi (eds), *Inovação em Serviços Intensivos em Conhecimento*, São Paulo: ed. Saraiva, pp. 483–502.
Saviotti, P.P. and J.S. Metcalfe (1984), 'A theoretical approach to the construction of technological output indicators', *Research Policy*, **13**, 141–151.
Schumpeter, J. (1934), *The Theory of Economic Development*, Cambridge, MA: Harvard University Press, (1st edn 1912).
Soete, L. and M. Miozzo (1989), 'Trade and development in services: a technological perspective', Maastricht Economic Research Institute on Innovation and Technology (MERIT), Working Paper no. 89–031.
Sundbo, J. (1994), 'Modulization of service production and a thesis of convergence between service and manufacturing organizations', *Scandinavian Journal of Management*, **10** (3), 245–266.
Sundbo, J. (1996), 'Development of the service system in a manual service firm: a case study of the Danish ISS', *Advances in Services Marketing and Management*, **5**, 169–191.
Sundbo, J. (1997), 'Management of innovation in services', *Service Industries Journal*, **3**, 432–455.
Sundbo, J., F. Orfila-Sintes and F. Soerensen (2007), 'The innovative behaviour of tourism firms: comparative studies of Denmark and Spain', *Research Policy*, **36** (1), 88–106.
Tether, B. (2005), 'Do services innovate (differently)? Insights from the European Innobarometer Survey', *Industry and Innovation*, **12**, 153–184.
Tether, B.S., C. Hipp and I. Miles (2001), 'Standardisation and particularisation in services: evidence from Germany', *Research Policy*, **30**, 1115–1138.

Tether B.S. and A. Tajar (2008), 'The organisational-cooperation mode of innovation and its prominence amongst European service firms', *Research Policy*, **37** (4), 720–739.
Toivonen, M. (2004), 'Expertise as business: long-term development and future prospects of knowledge-intensive business services', Doctoral dissertation, Helsinki University of Technology.
Van der Aa, W.T. and T. Elfring (1993), '*A typology of innovations in service firms*', conference on Managing Innovations in Services, Cardiff Business School, 5–7 April.
Van Der Wiel, H. (2001), 'Innovation and productivity in services', 2nd International Eindhoven Center for Innovation Studies (ECIS) Conference on The Future of Innovation Studies, Eindhoven, 20–23 September.
Windrum, P. and M. Garçia-Goñi (2008), 'A neo-Schumpeterian model of health services innovation', *Research Policy*, **37** (4), 649–672.

2 Innovation in services: a new paradigm and innovation model

André Barcet

2.1 Introduction

Empirically speaking, services have taken on an increasing economic role and today play an important role in the development of change; innovation in services can therefore be seen as a new factor in economics. However, this innovation appears to lack recognition in terms of economic policy; large programmes of innovation and investment are far more likely to be directed towards technology and manufacturing. Innovation in services also holds a relatively ambiguous place in economic writing, although research in the area is increasing rapidly. In particular, there is some confusion between innovation in services and innovation that takes place within a service company. Does a technological innovation created by a service company qualify as an innovation in services? Is the field of innovation in services restricted to companies that provide services? Manufacturing companies often say that they are innovating when they provide new services, which raises the question as to what innovation in services actually is. What are its features and what are the conditions for the design and success of innovations in services?

My reflections aim to show that innovations in services exist and cannot be mistaken for innovations in the service industry. However, the main point is without a doubt the fact that innovation in services implies a break with the dominant model of innovation based on an industrial paradigm. I will therefore begin by setting out my hypothesis of a shift from a paradigm qualified as industrial to a paradigm that corresponds to innovation in services, and describe the most significant features of this 'service-based'[1] paradigm of innovation. This paradigm shift must be analysed as a significant change in the forces at work in our economies. Innovations in services must be produced explicitly, which leads me to present a four-layered conceptual model of innovation in services to enable identification of the central issues to be resolved when looking to design an innovation. This conceptual model notably leads me to specify the service commitments set out by the service producer. Finally, the new paradigm enables me to demonstrate that there are conditions, specific to services, which must be fulfilled to ensure the success and durability of the innovation.

Some conditions relate to the innovating organization while others are related to the macrosociological system as a whole.

2.2 Innovation in services: paradigm change?

Like economic policy, economic analysis has historically been dominated by an 'industrial' approach to innovation and the innovation process, which leads to a dominant paradigm for innovation that Howells (2000a) qualified as based on the manufacturing sector. Innovation in services has been looked at only recently and often in a very incomplete manner. We must understand current changes in society and then analyse the consequences in terms of innovation 'design'. The paradigm change requires us to identify what innovation in services consists of and then set out its main features to account for the current transformation.

2.2.1 Innovation in services and the industrial paradigm

At least since Schumpeter, innovation has been considered an essential feature of long-term economic dynamics. Although Schumpeter's framework[2] (1939) was wider in scope and more complex (Fagerberg, 2003), dominant thinking has long seen innovation as essentially led by the manufacturing industry and produced by technological evolution. Technology is at the root of innovation in both processes and products and the degree of technological innovation is ultimately the central argument in defining the degree of innovation in products or procedures. Radical innovation therefore implies a technological breakthrough and the main issue is to manage technological trajectories (Dosi, 1982; Pavitt, 1984). As a result, innovation indicators are mainly based on research and development expenditure and patents submitted.

This view leaves services as the poor relation of innovation. On the one hand, services can only innovate when their providers bring in new technologies, which leads us to Barras's reverse cycle model (Barras, 1986). In these conditions, information and communication technologies play a central role and are the precondition for innovation in several service industries (telecommunications, banking, insurance, retailing, transport etc.). On the other hand, where innovation in services does exist, it is always considered as incremental innovation and therefore has very little influence on economic dynamics. In addition, it is easy to imitate and cannot be patented.

In the 1990s, research into innovation in services made significant progress, most notably in attempting to identify types of service innovations or the vectors on which innovations in services are based (see the work of Gallouj, 2002; Barcet, 1996; Gallouj and Weinstein, 1997; Sundbo, 1997; Howells, 2000b; Miles and Metcalfe, 2000; Djellal and Gallouj,

2002; Miles, 2006). This body of work produced very different results that are sometimes contradictory. While service companies are now recognized as innovators, and are, according to some, the main place where current innovation takes place[3] (OECD, 2007), the different approaches give rise to the question: 'What is innovation in services?'

2.2.2 Innovation in services: what does it mean?

I believe that understanding the challenges related to innovation in services requires a precise vision of what precisely is meant by the term in order to identify where significant economic changes are currently developing. Opinions are therefore justifiably very different depending on the issue addressed by each study.

It is important to distinguish the notion of innovation in companies that provide services from that of innovation in services. Companies that provide services, like manufacturing companies, have a certain view of the innovative process and this affects the skills enlisted, the internal organization of the different activities and functions, the techniques and technologies used and the result of the process (the output) that is sold or provided to an economic actor. The strong trend towards industrializing service activities, particularly the relatively significant development of back-office in comparison with front-office activities, undoubtedly plays a role in the convergence between manufacturing companies and service providers. This reduces the distinction between manufacturing companies and companies providing services when it comes to identifying their internal innovation processes. Neither would making a distinction between market and non-market companies be particularly relevant when it comes to analysing the various innovation processes at work within each organization.

Therefore, my research does not involve focusing attention on the different features that innovation may take in different organizations. Rather, I will define an innovation in services as an innovation with regard to the result obtained by the client (group of clients) or user (group of users), as the result introduces something new into the way of life, organization, timing and placement of what can generally be described as the individual and collective processes that relate to consumers. This innovation in services, which I call 'service-based' innovation, is not restricted to companies working in what is commonly known as the service industry. It potentially concerns all companies, regardless of whether they have historically been classified as belonging to the primary, secondary or tertiary sector. 'Service-based' innovation is therefore a specific, identifiable innovation. If there is a paradigm shift, as Howells's analysis (2000a) suggests, it is to this dimension that it relates.

Neither does the change regard what could be described as the degree of innovation, particularly on knowing whether the innovation will be radical or merely incremental or intermediate. The new paradigm is first and foremost a change in perspective, which to a certain degree implies a new way of looking at the roles of the actors and mechanisms that create value, followed by a change in the identification of results and finally of quantity. This new paradigm can, of course, be seen as a new trend when compared to the old paradigm.

2.2.3 'Service-based' innovation

This kind of innovation contains a variety of complementary dimensions. The first regards the mechanism of creating value and the need to return to the historical concept of use value as proposed by classical theorists (Smith and Ricardo), or the concept of 'utilization value' proposed by Giarini (1978). The idea of use value differs from the idea of usefulness value developed in neoclassical theory in that usefulness is the result of an individual and psychological preference that a consumer has for a product, whereas use value goes back to the idea that the process of utilizing or using a product creates value for the person using it. Just as there is a production process for a product that is carried out by a company, there is an utilization process carried out by a 'user' or group of users. This is the process that creates the value. The client, or to be more precise the user, is thus an actor in creating this value. Therefore, the process of utilization reveals some of the positive or negative effects produced by the product or service offered, or a combination of product and service (Barcet and Bonamy, 1998, 1999). These effects are never one-dimensional. Neither can they be fully individualized and appropriated by a single person; they contain collective and social dimensions (Gadrey and Zarifian, 2002).

The second complementary dimension regards the analysis of the utilization process itself. This process implies an analysis of the act of consumption as an economic process in its own right, whether carried out by a single person, a family, a group of people or an organization. The process thus takes place over a period of time and in a space; it involves actors, implies monetary and non-monetary costs, and produces results. Being a process in which the actor brings about a result provided by the supplier, it is necessarily individualized. Taking this process of utilization into account therefore results in a different perspective from that held by the producer. The producer attempts to obtain a result (the production output), which becomes a resource in the utilization process. In that sense an innovation in services is therefore an innovation on the effects obtained by the utilization(s). This requires the supplier to identify with the process that the result of their production will become a part of.

The third dimension relates to a general change in the economy, qualified as a function or functionality-related economy (Barcet and Bonamy, 1999; Zacklad, 2005; Buclet, 2005). This regards the substitution of a supply of skills and service flows for what used to be a product. In other words, by dealing with the client or user's problem in a new and more precise manner, the supplier tries to provide the client with a solution. There can be no doubt that this is the service sector, as it is possible to define the service as 'a solution to an identified problem'. The problem is obviously on the client's side with all the features of the solution provided by the supplier and being, to different degrees and in a variety of different ways, implemented by and for the client. This is not exclusive to companies classified as belonging to the service industry. It potentially concerns all companies but presumes that rather than the physical product produced by the company, the central dimension regards the totality of the effects produced by the solution on offer. As Zacklad (2005) noted, this is particularly relevant to environmental issues, which are continually increasing in significance.

The three dimensions that we have mentioned are different views of the same reality. Paradigm change mainly means an increased focus on the problem, not in terms of the immediate result of the provision and each service provider, but on the future effects of the implementation of the result. Designing and creating an innovation in services therefore means defining and creating an effect-producing process and new effects that respond to identified constraints or problems.

One of the consequences of this new approach regards measurement and evaluation, as evaluation is no longer limited to the product as a result of its production, but regards the product in its capacity to enable the production of positive and usually multidimensional effects from the point of view of the client.

2.2.4 The forms taken by 'service-based' innovations

Identifying this issue implies defining the precise ways in which the service-based innovation will be implemented. I would suggest that this kind of service-based innovation can be implemented on two different levels.

The first is easier to identify. I have qualified it as the goods and services integration process (Barcet and Bonamy, 1999). It has also been analysed using the concept of 'encapsulation' (notably by Howells, 2004). The observation is relatively simple: the supplier of the good or technology surrounds the product with a range of services that enable the purchase to be financed, maintained and updated; to create and transmit knowledge or learning processes, evaluate the results and then upgrade the entire product. The innovation as a whole therefore offers a solution that is adapted to a specific problem and/or a particular type of client.

This framework is particularly significant now, most notably in the link between information and communication technologies (ICTs) and associated remote services (Debonneuill, 2007).

The second level corresponds to a situation in which the technology or product does not play a central role, which could be qualified as 'pure service'. This situation can be observed in service innovations that are not based on high levels of technology or knowledge. Innovations linked to personal services, innovations with regard to looking after patients at home, innovations in urban mobility management (such as 'Vélo'v' in Lyons), in personnel recruitment mechanisms, in facilities management and innovations in the implementation of economic tourism projects by museums or tourist destinations, are all good examples of this kind of innovation. These situations require the precise identification of the effects to be obtained and the implementation of the processes necessary to carry out activities that will produce the positive effects.

This does not mean that the issue does not concern technologically or knowledge-intensive services (KIBS[4]), but that the technological dimension in these cases often tends to hide the service-based dimension.

To conclude, the paradigm change linked to current economic developments implies that relationships with clients (users) and their consumption processes are becoming central to our concept of innovation. This requires me to highlight a particular way of thinking about innovation, which I will call a layered model of innovation.

2.3 A layered model of innovation

The layered model of innovation is designed in a similar manner to the networks and communication systems' design model. The seven layers of an OSI[5] system are organized into two subsets: one, made up of four layers, corresponds to the utilization level; the other, with three layers, to the applications. What is particularly interesting in the idea of layering is that each layer corresponds to a particular issue, but the whole can only exist and be effective if all the layers are linked and form a coherent whole.

The framework in Figure 2.1 therefore presents a model of innovation[6] in services made up of four distinct but interacting layers. Each layer enables the identification of relevant questions as well as those actors (whether individual or collective) which the questions particularly concern.

The first layer leads to a definition of the use and usefulness that the service as a whole is likely to provide a client or group of clients. The essential questions regard the effects of the service, the sustainability of its effects, additional costs linked to the utilization process, the questions of learning while using and the question of the client being able

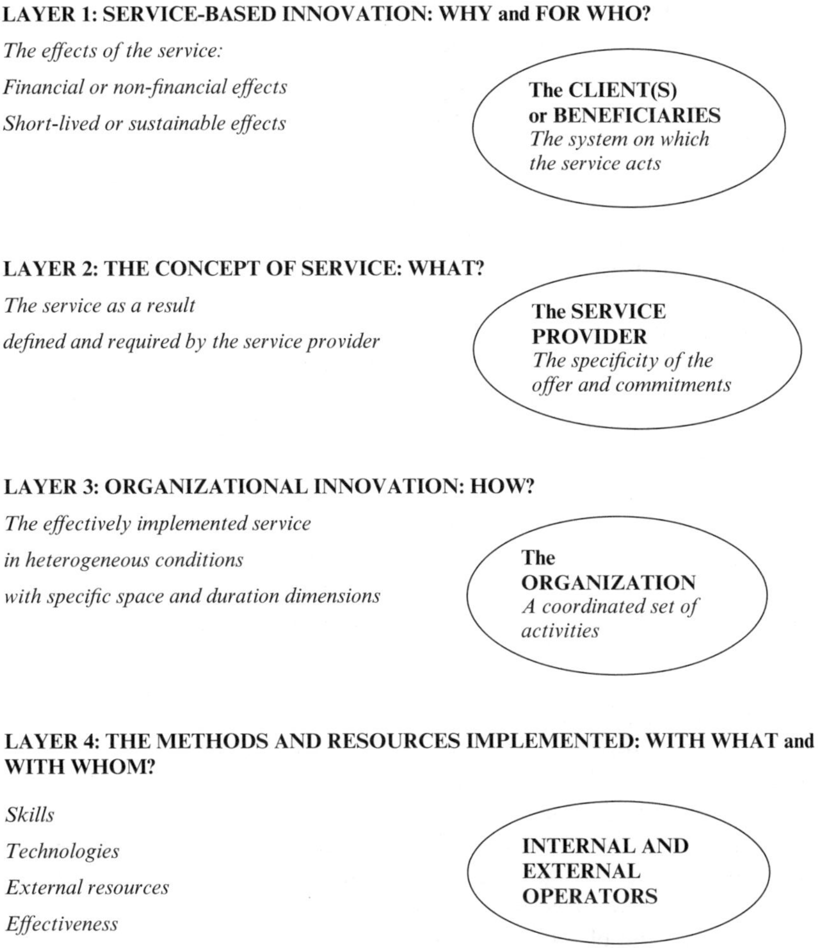

LAYER 1: SERVICE-BASED INNOVATION: WHY and FOR WHO?

The effects of the service:
Financial or non-financial effects
Short-lived or sustainable effects

The CLIENT(S) or BENEFICIARIES
The system on which the service acts

LAYER 2: THE CONCEPT OF SERVICE: WHAT?

The service as a result
defined and required by the service provider

The SERVICE PROVIDER
The specificity of the offer and commitments

LAYER 3: ORGANIZATIONAL INNOVATION: HOW?

The effectively implemented service
in heterogeneous conditions
with specific space and duration dimensions

The ORGANIZATION
A coordinated set of activities

LAYER 4: THE METHODS AND RESOURCES IMPLEMENTED: WITH WHAT and WITH WHOM?

Skills
Technologies
External resources
Effectiveness

INTERNAL AND EXTERNAL OPERATORS

Figure 2.1 The layers of innovation in services

to compare effects. Methodologies to look into this aspect are relatively undeveloped, because it requires anticipation of all the effects of the service and the necessary client or beneficiary[7] behaviours. At the same time as defining 'why', one must also set out 'for whom'; i.e. specify the clients or beneficiaries for whom the service is intended. An important aspect in the case of services is the need to create a certain 'construction' of the client; i.e. to define the actions and behaviours necessary to produce the desired effects, design the learning process to achieve the required concept and define the process of generating and harnessing any information held by the client in order for the informative resources to be available as part of the service supply system. This first layer could equally be analysed as the perception of innovation opportunities, potentialities or client expectations, whether explicit or otherwise. These expectations are perceived as unresolved or poorly resolved issues, such as constraints or poorly formulated desires. At this level, the essential questions are 'Why?' and 'For who?'.

The second layer is a definition of the product on offer; i.e. the concept that the provider wishes to present to the market. Innovation is always a response from a provider. The response is created by the provider who must present it to and, to a certain extent, impose it upon clients as a possible response to problems, constraints or desires that may be more or less well-defined. The response is always one among many; there is rarely only one possible solution. The provider's set-up will dictate the precise form that the response takes and, if successful, will ensure that the innovation has a certain durability. At a design level, the provider must also ensure that their concept is clearly differentiated from other existing concepts. In other words, the questions that are raised at this level regard: the identification of the offer, the offer's positioning in relation to competing offers and the definition of the main features of the service, particularly those that make it original. The concept is therefore a summary of the potentialities that emerge from the clients' system and the opportunities to differentiate the offer from its competitors.

One of the main questions with regard to defining the concept of 'offer' regards the level of precision and nature of the commitments that the service provider takes on with regard to his or her clients. The specificity and precision of each commitment have significant consequences on the confidence to be built up, the judgement of the quality of the service offered and the comparison with competing offers. This central question is sometimes completely neglected or incomplete, which can slow the development of service activities.

Another dimension is to specify the values on which the service offer is based. Like goods, services carry psychological, symbolic, artistic and social values. When designing an innovation it is essential to have a certain coherence between what is promised and the final product; between the values advertised by the proposer and the service as a whole. The price of the service must also be specified at this level, for a transaction to take place. At this level, the essential questions regard 'what to offer'.

The third layer responds to the question of designing a general system of service provision. This level requires a specification of the steps and phases necessary for the service's implementation. The challenge is to organize the different activities required for the service to be implemented.

This idea of the process must integrate the heterogeneous dimensions of the conditions for implementation of the service, knowing that a feature of many services is that they are never implemented in controlled places or environments. The provision of the service is therefore frequently subject to risks due to variations in locations. It is also linked to the individual and collective behaviour of the actors (particularly of clients), which can never be fully controlled. The essential questions therefore regard 'how', at the organizational level. To a large extent, designing the process in its entirety also implies designing the ways in which the organization's operators will coordinate matters among themselves.

In this layer, the question of 'making the service available' raises itself; i.e. the explicit forms taken by the connection between service providers and interested clients. It is here that the client's roles will be defined in terms of the information, resources and actions that must be accomplished. The methodology, procedure and standards dimensions are particularly important parts of this layer. In general, this layer can be analysed as 'organizational technology', which implies that every organizational technology corresponds to a different service.

The fourth layer specifies the necessary means and resources. Every activity requires resources, which have a variety of dimensions: financial, technological or information, skill or knowledge-based. These resources are obtained internally and/or externally. In service activities, information, skill or knowledge-based resources may be provided by the client. In some cases, the client becomes the co-producer of the solution, which can lead to intellectual property issues. Many services require very varied skills and therefore the involvement of partners, each of which will contribute. Specifying the skills of partners must take place at this level. There are multiple skills involved; the personal relationships dimension of services gives particular importance to interpersonal skills and 'knowing how to be' alongside 'know-how'. Also at this level, the historical development of certain technologies can be seen as the condition that enables certain innovations to appear. At this level information technologies provide new resources that enable the activation of new services. This layer is therefore dominated by the questions 'With whom?' and 'With what?'.

The design of a new innovation in services must specify the choices made at each level. The answers that the innovation's designer gives to the questions at each of the four levels are neither linear nor independent. The process is one of constant interaction. The impossibility of responding to certain questions in some of the layers in a satisfactory manner could obviously lead to a re-examination of all the choices made and a redefinition of the concept. Linking the layers therefore appears to be a necessary condition of an effective innovation design.

My observations, as well as my monitoring of concrete service innovation processes, appear to show that the second layer is the most delicate point for service companies. In other words, there may well be two aberrations in a process qualified as an innovation in services by its designers.

The first is an anticipation of the potentialities provided by a technology and the assumption that the implementation and control of the technology is sufficient to define the concept of service without worrying too much about the commitments that must be taken vis-à-vis the clients, the values that the service should achieve and the actions or questions that the client must resolve. This aberration, which can be qualified as technological, may lead to resounding failures, as one has effectively confused the means with the service to be provided.

The second aberration concerns the innovation designers' anticipation of a need, expectation or opportunity for clients and their way of life and consumption. The designers then rapidly create the service to be offered, without specifying the precise nature of the commitments taken and the values that underpin their service. As a result, the intention of the service may be commendable or justified, but the imprecision or vagueness of the service concept is a significant handicap in building trust.

The second layer, which regards setting out a precise concept, appears to be the crux of innovation in services. By specifying their commitments and defining the values on which the service is based, the designers of the innovation provide themselves with the means to build confidence, because clients can check that providers are fulfilling their promises. Any offer of a service begins as potential and a promise. Trust and loyalty-building emerge from the effective implementation of this promise. Continuing demand is not the only dimension that makes innovations in services sustainable.

2.4 The conditions for an 'effective' innovation in services

Examining the way the new innovation in services paradigm works enables the identification of a number of conditions that must be fulfilled to ensure the effectiveness of an innovation. My brief analysis will distinguish between conditions that are essentially microeconomic and those that are more systemic or macrosociological.

2.4.1 Microeconomic conditions

The main conditions for a successful innovation regard the importance of organization, the role of production and the way knowledge circulates in the coordination process, and the implementation of skills, interpersonal as well as technical, in the service offer.

BOX 2.1 WHAT IS VÉLO'V?

Vélo'v is a short-term bicycle rental service that aims to boost people's mobility in a given urban area. The Vélo'v innovation was piloted in the town of Lyons through a partnership between Greater Lyons council and a private operator. The objective was to persuade people not to use their cars to travel within the town and to ensure a smooth and easy transition between two points or two means of transport for a relatively cheap price. The success of the experiment has led it to be repeated in other towns.

The organization The layered model described above gives great importance to the question of organizing the various workflows and activities to ensure that the promised service is effective and recognized as such. A service's primary technology is therefore organizational, with the result that an innovation in services can also be defined as an architectural innovation (Henderson and Clark, 1990). The elements used in the innovation do not themselves present any innovative aspect; the innovation lies in the way in which the elements combine and the organization of workflows, resources and skills. The service provision should set out specific links between the different actors and resources used. It is these links that define the innovation. Vélo'v provides a perfect example (Box 2.1).

Typically, a new service is offered. The effectiveness of the innovation depends far more on the organizational framework than the technologies used. Bicycles are not high-tech but were reconfigured for the particular solution. Information and communication technology (ICT) plays an important role in managing information and payments, but the essential issue regards the organization of transfers (transporting bicycles from points where they are present in large quantities to points where they have run out or are low in number, whenever they may be required and particularly at peak times), and maintaining the bicycles and storage points. The challenge is to create a convenient and reliable service that can provide users with a bicycle wherever and whenever they need it. The effectiveness of the innovation is based on the probability of a bicycle being available at the exact time and location that the client needs it.

Coordination and the role of producing and distributing knowledge Producing the service implies coordination between different activities and different actors. This coordination is, of course, partially planned and organized in advance as part of the design of the entire service in the third layer of the framework above. It translates into an organizational timetable for tasks and activities and a definition of the procedures to be carried out to obtain the final result. However, this capacity for anticipating the way the process will develop is often not sufficient to control the provision in its entirety, because the fact that the actual situations in which the service is provided are different makes provision more complicated. Whereas in an industrialized service process, particularly in the back-office part of the service, it is possible to create a standard and stable process, in numerous service situations part if not all of the provision takes place in environments that are not under the service provider's control. Actors must therefore continually adapt themselves to changing circumstances. In certain kinds of service, providing the service always requires an adaptation of the offer to the specific requirements and individual characteristics of the clients, as well as to the interclient relationships present at the time. Taken as a whole, these features make it very difficult to anticipate and control situations. Finally, many service processes involve different actors and professions, who need to interact to enable the service to be produced. In these conditions, real-time reactive coordination becomes essential. This active and permanent coordination function is the key to effective service production.

In order for the coordination function to operate correctly it needs to define a process for the real-time production and distribution of information. This implies systematically setting up information 'sensors', which can simply consist of the actors providing the service, but can also be people whose main function is information gathering. It is then necessary to implement a framework to provide this information to the coordination function and then, pursuant to decisions taken, which will depend on the questions to be resolved, to activate other actors and systems to resolve any issues that may arise.

Information and communication technologies have a central and specific role to play in producing services, which is different from technology's role in producing goods in the sense that they do not actually produce the service, but are diagnostic tools to assist with decision-making. There is no process by which human labour is substituted for technology, as is the case in some industries. One of the consequences of these coordination issues regards the role collective work and the creation of working communities play in many service innovations (Barcet et al., 2004). The example of keeping patients at home[8] is a particularly revealing service innovation with regard to this question (Box 2.2).

An innovation of this nature requires the coordinated intervention of an entire range of professions (doctors, nurses, laboratory analysis

BOX 2.2 KEEPING PATIENTS AT HOME: WHAT DOES IT INVOLVE?

Keeping patients with long-term illnesses at home is a service innovation that has been tested during recent years. The idea is to enable a patient to live at home while receiving high-quality treatment under the permanent supervision of a doctor at the hospital responsible for the medical and surgical treatment, with the possibility of being readmitted to hospital if it should prove necessary. This also implies that the family environment is one in which all the financial, organizational and technical requirements linked to keeping the patient at home can be met. This innovation has two objectives.

The first regards hospital management (patients receiving long-term in-hospital treatment occupy beds; given the relative lack of beds, keeping patients at home enables beds to be freed up and the flow of patients to be managed under improved conditions).

The second objective regards the patients themselves (effective treatment at home means more favourable psychological and emotional conditions for the patient's well-being).

technicians, physiotherapists, people who hire medical equipment, data processors, equipment maintenance, meal delivery, psychological assistance staff or association and support staff – notably as occasional replacements for family members). Some of these interventions take place on a regular basis and can be timetabled, while others are provided on an ad hoc basis if required. Others will only be called upon in an emergency situation. In addition, the intervention of some actors assumes the transfer of information between contributors: information on the patient's state of health, on how the illness is evolving, information on the actions carried out by each contributor and the consequences of these actions on the future actions of other contributors, and information on the functioning of technological equipment at the patient's home. Finally, patients kept at home are subject to permanent monitoring from the hospital where they receive their treatment, which enables decisions on modified treatments or an eventual rehospitalization to be taken.

For this innovation to be a success a coordination 'unit' must be set up; a hub for information and decision-making that enables each actor and service flow to be activated in order to achieve the objective. An unexpected consequence of this innovation is that the information distributed to each actor creates new conditions for learning. This is because the transfer of knowledge between actors from different professions modifies the way each person acts, making them more likely to incorporate the consequences of their actions on other professions and therefore adapt their own practice. This creates 'real communities' that significantly affect the effectiveness of the whole, as well as that of each individual. Similar examples can be observed in several service innovations, every time the service in question requires actors and workflows to be coordinated in heterogeneous and fluctuating situations.

Service relationship skills Designing a service innovation requires the identification of the specific skills necessary to provide the service. Some of these skills are technical in nature, but the specific dimension linked to the 'service-based' paradigm above regards skills in the area of service relations (De Bandt and Gadrey, 1994). As this specificity is highlighted by many writers, a brief explanation will suffice here. It implies particular staff qualities, learning processes and attitudes. The service relation does not only consist of psychological qualities and an appropriate attitude; it also requires a certain empathy between service staff and clients. It also implies a capacity to listen to, analyse and understand concrete situations, features that are specific to every client and group of clients and implicit and explicit expectations, in order to be able to react or adapt the service to individual situations. The development of personal services has often demonstrated the difficulties that innovators in services face in selecting staff with these qualities and training them. One solution – which is, however, time-consuming – is to provide the employees concerned with opportunities to interact with regard to their practice and client relationships. This enables them to understand client behaviour, particularly when clients may appear to be criticizing the work of the operator or when acts of aggression occur.

2.4.2 *Macrosociological or systemic conditions for innovation in services*

Innovation in services as defined above can obviously take place at different levels. The innovation can be uncommon, isolated or limited to a given problem. However, most innovations in services form part of a series of changes taking place in society. These changes do not merely regard one specific kind of actor or one type of problem. Rather, they concern the way people live their lives, the way companies and organizations function and the search for an improvement in all-round relationships between people, institutions and places where people work and socialize. In other words, an innovation in services very often has significant macrosociological effects, as it affects the environment of every actor concerned. There are sociological conditions that affect an innovation's success and, at the same time, the innovation participates in transforming the system into which it fits. Designing and implementing an innovation in services implies analysing, understanding and becoming involved in the client's system as a whole. The intervention does not only affect the object of the transformation (a physical object, a group of objects and techniques, some information, a physical person, a group of people, a company or an organization), but also all the relationships built up between people, objects and techniques, which means that the system exists and has its own dynamic. This tends to lead to a systemic vision. One initial consequence of this situation regards the fact that the effects of the innovation are not one-dimensional; they are not only present for the innovation's direct beneficiary or client. They cannot merely be appropriated in an exclusive manner; the effects are also collective. In other words, a variety of purposes are at work, which are private, collective and territorial, as shown by Du Tertre with regard to the development of personal services (Du Tertre, 1999). The development of an innovation in services therefore only fully makes sense as part of this overall or systemic vision of reality. As a result, the effectiveness and sustainability of an innovation assume the presence of macrosociological conditions. We will briefly look at three sets of conditions: conditions under institutional control, conditions regarding the links between different services to obtain an 'overall service', and conditions of appropriation and distribution.

Conditions under institutional control Some service innovations may take on their full scope if significant changes, particularly regulatory changes, take place in society. This issue can be illustrated using an important and recurring theme in today's society: the question of time, particularly free time. One of the challenges that current service innovations face is that of enabling each individual, social group or organization to manage their time better, with the aim of overcoming certain time constraints, freeing up segments of time to develop new activities or enriching their free time. Industrial society has tended to allocate each segment of time to a specific activity, whether part of a day, a year or even a whole lifetime. Historically, this allocation produced relatively rigid organizational frameworks. While some service activities can enable time constraints to be overcome, society as a whole also needs to reconsider and reorganize the way it thinks about time, particularly as time that is freed up for some produces constraints for others. It is therefore necessary for the constraint to be shared so that it is not always the same actors who obtain an extra 'gain' or 'constraint'.

During the current phase, the issue of the link and balance between working time and free time is clearly expressed. The difficulty in resolving the issue can be considered as slowing innovation in services. This kind of change comes up against legal and regulatory constraints on working hours and schedules on the one hand and the varying organization of different kinds of time on the other. An innovation consisting of opening a crèche 24 hours a day, seven days a week would require specific regulations adapted to the situation in question. If some employees produce services while others consume, a variety of services (e.g. public, leisure, transport, health and education services) would need to be open during the former's free time as well.

Conditions regarding connections between services In some service innovations, a single service can only partially fulfil its role and will therefore only reach an incomplete market. This is because the attractiveness of the service depends partly on the functionalities that the service provides, but also on the presence and functionalities of other complementary services, sometimes directly associated and sometimes merely favourable to the former. As a result, at the social system level there is a need to design a 'global service', or at least a basket of services enabling each service to fulfil its role and have positive effects on the development of other services. In other words, an innovation in services will often have more positive effects if it has been designed and organized in a networked economy.

Several examples illustrate these issues. The example of 'competitive clusters', associating numerous actors and companies, is particularly relevant. In the tourism sector, the value of a ski resort is commensurate not only with the quality and size of its ski slopes, but also with the totality of services it can provide: reception, accommodation, food, leisure and other sporting activities. Some resorts have realized that making several dimensions fit together is essential and that clients are increasingly more likely to judge the service from an overall perspective. This obviously requires an actor (and sometimes several), who can conceptualize and implement this overall service. Institutional control certainly has a role to play in ensuring services' effectiveness and interconnectedness.

The conditions for distribution and appropriation One of the recurring issues with regard to innovations in services is the fact that they can very rarely be patented; often they can be copyrighted at most. The direct result is that innovations in services can be copied very easily, with companies that innovate in services being less protected than those innovating in technologies or goods and not having their income guaranteed over a certain period of time in order to recover the resources invested in designing and implementing the innovation. This in turn means that there are few incentives to innovate in services, which could explain why innovators are more inclined to innovate in technologies even if they believe the result may be a service innovation. The protection on the enabling technology provides a kind of indirect protection on the service innovation.

Nevertheless, the service sector does have certain ways of managing the distribution of innovations and the conditions under which innovations are appropriated. The role of brand names, which emerged from the world of industry, is one. Brand names have a significant intangible dimension that can be applied to services. The development of franchises is another tool that is relatively well adapted to services. Nowadays, the precise definition of a service concept, the affirmation of an intellectual property right with regard to the concept and the company's respect for its commitments are also important conditions that protect a company. A company that is capable of setting out and fulfilling specific commitments vis-à-vis its clients has an advantage that can often be decisive. It is by consistently improving the process and taking account of new client constraints and expectations that a service company can maintain its competitive advantage. A constantly renewed innovation is the standard by which an advantage maintained over a certain period can be measured. The innovation thus becomes a regular process. In other words, it is by paying attention to and constantly analysing developments in its client system(s) that a company can renew its innovative potential.

Finally, while it is easy to copy an innovation in services, this is not necessarily a complete disadvantage for service companies. Just as an isolated service innovation lacks effectiveness, when trying to affect an entire social system the presence of several companies that innovate in close proximity can be an effective tool for social transformation and therefore a condition for the success of a specific innovation.

2.5 Conclusion

My reflections have led me to propose the existence of specific characteristics related to innovation in services whose objective is to integrate new effects into the solutions implemented as part of clients' or beneficiaries' processes. The levels at which they do this may vary. This appears to me to imply a new paradigm in the way innovations are designed. Innovation in services, which I have qualified as 'service-based' to distinguish it from technological or product innovation, corresponds to a change in the relationship between supply and demand. In this sense, designing a new offer implies both anticipating client questions and constraints, and anticipating the active role that the client will play in a process that will use the resources gained from the provision of the service to produce the positive effects required. As the innovation regards the individual and collective consumption process it is always, to differing forms and degrees, multidimensional. It produces both private and collective effects at one and the same time, participating in the evolution of ways of life, the ways in which actors relate to each other and therefore the formation of a social system.

This new paradigm may appear either to oppose or to be in direct confrontation with the preceding industrial paradigm, as the old paradigm continues to exercise its influence in several sectors and on several innovation designers. However, it would be more correct to see the new paradigm as integrating and going one step further than its predecessor. In this sense, the 'service-based' model of innovation that we have presented will also have to define the immediate result that the service provider must consider and select the technologies necessary to provide the service in question.

Notes

1. The concept of 'service' is still a little problematic; we are forced to create terminology in order to distinguish between a service that produces effects for the client (dimension that we qualify as 'service-based'), a service as the result of the act of providing (the supply dimension) and services as activities or organizations.
2. It should be remembered that Schumpeter had a very wide view of innovation. He identified five types: product innovation, procedure innovation, market innovation, innovation in a new input and organizational innovation.
3. The OECD study (2007) notably demonstrates that the average intensity of innovation in European companies is particularly high in the business service sector.
4. KIBS stands for knowledge-intensive business services.
5. OSI stands for Open System Interconnexion, a reference model developed as part of the ISO standardization framework (Zimmermann, 1980).
6. This model was first presented in a summary of a collective study on innovation in services (Barcet and Bonamy, 1999). It was used as a basis for structuring service innovation management work (Flipo, 2001), and could also be used as part of an operational consulting approach to innovation in services.
7. In a service relationship the person receiving the effects of the services is not always the one who pays, which leads to a distinction between the client (with a financial dimension) and the beneficiary.
8. This innovation was described in Barcet et al. (2003).

References

Barcet, A. (1996), *Fondements Culturels et Organisationnels de l'Innovation dans les Services*, Research Report for French Ministry of Higher Education and Research, Cedes-CNRS, Lyon, May.

Barcet, A. and J. Bonamy (1998), 'La Valeur d'Utilisation au Coeur de l'Intégration des Biens et des Services', *INSEE Méthodes*, **87–88** (21), 41–66.

Barcet, A. and J. Bonamy (1999), 'Eléments pour une Théorie de l'Intégration Bien et Service', *Economies et Sociétés*, *EGS*, n°1, 197–220.

Barcet, A., J. Bonamy, L. Esnault, Y. Lespagnol and R. Zelliger (2004), 'Communautés de pratiques, management de la connaissance et production de service', XIVth International Conference of RESER, Information Communication Technologies and Service Relationship in the Global Economy: A Challenge for Enterprises and Societies in Europe.

Barcet, A., J. Bonamy and M. Grosjean (2003), 'Une Innovation de Service par la Mise en Réseau de Service', *Economies et Sociétés, EGS*, n°5, 1897–1916.

Barras, R. (1986), 'Towards a theory of innovation in services', *Research Policy*, **15**, 161–173.

Buclet, N. (2005), 'Vendre l'Usage d'un Bien plutôt que le Bien lui-même: Une piste pour concilier meilleure prise en compte de l'environnement et rentabilité des entreprises', XVe Conférence Internationale du RESER, Granada, Spain, 22–24 September.
De Bandt, J. and J. Gadrey (eds) (1994), *Relations de Service, Marchés de services*, Paris: CNRS Editions, Collection Recherche et entreprise.
Debonneuil, M. (2007), *L'Espoir Economique*, Paris: Bourin Editeur.
Djellal, F. and F. Gallouj (eds) (2002), *Nouvelle Economie des Services et Innovation*, Paris: L'Harmattan.
Dosi, G. (1982), 'Technological paradigms and technological trajectories', *Research Policy*, **11** (3), 147–162.
Fagerberg, J. (2003), 'Schumpeter and the revival of evolutionary economics: an appraisal of literature', *Journal of Evolutionary Economics*, **13** (2), 125–159.
Flipo, J.P. (2001), *L'Innovation dans les Activités de Service*, Paris: Editions d'Organisation.
Gadrey, J. and P. Zarifian (2002), *L'Emergence d'un Modèle de Service: Enjeux et Réalités*, Paris: Liaisons éditeur, Collection Entreprise et carrières.
Gallouj, F. (1994), *Economie de l'Innovation dans les Services*, Paris: L'Harmattan, Logiques économiques.
Gallouj, F. (2002), *Innovation in the Service Economy: The New Wealth of Nations*, Cheltenham, UK and Northampton, MA, USA: Edward Elgar.
Gallouj, F. and O. Weinstein (1997), 'Innovation in services', *Research Policy*, **26** (4–5), 537–556.
Giarini, O. (1978), 'Valeur d'Echange, Valeur d'Usage', *Futuribles*, **5**, 289–298.
Henderson, R.M. and J.B. Clark (1990), 'Architectural innovation: the reconfiguration of existing product technology and the failure of established firms', *Administrative Science Quarterly*, **35** (1), 9–30.
Howells, J. (2000a), 'Innovation and services: new conceptual frameworks', Centre for Research on Innovation and Competition, University of Manchester, CRIC Discussion Paper no. 38.
Howells, J. (2000b), 'The nature of innovation', OECD Workshop on Innovation and Productivity in Services, Sydney, 1–2 November.
Howells, J. (2004), 'Innovation, consumption and services: encapsulation and combinatorial role of services', *Service Industries Journal*, **24** (1), 9–36.
Miles, I. (2006), 'Innovation in services', in J. Fagerberg, D.C. Mower and R. Nelson (eds), *The Oxford Handbook of Innovation*, Oxford: Oxford University Press, pp. 433–458.
Miles, I. and S.J. Metcalfe (2000), *Innovation Systems in the Service Economy*, London: Kluwer Academic Publishers.
OECD (2007), *Globalisation and Innovation in The Business Services Sector*, Paris: OECD.
Pavitt, K. (1984), 'Sectoral patterns of technological change: towards a taxonomy and a theory', *Research Policy*, **13** (6), 343–373.
Schumpeter, J. (1939), *Business Cycles*, New York, USA and London, UK: McGraw-Hill.
Sundbo, J. (1997), 'Management of innovation in services', *Service Industries Journal*, **17** (3), 432–455.
Tertre, C. Du (1999), 'Les Services de Proximité aux Personnes: vers une Régulation Conventionnée et Territorialisée', *L'Année de la Régulation*, **3**, 203–236.
Zacklad, M. (2005), 'Innovation et Création de Valeur dans les communautés d'action', in R. Teulier and P. Lorino (eds), *Entre Connaissance et Organisation: l'Activité Collective*, Paris: Maspéro, pp. 285–305.
Zimmermann, H. (1980), 'OSI Reference Model – The ISO Model of Architecture for Open Systems Interconnection, Communications', *IEEE Transactions*, **28** (4), 425–432.

3 Services and innovation and service innovation: new theoretical directions

Jeremy Howells

3.1 Introduction

There have been numerous reviews and overviews of the conceptualisation and theoretical development of services and innovation over the last few years (see, for example, Miles, 2005; Hipp and Grupp, 2005; Howells 2007; Djellal and Gallouj, 2007; Chapter 1 in this volume). This has been paralleled by the increasing policy recognition in the growth and significance of services and in particular how innovation in services and service innovation is of increasing importance to national and local economies. The aim of this chapter is, therefore, not to review such developments, which are provided elsewhere in this volume. Instead, the objective of this review is to take a forward look at potentially new theoretical perspectives and promising avenues for future research in this field, as well as major gaps in our knowledge relating to service innovation.

This is no easy task. Although research into innovation in services and service innovation dates back over 40 years, an accepted and consolidated view of what theoretical trends have emerged over this period still has not been recognised. Thus, even the naming, definition and sequencing of innovation concepts and models has not been agreed, whilst much of the literature on service innovation remains isolated and fragmented from much of the wider services literature relating to organisational change and behaviour, marketing, consumption or new product development. This 'siloing' of service innovation from other parts of service and organisational research has undoubtedly slowed theoretical development on the topic of service innovation and its subsequent synthesis and agreement. Trying to make statements about new and future theoretical developments on service innovation when a 'received' view of such developments has not been accepted is therefore at best challenging if not still highly contestable. However, as a springboard for undertaking this endeavour it is important to outline where there is now agreement: these relate to what might be termed as service innovation axioms. These are outlined below.

3.2 Service innovation axioms: platforms for further development

There is now a well-established axiom of development phases of concepts and theoretical perspectives of services and innovation, although the titles of the phases are still not agreed upon. These conceptual frameworks can be grouped around three perspectives on service innovation centred on how innovation in services is perceived in relation to the general economy and in terms of innovation theory more specifically (see Chapter 1 in this volume). As such, this relates to: the extent that innovation is seen as endogenous or not to the sector overall and, related to this, how far it is based on embodied technological change related to the adoption and use of capital equipment (especially information technology, IT); and how similar, or dissimilar, the characteristics of the service innovation processes are to existing innovation models and especially those associated with manufacturing industry. On this basis, three main conceptual approaches have emerged (see Coombs and Miles, 2000; Gallouj, 2002a; Tether, 2003; Miles, 2000; Howells, 2007): the 'technologist' approach, the 'service-oriented' (or demarcation) approach and the 'integrative' approach.

Studies that have adopted the 'technologist' approach of innovation in services have seen the main driving force and shaping of service innovation being derived from the external, non-endogenous adoption of technologies and systems from outside the sector, in particular computers and other IT equipment, by service firms and organisations (Collier, 1983; Pilat, 2001; Pilat and Lee, 2001; Agrawal and Berg, 2008). This approach can, in turn, be linked to the supplier-dominated perspective associated with earlier sectoral taxonomies of technological activities (Pavitt, 1984; and Miozzo and Soete, 2001; although some such as Evangelista, 2000, suggest that his revised taxonomy confirms adopting a more integrationist view; see below). This traditional view of innovation in services is strongly held by many outside the services research community (Howells, 2001) and still remains perhaps the most numerous in terms of empirical analysis (Gallouj, 2002a, 2), although their influence has been waning in recent years.

As a reaction to this traditionalist approach and as a means to assert the distinctiveness of innovation in services saw the emergence of the 'service-oriented' approach largely in the late 1990s (Gallouj and Weinstein, 1997; Gadrey and Gallouj, 1998; Sundbo, 1998; Djellal and Gallouj, 2001; Sundbo and Gallouj, 2000; den Hertog, 2000; Gallouj, 2002b). Studies based on this approach have sought to move away from what might be seen as merely adapting manufacturing centred innovation models to focusing instead on the peculiarities of service innovation and how this might lead to new conceptualisations of innovation processes in relation to service activity. This set of scholars and their studies have emphasised the 'peculiarities of services' (Barras, 1986), and how services differ from archetypal manufacturing (see, however, more recently Hipp, 2008). For example, services are frequently intangible, and are often (but not always) produced and consumed at the same time, often with the direct involvement of the consumer (i.e. services are co-made or co-produced by the provider and user working together). This means it is much more difficult to define a service product and observe a moment at which the service product changed significantly. Research adopting this perspective has sought to identify distinctive endogenous innovation practices and patterns that are very different from the traditionalist supplier- and technologist-dominated models of service innovation.

As such, such studies have sought to highlight that:

1. much of service innovation has remained hidden because in the past it has not been properly conceptualised and therefore measured (Evangelista and Sirilli, 1995);
2. because of this, service innovation is more common and important than previously supposed;
3. its significance moreover does not simply reside within the service sector but it is important to all parts of the economy; and, lastly and above all,
4. service innovation is different from existing manufacturing- and artefact-driven models of innovation because of its:
    a. frequently intangible nature;
    b. emphasis on new organisational practices and routines;
    c. reliance on close client interaction and sometimes co-production; and,
    d. because of this 'simultaneity' of production and consumption, the inability to store service products; i.e. a high degree of perishability.

Much of the driving force behind this service-oriented approach has been through studies that have sought to analyse knowledge-intensive business services (KIBS) (Miles et al., 1995; Miles, 1994, 2000; O'Farrell and Moffat, 1991; O'Farrell and Wood, 1999; Bettencourt et al., 2002) in relation to client-oriented service innovation. However, a criticism of this approach is that by adopting the view of distinctiveness and differentiation of innovation in services the approach goes too far in emphasising the highly differentiated, bespoke nature of service provision with its close, knowledge-intensive links with clients. Although such a perspective has been guided by specific, in-depth studies of these specialised, knowledge-intensive links which has provided a rich insight into new dimensions in innovative activity, it ignores the more routine, standardised and low-technology nature of many service outputs. Moreover, even in these more knowledge-intensive business sectors, some more recent studies have tended to indicate many of the similarities with high-technology manufacturing sectors rather than the differences (Tether, 2005).

The emergence and rise of the 'service-oriented' approach and its focus on what might be termed the 'high peaks' of service activities in terms of both knowledge intensity, innovativeness and client-intensive relationships is however understandable in a Kuhnian sense. Thus it can be seen as seeking to create a major disruptive break that seeks comprehensively to reject the previous traditionalist paradigm, which saw services as simply being passive, supplier-dependent and non-innovative sectors (Howells, 2007). The development of the service-oriented paradigm has therefore been as much (if not more) about rejecting the traditionalist paradigm, than in creating a necessarily coherent new one (Kuhn, 1962). Thus, in seeking to cast out the traditionalist paradigm, proponents of the new paradigm were almost obliged to focus on selecting firms, activities and sectors that readily and most starkly contrasted with previous notions of what services and innovation were about. The privileging in the 'service-oriented' paradigm of co-production and client-intensive arrangements in service innovation may also have been influenced by the substantial set of research by geographers and others (Tordoir, 1994; Quinn and Dickson, 1995; Cornish, 1996; Wood, 2002; Bettencourt et al., 2002) in this field. This work has highlighted the importance of co-production between service providers and users, associated with its simultaneity and perishability, in relation to location and spatial trends (as well as the impact of telecommunications in decoupling the need for co-location of service production and use) (Illeris, 1994).

The third and last approach, the 'integrative' approach, comes from two closely related positions. Firstly, that there is already a great deal of similarity between manufacturing and service industries, but it is not being adequately conceptualised or measured properly. Indeed this applies just as much to manufacturing environments as it does to service environments; much remains 'hidden' because it is not recognised and therefore not measured and analysed properly. The second reason is that there is recognition that there actually has been a change in the fundamental operation of the economy. In particular, there has been a convergence and intertwining of goods and services in both their production and consumption (see, for example, Hill, 1977; Howells, 2001) and because of this there has been a need to develop an integrated theory to cover both segments of the economy. This includes the increasing bundling of services and manufactured goods into 'solutions' (Howells, 2004). This approach also recognises the major changes that have occurred in managerial practice, and the shift away from 'manufacturing' versus 'service' firms, towards organisations focused on the realisation of value. This has moved the focus of research away from technologies to knowledge, and away from individual firms towards understanding value chains or networks, locating service and manufacturing in a set of interrelated activities.

Both strands stress the often close links between manufacturing and services in provision, consumption and strategy (Howells, 2001, 2004; Uchupalanan, 2000; Tether et al., 2001; Tether, 2003; Daniels and Bryson, 2002; Bryson and Monnoyer 2004; Gallouj and Weinstein, 1997; Gallouj, 2002a). As a result, it was felt that there was a need for a single innovation theory to acknowledge this and to gauge accurately the true dimension of innovation in the modern economy (Gallouj and Weinstein, 1997) and therefore is the most desirable approach for service innovation studies to adopt (Gallouj 2002a, 27). Because such approaches are so relatively new and sparse, they are therefore understandably difficult to articulate fully. However, there a number of recent empirical studies which have sought to highlight many of the similarities between manufacturing and service sectors in terms of the association between innovation and economic performance (Cainelli et al., 2006, 454–5).

A final approach, which has emerged over recent years, is what might be termed as a 'segmentalist' service approach to innovation, that partially covers also the first two approaches – the technologist and service-oriented approaches. The argument for this perspective on service innovation is based on the fact that services are simply too big a 'sector' to study in any meaningful or coherent form en masse. Services are far from homogeneous and therefore, so the argument goes, it would be foolish to try and create a standard model or perspective that would cover the whole of the service economy. Different manufacturing sectors are treated differently in many innovation studies so why should we expect the much larger service sector to be treated differently? As such, with such size and diversity, it is perhaps not surprising that at least some researchers have sought a more segmented route. This approach is implicit, if not explicit, in the interesting study by Horgos and Koch (2008) on KIBS firms within Germany.

3.3 Avenues for future research

Regardless of which phase in service innovation research is accepted perhaps the biggest effect of service innovation will be its wider engagement, and impact upon, innovation conceptualisation and model building more generally. In reviewing the above axioms it is now useful to outline what are the key avenues for future research in the field of innovation in services and service innovation. What can be conceived or imagined here in terms of future developments? On this basis, there are a number of research avenues, which seem to be important foci for future research in the field of service innovation and which also relate to existing major gaps in our understanding of the pattern and process of service innovation.

The first relates to the changing contribution of service innovation research to the development of a wider notion of an integrated and holistic theory of innovation. This will in a sense see services research 'coming of age' in its ability to start influencing, rather than be influenced by, the wider constructs of innovation models and can be viewed as a further evolution of the integrative approach outlined earlier in section 3.2. Thus, although this may not mean the end of service innovation studies, potentially it will be the end of service innovation-specific models. This is not to mean that service innovation will no longer be of use or significance, rather to the contrary, but that such work and findings will be incorporated more directly and strongly into innovation-wide concepts and models. Indeed potentially much of value arising from the research into innovation studies will be generated from studies exploring the previously hidden or neglected areas of innovation research around services, low-technology sectors (Von Tunzelmann and Acha, 2005; Hirsch-Kreinsen, 2008), non-tangible forms of change and adoption and diffusion processes. This process also implies that much of what is happening in innovation in services will have applications and merit in primary (agriculture and extraction) and secondary (manufacturing) industries and in the utilities. Service innovation research will, therefore, further move away from being a research 'silo' and will continue to become more integrated in the wider innovation research community. How this will affect its long-term identity remains to be seen, but its importance as a focal point for research in innovation will undoubtedly remain and indeed be heightened by the need to understand the innovation process within a much wider economic and social setting.

Another major avenue of research in service innovation lies in investigating interaction in the innovation process. Service innovation research has done much to push the boundaries in terms of the interactive nature of innovation and the role of the consumer. However, there are still great opportunities for undertaking further work in a genuinely novel way in this area. What does this mean? It involves not looking at just a supplier (push) or a user (pull) view, but means that we should start to seek to explore the genuine interaction between suppliers and users in a two-way, interactive process. Some service researchers have already highlighted this area (den Hertog, 2000) but we have only just begun the process and clearly more empirical-based studies need to be undertaken to expand, develop and refine the basic concept. With the growth in user-led studies of innovation back in the 1970s centred on von Hippel's work (for example, see von Hippel, 1978, 1986, 1988; Parkinson, 1982; Shaw, 1985; Foxall, 1987; Callahan and Lasry, 2004; Lüthje, 2004) we have slowly started to move away from taking a supplier-dominated paradigm and perspective of innovation.

This remained largely a bipolar (producer or user) view until the notion of innovation 'co-production' started to appear in services research. The concept of the co-production of innovations by both suppliers and customers/users has been one that has originated, and been developed, from within the service innovation field (see, variously, Normann, 1984; O'Farrell and Moffat, 1991; Miles et al., 1995; Sundbo, 1997; Tordoir, 1996; O'Farrell and Wood, 1999; den Hertog, 2000; Bettencourt et al,. 2002; van der Aa and Elfring, 2002; Magnusson, 2003; Magnusson et al., 2003; Kuusisto and Viljamaa, 2004; Martinez-Fernandez and Miles, 2006). In turn, these studies have provided a conceptual centre ground, which has moved away from a simply supplier-, or user-, dominated view of innovation and emphasised one of genuine cooperation and development. Perhaps we should however move further to one of seeing a whole series of gradations of involvement and support between suppliers and users and one that also takes into account more multiple engagement patterns between sets of actors within wider collaborative teams or networks. Indeed, although the general consensus is that consumers have a more important role to play in service innovation, Freel (2006, 355) found that this is not always the case.

There also needs to be a clearer notion of what we mean by demand-, user- or consumer-'led' innovation. When discussion turns to user-, or consumer-, 'led' innovation, this can mean anything from a supplier picking up ideas from a customer and then doing virtually all of the work itself to the actual consumer/user generating the idea, developing it and then actually producing itself. We also need to be clearer about what types of customers we are analysing. Most of the discussion has been around other businesses (B), but final consumers (C) can also be important here (see, for example, Alam, 2006). Thus, business-to-business (B-to-B) relations in service innovation, not surprisingly, are dominant in many of the business service fields, where much of the focus on innovation in services has been, but business-to-customer (B-to-C) and even customer-to-customer (C-to-C) interactions are also important in many of the new Internet- and social-based areas of service innovation. A clearer articulation of the types of engagement and how engagements are maintained during these interactions still remains an important task here for future research.

Equally, and related to the above, we need much greater detail about the whole cycle of the processes and interactions involved in the service innovation process; i.e. the individual components in the innovation processes and the flows linking them together. Thus, Clark (1985) defined (from largely a design perspective) a component as a physically distinct portion of a product that embodies a core design concept and went on to articulate the notion of innovation components from within this manufacturing and production centred way. Components and the way they are integrated and configured in a wider structure (or 'architecture'; Henderson and Clark, 1990), to form collectively the wider product or 'system' is crucial here.

By contrast, there has been little research in services in identifying, defining and articulating different component elements (in contrast to wider systemic analysis and development; see, for example, Davies, 2003), in particular in a way that seeks to create generic definitions of components that can then be compared with one another in different service innovation frameworks. This relates in large part to the ability, or rather inability, to identify distinct service innovation components and in part this may be because they are more likely to be continuous in nature. There have been some attempts within 'new service development' (NSD; see below) research to granulate service activity (Silvestro et al., 1992) but these have remained general and have been focused more on stages of the development process rather than on generating identifiable components and processes. As such, there is a need for much more granulation but also comparability in the components that constitute service activity and service innovation. Without such developments, not only will detailed architecture of service innovations remain lost, but making generalisations and comparisons between different service innovations will remain limited.

If there is a lack of articulation over the components of service innovation, so too is there a lack of adequate description and measurement of the processes, flows and transactions associated with service innovation. This includes interactions right through the innovation cycle, covering not just the front-end, fuzzy idea generation phase, but also patterns of co-working during the process of developing the innovation and right through to the final outcome of the innovation itself. The problem here is that these stages and interactions are fairly clear when in relation to manufacturing, because of the tangible nature of at least some of the processes and components (see Armbuster et al., 2008), and this is true in some service sectors that involve dealing in tangible process or outputs (such as retailing or transport services). However, for most services gaining a complete understanding of the processes and interactions over time, producing an accepted definition of the elements and stages in each, and then enabling this to be applied generically to all or at least parts of the service sector, remain elusive. Progress in service innovation research is likely to remain constrained without a more detailed articulation and elaboration of these intangible processes and the production of an accepted terminology that is derived from them.

Although academics have been slow in identifying and charting new developments in this area, one area which led the field was in the adaptation of new product development concepts to fit or be measured against the service realm (Easingwood, 1986; Zeithaml et al., 1985, see also Chapters 10 to 13 in this volume) and from this generating the concept of 'new service development' (NSD) which originally grew out of the marketing literature (De Brentani, 1989, 1995, 2001; Storey and Easingwood, 1993, 1996; Hart and Service, 1993; Edvardsson and Olsson, 1996; Johne and Storey, 1998; de Jong and Vermeulen, 2003; Dolfsma, 2004). More recently there has been a new set of researchers seeking to develop a new conceptual framework and agenda in this area (see, for example, Smith et al., 2007). However, again, there remain few links here with the wider service innovation literature (see Menor and Roth, 2007).

A more specific element of this is how service firms and agencies generate, use and diffuse knowledge in their organisations. How firms tap and use knowledge appears to have a significant effect on their innovative capabilities (Leiponen, 2006, 250–4), but we have only a very partial view of how these knowledge processes actually work. Increasingly this requires an interorganisational perspective as knowledge collaboration involves interorganisational exchange and sharing around particular knowledge sets (Consoli, 2007; Consoli and Ramlogan, 2009). Dougherty (2004), for example, has explored the way routines are used in practice to create and share knowledge within a sample of firms in professional and utility-based service operations. In particular, service firms were seeking to create new forms of knowledge organisation with, for example, the creation of corporate research and development (R&D) based around principles of practice rather than that of basic science or technologies (see also Carlile, 2002). As such, service firms are creating new organisational forms and routines that are different from traditional manufacturing-based concepts to generate, manage and transfer their knowledge. This is perhaps at its most difficult when involved with such knowledge exchange processes between different firms and organisations and/or between units located overseas. Examples of service outsourcing and offshoring are perhaps best examples of such complexities (see, for example, Massini and Miozzo, Chapter 20 in this volume).

If we need a clearer conceptual framework on components and flows in the innovation process, we also need a clearer notion of the organisational routines that link them together. Moreover these new organisational routines are forms of innovation in themselves as organisational innovations (although some still believe that innovation should not be extended to include all types of organisational change; Drejer, 2004). Strangely, given the undoubted significance of disembodied, organisational forms of innovation in services, there remains a significant gap between service innovation literature and organisational innovation literature. Research by van der Aa and Elfring (2002) has sought to bridge this gap and in their work they have identified three types of organisational innovation: the multi-unit organisation, new combinations of services, and redefinition of roles and relationships with customers. The multi-unit organisation has links to the co-production and co-location concepts noted earlier; namely, because of the simultaneity of production and consumption, the growth of the business in any one location is limited and therefore service firms need to develop reproduction formulas and new organisational forms that seek to balance standardisation with customisation. New combinations of services are associated with creating new bundles of services and this in turn requires new capabilities in the integration and recombination of service operations. The third form of organisational innovation van der Aa and Elfring (2002) identified is related to the redefinition of roles and relationships with customers; in particular here changing forms of co-makership associated with new service concepts, new delivery systems and new markets.

Innovation in public services, although long highlighted as an area for future research, will surely come of age over the next ten years in service innovation research. Although the importance of innovation in public services and its potential for leading to improvements in quality of service and efficiency savings has been recognised for some time (at least from the public service and administration side; see, for example, Zegans, 1992; Joyce, 1998; Glor, 1997, 2008; Moore, 2005), in terms of service innovation research there have been few, if any, studies up until recently. The major PUBLIN research project has done much to highlight some of the key issues in Europe (Halvorsen et al., 2005) and it has been taken up more specifically by resent research by Windrum (2008; Windrum and García-Goñi, 2008, see also Chapter 6 in this volume) in the health sector. These more recent studies have undoubtedly highlighted some of the key differences in the innovation environment between public and private realms, although finding commonalities will also be of value and interest. Thus, the crucial role of the client (Kuusisto and Viljamaa, 2004) and also the issue of absorptive capacity of the organisation in adopting new innovation practices are themes, which have clear parallels with the private sector. However, in terms of the impact of service innovation researchers on the field of the management of public service organisations, it is salutary how there remains little or no acknowledgment of innovation research (see, for example, Greenhalgh et al., 2004). Nonetheless, over the next ten years this will undoubtedly be a major new area of service innovation research as improvements in the efficiency, quality, performance and client responsiveness of public services becomes an ever more important issue for public policy and administration.

Lastly innovation research has tended to focus on narrow, one-dimensional, 'single-issue' aspects, that have tended to highlight certain types of innovation such as product innovation or process innovation, in isolation, such that these different types of innovation are typically viewed as being independent of one another (Damanpour and Evan, 1984, 406; Reichstein and Salter, 2006). Even when innovations are treated together this tends to be confined to the firm or organization, rarely in relation to interorganisational types of innovation. By contrast, it should be considered that (perhaps especially in services) different types of innovation often interact with each other in complex, interdependent and complementary ways. Thus Howells and Tether (2004) sought to outline the more 'outward-looking' nature of innovation associated with how service firms externally interact with other actors, in particular customers in services compared to the more 'inward-looking' nature of manufacturing activity, which is associated with more harnessing in-house activities and resources. Recent studies by Tether (2005), Leiponen (2005) and Mansury and Love (2008) tend to support his view of innovation in service firms being more oriented towards the external sourcing of knowledge, interorganisational collaboration and customer interaction and networking than manufacturing firms, which are more oriented to in-house capabilities and knowledge resources.

3.4 Conclusion

Service innovation research has come a long way from its origins in the early 1970s. This should be seen not so much in terms of the study of innovation in services itself, but in terms of its increasing influence on the wider innovation research agenda and more importantly its impact on conceptualisation and theoretical developments. However, in terms of theory development itself, research in service innovation will only effectively come of age, once the concepts and models that have been developed have a predictive element to them; i.e. they should not be able to describe and understand current patterns, but should also be able to provide predictions about future patterns of change. Without providing reliable prediction of events or patterns, we remain fixed in a realm of only partial theories and models (Braithwaite, 1960, 1). We should therefore seize opportunities along a wide conceptual front. This includes not only formal model building and empirical testing but also from other methodological frameworks. Thus foresight and horizon scanning models have a role to play, although in a more policy oriented context (Toivonen, 2004).

Lastly, research in service innovation should come of age and realise the great potential it has for developing innovation research overall. The challenges it faces in terms of intangibility and simultaneity in many of the service processes and components it seeks to identify and measure are great, but so are the rewards in terms of helping to formulate and develop a truly integrative and holistic view of the innovation process and its theory.

References

Agrawal, G.K. and D. Berg (2008) 'Role and impact of "technology" in the service development process: a research study', *International Journal of Services Technology and Management*, **9**, 103–121.

Alam, I. (2006), 'Removing the fuzziness from the fuzzy front-end of service innovations through customer interactions', *Industrial Marketing Management*, **35**, 468–480.

Armbuster, H., A. Bikfalvi, S. Kinkel and G. Lay (2008), 'Organizational innovation: the challenge of measuring non-technical innovation in large-scale surveys', *Technovation*, **28**, 644–657.

Barras, R. (1986), 'Towards a theory of innovation in services', *Research Policy*, **15**, 161–173.

Bettencourt, L.A., A.L. Ostrom, S.W. Brown and R.J. Roundtree (2002), 'Client co-production in knowledge-intensive business services', *California Management Review*, **44** (4), 100–128.

Braithwaite, R.B. (1960), *Scientific Explanation*, New York: Harper.

Bryson, J.R. and M.-C. Monnoyer (2004), 'Understanding the relationship between services and innovation: the RESER review of the European service literature on innovation, 2002', *Services Industries Journal*, **24**, 205–222.

Cainelli, G., R. Evangelista and M. Savona (2006), 'Innovation and economic performance in services: a firm-level analysis', *Cambridge Journal of Economics*, **30**, 435–458.

Callahan, J. and E. Lasry (2004), 'The importance of customer input in the development of very new products', *R&D Management*, **34**, 107–120.

Carlile, P. (2002), 'A pragmatic view of knowledge and boundaries: boundary objects in new product development', *Organization Science*, **10**, 442–455.

Clark, K.B. (1985), 'The interaction of design hierarchies and market concepts in technological evolution', *Research Policy*, **14**, 235–251.

Collier, D.A. (1983), 'The service sector revolution: the automation of services', *Long Range Planning*, **16**, 10–20.

Consoli, D. (2007), 'Services and systemic innovation: a cross-sectoral analysis', *Journal of Institutional Economics*, **3**, 71–89.

Consoli, D. and R. Ramlogan (2009), 'Scope, strategy and structure: the dynamics of knowledge networks in medicine', Manchester Business School Working Paper No. 569, University of Manchester.

Coombs, R. and I. Miles (2000), 'Innovation, measurement and services: the new problematic', in J.S. Metcalfe and I. Miles (eds), *Innovation Systems in the Service Economy: Measurement and Case Study Analysis*, Boston, MA: Kluwer Academic Publishers, pp. 85–103.

Cornish, S.L. (1996), 'Marketing software products: the importance of "being there" and the implications for business service exports', *Environment and Planning A*, **28**, 1661–1682.

Damanpour, F. and W.M. Evan (1984), 'Organizational innovation and performance: the problem of organizational lag', *Administrative Science Quarterly*, **29**, 392–409.

Daniels, P.W. and J.R. Bryson (2002), 'Manufacturing services and servicing manufacturing: knowledge based cities and changing forms of production', *Urban Studies*, **39**, 977–991.

Davies, A. (2003), 'Integrated solutions: the changing business of systems integration' in A.

Prencipe, A. Davies and M. Hobday (eds), *The Business of Systems Integration*, Oxford: Oxford University Press, pp. 333–368.
De Brentani, U. (1989), 'Success and failure in new industrial services', *Journal of Product Innovation Management*, **6**, 239–258.
De Brentani, U. (1995), 'New industrial service development: scenarios for success and failure', *Journal of Business Research*, **32**, 93–103.
De Brentani, U. (2001), 'Innovative versus incremental new business services: different keys for achieving success', *Journal of Product Innovation Management*, **18**, 169–187.
de Jong, J.P.J. and P.A.M. Vermeulen (2003), 'Organizing successful new service development: a literature review', *Management Decision*, **41**, 844–858.
den Hertog, P. (2000), 'Knowledge intensive business services as co-producers of innovation', *International Journal of Innovation Management*, **4**, 91–528.
Djellal, F. and F. Gallouj (2001), 'Patterns of innovation organisation in service firms: postal survey results and theoretical models', *Science and Public Policy*, **28**, 57–657.
Djellal, F. and F. Gallouj (2007), 'Innovation and employment effects in services: a review of the literature and an agenda for research', *Services Industries Journal*, **27**, 193–213.
Dougherty, D. (2004), 'Organizing practices in services: capturing practice-based knowledge for innovation', *Strategic Organization*, 2, 35–64.
Dolfsma, W. (2004), 'The process of new service development: issues of formalization and appropriability', *International Journal of Innovation Management*, **8**, 319–337.
Drejer, I. (2004), 'Identifying innovation in surveys of services: a Schumpeterian perspective', *Research Policy*, **33**, 551–562.
Easingwood, C.J. (1986), 'New product development for service companies', *Journal of Product Innovation Management*, **3**, 264–275.
Edvardsson, B. and J. Olsson (1996), 'Key concepts for new service development', *Services Industries Journal*, **16**, 140–164.
Evangelista, R. (2000), 'Sectoral patterns of technological change in services', *Economics of Innovation and New Technology*, **9**, 183–221.
Evangelista, R. and G. Sirilli (1995), 'Measuring innovation in services', *Research Evaluation*, **5** (3), 207–15.
Foxall, G.R. (1987), 'Strategic implications of user-initiated innovation', in R. Aothwell and J. Bessant (eds), *Innovation: Adaptation and Growth*, Amsterdam: Elsevier, pp. 25–36.
Freel, M. (2006), 'Patterns of technological innovation in knowledge-intensive business services', *Industry and Innovation*, **13**, 335–358.
Gadrey, J. and F. Gallouj (1998), 'The provider–customer interface in business and professional services', *Services Industries Journal*, **18**, 1–15.
Gallouj, F. (2002a), *Innovation in the Service Economy: The New Wealth of Nations*, Cheltenham, UK and Northampton, MA, USA: Edward Elgar.
Gallouj, F. (2002b), 'Innovation in services and the attendant old and new myths', *Journal of Socio-Economics*, **31**, 137–154.
Gallouj, F. and O. Weinstein (1997), 'Innovation in services', *Research Policy*, **26**, 537–556.
Glor, E. (1997), 'What is public sector innovation?' *Innovation Journal*, **2** (**1**), article 1.
Glor, E. (2008), 'Toward development of a substantive theory of public sector organizational innovation', *Innovation Journal*, **13** (3), article 6, 1–28.
Greenhalgh, T., G. Robert, F. Macfarlane, P. Bate and O. Kyriakidou (2004), 'Diffusion of innovations in service organisations: systematic review and recommendations', *Milbank Quarterly*, **82**, 581–629.
Halvorsen, T., J. Hauknes, I. Miles and R. Roste (2005), 'Innovation in the public sector: on the differences between public and private sector innovation', Publin Report No. D9, STEP, Oslo.
Hart, S.J. and L.M. Service (1993), 'Cross-functional integration in the new product introduction process: an application of action science in services', *International Journal of Service Industry Management*, **4**, 50–66.
Henderson, R.M. and K.B. Clark (1990), 'Architectural innovation', *Administrative Science Quarterly*, **35**, 9–30.

Hill, P. (1977), 'On goods and services', *Review of Income and Wealth*, **4**, 315–338.
Hipp, C. (2008), 'Service peculiarities and the specific role of technology in service innovation management', *International Journal of Services Technology and Management*, **9**, 154–173.
Hipp, C. and H. Grupp (2005), 'Innovation in the service sector: the demand for service-specific innovation measurement concepts and typologies', *Research Policy*, **34**, 517–535.
Hirsch-Kreinsen, H. (2008), '"Low-tech" innovations', *Industry and Innovation*, **15**, 19–43.
Horgos, D. and A. Koch (2008), 'The internal differentiation of the KIBS sector: empirical evidence from cluster analysis', *International Journal of Services Technology and Management*, **10**, 190–210.
Howells, J. (2001), 'The nature of innovation in services' in OECD (eds), *Innovation and Productivity in Services*, Paris: OECD, pp. 55–79.
Howells, J. (2004), 'Innovation, consumption and services: encapsulation and the combinatorial role of services', *Services Industries Journal*, **24**, 19–36.
Howells, J. (2007), 'Services and innovation: conceptual and theoretical perspectives', in J.R. Bryson and P.W. Daniels (eds), *The Handbook of Service Industries*, Cheltenham, UK and Northampton, MA, USA: Edward Elgar, pp. 34–44.
Howells, J. and B.S. Tether (2004), *Innovation in Services: Issues at Stake and Trends* Inno Studies Programme (ENTR-C/2001), Brussels: Commission of the European Communities.
Illeris, S. (1994), 'Proximity between service producers and service users', *Tijdschrift voor Economische en Sociale Geografie*, **85**, 294–302.
Johne, A. and C. Storey (1998), 'New service development: a review of the literature and annotated bibliography', *European Journal of Marketing*, **32**, 184–251.
Joyce, P. (1998), 'Management and innovation in public services', *Strategic Change*, **7**, 19–30.
Kuhn, T.S. (1962), *The Structure of Scientific Revolutions*, Chicago, IL: Chicago University Press.
Kuusisto, J. and A. Viljamaa (2004), 'Knowledge-intensive business services and co-production of knowledge: the role of the public sector?' *Frontiers of E-Business Research*, **2004**, 282–298.
Leiponen, A. (2005), 'Organisation of knowledge and innovation; the case of Finnish business services', *Industry and Innovation*, **12**, 185–203.
Leiponen, A. (2006), 'Managing knowledge for innovation; the case of business-to-business services', *Journal of Product Innovation Management*, **23**, 238–258.
Lüthje, C. (2004), 'Characteristics of innovating users in a consumer goods field: an empirical study of sport-related product consumers', *Technovation*, **24**, 683–695.
Magnusson, P.R. (2003), 'Benefits of involving users in service innovation', *European Journal of Innovation Management*, 6, 228–238.
Magnusson, P.R., J. Matthing and P. Kristensson (2003), 'Managing user involvement in service innovation: experiments with innovating end users', *Journal of Service Research*, **6**, 111–124.
Mansury, M.A. and J.H. Love (2008), 'Innovation, productivity and growth in US business services: a firm-level analysis', *Technovation*, **28**, 52–62.
Martinez-Fernandez, M.C. and I. Miles (2006), 'Inside the software firm: co-production of knowledge and KISA in the innovation process', *International Journal of Services Technology and Management*, **7**, 115–125.
Menor, L.J. and A.V. Roth (2007), 'New service development competence in retail banking: construct development and measurement validation', *Journal of Operations Management*, **25**, 825–846.
Miles, I. (1994), 'Innovation in services', in M. Dodgson and R. Rothwell (eds.), *Handbook of Industrial Innovations*, Aldershot, UK and Brookfield, VT, USA: Edward Elgar, pp. 243–256.
Miles, I. (2000), 'Services innovation: coming of age in the Knowledge-Based Economy', *International Journal of Innovation Management*, **4**, 371–389.

Miles, I. (2005), 'Innovation in services', in J. Fagerberg, D.C. Mowery and R.R. Nelson (eds), *The Oxford Handbook of Innovation*, Oxford: Oxford University Press, pp. 433–458.
Miles, I., N. Kastrinos, K. Flanagan, R. Bilderbeek, P. den Hertog, W. Huntink and M. Bouman (1995), 'Knowledge-intensive business services: users, carriers and sources of innovation', Luxembourg: EIMS Publication No. 15, Innovation Programme, Directorate General for Telecommunications, Information Market and Exploitation of Research, Commission of the European Communities.
Miozzo, M. and L. Soete (2001), 'Internationalization of services: a technological perspective', *Technological Forecasting and Social Change*, **67**, 159–185.
Moore, M.H. (2005), 'Break-through innovations and continuous improvement: two different models of innovative processes in the public sector', *Public Money and Management*, **17**, 43–50.
Normann, R. (1984), *Service Management, Strategy and Leadership in Service Businesses*, Chichester: John Wiley & Sons.
O'Farrell, P.N. and L.A.R. Moffat (1991), 'An interaction model of business service production and consumption', *British Journal of Management*, **2**, 205–221.
O'Farrell, P.N. and P.A. Wood (1999), 'Formation of strategic alliances in business services: towards a new client-oriented conceptual framework', *Services Industries Journal*, **19**, 133–151.
Parkinson, S.T. (1982), 'The role of the user in successful new product development', *R&D Management*, **12**, 123–131.
Pavitt, K. (1984), 'Sectoral patterns of technical change: towards a taxonomy and a theory', *Research Policy*, **13**, 343–373.
Pilat, D. (2001), 'Innovation and productivity in services: state of the art', in OECD (eds) *Innovation and Productivity in Services*, Paris: OECD, pp. 17–54.
Pilat, D. and F.C. Lee (2001), 'Productivity growth in ICT-producing and ICT-using industries', OECD Working Papers No. 2001/4, Paris: OECD.
Quinn, J.J. and K. Dickson (1995), 'The co-location of production and distribution: emergent trends in consumer services', *Technology Analysis and Strategic Management*, **7**, 343–354.
Reichstein, T. and A. Salter (2006), 'Investigating the sources of process innovation among UK manufacturing firms', *Industrial and Corporate Change*, **15**, 653–682.
Shaw, B. (1985), 'The role of the interaction between the user and the manufacturer in medical equipment innovation', *R&D Management*, **15**, 283–292.
Silvestro, R., L. Fitzgerald, R. Johnston and C. Voss (1992), 'Towards a classification of service processes', *International Journal of Service Industry Management*, **3** (3), 62–75.
Smith, A.M., M. Fischbacher and F.A. Wilson (2007), 'New service development: from panoramas to precision', *European Management Journal*, **25**, 370–383.
Storey, C. and C.J. Easingwood (1993), 'The impact of the new product development project on the success of financial services', *Services Industries Journal*, **13**, 40–54.
Storey, C. and C.J. Easingwood (1996), 'Determinants of new product performance: A study in the financial services sector', *International Journal of Service Industry Management*, **7**, 32–55.
Sundbo, J. (1997), 'Management of innovation in services', *Services Industries Journal*, **17**, 432–455.
Sundbo, J. (1998), *The Organization of Innovation in Services*, Roskilde: Roskilde University Press.
Sundbo, J. and F. Gallouj (2000), 'Innovation as a loosely coupled system in services', *International Journal of Services Technology and Management*, **1**, 15–36.
Tether, B.S. (2003), 'The sources and aims of innovation in services: variety between and within sectors', *Economics of Innovation and New Technology*, **16**, 481–506.
Tether, B.S. (2005), 'Do services innovate (differently)? Insights from the European Innobarometer survey', *Industry and Innovation*, **12**, 153–184.

Tether, B.S., C. Hipp and I. Miles (2001), 'Standardisation and particularisation in services: evidence from Germany', *Research Policy*, **30**, 1115–1138.
Toivonen, M. (2004), 'Foresight in services: possibilities and special challenges', *Services Industries Journal*, **24**, 79–98.
Tordoir, P.P. (1994), 'Transactions of professional business services and spatial systems', *Tijdschrift voor Economische en sociate Geografie*, **85**, 322–332.
Tordoir, P.P. (1996), *The Professional Knowledge Economy: The Management and Integration of Professional Services in Business Organizations*, Dordrecht: Kluwer.
Uchupalanan, K. (2000), 'Competition and IT-based innovation in banking services', *International Journal of Innovation Management*, **4**, 455–489.
van der Aa, W. and T. Elfring (2002), 'Realizing innovation in services', *Scandinavian Journal of Management*, **18**, 155–171.
von Hippel, E. (1978), 'Successful industrial products from customer ideas', *Journal of Marketing*, **42**, 39–49.
von Hippel, E. (1986), 'Lead users: a source of novel product concepts', *Management Science*, **32**, 791–805.
von Hippel, E. (1988), *The Sources of Innovation*, Oxford: Oxford University Press.
Von Tunzelmann, N. and V. Acha (2005), 'Innovation in "low tech" industries', in J. Fagerberg, D.C. Mowery and R.R. Nelson (eds.) *The Oxford Handbook of Innovation*, Oxford: Oxford University Press, pp. 407–432.
Windrum, P. (2008), 'Innovation in public services', in P. Windrum and P. Koch (eds), *Innovation in Public Services: Management, Creativity, and Entrepreneurship*, Cheltenham, UK and Northampton, MA, USA: Edward Elgar, pp. 3–20.
Windrum, P. and M. García-Goñi (2008), 'A neo-Schumpeterian model of health services innovation', *Research Policy*, **37**, 649–672.
Wood, P.A. (2002), 'Services and the "New Economy": an elaboration', *Journal of Economic Geography*, **2**, 109–114.
Zegans, M.D. (1992), 'Innovation in the well-functioning public agency', *Public Productivity and Management Review*, **16**, 141–156.
Zeithaml, V.A., A. Parasuraman and L.L. Berry (1985), 'Problems and strategies in services marketing', *Journal of Marketing*, **48**, 33–46.

4 The two-sided cost disease and its frightening consequences

William J. Baumol

Since rates of labor-saving productivity growth are uneven, the growth in some activities *must* be below average (Baumol's Fourth Tautology).

[The AK-47] . . . has become the world's most prolific and effective combat weapon, a device so cheap and simple that it can be bought in many countries for less than the cost of a live chicken. (Kahaner, 2006)

4.1 Introduction

I have repeatedly argued that the rising real prices that constitute the cost disease that is named in my honor cannot force society to give up the patterns of consumption to which it is habituated and that it prefers now, or used to. Neither health care nor education are condemned to deteriorate in quality and decline in quantity by their rising real prices. For the nearly universal phenomenon of rising productivity means we can afford them, indeed, that we can even afford steady expansion in the amounts supplied and consumed, despite their disturbingly persistent and substantial rates of cost increase.

But this does not mean that society is unaffected by the cost disease. Over the years, general standards of living have increased, and our material possessions have multiplied. But at the same time, our communities have experienced a decline in the quality of a variety of public and private services, arguably the result mainly of their rising costs. Not just in the United States, but throughout the world, as the wages of the required workmen have risen, streets and subways have grown increasingly dirty. Bus, train and postal services have all been cut back. Amazingly enough, in the 1800s in suburban London, there were 12 mail deliveries per day on weekdays and one on Sundays. Today the UK mail service is hardly a subject of admiration any more.

Parallel cutbacks have occurred in the quality of private services. It used not to be necessary to push five buttons sucessively on the telephone to get to speak to a human being at the bank. Another example, although undoubtedly a matter for less general concern, is the quality of food served in restaurants. Even some of the most elegant and expensive restaurants serve frozen and reheated meals – charging high prices for what amounts to little more than TV dinners. And there is much more, as this chapter will show. We are indeed experiencing what seems likely to prove to be a foretaste of the future.

I must be excused for offering yet another (mercifully brief) explanation of the cost disease, one that is not complicated by reliance on the behavior of wages – the assertion that when wages in one sector of a competitive economy change, they are driven in the same direction in the remainder of the economy. Though wages continue to play an implicit role, the source of that disease can be made clear without attention to them.

The mechanism this time will be described by analogy with a deploring newspaper headline that I saw when I first arrived in England in 1946. As I recall, it sought to shock readers by reporting that 'Nearly half of UK student grades perform below average'. The cost disease is just like that. Here we need merely keep in mind that any index of the overall price level (the figure that serves as the measure of the economy's rate of inflation) is just an average of the prices in the economy. It follows immediately and unequivocally that if the prices of all commodities are not rising at the same pace, then some must be increasing at a rate above average (that is, their relative costs and 'real prices' must be rising), while others must be characterized by real prices that are falling. That rudimentary tautology is almost all there is to the cost disease. Its only additional element is the observation that the set of those items whose real prices are rising is roughly unchanging, decade after decade, and the same appears to be true of those that are falling. It is this relatively unchanging characteristic of the behavior of the two sectors with the passage of time that requires empirical evidence. The remainder of the story, as we now see, is true by definition, that is, the definition of an average.

The explanation of the persistence of these paths is not difficult and was given above. The items in the rising-cost (price) group generally have a handicraft element in their production process, whose labor content is therefore difficult to cut down. Those in the other group are predominantly manufactures that are more easily automated and whose consequently falling labor content is just the other side of their steadily rising labor productivity.

The same story can obviously be extended to the remainder of the analysis. Not only prices but also rates of growth of productivity must have the same relationship to their averages for the economy as a whole. If these growth rates differ from industry to industry then some must evidently be below the average and some above it.

4.2 The triumph of do-it-yourself

One of the main responses to rising prices in the stagnant sector of the economy is the transfer of many products with sharply rising prices, from commercial production to do-it-yourself. We all recognize that we do our own house cleaning and carry out other household tasks ourselves now that the professions of maid and butler have shrunk to insignificance. They were gone by the time I obtained my first teaching position, some six decades ago. But my older colleagues of that period had previously employed maids even when they were only assistant professors.

Today, even in the rare cases where such household help still is employed, it isn't what it used to be. Once when one of my former professors and his wife were staying at our house, they received a lunch invitation from a former student, one of the world's wealthiest men, at his out-of-town estate. The chauffeured car was sent to our home to pick up the guests. But for the after lunch return to our apartment, the host had to drive them himself because, as he explained, it was the chauffeur's afternoon off.

In my childhood, the milkman, with his horse-drawn vehicle would deliver dairy products from house to house. There was also an iceman who did the same, since refrigerators were a rarity.

A change that is even more striking is in the doctor's activity. I remember that when I came down with a fever, a house call, with the physician carrying his little black bag, was the norm. Office visits were a rarity, employed only for routine examinations.

For academics like myself the big change is the disappearance of secretaries. I used to write my articles by hand and then give them to my secretary. She would type six copies with carbon paper. Any error would have to be erased individually on all six copies. My secretary was a superb typist who loved to work with math and who was dedicated to her work. Still, it often took three weeks for her to decipher my handwriting and type out and correct the obscure hieroglyphics that came from my pen.

Once, when I headed the committee that allocated some research grants to our colleagues, we found that most of the applicants had requested funding for secretarial and research assistant help. As I recalled it, this amounted to some $10,000 per applicant for part-time help. It occurred to me that there was an alternative. We offered each applicant the choice between the requested assistants and a new computer. Every applicant selected the computer, saving us some $7000 per grant.

But perhaps the most striking change is the advent of the disassembled product which is available in a large carton on which is stamped in large letters the terrifying message 'easy to assemble'. Except for the few of us who are delighted with such challenges, we know that one or more days that could have been better spent may be eaten up in the process. We are apt to run into dead ends, endure breakage and be subject to the humiliation of having to turn to our children or grandchildren for help.

4.3 The throw-away society

I have often heard the complaint that people nowadays lack social conscience and that one of the extreme manifestations of this is the throw-away society. We simply dispose of valuable products that are essentially still functional but have run into some problem. Not only is that a waste of the product itself, but it also exacerbates the community's waste disposal problem which threatens to destroy the attractiveness of the neighborhood, and worsens the scarcity of space that is available for use as disposal. It may also contribute to pollution and deterioration of the air and waterways.

But this is largely misunderstanding. I still own my father's gold-filled pocket watch, a Waltham dated, I believe, 1915. It is a beautiful mechanism. But to have it run accurately, he would send it annually for cleaning at a cost, I recall, of $10. Anyone who has ever disassembled such a watch, cleaned the parts and reassembled it (as I have done, sometimes even successfully) knows that several hours of labor (it should be skilled labor) is entailed. Today such a thorough cleaning job, with complete manual disassembly, would probably cost hundreds of dollars. But I also own a wristwatch that cost me less than $7 new, and that has kept amazingly accurate time now for more than three years. When the battery finally expires and replacement undermines the water resistance of the watch, it will simply be logical for me to replace it with a new one.

There is a general point here. Maintenance and repair of products is a process that inherently resists automation. A mechanical watch that has been dropped and no longer runs can suffer from many different forms of damage. As the doctor must examine each sick patient individually to determine just what the problem is (and often fails at the task), so the person who repairs an antique watch or fountain pen must personally disassemble the item and inspect it with care. Even if one malfunction is identified, this does not preclude the possibility that there are others and that an ostensibly repaired product therefore still will not work. The fact is that it is typically far too expensive to arrange for the handicraft task entailed in repair of such an item rather than just throwing it out and buying a replacement.

This is all readily verifiable by example. Figure 4.1 shows 40 years of data on the cost of automobile repairs and insurance, as compared with the overall price level as given by the Consumer Price Index (CPI) of the US Department of Labor. It is clear that the speed of increase of auto repair prices significantly exceeds the economy's overall rate of inflation. Problems that need repair do not come in perfectly uniform patterns, particularly if they are the result of an accident. So they cannot just be put on an assembly line along with many other broken items and their

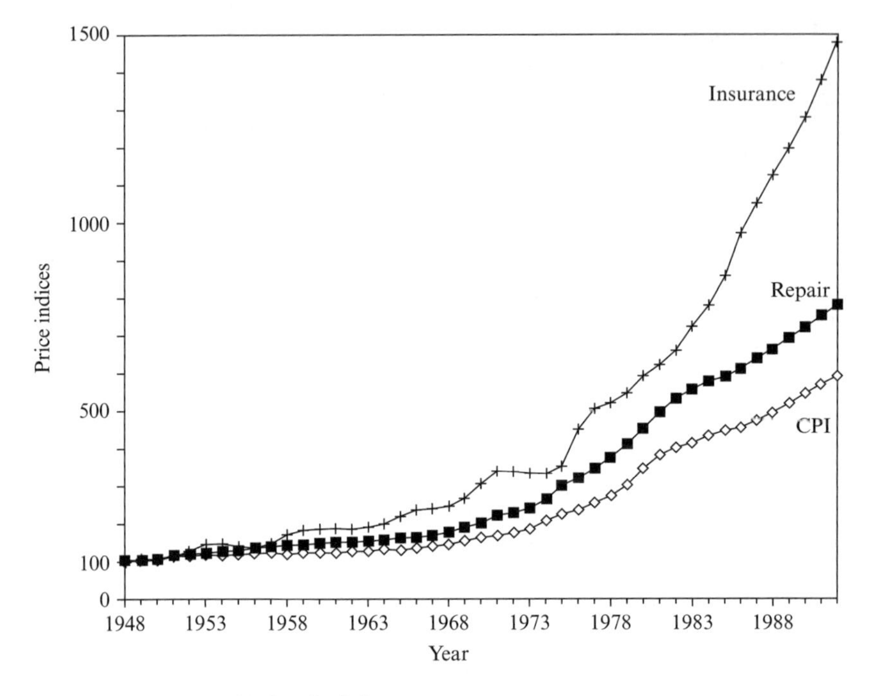

Source: US Bureau of Labor Statistics.

Figure 4.1 Price trends: auto repair and maintenance vs. auto insurance vs. CPI, 1948–1992 (1948 = 100)

restoration automated. There is no easy way to cut significantly the labor time required for the task.

The price of insurance has gone up faster still, as the graph shows, because it covers not only the cost of auto repair but also the medical cost of accident victims, who also require 'repairs' that are hardly standardized and homogeneous.

4.4 The contribution of automation

The cost-disease's rise in prices of handicraft activity has also served as an incentive for labor-saving innovation, some of which simply makes it possible to transform a task into a do-it-yourself activity. A straightforward example is the disappearance of the elevator operator. That is, the disappearance is not yet complete. It remains as an example of what Veblen called 'conspicuous waste'. It entails spending carried out simply to show to the world that the person making the outlay is very wealthy and can afford the useless expenditure. For this reason, places such as very elegant hotels that want to advertise their ostensible luxury, retain uniformed elevator operators wearing handsome gloves to run the machinery.

The marvelous machinery that is now found in every kitchen of high quality is also designed to transform common activities into do-it-yourself exercises. The dishwasher, the garbage disposal machine in the sink, the automatic toaster, the clothes washer and dryer, and much more, are evidently intended to serve this purpose and help to facilitate replacement of the labors of the hired cook with the resident's own efforts.

Outside the home one can find many other illustrations. The automated teller machine (ATM) is evidently designed to have users do for themselves what used to be carried out by bank tellers. The obvious insult entailed in the arrangement is the charge for its use. Bank depositors are expected to pay for the privilege of doing their own work in depositing cash or withdrawing it.

Equally widespread is the role of television and news websites, which are replacing newspapers, saving the cost of the human activity expended in delivery to news-stands. Production of the newspaper itself has been modified in ways that serve to combat the cost disease. Setting of type certainly is no longer done by hand or even with the aid of the labor-saving linotype machine. The electronic controls make typesetting and correction far less labor-intensive. Yet the price of a newspaper such as the New York Times has nevertheless exploded from 2 cents at the time of my childhood to $1 on weekdays and $3.50 on Sundays.

Automation is spreading even into realms that are suggestive of science fiction. Indeed, Steven Spielberg has reportedly been driven to comment that: 'there is no longer any science fiction'. Health care provides many startling examples. One, that was recently described at a meeting of the American Philosophical Society, involves robotics and X-ray technology. The patient undergoes an advanced form of radiology far more sophisticated than a CAT (computerized axial tomography) scan. The output, that has the appearance of a credit card, contains a set of photographs of the patient's innards, layer by layer, from the skin to the skeleton. Then, if the patient is in need of surgery, this can be carried out by a robot, preprogrammed by computer for the task, and guided by the credit card. Several dry-run surgeries can be carried out to ensure that the robot has it right. Then the actual operation is carried out, with any medicines, blood or other materials used in the process replaced immediately and automatically. A surgeon is in attendance while the operation is under way, but is there only for emergencies. This process has already been carried out upon a live patient, who was far from the location of the supervising human surgeon. Here the primary purpose, clearly, was not saving of human labor. But it suggests the imaginative forms that pursuit of that goal can take.

4.5 Surviving stagnant services

The foretaste of the future, as affected by the cost disease, has focused upon consequent changes in the ways in which these services are provided. Do-it-yourself procedures and the substitution of throw-away and replacement for repair are two examples in which the supply of the products at issue may or may not have changed considerably, but only the way in which they are obtained. But one of the issues raised by the ever-increasing prices that are analyzed by the cost disease is whether the future will lead to reduced demand and supply of those items. We have already noted that this need not happen, because we can afford not merely the quantities of those items that society obtains now, but we can have at our disposal ever-rising quantities of those items and often with ever-enhanced quality. Thus, there is no need in the future for reductions in the provision of health care, education or the performing arts. And here, too, the future is already with us. In much of the world in recent decades, health care has been improving substantially and education, at both the elementary and advanced levels, has become ever more common.

The upshot is that the cost disease may conceivably end up by constricting such activities, as it has done with household servant labor, for example, but it need not do so in the foreseeable future. For commodities whose demands are persistent and resistant to price increases[1] but expand when rising productivity enhances the purchasing power of the representative power, then despite appearances, the future will promise more and better health care and education. That, then, is not the real danger that the cost disease threatens to make unavoidable.

4.6 Evolution of military technology and the real threat of the cost disease

The steep decline, since the onset of the Industrial Revolution, in the cost of manufactured goods to the purchaser is spectacular, particularly if we use what may be considered the truest measure of the price of the new products in the cornucopia that has become available to consumers – the number of hours or minutes that a representative individual must work in order to earn enough to purchase those products. While no medieval peasant could have dreamed of possessing a vehicle such as was available to the lord's family, today most American, Western European and Japanese workers own at least one automobile that is far faster and more comfortable than any medieval vehicle and, besides that, most now possess a television, a microwave oven and much more. Further, consider the following. In 1919, the average US worker had to labor nearly an hour to buy a pound of chicken. At today's wages and poultry prices, less than five minutes of labor is required for the purpose. Food is not the only item that has become much less costly in terms of the labor time needed to pay for it. There have also been great cost reductions in various types of electronic equipment – a cut of 98 percent in the cost of a color TV between 1954 and 1997, 96 percent in the cost of a VCR between 1972 and 1997, and 99 percent in the cost of a microwave oven between 1947 and 1997. Of course, the most sensational decrease of all has been in the cost of computers. Computer capability is standardized in terms of the number of MIPS (millions of instructions per second) that the instrument is capable of handling. These days, it costs under 27 minutes of labor per one MIPS capacity. In 1984, it cost the wages of 52 hours of labor; in 1970, the cost was 1.24 lifetimes of labor; and in 1944, the price was a barely believable 733,000 lifetimes of labor (Cox and Alm, 1997).

Military equipment, or at least some of it, is no different. While the cost of the latest fighter jet goes heavenward and the military budgets of the world's major powers impose damaging deficits upon their governments, bargain-basement equipment has also made its appearance (note the quotation from the *Washington Post* at the head of this chapter). The terrorists and guerillas of the world have demonstrated beyond doubt how effectively these kinds of products can be put to use, and have proven how effective such weaponry can be in stymieing the most determined counter-efforts of major powers equipped with extensive manpower, organized forces and every device that money can buy. For perhaps the first time in history, vastly superior wealth, utilized with thought and determination (as, for example, occurred in the case of Ulysses S. Grant's destruction of the Confederate Army of the evidently cleverer General Robert E. Lee), no longer assures military success. Wealth is no longer the near guarantor of military success, as it apparently used to be, if the quotations at the beginning of this chapter are right.

The technology of warfare clearly has continued to evolve, arguably outpacing the incredible and utterly unprecedented pace of innovation in general during the past two centuries. There have been two major consequences critical for my topic. First, of course, is the plain fact that humanity now has it in its power to commit suicide, finally and completely, via nuclear holocaust. And, as just noted, developments in military technology have produced an outpouring of powerful and often bargain-basement-priced products, with a number of doomsday weapons among them – be they biological, chemical or nuclear.

One of the most chilling prospects, in the wake of the events of 11 September 2001, is the possibility that the perpetrators could strike again, this time with nuclear weapons, including so-called dirty bombs (in which the waste products of nuclear reactors are wrapped in conventional explosives), or a terrorist attack on a commercial nuclear power plant utilizing a commercial jet or heavy munitions, or the possibility that terrorists could build or obtain an actual atomic bomb and detonate it in a city. While it is less likely that 'fly-by-night' terrorist groups could produce sophisticated nuclear devices, a greater threat is a nationally supported program under the sponsorship of a malevolent rogue regime that provides the necessary resources and facilities. With more nations 'going nuclear' or wanting to go nuclear, the possibilities are all too frightening.[2]

The bottom line is that the cost disease is a disease indeed. But its frightening consequences are very different from what is most easily imagined. And a search for a cure is surely a matter of the highest priority.

Notes

1. That is, if they are what we economists call 'price-inelastic' (and 'income-elastic'); that is, demands are not reduced much by rising prices but are raised by growth in purchasing power.
2. A recent *New York Times* article reported that North Korea's successful atomic test in 2006 'brought to nine the number of nations believed to have nuclear arms. But atomic officials estimate that as many as 40 more countries have the technical skill, and in some cases the required material, to build a bomb' (Broad and Sanger, 2006).

References

Broad, William J. and David E. Sanger (2006), 'Restraints fray and risks grow as nuclear club gains members', *New York Times*, 15 October, available at http://www.nytimes.com.
Cox, W. Michael and Alm, Richard (1997), 'Time well spent: the declining *real* cost of living in America', *1997 Annual Report*, Federal Reserve Bank of Dallas, pp. 2-24.
Kahaner, Larry (2006), 'Weapon of mass destruction', *Washington Post*, 26 November, available at http://www.washingtonpost.com.

5 The environmental crisis and the economics of services: the need for revolution

Jean Gadrey[1]

The Earth is not a gift from our parents. It is on loan to us from our children. (Native American saying)

5.1 Introduction

With a few exceptions, the economics of services, as currently constituted, takes little account of environmental or social considerations. Except in the case of transport, little attention has been paid to the links between services and the environment and scarcely any more to inequalities, whether in access to services or those caused by services (a few analyses of public services or of the dualism of service-sector employment notwithstanding). This situation is set to change, and probably quickly, in the coming years. However, the environmental and social revolution in the economics of services will come up against various obstacles, some of which are not specific to services. Four are outlined below, although this is by no means an exhaustive list.

5.1.1 Obstacle 1: delayed awareness, for a number of reasons

A historically unprecedented environmental and social crisis is looming, one that might even lead to the 'collapse' of human life across the globe. However, awareness has been slow in coming, because the manifestations of this crisis are still limited, particularly for the dominant groups and in the rich countries. However, it has been delayed also because the solutions, which exist, conflict head-on with short-term private interests, with the similarly short-term principles driving financial globalisation and with a system that produces consumerist greed by creating superfluous needs, as reflected in, among other things, the exorbitant amount spent on advertising across the world, with total expenditure of almost $500 billion forecast for 2008.[2]

5.1.2 Obstacle 2: environmental economics is in its infancy and not familiar to economists

It is difficult to conceptualise an environmental and social economics (including an economics of services), given the need to make projections well beyond the 'long term' as defined by economists (and even beyond Kondratiev's long cycles) in order to incorporate environmental and human forecasts for a century ahead at least, as the Stern Review (2006) does, for example. A further difficulty is the need to take account, as the Stern Review does, of the contributions of a number of disciplines hitherto ignored by economists, including climatology, agronomy and the natural and life sciences, to say nothing of the other social sciences, of course.

5.1.3 Obstacle 3: services are ignored by political environmentalism, while environmentalism is neglected by the economics of services

Political environmentalism and environmental economics, which both have rich research traditions, have paid little attention to services, with the exception of transport. On occasions, they have actually considered services (or market services to households at least) as inglorious, even servile activities. Many of their key concepts were developed with reference to manufactured and agricultural goods rather than to services. For its part, the economics of services has practically ignored environmental issues. Now it can be shown that the environmental crisis will not be resolved without close attention being paid to services and the economics thereof (services now account for three-quarters of total employment in the rich countries, and for almost the same share of value added) and that the future of services will be determined by their relationship with the natural world, to the same extent as other activities, albeit in a somewhat different way.

5.1.4 Obstacle 4: internalising externalities

Environmental externalities will lie at the heart of an environmentalist economics of services rather than on the periphery which, as their name indicates, is where they are positioned in non-environmentalist (in fact anti-environmentalist) economics. Externalities will have to be internalised, not only by being incorporated into performance and cost evaluations but also by being regarded in theoretical terms as linked to common assets as essential as the 'internalities' – i.e. end products and services – from which they are derived. When the long-term collateral damage can be equal in magnitude to the benefits (in short-term well-being) derived from the production of goods or services, then analysis of the former is as vital as that of the latter.

5.2 Prologue: conflicting findings

One starting point for investigation of the role of services in terms of their sustainability (here in the environmental sense of the term) would be to compare two types of figures, which result in radically opposed diagnoses.

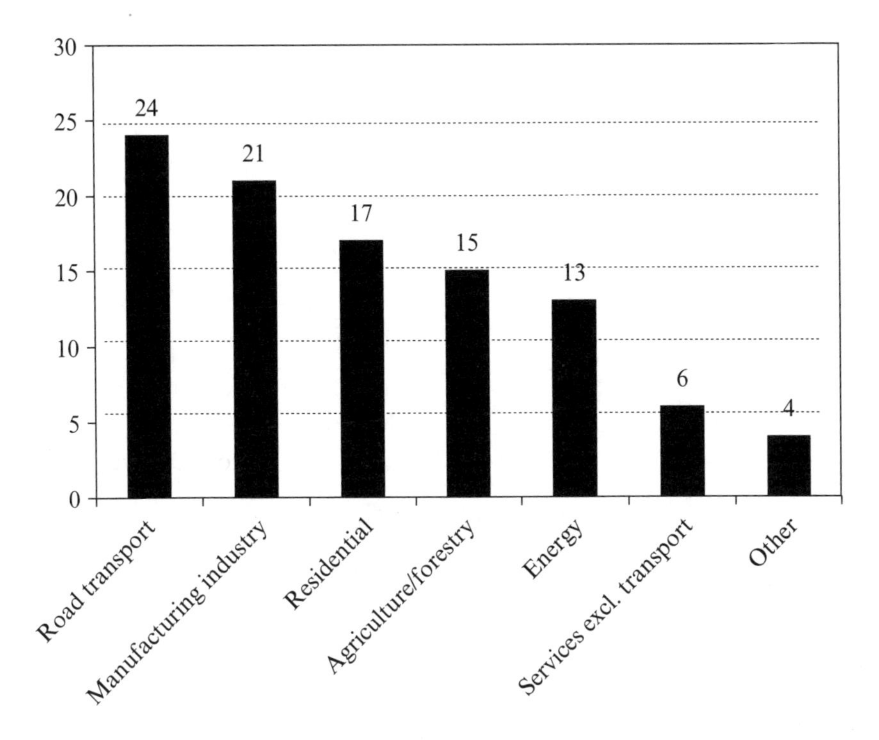

Source: Citepa (2007), http://www.citepa.org/emissions/nationale/Ges/Emissions_FRmt_GES.pdf.

Figure 5.1 *CO_2 emissions by sector, France, 2005 (%)*

According to the first type, services are astonishingly 'green' activities; while according to the second type, service economies are the heaviest consumers of natural resources and the most polluting.

The first type of figures – in France, data on CO_2 emissions, which form one of the major indicators of the environmental pressure associated with the risks of climate change – are published for different sectors of human activity. The results shown in Figure 5.1 are obtained.

A graph of this kind seems to give a decisive advantage to services (excluding transport). In 2005, they accounted for 71.5 per cent of employment in France but for only 6 per cent of emissions, which constitutes an apparently remarkable environmental performance compared with the other sectors of activity (the 'residential sector' being a sector of consumption rather than of activity). According to these figures, manufacturing industry, which accounts for 21 per cent of emissions but only 13.3 per cent of employment, is 19 times more polluting per job than services,

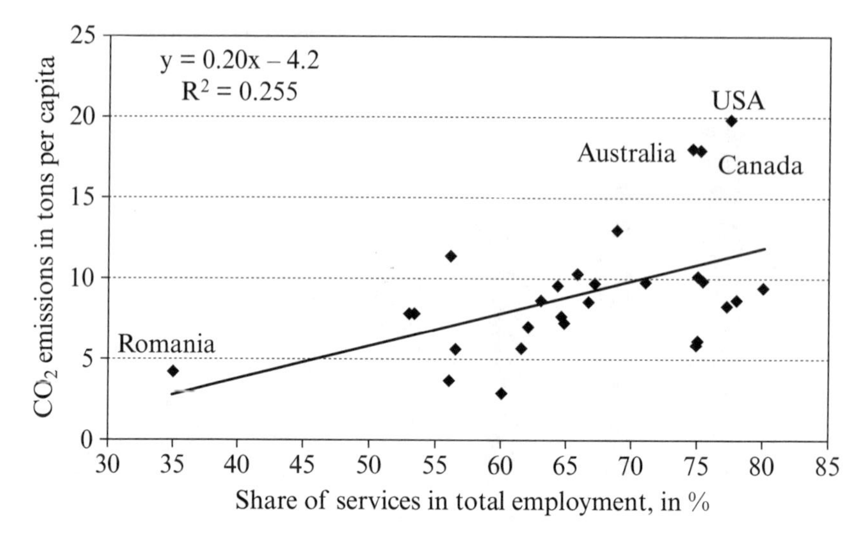

Sources: Employment in services: OECD (2008) and Eurostat (various years); emissions: UNPD (2006).

Figure 5.2 28 OECD and EU countries: share of services in employment and CO_2 emission, per capita, 2003

agriculture 50 times (for 3.6 per cent of employment), road transport 90 times (for 3.2 per cent of employment) and the energy sector 190 times more polluting (for just 0.8 per cent of employment).

It is on the basis of such figures that some commentators have come to regard 'service sector expansion' as a factor favourable to 'sustainable development'.

Regarding the second type of figures: if services are environmentally very 'light', it might be assumed that those countries whose economies are the most dominated by services perform better in terms of their environmental pressure per inhabitant. Taking CO_2 per inhabitant as the variable once again, and combining it with the share of services in total employment, the graph in Figure 5.2 is obtained for 28 Organization for Economic Co-operation and Development (OECD) and EU member states (some countries are missing because of an absence of data).

Although it is not very strong, the correlation is significant, and the trend is clear: the most service-oriented countries are the most heavily polluting according to this criterion. The league leaders are the three countries that have refused to sign the Kyoto Protocol, including the USA. Australia, which has been ravaged since 2005 by drought and fire, went back on this refusal in 2007.

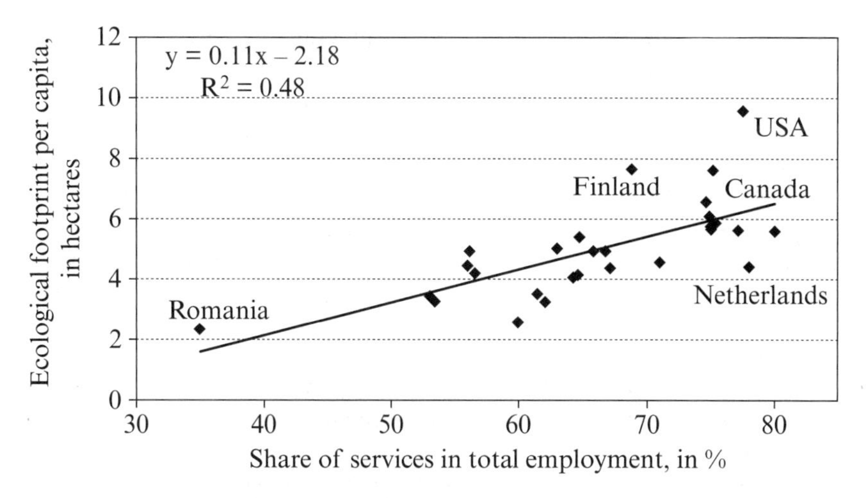

Source: for ecological footprint in 2003, WWF (2006).

Figure 5.3 28 OECD and EU countries: share of services in employment and ecological footprint per capita, 2003

The result is the same, with a stronger correlation, if the CO_2 emissions variable is replaced by the ecological footprint per person[3] (Figure 5.3).

These contradictory empirical findings are an inducement to investigate in greater depth the question of the environmental pressure that services exert and how to measure it (section 5.3). I will then go on to look at the future of service-sector employment under environmental constraints (sections 5.4 and 5.5).

5.3 The materiality of services and their externalities

The idea that 'the economy is becoming dematerialised', as a result in particular of the expansion of the service sector, is a falsehood. A service may be, as has jokingly been said, 'a product that doesn't hurt if it falls on your feet', but the production and consumption of services sometimes do as much damage to the planet as other products.

If this is true, and we shall see that it is, then services are not the miracle solution to the problem of the environmental sustainability of the economies of the future. It is true that, on average, less energy and fewer natural resources are consumed in producing them (per job, for example) than in agriculture, energy generation or manufacturing industry. Firstly, however, this average actually encompasses very different situations, ranging from road and air transport, which beats all the records for greenhouse gas emissions, to local relational services, which exert relatively little, though by no means negligible pressure on the environment. Secondly, and above all, it is based on a partial and misleading calculation that ignores one essential 'material' characteristic of most services, namely the simultaneous physical presence of the providers and users who 'co-produce' the service.

Thus it is not sufficient to calculate the environmental pressure exerted by services by taking into account only the material factors of production – in the strict sense of the term (buildings and heating, machinery, computers, energy etc.) – used to produce them. Account also has to be taken of the material factors used in the process of co-production (section 5.3.1). The second unfortunate oversight concerns the main 'factor of production' in services, that is the human factor, which is not included in the usual calculations of the environmental pressure that services exert. It has to be reintroduced, here as elsewhere (section 5.3.2). I will then – from a critical perspective – touch upon a third erroneous argument, namely that the reduction in the consumption of materials per unit of gross domestic product (GDP) is evidence of the 'dematerialisation' of the economy (section 5.3.3). This argument is not specific to services but does concern them indirectly. We will then see that, for the moment at least, the development of new information and communication technologies (NICTs), which has been particularly significant in services, has not in any way been accompanied by dematerialisation or by any reduction in environmental pressure (section 5.3.4).

5.3.1 The service relationship and its material environment: the need for a comprehensive environmental analysis

The 'service relationship' and 'co-production' are two of the characteristics of services that have, quite rightly, attracted considerable attention from socio-economists interested in services and been the subject of many studies. However, these studies have tended to disregard the material nature of these 'encounters' and instead have concentrated on analyses of verbal, cognitive and contractual exchanges, often inspired by interactionist sociology.

The following are the three main components of the materiality of the service relationship:

1. The travelling that is essential to provision of the service: journeys by service providers in some cases (home services, consultants etc.) and by users/customers in others (by far the most numerous: pupils and students, patients, customers of shops or of hotels and restaurants etc.) and, sometimes, both groups of actors in the relationship (tour guides, taxi and bus drivers etc.), as well as journeys made by service-

sector employees to government offices or the places where service organisations deliver their services. No services can be provided without a great deal of travelling, which tends to be forgotten in analyses confined to the standard 'factors of production'. And the materiality of this travelling lies not simply in the flows (emissions from private vehicles and public transport, energy consumed) but also in the stocks of vehicles and in the relevant infrastructures, to the extent that they are used in the production and consumption of services.

2. The materiality of the spaces where the service relationship takes place: offices, classrooms and lecture theatres, hospitals, bank counters, etc. Natural resources are required to build them (stocks) and others (regular flows) are used in maintaining, heating and lighting them.
3. The materiality of the technical tools deployed in support of the service relationship: at bank counters, in shops and, even more so, in hospitals, as well as the materiality of the 'back office', where both stocks and flows may be particularly plenteous.

The main explanation for the considerable advantage that services enjoy in the data on CO_2 emissions by sector (Figure 5.1) is that the figures include only the flows of emissions associated with heating and lighting buildings and, where applicable, other uses of energy for the technical tools. This is only a small fraction of the material resources used in producing and consuming services.

For example, in the carbon assessment carried out by the French Environment and Energy Management Agency (ADEME) for the Casino supermarket chain, almost 40 per cent of the CO_2 emissions in all the categories taken together were attributable to customers' journeys. Would this figure be very different in the case of a university campus or a hospital, given the volume of daily journeys induced by 'immaterial' activities such as teaching, research and the practice of medicine? Would it be very different in the case of variety shows and concerts (which, according to Baumol, are typical examples of 'stagnant services', see section 5.4)? Are the academics who are so keen on organising 'high-level', preferably international conferences aware of the contribution their 'intangible' production of knowledge is making to global warming?

5.3.2 Environmental assessments of service industries: essential to include service providers' consumption of natural resources

There is a tendency to forget a major fact: the growth in services has been possible only because, on the one hand, service providers have been able to enjoy a standard of living comparable to that of other economically active groups (this is the wage equality hypothesis in Baumol's model, see section 5.4) and, on the other, because consumers have been able to spend increasingly less (as a share of their budgets) on basic material goods and become purchasers of services (this is what is sometimes called 'Engel's law'). This has come about because the material goods that form the basis of this standard of living (those required to feed, house and clothe oneself) have been the object of fantastic gains in 'raw' productivity, which have been obtained, particularly since the early 1950s, by overexploiting natural resources and producing waste and greenhouse gas emissions at a level beyond that at which nature can regenerate itself at its own pace. To put it very provocatively, the service society, at least as it has developed – by relying on an 'unsustainable' form of productivism applied in other sectors but also sometimes in services as well – has unwittingly become a hyper-industrial, hyper-material, anti-environmental society.

Let me add another argument in support of my view that contemporary service societies can be described as 'hyper-material'. If we look just at employment figures, it is clear that, regardless of economic circumstances, there has been a considerable increase both in total employment in services and in its relative share in employment, to the detriment of agriculture and manufacturing industry. In the case of France, and confining myself to the last 30 years and focusing on the growth in employment (index 100 in 1978), Figure 5.4 is obtained.

The picture is very different when it comes to the quantities produced or consumed ('volumes' in the statistical sense). For example, taking the principal items in household final consumption (excluding public non-market services), the graph in Figure 5.5 is obtained.

Since 1959, the volume of industrial goods consumed by households has increased by a factor of more than 4.5, just as that of market services has. As for the volume of agricultural products consumed, the increase is much lower, it is true, but it has more than doubled, nevertheless. Consequently, even those who would explain the increased materialisation of the economy by pointing to the increased consumption of industrial and agricultural products should understand that a graph of this kind supports the idea that 'service economies' (i.e. those in which services account for the largest share of employment and value added) are increasingly 'material' economies. In 'service economies', increasing volumes of 'material goods' are being produced and consumed.

5.3.3 The reduction in emissions and in consumption of materials per unit of GDP has nothing to do with the dematerialisation of the economy

The notion that the economy is becoming 'dematerialised' is sometimes based on the observation that the evolution of GDP (in volume terms) is becoming decoupled from indicators of environmental pressure (use

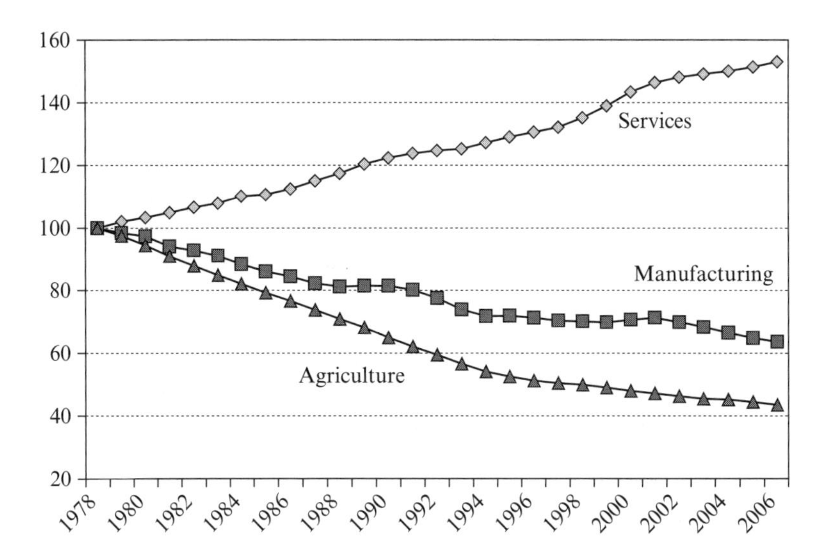

Figure 5.4 Index of employment variations by sectors in France (1978 = 100)

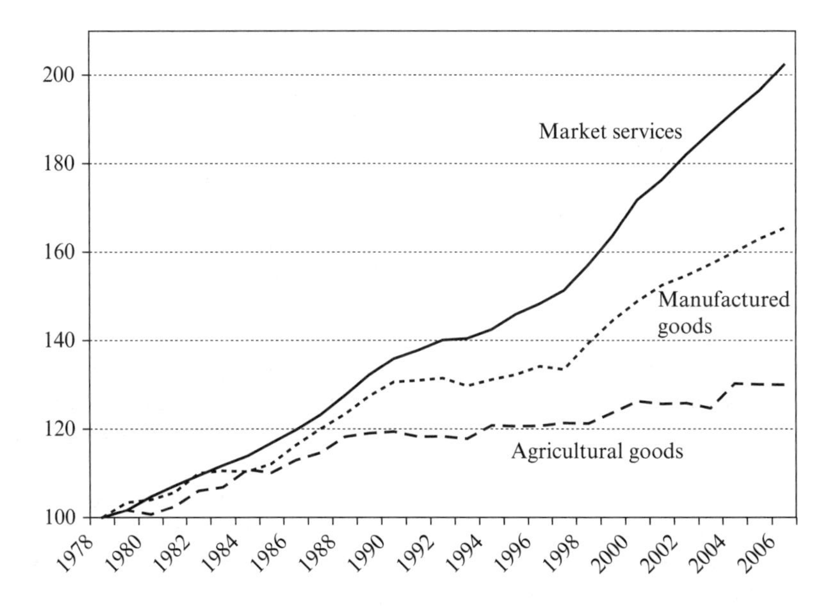

Source: Insee, French National Accounts.

Figure 5.5 Household consumption index, France (1978 = 100)

of resources, emissions and pollution) or – and this amounts to the same thing – on figures that show reductions in the volumes of materials used and pollution produced per unit of GDP. Such observations are not without interest, but they completely bypass the question of the finiteness of renewable natural resources and the exhaustion of non-renewable resources. The only thing that counts in assessing whether we live in societies whose 'objective materialism' is increasing or declining under the constraint of finite resources is the absolute total volume of appropriations and of waste. Both are continuing to increase and have reached levels considerably greater than nature's capacities (Figure 5.6). This criticism of the dematerialisation argument does not apply to services specifically, but it does concern them indirectly, since those economies most dominated by services are also the ones that make the most intensive use of natural resources.

5.3.4 Nor are NICTs, which are useful in some respects, 'the' solution to ensuring the sustainability of services

I will restrict myself to two indicators. The first concerns the materiality of computer production (there are said to be about 1.5 billion computers in operation around the world today), while the second relates to the comparative evolution of information flows and transport flows since 1800.

Computers and their networks The materiality of computers has been the object of various studies, some of which were carried out under the aegis of the United Nations. What matters is not so much the quantity of materials present in computers as finished products, or even the energy flows required to operate them, although they are becoming not insignificant (see below), but rather the flows required to produce them. The available orders of magnitude are as follows. The CO_2 emissions linked to the production and transport of a desktop computer with a flat screen manufactured in Asia amount to 1.3 tonnes, half of which are accounted for by air transport. Remember that, in a perfectly equal world, with its current population, each person's 'sustainable emission right' would be 1.8 tonnes of CO_2 per year. This is what nature can absorb without worsening the greenhouse effect.

However, other uses of material resources have also to be considered. It takes about 1.8 tonnes of various materials to produce a desktop computer, including 240 kilos of fossil fuels and 1500 litres of water.[4] 'The water footprint' of a 32 megabyte memory chip weighing just 2 grams (also known as the 'virtual water' contained in the chip) is approximately 32 litres.[5] Taking these various factors into consideration, an everyday office computer has a high environmental impact, which brings to mind another

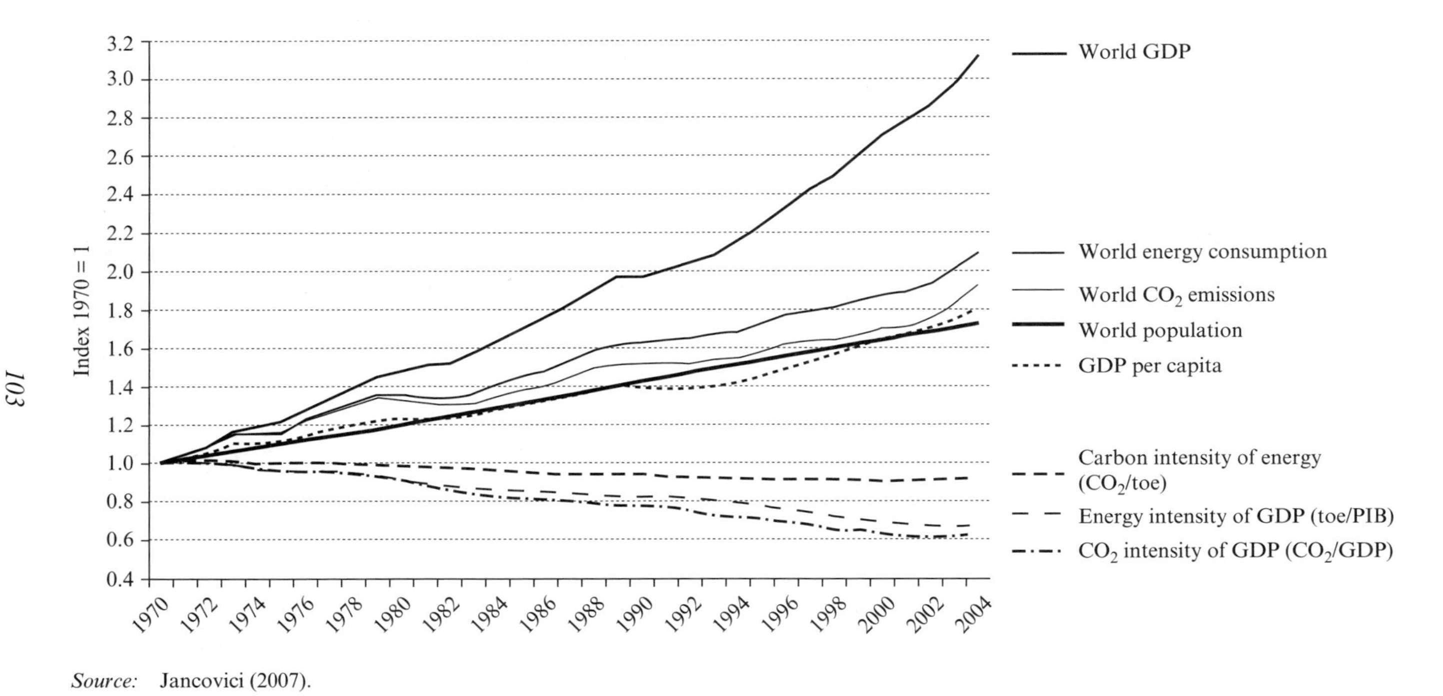

Source: Jancovici (2007).

Figure 5.6 Evolution of GDP and of indicators of environmental pressure (1970–2004)

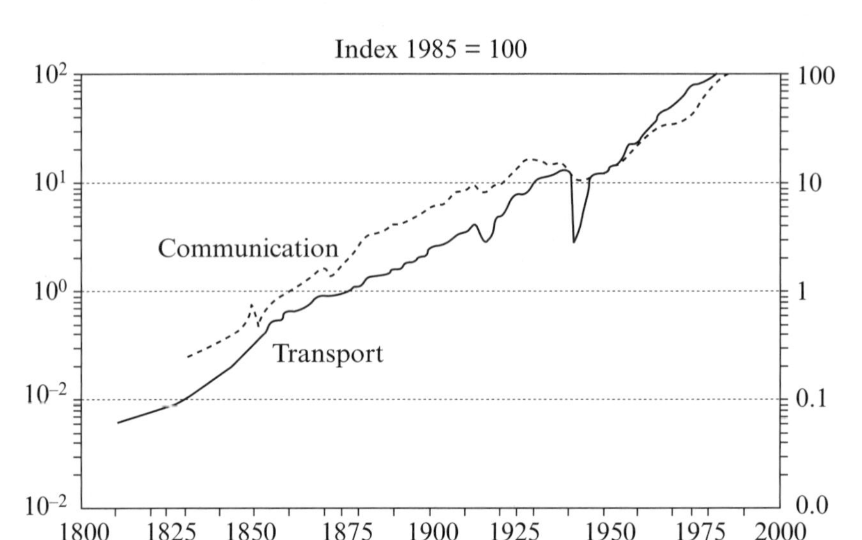

Source: Grübler (1990), quoted in IPCC (2001).

Figure 5.7 The evolution of transport and communication (1800–2000)

aspect of the impression of dematerialisation that might be created by superficial observation of modern service economies, namely the miniaturisation of the components and technical tools used in many services. In reality, what previous studies have highlighted is a form of 'decoupling' that is scarcely environmentally friendly. The more a product is miniaturised, the more the ratio between the volume of materials 'that remain in the end product' and the volume of materials used to produce it is reduced. In weight terms, this ratio is 1:2 for a car and 1:630 for an electronic chip (same source).[6]

If the energy flows required to operate computers and other information technologies are added to these material assessments, the picture darkens a little further: 'The energy footprint of information technologies is emerging as a major issue. According to the International Energy Agency, electronic equipment and standby settings have become one of the principal items in household electricity consumption. They have now overtaken the traditional "refrigeration" element and, in households so equipped, are only just behind electric heating.'[7]

The growth in information flows did not reduce the circulation of material flows in the past In the past at least (between 1800 and 1985), both these types of flows increased in parallel (Figure 5.7). It is possible to decouple them, and since the early 1990s indeed this is highly likely to have been the case, but it has not happened to the extent of achieving a partial substitution that would lead to a decrease in goods transport. I will return later to the case of transport and its rapid expansion in recent years.

Similar observations could be made, for the period from the 1980s onwards, with regard to the relationship between computer use and paper consumption,[8] or even the worrying increase in computer waste.

5.4 The trend towards increased service-sector employment will be affected or even reversed

This hypothesis constitutes a very serious challenge to a 'productivist' and 'non-environmentalist' theory that I, like all specialists in the economics of services, supported from the mid-1980s until very recently.

The observable historical trend (from at least the nineteenth century onwards) towards an increase in the share of services in total employment and value added will be modified and possibly reversed in developed countries, as will the trend towards economic growth and productivity gains, as they are currently defined and measured. This possible historical reversal, which would be nothing short of a revolution, can be analysed, initially, by asking what would become of the economic theory now accepted as authoritative in explaining the growth in services' share in employment if environmental externalities were incorporated into it (section 2.1). We will then see (section 5.4.2 *et seq.*) that we have to go further and change the conceptual foundations of these arguments, starting with the notion of productivity gains, the current definition of which is most inadequate when it comes to forecasting environmental sustainability and the future place of services in employment.

5.4.1 Revisiting Baumol's law: the relative prices of 'stagnant' services are not bound to rise continuously if externalities are internalised

So let us take as our starting point the approaches that have been standard since Clark (1939), Fisher (1935) and Fourastié (1949), in which there is a structural gap in productivity gains between manufacturing industry (and agriculture) and most services. From 1965 onwards, these approaches took the form of a simple model that has been universally applied under the name of Baumol's Law. Now this law should be revisited because the main underlying idea, namely labour productivity, is unable, at least as currently defined, to take account of externalities, particularly environmental ones. It will, however, be essential to internalise these externalities in order to deal with the main challenge of the twenty-first century, namely the increasing scarcity of the natural resources that first the 'industrial society' and then the 'service society' tapped and spewed out again, regarding them either as free (for example forests as 'carbon sinks') or available at low cost (oil, water). The way in which Baumol's model is affected can be demonstrated, in a non-formalised way, as follows.

Baumol's model rests on three hypotheses: 1) labour productivity gains are on average high in most manufacturing and agricultural activities (the 'progressive sector') and low or non-existent in services, or at least in many of them (the 'stagnant' sector'); (2) wages in these two sectors tend to converge. It follows from these two hypotheses that the relative price of services in the stagnant sector will rise continuously compared with those of manufactured or agricultural products. A third, bolder hypothesis then comes into play, based on certain empirical observations:[9] the relative demand for services does not decline compared with those for manufactured or agricultural products, despite the increase in their relative price. From this it is inferred, quite simply, that the share of employment in those services in which productivity is said to be stagnant will inevitably grow, with the only limit being that of the validity of the hypotheses.

This line of argument will have to be revised from its first two stages onwards if significant environmental externalities are internalised (by means of taxes or other political mechanisms), since the result will be increases in the unit cost of goods in the progressive sector depending on the 'materials' used and emissions produced per product unit, regardless of the evolution of labour productivity.

For example, if labour productivity continues to rise rapidly in manufacturing industry, then the 'labour value' (production cost as measured by the amount of labour required) contained in each unit of industrial product will certainly evolve in inverse proportion to the productivity gains, but the 'environmental value' (environmental cost, now internalised) of each unit will remain the same (the credibility of this hypothesis is examined below). It is as if a fixed proportion of 'stagnant' service had been incorporated into the production process – in this case a sort of 'service' provided by nature, which obstinately refuses to increase the productivity of its 'labour' (by absorbing more and more CO_2 per hectare of forest, for example), which is, after all, the cause of our concerns. In consequence, however, the first of Baumol's findings will have to be revised, since the relative price of manufactured goods relative to that of 'stagnant' services is no longer bound to tend towards zero. On the contrary, indeed, it will tend asymptotically towards a non-zero limit because of the fixed environmental unit cost of the manufacturing process. The conclusion remains the same even if the fact that stagnant services themselves also have environmental production costs is taken into account.

This questioning of the model would be unjustified if it could be assumed that, in the progressive sector (here manufacturing industry), the political costs (or prices) of 'nature's services' per unit produced, rather than being more or less fixed as has just been assumed, also tended towards zero (relative to the cost of stagnant services). This could happen if, firstly, the savings on materials, energy and emissions (i.e. the reductions in environmental pressure per unit produced) increased continuously and, secondly, the cost of nature's services (for example the politically fixed cost of absorbing a tonne of carbon) did not increase or if it increased less rapidly than the savings on natural resources and environment pressure per unit produced.

Now, the tendency in future decades will certainly be towards reducing the natural resources used and the emissions created per unit produced. On the other hand, given the constraints of increasing scarcity and the need to prevent long-term damage, it is very likely that the relative price (or 'tariff') of 'nature's services' will increase gradually, whether it be those that companies and governments already pay for by purchasing energy (including oil), water, etc. in regulated markets or those politically determined ones that they are going to pay for (as they already do in some cases) when externalities are internalised: taxes, public constraints, emission rights markets, etc. Thus it seems highly unlikely that the costs of nature's services per unit produced in manufacturing industry (or in agriculture) will be following a continuous downward trajectory in future. For the decades of transition, the constant price hypothesis that was initially considered is, in reality, 'optimistic'.

Even though manufacturers are succeeding in dematerialising their products to a significant extent (in terms of their natural resource content and emissions), for example through the reuse and recycling of materials, this scenario is all the more likely since such opportunities are virtually non-existent in agriculture, forestry and fishing, except with regard to emissions and pollution. A baguette cannot be dematerialised from the point of view of the quantities of wheat or flour required, even though it is well known that organic farming methods and an emphasis on local products can more than halve the associated emissions of greenhouse gases.

An objection The preceding argument rests on a hypothesis about the 'prices' of nature's services. Now if environmental policies succeed in re-establishing satisfactory ecological balances throughout the world after a certain transition period, there should no longer be any need for taxes, or at least tax rates should fall after this period. Under these circumstances, the conditions under which Baumol's law becomes valid would exist once again.

This objection is a fair one. However, nobody today knows how long the transition period during which high taxes on nature's services will be required is likely to last. Given the size of the current gap between the global use of these services and what nature can withstand, given the rates at which the Chinese and Indian economies are currently growing and given the persisting heavy investment in infrastructure projects, urban development, coal-fired power stations, etc., I would assume that it will take more than 50 years to achieve this goal, and in all likelihood the whole of the twenty-first century.

5.4.2 Beyond the concepts of productivity and growth

The preceding argument does not permit us to draw any conclusions as to the future of employment in the various sectors. After all, if the relative prices of agricultural products, for example, remain obstinately unchanged (or even rise), under the influence either of environmental taxes (water, energy, transport, biodiversity) or of inadequate supply (land use disputes, desertification, etc.), this will not for all that be a factor likely to cause the share of agricultural employment to rise. In themselves, taxes do not lead to any increase in value added net of taxes, nor in the volume of output or the level of demand, which is decisive for employment. As for price rises caused by inadequate supply relative to demand, they do not have this characteristic of stimulating employment either, if we remain locked into a productivist logic. They may very well lead to the emergence of agricultural rents (comparable to the current oil rent), without any positive employment effect.

Thus this argument has to be taken further by considering the following hypothesis from the point of view of the supply side: many necessary changes in production (of goods or services) will be accompanied by a reduction in labour productivity, as it is currently measured. Possible alternative measures are not to be excluded, but they do not yet exist. And although all sectors are affected, agriculture, energy, construction and parts of manufacturing industry will be in the front line. This will strengthen the tendency towards a halting of the decline in their share in total employment, particularly since some of the needs they exist to satisfy are basic ones, which will help to keep demand strong. Their share should even increase significantly in some 'material production' sectors.

The explanation can be summarised as follows. Sustainable, 'clean' or 'green' products (organic, recyclable, using less transport, energy, etc.) require, and will continue to require, more labour per unit produced than the polluted or polluting products derived from productivist procedures. Not overexploiting nature and managing its 'services' prudently is good for employment (at least in some sectors where demand should not weaken) but bad for 'raw' labour productivity, the measure that does not reflect the gains or losses of environmental quality caused by making the products in question.

All in all, on the basis of the hypotheses I am going to explore and for reasons that go beyond those I highlighted in revisiting Baumol's model, services will no longer be the only major sources of jobs in the future, even though some of them will function in this way in response to local social needs that will be the objects of innovations in sustainability. In order to make forecasts about sustainable employment, we will have to stop thinking in terms of the major (primary, secondary and tertiary) sectors of the economy and focus rather on the fairly disaggregated levels of individual industries and needs, taking into account not only the environmental sustainability of production and consumption but also the social issues that will emerge from this 'great transformation' towards an ecological economy.

Let us begin by enlarging on one major example. Employment should grow overall in one of the major sectors that, according to all the standard arguments, including Baumol's, is doomed to inexorable decline as a result of productivity gains, namely agriculture.[10] This fairly lengthy detour via agriculture in an analysis devoted to services may appear somewhat surprising. Firstly, however, the argument presented here can be applied to other areas of production and forces us to consider the way in which measures of 'raw' productivity act as blocks on progress. Secondly, economists specialising in services will have to rediscover 'a feeling for the land' and learn that service providers throughout the world will first have to be able to feed themselves within a framework of 'food sovereignty' that will have to be established. If nothing changes, we are heading straight for a world food crisis that will affect both the South and poor people in the rich countries. This would not be good for services.

5.4.3 Agriculture: an employment-rich green revolution to counter the world food crisis while respecting the environment

In the past, the main way of increasing the volume of economic output (as currently measured) was to obtain labour productivity gains, which can be defined as producing more of the same things with the same input of labour. Jean Fourastié (1949) showed, for example, that if around 200 hours' work were required to produce a quintal (100 kg) of wheat from the year 1000 until the eighteenth century, with considerable fluctuations from year to year, no more than 30 hours were required by about 1950. Today, that figure is down to 2.5 hours with the most 'productive' techniques, including the hours of work required to make the machines and 'inputs' required to produce the wheat.[11] What a remarkable contribution to growth. In the 'rich' countries, 2 per cent to 3 per cent of the economically active population are required to feed a country's entire population, and sometimes more if the trade balance is positive. All this is true, but such figures paint at best an incomplete picture.

Fourastié's argument, taken up by other economists, rests on one basic hypothesis, namely that a tonne of modern wheat is the 'same thing' as a tonne of wheat from the past. At first sight, this is indeed the case, and it would probably even be possible to identify certain physico-chemical properties that demonstrate the superiority of modern methods, if only the grains of wheat or the flour obtained from them were taken into consideration.

However, modern wheat is the product of an increasingly intensive form of agriculture that generates considerable negative externalities as well as the wheat. Intensive agriculture makes massive use of chemical inputs – pesticides as well as other chemicals – whose negative impacts on health, ecosystems and the pollinating insects that are essential for many other crops are beginning to be assessed. It is one of the factors contributing to the drying up and pollution of water tables and to increased desertification, both of which are reducing the two main resources required in agriculture, namely the land available for arable crops and water. It has replaced human energy with oil-based 'machine' energy, and it is this substitution of capital (and oil) for human labour that explains most of the productivity gains achieved in the sector. The largest tractors or combined harvesters are powered by 10- or 12-litre engines producing 500 hp. All this gives rise to considerable CO_2 emissions, and oil is going to be an increasingly scarce commodity. Finally, as the distance grows between the sites of large-scale agricultural production and the places where crops are processed and consumed, modern wheat or flour require transport, which is itself heavily polluting.

Thus intensively farmed wheat 'incorporates' a whole series of characteristics that impact negatively on sustainability. These negative characteristics are ignored in standard measures of productivity and growth. In order to take account of them, the estimated value of the extensive collateral damage caused by the production and transport of this wheat would have to be deducted from the value of this 'polluting' wheat, and the 'option value' (the value attributed to future uses) of the non-renewable resources used (the oil in particular) would have to be taken into account. This is difficult, but feasible. What would the result be? In the regrettable absence of any efforts in this direction, it is difficult to say, but taking as a starting point the price differences between intensive farming and organic farming, it can be estimated (very roughly) that the productivity of intensive agriculture, if adjusted in this way, would be cut in half, or even further, relative to its raw value, depending on the method used to assess the environmental damage and the oil depletion. The actual productivity gains achieved in the recent period, when the most serious environmental damage has been done, would probably be very low or even non-existent. More generally, according to the existing 'green' GDP indicators, the 'adjusted growth' would have been zero since 1970 in the USA and the fantastic growth rates in the Chinese economy could be divided by two or even more.

If intensive agriculture were to be converted in stages into a sustainable system, this would not mean a return to the Middle Ages. Such a conversion could even be described as a 'modernisation', requiring innovations, although one unexpected turnaround would have to be considered. Employment in agriculture, which has declined continuously since the Industrial Revolution (it now accounts for scarcely more than 3 per cent of total employment in France, compared with 27 per cent in 1954), should rise in order to meet sustainable demand, with 'sustainable' (higher) prices that incorporate the new requirements. The same would apply to value added in this sector: its share in GDP would increase significantly. This hypothesis about output rests implicitly on a second hypothesis concerning demand, namely that households would increase their 'budgetary coefficient' for these food products, which would have become both more expensive and sustainable; this in turn would require us to investigate the notion of a 'sustainable purchasing power' and to consider ways of reducing income inequalities in order that these sustainable products can be accessible for all.

The outcome of such a shift would be a sharp reduction in productivity, as measured by the current methods, but employment would increase, for a given output in raw volumes (for example, quintals of cereals). Zero growth (in this sector), but employment up? This seems difficult to believe. And yet it is possible if we take into account the necessary gains in quality and sustainability, which will constitute major new sources of jobs linked to 'sustainable development'. In reality, if the gains in quality and sustainability could be incorporated into measures of variations in prices and productivity, the probable result would be not a decline in productivity but rather gains.

In France, a sustainable, local form of agriculture, the aim of which would be to create a high degree of food sovereignty, which does not mean complete protectionism, could in the long term provide between 1.5 and 2 million jobs, compared with less than 1 million in 2007.[12] Only 1.7 per cent of French farms are organic, compared with 9 per cent in Austria. This figure will have to be increased ten- or twenty-fold. Furthermore, the prudent and sustainable use of biomass (one of the many sources of renewable energy, which includes dendroenergy) could by itself require the creation of 150,000 new jobs in agriculture.

5.4.4 Second example: renewable energy and energy-saving activities, supported by a wave of innovation, should create hundreds of thousands of jobs

Good-quality scenarios concerning the growing importance of renewable energies in France and in many other countries are beginning to become available. The 2006 scenario for France produced by the négaWatt association[13] (110 experts and practitioners) combines: (1) the growing use of clean energy technologies (renewables); (2) the drive for energy efficiency (in buildings and equipment, the aim being to consume fewer kWh for the same function, for example fewer kWh to maintain a given temperature in buildings); and (3) energy conservation (reduction of energy waste). This is far from being a decline scenario: between 2000 and 2050, electricity use would double with a stabilised level of kWh consumption (efficiency and conservation) and mobility (transport) would increase by 15 per cent (through a doubling of bus and train travel). Greenhouse gas emissions linked to the production and consumption of energy would be reduced fourfold compared with 2000.

In this scenario, employment would increase significantly in construction (particularly in renovation), production of components and materials, technico-economic research, energy services and maintenance, particularly if these emerging industries were assisted by public investment and tax incentives on the demand side.

The housing renovation programme would create more than 100,000 full-time jobs, wind energy would create more than 200,000 jobs between 2006 and 2050 and photovoltaic technology a further 150,000, while the prudent use of biomass would provide employment for 150,000 agricultural workers. These jobs, most of which cannot be relocated, would contribute to local development. This amounts to around 600,000 additional jobs.

But what happens in this scenario to gains in productivity and growth, as they are currently measured? The growth in employment and the stagnation of the amount of kWh produced will lead to a net decline in productivity and zero growth in the volume of output, as in the previous scenario for agriculture. However, such calculations ignore two essential factors. The first is that twice as much use (or 'end service') would be obtained for each kWh consumed. Now it is this latter use, and it alone, that counts in any assessment of the 'material well-being' linked to energy. The second is that a 'clean' kWh is not a 'dirty' kWh: the former does not produce the negative externalities the latter does. In order to be able to reflect what really matters in terms of sustainable well-being (the sustainable use value), the measure of productivity gains (and hence of unit prices, of purchasing power, etc.) would have to be capable of capturing the 'volume' of final utility as well as the gains in environmental sustainability (for example, the costs of the long-term damage that has been avoided). This is something to be gone into more deeply. Some hedonic indicators do exist; they are no doubt very imperfect, but they can be used to take account of the fact, for example, that a current computer 'gives its owner more' (uses, functions, power, speed, etc.) than a ten-year-old computer. Attempts could be made to subvert this method, provided agreements could be reached, so that it would be possible to make a dirty kWh and a clean kWh genuinely comparable, along with many other products and services that need to be 'greened' and better used: polluting agricultural products vs relatively non-polluting products, clean and unclean passenger miles in the transport sector, etc.

Since energy and agriculture are not isolated cases, it can reasonably be assumed that, among the main sources of useful jobs for the future, there will be improvements in quality, efficiency of use and environmental sustainability in all sectors (i.e. a reduction in negative externalities). In such cases, the current measures of productivity gains and growth will present a biased picture of 'progress' and may even indicate a decline that would not exist if account could be taken of all the very real and assessable components of the valued added created by the sustainable use of goods and services.

5.4.5 Local production as a source of sustainable jobs

It is now clear to all those concerned about the current and future state of natural resources that we have to prioritise local economic activities as much as possible. Their transport requirements are lower and they foster richer social ties, since they reduce the physical distance not only between production and consumption but also between producer, distributor and consumer networks by encouraging regionalised cooperation. This is what some people describe as 'relocalisation' and it applies not only to agriculture but also to manufacturing industry, construction, energy and most services as well. We will see that services can play a role of the utmost importance in this regard: unlike those approaches that have made globalisation the engine of expansion for services, it is their relocalisation and proximity that will undoubtedly be their main assets.

5.5 Back to service-sector employment: winners and losers

What have the observations in the previous section to do with a future-oriented analysis of services? Firstly, and fundamentally, if some major sectors that have been in continuous decline are once again to increase their share of total employment, this will, in purely arithmetic terms, leave less space for service-sector employment. However, this argument is not sufficient, since in both the primary and secondary sectors, there are industries that will develop, it is true, but also others whose decline in the long run is virtually certain. In the primary sector, fishing will decline (simultaneous scarcity of fish and of oil), as will cattle raising, among others. Other activities, both manufacturing and services, that will decline are those linked to the 'car, truck and plane society' and the corresponding infrastructures.

Thus the future of employment in services will be shaped by: (1) the sustainability of primary- and secondary-sector activities, reorganised on a non-productivist basis; (2) extensive innovations in services themselves, aimed at reducing their environmental footprint and developing local services that are both environmentally and socially useful; and (3) a decline in all forms of production and consumption that devour energy and ecological space for each unit of final service and produce high volumes of emissions, including greenhouse gases. This is what now has to be investigated, focusing on individual industries, sectors and distribution channels in order to identify the winners and losers among the various service activities.

5.5.1 The winners

We have seen that technological and social proximity is likely to become an essential element in the social organisation of sustainable production and consumption. Thus it would seem that services (excluding transport) enjoy a considerable comparative advantage in this regard, since most of them are provided as part of a 'service relationship' that requires the joint presence of providers and users. However, I have also seen fit to criticise this slightly idyllic vision from an environmental point of view, on the grounds that it tends to ignore, among other things, the travelling that is required, as well as the materiality of service providers' labour force and their environmental footprint.

It is still the case, nevertheless, that there is, on average, a clear difference between the environmental pressure exerted by local services and that exerted by the production of industrial and agricultural goods, the inputs for which and the outputs of which travel thousands of kilometres before arriving at the locations where they are handed over and used, even though there is a shortage of accurate figures, particularly on the extent of the travelling linked to services. In reality, regionalised services are very diverse, and the travelling and negative externalities to which they give rise depend on the organisational and locational choices that are made.

Large-scale retailing does not have the same environmental footprint as small neighbourhood shops organised into cooperatives linked to local producer cooperatives. Public services which, driven by the 'imperatives of productivity', shut down local offices or branches in order to concentrate their activities in amalgamated centres, increase pressure on the environment while at the same time reducing social utility. Home help services for elderly persons give rise to many more journeys than local retirement or nursing homes. Thus the spatial and social organisation of such services plays a more important role in determining their environmental impact than their shared description as 'local or neighbourhood services' would suggest.

However, the future of employment in services also depends, to a very large extent, on evaluations of future needs, with judgements being made as to the social and environmental utility of those needs, their universality and alternative ways of meeting them. There is no doubt, for example, that there are enormous legitimate needs pertaining to elderly people and young children, even though the responses to those needs (for example whether they are home-based or not) do not have the same sustainable societal value and depend on individual preferences and on the quality of the available options, including the human quality. The same applies in many public services. If France were to have the same density[14] of 'social services' (education, health, social work, services for the elderly and children, public authorities) as Sweden, 3 million additional jobs (many more than the number of unemployed people), of much better quality, would be required. This figure does not mean that other national models can be simply imitated, nor that the Swedish services in question are all environmentally viable, but it does indicate that there is scope for creating useful jobs, the environmental impact of which should be investigated.

At this point, the question arises of access to those very many services that can be associated with universal rights, either those already in existence or those still to be gained, subject to environmental constraints (another right). If the density of 'social services' is greater in the Scandinavian countries, it is also because it is believed in those countries that every citizen must be able to access them, even if that means they have to be provided free or at very low cost to people on low incomes. Thus the future of employment in such services also depends (still subject to environmental constraints) on choices between competing market solutions, access to which is determined by income, and 'universal' solutions, whether they are provided directly by the state or local authorities or whether they take the form of partnerships with non-profit associations or companies.

It is estimated, for example, that state provision of extensive early childhood services would require the creation in France of 1 million places from 2009 to 2025 and thus of an additional 150,000 jobs. If the significantly greater needs pertaining to the elderly (home help services, as well as – and perhaps above all – convivial retirement housing well integrated into the city, district and region) are added, there are no estimates of what the affirmation of a universal right would produce, but it is certain that it would amount to several hundreds of neighbourhood jobs. The Direction de l'animation de la recherche, des etudes et des statistiques (DARES, 2005) projections by groups of occupations are for 240,000 additional jobs created in ten years. These projections are not based on the assumption of universal services for young children and the elderly. If they were, this figure could be multiplied two- or threefold.

Other public or universal local services associated with rights that have yet to be asserted should be considered, particularly in the case of public housing construction and maintenance (where many jobs could be created), subject to environmental assessments of both stocks and flows. More generally, public services have helped (not without shortcomings) to sustain the regions. The weakening of public services and the disappearance of local services, quite apart from their adverse environmental impact, will drain some of the life out of rural regions (which will play a vital role in sustainability strategies) and urban fringe areas that are already experiencing social crisis and further reinforce the dual tendency toward polarisation, with a concentration of activities in certain areas and the dereliction of others.

Finally, against the background of an unfavourable public finance situation, employment in the cooperative, mutual and non-profit sector, which is essentially a service economy (and overlaps to some extent with local personal services), has increased, in France, by 15 per cent since 2000 and in 2008 accounted for 10 per cent of total employment, no longer very far behind manufacturing industry, with its 13.8 per cent share. Job quality is uneven and often low and the sector's environmental impact is little investigated, but its social utility is undoubted in most cases.

5.5.2 The losers

Heading the list of losers are road and air transport (and, to a now not insignificant extent, sea transport, emissions from which have recently been reassessed[15]), along with many other agricultural and industrial activities and services that depend heavily on them. There is no need to explain the causes of this decline; it is sufficient merely to mention the soaring price of fuel, the forthcoming taxes on energy and the absence, at least in the foreseeable future, of alternative technologies that reduce greenhouse gases by a sufficient amount. Considerable uncertainties exist, but they will not alter matters much, and even the European Environment Agency (EEA) has just admitted as much[16] by recognising: (1) that the explanations for the failure to reduce the environmental damage caused by transport are linked to the organisation of the sectors upstream that create transport needs (these include services, as well as urban planning) and to the political choices made in their regard; and (2) that it will be necessary to plan for their decline. This report notes that from 1995 to 2004, air passenger traffic increased by 49 per cent and private car use by 18 per cent, with the share of journeys made by car in the EU having risen to 74 per cent in 2004.

Among the services that are most heavily dependent on transport, international tourism and business travel based on air transport are going to founder, dragging down with them some economies that have made them the pivot of their development, unless of course some rapid restructuring can be achieved. Other sectors will be affected, however, including postal services, some distribution activities with large carbon footprints, etc. The international organisation of many activities (including academic research) will also be affected by these constraints. On the other hand, the future is more promising for less polluting forms of transport, including rail transport for both passengers and freight.

5.5.3 The long-term employment prospects: a synoptic table

Table 5.1 summarises my hypotheses about the future of service-sector employment in developed countries over the coming decades, from the point of view of sustainability and equity. It does not follow the standard classifications exactly. Primary- and secondary-sector activities are mentioned but not treated at any greater length. Some categories in the left-hand column overlap with each other. As it stands, this table cannot be used as a basis for an overall forecast of employment in services, even though it would seem that, between the winners and losers in the service sector and taking account of the other two sectors, we should not expect the share of services to rise in the long term.

5.6 Conclusion: the 'social issue' of the twenty-first century and services

The most important 'social issue' of the twenty-first century will be the challenge of dealing with the global social effects of the environmental crisis. This is what is shown by the 2007–08 United Nations Development Programme (UNDP) report (December 2007) on climate alone (see Box 5.1). However, an important earlier report (2005 Millennium Ecosystem Assessment[17]) examined the human effects of the destruction of ecosystems.

When it comes to services, the two strategic issues as far as equitable

Table 5.1 The future of service-sector employment in developed countries: winners and losers

Sectors	Challenges	Employment trends between now and 2050
Agriculture, forestry, fishing	Local organic farming as a major focus of development. Sustainable exploitation of forests and biomass. Sustainable fishing.	Overall growth in agriculture and forestry, with differences depending on environmental assessment of particular types of production. Sharp decline in fishing.
Manufacturing industry	Considerable reduction in volumes of materials and emissions per unit of final service. Partial relocalisation, advantage for local SMEs. Emergence of new sectors based on innovations in product sustainability.	Reduction in employment halted by restrictions on the productivism and organising principles of global firms. Restructuring and conversion possible in certain cases.
Energy	Proactive transition to renewable energy sources. Social pricing depending on income. Compromise between renewable energy sources with a large spatial footprint (biomass) and others that are more materials-intensive (photovoltaic cells, tidal barrages, wind power). Transition to the oil-free society. Energy conservation and efficiency.	Strong growth between now and 2050 in this sector and in all those contributing to energy savings, such as construction, heating, etc.
Construction and urban development	Savings on energy and materials, sustainable urban development, renovation, 'slow towns', focus on neighbourhood.	Growth during the decades of transition (renovation in the first instance).

Table 5.1 (continued)

Sectors	Challenges	Employment trends between now and 2050
Transport	Organised rundown of greenhouse gas-emitting modes of transport and decrease in other forms of pollution and production associated with the car, truck and plane society. Question need for speed. Increasing focus on less polluting, slower forms of transport. Investigation of factors generating transport needs.	Sharp decline in employment, except in the 'greenest' forms of public transport that will take over to some extent. It might be possible not to reduce travelling overall.
Commerce	Priority for local neighbourhood businesses based on cooperation with local producers. Cooperative groups. Scaling back of supermarkets/ hypermarkets and their unsocial opening hours.	Probable growth in overall employment, in order to encourage less productivist and concentrated modes of distribution.
Repairs, recycling, maintenance	These activities are some of the most important contributors to sustainability. Some could contribute to industrial redevelopment and conversion programmes ('service-oriented' manufacturing industry). Recycling of solid waste rather than incineration.	Strong growth for those repair, recycling and maintenance activities that remain affordable (since the cost of some may be prohibitive).
Goods rental	Establish an environmentally efficient distribution of use of goods.	Growth.

Table 5.1 (continued)

Sectors	Challenges	Employment trends between now and 2050
Craft enterprises	Important role in neighbourhoods: repairs, maintenance, recycling, etc.	Growth.
Postal services	How can (1) emissions per unit of mail or package; (2) flows be reduced? Impact on service quality, on mail order businesses, etc. Advantage for e-mail.	Net reduction (already under way, but on a productivist basis with multiple negative externalities).
Telecommunications and computer networks	Environmental footprint (materials, emissions) and water footprint of terminals and networks, value of sustainable use. Research into damage to health. Innovations designed with a view to 'greening' products and infrastructures.	Slowdown of growth in rich countries, increase elsewhere, subject to environmental constraints.
Banking, insurance and estate agency	Financial and housing crises will foster forms of public control or ownership and give rise to the need to regain control of real estate. Local institutions. New local currencies.	Reduction.
Business and administrative services (excl. post and telecoms)	For 'operational services' (cleaning, security, catering etc.), environmental and social challenges. For intellectual services, reduction of travelling by individuals, subsidy for local services, new sustainable development consultancy services, eco-design.	Slow growth in neighbourhood services, decline in advertising and marketing.

Table 5.1 (continued)

Sectors	Challenges	Employment trends between now and 2050
Hotel and catering, tourism	Sector adversely impacted by crisis in long-distance motorised transport, but advantage for local tourism and restaurants, subject to environmental supply constraints.	Probable reduction in employment, sharp decline in 'long-distance' tourism. Advantage for neighbourhood restaurants linked to local suppliers.
Services for young children and the elderly	Shift to universal solutions associated with rights, subject to environmental constraints. Assessment of relative advantages and disadvantages of home-based vs external solutions. Professionalisation.	Strong growth based on public funding.
Local government	Will be increasingly crucial for sustainability. Return to public control of many services subcontracted to the private sector under unsatisfactory conditions, both socially and environmentally.	Growth.
National government	Return to local offices and establishments, decentralisation but with social and environmental quality standards. With regard to health and education, debate on the excesses of the drive for formal qualifications and hypermedicalisation, but demands for equality based on rights. Same for justice.	National–local division will change, but the share of government departments and services should increase. Slight increase in education and health.

BOX 5.1 THE 2007–08 UNDP REPORT

Global warming is already, as of today, one of the main factors likely 'to stall human development', as a result of its effects on agriculture and food security (drought), water shortages, the disappearance of vital ecosystems, the exposure of coastal regions to flooding, increased health risks (malaria, etc.) and the strengthening of social inequalities.

What is already observable will get worse: 'climate change could leave an additional 600 million facing malnutrition by the 2080s . . . result in 330 million people being displaced through flooding . . . (and) an additional 1.8 billion people could be living in a water scarce environment'; 'For the poorest 40 per cent of the world's population – around 2.6 billion people – we are on the brink of climate change events that will jeopardize prospects for human development' (UNDP, 2007–08).

and sustainable human development is concerned are: firstly, the need to subject them to a comprehensive environmental analysis (see section 5.3 above); and, secondly, the question of universal access to essential services associated with rights that have already been recognised or are still to be asserted. However, neither of these two questions can be posed if services are considered in isolation from other activities and the way they are organised, both materially and spatially. For example, there will not be any sustainable services in unsustainable towns and cities, with unsustainable transport systems and buildings, and with providers whose standards of living and ways of life are not sustainable either.

The sustainability requirement brings with it an equality requirement, which concerns services as well. However, this double requirement will be the source of major conflicts.

The demand for equality will be expressed with regard to the notion of the available ecological space or footprint per person. The affluent classes and elites will do everything they can, in the name of their allegedly greater social utility, to retain their environmentally extravagant spaces by reducing as much as possible the environmental space used by people positioned at the bottom of the ladder. However, they will be in difficulty nevertheless: the rich could advance the argument, which is present for example in Adam Smith's work and has been constantly updated ever since by the elites, that rich people's wealth benefits everybody by sustaining business dynamism. This argument is based on a vision of material wealth that can be increased without limit. However, as soon as that wealth contains an absolutely vital element of natural finite resources, the argument no longer holds up: the environmental wealth of the rich (reflected, for example, in their ecological footprint) clashes head-on with the sustainable well-being of the greatest number. The struggle for spaces giving access to ever scarcer natural resources will be combined with a struggle between the classes for the control and distribution of economic output. This will lead to the re-emergence of the problems investigated by Fred Hirsch in *Social Limits to Growth* (1976).

Services will be directly affected by these egalitarian demands, since many of them are not only associated with universal rights (health, education, mobility, justice, housing, energy, space, etc.) but are also, in some cases, dependent on vital natural resources (water, energy, space, etc.). This will focus attention on the spontaneous tendency of competitive markets to produce or reproduce inequalities, exclusion and environmental damage. The 'competitive marketisation' of services, driven in particular by GATS (the General Agreement on Trade in Services), will be called into question. Although the current trend is towards the 'deregulation' of services, the environmental crisis could well require that they be re-regulated and that severe restrictions be placed on the current short-term competitive principle, which leads to social and environmental dumping. Shared environmental goods, or 'nature's services', will also have to acquire a legal status protecting them from the profit-driven market principle. This status will be a locus for major social challenges, since the removal of such goods from the competitive marketplace will not in itself be sufficient to guarantee equity of environmental sustainability.

In coming decades, the ecological economy in general, and the ecological service economy in particular, will require a higher level of public intervention and increased public funding, particularly in services, for three reasons: (1) those that have already been mentioned, namely the reduction in inequalities of access to nature's services and to the basic services that promote social cohesion and well-being; (2) in order to launch new investments that will be to the transition towards the ecological economy what the investment in infrastructure and the major network-based public services were to the economy of the industrial revolution that developed in the three decades after the Second World War; and (3) because the behaviour of consumers and of firms will have to be redirected by using, among other tools, much stronger incentives, norms and taxes, at both local and global level. This will raise another social question, namely that of knowing who, throughout the world and in each country, will support the conversion efforts, on what basis they will do so and with what level of commitment. Not to recognise the environmental debt owed by the 'rich', whose wealth has been accumulated largely by overexploiting or pillaging the South's natural resources, to the social groups and peoples who have contributed least to driving us across the thresholds of non-sustainability, but who are already the ones paying the highest price, would be paradoxical indeed.

Notes

1. I am deeply grateful to William Baumol for his remarks and suggestions about a previous version of this chapter. The same goes for the organisers of and participants at the international conference on Services, Innovation and Sustainable Development, Poitiers (France), March 2008. This chapter was carried out within the FP7 ServPPIN Project (The Contribution of Public and Private Services to European Growth and Welfare, and the Role of Public–Private Innovation Networks).
2. ZenithOptimedia, http://www.lesechos.fr/info/comm/4657216.htm.
3. A population's ecological footprint is defined as the area of the planet, measured in hectares, that that population requires, given its current lifestyle and technologies, to satisfy its needs for products of the soil (agriculture and forestry), for fishing grounds, for built-up or developed land (roads and infrastructure) and for forests capable of recycling its CO_2 emissions.
4. Greenpeace and the Planète Urgence website: http://www.infosdelaplanete.org/1900/combien-de-petrole.html.
5. See: http://www.pcinpact.com/actu/news/Un_m_deau_pour_1Go_de_RAM_.htm, citing researchers from the UN University in Tokyo: http://search.japantimes.co.jp/member/member.html?fe20030123sh.htm.
6. See United Nations in Tokyo, http://searchjapantimes.co.jp/member/member.html?fe20030123sh.htm.
7. See (in French) http://www.mediaterre.org/international/actu, 20070126161538.htm.
8. See: http://www.lepapier.fr/cestlavie.htm.
9. These observations are concerned in particular with the performing arts, education and health sectors.
10. There is one qualification. Some areas of agriculture are not likely to follow this average trend. For example, for reasons linked to its large environmental footprint, cattle raising should decline sharply in the long term, since it produces high levels of greenhouse gas emissions.
11. http://www.jean-fourastie.org/temoin1.htm.
12. Objection: does the gradual replacement of intensive agriculture by sustainable agriculture, which would be 'employment-richer' and geared to supply more local markets, not conflict with the imperative of feeding a growing world population? No, on the contrary. Firstly, we do not really have any choice, because without such measures, the world population will suffer catastrophes caused by disruptions to the environment, and the poorest people will be in the front line. Secondly, the Food and Agriculture Organization (FAO) itself states that sustainable agriculture has an essential role to play in feeding mankind and in providing local employment opportunities (http://www.fao.org/newsroom/fr/news/2007/1000550/index.html). This is confirmed by Bruno Parmentier's splendid book (*Nourrir l'humanité*, 2007), which advocates food sovereignty as a right for all peoples, in opposition to globalised productivism.
13. See: http://www.negawatt.org/.
14. In terms not only of their share in employment but also of their distribution across the whole country and accessibility for all.
15. According to a 2008 report by the IMO (International Maritime Organization), which was covered in the *Guardian* newspaper, sea transport emits 1.12 billion tonnes of CO_2 per year, which is 4.5 per cent of total human emissions, a figure that could increase by a further 20 per cent between 2008 and 2020. This figure is virtually three times the estimate adopted hitherto by the Intergovernmental Panel on Climate Change (IPCC)

and almost double the emissions produced by air transport. Estimations of the other pollutants released into the atmosphere by ships (sulphur, etc.) are increasing even more rapidly: the heavy fuel oil that is generally used is one of the worst combustible pollutants in existence. However, the Kyoto Protocol does not take sea transport into account. See *Guardian* (2008).

16. See: http://reports.eea.europa.eu/eea_report_2008_1/en.
17. See: http://www.millenniumassessment.org/en/index.aspx.

References

Ademe (2007), 'La démarche Bilan Carbone chez Casino', http://www2.ademe.fr/servlet/getDoc?cid=96&m=3&id=48543&p2=14134&ref=14134&p1=1

Baumol, William J. (1967), 'Macroeconomics of unbalanced growth', *American Economic Review*, **2**, 415–26.

Clark, Colin (1939), 'The conditions of economic progress', http://www.questia.com/library/book/the-conditions-of-economic-progress-by-colin-clark.jsp

DARES (2005), 'Les métiers en 2015', Premières synthèses, n° 50.1.

Eurostat (various years), Labour Force Surveys.

Fisher, Allan G.B. (1935), *The Clash of Progress and Security*, London: Macmillan.

Fourastié, Jean (1949), *Le grand espoir du 20ème siècle*, Paris: Presses universitaires de France.

Grübler, Arnulf (1990), *The Rise and Fall of Infrastructures*, Laxenburg, Austria: IIASA.

Guardian (2008), 'True scale of CO_2 emissions from shipping revealed', http://www.guardian.co.uk/environment/2008/feb/13/climatechange.pollution, 13 February.

Hirsch, Fred (1976), *Social Limits to Growth*, Cambridge, MA, Harvard University Press.

IPCC (Intergovernmental Panel on Climate Change) (2001), *Climate Change 2001*.

Jancovici, Jean-Marc (2007), 'La dématérialisation de l'économie: mythe ou réalité?', *La jaune et la rouge*, August–September.

OECD (2008), *Labour Force Statistics*, 1987–2007, Paris: OECD.

Parmentier, Bruno (2007), *Nourrir l'humanité*, Paris: La découverte.

Stern (Lord), Nicholas (2006), 'The Stern Review', http://www.occ.gov.uk/activities/stern.htm

UNDP (various years), *Human Development Reports*.

WWF (2006), *Living Planet Report*, Gland, Switzerland: WWF.

PART II

THE NATURE OF INNOVATION IN SERVICES: SECTORAL ANALYSES AND CASE STUDIES

6 Innovation in public health care: diabetes education in the UK

Paul Windrum, Manuel García-Goñi and Eileen Fairhurst[1]

6.1 Introduction

This chapter presents an application of the Windrum and García-Goñi (2008) model of public sector innovation. The case study that is being used to test the model is the provision of new education services for patients suffering Type 2 diabetes. The case study provides a useful introduction, and overview, of the particular issues that are raised by studies of public sector service innovations for innovation scholars. In particular, it is necessary to include explicitly the influence of public sector organisations and policy-makers within the analysis. Also, one needs to consider the influence of heterogeneous user preferences and needs on the innovation process. Each has, by and large, been omitted in past theoretical and empirical studies of innovation in (private sector) services and manufacturing. Building on the work of Saviotti and Metcalfe (1984) on manufacturing innovation, and Gallouj and Weinstein (1997) on innovation in private sector services, the Windrum and García-Goñi model is a multi-agent model that explicitly considers the way in which interactions between service providers, patients and government policy-makers shape the development and diffusion of innovations within public sector healthcare.

The Windrum and García-Goñi (2008) model is neo-Schumpeterian in two senses. First, one can use it to consider the five types of innovation discussed by Schumpeter (1934, 1943): product innovation, process innovation, organizational innovation, market innovation and material innovation. Second, it is neo-Schumpeterian in the sense that it is a generic model that can, in principle, be applied to studies of service innovations in both public and private sectors; and to innovations in manufacturing as well as in service sectors. This reflects its authors' advocacy of the neo-Schumpeterian perspective, the so-called 'synthesis' position. The objective of the synthesis approach is to take the insights of research on (private sector) services innovation and the insights of research in manufacturing innovation, and integrate these within a single, unifying framework. According to the synthesis view, manufacturing-based innovation studies privileged product and process innovation at the expense of organisational, market and input innovation, while services-based innovation studies have (re)invigorated research in these other dimensions. The aim of the neo-Schumpeterian synthesis approach is not merely to add one to another, but to develop an integrated account that is applicable to both services and manufacturing, and which covers all aspects of innovative activity (see Metcalfe, 1998; Gallouj, 2002; Drejer, 2004; Windrum, 2007; and Chapter 1 in this *Handbook*).

The application of the model to diabetes education is part of an ongoing programme of model testing and development. The original 2008 paper applied the model to ambulatory surgery. Ambulatory surgery is a radical process/organisational innovation that initially focuses on the restructuring of back-office activities of a public hospital, and the development of alternative procedures that facilitate patient recuperation at home rather than in a hospital bed. The latter involves the development of new home-provision services. By contrast, the diabetes education case study provides an example of a radically new health service product based on the application and development of a radically new education principle.

Diabetes is a chronic illness, i.e. there is no medical 'fix' (an operation or a pill), and the sufferer will die from the condition. In addition to shortening life expectancy, the condition directly impacts the quality of their remaining life. Diabetes can lead to blindness, amputations (particularly feet and hands) and coronary disease. Successful control of Type 2 diabetes positively impacts on life expectancy and on the quality of life. This requires a change in the lifestyle of the sufferer: notably, a change towards a healthier diet and exercise regime. The first step towards this change is education. First, the acceptance that one has the condition – newly diagnosed patients commonly enter into a state of denial, which is not helped by the fact that they may not exhibit symptoms of the illness; and, second, knowledge and information about the condition and how to change personal lifestyles.

Dealing with chronic diseases, such as Type 2 diabetes, represents a radical departure for the services and working practices of public sector providers. Public sector health providers have been built on the treatment of acute illnesses, i.e. illnesses which could be cured through surgical intervention or the use of pharmaceutical products. The dramatic rise in chronic illnesses since the mid-1990s demands a radical change not only in the services offered by practitioners, but also in their organisational practices, their competences and their conceptual outlook. Similarly, it requires a radical change in health policy. The rise of chronic illnesses is directly related to increasing life expectancy (people are living longer) combined with rapidly rising obesity levels. Hence, the old policy assumptions and frameworks, premised on the assumption that the objective was to treat acute illnesses and thereby extend life expectancy, are no longer sufficient. Diabetes education is thus a radical innovation; it is a new product service whose development and diffusion requires changes in the preferences and competences of patients, public sector practitioners and government policy-makers.

The case study presented in this chapter focuses on the development of a novel, patient-centred education programme that was developed and implemented in the Salford Primary Care Trust (PCT)[2] between 2002 and 2005. Two public sector organisations were involved: the Salford PCT Community Diabetes Team and a group of education experts from Manchester Metropolitan University. Hence, this was a purely public sector innovation, driven by a team of public sector experts in education and medicine who were interested in experimenting with new, alternative models of diabetes education. Prior to this innovation, the Diabetes Team had been delivering education in the traditional, didactic manner, i.e. the medical practitioner imparted information to a passive recipient (the patient) within a unidirectional patient–practitioner relationship.

The concept of patient-centred health is very different to this traditional didactic patient–practitioner relationship. In the traditional relationship, the health care provider imparts information, provides a drug, or operates upon, a 'passive' patient. The development of this relationship was intimately tied to the development and expansion of health services that dealt with acute treatments. By contrast, the patient-centred model is better suited to treating chronic diseases, such as Type 2 diabetes. In this relationship the health care practitioner and the patient enter into an open, ongoing relationship. In terms of developing and delivering effective education, this means the health care practitioner is the mediator of an open discourse.

The goals are threefold. First, to ensure that the newly diagnosed patient believes that they have the condition and so starts to change their lifestyle. Second, to impart essential information on the condition, on diet and ways to improve diet, and on exercise and ways to improve the patient's exercise regime. Third, to build patients' self-confidence; changing lifestyles requires both informed and self-confident patients. In a patient-centred education programme, the practitioner supports an open, two-way discussion, asks open questions, and helps patients consider alternative options for taking control of their diabetes, and the likely consequences of patients' decisions. Hence, there is an intensive interaction, and collaboration, between health providers and patients in a patient-centred education programme.

This case study represents a useful test case of the Windrum and García-Goñi model. The original 2008 paper drew upon a case study of the diffusion and development of ambulatory surgery in a Spanish hospital. This was a major process and organisational innovation, and its effect on innovation within that hospital was very much in line with Barras's (1986) discussion of back-office innovations. This case study involves the development of a radically new service product that requires the development of radically new client-facing (front-office) competences. In this sense, it is more akin to Saviotti and Metcalfe's (1984) discussion of the introduction and development of a radically new product innovation. Another aspect of interest is the interaction between the Salford PCT Community Diabetes Team and Manchester Metropolitan University. The role and services provided by the education experts at Manchester Metropolitan University in this collaboration have clear parallels with Gallouj and Weinstein's (1997) discussion of knowledge-intensive service firms. Hence, the case study provides an important test of the robustness of the Windrum and García-Goñi model of health service innovation.

The remainder of the chapter is organised as follows. Section 6.2 provides an overview of the Windrum and García-Goñi model. Section 6.3 provides the reader with an understanding of the medical condition of diabetes mellitus, explains the very different conceptual basis of didactic and patient-centred education, and describes the innovative patient-centred education programme that was developed and rolled out in the Salford PCT in the period between 2003 and 2005. Section 6.4 applies the case study to the Windrum and García-Goñi model. Using the model, we examine how interactions between providers, patients, and policy-makers shaped the innovation. The model is then used to identify the different types of innovation present in the case study, and to ascertain whether this is a radical or an incremental innovation. Additionally, the model provides a framework for comparing and evaluating the clinical benefits of alternative didactic and patient-centred education programmes that were delivered in the Salford PCT region. Section 6.5 concludes.

6.2 The Windrum and García-Goñi model of innovation

Here we briefly review the neo-Schumpeterian model of health services innovation developed by Windrum and García-Goñi (2008). The model is set firmly within the characteristics approach to modelling innovation. This approach stems from the work of Lancaster (1966, 1971). He observed that all types of products (both material goods and immaterial services) can be described as a bundle of 'service characteristics' (or attributes) that a good/service embodies. These are the stream of services, provided by a good/service, which the purchaser consumes over the lifetime of the purchased product. As Lancaster observed, consumers choose between alternative bundles of service characteristics and prices that are offered on the market by producers.

Lancaster's consumer theory was first translated into a model of innovation by Saviotti and Metcalfe (1984). Product and process innovation is the means by which firms compete, offering consumers more attractive combinations of service characteristics and prices than their rivals. Technological innovation, i.e. the development and introduction of radically new technologies, is a key aspect of the competitive process because the service characteristics of competing products are directly related to the set of technologies that underpin rival products. In other words, there is a direct mapping between the 'service characteristics' that consumers desire, the 'technical characteristics' of the artefact product, and the 'process characteristics' of the technologies and firm routines used to produce the product. Product innovations alter the vector of technical characteristics and, hence, quality of service characteristics; while process innovations alter the vector of process characteristics, enabling the innovating firms to reduce production costs and, hence, price.

The focus of Saviotti and Metcalfe was innovation in manufacturing sectors. Gallouj and Weinstein (1997) subsequently adapted the model in order to investigate innovations in services, or, to be more accurate, private sector services. Of particular interest was the interaction between knowledge-intensive business services (KIBS) and their client firms in the production of new service products. KIBS are specialist producers of intermediate inputs, that are either a component of the final good/service (i.e. enter the vector of technical characteristics), or else are part of the final supplier's production process (i.e. enter the vector of process characteristics). This is the basis of the 'co-production thesis'. KIBS firms, it is argued, not only provide higher-quality intermediate inputs; they also engage with their clients in the co-production of new knowledge and material artefacts. For the client and the KIBS provider to interact in this way, there must be a semi-permeable boundary between the organisations. Developments in the skills and competences of one will affect a change in the skills and competences of the other (Gallouj and Weinstein, 1997). Potential productivity gains for the client lie in the KIBS provider acting as an interface between the client's tacit knowledge base and the wider knowledge base of the economy, through the provision of high-quality information on new business opportunities, market trends and new technologies, and through new business models (Gallouj and Weinstein, 1997; Antonelli, 1998; Sundbo, 1998; Preißl, 2000).

In order to examine public sector services, the Windrum and García-Goñi model develops the Gallouj and Weinstein model in three key respects. First, it is a multi-agent framework that includes policy-makers

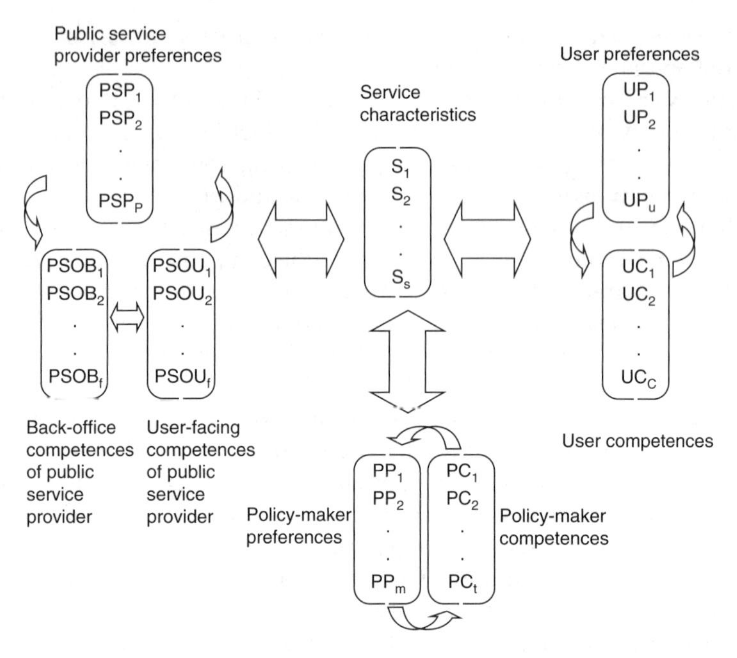

Source: Windrum and García-Goñi (2008).

Figure 6.1 Multi-agent framework of co-evolving service characteristics, competences and preferences

and public sector service providers, as well as firms and consumers (Figure 6.1). This allows one to investigate interactions between economic, social and political spheres. This is essential for understanding innovations in service sectors, such as health, that have a significant public sector component. The upshot is that the discussion of innovation is no longer confined to private sector firms. The focus is now extended to include public and private sector organisations that influence and shape the innovation process.

The second notable feature of the Windrum and García-Goñi model is that it considers agents' competences and interests, and how these co-evolve through the alternative technology products that developed over time. Different agents define their competences and preferences with respect to alternative technologies and, thereby, reveal their interests. In turn, the development and diffusion of radically new innovations alters the competence bases and preferences of consumers, firms, and policy-makers over time.[3] Following Barras (1986, 1990), a distinction is drawn between user-facing 'front-office' competences of service providers, and their 'back-office' competences. User-facing competences include all the tangible and intangible skills, know-how and technologies that are used to produce service characteristics. These outward-facing competences are mobilised by service providers when interacting with their users in order to define, produce and deliver the final service. For example, in diagnosing and treating an illness such as cataracts, medical practitioners must have knowledge and skills in a set of interrelated hardware and software technologies in order to provide the final medical service to the patient. Back-office competences are the skills and administrative activities such as payroll and patient booking systems that support these user-facing competences and activities.

The third notable feature of the model is that it captures all five types of innovation discussed by Schumpeter (1934, 1943): organizational, product, market, process and input innovations. A radical innovation alters not only the dimensions of the service characteristics vector but also the competences and preferences of the agents involved in the innovation process. Hence, the model captures the central neo-Schumpeterian message of long-run change. The multi-agent system changes qualitatively over time as new medical technologies facilitate the development of new services with new service characteristics, and the preferences and competences of agents change as these radical innovations diffuse.

The definitions of radical and incremental innovation in this model differ to those found in the Saviotti–Metcalfe and the Gallouj–Weinstein models. Definitions of 'radical' and 'incremental' are important, not only for our understanding of the sources and drivers of innovation but also for the measurement of innovation. In the Windrum and García-Goñi model, radical innovations change to the dimensions of the service characteristics vector S because they offer one or more new service characteristics. In addition to altering the dimensions of the service characteristics vector (S), radical medical innovations bring about changes in the competences and preferences matrices of patients/service users, providers, and policy-makers (*UC*, *UP*, *PSCB*, *PSCU*, *PSP*, *PC*, *PP*). An incremental innovation, by contrast, does not alter the dimensions of the service characteristics vector. An incremental innovation in this model involves the ongoing development of an existing medical technology service, i.e. the incremental improvement in the performance of one or more existing service characteristics S.

In the Gallouj–Weinstein model, all service characteristics must be different in order for the innovation to be considered a radical innovation, i.e. the entire set of elements within S must be new. By contrast, we expect that a radically new service innovation will share at least some characteristics with an existing service product because most medical innovations involve new medical services substituting existing medical services. At a minimum, a new service must differ from an old service by offering at least one new service characteristic. We suggest that applying the Gallouj–Weinstein definitions in medicine would seriously underestimate the number of radical innovations that occur.

The Saviotti–Metcalfe model also equates radical innovation with the introduction of a new service product. Their definition is even stronger than Gallouj–Weinstein's because it requires the introduction of a new product that is underpinned by a new, distinct technology. This discounts radical organisational and process innovations, such as the ambulatory surgery study discussed by Windrum and García-Goñi (2008). This fundamentally restructured the organisation of hospital services, led to the development of new outpatient home care services, and facilitated the development and expansion of other treatments within the hospital in which it was developed. In other words, it altered the dimensions of service characteristics and the dimensions of service providers' competences, led to changes in the dimensions of provider preferences, the competences and preferences of users, and the competences and preferences of policy makers. Yet innovations such as ambulatory surgery would not count as a radical innovation according to the Saviotti–Metcalfe definition.

6.3 The case study

Having discussed the key elements of the Windrum and García-Goñi model, we next consider the innovation case study. This innovation involves the development and introduction of a new, 'patient-centred' education programme for Type 2 diabetes patients in Salford, in Greater Manchester. This new education programme replaced a 'traditional' didactic education programme in which health professions imparted information to passive recipients in a unidirectional manner.[4]

6.3.1 Background

Diabetes mellitus is a chronic long-term condition for which there is, as yet, no cure. It is the fifth leading cause of death by disease in the US (Hogan et al., 2003), and is now prevalent in Australia (Dixon, 2005) and Canada (Ohinmaa et al., 2004). In England, there are 1.4 million people diagnosed with diabetes, and its incidence is rising. This is a direct consequence of an ageing population (more than 10 per cent of people over 65 are diabetic) and an increasing incidence of obesity.

The vast majority (85 per cent) of people living with the disease in England are diagnosed with Type 2 diabetes. Patients with Type 2 diabetes are able to produce some insulin but the levels are not sufficient to properly control their blood sugar levels. However, it is estimated that 1 million people in the UK have the condition but are currently undiagnosed due to a lack of screening. Type 2 diabetes tends to run in families and is particularly common amongst people of African, Caribbean and Asian origin. At the moment, the average age for developing the disease in the UK is 52 years old. Nevertheless, the average age is falling, and some very overweight children are starting to become affected.

Chronic conditions such as diabetes are of concern because they lead to higher rates of morbidity, and place significant demands on health service provision over the period of that provision (Ramsey et al., 2002). The complications associated with diabetes are very serious. These include blindness, heart disease, foot amputation and (in men) erectile dysfunction. Actions taken to minimise these complications directly improves patient welfare and reduces health expenditure. Through management of their lifestyle and daily care control, Type 2 patients can control and reduce the impact of complications. The key aspects are a healthy food regime, regular physical activity, the monitoring of blood sugar levels and (if necessary) the use of drugs to control blood sugar levels. For some, changing exercise patterns may be all but impossible in practice (i.e. for those who are badly overweight and suffering from arthritis), so the focus tends to fall on diet.

The health literature has devoted efforts to different actions undertaken towards the provision of diabetes care, and to the relationship between the health status of diabetes patients and their need for health service provision. Different clinical trials provide evidence, for example, on how lowering blood glucose levels has a positive effect in delaying the onset, and slowing the progression of, diabetic retinopathy, nephropathy and neuropathy in patients with insulin-dependent diabetes mellitus (Diabetes Control and Complications Trial Group, 1993). Intensive diabetes therapy has long-term beneficial effects on the risk of cardiovascular disease in patients with Type 1 diabetes (Nathan et al., 2005). Intensive blood-glucose control for adult patients with Type 2 diabetes decreases the risk of microvascular complications (UK Prospective Diabetes Study Group, 1998). Beaulieu et al. (2006) review the literature on alternative, cost-effective diabetes disease management programmes. They highlight, however, that even if diabetes disease management programmes are potentially beneficial, less than 40 per cent of adults with diabetes achieve guideline-recommended levels of medical care in the US. This finding suggests there is evidence of adverse selection in the health provision of diabetes care, given the payment system in operation in the US. Further literature reviews of economic studies of diabetes intervention are provided by Klonoff and Schwartz (2000) and Zhang et al. (2004).

6.3.2 Traditional and patient-centred education models

The patient-centred diabetes education programme developed by the Salford PCT Community Diabetes Team and Manchester Metropolitan University is based on the concept of patient-centred health. Patient-centred health is a new concept whose introduction and development marks a major departure from the traditional didactic health care model. The traditional model has proved highly effective for the treatment of acute illnesses. Health professionals are in overall charge of care, which includes diagnosis, treatment decisions and ensuring that treatment is carried out as prescribed. The person receiving treatment, meanwhile, is a passive patient who accepts medical decisions, complies with instructions, and is dependent on the health care professional. In other words, under the traditional didactic model there is no interaction between the health provider and the patient.

With patient-centred health, the patient is the central focus of health care, and engages in a partnership with the health care professional. In addition to the physical aspects of a condition, the knowledge and skills of the patient are important, as are their psychological, emotional and behavioural states. Within the patient–practitioner partnership, each side brings their knowledge and experience to the table, engages in negotiations, and makes decisions about areas for improvement in the patients' self-care and daily choices. Thus interactions between health providers and patients are crucial to the development of health service provision. Translating the concept to a diabetes education programme, health care professionals bring knowledge about the illness, treatment options, preventative strategies and prognosis. Diabetes patients bring their knowledge and personal experience of living with the condition, their values and beliefs, their social circumstances, habits and behaviour, and attitudes towards risk taking (Coulter, 1999). The health professional adopts a facilitating role, asking open questions, helping individual patients to look at alternatives and assisting in the setting of behaviour-based goals.

It is argued that a patient-centred health model is better suited to and, hence, more effective for chronic diseases such as diabetes (Hampson et al., 1990; Funnell et al., 1991; World Health Organization, 1998; Ashton and Rogers, 2005). This is because chronic patients, whose condition lasts for the remainder of their life (hopefully, a long period), need to learn how to manage the condition in order to minimise the medical complications, and to extend life expectancy. In the case of Type 2 diabetes, patients need to self-manage their condition by self-testing their blood glucose levels, eating healthily, caring for their feet, balancing their food intake against exercise and, where necessary, taking medication.

Finally, successful management is in the interests of both patients and the National Health Service (NHS). Patients can avoid the worst of the implications of the disease, e.g. debilitating heart attacks and amputations of feet and hands, while the NHS saves on the cost of surgical operations and other services.

6.3.3 The innovation

As noted, this innovative education programme was the outcome of collaboration between two public sector institutions interested in experimenting with new models of diabetes education: the Salford PCT Diabetes Education Unit (hereafter 'SCDT') and a group of education specialists at Manchester Metropolitan University (hereafter 'MMU'). The SCDT comprises eight staff members: seven are medical practitioners (two podiatrists, two dieticians, three specialist diabetes nurses), and there is one full-time administrator. They deliver diabetes education services to around 80 general practitioner (GP) practices in the local Salford PCT region. The education services are delivered in four venues: one local public library, and three large GP practices to which patients have easy access.

The SCDT–MMU programme was developed and rolled out in the period 2003 to 2005. The programme was one of a number of patient-centred programmes developed around England, prompted by the actions of two key policy agencies in the NHS: the National Service Framework (NSF), and the National Institute for Clinical Excellence (NICE). Together, these determine the minimum standards of care and targets for diabetes education services to be delivered by Primary Care Trusts (PCTs) in England. The set of NSF standards and NICE guidelines issued in 2002 deliberately set up a window of opportunity for groups to experiment with new, patient-centred education programmes. In 2005 a review of the alternative programmes was conducted, and a national standard established. Given that patient-centred diabetes education was a radically new concept, with no precursors, programme designers faced great uncertainty with respect to issues such as what type of content was most suitable, the best methods for delivering material, and the very notion of what 'patient-centred' education actually means. Hence, the programmes that were developed varied significantly in terms of their content, design and delivery.

The SCDT–MMU programme targeted key information and the development of key types of patient knowledge that maximise the probability of patients successfully changing their lifestyles. First, one needs to address the acceptance of the condition by newly diagnosed patients who may (as yet) not show any of the symptoms of the condition. Having accepted the condition, patients need to acquire a basic understanding of the illness itself. This understanding is the platform for seeking guidance on diet and exercise. Guidance on diet and exercise needs to be targeted to individual patients. Finally, the course addressed self-motivation; essential if better diet and improved exercise regimes are to be achieved and sustained.

Any patient-centred diabetes education programme needs to address a set of fundamental conceptual issues and constraints. First, it must operationalise the interrelated concepts of patient-centred health and empowerment. The goal of the patient-centred education is the development of empowered and self-motivated patients. This must take into account, and positively exploit, the growth of alternative sources of medical information, such as the Internet. This provides an opportunity for people to go beyond their GP and gain information about conditions and treatments when they want, and from many sources (Cumbo, 2001; Department of Health, 2004). Second, a programme must negotiate the fundamental economic trade-off between maximising service quality (given potentially unlimited demand for services) and highly expensive, extremely scarce NHS resources.

Two complementary vehicles were developed to achieve these goals. One was the introduction of mediated learning between patients and health providers on key areas of diabetes. In the Salford programme, one health practitioner would mediate a group of patients in open discussion, asking open questions and helping patients to consider alternative options for taking control of their diabetes, and the likely consequences. This group format enabled patients to compare personal experiences, opening up a forum in which they could share their initial fears and confusion, learn from one another, and build confidence in each other's experiences over the course of the programme. This is very far removed from the traditional didactic teaching model in which the expert (the health care professional) imparts information to a passive recipient (the patient).

This reflective group learning was underpinned by a specially designed 'Education Pack', comprising a number of sections of education material. Each section of the pack relates to a different topic area covered by the course. Prior to an education session, patients are given the supporting material as a hand-out. This provides patients with the material they will discuss in the next education session. The course requires that patients read the relevant section of material before attending each session. In this way, key content and messages are imparted, through a combination of prior reading and through discussion amongst group members during the education sessions. As the patient proceeds through the course, so the material in the Education Pack is built up. The material is held in a specially designed folder. The folder and all materials are provided to patients free of charge.

Prior to the development of this Education Pack, the SCDT had provided leaflets and other freely available printed materials published by the

NHS, Diabetes UK and other sources. This was very much in line with what patients received in other PCTs across England.

As well as underpinning the course content, the Education Pack facilitates ongoing learning. This takes two forms. First, it provides suggestions and references for future reading and learning, these include a number of useful Internet sites. Second, more detailed discussions of the material covered in the course are provided in an extended Appendix. The initial scoping exercise conducted by the project found that clear images combined with short, concisely written messages had greatest impact when patients were presented with information for the first time. This is particularly the case for patients with poor reading abilities (due to a combination of age and education). They are more likely to access, understand and remember information that is presented through clearly structured images. The scoping exercise found that more detailed and longer texts were effective when the same ideas were presented the second or third time around. What is more, a minority of patients, comfortable with extensive texts, were frustrated by the fact that these had not previously been made available. An effective programme must take on board the diversity of preferences and learning abilities of the local patient population in Salford.

A key consideration of the programme design was to maximise the number of 'learning points' while keeping delivery costs low. The learning material was organised and delivered across three group sessions. By designing the Education Pack in such a way that patients read the supporting material prior to attending each session, the education sessions plus the learning materials of the Education Pack together provide six 'learning points'. This compared very favourably with the two 'learning points' (taking the form of two, one-hour lecture sessions) that had previously been delivered. Of course, moving to this six-learning-point model places a great onus on the quality of the education material being provided.

The process does not end with the third education session. Patients return to their local GPs or diabetes nurses where they continue along their diabetes pathway. The final activity in the education programme is the drawing up of a patient contract, titled 'My Action Plan', between the SCDT and the patient. This contract is in part a certificate of achievement for having attended the course, and in part an agreement (of good intentions) on diet, exercise etc. that the patient takes back to their local GP. In this way, a link is made between the SCDT education programme and the local GP practice.

6.4 Analysing the case study

Section 6.2 discussed our long-term goal: to develop a model that captures the innovation processes in public sector services, such as health. In addition to providing theoretical insight and understanding, the model must be empirically operational. This enables one to evaluate the clinical and economic benefits of heath innovations. As part of this agenda, we here test the Windrum and García-Goñi model by applying it to the case study of patient-centred education. First, we consider its capacity to explain how interactions between providers, patients and policy-makers shaped the innovation. Second, the model assists us in defining the types of innovation that are present in the case study, and the location and impact of the innovation on agents' preferences and competences. Third, the model enables us to determine whether this is a radical or an incremental innovation. Fourth, using the model to frame a comparative analysis between the old and the new education programmes, we compare the benefits of the patient-centred diabetes programme with those of the didactic education programme delivered in the Salford PCT region.

6.4.1 Interaction between agents

In order to map the education innovation, we need to make one extension to the generic model. As discussed in section 6.3, the generic model can be extended in a number of ways. For example, it can account for heterogeneity in user (patient) preferences, i.e. multiple user niches. It can be extended to take into account the existence of multiple policy-makers (e.g. different government departments). It can also be extended to take into account interactions between multiple service providers. In the latter case, this can be the provision of drugs, medical instruments or equipment by private sector providers which form part of the final service that is delivered to patients by a public sector organisation. It may also be an interaction between a public (or private) sector service provider, providing a specialist input to the front- or back-office activities of the public service provider. This is the case here. The MMU team of education specialists entered into a collaborative research and development (R&D) project with Salford SCDT (the final service provider), in order to design and deliver a new, patient-centred education programme in the Salford PCT region. This is represented in Figure 6.2.[5]

The MMU team of education specialists are a public sector knowledge-intensive services (KIS) provider that co-produced the novel patient-centred education programme with the SCDT team of health practitioners. In designing the Education Pack, and training the SCDT team in techniques of mediated education learning, the client-facing competences of the MMU team (*KSCU*) directly interacted, and altered the patient/user competences of the SCDT (*PSCU*). This collaboration co-produced a new service product – a patient-centred education programme – with a distinct and new set of service characteristics. This co-production of a new, public sector service shares a number of the features described by Gallouj and

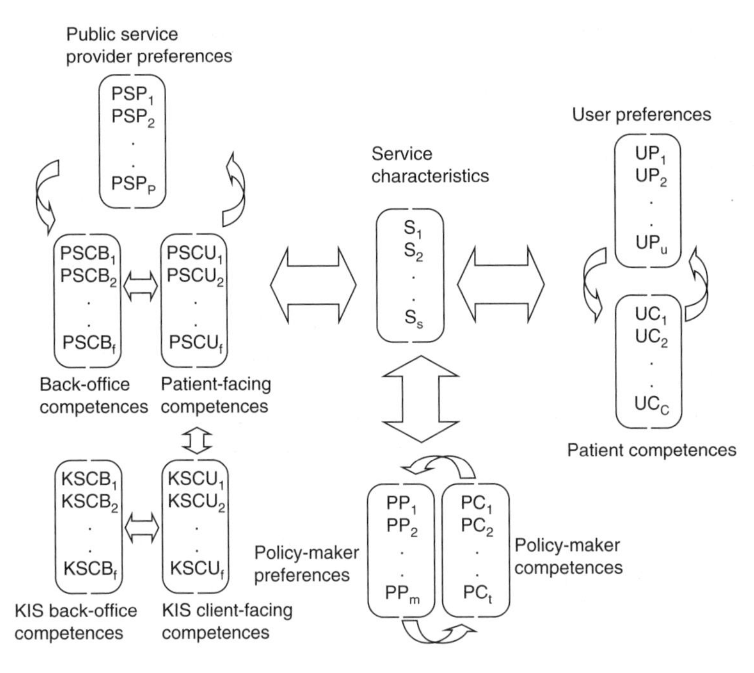

Figure 6.2 Interaction between SCDT and MMU in the co-production of the patient-centred education programme

Weinstein (1997) in their model of private sector KIBS-business client interactions (see section 6.3, above). In this case, the MMU team of education experts is a KIS producer of a specialist knowledge that is used to co-design the new, patient-centred education service (*S*), and trained the SCDT team in mediated education (*PSCU*).

With regards to the other agents in the model, the objective of an education programme for Type 2 diabetes patients is to affect a change in their competences (*UC*) and their preferences (*UP*). These are necessary conditions for the change in lifestyle that is required in order to control the disease effectively. The shift to a patient-centred health policy represents a radical change for policy-makers. They must also develop new competences (*CP*) if the policy is to be successful. Through policy learning, the preferences (*PP*) of policy-makers also change over time.

6.4.2 Radical innovation

New service characteristics The new education programme is a radical innovation, according to the definition used in the Windrum and García-Goñi model. First, patient-centred education is a radically new approach that provides new service characteristics, previously unavailable with a didactic education model. Most notably, the interaction between patient and practitioner was not a characteristic of the traditional model education programme, but is central to a patient-centred programme. These new service characteristics alter the dimensions of vector *S*. Furthermore, aspects of the programme address building confidence and empowering patients in order to become independent, self-motivated individuals. These were not considerations for the previous, didactic programme. In addition to adding new service characteristics, the patient-centred programme offers higher quality in the service characteristics that are shared with the old, didactic programme. Notably, the specially designed Education Pack delivers higher-quality, targeted information on the condition, its symptoms, the medical implications of diabetes, and on how to manage the condition through lifestyle changes.

Change in competences and preferences In addition to altering the vector of service characteristics *S*, a radical innovation alters the competences and/or preferences of one or more agents. In this particular case study, important changes occur in client-facing competences (*PSCU*) and preferences (*PSP*) of the SCDT, and in the competences (*PC*) and preferences (*PP*) of NSF policy-makers. NSF policy-makers were engaged in a novel, new experiment in setting standards and in policy learning (Windrum, 2008). In order to deal with the asymmetry of information between the standards setter, in this case the NSF, and the health care professionals who were able to develop innovative education services, a pseudo market for innovation was created. This encouraged local experiments in PCTs (some centrally funded, others self-funded by PCTs) for a pre-specified three-year period. At the end of this period, the results of the various local experiments were evaluated and a binding set of standards introduced.

Windrum (2008) argues that this circular process of policy learning is, of itself, a radically new policy innovation. It combines both top-down and bottom-up learning. A top-down induced opportunity for radical innovation (a window of opportunity) was created at the NSF policy level. An initial set of loosely set standards was deliberately introduced, providing a space for local initiatives. At the local level, highly innovative programmes were experimented with. Important differences existed between the programmes as different groups explored what a patient-orientated education programme actually is, what it looks like, and how effective it can be compared to more traditional, didactic education programmes. These local experiments provided the learning inputs for effective standards-setting. Once the evaluation process had taken place, the next part of the process involved the introduction of a more specific and prescriptive set of guidelines issued to PCTs, and a clear set of formal standards and targets that needed to be met. This learning-by-doing process altered the understanding and preferences of policy-makers with respect to effective health policy in diabetes, and changed the ways in which they sought to manage diabetes health policy. This affected a change in their competences (*PC*) and in their policy preferences (*PP*).

Turning to service provider competences and preferences, patient-centred education requires the development of a very different set of skills and knowledge in order to support self-motivated patient learning, and patient empowerment. These skills are not taught during the training of NHS GPs and nurses. The SCDT team therefore faced a serious challenge in developing these new skills, and in changing to the radically different ethos of patient-centred learning. Indeed, Windrum and Senyucel (2006) report that both practitioners and patients experienced difficulties when initially engaging in the concept of patient-centred education programmes. There is a continual temptation to revert back to the old didactic mode in which the SCDT team members were trained. These findings echo those of Roisin et al. (1999). They found that practitioners continually fell back on didactic education methods and practices in which they were trained. In the case of the SCDT, this team of health practitioners did manage to overcome their initial difficulties and successfully switched to the new education regime (Windrum and Senyucel, 2006).

In one sense, it is trivial to state that patients attending the course undergo a change in preferences (*UP*) and a change in competences (*UC*). After all, the explicit purpose of diabetes education is to affect a change in the preferences and competences of Type 2 diabetes patients in order to affect a change in their lifestyles and, hence, better to control their diabetes. However, one should not underestimate the conceptual challenge faced by patients in the shift towards patient-centred education. Like practitioners, diabetes patients can find it difficult to switch away from the traditional practitioner–patient relationship and buy in to the new paradigm. Not only may individual patients lack the necessary skills and knowledge to be self-motivated, independent managers of their own disease, but the concept itself is radically new. Their previous experience of interaction with professionals is invariably under the traditional medical model. Indeed, they continue to experience the traditional model in the majority of their interactions with the NHS, including aspects of their diabetes such as visits to hospitals for annual check-ups, visits to foot specialists, and surgery.

Ultimately, of course, what really matters is whether a patient-centred

Table 6.1 Attendance on the two education programmes

	Traditional programme			Patient-centred programme		
	Total	Men	Women	Total	Men	Women
# patients contacted with appointment	244	141	103	184	88	96
# patients who attend 1st session	119 (44%)	65 (46%)	54 (52%)	116 (63%)	52 (59%)	64 (67%)
# patients who complete the course	95 (80%)	48 (74%)	47 (87%)	111 (96%)	50 (96%)	61 (95%)
Drop-out rate	20%	26%	13%	4%	4%	5%

education programme is more successful in providing the information, and building the confidence, needed for patients to start changing their lifestyles. It is to this issue that we next turn.

6.4.3 *Clinical benefits*

Initial engagement and retention Data on the initial engagement of patients to an education programme, and retention rates over the course, are important success indicators. Unless a course can attract and retain newly diagnosed Type 2 patients, they will not access the information and build the self-confidence and skills necessary to affect a change in lifestyle and, hence, the control of their diabetes.

The data reported in Table 6.1 indicate that the proportion of patients attending the first session of the patient-centred education programme is significantly higher than for the traditional model: 63 per cent compared to 44 per cent. What is more, male attendance of education programmes is notoriously poor. The patient-centred concept appears attractive to men; the initial attendance amongst men being 59 per cent. This contrasts sharply with an initial male attendance on the traditional didactic programme of less than 50 per cent.

Table 6.1 indicates a remarkable difference in drop-out rates on the two education programmes. The drop-out rate on the traditional didactic programme run by the SCDT was 20 per cent. For the patient-centred programme, the drop-out rate was just 4 per cent. In particular, there was a noticeable difference in the drop-out rate of men: 26 per cent on the traditional didactic programme (compared to 13 per cent for women). On the patient-centred programme, the drop-out rate for men was approximately the same as for women – both were around 4 per cent.

Table 6.2 Statistical data on HbA1c scores

	Group	N	Mean	Std. deviation	Std. error mean
HbA1c score at the time of the programme	Traditional model	108	7.739	1.6165	.1534
	Patient-centred model	94	7.745	1.6193	.1661
HbA1c score 1 year after the programme	Traditional model	108	7.172	1.0099	.0972
	Patient-centred model	94	6.838	.8599	.0887

Diabetes control: HbA1c data The ultimate medical test of clinical benefit is the health status of patients. The primary test for successful diabetes control is the HbA1c blood test. The HbA1c blood test, which records the blood glucose levels of patients, is the best biomedical indicator of whether patients have changed their lifestyles and brought their diabetes under control. The HbA1c test is conducted every 2–3 months by a patient's GP. The test score should be below 7 per cent. For patients belonging to the test and experiment groups, data were collected on individual patients' HbA1c scores at two time points. The first recorded score was immediately following the education programme. The second score was approximately 12 months after the individual completed their education programme. Together, these scores give us a good indication of how well patients in the trial and control groups are controlling their diabetes, and the scores of each group can be statistically compared. There are complete data for 108 of the 120 patients in the control group, and data for 94 of the 116 patients in the trial group (80 per cent and 81 per cent samples respectively). Missing data are explained by the patients passing away, relocating away from the area, or their local GPs not recording data on the central information system.

Table 6.2 presents the average mean scores and standard deviations of patients following both types of education programmes at the time of the education programme and one year later. The estimated mean of both groups is very similar at the time of the education programme (the control group is 7.739 for patients following the traditional programme and 7.745 for patients in the patient-centred programme) with a standard deviation of 1.61 (both groups). This indicates patients in both groups are starting from the same base. By comparison, 12 months later the HbA1c results are statistically different. The average blood score for patients following the traditional programme is 7.172, while the average score for patients in the patient-centred programme is 6.838. Therefore, the HbA1c blood data indicates a marked contrast in diabetes control between the two groups one year after the education intervention, with patients who attended the patient-centred education programme controlling their diabetes far better than patients who attended the traditional education programme.

Running the test for equality of variances confirms this result. Immediately after receiving an education programme, the distribution of the average blood scores for the pooled and separate populations is the same (the difference in the variance is not statistically significant). However, one year after individuals receive their respective education programmes, there are statistically significant differences between the means of the distributions, as presented in Table 6.3.

6.5 Conclusion

The objective of the chapter has been to test, and further develop, the Windrum and García-Goñi model. This has been done by selecting a health innovation that is qualitatively different to the ambulatory surgery case that was originally used in the development of the model. Ambulatory surgery is a radical process/organisational innovation that initially focuses on the restructuring of back-office activities of the public sector provider (in this case, the *PSCB* of a hospital), and the development of alternative procedures that facilitate patient recuperation at home rather than in a hospital bed. The latter involves the development of new services, altering the vector of service characteristics (*S*). This has obvious parallels with Barras's (1986, 1990) discussion of the pattern of innovation that follows the adoption and diffusion of new information and communication technologies (ICTs) in private and public sector service organisations.

By contrast, the patient-centred education programme is a new service product based on the application and development of a radically new education principle. This is much closer to the examples drawn from manufacturing that are considered by Saviotti and Metcalfe (1984). There, as we have discussed, the focus is on product innovation; a radical product innovation being the first commercial application of a new scientific or engineering discovery. The case study also contains aspects of the private sector services model of Gallouj and Weinstein (1997). Notably, the MMU–SCDT collaboration was a partnership between a knowledge-intensive service provider (KIS) and a final service provider. The important difference here is that both MMU and the SCDT are public sector organisations, whereas the empirical cases studied by Gallouj and Weinstein involved private sector organisations.

Table 6.3 Independent samples test for HbA1c scores

		Levene's test for equality of variances		t-test for equality of means						
		F	Sig.	T	Df	Sig. (2-tailed)	Mean difference	Std. error difference	95% confidence interval of the difference	
									Lower	Upper
HbA1c score Month1	Equal variances assumed	.317	.574	–.029	200	.977	–.0065	.2261	–.4524	.4393
	Equal variances not assumed			–.029	199.014	.977	–.0065	.2261	–.4525	.4394
HbA1c score Month12	Equal variances assumed	2.780	.097	2.510	200	.013	.3339	.1330	.0716	.5963
	Equal variances not assumed			2.538	199.910	.012	.3339	.1316	.0745	.5934

In future research we will examine the relationship between these different examples. For the moment we can report that the empirical testing of the Windrum and García-Goñi model that has been conducted here indicates the model can encompass different types of public health service innovations. Further, the model appears to be robust, in that one does not need to make ad hoc assumptions or reformulate the model in order to apply it to a new case study.

Applying the model to the case study of Type 2 diabetes education provides a number of important insights. First, the model reveals the importance of interactions between interests of different agents within the innovation process. These shape the nature, direction and timing of innovations and their subsequent diffusion. In particular, the interaction between policy-makers and health practitioners in the development of new health standards for diabetes education is essential, as is the interaction between practitioners and patients within patient-centred health programmes. Second, the model helps us identify the drivers, interests and relative power of different agents within the provision of this new health service. Third, the model enables one to identify the location of the innovation and whether it is radical or incremental in nature. Patient-centred education is a radical new service product that is underpinned by a new concept of the practitioner–patient relationship. The development of this radically new service requires the development of a new set of competences and preferences on the part of professional health practitioners. The diffusion of patient-centred education across the NHS requires a radical change in policy. For this policy change to occur, policy-makers need to develop new competences and, through policy exploration and policy learning, the development of new preferences. Finally, the goal of diabetes education is to affect a change in patient attitudes and lifestyles through the provision of high-quality information on the condition, and on healthier diet and exercise regimes, and to build self-confidence in patients. These are the necessary prerequisites for a change in patient competences and preferences. Using the model, we can evaluate the relative success of the programme in terms of its attendance and in its clinical outcomes.

To summarise, this case study has provided an important test of the robustness of the Windrum and García-Goñi model of health sector innovation. The model captures the key drivers of this case study and its evolutionary dynamics. It predicts that the successful development and diffusion of radical health innovations depend on a coalition of interests between key actors in the innovation network. In the case of patient-centred diabetes education, it has been shown that key interactions and coalitions between policy-makers, practitioners and patients are essential for successful innovation. Looking forward, the agenda is to test the model further using other empirical case studies. In line with the neo-Schumpeterian synthesis approach, the ultimate goal is to develop an integrated model that is applicable to innovation in both services and manufacturing.

Notes

1. This research was funded through the Innovation in the Public Sector (PUBLIN) project, part of the EU Programme for Research, Technological Development and Demonstration on Improving the Human Research Potential and the Socio-Economic Knowledge Base, 1998–2002 under the EU 5th Framework Programme.
2. Salford is a district within the Greater Manchester area, in the UK.
3. Following Gallouj and Weinstein (1997), the vector of technical characteristics is replaced by a vector of producer competences. This is a useful approximation where the production of services is predominantly based on human capital. This is the case for innovations in knowledge-intensive health services, such as the one examined in this chapter. The approximation may be less useful for studying innovations in service sectors that are highly dependent on physical capital, or in large-scale manufacturing.
4. For more information, see the report by Windrum and Senyucel (2006).
5. The representation of the KIS provider is deliberately simplified in order to focus on the interaction between the KIS provider and the public sector provider. For this reason the vector of KIS preferences (PKP), and interactions between the preferences of the KIS and its back-office and client-facing competences, have been omitted.

References

Antonelli, C. (1998), 'Localized technological change, new information technology and the knowledge-based economy: the European evidence', *Journal of Evolutionary Economics*, **8** (2), 177–198.

Ashton, H. and J. Rogers (2005), 'A health promoting empowerment approach to diabetes nursing', in A. Scriven (ed.), *Health Promoting Practice: The Contribution of Nurses and Allied Health Professionals*, Basingstoke: Palgrave Macmillan, pp. 45–56.

Barras, R. (1986), 'Towards a theory of innovation in services', *Research Policy*, **15**, 161–173.

Barras, R. (1990), 'Interactive innovation in financial and business services: the vanguard of the service revolution', *Research Policy*, **19**, 215–237.

Beaulieu, N., D.M. Cutler, K. Ho, G. Isham, T. Lindquist, A. Nelson and P. O'Connor (2006), 'The business case for diabetes disease management for managed care organizations', *Forum for Health Economics and Policy*, **9** (1), Frontiers in Health Policy Research, available at www.bepress.com/fhep/9/1/1.

Coulter, A. (1999), 'Paternalism or partnership?' *British Medical Journal*, **319**, 719–720.

Cumbo, J. (2001), 'Better-informed patients question bedside manners', *Financial Times*, February 21.

Department of Health (2004), *Choosing Health: Making Healthier Choices Easier*, London: Stationary Office.

Diabetes Control and Complications Trial Research Group (1993), 'The effect of intensive treatment of diabetes on the development and progression of longterm complications in insulin-dependent diabetes mellitus', *New England Journal of Medicine*, **329** (14), 977–986.

Dixon, T. (2005), 'Costs of diabetes in Australia, 2000–01', Bulletin 26, *Australian Institute of Health and Welfare*. Cat. No. AUS 59. Canberra: AIHW.

Drejer, I. (2004), 'Identifying innovation in surveys of services: a Shumpeterian perspective', *Research Policy*, **33** (3), 551–62.

Funnell, M.M., R.M. Anderson, M.S. Arnold, P.A. Barr, M.B. Donnelly, P.D. Johnson, D. Taylor-Moon and N.H. White (1991), 'Empowerment: an idea whose time has come in diabetes education', *Diabetes Educator*, **17**, 37–41.

Gallouj, F. (2002), *Innovation in the Service Economy*, Cheltenham, UK and Northampton, MA, USA: Edward Elgar.
Gallouj, F. and O. Weinstein (1997), 'Innovation in services', *Research Policy*, **26**, 537–556.
Hampson, S.E., R. Glasgow and D.J. Toobert (1990), 'Personal models of diabetes and their relations to self-care activities', *Health Psychology*, **9**, 632–646.
Hogan, P. T. Dall and P. Nikolov, American Diabetes Association (2003), 'Economic costs of diabetes in the US in 2002', *Diabetes Care*, **26** (3), 917–932.
Klonoff, D.C. and D.M. Schwartz (2000), 'An economic analysis of interventions for diabetes', *Diabetes Care*, **23**, 390–404.
Lancaster, K.J. (1966), 'A new approach to consumer theory', *Journal of Political Economy*, **14**, 133–146.
Lancaster, K.J. (1971), *Consumer Theory: A New Approach*, Columbia, NY: Columbia University Press.
Metcalfe, J.S. (1998), *Evolutionary Economics and Creative Destruction*, London, UK and New York, USA: Routledge.
Nathan, D.M., P.A. Cleary and J.Y. Backlund (2005), 'Intensive diabetes treatment and cardiovascular disease in patients with type 1 diabetes', *New England Journal of Medicine*, **353**, 2643–53.
Ohinmaa, A., P. Jacobs, S. Simpson and J. Johnson (2004), 'The projection of prevalence and cost of diabetes in Canada: 2000 to 2016', *Canadian Journal of Diabetes*, **28** (2), 1–8.
Preißl, B. (2000), 'Service innovation: what makes it different? Empirical evidence from Germany', in J.S. Metcalfe and I. Miles (eds), *Innovation Systems in the Service Economy: Measurement and Case Study Analysis*, Boston, MA: Kluwer, pp. 125–148.
Ramsey, S., K.H. Summers, S.A. Leong, H.G. Birnbaum, J.E. Kemner and P. Greenberg (2002), 'Productivity and medical costs of diabetes in a large employer population', *Diabetes Care*, **25**, 23–29.
Roisin, P., M.E. Rees, N. Stott and S.R. Rollnkick (1999), 'Can nurses learn to let go? Issues arising from an intervention designed to improve patients' involvement in their own care', *Journal of Advanced Nursing*, **29**, 1492–1499.
Saviotti, P.P. and J.S. Metcalfe (1984), 'A theoretical approach to the construction of technological output indicators', *Research Policy*, **13**, 141–151.
Schumpeter, J.A. (1934), *The Theory of Economic Development: An Inquiry into Profits, Capital, Credit, Interest and the Business Cycle*, Cambridge, MA: Harvard University Press.
Schumpeter, J.A. (1943), *Capitalism, Socialism and Democracy*, New York: Harper & Row.
Sundbo, J. (1998), *The Organisation of Innovation in Services*, Cheltenham, UK and Lyme, NH, USA: Edward Elgar.
UK Prospective Diabetes Study (UKPDS) Group (1998), 'Intensive blood-glucose control with sulphonylureas or insulin compared with conventional treatment and risk of complications in patients with type 2 diabetes (UKPDS 33)', *Lancet*, **352**, 837–853.
Windrum, P. (2007), 'Services innovation', in H. Hanusch and A. Pyka (eds), *The Edward Elgar Companion to Neo-Schumpeterian Economics*, Cheltenham, UK and Northampton, MA, USA: Edward Elgar, pp. 633–646.
Windrum, P. (2008), 'Patient-centred diabetes education in the UK', in P. Windrum and P. Koch (eds.), *Innovation in Public Services: Management, Creativity, and Entrepreneurship*, Cheltenham, UK and Northampton, MA, USA: Edward Elgar, pp. 216–244.
Windrum, P. and M. García-Goñi (2008), 'A neo-Schumpeterian model of health services innovation', *Research Policy*, **37** (4), 649–672.
Windrum, P. and Z. Senyucel (2006), *Salford PCT Patient-Centred Diabetes Education Programme Report*, Manchester: Manchester Metropolitan University Business School.
World Health Organization (1998), *Health Promotion Glossary*, Geneva: WHO.
Zhang, P., M.M. Engelgau, S.L. Norris, E.W. Gregg and K.M. Venkat Narayan (2004), 'Application of economic analysis to diabetes and diabetes care', *Annals of Internal Medicine*, **140**, 972–977.

7 The economics of knowledge interaction and the changing role of universities

Cristiano Antonelli, Pier Paolo Patrucco and Federica Rossi

7.1 Introduction

The key role of interactions in order to understand the dynamics of economic systems is increasingly appreciated. Interactions among agents are at the origin of the endogenous change of both preferences and technologies (Lane, 1993; Lane and Maxfield, 2005; Durlauf, 2005).

Within advanced economies, based upon the production and use of services, the organization and implementation of interactions between a variety of business partners and institutions becomes a central issue in the generation and dissemination of knowledge. Within economic systems, agents do more than exchange and trade: they interact, in that they share and barter tacit knowledge and specific competencies. Such knowledge interactions take place vertically in the context of user–producer transactions that parallel market transactions, horizontally among firms engaged in competitive relationships, and diagonally among firms and other institutions. The intentional pursuit of qualified interactions, their organization and exploitation, are increasingly seen as effective innovative strategies that enable the generation of new knowledge by allowing access to external complementary knowledge (Antonelli, 2008b).

The new understanding of the dynamics of knowledge generation parallels major institutional and organizational changes in the universities' characteristics and modes of operation. The traditional 'open science' and 'knowledge Mode 1' models are being challenged by new organizational forms of the university, as described by the 'Mode 2', the 'entrepreneurial science' and the 'triple helix' models. These models have revisited an array of elements that typically characterize different systems of scientific knowledge creation and distribution: (1) the characteristics of knowledge flowing from universities (i.e. knowledge as a public good vs knowledge as a private good); (2) the nature of the research activity itself (i.e. basic research vs applied and contract research); (3) the processes through which knowledge is created and distributed (i.e. publication and teaching vs patenting, consulting, scientific entrepreneurship and more generally 'third-stream activities'); 4) the organizational and governance forms through which knowledge is created and disseminated (academic self-governance vs university–industry interactions).

Precisely, the current shift in the organization of science and knowledge production from Mode 1 (Bush, 1945) and 'open science' (Dasgupta and David, 1994) to Mode 2 (Gibbons et al., 1994; Nowotny et al., 2001), triple helix and entrepreneurial science (Etzkowitz, 2002; Etzkowitz and Leydesdorff, 2000) and university–industry networks (Lawton-Smith, 2006) seems to support the idea that scientific production benefits from agglomeration and concentration of different research organizations and from interaction between firms and universities. In Mode 1 and the 'open science' model, research was conducted in an individualistic way, within a single organization, within the boundaries of a single discipline, with few if any collaborations with industry. The scientific output of such research activity was then possibly and subsequently applied to the productive activities of industries and firms, so that the innovation process itself was conceptualized as a linear process. This model had its day between the 1950s and the 1980s, but it became less efficient following the decline of the innovative model based on large corporations. On the contrary, in Mode 2, the triple helix model and entrepreneurial science, research involves wider connections and collaborations across institutions, scientific fields, industrial sectors and countries. This supports the emergence of multidisciplinary science, vertical and horizontal integration across institutions, and scientific diversification. This also favours the view that the organization of scientific activity and the growth in knowledge production may benefit from consolidation and collaboration not only within the academic system but also, and especially, between universities and firms. This model has acquired increasing relevance both in the literature and in practice since the mid-1980s.

The objective of this chapter is to understand, both analytically and empirically, this shift in the organization of knowledge production. The integration between information economics, the economics of knowledge and the economics of interactions provide the basic tools to elaborate an appropriate framework to understand the closer relationship between university and industry. In this perspective, the economics of interactions and social networks is emerging as a fruitful field of analysis that qualifies communication among agents within economic systems as an essential determinant of knowledge dynamics. University–industry relationships gain relevance in the economics of innovation as an interesting case where the market provision of knowledge-based services and the sharing of localized competence and idiosyncratic expertise among heterogeneous agents with distinct and limited competencies provide the foundations for the generation of further new knowledge. Elaborating upon data for Italian regions, we provide evidence of a positive relationship between scientific production in universities and private research and development (R&D) performed by firms. Our argument supports the increasing emphasis given to university–industry collaborations as appropriate strategies aiming at the exploitation of the positive feedbacks between academic research and industrial R&D.

This contribution is structured as follows. Section 7.2 analyses the characteristics and implications of the shift in the organization of scientific production, from the traditional open science model to the new understanding based on the entrepreneurial activity of scientists and the collaboration between university and industry. Interactions support the emergence of quasi-markets for knowledge-based services, where universities are new entrepreneurial players. Section 7.3, elaborating upon the Italian case, provides preliminary and descriptive evidence for the positive relationship between scientific publications and private R&D activity. The conclusions in section 7.4 summarize and put the main results into perspective.

7.2 The new organization of scientific production: from open science, to university entrepreneurship, markets for science and university–industry interactions

Institutional and organizational characteristics of universities are at the centre of the economic analysis of the academic system. In particular, the effect that different organizational and governance forms have on the quantity, quality and efficiency of research activity has been an object of analysis in the economics of science (see, for instance, Bonaccorsi and Daraio, 2007; Geuna et al., 2003; Von Tunzelmann et al., 2003).

However, while it has long been appreciated that the distribution of scientific production across individuals as well as institutions is by no means normal but is well described by a Pareto distribution, where the largest proportion of output is accounted for by very few researchers and institutions (Lotka, 1926; Katz, 1999; Merton, 1968), the large body of empirical and theoretical literature investigating whether specific organizational and institutional forms have positive effects on scientific production has generated controversial results. Results are nuanced, with a set of studies that support the idea of positive effects of industry–university interactions, agglomeration of R&D and the commercialization of science on scientific production, but only under precise specifications and assumptions; while a different set of studies are more critical, finding that 'third-stream activities' are not only not relevant to explain the amount of scientific production, but that they are negatively related to its quality.

In particular, the relationship between publicly and privately funded science, the progressive commercialization of research, and the effects that contract research has on scientific production are the object of ongoing debate. On the one hand, criticisms of contract research are based on the idea that scientists would be less and less committed to publication activities and would substitute the creation of public knowledge and basic research with applied research tailored to the needs of individual firms. In turn, this would harm the traditional academic ethos based on the publicity of scientific results, the circulation of information and the sharing of knowledge among colleagues, who would instead be seen as competitors in the market for private consulting and contract research (see, e.g., Nelson, 2004). Moreover, according to this view, not only the rise of entrepreneurial activity at universities is detrimental to the traditional academic ethos and culture, but the introduction of intellectual property rights and commercial exploitation of basic research also undermine the transfer of knowledge from university to industry, by restricting the upstream diffusion of knowledge (Mowery and Ziedonis, 2002; Sampat, 2006). This argument is used to support the claim that the increasing commercialization of science may even hamper the economy's overall rate of innovation (Florida, 1999).

On the other hand, however, it has been stressed that the generation of public science and contract research are complementary rather than substitutes, and that the production of basic research would benefit from scientists closely interacting with firms.

In this respect, the seminal study by Mansfield (1995) points to the idea that the academics' scientific production benefits from their interactions with industrial partners. Moreover, in some sectors like biotechnology, higher levels of scientific knowledge production often result from the presence of social networks between 'star scientists' and industry (Zucker et al., 1998; Zucker and Darby, 2001). Van Looy et al. (2004) and Lowe and Gonzalez-Brambila (2007) find that the combination of basic research, patenting and entrepreneurial activity by academic scientists is beneficial for the intensity of publication.

The specific characteristics of different scientific fields and disciplines seem to matter in explaining to what extent 'third-stream activities' and university–industry interactions may or may not favour scientific productivity. For instance, applied sciences seem to benefit from the agglomeration of and the collaboration between public and private R&D organizations, because of the greater endowment of technical equipment and the need to rely upon external technological resources in order to develop experimental research activities. The benefits seem to stem also from easier access to financial resources and the sharing of equipment costs among a larger number of partners. These, in turn, generate greater efficiency in scientific production (Bordons et al., 1996; Bordons and Zulueta, 1997). Similarly, Van Looy et al. (2004) and Lowe and Gonzalez-Brambila (2007) find specific differences between, for instance, engineering, where entrepreneurial activities of academic scientists exert a stronger positive effect on scientific productivity, and chemistry and biomedicine where such effects are much weaker. Their results seem to point to the fact that in certain disciplines, the setting up of university–industry collaborations and the development of university entrepreneurship are more time-consuming and less mutually advantageous. This produces a misallocation of resources and efforts in favour of 'third-stream activities' that is detrimental to scientific production.

7.2.1 The open science model

In the traditional 'open science' model (Dasgupta and David, 1994), the academic system provides the institutional context appropriate to combine the incentives to both the creation of new knowledge and its dissemination. Publication activity is the keystone of this model. Researchers compete for collective reputation within the international scientific community through peer review and the process of selection. On the basis of the reputation achieved internationally, academics are rewarded in both hierarchical and monetary terms. At the same time, clearly, publications are the main channel through which knowledge can be created and disseminated. In such context, well-described by the famous metaphor of the 'ivory tower', research is generally conducted in an individualistic way, within a single organization, within the boundaries of a single discipline, with few collaborations with industry. Interactions between university and industry are possible but limited to large firms able to undertake large R&D projects in their laboratories, and also to hire young PhDs and scientists. The scientific output of such research is applied to the productive activity of industries and firms in a 'linear' way.

The functioning of the system, and more specifically the possibility of interaction between universities and large firms performing R&D internally, is possible only as long as the state intermediates between universities and firms. Such an indirect relationship between the business sector and the academic system is based upon the following circular scheme: (1) firms agree to pay taxes that the state reallocates to the funding of universities; (2) the academic system assesses the quality of scientific publication and the creativity of scientists, on the basis of the peer-review mechanism, and provides them with the appropriate rewards by financing chairs and tenured positions using state funds; (3) academics create and disseminate scientific knowledge by both publishing and teaching; (4) firms access knowledge produced externally by universities through the hiring of highly educated workers and PhDs able to absorb and build upon the scientific contents of publications.

The basic tenets of the economics of knowledge as they have been put forward by Kenneth Arrow and Richard Nelson provide the conceptual tools to analyse the characteristics, processes and institutional forms that qualify scientific knowledge production in the open science system.

Knowledge created in such a system is public in nature and its characteristics are consistent with the notion of information as a typical public good (Arrow, 1962): it is non-rival, since more than one person can use it at the same time, and non-exclusive, since it can be shared easily and rapidly, and it is difficult to prevent potential free-riders from accessing it. Non-rivalry and non-excludability imply that information cannot be appropriated – or at least it cannot be appropriated completely – by the agents that have invested resources in order to produce it. Information is moreover indivisible and there is a fundamental asymmetry in the assessment of its content: the potential buyer cannot appreciate the value of information without knowing its content, but if the content is disclosed the buyer no longer needs to purchase it. Scientific knowledge shares many of the economic properties of information, and in particular it has the character of a durable public good, since: '(i) it does not lose validity due to use or the passage of time per se, (ii) it can be enjoyed jointly, and (iii) costly measures must be taken to restrict access to those who do not have a "right" to use it' (Dasgupta and David, 1994, 493).

Focusing more specifically on the production of scientific knowledge, Nelson (1959) pointed out that the amount of basic research activities performed by private agents competing in a market setting is likely to be inferior to the socially optimal amount. This is due to several features of basic research. First, the outcomes of basic research are characterized by fundamental Knightian uncertainty (there is no known probability distribution over their attainment); and even when scientific discoveries are made, the realization of economic pay-offs may require a very long time. As a consequence, the economic value of basic research is difficult to quantify. Second, the discoveries that stem from basic research tend to produce large externalities: results and applications may be obtained that are far from those that were expected *ex ante* ('serendipity') and hence they may benefit economic agents that are different from those that have invested in their production. Therefore, social returns to basic research are larger than private returns, and this divergence causes a systematic market failure which, in the absence of remedial actions, would result in private underinvestment in science: in order to guarantee that the socially optimal amount of basic research is performed, public investment becomes necessary. This 'market failure' argument has constituted the main economic rationale for public intervention in stimulating scientific production since it was first formulated (Mowery, 1983): it has provided economic justification either for direct public funding of research, or for the design of appropriate incentives and constraints able to induce individuals to behave in ways that lead to globally efficient solutions.

In this context, one of the most fruitful applications of the Arrovian economics of information and knowledge to the analysis of university activities relates to the study of the norms that govern the production and transmission of academic knowledge. An early influential account of the incentives and norms that guide the behaviour of research scientists was provided by Merton (1973), who identified the main institutional goal of science as 'the extension of certified knowledge', and described four interrelated norms that govern its production: scientific findings are the product of social collaboration and should be made available to the scientific community ('communalism'); the truthfulness of claimed observations is to be determined on the basis of impersonal criteria without regard to the identity of the scientist who makes the observation ('universalism'); scientists should be seeking truth, rather than seeking to further their own interests by advancing unfounded claims ('disinterestedness'); the scientific community should subject the claims and beliefs of its members to empirical scrutiny before accepting them ('organized scepticism').

These norms imply the autonomy of science in setting its own goals and in pursuing objective knowledge without outside pressures. This was at the basis of the well-known 'university of culture' model, which developed the ideas of Humboldt and the German idealists, and which was closely interlinked with the 'open science' model of scientific production. The norm of self-governance and control over the research agenda exercised by the scientific community relies precisely on this view. This norm is justified on the basis of the asymmetric information problem due to society's inability to appreciate the quality of scientists and of their publications, as well as to identify the most promising directions of research (Cowan, 2006).

However, as Antonelli (2008a) pointed out, the system works only if the inventor is rewarded appropriately, so that they are induced to make the results of their work public, if scientific publication is an effective form of knowledge dissemination, and if some form of economic compensation is granted also to the losers in the priority race. In the absence of such rewards no individuals would be encouraged to undertake uncertain and costly research activities. In this respect, the great ingenuity of the academic system is that it is based on a two-part payment schedule. This consists of a flat salary for entering science, supplemented by rewards to winners of scientific competitions. The flat salary is paid even in the absence of research activity, but it is economically justified because it is tied to a complementary productive activity, teaching. The reward for academic priority is instead granted by indexing career advancements and/or wages to publication performance.

Over time, a set of external and internal forces has induced important changes in the academic system, pushing universities towards new organizational configurations. Firstly, the decline in the amount of R&D funded and performed directly within large corporations has been paralleled by the increased division of innovative labour between specialized and often small firms, often able to command high technologies and scientific knowledge, and to combine these with more tacit and practical skills. This decline in the well-established innovation model – based on large corporations performing R&D in their own labs and receiving scientific knowledge from universities – alters the conditions that justify the funding structure of the open science model.

Secondly and relatedly, following the Anglo-Saxon model,[1] evaluation and accountability schemes have been progressively introduced in the majority of academic systems in developed countries in order to assess the scientific and organizational performances of academic institutions. These are now compelled to achieve well-defined goals in terms of scientific productivity and administrative efficiency.

Thirdly, some inefficiencies of the 'open science' system itself have progressively become apparent. The model appears to provide poor guidance to identify both the correct amount of public funds and the criteria to distribute these funds among disciplines and among institutions. The reward system sets up a tension between compliance with the norm of full disclosure and the competition for priority. Competition among researchers may encourage rival teams to undertake a too risky set of research projects within a given programme. It may also induce them to choose too similar projects within the programme; and if the programme involves projects that do not display large fixed costs, too many research teams may be attracted to a given research area, to the possible neglect of other areas.

In sum, it seems clear that the lack of actual communication between the generation of knowledge and its usage lies at the heart of the demise of the 'ivory tower' model. In the 'ivory tower' model, the university is expected to perform the function of issuing 'knowledge signals' without paying any attention to their actual reception on the business side. At the same time universities are not expected to listen to the knowledge signals emitted from the business community. In the new model, industrial R&D feeds academic research and vice versa, both with the bilateral provision of intermediary knowledge inputs and with better signals about the emerging direction of the needs and opportunities of both parties.

7.2.2 The university–industry interaction model

The interpretation of technological knowledge as a collective activity provides the foundations to appreciate the role of knowledge interactions, as well as to qualify the shift in the role and functions performed by universities within economic systems. In this approach, in fact, knowledge is regarded as fragmented and dispersed among a variety of heterogeneous agents where each possesses complementary bits. Their communication by means of both knowledge transactions and knowledge interactions, and their eventual integration, enables the generation of new knowledge (Antonelli, 2008b). The generation of scientific knowledge is not an exception to this general view. Both scientific and technological knowledge are the result of qualified knowledge communication among agents that have access to dispersed fragments of the general knowledge. For communication to take place it is clear that both parties need to take an active role: both the emission and the reception of signals are necessary. The analysis of university–industry relationships provides an interesting perspective from which to understand the role of both knowledge-based services transactions and interactions as vehicles for knowledge dissemination and as central sources in the innovation process. Interaction and transactions are complementary aspects of a broader process of communication. Communications benefit from the concentration of organizations that are heterogeneous in terms of scientific and technological domains. In this context geographical and institutional proximity are seen as means for more efficient knowledge-based interactions.

Two distinct strands of literature contribute to this approach. On the one hand the analysis of limited knowledge appropriability (Arrow, 1962) has led to the identification of knowledge spillovers (Griliches, 1992) and eventually of knowledge absorption costs (Cohen and Levinthal, 1990). This line of enquiry has made it possible to identify knowledge transactions as sources of knowledge creation and dissemination because they provide economic actors with the opportunity to access external knowledge at costs that are below equilibrium levels because of the well-known effects of pecuniary knowledge externalities (Antonelli, 2008c). On the other hand the appreciation of the role of user–producer interactions – which were originally understood within the limited context of market transactions between customers and suppliers of well-identified goods – has progressively led researchers to identify and emphasize the broader flows of knowledge interactions including those between firms and universities (Lundvall, 1985; Russo, 1985; von Hippel, 1988, 2005). The merging of these lines of enquiry into an integrated framework has stressed the key role of knowledge communication, consisting both of knowledge transactions and knowledge interactions, as a key source of the generation of new knowledge. Knowledge transactions and knowledge interactions are strictly intertwined and parallel each other. They are often complementary: knowledge interactions add on and qualify knowledge transactions.

The identification of the interaction content of transactions becomes crucial in this approach.

In this perspective, universities are progressively emerging as new and crucial partners of business firms in the innovation process, since they are performing an increasingly important role as providers of knowledge-based services supplied in the marketplace and as actors in the organization and implementation of qualified interactions with firms.

It is generally acknowledged that the substantial shift towards both new organizational forms and new ('third-stream') activities performed within universities has been supported by four complementary processes: (1) the introduction of accountability and assessment criteria for academic scientists and their research outcomes; (2) the approval on the part of governments of legislative measures aimed at facilitating the commercialization of university research and at fostering industry–university collaborations, of which the most important and the most questioned was the Bayh–Dole Act (1980) approved by the US Congress; 3) the emergence of biotechnology, a scientific discipline that, more than other disciplines, produces results for which commercial applications can be found quite rapidly and profitably; (4) the assignment to universities, in addition to their traditional missions of research and teaching, of an economic development mandate, with a particular emphasis on the generation of benefits for firms' R&D performances, especially in the case of small and technology-based firms. In particular, university–industry interactions as strategies through which universities exert entrepreneurial activity have recently experienced a dramatic rise through increasing patenting, licensing, research joint ventures with private firms, university spin-offs and technological consultancy (Rothaermel et al., 2007).

A visible qualitative effect of the increased interaction with industry on the part of academia has been, in the 1980s and 1990s, the flourishing of new organizational forms based on both formal and informal university–industry interactions and communication of scientific knowledge. While traditional university–industry interactions and the commercialization of scientific results were based upon licensing (Siegel et al., 2003), universities are now progressively enlarging the range of strategies devoted to transferring and marketing their scientific outcomes. In this respect, formal interactions and university technology transfer mechanisms rely upon academic–industrial liaison and technology licensing offices, industry–university joint research centres and, more recently, university spin-offs, and generally result in a legal instrument such as a patent, licence or royalty agreement. Informal interactions are instead based on the transfer of commercial technology, joint publications and industrial consulting (Link et al., 2007).

Firms create various linkages with universities and other R&D organizations, with clear benefits for technology-based firms, especially in terms of firm productivity, R&D capability and R&D output (Medda et al., 2005). The involvement in technology and innovation platforms, through various channels such as formal and informal collaborations, facility sharing, joint R&D projects and the development of university incubators (Zucker and Darby, 2001; Rothaermael and Thursby, 2005a, 2005b) enable interaction and coordination processes between firms, public laboratories and universities and support the development of scientific clusters, such as in the case of biotechnology (Robinson et al., 2007). Technology and innovation platforms emerge as directed governance forms for the provision of new knowledge-intensive activities based on the interactions between different organizations, and as institutions that are distinct from the spontaneous organization of economic activities such as in the traditional notion of markets for knowledge (Consoli and Patrucco, 2008).

In the models based on the interactions between university and industry, such as Mode 2, the triple helix and the academic entrepreneurship models, research is the outcome of collaborations across organizations, disciplines, sectors and technologies (Mowery et al., 2004; Siegel, 2006). The importance given to the implementation of vertical and horizontal linkages across institutions supports the idea that the organization of scientific production benefits from agglomeration effects. Consistently with the 'triple helix' and 'entrepreneurial science' approach to innovation processes, the spatial concentration of technology centres, R&D laboratories and of academic infrastructures (Abramovski et al., 2007) – which characterizes successful regional innovation systems – provides the suitable endowment to generate opportunities for co-localized firms to take advantage from the diversity of science- and technology-based knowledge, as well as from better opportunities to transfer efficiently both codified and tacit knowledge. A well-established empirical literature confirms that the local diffusion of scientific and technological complementary knowledge bases is increased via the knowledge externalities which stem from human capital in university and R&D laboratories, e.g. by means of postgraduates, researchers' mobility and personal contacts among them (Audretsch and Feldman, 1996; Audretsch and Stephan, 1996; Feldman and Audretsch, 1999).

Here, regions can be loci where effectual industry-related R&D infrastructures are built around new activities and functions of the academic system, such as patenting and licensing, consulting, research outsourcing, and scientific spin-offs. Small and medium-sized firms, especially those located on the innovation frontier, can also benefit from academic research.

Universities and the set of other actors involved in R&D (i.e. R&D centres, technology experts and consultants, regional agencies for innovation and research) provide new knowledge to firms, and in turn contribute industrial innovation processes, in three major ways. Firstly, firms can receive new inputs in terms of codified knowledge through individuals, both in the form of highly and formally educated human capital and in the form of scientists and senior researchers. Secondly, the academic system diffuses new knowledge that can be used in the industrial process of knowledge creation through publications. Thirdly, cooperative R&D projects focused on the development of specific technological applications for industrial needs are more and more characterized by the strategic presence of universities (Geuna, 1999).[2] The university emerges as a first interface among the variety of knowledge bases, not only favouring the effective introduction of generic and scientific knowledge in the activities of business firms, but also creating the conditions that support the application of new knowledge in different contexts.

Formal and informal university–industry interactions are based also on the personal relationships and social networks between academics, industry scientists, and managers and entrepreneurs in local firms. Social networks in fact can account for a better local exploitation of the most excellent skills of each academic vintage. Academics and scholars can provide appropriate consultancy and research support for local firms, even benefiting from close interaction with firms in terms of reputation, new chances and stimuli for academic research, and also in terms of the diffusion of knowledge through the creation of students' job opportunities and placement on the local labour markets. From the firms' point of view, pursuing their research processes in collaboration with universities may be preferable to funding them all internally for several reasons: easier access to a wider range of already existing competences, lower costs of personnel training and internal competence creation, greater cognitive heterogeneity, and, in an increasingly fast-paced economic system, more opportunities for shaping their environment by at least partially influencing the actions of potential competitors in research. Firms can also benefit from the access to new knowledge in the form of infrastructures (such as laboratories and databases) and from the opportunity to temporarily post researchers and scientists in academic infrastructures, establishing new chances for learning and research.

In this perspective, university-based spin-offs might be a most appropriate entrepreneurial strategy. University spin-offs can benefit from knowledge transfer embodied in academic scientists and researchers, eventually leading to the development of university-based technology and science parks where small firms and especially new technology-based firms (NTBFs) are not simply tenants or mechanisms to generate financial returns from academic intellectual property. NTBFs can be an effective mechanism to foster innovation and the generation of technological knowledge in the region when they can benefit from a common pool of scientific and technological knowledge and from social networks, in turn increasing the rate of survival of new, technology-based firms and supporting the persistence of their innovative activity (Murray, 2004). Financial institutions at large and venture capitalists in particular can foster the creation of new technology-based entrepreneurship. Interactions between venture capitalists, new high-tech firms often resulting from academic spin-offs, their clients and prospective investors, enable a more effective screening and a better assessment of the value and reliability of new ventures' knowledge (Antonelli and Teubal, 2008). Venture capitalism supports, on the one hand, the growth of new business ideas and academic entrepreneurship through the funding of technology-based and university spin-off companies and, on the other, the diffusion of successful knowledge.

Also in the new approach, scientific reputation based upon publication is a key element in the implementation of effective university–industry interactions (Antonelli, 2008a). As a matter of fact, academic scientists need to publish so as to reinforce their reputation, signal their competencies in the markets for research services and therefore attract new funds for their activity. Scientific reputation built upon publication now engenders rewards that can be earned in the market for research and professional activities. The interplay between epistemic communities and local communities of practice as new markets for scientists is most important in this context, and the direction of scientific and basic research is strongly affected by developments in existing technologies and products. Very often, basic research provides scientific explanations of technological artefacts that are already in use (Rosenberg, 1994). The greater the overlapping between scientific recognition and the reputation in the professional community, and the closer the interplay between local academic and professional communities, the more efficient is the role of publication as a system of incentives (Knorr-Cetina, 1999; Feldman and Desrochers, 2004).

In this context, the boundaries between knowledge as a public good and knowledge as a private good are blurry. Knowledge flowing from universities can be seen as a quasi-public good, to some extent similar to knowledge flowing from national laboratories where, such as in the case of the development of standards and technical measurement and methods, it involves close collaborations with industry. Knowledge is now understood as an essential facility that engenders the creation of further scientific and technological developments (Antonelli, 2007).

In this new context, the more intensive is the communication between business and academic research, implemented by means of both knowledge transactions and knowledge interactions and favoured by geographic proximity and organizational interfacing, and the larger the productivity of the resources invested in research is expected to be.

7.3 Preliminary evidence on university–industry interactions in Italy

According to the interactionist approach developed so far, scientific activity within universities and R&D investments by business firms are interdependent elements in the innovation process. Collaborations between academic scientists and industrial researchers, for instance in the forms of technological consultancy or joint publications, are crucial means through which external knowledge can be accessed and knowledge-based services provided. The creation of scientific knowledge in universities and the undertaking of R&D within private firms are complementary and closely related, rather than substitutes, and reinforce each other in the creation of further new knowledge.

This leads us to advance the hypothesis that the intensity of knowledge transactions and interactions between universities and firms should increase the amount of technological and scientific knowledge that can be generated with a given level of resources. The preliminary empirical analysis presented in this section will show the distribution of both academic scientific production and private R&D investments for Italian regions so as to provide some tentative evidence about the positive relationships between the two.

This section elaborates upon an original database made of 2673 Italian researchers (assistant, associate and full professors), distributed across 61 universities, active in the fields of chemistry (physical chemistry; general and inorganic chemistry; organic chemistry), engineering (metallurgy; material engineering; electronics measurement), earth sciences (petrology) and physics (theoretical physics). These scientific fields were chosen first and foremost because in these fields, with the exception of physics, Italy has a scientific impact higher than the European average. Such fields can be thought of as 'best practices' or 'scientific champions' in Italian science. Secondly, they mirror quite well the traditional distinction between theoretical science (physics), applied science (chemistry) and technical or technology-oriented science (petrology and engineering).

This database has been implemented using data from the Italian Ministry of University, Research and Technology (MIUR), and the 2673 researchers included in the database represent the universe of the researchers in those fields. For each researcher, the database provides information on their position (assistant, associate and full professor), the institution in which they are employed (university level), and the region in which the university is located.

In order to analyse scientific production for the different fields, two indicators for output will be used, namely the number of publications cited by articles published in ISI journals, and the number of citations received by such publications. Such output indicators will then be correlated with an input indicator represented by the number of researchers per university and region. In particular, in order to account for the variability of output due to scale effects, we used the number of researchers as a proxy for the dimension of the resources available in a given university. The number of publications and citations is referred to the period 1990–2004 and it is based on the Science Citation Index elaborated by the Institute for Scientific Information (ISI).

Table 7.1 shows the distribution of researchers, publications and citations across the different scientific fields (Patrucco, 2006). Here, the larger fields are those in chemistry, which represent 25 per cent, 23 per cent and 18.5 per cent of the population of researchers, respectively. On the contrary, engineering sectors are the smallest, covering 8 per cent, 4 per cent and 3 per cent of the total researchers. These relative weights are, quite obviously, reflected also in the shares of publications and citations. More interestingly, when considering simple productivity measures such as the number of publications per researcher, researchers in physics are the most productive, with 62 publications per researcher. When looking at the citation impact (No. citations/No. publications), chemistry, and in particular the field of general and inorganic chemistry, has the highest impact (6.79).

Table 7.2 shows instead the distribution of researchers, publications and citations by region. Emilia Romagna is the region with the largest proportion of researchers (14 per cent), publications (15.76 per cent) and citations (17.69 per cent), and also with the highest citation impact (6.76). in terms of scientific productivity, that is to say the ratio between publications and researchers. However, Tuscany and Piedmont are, among the largest regions, the most productive ones, with 62.55 and 60 publications per researcher respectively.

Table 7.3 shows the distribution of total, public and private R&D expenditures across Italian regions and compares these with the number of publications.

Figure 7.1 shows the relationship between the intensity of private R&D expenditures (see Table 7.3) and the effectiveness of academic research activity in each region. The log specification of both variables implies the existence of a power functional relationship. The regression line fits the data quite well, supporting the idea that there is a mutually reinforcing

Table 7.1 The distribution of researchers, publications and citations across the different scientific fields

Scientific field	No. res	% res	No. publ	% publ	No. cit	% cit	Publ/res	Cit/res	Cit/publ
Metallurgy	88	3.29	2,365	1.68	9,874	1.17	26.88	112.20	4.18
Material engineering	220	8.23	7,375	5.24	31,680	3.74	33.52	144.00	4.30
Electronics measurement	108	4.04	1,890	1.34	4,949	0.58	17.50	45.82	2.62
Petrology	117	4.38	3,889	2.76	18,860	2.23	33.24	161.20	4.85
Physical chemistry	493	18.44	27,715	19.70	157,915	18.63	56.22	320.31	5.70
General and inorganic chemistry	621	23.23	37,598	26.73	255,292	30.12	60.54	411.10	6.79
Organic chemistry	670	25.07	37,871	26.92	234,397	27.66	56.52	349.85	6.19
Theoretical physics	356	13.32	21,956	15.61	134,494	15.87	61.67	377.79	6.13
TOTAL	2,673	100.00	140,659	100.00	847,461	100.00	52.62	317.04	6.02

Table 7.2 The distribution of researchers, publications and citations across the Italian regions

Region	No. res	% res	No. publ	% publ	No. cit	% cit	Publ/res	Cit/res	Cit/publ
Emilia-Romagna	376	14.07	22,167	15.76	149,955	17.69	58.95	398.82	6.76
Lombardia	346	12.94	18,634	13.25	116,200	13.71	53.86	335.84	6.24
Toscana	248	9.28	15,513	11.03	100,052	11.81	62.55	403.44	6.45
Lazio	256	9.58	13,229	9.41	77,500	9.14	51.68	302.73	5.86
Campania	232	8.68	11,050	7.86	63,558	7.50	47.63	273.96	5.75
Veneto	182	6.81	9,483	6.74	55,365	6.53	52.10	304.20	5.84
Piemonte	152	5.69	9,143	6.50	55,815	6.59	60.15	367.20	6.10
Sicilia	207	7.74	7,748	5.51	45,357	5.35	37.43	219.12	5.85
Puglia	119	4.45	5,460	3.88	25,164	2.97	45.88	211.46	4.61
Friuli	89	3.33	5,436	3.86	34,765	4.10	61.08	390.62	6.40
Umbria	75	2.81	4,823	3.43	25,439	3.00	64.31	339.19	5.27
Sardegna	97	3.63	4,351	3.09	25,739	3.04	44.86	265.35	5.92
Liguria	80	2.99	3,621	2.57	16,236	1.92	45.26	202.95	4.48
Marche	63	2.36	3,396	2.41	16,752	1.98	53.90	265.90	4.93
Calabria	54	2.02	2,116	1.50	11,115	1.31	39.19	205.83	5.25
Trentino-A.A.	30	1.12	1,709	1.21	10,775	1.27	56.97	359.17	6.30
Abruzzo	32	1.20	1,656	1.18	10,505	1.24	51.75	328.28	6.34
Basilicata	28	1.05	887	0.63	5,771	0.68	31.68	206.11	6.51
Molise	7	0.26	237	0.17	1,398	0.16	33.86	199.71	5.90
Total	2,673	100.00	140,659	100.00	847,461	100.00	52.62	317.04	6.02

Table 7.3 R&D expenditures and publications across Italian regions (000 euros; constant prices 1995; 1990–2001 mean values)

Region	No. Publications	Total R&D	Private R&D	Public R&D
Emilia-Romagna	22,167.00	711,752.58	379,880.17	331,872.67
Lombardia	18,634.00	2,329,485.25	1,808,760.58	520,724.58
Toscana	15,513.00	558,687.58	206,042.25	352,645.25
Lazio	13,229.00	1,904,128.17	606,956.08	1,297,171.75
Campania	11,050.00	500,306.83	191,737.92	308,569.00
Veneto	9,483.00	423,185.33	213,965.83	209,219.25
Piemonte	9,143.00	1,567,384.17	1,367,001.75	200,382.25
Sicilia	7,748.00	311,484.67	59,695.00	251,789.50
Puglia	5,460.00	205,285.08	68,893.08	136,392.25
Friuli	5,436.00	222,353.50	119,898.08	102,455.42
Umbria	4,823.00	84,283.17	17,613.50	66,669.92
Sardegna	4,351.00	117,888.08	17,461.33	100,426.75
Liguria	3,621.00	332,267.25	166,293.00	165,974.17
Marche	3,396.00	96,425.67	31,421.08	65,004.17
Calabria	2,116.00	49,656.25	4,463.67	45,333.92
Trentino-A.A.	1,709.00	75,325.58	31,990.83	43,334.50
Abruzzo	1,656.00	150,082.92	84,504.83	65,578.00
Basilicata	887.00	37,495.42	9,772.33	27,723.08
Molise	237.00	10,239.17	2,067.00	8,171.92
Italy*	140,659.00	509,879.82	283,600.96	226,286.23

Note: *Mean values for R&D expenditures.

relationship between private and academic research. The position of each region with respect to the regression line reflects the extent to which they are above or below average. Out of the virtuous regions, Piedmont is located at the top right in the diagram, showing the highest combination of the two variables. Then one can find Lombardy, followed by some north-eastern regions like Friuli and Emilia-Romagna. Finally, two central regions, i.e. Lazio and Abruzzi, deserve to be mentioned as contexts in which industrial and academic research are mutually reinforcing. Moreover, it is hardly surprising to note that most of the less-developed southern regions are in the bottom left part of the box.

The preliminary evidence gathered confirms that the productivity of research activities is larger where there is an agglomeration and concentration of different research organizations and, hence, a closer interaction between firms and universities.

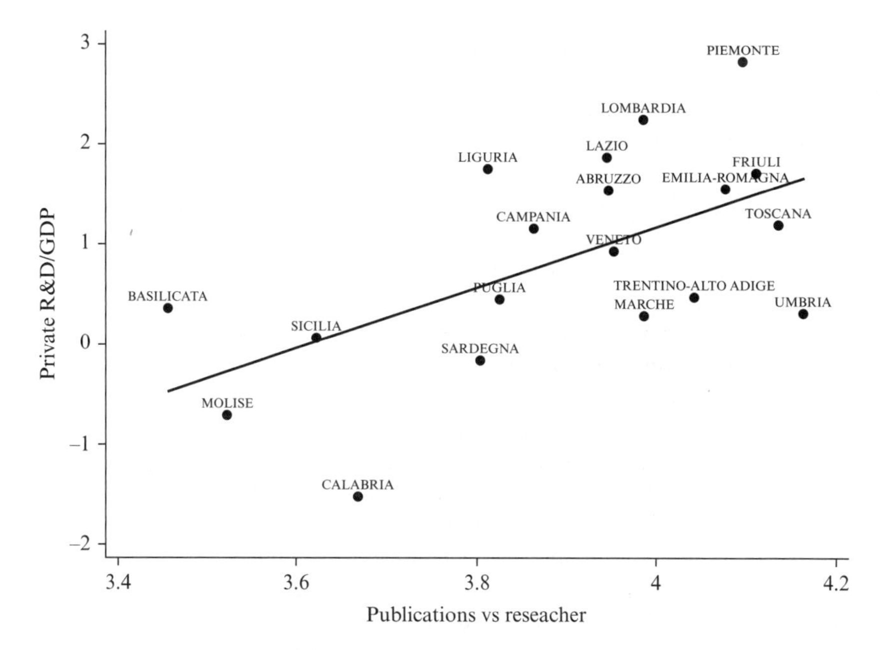

Figure 7.1 The relation between private R&D expenditures and scientific productivity

7.4 Conclusions

The academic system is undergoing a deep transformation in the way in which it creates and disseminates knowledge, as well as in the characteristics of the knowledge it produces. The new academic system emerging from such transformation seems to be able to fill the gap between the two extreme and traditional cases of knowledge production: on the one hand, the public provision of knowledge through basic research, and on the other, the private provision of knowledge as a (quasi) proprietary good on the part of firms.

Interorganizational and qualified interactions are an intermediate knowledge governance mode between the extremes of pure market transactions and vertical integration. Knowledge interaction consists of the intentional implementation of selective and preferential communication between providers of interdependent knowledge-based services. Because of the peculiar characteristics of knowledge as partially appropriable, excludable and indivisible, in a context where transactions are not completely efficient, knowledge interactions emerge as the appropriate strategy pursued by innovators in order to endogenize the effects of positive knowledge externalities and complementarities.

The emergence of a (quasi) market for scientific knowledge and research-based services and the ensuing flows of knowledge interactions are the result of the interdependence among numerous elements: (1) the typical non-exclusivity that characterizes academic employment and the freedom to enter professional markets traditionally accepted for academics; (2) the enhanced knowledge transactions made possible by increasing knowledge appropriability and tradability through licensing, patenting and consultancy; (3) the rise of venture capitalism and of dedicated financial markets for knowledge and innovation; (4) the development of a complementary knowledge-intensive business services (KIBS) sector able to intermediate between universities and firms, especially new ones; (5) the improved mutual understanding between academics and firms with respect both to the demand for knowledge and to the identification of local pools of knowledge characterized by high levels of complementarity; (6) the implementation of technology infrastructures, such as research consortia and technological platforms as specific forms of qualified interactions that support technology and knowledge-sharing between firms and universities; (7) the faster pace of innovation and the increasing uncertainty confronted by firms, which are therefore increasingly incentivized to liaise with external agents in order to attempt to 'control' their environment.

It has been highlighted that there is an inherent conflict between the set of incentives and rewards in the traditional open science system, based on peer-reviewed publication of basic research, and the incentives and rewards that are at the core of the new entrepreneurial university model, focused on the revenue generated from commercial application of basic research.

However, the two different approaches are not only and not merely historically specific, divergent and substitute. The two approaches can be understood also as localized and more or less appropriate according to the nature of the knowledge (e.g. tacit vs codified) and of the innovations (disruptive vs incremental) that are being created, or according to the relationship between technological and scientific knowledge that characterizes different fields (Nelson and Rosenberg, 2004). For instance, while 'knowledge mode 1' and 'ivory-towerism' can be useful for the creation of codified knowledge in the case of radical innovations, interactions and close collaboration between university and industry can be appropriate in the case of more applied, tacit and incremental innovations. The integration between the economics of information, the economics of knowledge and the interactionist approach provides an analytical framework able to clarify the variety of organizational and governance systems in the research sector. In this framework, open science and university–industry interactions coexist as two complementary modes of knowledge production whose synergies are beneficial to the overall innovative and scientific development (Rossi, 2008). Open science enables the rapid advancement of the scientific frontier and the opening up of a wide range of scientific possibilities and opportunities. Interactions between firms and universities allow for such findings not only to find useful applications, but also to be constantly ameliorated and adjusted according to users' needs.

In this perspective, knowledge is increasingly viewed as a collective good based upon the integration of external resources by means of interactions and communication, where the division of knowledge renders networks and platforms appropriate coordinating forms. The importance progressively given to intentional interactions and communication in the new economic understanding of knowledge production, requires consideration of: (1) the way in which different knowledge bases require different institutional patterns especially in terms of learning and communication norms; and (2) the way in which these patterns and norms imply different forms of organization and governance of knowledge production (Patrucco, 2008, 2009).

The economics of interactions and social networks provides the elementary principles to understand the role of external resources in the overall process of innovation and knowledge creation. Knowledge transactions and knowledge interactions are often complementary: knowledge transactions add to and qualify knowledge interactions. Connections and interactions between actors emerge as a crucial institutional element to understand the dynamic properties of innovation systems and the governance of knowledge creation and dissemination. The growing array of relationships between university and industry are only a specific, and yet increasingly important, case of the more general properties recognized by the economics of interactions and social networks.

In sum, the economics of social interactions provides the economics of knowledge with a powerful tool of analysis to understand and guide the evolution of the organization and governance of knowledge generation. Specifically, the grasping of the key role of knowledge interactions provides basic guidance in understanding the evolution of the organization of the academic system. Much work is necessary in order to qualify the actual amount of knowledge communication and knowledge generation that transactions and interactions actually produce, and to qualify the characteristics of the context into which they take place most effectively. The identification and valorization of knowledge transactions that are rich in knowledge interactions and the sorting of transactions with a low interactionist content become central issues in the organization of effective knowledge governance mechanisms.

Notes

1. In Europe, for instance, the UK introduced the first research assessment exercise in 1986, while the Netherlands first developed a research evaluation method in 1983, which was then upgraded into a systemic assessment exercise in 1993; in Germany the first attempts to link funds to research evaluation date back to 1990, and were developed to a fuller extent in 1998. Finland and Denmark carried out their first systemic national research evaluations in 1994, although in Finland measures of scientific assessment were present already in the early 1980s (Geuna and Martin, 2003). In Italy, the first national research assessment exercise has been implemented in 2003, on a three-year basis.
2. While the first two features are common to both the open science and university–industry interaction models, the third qualifies the difference between the two approaches.

References

Abramovski, L., R. Harrison and H. Simpson (2007), 'University research and the location of business R&D', *Economic Journal*, **117** (3), 114–141.

Antonelli, C. (2007), 'Technological knowledge as an essential facility', *Journal of Evolutionary Economics*, **17** (4), 451–471.

Antonelli, C. (2008a), 'The new economics of the university: a knowledge governance approach', *Journal of Technology Transfer*, **33** (1), 1–22.

Antonelli, C. (2008b), *Localized Technological Change: Towards the Economics of Complexity*, London: Routledge.

Antonelli, C. (2008c), 'Pecuniary knowledge externalities: the convergence of directed technological change and the emergence of innovation systems', *Industrial and Corporate Change*, **17** (5), 1049–1070.

Antonelli, C. and M. Teubal (2008), 'Knowledge-intensive property rights and the evolution of venture capitalism', *Journal of Institutional Economics*, **4**, 163–182.

Arrow, K.J. (1962), 'Economic welfare and the allocation of resources for invention', in R.R. Nelson (ed.), *The Rate and Direction of Inventive Activity: Economic and Social Factors*, Princeton, NJ: Princeton University Press for NBER, pp. 609–629.

Audretsch, D.B. and M.P. Feldman (1996), 'Spillovers and the geography of innovation and production', *American Economic Review*, **86** (3), 630–642.

Audretsch, D.B. and P.E. Stephan (1996), 'Company–scientist locational links: the case of biotechnology', *American Economic Review*, **86** (3), 641–652.

Bonaccorsi, A. and C. Daraio (2007), *Universities and Strategic Knowledge Creation: Specialization and Performance in Europe*, Cheltenham, UK and Northampton, MA, USA: Edward Elgar.

Bordons, M., I. Gomez, M.T. Fernandez, M.A. Zulueta and A. Mendez (1996), 'Local, domestic and international scientific collaboration in biomedical research', *Scientometrics*, **37** (2), 279–295.

Bordons, M. and M.A. Zulueta (1997), 'Comparison of research team activity in two biomedical fields', *Scientometrics*, **40** (3), 423–436.

Bush, V. (1945), *Science: The Endless Frontier: Report to The President on a Program for Postwar Scientific Research*, Washington, DC: Department of Defense.

Cohen, W.M. and D.A. Levinthal (1990), 'Absorptive capacity: a new perspective on learning and innovation', *Administrative Science Quarterly*, **35** (1), 128–152.

Consoli, D. and P.P. Patrucco (2008), 'Innovation platforms and the governance of knowledge: evidence from Italy and the UK', *Economics of Innovation and New Technology*, **17** (2), 701–718.

Cowan, R. (2006), 'Universities and the knowledge economy', in B. Kahin and D. Foray (eds), *Advancing Knowledge ad the Knowledge Economy*, Cambridge, MA: MIT Press, pp. 135–149.

Dasgupta, P. and P.A. David (1994), 'Towards a new economics of science', *Research Policy*, **23** (5), 487–521.
Durlauf, S. (2005), 'Complexity and empirical economics', *Economic Journal*, **115** (504), 225–243.
Etzkowitz, H. (2002), *MIT and the Rise of Entrepreneurial Science*, London: Routledge.
Etzkowitz, H. and L. Leydesdorff (2000), 'The dynamics of innovation: from national systems and "Mode 2" to triple helix of university–industry–government relations', *Research Policy*, **29** (2), 109–123.
Feldman, M.P. and D.B. Audtretsch (1999), 'Innovation in cities: science-based diversity, specialization and competition', *European Economic Review*, **43** (2), 409–429.
Feldman, M.P. and P. Desrochers (2004), 'Truth for its own sake: academic culture and technology transfer at Johns Hopkins University', *Minerva*, **42** (2), 105–126.
Florida, R. (1999), 'The role of the university: leveraging talent, not technology', *Issues in Science and Technology*, **15** (4), 67–73.
Geuna, A. (1999), *The Economics of Knowledge Production*, Cheltenham, UK and Northampton, MA, USA: Edward Elgar.
Geuna, A. and B. Martin (2003), 'University research evaluation and funding: an international comparison', *Minerva*, **41** (4), 277–304.
Geuna, A., A. Salter and E. Steinmueller (eds) (2003), *Science and Innovation: Rethinking the Rationales for Funding and Governance*, Cheltenham, UK and Northampton, MA, USA: Edward Elgar.
Gibbons, M., C. Limoges, H. Nowotny, S. Schwarzman, P. Scott and M. Trow (1994), *The New Production of Knowledge: The Dynamics of Research in Contemporary Societies*, London: Sage Publications.
Griliches, Z. (1992), 'The search for R&D spillovers', *Scandinavian Journal of Economics*, **94**, 29–47.
Katz, J.S. (1999), 'The self-similar science', *Research Policy*, **28** (5), 501–518.
Knorr-Cetina, K. (1999), *Epistemic Cultures: How the Sciences Make Knowledge*, Cambridge, MA: Harvard University Press.
Lane, D.A. (1993), 'Artificial worlds and economics, part II', *Journal of Evolutionary Economics*, **3** (3), 177–197.
Lane, D.A. and R.R. Maxfield (2005), 'Ontological uncertainty and innovation', *Journal of Evolutionary Economics*, **15** (1), 3–50.
Lawton-Smith, H. (2006), *Universities, Innovation and the Economy*, London: Routledge.
Link, A.N., D.S. Siegel and B. Bozeman (2007), 'An empirical analysis of the propensity of academics to engage in informal university technology transfer', *Industrial and Corporate Change*, **16** (4), 641–655.
Lotka, A.J. (1926), 'The frequency distribution of scientific productivity', *Journal of the Washington Academy of Sciences*, **16** (12), 317–323.
Lowe, R. and C. Gonzalez-Brambila (2007), 'Faculty entrepreneurs and research productivity: a first look', *Journal of Technology Transfer*, **32** (3), 173–194.
Lundvall, B.A. (1985), *Product Innovation and User–Producer Interaction*, Aalborg: Aalborg University Press.
Mansfield, E. (1995), 'Academic research underlying industrial innovations: sources, characteristics and financing', *Review of Economics and Statistics*, **77** (1), 55–65.
Medda, G., C. Piga and D.S. Siegel (2005), 'University R&D and firm productivity: evidence from Italy', *Journal of Technology Transfer*, **30** (1/2), 199–205.
Merton, R.K. (1968), 'The Matthew effect in science: the reward and communication systems of science are considered', *Science*, **5** (159), 56–63.
Merton, R.K. (1973), *The Sociology of Science. Theoretical and Empirical Investigations*, Chicago, IL: University of Chicago Press.
Mowery, D. (1983), 'Economic theory and government technology policy', *Policy Science*, **16** (1), 27–43.
Mowery, D.C., R.R. Nelson, B. Sampat and A.A. Ziedonis (2004), *Ivory Tower and Industrial*

Innovation. University–Industry Technology Transfer Before and After the Bayh–Dole Act, Stanford, CA: Stanford University Press.
Mowery, D.C. and A.A. Ziedonis (2002), 'Academic patent quality and quantity before and after the Bayh-Dole Act in the United States', *Research Policy*, **31** (3), 366–418.
Murray, F. (2004), 'The role of academic inventors in entrepreneurial firms: sharing the laboratory life', *Research Policy*, **33** (4), 643–659.
Nelson, R.R. (1959), 'The simple economics of basic scientific research', *Journal of Political Economy*, **67** (3), 297–306.
Nelson, R.R. (2004), 'The market economy and the scientific commons', *Research Policy*, **33** (3), 455–471.
Nelson, R.R. and N. Rosenberg (2004), 'American universities and technical advance in industry', *Research Policy*, **23** (3), 323–348.
Nowotny, H., P. Scott and M. Gibbons (2001), *Re-Thinking Science: Knowledge and the Public in an Age of Uncertainty*, Cambridge: Polity Press.
Patrucco, P.P. (2006), 'The production of scientific knowledge in Italy: evidence in theoretical, applied and technical sciences', European University Institute Working Papers RSCAS No. 2006/12.
Patrucco, P.P. (2008), 'The economics of collective knowledge and technological communication', *Journal of Technology Transfer*, **33** (6), 579–599.
Patrucco, P.P. (2009), 'Collective knowledge production costs and the dynamics of technological systems', *Economics of Innovation and New Technology*, **18** (3), 295–310.
Robinson, D.K.R., A. Rip and V. Mangematin (2007), 'Technological agglomeration and the emergence of clusters and networks in nanotechnology', *Research Policy*, **36** (6), 871–879.
Rosenberg, N. (1994), *Exploring the Black Box*, Cambridge: Cambridge University Press.
Rossi, F. (2008), 'The economics of knowledge and the governance of universities' third stream activities', DIME-1PR Working Paper n. 87, http://www.dime-eu.org/files/active/0/WP87-1PR.pdf.
Rothaermel, F.T., S.D. Agung and L. Jiang (2007), 'University entrepreneurship: a taxonomy of the literature', *Industrial and Corporate Change*, **16** (4), 691–791.
Rothaermel, F.T. and M.C. Thursby (2005a), 'University-incubator firm knowledge flows: assessing their impact on incubator firm performance', *Research Policy*, **34** (3), 305–320.
Rothaermel, F.T. and M.C. Thursby (2005b), 'Incubator firms failure or graduation? The role of university linkages', *Research Policy*, **34** (7), 1076–1090.
Russo, M. (1985), 'Technical change and the industrial district: the role of interfirm relations in the growth and transformation of ceramic tile production in Italy', *Research Policy*, **14** (6), 329–343.
Sampat, B.N. (2006), 'Patenting and US academic research in the 20th century: the world before and after Bayh–Dole', *Research Policy*, **35** (6), 772–789.
Siegel, D.S. (ed.) (2006), *Technology Entrepreneurship: Institutions and Agents Involved in University Technology Transfer*, Vol. 1. Cheltenham, UK and Northampton, MA, USA: Edward Elgar.
Siegel, D.S., D. Waldman and A.N. Link (2003), 'Assessing the impact of organizational practices on the relative productivity of university technology transfer offices: an exploratory study', *Research Policy*, **32** (1), 27–48.
Van Looy, B., M. Ranga, J. Callaert, K. Debackere and E. Zimmerman (2004), 'Combining entrepreneurial and scientific performance in academia: towards a compound and reciprocal Matthew-effect?', *Research Policy*, **33** (3), 425–441.
Von Hippel, E. (1988), *The Sources of Innovation*, Oxford: Oxford University Press.
Von Hippel, E. (2005), *Democratizing Innovation*, Cambridge, MA: MIT Press.
Von Tunzelmann, N., M. Ranga, B. Martin and A. Geuna (2003), 'The effects of size on research performance: a SPRU review', SPRU, University of Sussex, Brighton.
Zucker, L.G. and M.R. Darby (2001), 'Capturing technological opportunity via Japan's

star scientists: evidence from Japanese firms' biotech patents and products', *Journal of Technology Transfer*, **26** (1–2), 37–58.
Zucker, L.G., M.R. Darby and M.B. Brewer (1998), 'Intellectual human capital and the birth of US biotechnology enterprises', *American Economic Review*, **88** (11), 290–306.

8 Innovation and creative services

Ian Miles and Lawrence Green

8.1 Introduction

Much of the 'creative economy' is located in the services sectors, and the creative industries have become the foci of attention in many countries. It has been recognised that sectors such as advertising, architecture, broadcasting, design, entertainment software, and many 'cultural' activities, are important contributors to economic output as well as quality of life. Recent attention to the 'creative class' also argues for the importance of these sectors in fostering innovative environments more generally. The nature of innovation in these sectors has attracted little analysis until recently – in many ways, the situation is reminiscent of that encountered when services innovation first began to emerge as an area for innovation studies to address seriously. Researchers have begun to dip their toes into the water. A number of studies are indicating useful approaches to, and providing informative results about, the management and broader organisation of innovation in creative sectors, and what types of innovation are under way here. This is highly relevant to attempts to develop policies for creative industries. This chapter will review the features of these creative industries and how they resemble and differ from other services, and explore the main points emerging from recent studies of innovation in creative sectors.

8.2 Who are the creatives?

In the UK, the idea took root in the 1990s that knowledge-based and creative activities were the most likely route to becoming a successful twenty-first-century economy. The Department of Trade and Industry (DTI) and the Department of Culture, Media and Sport (DCMS) have made several efforts to identify and explore the creative economy. The DCMS's broad definition of the 'creative industries' has been influential in many countries, as well as guiding UK debates. It proposed that these are: 'those industries which have their origin in individual creativity, skill and talent and which have a potential for wealth and job creation through the generation and exploitation of intellectual property.' (DCMS, 2001, 4).

This is only of limited value as a starting point. Depending on how one reads 'individual creativity, skill and talent', we could be talking about practically any non-routine economic activity. What is intended becomes clearer when we consider the sectoral classification proposed by the DCMS. In 2001 it saw the following broad sectors of the economy as forming creative industries (the number after each sector is the estimated UK employment it involved at that time, and the list is organised in terms of employment size):

- Software and computer services: 555,000.
- Publishing: 141,000.
- Music: 122,000.
- Television and radio: 102,000.
- Advertising: 93,000.
- Design: 76,000.
- Performing arts: 74,000.
- Film and Video: 45,000.
- Art and antiques market: 37,000.
- Crafts: 24,000.

The SIC(92) classes with which these activities are associated are displayed in Box 8.1. Note that there are no SIC codes reported for 'crafts' (fortunately the smallest sector).[1]

The set of industries has some problematic features. If publishing and broadcasting are included, why not (tele)communications? To be sure, person-to-person conversations are not always very creative, but delivery of electronic content via telecommunications is an activity very like publishing and broadcasting. Why is software included here? Perhaps videogames and websites have many aesthetic components, but much software is developed for operating equipment rather than interfacing with end-users, and much software that does interface with end-users is a matter of work functions rather than entertainment or culture. Research and development is excluded (though it is mentioned in some DCMS texts) while it has many features of individual creativity – as, we might add, do many professional services (indeed, some legal and consultancy activities, for example, involve a great deal of performance).

What seems key to the DCMS definition – and the industries chosen – is that these are to a large extent commercial activities that are focused on end-user experience (often through the creation of creative content[2]). Some 'cultural industries' like education and museums are excluded, because they are in large part non-commercial. While the dominant broadcaster in the UK is funded by licence fees, there is strong competition from commercial broadcasters, and there are large amounts of material from independent producers carried by the BBC. Software is still in many ways an anomaly – much software activity is customising the functional

BOX 8.1 DCMS CREATIVE SECTORS (2001), CODED INTO SIC(92) CLASSES

Film:
22.32 reproduction of video recording (+ 25)
92.11 motion picture and video production
92.12 motion picture and video distribution
92.13 motion picture projection

Music and the visual and performing arts:
22.14 publishing of sound recordings
22.31 reproduction of sound recording (+ 25)
74.81 photographic activities (+ 25)
92.31 artistic and literary creation and interpretation
92.32 operation of arts facilities
92.34 other entertainment activities not elsewhere classified (+ 50)
92.72 other recreational activities not elsewhere classified (+ 25)

Architecture:
74.20 architectural and engineering activities and related technical consultancy (+ 25)

Publishing:
22.11 publishing of books
22.12 publishing of newspapers
22.13 publishing of journals and periodicals
22.15 other publishing (+ 50)
92.40 news agency activities

Computer games, software, electronic publishing:
22.33 reproduction of computer media (+ 25)
72.20 software consultancy and supply

Radio and TV:
92.20 radio and television activities

Advertising:
74.40 advertising

Designer fashion:
17.71, 17.72, 18.10, 18.21, 18.22, 18.23, 18.24, 18.30, 19.30 (these codes cover manufacture of clothing and footwear, within which are designer fashion activities) (+ 5)
74.84 other business activities not elsewhere classified (+ n/a)

Art/antiques trade:
52.48/9 retail sale in specialised stores not elsewhere classified (+ 5)
52.50 retail sale of second-hand goods in stores (+ 5)

Note: The + symbol indicates that not all firms in the class in question are believed to contribute to Creative Industries; the numbers following this sign are an estimate of the proportions that do so.

Source: DCMS (2001).

structure and content of databases and the like, rather then the design of interfaces. Much industrial design is precisely about the aesthetics and usability of industrial products, but there is also a large industrial process design subsector. Operational definition of the creative industries is liable to be the source of challenge and controversy for a long time.

In the more recent DCMS 'Evidence Toolkit' (2004) a more elaborate effort was made to classify creative (and cultural) industries in terms of: (1) their function or product (audio-visual, performance arts, sport, etc.); and (2) their role or activity in relation to this type of product (creation, making, dissemination, exhibition/reception, archiving/preservation, and education/understanding activities). This latter classification is rather helpful, since the 'making' category tends to capture many of the manufacturing activities caught up in the creative industries, such as publishing and printing of books, magazines (and audio and video recordings), while the other categories tend to be dominated by services.[3] It resembles the classifications used by Aksoy (1992) and Locksley (1992), who differentiated four information activities: (1) production, innovation, creation of 'content'; (2) packaging, publication, reproduction; (3) distribution, transmission, diffusion; and (4) marketing and servicing. These authors created a quadrant framework, with producers, packagers, distributors and users being linked in a chain, and Aksoy went on to plot industrial sectors in this framework. The DCMS study also sets out to map its categories in terms of SIC codes, though many of the activities involved – especially archiving/ preservation, and education/understanding – were regarded as in need of further work.

8.3 Some statistical analyses

The Community Innovation Survey (CIS) is the most widely used instrument for statistical assessment of innovation. It is, however, known to be incomplete where it comes to sampling services. Consider the fourth round of the CIS (CIS4) conducted in 2004. In the UK sample, at least, the public sector is absent – as are private or charity-run community and, health and education services, etc. Also missing are personal services, and some other subcategories that may be liable to be quite innovation-active. For example SIC 91[4] is 'Activities of Membership Organisations Not Elsewhere Classified'. This includes 91.1 'Activities of business, employers and professional organisations'; 91.11 'Activities of business and employers organisations'; the latter may be significant in that some such organisations are believed to be important sources of innovative information for some firms and sectors. Quite possibly, some associations that play roles in diffusing innovation, setting standards, etc. in creative industries will be present here. Another serious omission is SIC 92, which covers 'Recreational, Cultural and Sporting Activities'. This group, missing from CIS4 in the UK, included 92.1 'Motion picture and video activities'; 92.2 'Radio and television activities'; 92.3 'Other entertainment activities' – including artistic and literary creation and interpretation,[5] live theatre, arts facilities, fairs and amusement parks, and a set of activities not elsewhere classified which covers, *inter alia*, circuses, puppet shows, rodeos, firework displays, and model railway installations; 92.4 'News agency activities'; 92.5 'Library, archives, museums and other cultural activities'; 92.6 'Sporting activities'; and 92.7 'Other recreational activities' (for example gambling and betting). The omission of sectors 92.1–92.3 is probably the biggest concern for analysis of creative industries, and the obvious recommendation is that future CIS surveys should be extended to cover as many of these sectors as possible. There is no clear rationale for their exclusion if we are concerned with assessing innovation across the economy. Indeed, an ICM survey (ICM, 2006) asking the question, 'Has your business ever developed a new product or service in order to generate greater commercial return?' (a broader question than that asked by the CIS about innovation in the last three years, though a narrower one in that greater commercial return is only one of the motives for innovation) found that 58 per cent of firms in the film business answered 'yes'. This was higher than was the case for the creative sector as a whole, including those industries that were included in CIS4.

Several studies have examined creative industry innovation as captured by CIS4. The DTI Occasional Paper No. 6 (DTI, 2006) notes the DCMS definition of creative industries cited above, and suggests that CIS4 sampled around two-thirds of these creative sectors in the UK, and presents some results based on this. It notes that the data are 'indicative only', not only because of the exclusion of creative industries with an SIC code beyond 74, but also because the classification covers a particular set of industries in a rather coarse-grained way. The sectors included in the classification we covered as a whole – but it is understood that only a proportion of the firms falling within these industry groups may be 'creative' in the sense that was earlier discussed. Some sectors were dropped from the analysis for this reason – fashion clothing, for example, which is lost within a much larger clothes manufacturing sector. In contrast, publishing (SIC division 22) was included – there is generation of creative content here, but there is also much aggregation and reproduction of content along with marketing, logistics and other more mundane activities. Computer services (72.2) were included, though as noted these are liable to include many functional activities with operational rather than creative content. Industries in division 74 which were included covered many that are widely agreed to belong to the creative economy (architecture, advertising, photography, design, etc.). However, note that Group 74.2 includes architecture and engineering services (among these design services). Several of the engineering activities – for instance, design of industrial processes, geological and prospecting activities, even weather forecasting[6] – seem much closer to R&D and technical services than to other creative activities. Group 74.4, advertising, includes provision of spaces for advertising and a number of similar activities; and Class 74.87 – like other 'not elsewhere classified' groupings – encompasses strange bedfellows (credit reporters and debt collectors are surely not part of the creative economy), among them 'specialty design'.

Putting these groups together, the DTI paper presented a series of interesting results based on this categorisation, among which we can highlight:

- The workforce in these enterprises has a high proportion of graduates – notably science and engineering graduates, and these businesses are significantly more innovative (about 70 per cent report innovation activity) compared to UK enterprises on the whole (c50 per cent).
- The creative sector enterprises attribute over half of their turnover to their product innovations. In terms of the impacts of innovation, these creative industries reported that product-orientated effects were highest; improved quality of goods or services were the most significant. Thus there is a focus on new products and better-quality products, rather than efficient processes, meeting regulations, etc.
- Over a fifth of creative businesses report having cooperation

agreements for innovation – this is nearly twice as many as other industries, where cooperation emerges as a fairly rare phenomenon.

- These creative businesses also reported being more active at protecting their innovations than did other firms – every category of intellectual property (IP) (confidentiality agreements; copyright, trademarks, patents, secrecy, registration of design, lead-time advantage on competitors, complexity of design) was reported as being used more. The DTI paper suggests that this was due in part to the greater levels of originality in these businesses; the result tends to confirm the DCMS definition of these as industries where IP is particularly important.
- These creative businesses also reported facing greater barriers to innovation than did businesses in general. (Though as the DTI notes that in general, more innovative organisations report higher barriers.) Lack of qualified personnel is reported more frequently as a problem than in other industries, while (interestingly) regulatory impediments are less frequently encountered.

We have explored the CIS4 data in some detail, though the small numbers of cases involved means that we cannot provide in-depth analysis of specific industries like advertising or design. What proved particularly interesting was to compare firms falling into the DCMS categories of creators, makers, distributors and marketers. In the CIS4 sample for the UK, some 2025 creative industry firms feature in the total sample of 16,446 enterprises; the largest group are 'creators' (1093 – large groups being software firms, engineering consultants and designers, and research and development (R&D) services[7]), followed by 'makers' (568 – almost half of these being printers), 'distributors' (359 – half of which are telecommunication firms[8]), and only five 'exhibitors'.

One striking result is apparent from the data on graduate employment. In many of these industries – especially, but not only, the 'creators' – the figures are very large as compared to other sectors. The high share of science and engineering graduates is not so surprising when one considers that the sets of industries noted above as being very prominent in the sample are typically technology-based ones. These 'creative industries' are by and large not cultural, entertainment or aesthetically focused firms. Only around 130 firms could be seen as falling into this category – mainly advertisers, speciality design, public relations and market research, and social science R&D firms – and these have particularly large shares of non-science and engineering graduates in their workforce (around 30 per cent of employees). 'Creators' – whether socially-focused, as above, or more technologically oriented (engineering, architecture, software, natural science R&D, software) were outstandingly innovative, and the 'maker' creative classes were also more innovative than firms in general. 'Distributors' vary between the generally innovative telecommunications firms and the relatively low-innovation trade and retail services featured here.

There are limits to how far one can go with this sort of analysis, due to sample size issues; it would be worthwhile to merge data for several EU countries to allow for a more fine-grained analysis. Also it would be valuable to extend CIS sampling to capture more of the creative services that are currently omitted (and in addition to missing sectors, the restriction of the survey to firms of ten or more employees means missing out many smaller organisations that characterise the 'long-tail' creative services).

8.4 The creative challenge

Few researchers have applied the insights of innovation studies to the creative industries in general. A few specific industries have attracted attention,[9] but until very recently it was rare to find innovation research applied to creative industries. We suspect this is due to two factors: first, that they were generally seen as 'frivolous' (but their heavy economic weight offsets this factor); second, because much novelty is seen to involve aesthetic issues, fashion trends, and the sorts of content discussed at length by media and cultural studies. Determining what is original in aesthetics and content is a minefield which innovation researchers are understandably wary of.

Two recent exceptions are Handke (2004a, 2004b) and Stoneman (2007). Handke contrasts 'content creativity' with 'humdrum innovation' – traditional technological innovation. He surveyed creative industry firms about their production of new content as well as about innovation in the production process – for example, asking music companies about the release of new CDs. This does not give us much of a handle on novelty, however, since old material can be re-released in various ways, and clearly some CDs feature established performers continuing to plough old furrows; while some albums may shock the industry and produce hundreds of imitators. Stoneman (2007) notes that alongside traditional technological innovation, creative industries innovation has more unusual features. He examines two types of 'soft innovation'. First, new products and new ways of producing products that are largely aesthetic in nature (he cites music, books, film), in other words the primary products of many of the creative industries.[10] Second, design innovation in industries that produce mainly functional products – he cites such cases as new designs of cars, new food products, and redesigned electrical products, often overlooked because of the focus in innovation studies on changes in the functionality itself, and dismissal of what is regarded as aesthetic or product differentiation.

But aesthetic innovation is certainly not the only form of innovation in creative industries. It is apparent that there are innovations under way, for instance, in the use of hand-held video-recorders, virtual design studios, techniques for mixing audio recordings, internet distribution of advertisements, and much more. Several other researchers have recognised this. Voss,[11] for example, suggests that there are five important areas in which innovation may be created in the design of 'experiential services': (1) physical environment; (2) service employees; (3) service delivery process; (4) fellow customers; and (5) back-office support. He stresses the role of customer insights, and importance of methods for collecting these, in the design process, arguing that 'experiential innovations' are typically customer-driven rather than technology-driven. In the burgeoning literature on the 'experience economy' we can find other contributions that touch on innovation strategies, such as Pine II and Gilmore (1998), whose set of principles for the design of experiences involves: (1) theming the experience; (2) harmonising impressions with positive cues; (3) eliminating negative cues; (4) ensuring the integrity of the customer experience; (5) mixing in memorabilia; and (6) engaging all five senses.

8.5 Case studies of creators

Miles and Green (2008) studied creative industries in one of a number of studies funded in the UK by NESTA (National Endowment for Science, Technology and the Arts) and exploring 'hidden innovation' (i.e. innovation missed or neglected by conventional instruments and policies). They explored cases from four 'creator' industries – advertising, broadcasting, product design and videogames development – examining the nature and management of their innovations, and how these industries are (or are not) linked to wider systems of innovation.

Practically every firm studied was associated with several different innovations. Fairly conventional technological innovation involving new information technology (IT) applied to products and processes was so common as to be almost universal. This is particularly true in videogames development, but is occurring across the creative industries. Associated with this the digitisation of content allowed for repackaging or 'repurposing' content into new products, and for delivering it in new ways and to new markets. Innovations in new business models and organisational structures were also commonplace, and the focus on the 'experience' of consumers led to distinctive styles and approaches. The products of many creative industries constitute services that are aimed at provoking particular responses from their users: the consumers co-produce the experiences in question (individually or collectively). This is not just a reflection of new technological opportunities: 'creators' report that consumers – both the public and other firms – are becoming more sophisticated and demanding, and are also sharing their views more readily among their peers, as well as with the creative service suppliers.

The 'hidden innovation' that was uncovered as common in the creative industries could be organised into four groups, identified in earlier NESTA work.[12]

Firstly, innovation that is closely similar to activities measured by traditional indicators, but that is excluded from measurement. Services in general are less prone to think of and organise their innovation activities under the rubric of R&D (Miles, 2007), and this is true of the creative services. There are some exceptions – in earlier work we noted that some broadcasting and software companies do use the concept, as did some of the product designers studied here. But research into people's tastes and preferences was important in shaping new products and services. Such activity is excluded from R&D surveys and tax credit systems, and is rarely labelled as R&D or organised in terms of full-time researchers in a specialised department.

Secondly, innovation in organisational forms or business models was also very common in our creative industries. The most important developments often involve moving toward more open and user-generated innovation, or using user inputs of creative content (ranging from social networking sites and discussion fora, to reality TV phone-ins and 'viral marketing'). Reorganisation of value chains (often with use of new communication systems) is also pervasive, with firms sometimes finding new locations for themselves in such chains. Many businesses have been outsourcing work overseas or even relocating abroad (for example, design services that need to be close to the manufacturing operations located in Asia). New approaches to generating revenue are apparent too, with advertising (embedded alongside or within the content of products) becoming a source of income for creative firms in videogames, in preparing broadcast content (though 'product placement' is restricted in the UK), in online publishing and many other online services (as an alternative to subscriptions), and in many other creative industries and service activities more generally.

Thirdly, innovation involving novel combinations of existing technologies and processes has already been mentioned, when creative industries use existing content for new purposes. For instance, TV programmes are repackaged for DVD or other media (even mobile phone displays or Internet downloads); music is repackaged in new compilations, made available in new formats for MP3 players, 'mashed-up' so as to create a different listening experience. New products are also being generated for new markets – for example, videogame firms moving into educational markets with products that combine game features with learning content, or manufacturing firms becoming service providers (which happens in many areas and not just creative industries).

And fourthly, as in many professional services, creative industries feature innovations that take place during the creation and delivery of services – which typically vary continually one from another, as new audiences are encountered or as old audiences require new content within established formats. Often these are 'ad hoc' and usually incremental innovations, that may not be recognised or replicated outside of the specific service encounter (or by anyone other than the particular service provider).[13] The creative industries demand innovative problem-solving, but many of the new solutions are one-offs. Businesses do not find it easy to reproduce such new approaches, though some technical developments (for example, useful lines of code in videogames) may be systematically archived, and new digital content will be stored and often reused. What may not be 'captured', however, is the innovation process which led to the generation of such products: the results may be re-used, but there may have been limited learning as to how to replicate or extend such creativity.[14] In some of the creative industries, the main way of learning to operate creatively seems to be an apprenticeship-type model of participating in brainstorming sessions, working as assistant to senior professionals, and the like.

Despite the pervasiveness of innovation, and the many forms that it takes, the firms we studied find it difficult to manage their innovation processes systematically. Innovation often remains spontaneous or ad hoc – type 4 'hidden innovation' in the categorisation above. Creativity tends to involve the ideas of charismatic senior professionals, with little formal R&D.[15] University links are limited for innovation, though graduates provide vital technical skills, and the inflow of graduates provides access to new methods, thinking, and fashion.[16] It is apparent that communities of practice – professional associations and more informal groups – are an extremely important source of new ideas. Creative industries vary considerably in terms of how far such communities are organised as professional groups, with awards, accreditation, and the like. Many entertainment and advertising industries are highly professionalised; many more 'artistic' activities – where even the term 'industry' is anathema to many practitioners – tend to be organised more through locality or culturally based informal networks. These characteristics resemble those of many service firms and KIBS more generally. They seem to have more to do with the nature of knowledge-based work (and its traditional craft-like form of labour) than specific features of 'creative' industries.

8.6 Reconceptualising innovation

The range of innovations uncovered in the research could be understood in various ways. In the earliest reports based on the study we suggested added 'content' dimensions to den Hertog's framework for understanding service innovations (in terms of service idea, user interface, service delivery and technological infrastructure).[17] Ultimately, we consider that innovation can take place in just about any business process undertaken by the firm, with different types of firm (creators, distributors, makers, etc., and firms of different size and capability) tending to focus on different areas at any one time. The areas identified are described below:

General administrative activities and financial management

Innovations in office automation and financial control systems, largely dependent on new IT, are common to firms in many sectors (not just creative industries); differences among firms relate especially to the size and network structure of the business.

Business model (more specifically, revenue model)

Innovations may involve how finance and profits are derived: the case mentioned above of publications and online services being funded through advertising revenues, rather than payment from readers, is a case in point.

Value chain location and positioning

Whether organisational repositioning itself is organisational innovation or some other category of organisational change may be argued about, but innovation in terms of new business routines, communication practices, and networking techniques is almost invariably associated with the strategies that are generally described as 'moving up the value chain' or 'focusing on core competences'. In these industries decisions about what parts of the creative product to produce and/or process, how to collaborate with partners in major creative products, and the like, have been and continue to be pressing ones – and many firms find themselves having to flexibly take different roles in different projects.

Communications beyond the boundary of the firm

Innovations involving use of communication networks have already been mentioned in the context of other strategies, and will be encountered below. Suffice it to say here that tools and techniques for relating to partners, customers, suppliers, regulators, industry bodies and others are all the locus of various kinds of innovation: websites and mobile services being cases in point.

Internal communications
And more generally, the management of human resources and work organisation within the enterprise. These are also activities where innovation is prevalent. Knowledge management systems are particularly interesting cases, since these may be designed so as to capture innovations or knowledge about where expertise is located in the firm. Ways of maintaining contact with staff in the field, new training systems, and other tools are also widely used.

Back-office and backstage production processes
Innovation concerning how the product or service is designed, scripted, rehearsed, prototyped, etc., is what is classically thought of as 'process innovation', though much of this remains hidden from conventional metrics. The particular forms of innovation adopted vary considerably according to the type of industry: the activities can be heavily dependent on skilled or unskilled labour, or on technology of various kinds (though IT is almost universally involved). The processes and 'architecture' of the service may even be rendered visible as part of the consumer experience. Innovation in service design and the new service development process is so significant that we are tempted to identify this as a distinctive category. Tools to aid the design process – from computer-aided design (CAD) through to virtual laboratories and brainstorming spaces – are already significant, and likely to become more so.[18]

Transactions
Innovations centring on the process of payment for access to product are common, with e-commerce and related systems for online bookings and reservations being widely used for consumer services, as are loyalty card and similar schemes. Innovations may also involve less technological novelty, such as various types of season ticket and membership scheme. Transactional innovations – often highly influenced by developments in the financial and retail sectors – may be more or less closely tied to marketing and customer interfaces.

Marketing and customer relationship management
'Marketing' is an area where the CIS has asked about 'changes introduced in the last 3 years' from CIS4, and where a great deal of innovation is common across creative industries (and many other service industries). Such innovations drawing on knowledge about customer segmentation (applying market research and other social science), eCRM (electronic customer relationship management) systems, data mining, and the like. The interactive nature of many creative products

means that innovations such as online 'fan clubs' are distinctive, even compared to the user groups for more conventional industrial products. One of the challenges for creative firms has been achieving accommodation with spontaneous communities arising from the consumer audience (sometimes with support from content creators but not from publishers).

Content and symbolic substance of product
As we have seen, the term 'content' may not be universally employed, but creative industries typically seek to influence user experience through deploying text, imagery and other symbolic substance. Unsurprisingly, this is the site of much routine, incremental and radical innovation. Innovations can range from the creation of completely new genres of content through to reframing of familiar content within a new context (for example a new production of a drama or piece of music).

Performance and front-stage production processes
Varying very much across different creative industries, another form of 'process innovation' (often hard to differentiate from 'product innovation') involves the performance of a creative work by artists, actors, musicians, etc. Some innovation surrounds ways of capturing such performance (recording, broadcasting, etc.), some involves the supporting technology and infrastructure, and the organisation of creative work in the course of performances and displays.

Product format
Innovations in the format and character of creative products vary according to the sorts of media and performance that are involved. They can involve new types of product (such as new media like DVDs) and improved features of existing products.

Delivery of product
Much service innovation has concerned delivery, the ways in which the information content of services ('creative content' in the case of creative industries), or the physical medium for such content, reaches the consumer (or how the venue for performance and display is constituted). Electronic delivery of information services and technological support for conventional performances are particularly important. Innovation can also involve the creation of new venues, the repurposing of existing venues (perhaps by introducing live music to a restaurant), and the restructuring of venues to provide new dimensions to the consumer experience.

User interface with product
Innovation can involve changing how the consumer engages with the product, their points of access to its content and functionality. For some creative industries, the interface may be electronic, but it may also or instead involve creative facilities and premises such as cinemas, theatres or galleries. Innovation in the design and use of such facilities, in how they are configured and rendered appropriate settings for specific types of experience is one vector of innovation. Physical media 'platforms' or 'carriers', such as TVs, PCs, phones, or even print media, also involve interfaces with their own functionality and capacity to generate experience through the control and entry points that are presented.

User interaction
Innovation here goes beyond just the interface innovation discussed above, and works with new types of consumer inputs, which may simply be used by the creative firm (for example feedback on the product) or to enhance consumer experiences by making interactions among users a constitutive part of these experiences. Co-production is an important feature of many creative services (along with other knowledge-intensive services). Much attention has been attracted by recent innovations collectively labelled 'Web 2.0', where users supply much of the content to websites; and to the innovation processes labelled 'open innovation' which build on user (and other) innovative inputs. Social networking and user content may be important in a range of creative industries as evidenced by the popular facilities offered by the BBC for their users to add their own reviews or comment to the views and information provided by other users. The BBC even lets users dream up plot lines for, and write jokes and poems around, soap operas. Interestingly, such facilities may be employed by retail firms – Amazon is a case in point. Creative content may flow from users into firms that do not themselves produce much of such content other than in their marketing.

User capabilities
A wide range of innovations that are only fractionally within the control of all but the most influential creative industry firms are very important for their innovations. These include new hardware platforms, for instance mobile phones that show live TV or can carry 'visual radio' alongside FM radio. We anticipate that there will be a vast range of new creative services emerging to make use of location-based PDA and phone systems in the coming years (for example jokes, music, cultural histories associated with the places one is passing through). Users themselves are evolving in terms of their skills, tastes and practices. Creative firms can participate in

these processes, sometimes in alliance with hardware firms and other creative firms in software, telecommunication or broadcasting services, and alliances will be an important part of the future landscape.

8.7 Some practical conclusions

Innovation management needs to be recognised as a key capability in creative firms; but the nature of innovation management is changing rapidly. Much creative industry innovation is based on 'co-production' with significant input from users; and networks, partnerships and collaborations are also important sources of innovation. This has substantial implications for management practices and skills. Conventional project and innovation management skills are bound to remain important, along with the aesthetic and cultural awareness of professionals in many creative industries. But managers will increasingly need interpersonal and communication skills – skills such as team-building, conflict resolution and problem-solving. These will be required for collaboration with professionals of various types and for engagement with consumers and other firms.

Policy-makers need evidence as to how policy might assist (or hinder) innovation in the creative industries. Though some research has been undertaken on this theme, more detailed evidence would underpin and guide the policy process. This will mean extension of CIS-type surveys, but also case study work on the impacts of training, intellectual property rights (IPR), cluster and a host of other policy initiatives on the industries. It is likely to be the case that targeted innovation support for creative industries will prove to be helpful, since they do not find general innovation support programmes (such as R&D Tax Credits) to be relevant to the sort of innovations they undertake. Finally, since new forms of innovation are emerging rapidly, we need mechanisms for monitoring change, exploring good practice, and sharing this intelligence with policy-makers.

8.8 Some research implications

Research is certainly required to inform policy, training and management strategy in the ways outlined above. It is important to encompass more creative industries and more forms of creative work, both through case study research and large-scale surveys. It will be important to integrate, or at least relate together, bodies of research and analysis that have been largely compartmentalised into poorly linked areas of design, new service development and innovation.

Just as research on service sectors has thrown up points for innovation research that apply to the service activities of all economic sectors, so research on creative services is liable to have implications for the creative work in all sectors. Indeed, since there are actually more people employed in 'creative professions' in the UK economy in sectors outside of the creative industries than within the creative sectors,[19] the implications for the wider economy, and for innovation more widely, are likely to be profound. Thus research into 'creative service' innovation should extend beyond, and is relevant to, more than these creative industries. It is not just a matter of media, entertainment and culture as narrowly understood.

Exploring innovation in creative industries has also suggested that there are many more forms of innovation than are readily captured by the received categories of product and process innovation, and that even newer notions and metrics of organisational innovation may be failing to observe many new ways of proceeding in business. We have no way of knowing whether the creative industries studied here are in the vanguard of such forms of innovation, or are quite unremarkable. (We suspect that in some of the cases cited, they are in the vanguard, but not in others.) Development of tools for assessing the incidence, management and impact of this diverse range of innovative activities, that can be applied widely across service sectors and the whole economy, is an important task. Case studies such as the ones focused on above are needed to establish the terminology that can be applied in various industries, the sorts of informant (or data capture) required to provide intelligence about empirical developments, and how far we can fruitfully apply constructs like incremental and radical innovation and assess the technology content of innovation. Such case study work will inform better metrics. Perhaps more significantly, it will help deepen understanding of just how the various, and evolving, ways in which the innovations that are transforming our daily experience are themselves being shaped and reshaped.

Notes

1. The branch 74.84 (other business activities not elsewhere classified) is renumbered 74.87 in the SIC(2003): this has some potential for confusion.
2. The term 'content' is rather problematic for some fine arts like sculpture, and for many antiques and interactive performance situations. These all carry substantial symbolic charge, if they are to be successful, but the distinction between form and content is far more blurred than in, say, a TV programme, a novel, an advertisement – though in all media we can find instances of boundary-pushing cases that defy easy form–content (and genre) distinctions.
3. And does not construction bear the same relation to architecture as publishing does to film, music or text creation?
4. SIC is the Standard Industrial Classification, now being revised under the acronym NACE, with some improvement of the treatment of services and creative services. Later runs of the CIS are intended to capture a wider range of NACE sectors.
5. Notably, this includes 'technical writing', which might have been classed among the technology-based knowledge-intensive business services.
6. This of course is widely believed to feature considerable creative content, not to mention wild fantasies.
7. We included these in the definition of creative sectors for reasons explained earlier.

8. As note 7.
9. For example, videogames – perhaps chosen because it is a new and high-tech industry with remarkable market growth – where researchers include Cohendet and Simon (2007), Grantham and Kaplinsky (2005), Tschang (2007). Less academically, but of considerable interest, there is even a website on the topic (http://www.gameinnovation.org/) and numerous blogs cover this.
10. Although just what constitutes novelty, or how much novelty there is, will be an issue.
11. For example, Voss and Zomerdijk (2007) focus on innovation.
12. E.g. NESTA (2007).
13. Successful new ideas as to content are, however, often picked up and emulated rapidly, where mass media are concerned. There are even supposed to be 'joke thieves' who pass on information about new content being produced by stand-up comedians.
14. See Dougherty (2004) for analysis of how KIBS are trying to capture ad hoc innovation.
15. See Miles (2007) for services' (apparently) low R&D performance.
16. See IoIR (2004) for services' links to universities.
17. Green et al. (2007); den Hertog (2000).
18. See, for example, Gann and Dodgson (2007).
19. See Higgs et al. (2008).

References

Aksoy, A. (1992), 'Mapping the information business: integration for flexibility', in K. Robins (ed.), *Understanding Information: Business, Technology and Geography*, London: Belhaven Press.

Cohendet, P. and L. Simon (2007), 'Playing across the playground: paradoxes of knowledge creation in the videogame firm', *Journal of Organizational Behavior*, **28**, 587–605.

DCMS (2001), 'Creative industries mapping document', Department for Culture, Media and Sport.

DCMS (2004), 'DCMS Evidence Toolkit – DET' (Formerly, The Regional Cultural Data Framework) London, Department for Culture, Media and Sport Technical Report August 2004.

den Hertog, P. (2000), 'Knowledge-intensive business services as co-producers of innovation', *International Journal of Innovation Management*, **4** (4), 491–528.

Dougherty, D. (2004), 'Organizing practice in services to capture knowledge for innovation', *Strategic Organization*, **2** (1), 35–64.

DTI (2006), 'Innovation in the UK: Indicators and Insights', London: Department of Trade and Industry, Occasional Paper No. 6, July.

Gann, D. and M. Dodgson (2007), '*Innovation Technology: How new technologies are changing the way we innovate*', London: NESTA Provocation at http://www.nesta.org.uk/assets/Uploads/pdf/Provocation/innovation_technology_provocation_NESTA.pdf.

Grantham, A. and R. Kaplinsky (2005), 'Getting the measure of the electronic games industry: developers and the management of innovation', *International Journal of Innovation Management*, **9** (2), 183–213.

Green, L., I. Miles and J. Rutter (2007), 'Hidden Innovation in the Creative Industries', London: NESTA Working Paper at: http://www.nesta.org.uk/assets/Uploads/pdf/Working paper/hidden_innovation_creative_sectors_working_paper_NESTA.pdf.

Handke, C.W. (2004a), 'Defining creative industries by comparing the creation of novelty', presented at WIWIPOL and FOKUS Workshop, Creative Industries: A Measure for Urban Development? Vienna, Austria, 20th March.

Handke, C.W. (2004b), 'Measuring innovation in media industries: insights from a survey of German record companies', mimeo, Humboldt-Universität zu Berlin / Erasmus Universiteit Rotterdam.

Higgs, P., S. Cunningham and H. Bakhshi (2008), 'Beyond the creative industries', London:

NESTA, http://www.nesta.org.uk/assets/Uploads/pdf/Research-Report/beyond_creative_industries_report_NESTA.pdf.

ICM (2006), 'Survey of creative businesses', London: NESTA, http://www.nesta.org.uk/assets/Uploads/pdf/Research-Report/Archive/creative_businesses_survey.pdf.

Institute of Innovation Research (2003), 'Knowing how, knowing whom: a study of the links between the knowledge intensive services sector and the science base', University of Manchester IOIR (now MIoIR), Report to the Council for Science and Technology, available at http://www.cst.gov.uk/cst/reports/files/knowledge-intensive-services/services-study.pdf.

Locksley, G. (1992), 'The information business', in K. Robins (ed.), *Understanding Information: Business, Technology and Geography*, London: Belhaven Press.

Miles, I. (2007), 'R&D beyond manufacturing: the strange case of services' R&D', *R&D Management*, **37** (3), 249–268.

Miles, I. and L. Green (2008), 'Hidden innovation in the creative industries', London, NESTA Research report HICI/13, http://www.nesta.org.uk/assets/Uploads/pdf/Research-Report/hidden_innovation_in_creative_Industries_report_NESTA.pdf.

Pine II, B. and J. Gilmore (1998), 'Welcome to the experience economy', *Harvard Business Review*, **76** (4), 97–105.

Stoneman, P. (2007), 'An introduction to the definition and measurement of soft innovation', London: NESTA Working Paper, http://www.nesta.org.uk/assets/Uploads/pdf/Working-paper/soft_innovation_working_paper_2_NESTA.pdf.

Tschang, F.T. (2007), 'Balancing the tensions between rationalization and creativity in the video games industry', *Organization Science*, **18** (6), 989–1005.

Voss, C. and L. Zomerdijk (2007), 'Innovation in experiential services: an empirical view', in DTI (ed.) *Innovation in Services*, London: DTI, pp. 97–134.

9 Social innovation, social enterprise and services

Denis Harrisson, Juan-Luis Klein and Paul Leduc Browne

9.1 Introduction

Organizations with a social mission such as La Boussole which helps people with a criminal record to integrate into the labour market or the group Information which provides labour market integration services to motivated people by offering workshops, workplace internships, assistance with job search and personalized follow-up, are examples of social innovation, that is, initiatives taken by social actors to respond to a need while being supported through public recognition. Indeed, these particular initiatives are supported by Emploi-Québec, a government agency that grants financial support to this type of organization. Social innovation in Quebec is also transmitted through the Beaubien movie theatre, a cultural non-profit organization (NPO), or TOHU, the City of Circus Arts in Montreal. The latter's mission also includes an environmental component involving the rehabilitation of a former waste disposal site located in an urban area and a social commitment to the citizens of the Saint-Michel district in Montreal by contributing to the development of this district, one of the most densely populated and most disadvantaged neighbourhoods in Canada.

The concept of social innovation has not been widely theorized (Harrisson and Vézina, 2006), except with regard to its managerial dimension. Research conducted on social innovation has barely started bringing out the main dimensions marking its boundaries. Several scholars have highlighted the eclectic and heterogeneous nature of social innovation (Cloutier, 2003; Goldenberg, 2004; Moulaert et al., 2005; Harrisson and Klein, 2007; Drewe et al., 2008). There is as yet no common interpretation of it. The latest edition of the *Oslo Manual* (OECD, 2005) refers to two new types of innovation that were not included in the two earlier editions – innovation in services and organizational innovation – while the concept of social innovation itself still does not figure in the manual. This chapter does not aim to put forward a general concept of social innovation, which remains to be constructed, or to conduct an exhaustive review of all the approaches to the development of a theory of social innovation in the scientific literature. Rather, our goal is to present one facet of social innovation in the service sector, that is, the contribution of the social economy, also referred to as the third sector or social enterprise sector.

It should be specified that this concept already existed in the nineteenth century through the creation of numerous support groups among the working class that was beset by the destructive effects of industrial capitalism. The solidarity ties that were a key characteristic of the mutual organizations, friendly societies and budding cooperatives were the cornerstone of social innovation (Petitclerc, 2007; Browne, 1996; Finlayson, 1994). Unlike capitalism, which was founded on the principles of profit maximization through competition among entrepreneurs, the social economy aimed at the social protection and the betterment of those who participated in it. Thus, friendly societies enabled members to defray the costs of their funerals or of the construction of a house; and cooperatives made consumer goods accessible at affordable prices, allowed members of the working class to borrow money at reasonable rates and gave workers the means to be their own bosses while collectively holding the means of production. However, innovation was not limited to the forms of collective entrepreneurship illustrated by these examples; it also covered the social transformation involved in these initiatives, in particular in terms of strengthening the social fabric through the creation of networks of relations as well as in terms of political action (for example, the importance of the cooperative movement in the birth of the Labour Party in Great Britain and the Co-operative Commonwealth Federation in Canada). Similarly, in the early twentieth century, urban planners suggested forms of planned territorial development aimed at bringing the work environment and the living environment together within the same territory (Howard, 1902). In all cases, these experiences mobilized civil society and the communities involved.

In its contemporary version, the social innovation that occurs in the social economy refers to the adoption of innovative solutions to complex economic and social problems. It is produced by social actors from civil society, either autonomously or with state support or by creating a partnership in service delivery. There are several forms of partnership, depending on the actors involved. Actors participating in social innovation can be community groups, social movements or social entrepreneurs who have new ideas and take creative initiatives in order to change the social environment and solve various types of problems. It is the local and community dimension of these social arrangements which makes it possible to get around or overcome institutional constraints and to come up with inventive courses of action that would be impossible with a centralized, unilateral and linear approach caught in the grip of the administrative silo. These civil society actors forge links, build networks and develop joint projects. Developing and implementing activities, initiatives, services and improved procedures serves to enhance economic and social responses to the problems faced by communities and individuals. The rest of the chapter is divided into sections as follows. Section 9.2 presents social innovation as a way of changing society through creativity serving the public interest and the community. Section 9.3 briefly reviews the different meanings of 'social' and examines the way in which social innovation represents a new relational form between social actors. Section 9.4 studies the theory which supports the creation of new forms of enterprise that include both economic and social aims. Social innovation and its links to services is analysed in section 9.5. The innovation is not in the service per se but rather in the way in which the service is created and the specific kind of community members to whom the service is addressed. Section 9.6 concerns the role of research in social innovation. Social innovation is not a result of research specifically designed to answer issues and problems identified by the actors. However, research is needed for the codification of knowledge of the process of both creation of social innovation and its diffusion toward the community and the society. A case analysis of social innovation and service is introduced in section 9.7 of the chapter. This example draws on the role played by civil society in the organization and the delivery of home care and home support services in Quebec. Section 9.8 concludes.

9.2 Social innovation: creativity serving the public interest

Social innovation occurs as the renewal of intervention strategies by civil society which reconstructs its various responses to economic and social crises, weakened social links (Putnam's theory, 2000) and the evolving welfare state (Evers and Laville's theory, 2004, and that of Esping-Anderson, 2002). It can generally be hypothesized that the form of the social economy (at least in its not-for-profit dimension) depends on the forms of emergence, growth and transformation of the state (Browne, 1996, 16). Lester Salamon noted more than 20 years ago that non-profit organizations in the health and social services sectors in the United States 'actually deliver a larger share of the services government finances than do government agencies themselves' (Salamon, 1987, 29–30). In Quebec, where the Quiet Revolution of the 1960s led to the construction of a rather centralizing welfare state, both the funding and the delivery of these services were largely integrated into the public sector and developed under its authority.

Although it is true that voluntary, not-for-profit or cooperative action has existed for a very long time, there is no doubt that the social economy has today acquired a new salience throughout the world, as shown by the increase not only in scientific journals and university programmes devoted to it, but also in the local, national and international public programmes seeking to promote it. A new industry sector has emerged, fostering social initiatives which belong to neither the public sector nor capitalist enterprise. These initiatives stem from voluntary organizations or cooperatives, and operate under several designations and legal structures. This third sector or social economy sector is made up of organizations and associations that develop, produce and offer services and resources which have effects on members' mutual interest or on the public interest.

This sector has been growing at a phenomenal rate, expressed in terms of the large number of organizations in both developed and developing countries. In Canada in 2003, non-profit organizations and those in the voluntary sector accounted for 6.8 per cent of the gross domestic product (GDP), including hospital centres, universities and colleges. If these organizations are withdrawn, in accordance with the practice of those who adhere to a firmer position on the social economy, they account for 4 per cent of the GDP (Hall et al., 2005). These organizations employ 12 per cent of the labour force in Canada. The service sector dominates among the activities of these organizations. Thus, 74 per cent of employees in this sector are engaged in activities which involve delivering direct services to the population, such as education, health and housing.

Although this sector is important in terms of its contribution to the GDP and national employment, it is mainly at the local level that the significance of its contribution can be felt. Recent empirical research has shown that organizations linked with this sector contribute substantially to improving living conditions and sustaining local communities by carrying out activities, delivering collective and individual services, and generating numerous collateral but nevertheless crucial effects, such as job creation. They mobilize a considerable resource – voluntary action – which they harness to serve disadvantaged people. However, they do not only mobilize voluntary action, but also obtain large revenues which they inject into local communities. Thus, although the services provided by these organizations target low-income population groups, their action can be felt by the entire local community because they increase the density of community networks, create jobs, spend locally and contribute to the local market, and because they foster a degree of social vitality which is expressed through participation (Klein et al., 2004).

This contribution makes up for the weaknesses of the state and the private sector. The state is struggling with reduced budgets and deficits, not to mention the difficulties of relying on more traditional policies to meet the challenges posed by today's complex society (Ackers, 2002; Breton et al., 2004). The private sector, on the other hand, is caught up in the profound transformation of the economy which entails lay-offs, offshoring, restructuring towards the service economy and the knowledge economy, and a transition to other forms of organization such as networks or partnerships (Harrisson and Laberge, 2002). Civil society has thus stepped in in Quebec by means of economic initiatives such as the Fonds de solidarité (solidarity fund), Fondaction (CSN investment fund) or the Technopôle Angus (business park) (Fontan et al., 2005) as well as social initiatives, such as daycare centres, work integration social enterprises, and the development of social enterprises providing home care services (Vaillancourt et al., 2003; Tardif, 2007). This entrepreneurial spirit has social aims. The modern version of social innovation emerged as early as the 1970s. According to Defourny, the contemporary third sector has been developing through two main approaches. The first approach consists in non-profit organizations (NPOs) whose association forms the basis of and is based on associative life as imagined by Alexis de Tocqueville; the second approach puts forward an economy based on cooperatives and mutual associations which support economic development (Defourny, 2001). Beyond the legal or organizational approaches, the normative question is also crucial since it deals with the values and principles guiding the voluntary sector and/or the social economy sector. These organizations aim to serve members of the community even before generating a profit. The decision-making process is most often democratic and priority is given to people and employees before capital and income distribution, even though funding social innovations is difficult. Moreover, private sector enterprises are involved in new and more recent issues such as corporate social responsibility and sustainable development (Lapointe and Gendron, 2004). In addition, new state configurations focus more on partnership with private enterprise as well as community organizations and social economy enterprises in order to intervene in a more targeted, responsible and legitimate manner (Klein, 1992). Social innovation is not exclusive to a single sector of economic activities but is spread among the three sectors in an asymmetrical way, however with striking differences in terms of the social purpose of initiatives. This is why innovations seem to be predominant in the social economy sector (Lévesque, 2006).

The organizations involved in the social economy show that, at the ontological level at the very least, social innovation comes in various dimensions, occurs in several settings, involves heterogeneous actors and brings solutions to various types of problems. Social innovation is guided first by the problems of practical life. The pragmatism displayed by actors who are engaged in these activities is rapidly confronted with the limits of these interventions when a more rigorous conception of the field is lacking.[1]

9.3 The meaning of 'social' in social innovation

The term 'social' used in 'social innovation' refers to different interpretations and opens up several approaches. First, it helps to identify the type of service for which innovation is carried out. This can be a service to members of the community who, if profits are generated, advise the promoters to reinvest in the community, either by improving service delivery or by broadening the field of intervention (for example, the daycare centres or consumer cooperatives). Second, 'social' can also refer to a service benefiting the whole community, as demonstrated by the Technopôle Angus which is beneficial for all citizens in the Rosemont district in Montreal (Fontan et al., 2004, 2005), or target groups, as demonstrated by the home help services for the elderly (Ducharme et al., 2004). Third, the term refers to the non-economic aspect of economic interventions. The services deriving from social innovation are activities that are necessary but not lucrative or not highly lucrative, as shown by the activities involving the social integration of disadvantaged people; or these activities can be lucrative when they are provided to a population that can pay for them, but are not lucrative when they are provided to disadvantaged groups (for example, the Restaurants du Coeur in France, a French charitable organization). Fourth, the term 'social' is also used in relation to the methods adopted by the organizations which rely on a participatory dynamic involving employees, users or partners. Social innovation in fact refers to a particular process involved in its implementation which relies on the participation and empowerment of the various stakeholders (Mendell, 2006).

The type of service offered through social innovation cannot always be distinguished from the services offered by public or private enterprises in the market sector. A service is a joint production of an intangible good, often consumed as soon as it is produced and, in some cases, is the result of a joint production between a provider and a recipient. 'Services are produced when an organization delivers the "right to use" a capacity or skills and produces useful effects on the person or these goods' (Gadrey, 2002: translation). The recipient can be associated with the production or creation of the service delivered. In the forms of social innovation discussed here, the creation starts beforehand through alliances between the various actors who exchange ideas and knowledge as well as material, financial and human resources, and who share mutual interests. The service delivered is thus the result of an alliance that creates the conditions conducive to innovation. The client or user has expectations and aspirations. The production is structured by the negotiation of the rules attributing the responsibilities of each party. In some cases, the service is based on solidarity and entails a production of social values through the delivery. A consumer who chooses to buy fair trade coffee is not adopting a better taste or better-quality roasting. Rather, his or her choice is motivated by the opportunity to change the rules of commercial production of coffee by promoting the values of equality and equity in the production chain that lead to decent working conditions. These values take precedence and influence the consumer's choice. As regards other areas, social exclusion, poverty and population ageing compel the community to find solutions and implement services as a result of interactions between providers and users. In addition, various actors are directly involved in the relationship in order jointly to produce spaces in the community which allow for the development of services for individuals.

Several contextual factors have contributed to the emergence of the contemporary version of social innovation. First, a crisis arose in most of the systems and institutions that could no longer satisfy the needs and aspirations of users and citizens, a situation that has been compounded by the emergence of newly expressed needs. Solutions to the problems have thus been considered in terms of the actors' values and the need to redesign a framework that is conducive to social cohesion and the affirmation of new democratic rights (Ackers, 2002). Social innovation has thus arisen as a form of interaction between agents who can move beyond the mere commodity relation between buyers and sellers.

Second, there are different forms of social innovation which thus require specific structural changes. Social innovations are complex in nature, because their emergence is based on the simultaneous presence of several conditions such as the transfer of (some but not all) public service provision to the third sector, the growth of non-governmental organizations, the role of civil society associated with new ideological and philosophical values and beliefs (associationism, solidarity, communitarism, social entrepreneurship, democratic citizenship, participation, autonomy and empowerment), the presence of strong mobilization networks and social movements, and the existence of institutions that can diffuse innovations.

Third, for a state which withdraws from certain fields of direct intervention, the transfer or sharing of responsibilities to or with other sectors is based on the belief in managerial efficiency, that is, that it can do better at a lesser cost; and in the community being able to integrate the innovation activities, that is, the provision of services is rooted in the community and thus becomes legitimate. Nevertheless, social innovations are not accepted right away by the institutions, but rather establish themselves gradually following tension, conflict and, finally, compromise.

The Grameen Bank in Bangladesh is a case in point. It was created by Mohammad Yunus because the banking institutions refused to offer microcredit to the most disadvantaged peasants and artisans. This social entrepreneur aimed to reduce poverty and help people become economically independent by offering them small loans meeting their economic needs and financial capacity, while trusting that these borrowers were capable of reimbursing their loans. Traditional banks could not engage in this type of loan because the socio-economic category of borrowers did not correspond to the type of loan offer. In Quebec, the Fonds de solidarité, a venture capital fund developed on the initiative of the Fédération des travailleuses et travailleurs du Québec (FTQ, Quebec Federation of Labour) and still under its control, manages to combine investment in profitable enterprises with a decision-making role on the part of employees. This is matched by an educational component which deals with the mechanics of an enterprise's economic operation and, in so doing, the fund manages to impose respect for social ethics on the enterprises in which it participates, as shown by its intervention in the Gildan enterprise. (Gildan is a sportswear manufacturer blamed for unethical conduct by taking advantage of low labour costs in Honduras. Since then the company has redressed the situation and adopted a code of ethics.) Private investors still only consider the first aspect mentioned above, that is, investing in profitable enterprises, while ignoring the second, non-economic, aspect. Just as in the case of microcredit, banks analyse loans from the viewpoint of profitability and rarely from the viewpoint of effects on social inclusion, social cohesion or financial autonomy. During the Fordist period, this role was played by the Keynesian state, whereas, following the crisis in this mode of regulation, other types of regulation have been experimented with.

Civil society has gradually contributed to establishing the conditions and structures conducive to the emergence of social innovations. Organizations of civil society are capable of taking initiatives, that is, of inventing and borrowing ideas, developing new knowledge, transferring knowledge, reorganizing existing structures, constructing analogies, and deconstructing the established standards. This innovating civil society engages in collaborative relationships and, if the institutional framework allows it, new alliances thus emerge.

9.4 Social innovation and its theoretical reference points

Social innovation is ambiguous, even contradictory. It simultaneously encompasses initiatives to consolidate social cohesion, movements protesting against the established order, managerialist approaches and approaches emphasizing the emergence of grassroots initiatives aimed at transforming living conditions. It links actors belonging to the different socio-economic classes in all sectors of economic and social activities. This can contribute to corporatist governance or the forming of anti-hegemonic alliances. Through some of its manifestations in the third sector, it participates in the creation of forms of social cohesion which strengthen the established social and political order. On the other hand, social innovation can be the beginnings of a society undergoing radical transformation and can challenge the existing order and the distribution of power and authority. Social innovation can offer an alternative strategy involving opposition to domination and hegemony through the creation of projects which authorize a transfer of political power to civil society. The participatory budgeting in Porto Alegre in Brazil is, in this respect, a good example of social innovation which entails an effort to democratize and institutionalize without social legal codification (Novy and Leubolt, 2005). A project's value is assessed based on the measure of satisfaction of basic needs and investments made in the sectors involved (home help, daycare services, local development). Thus, the actors learn new participatory rules and create a space in the community, thereby linking the public space with networks of civil society actors and social movements.

Some approaches consider social innovation uniquely in its instrumental dimension, that is, as a backup sector which meets needs that are disregarded by both the private and public sectors. This suggests the theories which explain the non-profit sector on the basis of failures of the market and the state (James and Rose-Ackerman, 1986; Douglas, 1983; these theories were reviewed by Salamon, 1987 and Laville, 2007). This approach, which stems from economics and the more managerialist trends, puts greater emphasis on the effectiveness dimension of social innovations associated with the service offered regardless of other dimensions such as democratic governance or the transformation of socio-economic rules. Numerous management schools in North America and Europe have now created programmes to train managers who will be able to make these organizations efficient while protecting social values. The difference between this instrumentalist approach and that of the social economy is not merely semantic, but reveals a division with regard to the sector's aim as a vector of social transformation.

The social economy combines two aspects of social life. The first is the economic aspect, which refers to the production of goods and services and thus to participation in the creation of collective wealth. The second is the social aspect, which refers to 'the promotion of values and initiatives involving individual and collective empowerment', and the contribution to developing participatory democracy and responsible citizenship (Chantier de l'économie sociale, 1996). Thus, the social economy contributes to improving quality of life and collective and individual well-being. Social economy enterprises serve their members and the community, and their management is autonomous and independent of the public and private sectors. Their decision-making procedures are based on user and employee involvement, and the 'principles of participation, empowerment, and individual and collective responsibility' (Chantier de l'économie sociale, 2001).

The Schumpeterian notion of entrepreneurship is important for one level of analysis of social innovation (Klein and Harrisson, 2007; Lévesque, 2007). It focuses on a creative process whose starting point is the social entrepreneur, that is, an individual capable of taking the initiative, who has access to multiple resources and can break with the institutional links in order to create a specific itinerary for their project. Innovation is based on this emblematic figure of a leader who knows how to persuade, create convincing arguments and get others interested in the project. The effectiveness and efficiency of the services delivered through social innovation are closely associated with the type of funding and the management of enterprises. In fact, the social entrepreneur develops links with various sources of funding coming from the private sector (banks), the public sector, through a variety of types of financial assistance (ranging from tax exemption to direct subsidy), and the third sector, in particular the NPOs (philanthropy or credit unions). The NPOs can make their assistance conditional, or require a periodic evaluation of the service offered using different measurable indicators of satisfaction of human needs, independently of modes of governance or effects on representative democracy. The representatives of these organizations will thus strive to demontrate the impact of their interventions by relying on the managerial methods borrowed from private enterprise, but adapted nevertheless to the non-market sector.

At another level of analysis, social innovation in the service sector, viewed in particular in terms of governance and empowerment (Mendell, 2006), can be distinguished through its process of emergence and consolidation. Social innovation involves the creation of a particular type of services but pertains above all to a specific process involving both the key role of social entrepreneurs and the forming of alliances and networks. The result is the creation of a system giving rise to a proliferation of social innovations whose effects on one another consolidate the basis. Local governance contributes to enhancing the knowledge of actors and a relatively formal framework gives rise to other creative and initiative-taking processes. In doing so, the actors learn the new participatory rules and create a space in the community, thus linking the public space with networks of civil society actors and social movements. Social innovation includes a social dimension which involves strengthening the social link, an economic dimension which involves producing wealth, and a political dimension involving demand-based or militant actions, the will to participate in and even democratize socio-economic life, and the quest for change.

This is why Benoît Lévesque, for example, has suggested that social innovation be analysed within the framework of theories on social development (Lévesque, 2007). Civil society actors create new organizations, in particular in the social economy. They also participate in including and integrating new initiatives in existing organizations and that penetrate into different realms of society, whether the labour market, education, health, housing, social and economic development, or social life in general. Social innovation represents access to social transformation although it takes action mainly at the local level and in organizations (Moulaert et al., 2005).

Social innovation is first of all a pragmatic approach to social change, the fight against poverty, and the creation of mechanisms for redistribution and equality between the social classes. Social innovation is a structured intervention in local environments or socio-territorial networks (Klein, 1992, 2008). This is why so many experts in local development refer to innovative environments as the distinctive feature of an intervention linked to and structured around a specific place, region, city, or even district, village, organization or network of organizations and associations. The impact and diffusion of social innovation are of varied scope, depending on the type and the place.

There are thus three main dimensions of social innovation (Harrisson and Klein, 2007; Lévesque, 2007; Malo and Vézina, 2004; Moulaert et al., 2005; Cloutier, 2003). First, social innovation responds to an unsatisfied human need by creating a service or good. This can also involve reproducing a service delivered by private or public enterprises, but in which social values predominate. For example, in Quebec City, a supermarket owner hires employees who have difficulty integrating into the job market. From the consumer's perspective, there is hardly any difference between this establishment and another similar business, except that the employees form a group of people who are vulnerable in the labour market and the storekeeper is aware of this fact and supports them. In this specific case, integration represents the service. However, this dimension on its own is not enough since there is also the issue relating to changes or innovations in relations and interactions between socio-economic groups and individuals in organizations, associations or communities. The relations are equal, equitable and participatory. This second dimension involves the governance of social innovations. The third dimension involves the transformation of representative democracy and the governance of democratic institutions, in particular when social innovation is diffused to a broader community and over a vast territory as will be seen below.

Social innovation is multidimensional and polysemic. It involves commitment on the part of the social actors who contribute to it. This is why the interactions which take place in it involve more than a simple service relationship evaluated according to market principles. Social innovations contribute to improving a situation and, consequently, they lead to greater organizational effectiveness, productivity and competitivity when deployed on a large scale, and they also contribute to the well-being of the community. It should be understood, then, that this well-being is the simultaneous result of three dimensions of social innovation: effective service, legitimate governance and valid empowerment. These are the specific characteristics of social innovation to which other components should be added. Then, social innovation possesses a local hold, from where it can be diffused to a larger community. Nevertheless, its diffusion also poses a number of problems, as will be seen further on. The density of social innovation, its rootedness in the community or the organization, and the degree of transformation in the standards and rules of conduct of actors should also be taken into account. Therefore, a change in any one of the components can lead to an innovation. To sum up, innovation can remain effective at the local level without being diffused or it can be diffused very widely, but it can also dissolve and disappear.

9.5 Social innovation and services

For Gallouj (2002b), the service relationship is a particular mode of coordination between economic agents. This relationship, however, poses a number of challenges in analytical terms because there is no product that can exist independently of this relationship. The service is thus an act, a process or a processing protocol that appears above all as a 'social construction'. The service delivered through social innovation is also the result of interactions and a process. This is why the École de Lille has developed this key concept of the service relationship, that is, the 'co-production' between producer and client.

However, some analysts are against the 'co-production' approach and in favour of a more holistic approach to innovation in the service sector. In fact, there can no longer be a universal 'one-size-fits-all' approach (Tether, 2003). The sources of innovation are diversified, the role of research is changeable and the goals of the innovation are variable. The participation of actors in cooperative arrangements is not compulsory either. For Tether, there is no use developing a theory of innovation in the service sector since several patterns co-exist. Rather, it is necessary to establish a taxonomy and a typology associated with sectoral analyses, thereby reducing the diversity of innovation trajectories. Since the sectors are also dynamic, innovation should be examined from the perspective of innovation systems, because organizations are variable, as are the activities generated by these organizations.

A service produced through social innovation combines these two approaches to innovation in the service sector since 'co-production' is clearly an essential dimension. However, social innovation is difficult to grasp because it reveals a great diversity of patterns which show the system of social innovation as the concept to be analysed. The organization is important here because it is a place of coordination which gives rise to networks of communication at the source of the innovation. Moreover, these networks spread if they are able to develop procedures resulting in imitations and thus diffusion (Nijssen et al., 2006). Innovations are the product of interfaces and interactions which involve exchanges of information, knowledge and emotions (Collins, 2004). These interactions vary in intensity and give meaning to 'co-production'.

For example, the Réseau québécois en innovation sociale (RQIS, Quebec Social Innovation Network) diffuses information on social innovations in Quebec. This network gathers proposals for action and encourages consultation and networking among key actors in order to consolidate the interest shown in social innovation and foster the emergence of regional strategies. This network encourages exchanges and debates and influences decisions on public policy measures and development policies, as well as directions for planning funding from public, regional and local funds together with private funds. This network reveals multiple paths drawn by actors between the different organizations, because social innovation involves all spheres of social life in public, private and social organizations and associations, as well as the domestic sphere. Different streams of social innovation can contribute to defining the scope of social innovation and emerge from the three sectors that are so different from one another. However, the social economy enterprises are the ones which offer a great variety of services in several areas, such as consumer cooperatives, financial services, home help, community services, help with homework, individual support and many other services. The social economy sector seems to be more conducive than other sectors to social innovation because the organizations in this sector are all places of creation pertaining to social critique. Protest movements mix with alternative movements that do not merely criticize while shifting the solution to other realms, but rather develop the solutions themselves.

9.6 The role of research

Research and development have often been considered to play an unimportant role in innovation in the service sector (Nijssen et al., 2006; Tidd and Hull, 2006). The particular characteristic of services is the role of the client or user. Thus, services are produced by interfaces and interactions which involve information exchanges that vary in intensity (Gallouj,

2002a). Services are also considered in terms of social rules that control relations between the agents involved. The notion of service is therefore socially regulated prior to the modes of coordination within the organization (Gadrey, 2002). With the emergence of social innovation networks, the system of transferable rules and procedures thus plays an important role. It is therefore at this level that social research can contribute to initiative-taking, and to the creation and diffusion of social innovations. Social innovation occurs in real time in practical settings, either in organizations or associations, without the benefit of prior laboratory experimentation.

It is difficult to count the number of social innovations or to measure their effects. Unlike innovation in the market sector which can be understood as a competitive advantage and is thus maintained without competition for the longest time possible, social innovation calls for imitation. Starting out at a strictly local level, social innovation has a short range which makes its diffusion difficult due to the lack of codification of its procedures on the one hand and, on the other hand, strong institutional barriers, despite the fact that there is no legal protection such as a patent. Collective ownership would appear to be more important for diffusing social innovations than it is for technological innovations whose patents impose duties to be paid by competitors, thus restricting their diffusion except through the market. The identification of innovation as being open and democratic is associated with the existence of various forms of social networks. Networks are thus essential to innovation because they implement different forms of collaboration, whether geographical clusters, strategic alliances or partnerships. The life cycle of a network is also important since the governance and management of networks must support the dual goal of creating and diffusing an innovation, otherwise these networks become defensive and cling to the results achieved rather than concerning themselves with progress. The key change in these networks often involves moving from an informal mode, limited over time and based on a single alliance, to a multi-alliance network governed as an entity serving a single goal or at least merged goals (Powell and DiMaggio, 1991; Abrahamson, 1991; Alter, 2003). This clearly shows that a system of innovations exists and, although effective, it makes it difficult to measure social innovation.

Another way to measure social innovation consists in developing indicators related to the role of social research. Research on social innovation is mainly conducted by the social sciences and humanities. Knowledge is a crucial resource that is reflected in the innovation process (Hipp and Grupp, 2005). New knowledge is generated by numerous actors through their learning processes. One challenge involves transferring knowledge to practical settings in order to facilitate the success of social innovation, in particular when the innovation stems from academic research. It is difficult for some practical settings to appropriate the strategies developed, but appropriation at the local level is a condition for success. Research plays a less important role in innovation in the service sector, in particular because services spread when they are innovative or, on the contrary, they disappear, based on the precepts of Schumpeterian theory. However, social innovations cannot merely rely on these voluntary strategies or let themselves be guided by the rules of the market. Although, initially, a social innovation can effectively do without research and give rise to an innovative service, its success remains local and it is in danger of disappearing if it does not receive public recognition which contributes to sustaining its existence over the long term. This recognition facilitates the codification of procedures and interactions which are typical of the service delivered as well as the social rules and conventions that authorize participation and democratic exercise. Some innovations are the product of a deliberate intention or of research results, but there is no standardized model of social innovation.

Research is conducted using several methods which involve a relationship with the actors of social innovation before and after the fact. Social innovation can be the result of interactions between the actors in a particular setting that bring about the creation of a service to the community. Aiming to transfer the knowledge derived from this particular process, research can *a posteriori* help the actors to codify all activities having led to the social innovation, in order to produce specific knowledge that can then be transferred to other experiences. Tacit knowledge is thus formalized. Actors involved in a new innovation process can benefit from and appropriate the research results in order to affect the new local process by relying on an already validated approach, although the latter may be transformed by the interactions of people with new expectations or intervening in another context. Another method involves the researchers intervening right at the beginning of the social innovation process. The actors focus on the various potential ways the action is conducted during the process while the researchers document some critical situations and feed various types of knowledge to the actors. This is the model of partnership-based research, in which the researchers are directly involved and become actors themselves. This second method links the researchers with networks of associated actors. This pluralist system links universities and organizations to the system of innovations (Gibbons et al., 1994). Associations and partnerships are encouraged by public policies on research which devote specific programmes and adequate funding to it. Research projects thus emerge from meetings between researchers and actors who jointly decide on the needs.

Social sciences and humanities research comes into play at various stages of the process. It can be at the beginning, at one of the stages along the way or at the end of the process. It can relate to the process of creation or to the process of diffusion. Its goal is to develop knowledge that is also linked to other external dimensions. Lastly, research becomes important for the diffusion of social innovation. This pertains to the hold of the innovation. In fact, diffusion is based on the principle of imitation (Rogers, 1983). It then becomes a practice which is assimilated to a social norm. It is through these practices codified by research that information is transmitted to the influential networks. Rogers showed, as early as 1961, that social networks play a fundamental role in the diffusion of innovations, provided that the innovation has a use value and that its usefulness is promoted by the inventors and first users. Social innovation, then, must promote new rules, new norms of interaction and service without the visibility enjoyed by technological innovation, since it is intangible and immaterial.

9.7 A case of social innovation

Through social economy enterprises, civil society in Canada and Quebec plays an important role in the organization and the delivery of home care and home support services. In Quebec, state recognition of the participation of community organizations has converged with the evolution of local and regional development policy (Tardif, 2007). A series of recent reforms in both the health and development policies have highlighted the community as a crucial actor and stakeholder (Lévesque and Vaillancourt, 1998, 4). The reforms have led to the creation of territorially based agencies and of specific funds with a view to financing social economy organizations and establishing partnerships with community organizations. In the context of Quebec, such partnerships between state agencies and social economy enterprises are considered to be a social innovation in the delivery of services, even though the latter in themselves are not very different from those delivered by the public or the private sector. It is rather what happens upstream that is considered innovative.

Community participation in the delivery of health services is mainly concentrated in certain types of local services. Most of those who receive these services are individuals with a variety of health problems. These citizens include the unemployed, people with mental or physical health problems, people with disabilities, senior citizens, alcohol or drug addicts, homeless persons and others (Klein et al., 2004). In producing these services, non-profit organizations seek to improve working conditions, as well as to promote solidarity, social cohesion and community empowerment.

These organizations get more than half of their funding by virtue of agreements with the public health sector that are crucial in ensuring their long-term existence. The rest comes from agreements with other levels of the public sector, such as municipalities or regional agencies. Private foundations and philanthropic organizations such as the United Way are also involved. In certain circumstances, the users themselves are required to pay for the services they receive (the most deprived users do not pay). However, they do not pay the market price for what they receive from the social economy enterprises.

Home care is one of the sectors where the social economy has particularly established itself in Quebec. It offers personal support, assistance with meals, psychological services, respite services and some forms of housework. In performing these functions, social economy organizations act in partnership with the Centres locaux de services communautaires (CLSC community health and social service centres), local public bodies set up by the provincial government in the late 1960s in the wake of a wave of community-initiated social innovation. Now a pillar of the Quebec health care system, CLSCs were originally community-based health and social-service clinics founded by civil-society actors in the 1960s (Jetté, 2008) and then absorbed into the public health care system in the 1970s. However, at the end of the 1980s, the public health care system took a new direction. A major reform recognized the role of community organizations as crucial stakeholders. At the same time that it promoted partnership with civil-society organizations, the government followed the global trend towards de-institutionalization and making citizens accountable for their own health.

Two types of social economy organizations offer home support services: non-profit organizations and cooperatives. In addition to having a role in the delivery of local services, the participation of community and social economy organizations in the field of home support has been part and parcel of a new institutionalization of economic activities that were rife with illegal work and moonlighting. In addition to enabling vulnerable people to stay at home, these projects have provided steady jobs with formal working conditions (Chantier de l'économie sociale, 2001, 14).

To some extent, these social economy organizations play a palliative role in partly compensating for the lack of beds in long-term care facilities and hospitals. However, it is widely recognized that home care is often not only a cheaper, but also a healthier alternative to placing senior citizens in institutions. Cooperative housing with social support is an interesting option here. Many cooperatives or non-profit organizations for independent and partly self-sufficient seniors have been founded. Some projects combine several types of cooperative organizations in one and the same place, for example a housing co-op with home care and health services delivered by social economy enterprises, often with government financial support and financial contributions by the residents. These projects develop roots in the community; tenants may participate in the management of their residences (Ducharme et al., 2004). This type of partnership in the process of making up a social innovation is also possible with other kinds of vulnerable categories of population such as people with mental disabilities.

The turn to government–third-sector partnerships is not without challenges and problems. The dissemination of such innovations throughout Quebec remains an issue. Not all communities have the same resources and there are disparities in the availability of services. This asymmetry could lead to serious inequalities between regions due to the absence of organizations strong enough to assume a partnership role (Bélanger, 2002). The literature on the third sector points to the latter's weaknesses with respect to equity, need, accessibility, accountability, resources, coherence and planning. As Josephine Rekart has pointed out, government involvement is indispensable: 'to provide universal and integrated planning, to set and monitor standards, to achieve equity and social justice through resource allocation, and to preserve democratic control' (Rekart, 1993, 21). Social economy enterprises' proximity to service users, and their capacity for innovation and flexibility, require the ballast of stable and adequate government funding, as well as strong government leadership in the areas of policy and regulation (Browne, 1996, 41–50; Ostrander, 1987).

Partnerships between different levels of government and community and social economy enterprises take different forms (Coston, 1998). Four of them appear to be essential: subcontracting, co-existence, subsidiarity and co-building. The first one, subcontracting, confines the community partner to an instrumental role. The second one, co-existence, shows a parallel evolution of both the public and the community-based sectors. The third one, subsidiarity, indicates that the social sector has a key role in the implementation of different programmes, but the stakeholders do not participate in their design. Finally the fourth one, co-building, is the most complete, since the social economy organizations operate as actors in the development, as well as in the implementation of social policies (Proulx et al., 2005). Social innovations in the sectors highlighted here demonstrate that services can be provided and delivered with a certain sort of empowerment and democratic participation. The challenge is to keep alive projects that promote community bonds as a final goal and not just as another utilitarian measure (Duperré, 2004).

9.8 Conclusion

Social innovations contribute to the transformation of capitalist society not through large-scale reforms, but through a recognition of relationship capitalism based on the benefits of collaboration, trust and social inclusion. Social innovation comes into play in this type of society which is constructed by a social market system, that is, democracy, partnership, participation, consultation, citizen and user involvement, and civil society. The state is always present since it has the responsibility to support social innovations without being their instigator. It formulates public policies and is often a partner integrated into a network of institutions and actors of civil society belonging to various associations and organizations. Public policy leads to more effective diffusion of social innovations, supported by regulation, legislation or investment of public funds.

At the local level, the main effect of the involvement of organizations which are representative of civil society (community organizations, social economy enterprises, voluntary associations, unions) in the delivery of services at both the individual and collective levels can be considered to relate to the production and reproduction of more viable and more inclusive living environments. Organizations of civil society forge links with the social sector in a specific way. In addition to contributing to the service sector, they link the social issues with the economic issues at the local level while causing innovative social arrangements to be formed. Moreover, owing to their integration into the national, international and global networks, they contribute to cross-linkage at various levels and, thus, to a connection with successful networks of communities which otherwise would be isolated from the new economy. These organizations thus link the private sphere with the public sphere while helping to connect the local level with the global level.

In social innovations, relationships are of utmost importance, even before the service or the product. Innovative practices without values are no longer social innovations. This is why actors in social innovation advocate transmitting values before dealing with practices.

Note

1. Pragmatism is defined here as a social philosophy developed by John Dewey who views experience as a source of knowledge that leads to action: 'Knowing is acting.' In its modern interpretation, pragmatism has given rise to social constructivism, one of the main theoretical sources of social innovation. However, these theoretical issues are not dealt with in this text.

References

Abrahamson, Eric (1991), 'Managerial fads and fashions: the diffusion and rejection of innovations', *Academy of Management Review*, **16** (3), 586–612.

Ackers, P. (2002), 'Reframing employment relations: the case for neo-pluralism', *Industrial Relations Journal*, **33** (1), 2–19.

Alter, Norbert (2003), *L'innovation ordinaire*, Paris: Presses universitaires de France.

Bélanger, Maude (dir. Paul R. Bélanger) (2002), *Rapport de recherche sur les entreprises d'économie sociale en aide domestique*. Cahier du CRISES n° ET0209.

Breton, R., N.J. Hartmann, J.S. Lennards and P. Reed (2004), *A Fragile Social Fabric? Fairness, Trust, and Commitment in Canada*, Montreal and Kingston: McGill-Queen's University Press.
Browne, Paul Leduc (1996), *Love in a Cold World? The Voluntary Sector in an Age of Cuts*, Ottawa: Canadian Centre for Policy Alternatives.
Chantier de l'économie sociale (1996), *Osons la solidarité*, Rapport du groupe de travail sur l'économie sociale, Sommet sur l'économie et l'emploi, Québec Montréal, www.chantier.qc.ca.
Chantier de l'économie sociale (2001), *De nouveau nous osons, Une démarche de rassemblement*, Montréal, www.chantier.qc.ca.
Chanial, Philippe (2001), *Justice, don et association*, Paris: La Découverte/MAUSS.
Cloutier, J. (2003), *Qu'est-ce que l'innovation sociale*? Montrëal: Cahier de recherche du CRISES, n° ET0314.
Collins, Randall (2004), *Interaction Ritual Chains*, Princeton, NJ, USA and Oxford, UK: Princeton University Press.
Coston, Jennifer (1998), 'A model and typology of government–NGO relations', *Nonprofit and Voluntary Sector Quarterly*, **27** (3), 358–382.
Defourny, J. (2001), 'Introduction: from third sector to social enterprise', in C. Borzaga and J. Defourny (eds), *The Emergence of Social Enterprise*, London and New York: Routledge Studies in the Management of Voluntary and Non-Profit Sector, Routledge, pp. 1–28.
Douglas, James (1983), *Why Charity? The Case for a Third Sector*, Beverly Hills, CA: Sage.
Drewe, P., J.L. Klein and E. Hulsbergen (eds) (2008), *The Challenge of Social Innovation in Urban Revitalization*, Amsterdam: Techne Press.
Ducharme, M.-N., M. Charpentier and Y. Vaillancourt (2004), 'Les OSBL et les coopératives d'habitation pour personnes âgées: des initiatives méconnues', in Y. Comeau (ed.), *Innovations sociales et transformations des conditions de vie*, Cahier du CRISES ES0418, pp. 85–100.
Duperré, Martine (2004), 'Innovation sociale et milieux innovateurs: un exemple de construction institutionnelle dans le domaine socio-sanitaire', in Y. Comeau (ed.), Innovations sociales et transformations des conditions de vie, Cahier du CRISES ES0418, pp. 55–69.
Esping-Anderson, G. (ed.) (2002), *Why We Need a New Welfare State*, Oxford: Oxford University Press.
Evers, A. and J.L. Laville (2004), *The Third Sector In Europe*, Cheltenham, UK and Northampton, MA, USA: Edward Elgar.
Finlayson, Geoffrey (1994), *Citizenship, state and social welfare in Britain, 1830–1990*, Oxford: Clarendon Press.
Fontan, J.-M., J.-L. Klein and D.-G. Tremblay (2004), 'Collective action in local development: the Case of Angus Technopole in Montreal', *Canadian Journal of Urban Research*, **13** (2), 317–336.
Fontan, J.-M., J.-L. Klein and D.-G. Tremblay (2005), *Innovation socioterritoriale et reconversion économique: le cas de Montréal*, Paris: L'Harmattan.
Gadrey, Jean (2002), *New Economy, New Myth*? London: Routledge.
Gallouj, F. (2002a), *Innovation in the Service Economy: The New Wealth of Nations*, Cheltenham, UK and Northampton, MA, USA: Edward Elgar.
Gallouj, Faïz (2002b), 'Innovation in services and the attendant old and new myth', *Journal of Socio-Economics*, **31**, 137–154.
Gibbons, Michael, Camille Limoges, Helga Nowotny, Simon Schwartzman, Peter Scott and Martin Trow (1994), *The New Production of Knowledge: The Dynamics of Science and Research in Contempory Societies*, London, UK, Thousands Oaks, CA, USA and New Delhi, India: Sage Publications.
Goldenberg, M. (2004), 'Social innovation in Canada: how the non-profit sector serves Canadians and how it can serve them better', Research Report w/25, Ottawa: Canadian Policy Research Networks.
Hall, M.H., C.W. Barr, M. Easwaramoorthy, S.W. Sokolowski and L.M. Salamon (2005), 'The Canadian nonprofit and voluntary sector in comparative perspective', Imagine Canada, Toronto, www.imaginecanada.ca.

Harrisson, Denis and Juan-Luis Klein (2007), 'Introduction: Placer la société au centre de l'analyse des innovations', in Juan-Luis Klein and Denis Harrisson (eds), *L'innovation sociale: Émergence et effets sur la transformation des sociétés*, Quebec: Presses de l'Université du Québec, p. 1–14.

Harrisson, Denis and Murielle Laberge (2002), 'Innovation, identities and resistance: the social construction of an innovation network', *Journal of Management Studies*, **39** (4), 497–521.

Harrisson, Denis and Martine Vézina (2006), 'L'innovation sociale: une introduction', *Annals of Public and Cooperative Economics*, **77** (2), 129–138.

Hipp, C. and H. Grupp (2005), 'Innovation in the service sector: the demand for service-specific innovation measurement concepts and typologies', *Research Policy*, **34** , 517–535.

Howard, E. (1902), *Garden Cities of To-Morrow*, London: Faber & Faber.

James, Estelle and Susan Rose-Ackerman (1986), *The Nonprofit Enterprise in Market Economics*, Chur, Switzerland: Harwood Academic Publishers.

Jetté, C. (2008), *Les organismes communautaires et la transformation de l'État-providence*, Quebec: Presses de l'Université du Québec.

Klein, J.-L. (1992), 'Le partenariat: vers une planification flexible du développement local?' *Canadian Journal of Regional Sciences/Revue canadienne des sciences régionales*, **15** (3), 491–505.

Klein, J.-L. (2008), 'Territoire et régulation', *Cahiers de recherche sociologique*, **45**, 41–58.

Klein, J.-L. and D. Harrisson (eds) (2007), *L'innovation sociale. Émergence et effets sur la transformation des sociétés*, Quebec: Presses de l'Université du Québec.

Klein, J.-L., C. Tardif, M. Tremblay and P.-A. Tremblay (2004), *La place du communautaire: évaluation de la contribution locale des organisations communautaires*, Montreal: Cahiers de L'Alliance de recherche université – communautésen économie sociale R-07-2004.

Klein, J.-L., D.-G. Tremblay, J.-M. Fontan and N. Guay (2007), 'The uniqueness of the Montreal fur industry in an apparel sector adrift: the role of proximity', *International Journal of Entrepreneurship and Innovation Management*, **7**, 298–319.

Lapointe, A. and C. Gendron (2004), 'Corporate codes of conduct: the counter-intuitive effects of self-regulation', *Social Responsibility; An International Journal*, **1** (1–2), 213–218.

Lapointe, P.-A., P. Bélanger, G. Cucumel and B. Lévesque (2003), ' Nouveaux modèles de travail dans le secteur manufacturier au Québec', *Recherches sociographiques*, **44** (2), 313–347.

Laville, J.-L. (2007), 'Éléments pour l'analyse du changement social démocratique', in J.-L. Klein and D. Harrisson (eds), *L'innovation social: Émegence et effects sur la transformation des sociétés*, Quebec: Presses de l'Université du Quebec, pp. 88–120.

Lévesque, B. (2006), *Le potentiel d'innovation et de transformation de l'économie sociale: quelques éléments de problématique*, Cahier du CRISES no ET0604.

Lévesque, Benoît (2007), 'L'innovation dans le développement économique et social' in Juan-Luis Klein et Denis Harrisson (ed.), *L'innovation sociale. Émergence et effets sur la transformation des sociétés*, Quebec: PUQ, 43–67.

Lévesque, B. and M. Mendell (2004), 'L'économie sociale: diversité des approches et des pratiques', Proposal for area CURA programme in social economy, Working Paper for the Chair Office of SSHRC, Montreal.

Lévesque, Benoît and Yves Vaillancourt (1998), *Les services de proximité au Québec: de l'expérimentation à l'institutionnalisation.* Copublication CRISES/ LAREPPS/ CRDC, Cahier du CRISES no ET9812.

Malo, M.C. and M. Vézina (2004), 'Gouvemance et gestion de l'entreprise collective d'usagers', *Économie et Solidarité*, **35** (1–2), 100–120.

Mendell, Margie (2006), 'L'empowerment au Canada et au Québec: enjeux et opportunités', *Géographie, économie, société*, **1**, 63–86.

Moulaert, Frank, F. Martinelli, E. Swyngedouw and S. Gonzalez (2005), 'Toward Alternative Model(s) of Local Innovation', *Urban Studies*, **42** (11), 1969–1990.

Nijssen, E.J., B. Hillebrand, P.A.M. Vermeulen and R.G.M. Kemp (2006), 'Exploring product and service innovation similarities and differences', *International Journal of Research in Marketing*, **23**, 241–251.

Novy, Andreas and Bernhard Leubolt (2005), 'Participatory budgeting in Porto Allegre: social innovation and the dialectical relationship of state and Civil Society', *Urban Studies*, **42** (11), 2023–2036.
Organisation for Economic Co-Operation and Development (OECD) (2005), *Oslo Manual Guidelines for Collecting and Interpreting Innovation Data*, 3rd edition, Paris: European Commission (Eurostat) and OECD.
Ostrander, Susan (1987), 'Toward implications for research, theory, and policy on nonprofits and voluntarism', *Nonprofit and Voluntary Sector Quarterly*, **16**, 126–133.
Petitclerc, Martin (2007), *'Nous protégeons l'infortunes', Les origines populaires de l'economie social au Québec*, Montreal: VLB editeur.
Powell, W.W. and P.J. DiMaggio (eds) (1991), *The New Institutionalism in Organizational Analysis*, Chicago, IL: University Press of Chicago.
Proulx, Jean, Denis Bourque and Sébastien Savard (2005), 'Les interfaces entre l'État et le tiers secteur au Québec', Cahier de l'ARUC no C-01-2005.
Putnam, Robert D. (2000), *Bowling Alone: The Collapse and Revival of American Community*, New York: Simon & Schuster.
Rekart, Josephine (1993), *Public Funds, Private Provision: The Role of the Voluntary Sector*, Vancouver: UBC Press.
Rogers, Everett (1983), *Diffusion of Innovations*, 3rd edn, New York: Free Press; 1st edn 1961.
Salamon, Lester (1987), 'Of market failure, voluntary failure, and third-party government: toward a theory of government–third-party relations in the modern welfare state', *Journal of Voluntary Action Research*, **16**, 29–49.
Salamon, L.M. and H. Anheir (1998), 'Social origins of civil society: explaining the non-profit sector cross-nationaly', *Voluntas*, **9** (3), 213–248.
Schumpeter, J.A. (1967), *The Theory of Economic Development*, Oxford: Oxford University Press.
Tardif, C. (2007), 'Les corporations de développement communautaire au Québec: processus d'institutionnalisation et trajectoires socio-territoriales spécifiques', Dissertation, Urban Studies. Université du Québec à Montréal.
Tether, Bruce S. (2003), 'The sources and aims of innovation in services: variety between and within sectors', *Economics of Innovation and New Technology*, **16** (6), 481–505.
Tidd, J. and F.M. Hull (2006), 'Managing service innovation: the need for selectivity rather than "best practice"', *New Technology, Work and Employment*, **21** (2), 139–161.
Vaillancourt, Y., F. Aubry and C. Jetté (2003), *L'économie sociale dans les services à domicile*, Ste-Foy: Presses de l'Université du Québec.

PART III

ORGANISATIONAL AND STRATEGIC PATTERNS FOR SERVICE INNOVATION

10 Different types of innovation processes in services and their organisational implications

Marja Toivonen

10.1 Introduction

Since the mid-1990s, innovation in services has aroused growing interest and studies on this topic are today accumulating rapidly. One of the observations confirmed in several studies is that innovation activities in service sectors and service firms are less systematic than in the industrial context. Researchers have usually linked this observation to the fact that service firms only rarely have research and development (R&D) departments for innovation activities. Rather, these activities are distributed within the firm; they are conducted, for example, in connection with strategic planning, training and market development (Coombs and Miles, 2000; Djellal and Gallouj, 2001; Preissl, 2000). Many researchers have emphasised that this finding should not lead us to conclude that service firms are less innovative than industrial firms. On the contrary, we should broaden our view about the organisation of innovation, and strive for a better understanding of other forms of innovation activities in addition to those concentrating on the conduct of R&D (Hipp and Grupp, 2005).

Three main approaches can be identified in studies that aim at revealing alternative forms of innovation – important in services but remaining hidden if the starting point is a manufacturing-based innovation paradigm and accompanying indicators. The first approach focuses on quantitative innovation surveys and in this context has tried to develop such new indicators that are better applicable in services than the earlier ones. Both input and output indicators have been suggested. As regards the former, investments in human resources have been highlighted in particular as an important form of innovation expenditure (Coombs and Miles, 2000). Concerning the latter, many efforts have focused on the modification and broadening of the categorisation of innovations (as outputs). The most generally used categories of product and process innovations have proved to be difficult to apply in services, as a service product constitutes a process in its basic nature (Evangelista and Sirilli, 1998). Thus, attention has been paid to other types of innovations based on the original Schumpeterian classification (Schumpeter, 1934) – in particular, the importance of organisational innovations has been emphasised (van der Aa and Elfring, 2002; Antonelli et al., 2000; Gadrey et al., 1995). New categories have also been introduced, the category of a 'delivery innovation' being one such (Coombs and Miles, 2000). In addition, direct questions about new or significantly improved services have been used to map service innovations in some studies (e.g. Tether and Hipp, 2000).

The second approach consists of service theories that consider alternative ways to carry out innovation activities successfully in an organisation without concentrating them into a specific department – the R&D department. Both one broad alternative and several alternatives have been suggested in this framework. Sundbo presented the broad alternative in 1996 in an article (Sundbo, 1996) where he argued that innovation activities can be organised in two main ways: as an expert system (R&D department) or as 'a balanced empowerment system'. The latter is a general innovation system involving most of the people in the firm. Like all social systems, firms also show a dual structure according to Sundbo: an informal, loosely coupled interaction structure among employees, and a formal management structure which expresses the official goals, norms and values of the organisation. It is the task of the management system to induce the evolution of innovations in the loosely coupled system on the basis of the firm's strategy. As the strategy expresses the production and market situation of the firm, it also ought to express the firm's objectives concerning innovation. It should inspire innovations, but at the same time define the framework within which the innovations should be kept. Broad empowerment is advantageous particularly in the case of innovations which emerge from practical experience. In services, where the ideas often develop from the interaction with the client, this system is more natural than an expert system where the firm establishes special resources for the purpose of innovation.

Later, Sundbo together with Gallouj has suggested several different 'innovation patterns' (Sundbo and Gallouj, 2000; see also Djellal and Gallouj, 2001). Besides the R&D pattern, they identify the organised strategic innovation pattern, the service professional pattern, the entrepreneurial pattern, the artisanal pattern and the network pattern. The organised strategic innovation pattern is highly similar to the above-described broad empowerment, and according to the authors it is the most common pattern in the service sectors. In the service professional pattern, found primarily in KIBS (knowledge-intensive business services), the non-repeatable (ad hoc) mode of innovation is dominant. The expertise of individual professionals plays an important role here, and the locus of innovation is the interface with the client. The entrepreneurial pattern characterises small, new service firms (e.g. in the information technology sector); these firms are based on a radical innovation and their main activity is to sell this initial innovation. The artisanal pattern is typical in operational services (cleaning, hotels, etc.) among small, conservative firms which have no innovation strategy. If innovation is present here, it is through improvement models and learning processes. The network pattern refers to a situation where a number of service firms have created a common firm whose purpose is to innovate on their behalf. This pattern has been found in tourism and in financial services, for example.

In the above-mentioned approaches, the focus of analysis has been a service firm. The suggestions for new categories of innovation outputs have also discussed the level of service products – yet, without a deeper theoretical analysis. The third approach that I want to bring up here fills this gap: some years ago, Gallouj and Weinstein (1997) developed a model whose aim was to lay grounds for a theory of innovation at the level of service products. The model applies the Lancasterian characteristics-based analysis of goods to services (Lancaster, 1966; Saviotti and Metcalfe, 1984). A service is described as a set of final characteristics (*Y*), technical characteristics (*X*) and competence characteristics (*C*). The final characteristics consist of the benefits provided to the client. They are obtained through a certain combination of technical characteristics, which describe the organisation's tangible and intangible systems used in the production of the service; they also include the process characteristics of the service. Each technical characteristic mobilises certain competences, which are composed of the individual skills of the service provider and the client. Gallouj and Weinstein define innovation as any change affecting one or more of the above described characteristics. They specify different types of innovations based on their model. Among others, improvement innovation is created when the value of a certain *Y* is increased by improving certain *X* or certain *C*. Innovation by addition or substitution occurs when one or more new elements are added to *X*, or one or more of its elements are replaced (*C* and *Y* change accordingly). Recombinative or architectural innovation takes place when a new service is developed either by combining characteristics of two or more existing services (bundling) or by splitting up an existing service (unbundling).

All three approaches summarised above have increased and structured our understanding of service innovation in a valuable way – each from a different angle. However, not one of them has tried to describe the procedure of innovation processes at a detailed level. The models that tackle this topic are still more or less based on a traditional industrial paradigm, which means that a service innovation process is described as a separate design project preceding the actual service practice. Within this framework, there are different views regarding the degree of linearity and systematisation of the process. What is common, however, is that the process starts from an idea, several preparatory steps are taken before the launch of a new service on the market, and the procedures after the launch are not included, or are discussed only in passing. Some researchers representing this thinking openly start from the models applied in manufacturing and aim to transfer the basic idea into services (e.g. Cooper and De Brentani, 1991; Cooper and Edgett, 1996). Others are researchers, who have studied profoundly the nature of services and pointed out their specificities, but regarding the development of new or improved services make conclusions that are highly similar to traditional R&D process (e.g. Edvardsson et al., 2000).

Those researchers who have emphasised the necessity of broadening the R&D-based view of innovation in order to reveal 'hidden innovation activities' in services are nowadays increasingly proponents of the so-called synthesis view as regards innovation in manufacturing and in services (Boden and Miles, 2000; Gallouj and Weinstein, 1997; Preissl, 2000; Sundbo 1994). Yet, their idea is not that manufacturing is more advanced in innovation and services should imitate it; rather, they argue that studies are gradually revealing new issues common to both manufacturing and services. Some of these issues have first come out in the context of service studies, as service innovation has not conformed to earlier thinking. Later, researchers have found that many phenomena thought to be 'peculiarities' of service manifest themselves in manufacturing, too (see Miles and Boden, 2000). The questioning of 'old truths' is gaining ground also in the context of industrial and technological innovations. The most narrow and linear views of innovation have been criticised since Kline and Rosenberg (1986). Today some topics that are at the core of service research are very actively discussed among the researchers of technological innovations. Users as innovators (Tuomi, 2002) and the so-called open innovation paradigm (Chesbrough, 2003) can be mentioned as examples. These topics are closely linked to a search for alternative innovation processes besides the R&D process.

This chapter analyses alternatives in the procedure of a service innovation process in more detail. My main argument is that a project-type process that precedes the service practice and can be clearly separated from it is not the only successful or the most successful way to create innovations in services. It is, however, one plausible alternative and works well in some cases. Both service innovation literature and my own empirical studies indicate that there are three basic ways in which a service innovation process can proceed. In addition to the above-mentioned model – which more or less resembles an R&D process – the other two ways are an innovation process integrated in service practice, and a process where the emergence of an innovation is recognised afterwards. In the following,

I first describe these three process types on the basis of the literature. Thereafter I use empirical examples from my own studies to illustrate in which kinds of situations each process type has been applied and what have been their specific benefits and challenges from the viewpoint of the organisation of innovation activities. In the final section, I summarise my conclusions and present some ideas for further studies.

10.2 Basic characteristics of three types of service innovation processes

Simplifying a little, we can say that any innovation process includes the following crucial elements: the emergence of an idea, the development of the idea and the implementation of the idea (see Heusinkveld and Benders, 2003). In the R&D-based innovation models, i.e. in the models where innovation is seen as a project separated from practice, the sequence of activities is from an idea through development to implementation. What I argue in this chapter is that service firms may also develop the idea hand in hand with implementation. I call this model, where the main part of the innovation process is integrated in service practice, the model of rapid application. Further, I argue that a significant part of innovations emerge without deliberate innovative efforts. Here the starting point is not an idea, but a new type of service practice developed as an answer to some perceived opportunity or challenge (most typically to clients' needs). The underlying new idea is found only afterwards in these cases. I call this alternative the practice-driven model.

10.2.1 Innovation as a project separated from practice

The 'orthodox' R&D model, which to a great extent coincided with the view of linearly proceeding innovation process, is left out of the discussion in this chapter. As mentioned in the introduction, hard critique has been presented against this model for more than 20 years now, and empirical studies have shown that it is very rare in services (see Sundbo and Gallouj, 2000). There are, however, modified versions which still see R&D as a core in innovation processes, but have supplemented the picture with other types of activities and highlighted the non-linear nature at least of some parts of the innovation process. In services, R&D is usually understood to be projects instead of more permanent R&D departments (Miles and Boden, 2000). In the following, I first summarise briefly the most important modifications made to the 'orthodox' R&D model in the manufacturing context. Thereafter I describe the ideas of those researchers who have transferred the model from manufacturing to services.

The proponents of the linear innovation model thought that the development of new products consists of a sequence of clear stages: idea generation, screening, commercial evaluation, technical development, testing and commercialisation (e.g. Saren, 1984). An alternative view gained ground when empirical studies showed that models of a spiral or circular type capture the nature of real-life innovation activity better. Innovation processes proved to be complex and recursive: the end of one innovation process is usually the beginning of the next, and the stage from which the process begins may vary. In the early stages of innovation it is often difficult to find any kind of structure; experimental activity which includes abundant sidesteps and iterations is characteristic here. The term 'fuzzy front end' has been adopted to describe the chaotic nature of these stages. Through the separation of the front end, researchers have pursued a synthesis between the view that emphasises creative problem-solving and the view that highlights rational planning: the unpredictable front end is like an 'introduction' to actual product development where logical and goal-directed activity with a project plan plays a central role. According to the current view, a sequential structure is not, however, the only possibility even here – the use of cross-functional teams enables the conduct of different product development tasks in parallel (Buijs, 2003; Koen et al., 2001).

In addition to the new notions concerning intra-firm innovation processes, studies have highlighted the role of clients as sources of innovation. The concept of value innovation has been introduced to point out that innovation should be anchored with the added value provided to the client (Kim and Mauborgne, 1999). Also, the search for unexpressed and future needs of clients has been emphasised (Slater and Narver, 1998).

In the technical and manufacturing context, the concept of new product development (NPD) has been generally used to refer to models where innovation has been analysed as a formal and planned process. Researchers who have transferred this basic idea to services often use the corresponding concept of new service development (NSD). The aim of these researchers has been to integrate the paradigms evolved from studies of manufactured goods with the results of service studies, usually with those gained in the framework of service marketing (De Brentani, 1991). A great part of studies under the NSD title have concentrated on the examination of the factors which cause success or failure in the development of new services. The application of a formal NSD process – with clear pre-planned stages – has been identified as one success factor (Cooper and De Brentani, 1991; Cooper and Edgett, 1996). On the basis of this finding, normative models for managing NSD processes have been built. The number of stages varies in these models, being quite high in some of them. For example, Scheuing and Johnson (1989) differentiated 15 developmental stages in their study of financial services. The general structure of the models is highly similar to those construed for the development of tangible products. In addition to clear stages, the idea of formal checkpoints or gates for making 'continue/stop decisions' at the end of each stage has usually been included.

Along with the additions and modifications made to NPD models, NSD models have also been supplemented. In particular the emphasis on clients has been visible in newer studies. Researchers have concluded that in services the client involvement often speeds up the development process: instead of separate market research exercises, ideas are received directly from the participating clients. Some new models have been built to show how the input from clients can be taken into account at every stage of the development process (Alam and Perry, 2002). The second new topic in NPD – the questioning of the linearity of innovation processes – has also impacted on NSD studies, but the most radical ideas of partially chaotic processes have not gained popularity. A pre-planned, sequential process is still the ideal; yet, the conduct of some stages in parallel in order to achieve time reduction is considered possible. The idea of cross-functional teams has been adopted as a new success factor from NPD (ibid.).

The concept of innovation has been used only in some NSD studies, and even in these a deeper discussion of the concept and its dimensions is lacking. Outside the NSD framework, Sundbo and Gallouj have included the R&D-based model in their analysis of typical service innovation patterns (see the introduction to this chapter). Their view is similar to mine in the respect that they regard the new, less linear and more customer-oriented variants of this model as one alternative in services – but not as a norm (Sundbo and Gallouj, 2000).

10.2.2 Innovation following the model of rapid application

In the context of product development, some researchers have questioned the idea that planning always occurs first and is followed at a later time by implementation. One motivation behind the search for alternative approaches has been the time-taking nature of R&D and stage-gate models. Eisenhardt and Tabrizi (1995) stated that there are basically two ways to accelerate the product development process: the compression model and the experiential model. In the compression model, planning and execution remain distinct but happen more quickly – the sequential steps of the product development are squeezed together in a rational process. If product development is a predictable path through well-known markets and technologies, then this model is relevant. On the other hand, if product development is an uncertain path through shifting markets and technologies, then the experiential model is more effective. The experiential model merges planning and execution and assumes a process which relies on real-time experience. The acceleration of product development is achieved here through rapidly building intuition and flexibility. Yet, it is also important to provide enough structure so that people will be confident enough to act in these highly uncertain situations. Eisenhardt and Tabrizi emphasise that iterations, testing, milestones and powerful leaders characterise successfully applied experiential models. Their empirical study was targeted at the computer industry and they found that the experiential model describes well the typical product development process of this industry.

Moorman and Miner (1998) discuss the alternative in which the construction of a product is closely connected to its application in the market. They agree with Eisenhardt and Tabrizi and make some additional contributions to their view. Moorman and Miner argue that it is general in organisations – in many types of settings – that the time gap between planning and implementation narrows so that, at the limit, these two events converge. They call this phenomenon 'organisational improvisation' and suggest that it is not only what organisations actually practise, but also what they should practise to flourish. Firstly, the convergence of planning and implementation can be an effective choice when a firm faces environmental turbulence that requires action in a short time-frame (an observation which corresponds to the main argument of Eisenhardt and Tabrizi). Unexpected stimuli, among other turbulence factors, may create the need for organisational action without providing time for planning. Secondly, this approach might be prompted when planning cannot provide all the details needed in implementation. Thirdly, a situation where much real-time information from the marketplace is available evokes immediate responses, and thus favours the convergence of planning and implementation in new product development.

Engvall et al. (2001) consider in more detail the second point mentioned by Moorman and Miner. They argue that R&D-based models have concentrated on the systematisation of the form of the product development process, but say very little about the content. However, it is just the content which is the main problem in product development: the idea included is still immature and difficult to express in words. Constructing a plan for something which is not well known and involves abundant tacit knowledge is not a reasonable approach. Much more effective is a strategy which enables the creation of shared experience of the object to be developed. This can be done by putting the idea into practice right from the beginning in a preliminary or small-scale form. Cooperation with a pilot customer is essential here: in this way, a common understanding is created between the producer and the customer of what customers really desire and need, and what the result should look like in order to fulfil this desire and need.

The examples presented by Engvall et al. come from the telecommunications sector and at least some of them primarily concern services. The authors also use the term 'rapid application development', i.e. the same term that I have selected to describe my second service innovation model. A deeper discussion about the topic of service innovation is, however, missing in their study. As far as I know, there are no other studies either that would have analysed the experiential or rapid application model in more detail in the service context. However, it is apparent that the conditions where this model has been suggested to work well are not specifically linked to goods production, but may exist in services as well. We can even argue that some of these conditions are highlighted in the service context. Planning all the details beforehand is particularly difficult in services due to their immaterial nature and the central role of tacit knowledge (see Antonelli et al., 2000). On the other hand, the specific strength of services – the close linkages to markets and the abundance of real-time information – favours the experiential approach. The third factor, environmental turbulence, varies by sector, by firm and from one time to another – here it is difficult to see any generally applicable difference between manufacturing and services.

It is important to note that all the above-cited authors discuss the experiential or rapid application approach as a conscious strategy. This is a central point also in my idea of the model and differentiates it from the practice-driven model. Before turning to the literature which analyses the latter, I supplement my present discussion with views that do not emphasise a rapid market entry, but are in line with the accompanying argument that the construction of a product or service is closely connected to its application in the market. I refer to the views where the role of clients and users in the innovation process is under the spotlight. These views are particularly important in services where the interactive process with the client is central and clients actually function as 'co-producers' of innovations (Gadrey et al., 1995; de Vries, 2006). However, client- and user-centric studies are becoming general in the industrial context, too. They have their roots in the classic writings of Abernathy and Utterback (1978) and von Hippel (1978). Abernathy and Utterback stated in their often-cited article that under conditions where market needs are ill-defined, and uncertainty is high regarding the relevant means with which to answer these needs, the role of users as innovators is central. Users often have a more intimate understanding of the requirements included, due to which they may play a major role in suggesting the ultimate form of innovation (in addition to the need).

In the newer user-centric views, it is often argued that the innovation process continues after the launch. Tuomi (2002) has highlighted this point in the context of new technology. According to him, new technologies are not completed and unchangeable artefacts, but are very often modified in use, and therefore include an element of reinvention. Technological novelties are also actively interpreted and appropriated by the users; one technological artefact can have different meanings for different user groups.

Sundbo (2008) has recognised the same phenomenon – which he calls 'after-innovation' – in the service context when studying knowledge e-services. He states that an innovation in these services is not completed when it is launched on the market, but adjustments must be made if it is to be successful. The innovation process is typically a series of small incremental innovations or improvements which continue after the official launch. The importance of after-innovation can be explained by the way in which customers can be involved in the innovation process. It is difficult to base innovations on customer knowledge before the launch because customers cannot say beforehand what they want, and they even have difficulties in assessing prototypes. They react by suggesting ideas for improvements when they use the service in practice. Thus, the knowledge service firm should emphasise after-innovation and establish channels to catch customers' ideas in the after-innovation process.

Sundbo raises customer-based after-innovation to the position of an ideal in e-service innovations. He emphasises that customer involvement should not be pursued in the early stages, but in the later stages of the innovation process. Here he seemingly looks for a solution which is at the other end of the innovation process compared with the idea of rapid application. However, because Sundbo does not make explicit what kind of process, and how long before the launch it typically is or should be, it is difficult to evaluate how far the ideas of after-innovation and rapid application are from each other. It also remains somewhat unclear whether after-innovation is a specific characteristic of the most modern services – knowledge services delivered via the Internet or via other information technology devices – or whether the phenomenon could be supposed to be more general. Anyway, Sundbo's view is linked to the search for such an alternative innovation model where actual operations in the market are included as an important part in the service innovation process. In this sense, his effort is very similar to my own.

10.2.3 Innovation following the practice-driven model

In both the 'modernised' R&D model and the model of rapid application, the innovation process is deliberate – something new is pursued consciously. In R&D-based models the deliberate nature of the process is evident due to the systematic steps included. In the model of rapid application, the experiential nature of the process makes the deliberateness less visible; yet, the model includes a clear goal which is just aimed for on the basis of an alternative approach. In my third model, the process does not start from a conscious aim to create new solutions. These solutions develop in the course of service practice and afterwards some of them are recognised as innovations or, which is more common, some element in them is recognised to include an opportunity for innovation when developed further. Actually, it is a well-known phenomenon that many technological innovations have also emerged by chance in the course of various processes. In services 'non-deliberate' innovation processes have, however, specific characteristics that deserve discussion. The relationship of innovation to tailor-made solutions, which play a central role in services, requires clarification in particular. As each service act is unique – a response to a specific problem in a specific environment – practitioners often argue that all of them are innovations (see Preissl, 2000). Confusion, or at least insufficient clarity, can be found also in research literature, and here it has usually been linked to discussions of innovations recognised afterwards.

Gallouj and his colleagues were the first researchers who clearly formulated the argument that a service innovation process does not always start with a priori planning, but can also start in the service practice. This argument is included, among other writings, in the article of Gallouj and Weinstein (1997) where they present their characteristics-based service model (see the introduction to this chapter). They call this innovation type an ad hoc innovation. The description of an ad hoc innovation as a 'solution to a particular problem posed by a given client' (Gallouj and Weinstein, 1997, 549) has justifiably aroused critique (Drejer, 2004). The difference between a tailor-made solution and an innovation is not clear in this definition, and the definition may even support the conclusion that every service act is an innovation. In some other writings – both earlier and later – Gallouj and his colleagues have specified the idea of ad hoc innovation in a way which overcomes the above-mentioned difficulty: an ad hoc innovation refers to that expertise and those elements which develop in connection with tailored solutions and can be transferred to new situations, even though the solutions as such cannot. Thus, ad hoc innovation involves reproducible elements, which could not be recognised before the actual service provision, but are perceived *a posteriori* (Gadrey et al., 1995; Gallouj, 2002).

This later specification of an ad hoc innovation is in line with Schumpeter's basic idea of innovation as an economic concept. One central criterion linked to this idea is that innovation has more than one specific application. A true innovation is not only beneficial for the company that has created it, but as a more viable way of doing things leads other companies to follow. Thus, innovation gets diffused through imitation and in this way promotes the development of the branch, even the development of the entire economy (Schumpeter, 1934; see also Drejer, 2004). The whole idea of this particular role of innovations as promoters of economic development loses its meaning if we accept client-specific solutions as innovations.

In his later texts, Gallouj uses interchangeably the concepts of ad hoc innovation and *a posteriori* innovation. I prefer the latter, because it is not linked to the above-mentioned inaccuracies and misunderstandings. In addition, when the focus is on the proceeding of the innovation process, the concept of *a posteriori* innovation is more illustrative. According to Gallouj's (2002) description, this concept refers to an innovation process which continues 'unconsciously' for a shorter or longer time – the purpose being just to fulfil the changing needs of clients; the tacit idea included in practical operations is recognised only afterwards. In their analysis of various service innovation patterns, Sundbo and Gallouj (2000) have discussed the concept even more clearly, specifying how the innovation process is finalised in the *a posteriori* model. According to them, the steps of production, selling and innovation take place simultaneously or are merged in this kind of a process. The client's problem is the starting point. Yet, Sundbo and Gallouj emphasise that the service production process, which *a posteriori* becomes an innovation process, ends with a formalisation step. This formalisation step is achieved without the client's participation. It aims at going through the problem and the innovating final solution again, and at formalising and modifying them in order to reappropriate some of their components and to capitalise them in the memory of the firm. Understood in this way, *a posteriori* innovation corresponds closely to my practice-driven model. It is particularly important that Sundbo and Gallouj make a clear difference between a tailor-made service process and an innovation process, and present how a genuine innovation process can grow out of individual processes with clients.

10.3 Empirical findings based on two case studies

The empirical findings presented in this chapter are based on two research projects carried out by me together with my colleagues in Finland. The first project focused on knowledge-intensive business services (KIBS) and was conducted 2005–07 (see Toivonen et al., 2007). Here I have the final results available. The second project is still continuing at the time of writing this chapter in summer 2008, and I have preliminary results from it for the first year. This project started in spring 2007 and will continue until 2010. Both projects have applied multiple case study methodology. Case studies are a recommendable approach when the topic under exploration is new (Eisenhardt, 1989). As shown in preceding sections, understanding the different types of innovation processes in services at a detailed level is such a topic.

In the first project, innovation was studied at three levels: the level of individual services, the level of the firm and the level of networks. The present chapter discusses only that part which concerns the level of individual services. In the second project, the main focus is at the level of individual services (some analysis at the firm level will be included in a later stage of the project in 2010). The results to be presented in this chapter describe the processes in which some new service is created or an existing service changed in an innovative way. The material of the first project was collected from four KIBS companies: an engineering company, an architect's office, an advertising agency and an accounting company. In the second project, nine companies are included: three KIBS (an engineering company, an architect's office and an ICT company), three more 'traditional' service companies (a wholesale company, a retail company and an insurance company), and three manufacturing companies that have started to provide or are planning to provide services as a part of their business (a metal products company, a warehouse trucks company and a mining machinery company). All companies involved in the two projects are relatively large compared to the average firm size in their own sector in Finland, and most of them have international activities.

In the first project, the innovation processes to be studied were selected in a preliminary interview of the company management. We asked managers to identify such services which they thought included some element that was new and valuable, not only from their own viewpoint, but also in the market where they operated. In addition, the novelty had to be replicable, i.e. tailor-made services were excluded. All services studied had been on the market for at least a couple of years during the time of the study; thus, their development process was construed retrospectively. Regarding each service, two or three persons who had been involved in its development were interviewed; the persons represented different management levels. The total number of interviews, conducted by a semi-structured face-to-face method, was 28. In addition to interviews, we used primary documents and memoranda of company workshops as supplementary data; six workshops were organised during the study. (The numbers of interviews and workshops describe only the project part concentrating on the level of individual services.) We originally selected 13 service cases for the analysis; however, we found later that in four of them no genuine innovation was identifiable. Three cases were just tailor-made services of high quality, and one case was unsuccessful, the service being no more on the market. Thus I consider in this chapter only the remaining nine cases.

In the second project, the selection process for cases has been similar to the first project except that we have sought innovation processes which are ongoing, preferably in an early stage. Our aim is to follow up their development, and for this aim we have adopted participatory observation as an additional method. In some cases, besides the service provider it has also been possible to interview clients. At the time of writing this chapter in summer 2008, we had preliminary material from six processes where innovative elements were clearly recognisable. We were also following up five other processes which were in such an early stage that it was difficult to say whether they would lead to some service innovation, or whether they were rather examples of the diffusion of innovations created by other companies. The number of interviews conducted until that point was 61, and the number of company workshops organised was seven. In the cases where the companies had specific development projects, we had the possibility to participate in the project meetings.

In the following I present our central findings regarding the nature of innovation processes in the selected cases. I start by describing results from KIBS, mainly on the basis of the first project, but supplemented with a few observations from the second one. Thereafter I discuss 'traditional' services and services linked to manufacturing in a common subsection. Because the first project was focused on KIBS, my understanding is deeper in these types of services. However, the material accumulated on the two other groups is also quite abundant and enables some important observations regarding them.

10.3.1 Innovation processes in knowledge-intensive business services (KIBS)

As the results from the first project describe services that have already been on the market quite a while, I can speak about them openly. Table 10.1 summarises the nine KIBS cases by presenting the main idea of the service, the innovations included, and my categorisation of the innovation process according to the three models identified in the literature analysis.

The table reveals that I did not exclude all tailor-made services from my analysis. Services in cases 2, 4 and 7 had been provided to one client only at the time of our study, but they included such clearly replicable elements that I interpreted them as innovation processes, differing however from the other cases in the respect that the process was unfinished. As regards my main research issue – the application of the three models – the table shows that in the vast majority of cases the innovation process had followed the model of rapid application. One 'finalised' innovation case – the follow-up and consultancy concerning the operation of building services systems – represented the model where innovation had been created in a project separated from practice. The practice-driven model was also found

Table 10.1 Summary of the examined innovation processes in KIBS

The main idea of the service	Innovations included	Type of the innovation process
Case 1 (engineering company) A web-based follow-up and consultancy aiming to optimise the operation of building services systems (heating/cooling, water supply, air conditioning and energy).	Several small innovations: data gathering is independent of the automation system and allows configurability; the service includes tools for the clients willing to make further analyses in-house.	A project separated from practice
Case 2 (engineering company) Application of project management experience gained in building construction in a new context: in the rail track renewals and track maintenance.	A tailor-made service provided to one client (the Finnish Rail Administration). Includes the possibility for the creation of a replicable service linked to PPP-models in the infrastructure development.	The practice-driven model
Case 3 (engineering company) A commodified consultancy service for the planning and developing of telecommunications systems in large firms and public organisations.	Innovativeness is linked to the formalisation of consultancy into a specified service product in a new area. The optimisation of the cost-efficiency of solutions results in savings for clients.	The model of rapid application
Case 4 (engineering company) A systematised model for the development of a firm's contingency plan. The purpose is to secure the continuity of business during crises and other exceptional conditions.	A tailor-made service provided to one telecom group; potential for replication. Innovative elements: contingency plan covers the whole value chain of the firm and is embedded in everyday activities.	The model of rapid application

Table 10.1 (continued)

The main idea of the service	Innovations included	Type of the innovation process
Case 5 (architect's office) Workplace – product: the planning of work-places is linked to clients' strategies. Means change in architects' practice – they enter the planning process in an earlier stage.	A new service in Finland in its time. The development started at the end of the 1980s. Later, the formalisation of the service into a specified product, focusing on consultancy, has been highlighted.	The model of rapid application
Case 6 (architectural office) Shopping centre – concept: a combination of feasibility studies, consultancy and interior design. Encourages broad participation in the planning process.	In this form the service is new in Finland and in the Baltics. Innovativeness is also manifested in the formalisation of the service into a specified product and in the multidisciplinary consultative approach.	The model of rapid application
Case 7 (advertising agency) Consultancy concerning promotion and category management. Aims at helping manufacturers to reconcile extensive sales and the maintenance of brand value.	A tailor-made service provided to one client; replication is possible. An example of the development of advertising agencies towards a greater responsibility for the actual sales, not only for visibility.	The model of rapid application
Case 8 (accounting company) Services for purchase price allocation and goodwill impairment testing in the context of acquisitions. Based on new international regulations.	Emphasises the valuation of immaterial property, which is new in the accounting practice. The necessities caused by regulatory changes have been linked with the promotion of the clients' business.	The model of rapid application

Table 10.1 (continued)

The main idea of the service	Innovations included	Type of the innovation process
Case 9 (accounting company) So-called 'people services', including international assignment management (training, contracts, social security advice etc.) and reward systems design.	Innovative elements are linked to the broadening of services of accounting companies into a new expertise area: human relations consultancy.	The model of rapid application

in one case: the 'potential' innovation included in the project management service that the engineering company provided to the Finnish Rail Administration. In the following, I first consider these two 'exceptional' cases in slightly more detail and thereafter analyse the dominant model, the rapid application.

The follow-up and consultancy concerning the operation of building services systems is a service tightly linked to the advancements of information technology. Its development project can be divided into four main stages:

1. Measurements, information systems and commissioning linked to building automation had become general by the end of the 1990s. The case company got the idea for their integration into a unified service product from one of its clients.
2. Development of the prototype was started together with another client. Competitive advantage was pursued through two incremental innovations: data gathering was made independent of the automation system, and data were gathered from smaller units instead of the building's main service meter in order to ensure configurability.
3. After several improvements (e.g. in the software used) and changes based on the feedback from the pilot client, the actual selling of the service was started. A similar package consisting of follow-up, reporting and consultancy was offered to all clients. However, some clients wanted to purchase only tools and carry out the follow-up themselves or use another consultant (due to a prior established relationship).
4. In order to satisfy the differing needs of clients, tools were added to the service offering and its structure was made modular. Nowadays, clients can purchase just one or two of the three modules – measuring

and reporting, tools for analysis, and consultancy – but they can also purchase all modules in a single package.

All stages in the development show the important role of clients: clients provided the basic idea, piloted the prototype and provided reasons for further development. The development project was partially funded by public R&D funding. Yet, the resources used for the development were quite small: one engineer worked full-time for the project and was supported by some colleagues and some students.

In the case which I have interpreted as an example of the practice-driven innovation model, the engineering company has used its project management experience gained in building construction in a new context: in rail track renewals and track maintenance. The development of the new services was linked to the initiation of tendering for these functions, which in Finland had been carried out by public organisations until the mid-1990s. The case company has helped the Finnish Rail Administration to develop new practices needed in the sphere of public–private cooperation. The development process has been long; at the time of our research project in 2006 it was 'in the final straight' after taking eight years. In addition to the development work, the case company has also operated as the manager in concrete track projects after the winning of the tender. When we started our study, both services were still tailor-made – planned for the purposes of the Rail Administration only. Later the engineering company realised that there were broader application possibilities for the basic elements included in the services, and when we finished our study it was concretely planning to enter new markets. As privatisation and different public–private partnership models are spreading in the field of infrastructure administration, the development of the service into a replicable form seems realistic.

As mentioned above, the majority of the innovation processes that we studied in KIBS followed the model of rapid application. In addition to the 'finalised' innovation cases, I have included in this group two services which still were tailor-made, but showed clear promise for replication (consultancy concerning a firm's contingency plan and consultancy concerning promotion and category management). In both cases, the starting point had been to develop a replicable service; the finding of suitable markets had just proved more difficult than expected. The model of rapid application was manifested in our study as follows: after the emergence of a new idea, it was almost directly brought to the markets and developed hand in hand with the actual delivery of the service. There was a short stage of 'desk-study', but no separate testing or piloting stage which the provider would make clear to the first clients. These clients may or may not be aware of the novelties and of the piloting function that inevitably is part of their role; in any case, the client–provider relationship was that of a normal service transaction right from the start.

According to our study, the main reason for the use of rapid application seemed to be the fact that a new service usually includes abundant questions that cannot be correctly posed – let alone answered – without operating in real markets. A separate planning stage provides a limited benefit here. An urgent need in the markets may function as an additional motivation for the use of this model. Case number 8 – services for purchase price allocation and goodwill impairment testing – illustrates well a situation which leads to rapid application. These services are based on internationally unified new accounting standards, concerning the valuation of property in the context of acquisitions. The anticipation of the future productive capacity instead of past returns, and a more detailed identification of immaterial items, have been the essential changes in these standards. Despite the general regulatory basis, the practical solutions can be made in many ways and the solution selected has a great impact on the development of the business of the client companies. The anticipatory element included in the valuation, and the broadening of the issues into the realm of immaterial values, make the subject highly complex. Thus, several incremental innovations were called for when the regulations were transformed into actual services. All these complexities together with an effective date, which in Europe was 1 January 2005, favoured the use of the model of rapid application.

The main challenge in the innovation process, which proceeds hand in hand with the service delivery, is the balancing of resources between developmental work and business practice. If this point is not carefully considered, the model leads to a greater or smaller amount of 'voluntary work' among the developers. Due to the nature of KIBS as an expert sector, the individual employees are often enthusiastic enough to tolerate these kinds of situations, as the extra work load is usually temporary. However, an innovation process which employees carry out in addition to their regular jobs without any special arrangements is not very efficient. We found in our case companies three main ways in which the resource problem had been tackled. Firstly, the company may relieve a group of employees from their everyday duties for a short period of time during which they make the 'desk-study', preceding the launch of the idea on the market. Secondly, the firm may select an employee to coordinate a specific development process, and they together with a group of other employees devote part of their time to the innovation process. Here the development work is somehow discernible also in its later stages. In a few cases in our study, the time that employees could devote to the development was exactly defined. Thirdly, the firm may recruit a new employee, known as an expert in the area that the firm wants to develop. Responsibility for the innovation process is their main duty. A risk included in this solution is that new employees may remain 'outsiders', and from this position it is not easy to cooperate with other employees, which would be necessary in order to bring the innovation process to fruition.

Preliminary results from the second project confirm the finding that innovation processes in KIBS often apply the model of rapid application. We have followed up two processes in the architect's office and in the engineering company participating in our study. These companies were originally separate, but merged very soon after the project's start. The first process is architecture-centred and aims at providing new types of office spaces for a client in the banking sector. The experience of end customers is carefully anticipated and new innovative solutions are sought by combining the expertise of architects and industrial designers. At the time of writing this chapter in summer 2008 the process was still tailor-made, but developing a replicable service product was a goal in the architect's office. Like in other examples of the model of rapid application, the service was 'piloted' with the first client in the context of a normal transaction. Also the second process started in the architect's office, but after the merge the engineering company has been involved, too. Here the aim is to improve the coordination of design projects – and hence the quality of construction – by strengthening the role of the partner who takes the main responsibility. The idea is to gather the tasks of the coordinator into the form of a replicable service product. The basics of this new service had been formulated several years earlier and some preparatory steps were taken for its implementation, e.g. the planning of necessary databases was started. However, the architect's office had difficulties in putting the idea into practice without partners. Merging with the engineering company changed the situation, and at the time of writing this chapter the idea was implemented in a project where all designers come from this merged company (the engineering company includes building services, civil engineering and construction management units). The innovation process here shows features from both separate planning (a long ideation phase with some preparatory steps) and rapid application (implementation directly in real markets).

10.3.2 Innovation processes in 'traditional' services and in services linked to manufacturing

Our preliminary findings concerning innovation processes in 'traditional' services come from the insurance company and the wholesale company. The findings that describe service innovation processes in the manufacturing context come from the metal products company and the warehouse trucks company.

In the insurance company we have been following up an innovation process in which a 'locking-in' service has been developed for individual customers and households. The main idea is to map the customer's insurance needs deeply and thoroughly, based on their specific life situation. Customers are met face-to-face, the time reserved for the needs analysis is about one hour, and the service is free of charge. Analysis is provided as a separate service, i.e. the purchasing of insurance products from this company is not a prerequisite for the service. Yet, an explicit purpose is to acquire new customers and keep the existing ones, maybe getting the latter to supplement their present insurance portfolio. The basics of this type of service are not new in Finland – all the biggest insurance companies provide some free analysis service as the starting point for their insurance selling. What is new is the strong emphasis put on this service in the case company: the company regards analysis as a part of its core services and is convinced that a carefully and individually conducted analysis is the best way to create new loyal customer relationships and increase commitment among existing customers. Thus, the analysis is not primarily linked to the selling of insurance here and now, but to the building of long-term customer relationships. A specific strength in the company is its expertise in public social security. This expertise enables the consideration of the customer's situation as a whole: a combination of public security and private insurances needed to supplement it.

Like other insurance companies,the case company had provided some analysis service earlier. However, as the service had never been systematised, its aim and content were not clear among employees. There was a wide variety in the ways of conducting the analysis, and at its worst a clear gap existed between the ideal and practice. Tackling the challenge of inconsistencies had been discussed within the company for quite a while, and when our study started, a specific four-month project had been established to systematise the analysis service in the above-described advanced form. The project group consisted of both managerial and grassroots staff, one unit manager being the responsible developer who prepared the material for group meetings. The group held six meetings and we participated in all of them. The meetings focused on the development of the service concept and on detailed analysis of the service process. The most concrete result at the end of the project was a process blueprint, on the basis of which training of personnel was planned as the next step. Specific activities regarding the marketing and launching of the renewed service were not planned at this stage – the aim was just to continue the provision of the service at a higher level of consistency and quality.

The wholesale company participating in our study is a technical trader in business-to-business markets. It has actively developed e-commerce, having an online shop and electronic information services, among others. The service whose innovation process we are following up is totally new in the company, and as far as we know other technical wholesalers competing in the Finnish markets do not have a corresponding service either. Due to the genuinely innovative nature of this service, I can describe it at quite a general level only. The main idea is essentially to broaden the existing service portfolio of the company, from actual wholesale activities towards advisory services and other services supporting the clients' business. Like in other activities of the company, the utilisation of the Internet plays a central role in this new service. Some parts of the service will be provided via partners.

When we started our study, the idea of the new service had been unofficially discussed for some time, and the discussion had led to the decision to commence practical actions. The company had not established a specific project for the innovation process, but the first steps in putting the idea forward had been taken, and the process was proceeding purposively and in quite a systematic way. The responsibility for the development was clearly stated and delegated to the marketing manager of the company. He and one of his colleagues have constituted the in-house resources devoted to the process. The first step in the process was hiring an outside consultant who made a preliminary description of the elements of the new service, the resources needed and the challenges to be faced. Thereafter the marketing department of the company sent to clients a questionnaire which indirectly mapped out the demand for the planned service. In order to reveal real needs instead of superficial opinions, the questionnaire focused on the ways in which the clients presently handle the tasks that the company's new service could take care of. At the time of writing this chapter in summer 2008, the innovation process was in a phase where the finding of partners was considered to be the necessary next step. For this task the company used the same consultant who had made the preliminary outline of the new service.

I now move to the description of service innovation processes in the metal products company and the warehouse trucks company. These cases are important because 'servicisation' of industrial companies is a growing trend, and yet service innovation in manufacturing is a rarely discussed topic. The process that we have been observing in the metal products company had its starting point in client segmentation, the aim of which was an essential improvement and diversification of services provided to key customers. The company's services have been focused on three types of offerings: prefabrication services, transport services and warehouse services. In all of these, the service has been linked to the goods provided. Now the company also wants to provide such services that support the use of goods in the client's processes or that support the client's business in a broader sense. The former means, among others, that the service provider takes more responsibility for securing the availability of goods and materials according to the nature of the client's process. The latter includes, for instance, the acquisition of financing for the client's own warehousing – a service which big manufacturers can provide through their partners.

The innovation process in the metal products company has included two sequential parts, and for both of them a specific project was established. The first project lasted three months, the project group being composed of five members who represented the development, management and marketing functions of the company. The second project lasted eight months, but only two permanent members participated in it; here the responsibility rested with the marketing department. We as researchers had the opportunity to participate in all meetings of the project groups (19 in total). The first project analysed the service portfolio of the company in detail and specified the service promise separately for the key, priority and standard customers. Important differences in the service promise given to these three customer groups are, among others, the rapidity and preciseness of the delivery. The project also familiarised itself in the business processes of key customers using both interviews and company visits as information sources. The second project focused on standard customers who order the company's products in small quantities. Broadening the service portfolio towards more value-adding offerings in the case of key customers is not possible if the company also continues serving the smallest clients to the present extent. A solution for this problem was sought in partnering arrangements. At the time of writing this chapter in summer 2008, some partners had been found and the implementation of the new ideas was going on regarding all customer groups. This shows that the company was well on the way towards services that support the use of their goods in the clients' processes. Yet, benefiting the clients' business in a broader sense was still at the idea phase.

The warehouse trucks company had made corresponding client segmentation, and before we started our study it had also modularised its services. Based on the modules, it had composed three service packages targeted to three customer groups. The number of modules grows when moving from standard customers to key customers. Besides the whole packages, customers can purchase single service modules. Traditionally the services of this company have focused on after-sales operations: maintenance and repair of trucks. In the new portfolio, a wider range of services is provided and securing the availability of well-working trucks is emphasised. For the key customers, the company provides total solutions, taking the responsibility for the maintenance of the customers' trucks fleet. Among individual services, that may belong as a module to the packages or can be purchased separately, particularly interesting is a service which aims at optimising the client's logistics fleet. The service consists of an analysis of the client's present fleet and operating environment, and of the development plan regarding the purchase, maintenance and use of trucks and other logistics equipment. As the service concerns even the issues of work safety and skills of personnel, it indicates that this company (like the metal company) is extending its business from services linked to goods to services that support the use of goods in the clients' processes.

In the innovation process that we have been observing, the company has elaborated the key customers' service package and developed the above-mentioned optimisation service that it had not provided earlier. Our research material consists of the observation of company workshops and smaller meetings (12 in total) as well as of a few interviews. The company established a specific project for the elaboration of the key customers' package and a sub-project for the optimisation service. The structure, tasks and schedule of the main project were clearly defined. The project consisted of a steering group, functional teams and cross-functional teams, which developed the key customers' services in workshops. The development work focused on the different stages in the service process: selling, mobilising and operating. The result was a service model, which was tested using a simulation method: all project members participated in the so-called 'workflow game'. At the time of writing this chapter in summer 2008, the company had prepared marketing material for the key customers' services together with the other two service packages, and their implementation was about to start. The optimisation service had also been outlined; an outside logistics consultant played an important role in the development of this service.

Summarising my description, I can say that the model of rapid application, which was the most usual way to create innovations in KIBS, does not manifest itself equally clearly in 'traditional' services and in services linked to manufacturing. In three of the four cases, there was a separate project established for the development of new or improved services. On the other hand, most of the projects were small regarding both time resources and personnel (regarding the latter, the involvement of marketing personnel was outstanding in both 'traditional' and industrial services). Most projects also included only the early stages of the innovation process. New or improved services were implemented in practice quite quickly, which means that the processes had features of rapid application, too. The only project which to a greater extent followed the normative stage-gate model was the project of the warehouse trucks company. Thus, the innovation processes in the above-described cases could be characterised as some kind of intermediate form between the separate project and rapid application. Due to the case study nature of this research, our results are naturally only indications that have to be tested later with bigger samples. However, they revealed an additional reason for rapid application: if the service to be improved is already in use, as in the case of the insurance company, it is more plausible to collect experience from real markets than to arrange a separate testing and piloting.

Finally, I briefly point out the nature of the services that the case companies were innovating. It turned out that different types of analysis services, that aim to consider the clients' situation broadly and thoroughly, are one possible arena for innovations in both the business-to-consumer and business-to-business contexts. In manufacturing companies, the development of analysis services shows that these companies increasingly also sell 'pure' services, i.e. services that can be purchased without purchasing a good. In addition, manufacturing companies supplement services linked to goods (transport, maintenance, etc.) with services supporting the client in the use of goods (selling availability, total solutions) (see Mathieu, 2001).

10.4 Conclusion

In this chapter I have analysed the nature of innovation processes in services. I have argued that despite the recent accumulation of studies on service innovation, the process aspect has been examined quite one-sidedly. The majority of studies have started from the view that planning always precedes practice and can be separated from it. Yet, some critical authors have pointed out that an explorative approach, where the idea of a new or improved service (or a good) is rapidly launched on the market and developed hand in hand with the actual production, may be more effective particularly in circumstances that involve abundant uncertainties. Other critical authors have added that an innovation process may also start in service practice, the novelty being recognised only afterwards.

My own empirical analysis confirms that all three types of processes exist in the context of service innovation. I have called them: (1) innovation as a project separated from practice; (2) innovation following the model of rapid application; and (3) innovation following the practice-driven model. These three models differ from each other at two points: whether an innovation is deliberately pursued, and whether the main part of the innovation process is carried out before the new or improved service is on the market. In a separate innovation project both statements are valid, whereas only the first statement describes an innovation process which is integrated in service practice; in an a posteriori recognised innovation process neither of the statements holds true.

My empirical material comes from two case study projects. Thus, I am not able to make statistical generalisations. On the other hand, in these studies I and my colleagues have had the possibility to analyse some service innovation processes deeply and thoroughly, using both retrospective and follow-up approaches. Thus, I believe that our material provides important indications that form bases for further studies. In this chapter, I have summarised findings from 15 innovation processes, most of them being from KIBS firms, which the first project concentrated on. I have, however, some preliminary results also from 'traditional' services and industrial services. The findings regarding the latter group are particularly interesting, because services linked to manufacturing have only rarely been studied from the innovation perspective. In one of the two manufacturing cases, the service innovation process followed the stage-gate model, where a project separated from practice plays a major role. The other case exemplified a model which has features from both a separate project and rapid application: the firm established a specific development project, but only for the early stages of the innovation process. This kind of an intermediate model was found also in the insurance company participating in our study. Thus, it seems that the service innovation process may, at least in some cases, be similar in manufacturing companies and in service companies.

Actually, my analysis indicates that it may be KIBS firms which more clearly differ from firms in other sectors: in the vast majority of our 11 KIBS cases, the idea was brought to the markets very rapidly after its emergence, i.e. the model of rapid application was perceivable in a 'pure' form. I can suggest two reasons for this finding. Firstly, KIBS provide consultancy-type services whose development often requires only a small amount of investment. Thus, it is possible to test ideas directly in the markets without fear of great economic losses (some discretion is naturally required in order to maintain the good reputation of the firm). Secondly, the characteristic of knowledge-intensity implies that KIBS offerings are usually answers to quite complex needs of clients (see Morris and Empson, 1998). A typical situation in KIBS thus corresponds well to the situation where an explorative approach has been found to be beneficial: planning can not provide all the details needed in the implementation of novelties.

Due to the difference in the number of cases between KIBS and other types of firms in our studies, additional research is needed in order to validate my perceptions. Based on the literature and my own studies I can, however, conclude that instead of the search for one generally applicable ideal model, it should be accepted that innovations in services (and probably also in manufacturing) emerge in different ways. Consequently, more emphasis should be put on identifying the strengths and vulnerable points in different types of processes, and on finding good managerial practices for them. Each type of the above-described innovation processes creates its own management challenges. As regards a separate innovation project, a central challenge is the difficulty in digging out all relevant issues before entering the real markets. Careful piloting and testing reduce the problem, but may lengthen the planning stage and result in slow response to market needs. A longer project also requires considerable resources, due to which this type of development is often linked with the chances of obtaining external funding. As regards the practice-driven model, our study shows that it can take a surprisingly long time before the company operating in a new way becomes conscious of the novelties. Thus, a central management challenge in this model is to develop mechanisms through which new, tacit ideas could be made explicit earlier and be linked to the other knowledge resources of the firm. In the model of rapid application, the main challenge is the organisation of the work in a way which prevents the subordination of development to immediate market needs. Clearly devoting part of the working time to development provides one solution.

In addition to those further studies that could validate and specify our findings with bigger samples, I want to point out three topics that I consider particularly important to be examined in the future. Firstly, my present chapter has focused on the nature of innovation processes targeted to the creation of novelties in service products. There are, however, important innovation processes also at the level of service firms: many organisational and market innovations take place at this level, and the processes leading to them are not well understood. Secondly, the specific characteristics of those innovation processes that do not aim at developing a new service but improving the existing ones have not been emphasised sufficiently. The case of the insurance company shows, however, that the prior existence of the service in the markets may essentially influence the nature of the innovation process: it may make the integration of planning and execution necessary at least to some extent. Thirdly, the model of rapid application can be interpreted as one possibility of carrying out user-based innovation. Due to the central position of the client interaction in services – the phenomenon of co-production (Sundbo and Gallouj, 2000) – the concrete forms in which the participation of clients is organised in the different types of services innovation processes is worthy of further study.

References

van der Aa, W. and T. Elfring (2002), 'Realizing innovation in services', *Scandinavian Journal of Management*, **18**, 155–171.

Abernathy, W.J. and J.M. Utterback (1978), 'Patterns of industrial innovation', *Technology Review*, June/July, 41–47.

Alam, I. and C. Perry (2002), 'A customer-oriented new service development process', *Journal of Services Marketing*, **16** (6), 515–534.

Antonelli, G., G. Cainelli, N. and R. Zoboli (2000), 'Structural change and technological externalities in the service sector: some evidence from Italy', in Metcalfe, J.S. and I. Miles (eds), *Innovation Systems in the Service Economy. Measurement and Case Study Analysis*, Boston, MA, USA Dordrecht, Netherlands and London, UK: Kluwer Academic Publishers, pp. 170–191.
Boden, M. and I. Miles (2000), 'Conclusions: beyond the service economy', in M. Boden and I. Miles (eds), *Services and the Knowledge-Based Economy*, London, UK and New York, USA: Continuum, pp. 247–264.
Buijs, J. (2003), 'Modelling product innovation processes, from linear logic to circular chaos', *Creativity and Innovation Management*, **12** (2), 76–93.
Chesbrough, H. (2003), *Open Innovation: The New Imperative for Creating and Profiting from Technology*, Boston, MA: Harvard Business School Press.
Coombs, R. and I. Miles (2000), 'Innovation, measurement and services: the new problematique', in J.S. Metcalfe and I. Miles (eds), *Innovation Systems in the Service Economy. Measurement and Case Study Analysis*, Boston, MA, Dordrecht, Netherlands and London, UK: Kluwer Academic Publishers, pp. 185–103.
Cooper, R.G. and U. De Brentani (1991), 'New industrial financial services: what distinguishes winners', *Journal of Product Innovation Management*, **8** (2), 75–90.
Cooper, R.G. and S.J. Edgett (1996), 'Critical success factors for new financial services', *Marketing Management*, **5** (3), 26–37.
De Brentani, U. (1991), 'Success factors in developing new business services', *European Journal of Marketing*, **25** (2), 33–59.
Djellal, F. and F. Gallouj (2001), 'Patterns of innovation organisation in service firms: postal survey results and theoretical models', *Science and Public Policy*, **28** (1), 57–67.
Drejer, I. (2004), 'Identifying innovation in surveys of services: a Schumpeterian perspective', *Research Policy*, **33**, 551–562.
Edvardsson, B., A. Gustafsson, M.D. Johnson and B. Sandén (2000), *New Service Development and Innovation in the New Economy*, Lund, Sweden: Studentlitteratur.
Eisenhardt, K.M. (1989), 'Building theories from case study research', *Academy of Management Review*, **14** (4), 532–550.
Eisenhardt, K.M. and B.N. Tabrizi (1995), 'Accelerating adaptive processes: product innovation in the global computer industry', *Administrative Science Quarterly*, **40** (1), 84–110.
Engvall, M., P. Magnusson, C. Marshall, T. Olin and R. Sandberg (2001), 'Creative approaches to development: exploring alternatives to sequential stage-gate models', Fenix WP 2001:17, available at http://www.fenix.chalmers.se.
Evangelista, R. and G. Sirilli (1998), 'Innovation in the service sector: results from the Italian Statistical Survey, *Technological Forecasting and Social Change*, **58**, 251–269.
Gadrey, J., F. Gallouj and O. Weinstein (1995), 'New modes of innovation: how services benefit industry', *International Journal of Service Industry Management*, **6** (3), 4–16.
Gallouj, F. (2002), 'Knowledge-intensive business services: processing knowledge and producing innovation', in J. Gadrey and F. Gallouj (eds), *Productivity, Innovation and Knowledge in Services: New Economic and Socio-Economic Approaches*, Cheltenham, UK and Northampton, MA, USA: Edward Elgar, pp. 256–284.
Gallouj, F. and O. Weinstein (1997), 'Innovation in services', *Research Policy*, **26** (4–5), 537–556.
Heusinkveld, S. and J. Benders (2003), 'Between professional dedication and corporate design: exploring forms of new concept development in consultancies', *International Studies of Management and Organization*, **32** (4), 104–122.
Hipp, C. and H. Grupp (2005), 'Innovation in the service sector: the demand for service-specific innovation measurement concepts and typologies', *Research Policy*, **34**, 517–535.
von Hippel, E. (1978), 'A customer-active paradigm for industrial product idea generation', *Research Policy*, **7** (3), 240–266.
Kim, W.C. and R. Mauborgne (1999), 'Strategy, value innovation, and the knowledge economy', *Sloan Management Review*, Spring, 41–54.
Kline, S.J. and N. Rosenberg (1986), 'An overview of innovation', in R. Landau and

Rosenberg, N. (eds), *The Positive Sum Strategy: Harnessing Technology for Economic Growth*, Washington, DC: National Academy Press, pp. 275–305.
Koen, P., G. Ajamian, R. Burkart, A. Clamen, J. Davidson, R. D'Amore, C. Elkins, K. Herald, M. Incorvia, A. Johnson, R. Karol, R. Seibert, A. Slavejkov and K. Wagner (2001), 'Providing clarity and a common language to the fuzzy front end', *Research-Technology Management*, **44** (2), 46–55.
Lancaster, K.J. (1966), 'A new approach to consumer theory', *Journal of Political Economy*, **14**, 132–157.
Mathieu, V. (2001), 'Product services: from a service supporting the product to a service supporting the client', *Journal of Business and Industrial Marketing*, **16** (1), 39–58.
Miles, I. and M. Boden (2000), 'Introduction: are services special', in M. Boden and I. Miles (eds), *Services and the Knowledge-Based Economy*, London, UK and New York, USA: Continuum, pp. 1–20.
Moorman, C. and A.S. Miner (1998), 'The convergence of planning and execution: improvisation in new product development', *Journal of Marketing*, **62** (3), 1–20.
Morris, T. and L. Empson (1998), 'Organisation and expertise: an exploration of knowledge bases and the management of accounting and consulting firms', *Accounting, Organizations and Society*, **23** (5/6), 609–624.
Preissl, B. (2000), 'Service innovation: what makes it different? Empirical evidence from Germany', in J.S. Metcalfe and I. Miles (eds), *Innovation Systems in the Service Economy. Measurement and Case Study Analysis*, Boston, MA, Dordrecht, Netherlands and London, UK: Kluwer Academic Publishers, pp. 64–85.
Saren, M.A. (1984), 'A classification and review of models of the intra-firm innovation process', *R&D Management*, **14** (1), 11–24.
Saviotti, P.P. and J.S. Metcalfe (1984), 'A theoretical approach to the construction of technological output indicators', *Research Policy*, **13**, 141–151.
Scheuing, E.E. and E.M. Johnson (1989), 'A proposed model for new service development', *Journal of Service Marketing*, **3** (2), 25–35.
Schumpeter, J. A. (1934), *The Theory of Economic Development: An Inquiry into Profits, Capital, Credit, Interest, and the Business Cycle*, Cambridge, MA: Harvard University Press.
Slater, S.F. and J.C. Narver (1998), 'Customer-led and market-oriented: let's not confuse the two', *Strategic Management Journal*, **19** (October), 1001–1006.
Sundbo, J. (1994), 'Modulization of service production and a thesis of convergence between service and manufacturing organizations', *Scandinavian Journal of Management*, **10** (3), 245–266.
Sundbo, J. (1996), 'The balancing of empowerment: a strategic resource based model of organizing innovation activities in service and low-tech firms', *Technovation*, **16** (8), 397–409.
Sundbo, J. (2008), 'Customer-based innovation of knowledge e-services: the importance of after-innovation', *International Journal of Technology and Management*, **9** (3–4), 218–233.
Sundbo, J. and F. Gallouj (2000), 'Innovation as a loosely coupled system in services', in J.S. Metcalfe and I. Miles (eds), *Innovation Systems in the Service Economy. Measurement and Case Study Analysis*, Boston, MA, Dordrecht, Netherlands and London, UK: Kluwer Academic Publishers, pp. 43–68.
Tether, B. and C. Hipp (2000), 'Competition and innovation amongst knowledge-intensive and other service firms: Evidence from Germany', in B. Andersen, J. Howells, B. Hull, I. Miles and J. Roberts (eds), *Knowledge and Innovation in the New Service Economy*, Cheltenham, UK and Northampton, MA, USA: Edward Elgar, pp. 49–67.
Toivonen, M., T. Tuominen and S. Brax (2007), 'Innovation process interlinked with the process of service delivery: a management challenge in KIBS', *Economies et Sociétés: Economics and Management of Services Series*, **3**, 355–384.
Tuomi, I. (2002), *Networks of Innovation: Change and Meaning in the Age of the Internet*, Oxford: Oxford University Press.
de Vries, E. (2006), 'Innovation in services in networks of organizations and in the distribution of service', *Research Policy*, **35**, pp. 1037–1051.

11 Service innovation: development, delivery and performance

Joe Tidd and Frank M. Hull

11.1 Introduction

In this chapter we develop and test a framework for the development and delivery of service products. A generic framework is potentially useful because goods and services are increasingly bundled; for example, many intangible service products have physical manifestations that require operations similar to those in manufacturing, and products include intangible services. A more robust model therefore offers an opportunity for adding value both to theoretical understanding of systems and to practical applications to a wider scope of activities within a greater variety of contexts. We test the framework using data from comparable studies of 108 service enterprises in the UK and USA (Tidd and Hull, 2003, 2006). The dependent variables include an index of product development plus service delivery, and explores the trade-offs between improving cost and time on the one hand, and innovation and quality on the other. The latter is important because the delivery process is sometimes tantamount to the service offering itself.

We know a great deal about the organization and management of new product development in the manufacturing sectors, but we know comparatively little about how applicable this is to the service sector (Miles, 2000; Tidd and Bessant, 2009). In this chapter we explore the extent to which a framework for organizing and managing new product development, derived from good practice in manufacturing, predicts variation in performance in service companies in the USA and the UK. The framework we developed is based on proven good practice in the manufacturing sectors, and adapted to encompass services as well as goods. A generic framework is potentially useful because large corporations increasingly offer and bundle products and services, for example field installation, after-sale upgrades and add-ons, support, and maintenance (Chase and Garvin, 1989; Chase and Hayes, 1991). Large, high-tech companies employ as many people in service as manufacturing jobs. Moreover, the proportion of non-manufacturing jobs is growing as many industrial leaders are diversifying into services (Wise and Baumgartner, 1999).

The framework presumes that the basic steps in new service development are broadly similar to those in manufactured goods with only minor differences in emphasis and execution. From a systems perspective, inputs, throughputs and outputs must meet customer needs regardless of the type of product and transformation process – physical, symbol or human. However, despite these similarities in the results of analysis for goods and services, some important differences remain. The cornerstone of service development and delivery is found to be early simultaneous influence (ESI) by all relevant functions on decisions throughout the development cycle, because many functions in serial development methods have input only after most critical decisions have already been made regarding key service features and the process of service delivery. For example, the involvement and engagement of potential users early in the development of services, in particular in the public sector, appears to be increasingly important (Bessant and Tidd, 2007; Flowers, 2008).

11.2 Theoretical framework

There is a well-established literature on the organization and management of new product development, and we will not provide yet another review of this here. Although the disciplinary perspectives and methodologies differ, there is a significant consensus regarding the factors that contribute to new product success (Brown and Eisenhardt, 1995; Trott, 2008; Tidd and Bessant, 2009). However, almost all of these factors are derived from studies of new product development in the manufacturing sectors. There are similarities between the organization and management of manufacturing and services (Meyer and DeTore, 2001), and some authors have argued that manufacturing good practice should be applied to services, but there is good reason to believe that researchers and practitioners should question the direct application of manufacturing good practice to a service context without significant adaptation (Fitzsimmons and Fitzsimmons, 2000). For example, service and manufacturing differ in terms of the tangibility of outputs, the relationship between processes and products, and the interaction with end-users. However, the organization and management of new product development in services is still not well researched or understood (Menor et al., 2002). In this chapter we test the applicability of manufacturing good practice in new product development (NPD) to services.

In the framework we have developed, the organization and management of NPD is summarized by five constructs: strategy, process, organization, tools/technology and systems – SPOTS (Figure 11.1). This framework is adapted and expanded from a model of NPD effectiveness tested in industrial corporations in the US (Collins and Hull, 2002), and validated during the course of conducting case studies of companies participating in a user group. Varied formulations of this framework are

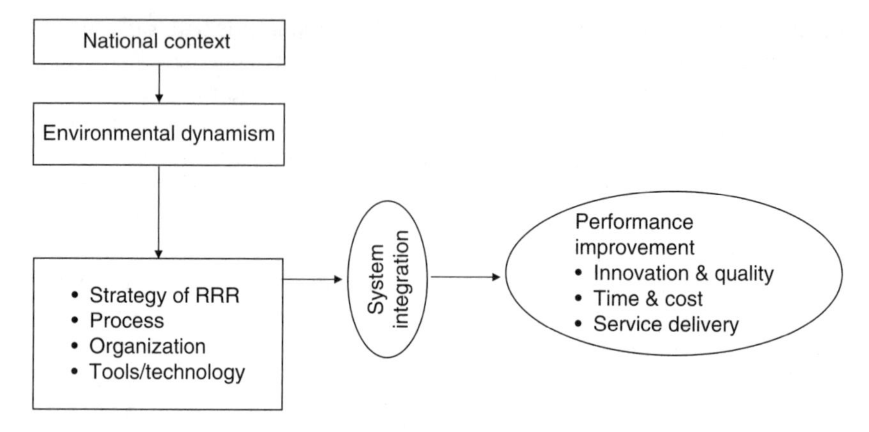

Figure 11.1 Performance, SPOTS model, environmental and national context

commonly used in the literature on concurrent engineering (Zirger et al, 1990; Susman and Dean, 1992), organization design (Hage, 1980; Daft, 1995) and as building blocks in models for continuous improvement (Bessant, 2003).

Each of the five constructs plays a different role in performance improvement. Strategy provides focus; process provides control; organization provides coordination of people; tools provides transformation and transaction capabilities, and system provides integration. Each can be expressed on a continuum ranging from a mechanistic bureaucracy to an organic-professional organization (Burns and Stalker, 1961; Lawrence and Lorsch, 1967; Mintzberg and Quinn, 1996; Damanpour, 1991; Leonard-Barton, 1992; Liker et al., 1999). In general terms, the mechanistic form is believed to be best for cost-efficiencies, and the organic for innovative product differentiation. However, simultaneously to achieve both kinds of generic competitive advantage, cost and innovation, organic and mechanistic design elements need to be combined in an integrated system (Tidd and Bodley, 2002; Tidd and Bessant, 2009). The need for combined organic and mechanistic practices is one reason why our constructs are defined below in ways that are relatively balanced in terms of mechanistic and organic compatibility (Dolfsma, 2004).

11.2.1 S – Strategy

We hypothesize a balanced strategy of RRR (rapid, reiterative, redevelopment) as being associated with higher service performance. This approach to product differentiation accumulates numerous incremental innovations instead of striving for a risky radical approach (Dewar and Dutton, 1986;

Mintzberg and Quinn, 1996). The RRR approach has enabled many companies to achieve greater cumulative novelty per unit of time than more radical approaches to innovation (Clark and Wheelwright, 1993; Tidd and Fujimoto, 1995). Short, repeat development cycles are hypothesized as making it easier for systems to maintain closely integrated relationships while making continual adjustments because multiple functions have opportunities for concurrent instead of serial input. The strategy of RRR is also hypothesized as improving performance because its focus on time compression and knowledge reuse provides a stimulus for both up and downstream functions to engage in frequent, constructive exchanges. Partly because the RRR approach entails conservatism and improvement, its strategic intent is to achieve both kinds of generic competitive advantage simultaneously, low cost and innovation.

11.2.2 P – Process

Processes for service development involve external benchmarking, structured methods for translating customer needs into requirements and specifications, setting standards for project and product performance, and systematic reviews (Garvin, 1995; Tidd and Bodley, 2002). By providing road maps and boundary conditions, cross-functional teams may be given more responsibility and stage-gates less rigidly enforced in design reviews. Such formal specific practices for service development are supported by more general methods of process improvement, such as mapping to identify the ways activities are conducted, reveal redundant steps, and reduce unnecessary handovers in development. These good practices are documented as standards for helping to keep product development projects on track and continuously updated. This approach to control is in stark contrast to the rigid, mechanistic use of static manuals of standard procedure, typical of ISO 9000 and Six Sigma approaches to quality. One reason is because cross-functional teams are often involved in the creation, maintenance and continual improvement of processes, thereby melding mechanistic and organic practices. The focus is on encouraging structured experimentation and supporting good practices, rather than controlling behaviors through formal process and measures (Hales and Tidd, 2008).

11.2.3 O – Organization

A cornerstone of NPD good practice is organizational, that is, the simultaneous integration of work performed by functions along a value chain of product development and delivery. The notion of organic team structure enables people to cut across bureaucratic barriers due to size and structural differentiation. Concurrent product development involves multiple functions simultaneously at early steps of decision-making to maximize opportunities and preclude downstream problems, such as difficulties with manufacturability, marketability and serviceability (Susman and Dean, 1992). Early simultaneous influence exerted by downstream functions fosters continuing cross-functional communication patterns among diverse people to develop better products that are well thought out from early stages, thereby avoiding costly late changes. Heterogeneous input typically results in better solutions to complex, dynamic problems (Isaksen and Tidd, 2006). The literature confirms the benefits of cross-functional teaming, especially to the extent that products are novel (Collins and Hull, 2002; Liker et al., 1999; Tidd and Bodley, 2002). One reason for its benefits, relative to serial approach where each specialist hands over to another, is that the group exerts some degree of control over individual behaviors, thereby melding some of the equivalents of mechanistic design elements with organic ones (Takeuchi and Nonaka, 1986).

11.2.4 T – Tools/technology

Despite their intangibility, many service products are knowledge-based and/or are heavily dependent upon ICT (information and communication technology). Significant improvements in ICT have increased the potential for conceiving of new kinds of products as well as for managing development and delivery processes, rather than simply focusing on the reduction of cost through automation (Tidd and Hull, 2003). Tools that were once hard to change and difficult to distribute are now soft, flexible and easily shared via electronic networks. ICT can be an enabler of continually updated processes and more rapid exchanges among cross-functional team members, regardless of distance. The mechanistic constraint of tools is thereby softened by the greater flexibility it permits in controls and increased frequency of communications. ICT tools are hypothesized as having a positive impact on system integration because of the speed and scope of data distribution. However, data are not information and ICT is presumed to be more effective to the extent that it supports mature processes and cross-functional organization (Crespi et al., 2006).

11.2.5 System integration

System integration is a property involving mutual adjustment among multiple functions and adaptive practices that have become institutionalized (Thompson, 1967). In terms of the framework presented here, a system has a balance of deployment of the sets of practices in the model. Each of these constructs offers a particular kind of advantage, not only solo, but also in concert. Synergy is lost if a single construct is overemphasized to the exclusion of others, for example novelty and cost (Tidd and Bessant, 2009). Such a system is more capable because reciprocal integration enables design, development, production, delivery and support to be mutually responsive to the needs of one another, and to align resources so as to respond more rapidly to customer needs. Examples common in service organizations would be business process re-engineering (BPR) and enterprise resource planning (ERP). However, BPR tends to optimize processes around an existing service offering, whereas ERP has a greater potential for promoting knowledge-exchange and new service development, although in practice both approaches have been associated with an increased emphasis on cost and time, but a reduction in innovation and quality (Trott and Hoecht, 2004).

11.2.6 Environmental context

Environmental dynamism is a contingency presumed to stimulate the adoption of concurrent practices and/or alternative initiatives for improving performance (Atuahene-Gima and Ko, 2001; Tidd, 2001; Tidd and Bessant, 2009). Measures of environmental competitiveness and turbulence were found to be correlated with the adoption of concurrency as defined by the framework in the USA data. These dynamics included increased technological complexity, faster rates of service product introduction, higher customization, globalization and demand for new and better services.

11.3 Research methods

11.3.1 Samples and methodology

Based on a review of the literatures on new product and new service development, and discussions with a small group of executives from prominent service companies, we developed a detailed checklist of practices. This checklist was used in formal interviews and workshops to help us to develop and refine our postal questionnaire into one that was appropriate for services. Many of the items had to be reconstructed at a more abstract, general level because of the intangibility and diversity of service products.

We discussed the initial case studies and questionnaire at two workshops for those involved in new service development, one in London and the other in New York. The 130 participants from 75 organizations provided further feedback on our proposed framework and questionnaire. A user group of 18 service organizations was formed from these events, and a total of 27 case studies were created from these 18 organizations. This user group and the more detailed case studies helped us to ground the formulation of hypotheses and to interpret better the statistical results from the postal surveys.

US sample Participants in the survey were companies in the New York metropolitan area. Large companies were targeted because they were more likely to have a broad experience of process and product innovation, and structural differentiation increases the need for integration provided by cross-functional organization and process tools. From the 100 largest employers in New York, 58 service companies were identified. Accounting, consulting, legal and non-profit organizations were not pursued. A total of 120 questionnaires were mailed, and the preferred respondent was someone in the service product development function, or alternatively persons in business development, quality management or productivity improvement. Respondents from 70 businesses in 51 corporations returned questionnaires. These included 11 of the 12 largest publicly held corporations, and all but four employed more than 500 people.

UK sample Respondents were drawn from the alumni network of contacts of a business school. The sample was one of convenience, and no claim is made for it being representative. As shown in Appendix Table 11A.1, the UK sample reflects the composition of service in London and the south-east of the UK, with a bias for financial services and consulting, similar to the sample in New York. In total, 100 questionnaires were mailed, and after a reminder 38 usable questionnaires were returned.

Measures and analysis The measures were adapted from a 200-page inventory of best practices based on 16 case studies and analysis of 100 companies. Many of the items had to be reconstructed at a more abstract, general level because of the intangibility and diversity of service products. The questionnaire consisted of 150 questions using seven-point Likert scales. Pilot surveys were conducted in the USA and the UK, and subsequent workshops in New York and London were used to help to refine the questions for the service context. Multiple regression analysis is used to predict variation in performance measures. The step-wise method was used to maximize variance explained.

11.3.2 Three studies

All three studies use the same framework (strategy, process, organization, tools and system) to predict variation in performance improvements in service products development and delivery processes. Performance is analyzed as a total index and as three subscales: (1) innovation and quality; (2) time compression in development and cost reduction in development and delivery, signature indicators of concurrent engineering (CE) effectiveness; and (3) service delivery. The first two factors roughly correspond to generic strategic alternatives, differentiation versus cost. The third factor is conceptually important because it distinguishes service delivery process from product features. Delivery processes often comprise a significant proportion of value added by services, especially if interpersonal exchanges are involved (Storey and Easingwood, 1999).

Study I explores the extent to which the framework transcends national boundaries by analyzing combined data from 70 USA with 38 UK service businesses. A dummy variable is used to assess the extent to which the USA and UK results are like and unlike. Study II assesses the extent to which practices as defined by model are like and unlike in the two nations. Study II replicates the methods used in the combined analysis for the US and UK subsamples for concurrency as defined by the framework. Study III analyzes the context within which NPD methods are deployed in the two nations. The product-focused strategy is compared with alternative process-focused interventions such as: TQM (total quality management), ISO 9000 certification, and BPR (business process re-engineering). The impact of each of these interventions on performance is compared in the context of environmental changes, such as increased technical complexity, globalization, etc.

11.3.3 Correlations and descriptive statistics

The performance subscales are significantly intercorrelated as shown in the first rows of Table 11.1 for relationships in the combined data. The intercorrelations among the performance indicators are stronger in the US than the UK subsample. Product innovation and quality has an insignificant negative correlation with time and cost in the UK subsample.

The constructs are significantly intercorrelated with one another in the combined data. The constructs are significantly intercorrelated with one another in the USA subsample. The correlation between process and organization is so strong in the combined and USA data that they are either added together or entered separately in multiple regression analyses. In the UK subsample, these intercorrelations are all positive and statistically significant except for two, strategy with organization and tools/technology. Each of the constructs is significantly correlated with all performance indices in the combined sample. However, the correlation between tools/technology and service performance is weak. In the US subsample, the constructs are significantly correlated with all performance indices. In the UK subsample, the correlations of constructs with performance are weaker, but statistically significant in all but two instances, tools/technology and system with time and cost.

The correlations of the dummy variable representing UK cases are consistently negative as shown in Table 11.1. A comparison of means show that these differences are significant for product innovation and quality,

Table 11.1 Correlation coefficients and descriptive statistics

	1.	2.	3.	4.	5.	6.	7.	8.	9.	10.	11.	12.	13.	14.	15.	16.
1. Performance overall	**1.0** 1.0 *1.0*															
2. Innovation & quality	**0.71**** 0.81** *0.54***	**1.0** 1.0 *1.0*														
3. Time & cost	**0.77**** 0.82** *0.61***	**0.43**** 0.73** *−0.13*	**1.0** 1.0 *1.0*													
4. Service Delivery	**0.90**** 0.91** *0.87***	**0.49**** 0.56** *0.36**	**0.52**** 0.58** *0.36**	**1.0** 1.0 *1.0*												
5. Strategy of RRR	**0.53**** 0.53** *0.54***	**0.53**** 0.60** *0.40***	**0.49**** 0.56** *0.34**	**0.39**** 0.38** *0.39***	**1.0** 1.0 *1.0*											
6. Process	**0.58**** 0.56** *0.64***	**0.51**** 0.56** *0.41***	**0.49**** 0.55** *0.36**	**0.48**** 0.45** *0.53***	**0.57**** 0.60** *0.50***	**1.0** 1.0 *1.0*										
7. Organization	**0.62**** 0.65** *0.52***	**0.53**** 0.56** *0.46***	**0.58**** 0.64** *0.43***	**0.50**** 0.55** *0.35**	**0.54**** 0.65** *0.25t*	**0.76**** 0.82** *0.59***	**1.0** 1.0 *1.0*									

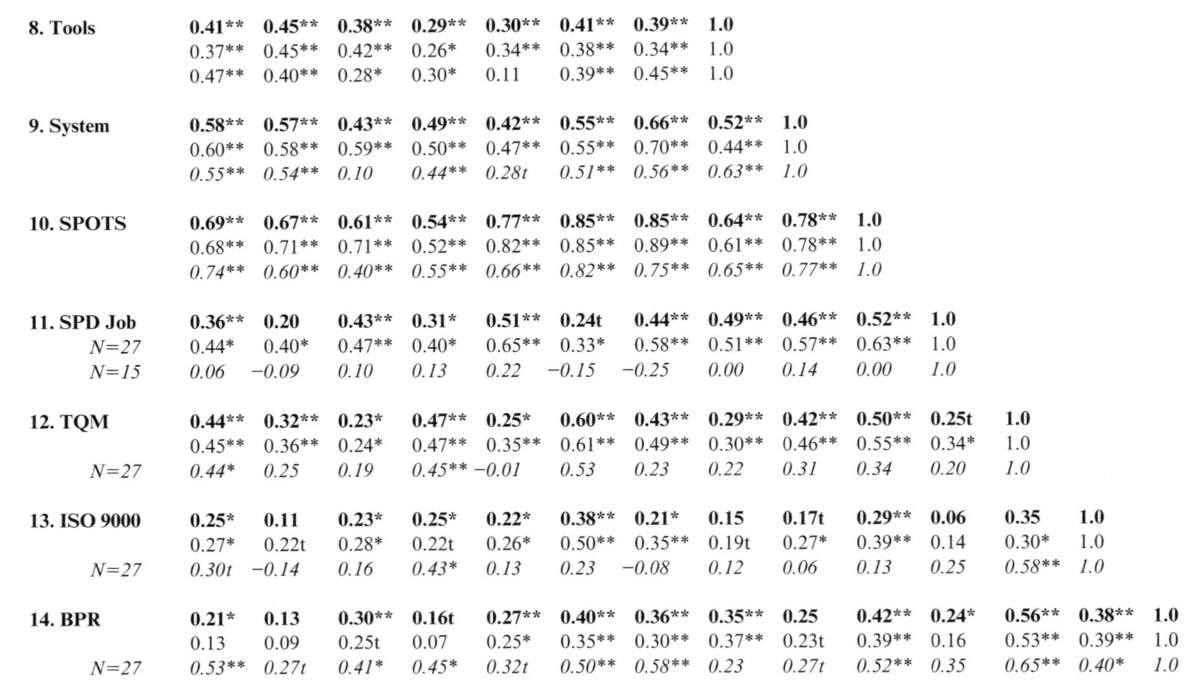

	1	2	3	4	5	6	7	8	9	10	11	12	13	14
8. Tools	**0.41****	**0.45****	**0.38****	**0.29****	**0.30****	**0.41****	**0.39****	**1.0**						
	0.37**	0.45**	0.42**	0.26*	0.34**	0.38**	0.34**	1.0						
	0.47**	0.40**	0.28*	0.30*	0.11	0.39**	0.45**	1.0						
9. System	**0.58****	**0.57****	**0.43****	**0.49****	**0.42****	**0.55****	**0.66****	**0.52****	**1.0**					
	0.60**	0.58**	0.59**	0.50**	0.47**	0.55**	0.70**	0.44**	1.0					
	*0.55***	*0.54***	*0.10*	*0.44***	*0.28t*	*0.51***	*0.56***	*0.63***	*1.0*					
10. SPOTS	**0.69****	**0.67****	**0.61****	**0.54****	**0.77****	**0.85****	**0.85****	**0.64****	**0.78****	**1.0**				
	0.68**	0.71**	0.71**	0.52**	0.82**	0.85**	0.89**	0.61**	0.78**	1.0				
	*0.74***	*0.60***	*0.40***	*0.55***	*0.66***	*0.82***	*0.75***	*0.65***	*0.77***	*1.0*				
11. SPD Job	**0.36****	**0.20**	**0.43****	**0.31***	**0.51****	**0.24t**	**0.44****	**0.49****	**0.46****	**0.52****	**1.0**			
N=27	0.44*	0.40*	0.47**	0.40*	0.65**	0.33*	0.58**	0.51**	0.57**	0.63**	1.0			
N=15	*0.06*	*−0.09*	*0.10*	*0.13*	*0.22*	*−0.15*	*−0.25*	*0.00*	*0.14*	*0.00*	*1.0*			
12. TQM	**0.44****	**0.32****	**0.23***	**0.47****	**0.25***	**0.60****	**0.43****	**0.29****	**0.42****	**0.50****	**0.25t**	**1.0**		
	0.45**	0.36**	0.24*	0.47**	0.35**	0.61**	0.49**	0.30**	0.46**	0.55**	0.34*	1.0		
N=27	*0.44**	*0.25*	*0.19*	*0.45***	*−0.01*	*0.53*	*0.23*	*0.22*	*0.31*	*0.34*	*0.20*	*1.0*		
13. ISO 9000	**0.25***	**0.11**	**0.23***	**0.25***	**0.22***	**0.38****	**0.21***	**0.15**	**0.17t**	**0.29****	**0.06**	**0.35**	**1.0**	
	0.27*	0.22t	0.28*	0.22t	0.26*	0.50**	0.35**	0.19t	0.27*	0.39**	0.14	0.30*	1.0	
N=27	*0.30t*	*−0.14*	*0.16*	*0.43**	*0.13*	*0.23*	*−0.08*	*0.12*	*0.06*	*0.13*	*0.25*	*0.58***	*1.0*	
14. BPR	**0.21***	**0.13**	**0.30****	**0.16t**	**0.27****	**0.40****	**0.36****	**0.35****	**0.25**	**0.42****	**0.24***	**0.56****	**0.38****	**1.0**
	0.13	0.09	0.25t	0.07	0.25*	0.35**	0.30**	0.37**	0.23t	0.39**	0.16	0.53**	0.39**	1.0
N=27	*0.53***	*0.27t*	*0.41**	*0.45**	*0.32t*	*0.50***	*0.58***	*0.23*	*0.27t*	*0.52***	*0.35*	*0.65***	*0.40**	*1.0*

Table 11.1 (continued)

	1.	2.	3.	4.	5.	6.	7.	8.	9.	10.	11.	12.	13.	14.	15.	16.
15. Environment	**0.41****	**0.37****	**0.31****	**0.32****	**0.23***	**0.20***	**0.17t**	**0.33****	**0.27****	**0.30****	**0.28***	**0.11**	**0.03**	**0.20***	**1.0**	
	0.41**	0.51**	0.32**	0.32**	0.19	0.23t	0.23t	0.44**	0.36**	0.35**	0.50**	0.07	−0.06	0.11	1.0	
	*0.46***	*0.18*	*0.31**	*0.40***	*0.36**	*0.22t*	*0.07*	*0.14*	*0.19*	*0.28**	*0.07*	*0.29**	*0.14*	*0.56***	*1.0*	
16. UK = 2; USA = 1	**−0.10**	**−0.15t**	**−0.01**	**−0.10**	**−0.08**	**−0.17***	**−0.13t**	**−0.24****	**−0.22***	**−0.20***	**−0.39****	**−0.11**	**0.20***	**−0.09**	**0.10**	**1.0**
Mean	**2.65**	**3.01**	**2.31**	**2.67**	**2.69**	**2.77**	**2.61**	**2.71**	**2.83**	**2.73**	**1.31**	**3.18**	**2.06**	**2.76**	**2.26**	**1.35**
Mean	2.70	3.09	2.31	2.73	2.75	2.86	2.68	2.83	2.93	2.81	1.44	3.25	1.89	2.83	−7.50	1.00
Mean	***2.56***	***2.86***	***2.29***	***2.56***	***2.58***	***2.60***	***2.48***	***2.49***	***2.64***	***2.56***	***1.07***	***3.00***	***2.44***	***2.60***	***.14***	***2.00***
S.D.	**0.66**	**0.76**	**0.86**	**0.80**	**10.00**	**0.71**	**0.73**	**0.67**	**0.65**	**0.59**	**0.47**	**10.05**	**10.24**	**10.16**	**0.95**	**0.48**
S.D.	0.74	0.71	0.91	0.86	10.04	0.74	0.78	0.69	0.65	0.62	0.51	1.05	1.17	1.15	0.99	0.00
S.D.	***0.51***	***0.83***	***0.76***	***0.68***	***0.91***	***0.64***	***0.62***	***0.58***	***0.61***	***0.49***	***0.26***	***1.04***	***1.33***	***1.19***	***0.81***	***0.00***

Notes:
Top=Combined Sample (N = 108) ◆ Middle = US (N = 70) ◆*Bottom= UK (N = 38)*
** p =0. 01, * p = 0. 05, t = p0.10 (two-tailed significance)

Table 11.2 Regression of system on predictors

	System
RRR strategy	**0.12t** 0.14 *0.13*
IDC Process & OTS organization	**0.40**** 0.46** *0.27**
CIT tools	**0.28**** 0.18* *0.46***
Nation (UK = 2, US = 1)	**−0.10t**
R2 Unadjusted	**0.42** 0.41 *0.44*
R2 Adjusted	**0.40** 0.39 *0.36*
F-ratio	**24.8**** 15.6** *8.9***

Notes:
Top = Combined Sample (N = 108) ◆ Middle = US (N = 70) ◆*Bottom = UK (N = 38)*
** p 0.01, *p 0.05, t = p 0.10 (two-tailed significance)

process, tools/technology, system and the good-practice index. This result suggests that the practices as measured by the framework are less implemented in the UK than the USA samples.

11.4 Study I: combined analysis

Strategy, process, organization, and tools explain about 40 percent of the level of system integration for the combined sample as shown in the top row of Table 11.2. In the top row of Table 11.3, variance SPOTS constructs explain in performance is shown as 0.47 for overall, 0.42 for innovation and quality, 0.31 for time and cost, and 0.30 for service delivery. These results are generally supportive of the SPOTS model as predictive of performance improvement in services.

Table 11.3 Regression of performance measures on SPOTS

	Performance	Innovation & quality	Time & cost	Service delivery
Strategy of RRR	**0.25****	**0.30****	**0.27****	**0.12**
	0.20*	0.28**	0.26*	0.07
	*0.38***	*0.26**	*0.25t*	*0.23t*
Process & Organization	**0.28****	**0.09**	**0.24***	**0.32****
	0.32**	0.15	0.18t	0.37**
	0.19	*0.02*	*0.30t*	*0.23*
Tools/Technology	**0.08**	**0.16***	**0.17***	**0.00**
	0.05	0.15t	0.14	-0.02
	0.19	*0.13*	*0.30t*	*-0.03*
System	**0.29****	**0.32****	**0.11**	**0.25***
	0.30**	0.27**	0.29**	0.24*
	0.24t	*0.38**	*-0.34t*	*0.25*
Nation (UK = 2, US = 1)	**0.03**	**0.01**	**-0.09**	**0.01**
R^2 Unadjusted	**0.50**	**0.45**	**0.35**	**0.33**
	0.50	0.50	0.47	0.34
	0.51	*0.36*	*0.29*	*0.31*
R^2 Adjusted	**0.47**	**0.42**	**0.31**	**0.30**
	0.47	0.47	0.43	0.30
	0.46	*0.29*	*0.16*	*0.22*
F-Ratio	**20.0****	**16.7****	**10.8****	**10.0****
	16.0**	16.5**	14.2**	8.3**
	*8.7***	*4.7***	*2.7**	*3.7*

Notes:
Top = Combined Sample (N = 108) ◆ Middle = US (N = 70) ◆*Bottom = UK (N = 38)*
** p=0. 01, * p = 0.05, t = p 0.10 (two-tailed significance).

The effects of different groups of constructs vary and interact. Strategy of RRR has significant relationships with all indicators of performance improvement in the combined data except for service delivery. Process summed with organization, has a strong main effect on three of four performance indicators, innovation and quality being the exception. If process is entered without organization, it explains significant variance in the overall performance index, time and cost, and service delivery (not shown in the table). Organization summed with process has at least weak effects on three of four performance indicators in the combined data, innovation and quality being the exception. If organization is entered without process, it explains significant variance in the overall performance, time and cost, and service delivery. Tools/technology has a significant effect on innovation and quality and time and cost. However, the coefficients are not as strong as for other constructs in the SPOTS model. System has at least significantly weak relationships with three of four performance indicators, time and cost being the exception. In summary, a strategy of RRR is particularly predictive of product innovation and quality and of time and cost. Process and organization are strongly predictive of time and cost and of service delivery. Tools/technology has significant relationships with product innovation and quality and with time and cost. System integration had strong significant effects on all performance indicators except time and cost.

11.5 Study II: comparison of the USA and UK

Nation as a category had little impact in any of the regression models shown in Table 11.3 despite several significant differences in means shown in Table 11.1. This suggests that differential implementation of practices as measured by the constructs in the SPOTS model explains significant differences in the two samples instead of unique cultural factors.

Strategy, process, organization and tools explain a little less than 40 percent of the level of system integration in both the USA and the UK samples as shown in the second and third rows of Table 11.2 respectively. However, strategy is insignificant in both. Process and organization in combination have roughly similar effects in both nations. If each is entered separately, the effects for process are almost identical. The effects for organization are strong in both, but more so in the USA data. The relative effect of tools is appreciably greater in the UK.

SPOTS constructs predict approximately half of the variance in the overall performance index in both nations. The regression coefficients with performance for the USA and UK samples are remarkably similar as shown in the second and third rows of Table 11.3. However, the significance levels are somewhat lower in the UK data because of the smaller sample size. Moreover, concurrency as measured by SPOTS explains somewhat less variance in the UK data.

The greatest variance explained in performance in both nations is for innovation and quality, 47 percent and 29 percent respectively. Variance explained in time and cost is much greater in the USA data, 43 percent. In the UK data it is only 16 percent. Variance explained in service delivery is relatively low in the USA and UK data, 30 and 22 percent respectively.

Strategy is the exception to parallel effects in the two nations. Strategy has a significant relationship with service delivery in the UK, but not the USA. Interestingly, strategy is the only one of the SPOTS constructs that is not significantly lower in the UK than the USA. Process summed with organization has roughly similar relationships with performance in both nations. If process is entered without organization in the USA data, it adds significant amounts of variance explained to the overall performance index and service delivery (not shown in the tables). If process is entered without organization in the UK data, it explains significant amounts of variance in the overall performance index, time and cost, and service delivery. Although the pattern of relationships for process is roughly similar in the two nations, it is relatively stronger in the UK than the USA data and includes an impact on time and cost. Organization entered without process in the USA data adds significant amounts of variance explained to the overall performance index, time and cost, and service delivery (not shown in the tables). If organization is entered without process in the UK data, it adds significant amounts of variance explained to the overall performance index, and time and cost. Thus, organization is relatively more important in the UK for predicting time and cost. It is relatively more important than process in the USA data for predicting overall performance and service delivery. Tools/technology has only two weak relationships, one with product innovation and quality in the USA, and another in the UK with time and cost (mostly the former). However, the coefficient for tools in multiple regression analysis may be artificially strengthened somewhat because of its correlation with system which has an insignificant correlation with time and cost. System has significantly positive effects in the US data with all performance indicators. In the UK data, the only strongly positive relationship of system integration is with product innovation and quality. By contrast, the relationship of system with time and cost is weakly negative partly because of its correlation with tools noted above.

11.6 Study III: context of service development and delivery

Study III compares methods of performance improvement in the two nations. Concurrency is a product-focused initiative. Alternatives to concurrency that are relatively more process-focused include: TQM (total quality management), ISO certification, and BPR (business process re-engineering). Each of these initiatives is significantly correlated with overall performance as shown in Table 11.1. Shown in Figure 11.2 are interrelationships among measures of the following concepts: environmental dynamism, the formalization of the product development function

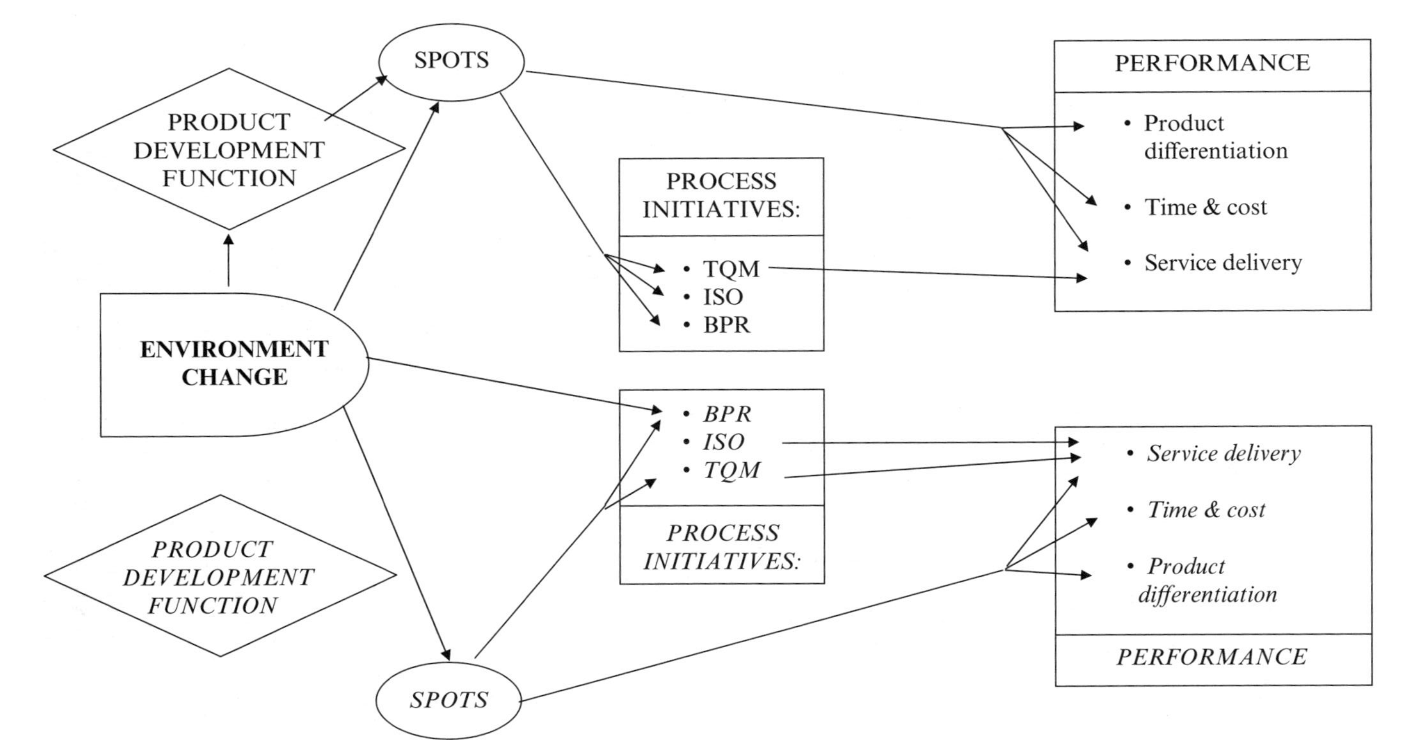

Figure 11.2 Context and performance: USA top, UK bottom

as a job, the deployment of concurrent methods as defined by SPOTS, and alternative process-focused methods of improving performance. Arrows depict statistically significant correlations in the USA data. Arrows in the bottom half depict relationships in the UK data.

The average level of SPOTS deployment in the USA was higher than in the UK. One reason may be that USA enterprises were more likely to formalize the product function as a job. Forty-four percent of companies in the USA subsample of 27 for which this question was asked had a job title for product development. Only 7 percent (a single company) in the UK subsample of 15 had such a job title. In the USA data, companies with this job title were significantly more likely not only to deploy concurrency as measured by the SPOTS index (0.52), but also to achieve high levels of performance overall (0.44), innovation and quality (0.40), time and cost (0.47), and service delivery (0.40) as shown in Table 11.1.

The formalization of the product development function appears to be pivotal in comparisons of the two nations. For example, controlling for job formalization of product development in analysis of the subsample of 42 companies for which this information is available reduces differences in practices between the two nations to insignificant levels. The deployment of concurrency in the USA data was correlated with all three process-based initiatives: TQM, ISO and BPR. By contrast, SPOTS in the UK data had a strongly significant correlation with BPR, but not the others. This suggests that USA enterprises were more likely to invest simultaneously in both product and process-based initiatives. By contrast, UK firms were significantly more likely than their USA counterparts to invest in ISO process standards.

The impact of concurrency as measured by SPOTS compared with more process-focused initiatives is shown in multiple regression analysis in Table 11.4. Results of these comparisons should be considered as exploratory because SPOTS is measured as an index of many practices while only a single scale measures each of the process initiatives. The SPOTS index has main effects on all performance measures in all samples when entered simultaneously with each process alternative controlling for environmental dynamism. TQM is positively related to overall performance and service delivery in the combined and USA samples. However, it has a weak negative relationship with time and cost in the USA sample. ISO has a weak positive effect in the UK sample for service delivery. BPR has negative effects in the total and USA samples for overall performance, innovation and quality, and service delivery. One may speculate that the implementations of BPR in the USA have been relatively more one technology-focused than in the UK. Comparing correlations of BPR in the two nations is consistent with this speculation. BPR is relatively

Table 11.4 Regression of performance measures on SPOTS and TQM

	Performance	Innovation & quality	Time & cost	Service delivery
SPOTS	**0.58****	**0.63****	**0.57****	**0.38****
	0.52**	0.58**	0.71**	0.31**
	*0.65***	*0.62***	*0.28t*	*0.47***
TQM	**0.21****	**0.07**	**−0.11**	**0.33****
	0.23**	0.11	−0.19t	0.40**
	0.13	*0.22*	*−0.11*	*0.15*
ISO 9000	**0.08**	**0.01**	**0.05**	**0.10**
	0.10	0.07	0.04	0.09
	0.10	−0.21	0.08	0.22t
BPR	**−0.21****	**−0.21****	**0.09**	**−0.26****
	−0.24**	−0.23**	0.06	−0.21**
	−0.09	*−0.18*	*0.21*	*−0.13*
Environment	**0.25****	**0.21****	**0.12**	**0.21****
	0.24*	0.32**	0.07	0.20*
	*0.28**	*0.05*	*0.14*	*0.25t*
Nation(UK = 2, USA = 1)	−0.02	−0.05	0.09	−0.05
R^2 Unadjusted	**0.56**	**0.49**	**0.38**	**0.38**
	0.55	0.59	0.48	0.40
	0.64	*0.41*	*0.22*	*0.44*
R^2 Adjusted	**0.54**	**0.46**	**0.34**	0.36
	0.51	0.56	0.44	0.35
	0.58	0.31	0.09	
F-Ratio	**21.6****	**6.3****	**10.2****	**11.1****
	15.4**	18.6**	11.6**	8.6**
	*11.2***	*4.4***	*1.8t*	*5.0***

Notes:
Top = Combined Sample (N = 108) ◆ Middle = US (N = 70) ◆*Bottom = UK (N = 38)*
** p = 0. 01, * p = 0.05, t = p 0.10 (two-tailed significance).

more correlated with tools in the USA, but more strongly correlated with strategy, process, organization and system in the UK. This inference is also consistent with observations in the qualitative studies in the USA that adopters of BPR typically focused on computer technology without including the people in the organization as part of initiatives to improve processes. TQM, and to a lesser extent ISO 9000, complement concurrency as measured by SPOTS in improving service delivery. The gap concurrency has in explaining service delivery performance suggests that its product focus may limit its applicability to delivery processes.

Environmental change is not significantly different in the two nations. In both it is significantly correlated with concurrency as measured by the SPOTS index. However, the relationship is slightly stronger in the USA than the UK. The correlation between environmental change and process-focused initiatives is insignificant in the USA data. By contrast, environmental change is strongly correlated with BPR in the UK data and more strongly so than for the SPOTS index. Response to environmental change appears to be relatively more process-focused than product-focused in the UK than in the USA data.

11.7 Conclusion

A more robust model of service development and delivery provides both a greater understanding of service systems, and application in a wide range of service contents (Tidd and Hull, 2006). Analysis of the combined data from our two samples is generally supportive of the framework we develop as being predictive of performance services. However, there are some interesting differences across the two samples, in terms of emphasis and execution of specific practices in service development and delivery.

Given that environmental dynamism in the two nations does not differ statistically, one may speculate on reasons for somewhat different responses to this stimulus for change. One possibility is that USA service companies are more focused on reducing lead and response time. This interpretation is consistent with the weak relationships of time compression with other variables in the UK data. A related possibility is that UK service firms lag in their adoption of the relevant practices, but are catching up. This view is consistent with the fact that environmental dynamism was significantly correlated with the adoption of a strategy of RRR in the UK and that this was the only one of the SPOTS constructs that was not significantly below the USA average. An alternative conclusion is that UK service companies pursue niche strategies and have adopted multiple kinds of systems to compete effectively in them, whereas their US counterparts adopt more large-scale, factory services (Berry et al., 2006). Consistent with this line of reasoning is that many more factors for practices and performances were observed in the UK data. For this reason, a separate analysis is reported in a separate paper that builds the analysis from a uniquely UK perspective without imposing the concurrency paradigm as defined by SPOTS onto the data (Tidd and Hull, 2006).

Limitations of the study include difficulty in matching the two national samples in terms of type of service and size of establishment. Although the nature of the service rendered was not found to affect results in the USA data, the possibility remains that differences in business type between the two nations is a factor. As size is a strong predictor of structuring, difference might partly account for lesser coalescence of formal product development practices in the UK data. Further research opportunities include extending the study to public sector services, as practice in this sector appears to lag that in the leading private sector service organization by many years, and is still focusing on ITC rather than other aspects of service development and delivery. Another interesting avenue would be to capture better the potential role of external inputs, consistent with the current interest in open innovation (Chesbrough et al., 2006): early user involvement in services, which is likely to be more critical and very different to the experiences and opportunities in developing manufactured products (Flowers, 2008); and of specialist outsourcing of knowledge-intensive inputs (Miles, 2000; den Hertog, 2002).

References

Atuahene-Gima, K. and A. Ko (2001), 'An empirical investigation of the effect of market orientation alignment on product innovation', *Organization Science*, **12** (1), 54–74.

Berry, L.L., V. Shankar, J.T. Parish, S. Cadwallader and T. Dotzel (2006), 'Creating new markets through service innovation', *MIT Sloan Management Review*, **47** (2), 55–63.

Bessant, J. (2003), *High Involvement Innovation*, Chichester: John Wiley & Sons.

Bessant, J. and J. Tidd (2007), *Innovation and Entrepreneurship*, Chichester: Wiley.

Brown, S.L. and K.M. Eisenhardt (1995), 'Product development: past research, present findings and future directions', *Academy of Management Review*, **20** (1), 343–378.

Burns, T. and G.M. Stalker (1961), *The Management of Innovation*, London: Tavistock.

Chase, R.B. and D.A. Garvin (1989), 'The service factory', *Harvard Business Review*, **57** (4), 61–69.

Chase, R.B. and R.H. Hayes (1991), 'Beefing up operations in service', *Sloan Management Review*, **33** (1), 15–26.

Chesbrough, H., W. Vanhaverbeke and J. West (2006), *Open Innovation: Research in New Paradigm*, Oxford: Oxford University Press.

Clark, K.B. and S.C. Wheelwright (1993), *Managing New Product and Process Development*, New York: Free Press.

Collins, P. and F. Hull (2002), 'Early simultaneous involvement in new product development', *International Journal of Innovation Management*, **1**, 1–26.

Crespi, G., C. Criscuolo and J. Haskel (2006), 'Information technology, organisational change and productivity growth: evidence from UK firms', The Future of Science, Technology and Innovation Policy: Linking Research and Practice, SPRU 40th Anniversary Conference, Brighton, UK, September.

Daft, Richard, L. (1995), *Organization Theory and Design*, St Paul, MN: West.

Damanpour, F. (1991), 'Organizational innovation: a meta-analysis of the effects of determinant and moderators', *Academy of Management Journal*, **34**, 555–590.
Den Hertog, P. (2002), 'Knowledge-intensive business services as co-producers of innovation', *International Journal of Innovation Management*, **4** (4), 491–528.
Dewar, D. and J.E. Dutton (1986), 'The adoption of radical and incremental innovations: an empirical analysis', *Management Science*, **32** (11), 1422–1433.
Dolfsma, W. (2004), 'The process of new service development: issues of formalization and appropriability', *International Journal of Innovation Management*, **893**, 319–337.
Fitzsimmons, J. and M. Fitzsimmons (eds) (2000), *New Service Design*, Thousand Oaks, CA: Sage.
Flowers, S. (2008), 'Special issue on user-centered innovation', *International Journal of Innovation Management*, **11** (3).
Garvin, D. (1995), 'Leveraging processes for strategic advantage', *Harvard Business Review*, **73** (5), 77–90.
Hage, J. (1980), *Theories of Organization*, New York: Wiley.
Hales, M. and J. Tidd (2008), 'The practice of routines and representations in design and development', *Industrial and Corporate Change*, **18** (4), 551–574.
Isaksen, S. and J. Tidd (2006), *Meeting the Innovation Challenge: Leadership for Transformation and Growth*, Chichester: Wiley.
Lawrence, P.R. and J.W. Lorsch (1967), *Organization and Environment: Managing Differentiation and Integration*, Boston, MA: Harvard Business School Press.
Leonard-Barton, D. (1992), 'Core capabilities and core rigidities: a paradox of managing new product development', *Sloan Management Review*, **13**, 111–25.
Liker, J., P. Collins and F. Hull (1999), 'Flexibility and standardization: test of a contingency model of product design-manufacturing integration', *Journal of Product Innovation Management*, **16**, 248–267.
Menor, L.J., M.V. Tatikonda and E.S. Scott (2002), 'New service development: areas for exploitation and exploration', *Journal of Operations Management*, **20**, 135–157.
Meyer, M.H. and A. DeTore (2001), 'Perspective: creating a platform-based approach for developing new services', *Journal of Product Innovation Management*, **18**, 188–204.
Miles, I. (2000), 'Special issue on innovation in services', *International Journal of Innovation Management*, December.
Mintzberg, H. and B. Quinn (1996), *The Strategy Process: Concepts, Contexts, Cases*, Upper Saddle River, NJ: Prentice-Hall.
Storey, C. and C.J. Easingwood (1999), 'The augmented service offering: a conceptualization and study of its impact on new service success', *Journal of Product Innovation Management*, **15** (4), 335–351.
Susman, G.I. and J.W. Dean Jr (1992), 'Development of a model for predicting design for manufacturability effectiveness', in G.I. Susman (ed.), *Integrating Design and Manufacturing for Competitive Advantage*, New York: Oxford University Press, pp. 207–227.
Takeuchi, H. and I. Nonaka (1986), 'The new product development game', *Harvard Business Review*, **64** (1), 137–46.
Thompson, J.D. (1967), *Organizations in Action*, New York: McGraw-Hill.
Tidd, J. (2001), 'Innovation management in context: environment, organization and performance', *International Journal of Management Review*, **3** (3), 169–197.
Tidd, J. and J. Bessant (2009), *Managing Innovation: Integrating Technological, Market and Organizational Change*, 4th edn, Chichester: Wiley.
Tidd, J. and K. Bodley (2002), 'The effect of project novelty on the new product development process', *RandD Management*, **32** (2), 127–138.
Tidd, J. and T. Fujimoto (1995), 'Work organization, production technology and product strategies of the British and Japanese automobile industries', *Current Politics and Economics of Japan*, **4** (4), 241–280.
Tidd, J. and F.M. Hull (2003), *Service Innovation: Organizational Responses to Technological Opportunities and Market Imperatives*, London: Imperial College Press.

Tidd, J. and F.M. Hull (2006), 'Managing service innovation: the need for selectivity rather than best-practice', *New Technology, Work and Employment*, **21** (2), 139–161.
Trott, P. (2008), *Innovation Management and New Product Development*, 4th edn, Englewood Cliffs, N: Prentice Hall.
Trott, P. and A. Hoecht (2004), 'Enterprise resource planning and its impact on innovation', *International Journal of Innovation Management*, **8** (4), 381–398.
Wise, R. and P. Baumgartner (1999), 'Go downstream: the new profit imperative in manufacturing', *Harvard Business Review*, **77** (September–October), 133–141.
Zirger, B.J., A. Modestom and B. Maidque (1990), 'A model of new product development: an empirical test', *Management Science*, **36** (7), 867–883.

Appendix

Table 11A.1 Types of companies in samples

Category	US	UK
Financial services	18	13
Retail banking	5	1
Credit card	3	–
Lending	2	2
Private banking	1	–
Investment services	7	10
Insurance	8	2
Consulting services	4	5
Construction	1	–
Distribution/logistics (*)	6	2
Education/training	1	0
Healthcare	8	4
Diagnostic services	4	2
Hospital	1	1
Pharmaceutical services	2	1
Manufacturing-related services (**)	4	–
Non-profit	3	1
Publishing	2	1
Retail	3	2
Travel/hotel	2	3
Telecommunications	5	2
Transportation	5	3
Total	70	38

Notes:
* Utilities, Engineering Services, Distribution of Product, etc.
** Credit, Risk, etc.

Appendix

Table 11A.2 Measures

		Total	US	UK
PERFORMANCE *To what extent have your service products changed during the past 5 years?*	*Alpha*	***0.90***	***0.94***	***0.77***
Product innovation & quality	*Alpha*	*0.86*	***0.84***	***0.88***
1. New features		2	1	1
2. Upgraded features		2	1	1
3. Higher quality		2	1	1
Time & Cost	*Alpha*	*0.83*	***0.88***	***0.71***
4. Shorter time from concept to test market of service product		3	1	4
5. Shorter time from test market to full-scale delivery of the service product		3	1	4
6. Reduced cost of service product development		3	1	3
7. Reduced cost of service product delivery		3	1	3
To what extent has the process of delivery of your service product changed during the past 5 years?		*0.87*	***0.91***	***0.75***
8. Shorter response time to order for existing service products		1	2	2
9. Shorter time for adjustments to complaints		1	2	2
10. Better after sales support services		1	2	2
11. Higher quality of delivery process, e.g., fewer customer complaints		1	2	2
12. Conformance with service product development process and procedures		1	2	2

Table 11A.2 (continued)

		Total	US	*UK*
STRATEGY *To what extent did your strategy for the past 5 years focus on:*	*Alpha*	***0.75***	***0.80***	***0.59***
1. Making major changes to existing service products		1	1	1
2. Making rapid changes to existing service products.		1	1	1
PROCESS *To what extent have you engaged in the following activities during the past 5 years in the development of service products?*	*Alpha*	***0.86***	***0.88***	***0.81***
Product development processes:	*Alpha*	*0.78*	***0.84***	***0.61***
1. Benchmarking best-in-class companies		2	2	2
2. Using structured processes for identifying customer needs and translating into requirements (QFD)		2	2	1
3. Setting performance criteria for projects		2	2	3
4. Setting standards for the performance of products		2	2	3
5. Institutionalizing systematic reviews for development projects		2	2	1
Continuous Process Improvement: *To what extent have you engaged in the following process initiatives during the past 5 years?*	*Alpha*	*0.84*	***0.86***	***0.80***
6. Mapping processes to reduce non-value activities		1	1	1
7. Improving documentation of processes		1	1	2
8. Measuring conformance with processes		1	1	2
9. Institutionalizing continuous improvement processes		1	1	2

ORGANIZATION *To what extent have you engaged in the following activities during the past 5 years?*	*Alpha*	***0.87***	***0.94***	***0.74***
Cross-functional collocation	*Alpha*	*0.75*	***0.80***	***0.66***
1. Cross-functional teaming		3	1	4
2. Collocation		3	1	4
Project-based organization	*Alpha*	*0.85*	***0.94***	***0.73***
3. Cross-training specialists		1	1	3
4. Rewarding project teams/groups		1	1	3
5. Strengthening the role of project managers		1	1	2
6. Increasing the influence of downstream functions in upstream decisions, e.g., customer service input in prod. dev.		1	1	2
7. Reorganization of jobs to reduce hand-offs		1	1	2
8. Flatter hierarchy in the organization chart		1	1	4
External relationships: *Have you changed your external relationships during the past 5 years by*	*Alpha*	*0.82*	***0.85***	***0.77***
9. Involvement of customers in decisions about service product development		2	2	5
10. Involvement of customers in decisions about delivery processes		2	2	1
11. Involvement of suppliers in decisions about service product development		2	2	1
12. Involvement of suppliers in decisions about delivery processes		2	2	1
TOOLS *To what extent have you emphasized the following kinds of activities during the past 5 years?*	*Alpha*	***0.76***	***0.83***	***0.62***

Table 11A.2 (continued)

		Total	US	*UK*
Internal technology	*Alpha*	*0.72*	***0.65***	***0.74***
1. Email		2	1	1
2. Management Information Systems		2	1	1
3. Expert Systems		2	1	3
Information distribution	*Alpha*	*0.75*	***0.80***	***0.71***
4. Distributed databases on-line to multiple functions		1	1	2
5. Common software for project management		1	1	2
6. Common software for process mapping		1	1	2
7. Building on-line databases with lessons learned and best practice templates		1	1	2
External linkages:	*Alpha*	*0.67*	***0.71***	***0.59***
1. Electronic data interchange with customers		3	2	3
2. Electronic data interchange with suppliers		3	2	3
SYSTEM *To what extent do you systematically use the following approaches?*	*Alpha*	***0.87***	***0.89***	***0.78***
1. Focus on achieving a balanced portfolio of competitive advantages for which customers are willing to pay, e.g., cost with novelty		1	1	2
2. Align competing product requirements by focusing on Voice of the Customer		1	1	1
3. View knowledge as a paramount competitive advantage to be gained from outside as well as inside the company		1	1	2

4. Transfer lessons learned from previous activities to succeeding people so that they build upon an existing base to reach ever higher future targets	1	1	2
5. Cultivate staff to provide holistic, system-wide thinking as well as specialized knowledge	1	1	1
6. Open communication channels to all functions and ranks in the organization	1	1	3
7. Act as a good partner with others, such as suppliers, external service providers, alliance partners and customers, in creating and maintaining mutual win/win scenarios	1	1	3
8. Involve customers early in the service product development process, pulling the product design in the direction of customer needs	1	1	1
9. Involve all functions throughout the development and delivery process with few hand-offs so that everyone works together reciprocally – sharing responsibility for the service product	1	1	3
SPD Function Do you have a job title for persons who are responsible for differentiating your service products from those of competitors?			
PROCESS–FOCUSED CHANGE INITIATIVES *Alpha*	*0.68*	*0.71*	*0.71*
1. TQM (Total Quality Management)	1	1	1
2. ISO 9000	1	1	1
3. BPR (Business Process Reengineering)	1	1	1

Table 11A.2 (continued)

		Total	US	*UK*
ENVIRONMENT CHANGE (sum of two sets of 6 items)	*Alpha*	*0.77*	*0.81*	*0.69*
Market Change: To what extent have your markets changed in the following ways+ absolute change regardless of direction	*Alpha*	*0.63*	*0.68*	*0.58*
1. Technological complexity of service products				
2. Rate of service product introduction				
3. Compatibility of service products with other products				
4. Customization				
5. Globalization				
6. Quality				
*Turbulence = absolute change regardless of direction (Alpha = 0.59)**	*Alpha*	*0.55*	*0.81*	*0.49*
7. Technological complexity of service products				
8. Rate of service product introduction				
9. Compatibility of service products with other products				
10. Customization				
11. Globalization (factor 2 in combined)				
12. Quality				

12 The toilsome path of service innovation: the effects of the law of low human multi-task capability

Jon Sundbo

12.1 Introduction

This chapter looks at the character of service innovation and the organisational and strategic system that produces innovations in services. The point the chapter wishes to make is that innovation in service firms is a social activity involving many actors and having many trajectories. It is a labile process that can go in many different directions and easily comes to a halt. Optimising the process (not driving it too hard, but not letting it come to a halt) is difficult as it demands extra effort from the management. The chapter discusses explanations of why this is so, and whether it needs to be so.

While this chapter is theoretical in nature, it is based on many empirical case studies of innovation in services which have been carried out since the 1980s (these include Gallouj, 1994; Andersen et al., 2000; Boden and Miles, 2000; Tidd and Hull, 2005; Toivonen et al., 2007), and the author's own contributions (e.g. Sundbo, 1996, 1997, 1998, 2008b; SIC, 1999).

Why are attempts to innovate in services apparently not very efficient, and why is radical innovation in services so rare (European Commission, 2004b)? These issues need investigating if we want to increase the beneficial effects of services in terms of growth and the radical strategic advantages of innovation to service firms (such as for example expressed in the 'blue ocean', Kim and Mauborgne, 2005). This chapter will discuss barriers to innovation, and it will be argued that the seemingly low efficiency of service innovation acts is caused by the fact that they are social processes with multiple actors and trajectories.

12.2 The character of services and service innovation

12.2.1 The classic service and its innovation process

A service is fundamentally an act which may, in varying degrees, use technology as a medium. Every human act is directed towards another human being who then will respond to the act, either by reacting or by creating an attitude towards the first actor (e.g. the customer thinks the service front employee is nice). Thus, a service is an interactive process in which a problem is solved (e.g. a dirty room is cleaned, or we gain information about how we should file our tax returns). Knowledge services are also interactive because in building our knowledge we relate ourselves to other people. Even the information we download from the Internet is elaborated through a social process. This understanding is also the core of the classical service management or marketing theory: a service is a problem-solution in which the service provider and the customer are involved (e.g. Eiglier and Langeard, 1988; Grönroos 2000).

To develop a new service is generally a complex act. It may include the act of producing a new service, the act of selling and delivering it, or it may be an organisational development which might include the front personnel behaving in a new way, new technology, new market behaviour, etc. This is the case when the service is developed within a pure service company, or in a manufacturing company. A service can change in the moment of its production and even afterwards, because it is a human act – a sociological phenomenon – in which several human beings are involved (for example the service front personnel, the manager, the customer, the customer's friends and family). Each human being has a personal will, they do not just follow a standard routine, this means that they may act in unforeseen ways during the service production and delivery, and in the quality assessment phase which takes place afterwards. The way individuals act may be the result of certain norms, attitudes or social habits gained from other spheres (such as an unconscious influence resulting from the individual's social or cultural capital; cf. Bourdieu 1994). Or it may be because the individual invents a new way of doing things. Human beings are creative – in contrast to things and animals.

Even sociological phenomena other than the ones which are of interest to classical economics – price and an objective quality (characteristics of the result of the service act that can be specified objectively and measured quantitatively) – can be involved in the service, such as power and aesthetics or ethics. Who should decide how the new bathroom should be made – me as a customer or the plumber? Who should decide our human resource management (HRM) strategy – me as the manager or the consultant? Aesthetics and ethics may also be issues that consumers contemplate concerning goods. In this respect, there may not be any difference between manufacturing and services. However, in manufacturing when the good has been designed it has found its final form. A service can be changed throughout the interactive production process and even after delivery (cf. Sundbo, 2008a). Therefore services are difficult to standardise.

The production of services is also a complex process of social interaction, as the service is, and it is close-knit with the product, namely the service. In manufacturing, which has standardised production processes, the production process and the product are separated. Thus the production is more social in services than in manufacturing.

In services there are rarely methods and traditions that can delimit, define and maintain the new service, unlike those that exist in manufacturing such as engineering methods and traditions for the exact description of the product. In manufacturing, the patent is an example of such an exact method that can maintain the focus of what the firm is attempting to innovate. By contrast, a service is more fluid and the firm may have difficulties in defining and maintaining what it is attempting to innovate. Maintaining focus is problematic for two reasons, namely the complex character of the service innovation and the large number of actors and knowledge trajectories that may be involved in developing it. This makes surveying and managing innovation difficult.

The universal-rational theories propounded by economics may be too narrow for explaining the development of services; even Nelson and Winter's (1982) evolutionary theory that emphasises routines may be too rigid to explain service production and innovation. Thus one may have to turn to more sociologically inspired interpretations of innovation such as those offered by Karl Weick which stress how actors attempt to create meaning with every act (Weick, 1995).

12.2.2 Patterns of service innovation

As stated above, the complex character of services and service innovation means that many actors and trajectories can be involved. Examples include:

- The users (customers) as a source of innovative ideas.
- The employees as a source of innovative ideas and acting as corporate entrepreneurs.
- Technological trajectories – i.e. new technology and technology suppliers.
- Service trajectories – new ideas about how to carry out the service activity.
- Management trajectories – new management ideas.
- New values and priorities in society.

The development of a new service could take its point of departure in each of these actors or trajectories. Other actors are involved later in the innovation process. This can lead to different patterns of innovation processes. These patterns have been described as (Sundbo and Gallouj, 2000):

1. The classic R&D pattern:
   a. Fordistic variant;
   b. neo-industrial variant.
2. The service professional pattern.
3. The organised strategic innovation pattern.
4. The entrepreneurial pattern.
5. The artisanal pattern.
6. The network pattern.

The innovation process may even change its pattern. New knowledge, interaction between employees, customers and managers, management decisions and individual interests can make the pattern change. An entrepreneur can give up or can emerge in the process. This makes service innovations even more complex.

12.2.3 Service innovation is a labile process

Analyses of service innovation nearly always observe that the process is not systematic; innovations arise out of practice, not from R&D; they are often small improvements and rarely radical; process and product innovations are close-knit with organisational and market innovations (Gallouj, 2002; Sundbo, 1997; European Commission, 2004b). Such innovations are often fluffy, and the development of an innovation is not structured. Many actors and knowledge resources are involved. This is not just to say that it is necessarily an open source process (cf. Chesbrough, 2006). Nor is it necessarily the kind of rational process that we know from engineering and military actions. The innovation process is often a labile social one.

In most cases, the economic consequences are unknown, for example service firms rarely use rational project management tools. Each of them has their set of, often differentiated, interests and problems. Service innovation is not an isolated, clearly goal-oriented process based on a plan of what to realise that can be expressed physically such as a manufacturing research and development (R&D)-based innovation can (at least in the view of some theoretical frameworks, e.g. Schmookler, 1966; Christensen, 1997).

The danger of service innovation is very often that the process stops, not because of a decision, but because the organisation and the management 'forget' that there was an innovation. This may happen because there is rarely a specific manager or department with the responsibility for maintaining innovation processes.

Often the employees and managers are supposed to undertake the innovation activities simultaneously with other activities without having any specific time allocated to innovation activities. It is to some degree up to the employee or the manager himself. The process may stop because the persons need to do other jobs, typically to solve a production problem. For example, if quality problems in the delivery to a customer arise and everybody must abandon their current activity to participate in solving that problem, then the innovation process dies for a period before being reinvigorated. Sometimes the employees and managers forget to reactivate the innovation process. Organisations may also observe lost opportunities if, for example, a competitor has launched a similar product or market conditions have changed. Sometimes organisations reactivate the process, but in the meantime new knowledge or new ideas have arisen and the innovation goes in another direction and may in some cases be almost completely changed, thus the final innovation is very different from the original idea. This is another reason for never being sure of what the outcome of a service innovation process will be – even if, in some firms and some situations, the innovation process may run as planned and with military precision. It cannot always be predicted.

Thus the service innovation process is labile, unpredictable and fluid; on occasions this is so much so that it is difficult to say whether or not there is an innovation at all.

12.2.4 The law of low human multi-task capability

This labile character of the service innovation process could be explained by a law of low human multi-task capability. The foundation for this law might be the fact that human beings are social animals who have difficulties in maintaining their focus on a task when they are in groups. The assumption underlying this argument is that a group has several processes running simultaneously and several interests. Other interests might include departmental power, individual prestige and position, short-termed production and delivery care, personal expressive tendencies such as friendship-making, becoming a popular colleague and hope of future rewards of a social or economic kind – all of these might detract the focus away from innovation.

Further, even individuals in the organisation, both employees as well as managers, have difficulties in maintaining their focus on specific tasks in a multi-task situation. Even if the human being is a very intelligent and creative creature, it has difficulties in multi-tasking. Thus, if innovation is one task amongst others, individuals have difficulties in balancing the different tasks – when to do what and with how much effort. The situation becomes even worse if several innovation projects are running simultaneously. Therefore, the social process form of innovation, which is the most common form in services, becomes very labile. Innovation is much easier when there is only one agenda such as in rational R&D-based innovations or entrepreneurship, where only one individual follows their goal and interests and ignores other processes and goals.

Innovation processes are thus much more focused and much more likely to be maintained when they are either based on one particular individual, the entrepreneur (whether they establish their own firm or operate within an organisation – an intrapreneur; cf. Pinchot, 1985), or incorporate a very planned and controlled process of the military kind such as the R&D-based development of industrial products (e.g. Cooper and Edgett, 1999). When the innovation process is a complex social one, as it usually is in services, it can very easily stop momentarily or permanently, or it may go unforeseen ways. This is caused by the law of low human multi-task capability. The human being is, as all living species, routine-oriented. The same attitudes, acts, structures and habits are repeated generation after generation. The functions and structures of social groups or societies are predictable.

People in organisations mostly follow their own interests and goals, though maybe not only them; they can also follow the interests and goals of the organisation or the firm, but they always have an element of self-interest. Further, individuals have difficulties in manoeuvring in new social relations as is the case in innovation processes because the innovation, even if it is a product innovation, will change social positions and relations. The individual cannot know what will be 'in it for me' and therefore has difficulties in deciding the right course of action.

However, modern human beings have left behind the simplicity of former societies and communities as a framework for their acts and attitudes. They are getting used to a changing world – even though it takes generations. Individuals are being transformed from being security-seeking and routine-oriented to orienting themselves in a labile world. Instead of striving for re-establishing old routines and structures, they strive for realizing an individual strategy (cf. Giddens, 1984) that can guide them through all the changes so that they themselves benefit in the end. Often they do not succeed, but they make attempts. If we want to understand service innovation processes, we must understand this situation.

If a change comes from outside (for example a natural disaster) or from inside (for example from an entrepreneur who can be compared to mutation of cells in nature), the structure, function and attitudes of a social group will change as a reaction to the change (e.g. Luhmann's, 1995, theory of social systems). The reaction can either be to re-establish the old, well-known and secure situation (as is often the case following natural disasters) or to follow the new tendencies (for example to come first to a new stage so as to get benefits from it). People can act in both ways in development of the innovation. As mentioned, there is natural resistance in organisations towards change (e.g. Hage, 1980; Weeks, 2004). Thus, people will often try to resist the innovation attempting to restore the old and well-known state of the art. They may also be followers of the innovation process, hoping that it will bring (social or economic) benefits to them. The first reaction is predictable and it is even possible for the management to influence and plan it. The latter very easily becomes unpredictable.

Here, I am talking of an innovation as a social process within an organisation. This form of innovation is promoted by open innovation (Chesbrough, 2006); however the specific problems of the typical service innovation are not directly caused by the open source. They are caused by the specific nature of social processes.

12.3 Peculiarities to services or a general tendency?

Is this toilsome way of innovating a specific characteristic of services? It is a characteristic of traditional service innovations while manufacturing innovations have been supposed – and observed – to follow a more unilinear, rational R&D-based innovation development (more the military kind of operation). The latter has, in theory, been supposed to be more efficient and economically beneficial and to lead to more radical innovations (Freeman and Soete, 1997; Christensen, 1997).

However, things are changing. Manufacturing industries are increasingly facing market conditions where the launching of radically new goods is not sufficient to obtain substantial competitive advantage. Markets are increasingly satisfied. It has become ever more uncertain and confusing what manufacturing firms should do to survive and grow. Goods must be combined with services and even experiences (cf. Pine and Gilmore, 1999). Process innovations, which have always been developed through more complex social processes under the law of low human multi-task capability, have become more integrated with product innovations. Since services and experiences are combined with goods, the goods-producing process cannot be separated from the process of developing service and experience innovations (cf. Sundbo, 2008a) and delivering services. The customers are increasingly becoming political, thus they are interested in how goods are produced and under what circumstances. This means that they are interested in the production process, which can no longer be separated from the product in the mind of the customers. Thus, even in manufacturing, process and product innovation are melting together. Employees and customers (or users) are increasingly being involved in the innovation process, which consequently becomes more toilsome.

Thus the issues discussed in this chapter are becoming increasingly relevant for understanding innovation in manufacturing, agriculture and other parts of the primary sector as well. However, it is neither the aim of this chapter, nor the *Handbook*, to discuss manufacturing innovation. Therefore, we just point to this interesting similarity here and continue with the main topic.

Services are also facing the problems of the toilsome innovation process and service firms are attempting to get more of a grip over the process. Managers have become interested in systematising, steering and rationalising the innovation process, which can be very expensive, often using much of the employees' and managers' time. This is not easy. It is not easy to systematise and steer the social processes or introduce other types of innovation processes in services. The complex process also has some advantages that the managers do not want to abandon. This problem and what it can lead to will be discussed in the following sections.

12.4 Creativity and employee and customer focus as the central elements of the complex social innovation processes

The complex, labile innovation processes in services is not as efficient as it ideally could be because too many resources are used in the slack and labile process. The innovations are very incremental and thus do not have a great effect on competitive advantage as do radical innovations. Service firms miss the benefits that systematic research could give. Are there any advantages to be gained from the complex, labile innovation processes in services? Or are they just a sign of human inability to do the right thing?

Both views are right. The research concerning the service innovation process demonstrates that firms are highly unable to systematise and steer it. The theoretical arguments have been presented above. However, case studies have also shown that there are some advantages to the complex social innovation process. These advantages could even be beneficial in other sectors such as the manufacturing and primary (agriculture, fishing, forest) sector, thus they are general advantages. They are twofold.

12.4.1 Creativity-in-the-process

The first advantage is 'creativity-in-the-process'. By this I do not mean the initial creativity that takes place in science that could be the basis for the R&D activities, or the first idea for a new service. These are, of course, also important creative activities, but they are part of the R&D model, creativity tools, or any other model of systematic innovation process. Here I am referring to the creativity involved in solving problems in the later stages of the innovation development. If the innovation process is not just a practical implementation of a given R&D result, but the development of an idea the details and market potential of which are not given, the innovative idea needs to be adjusted and renewed during the process. An example of an idea of a service innovation could be a new type of travel insurance (health, lost luggage, etc.) combined with a travel agency, experience-oriented information (for example via the Internet) about tourist destinations, cafes with music, food and other authentic experiences gained from the tourist destinations. The idea is imaginative. There may even be some R&D behind the idea, for example legal considerations and calculations of risk probability.

However, in the process of realising the idea, many unknown problems and issues emerge. For example: should the insurance company establish a travel agency in-house or outsource that part of its activities? How should the story about this concept be told to the market to guarantee market success? How do we cook the food and play the music from, for example, the Seychelles? (Do they have any authentic music at all? Who knows about it?) It might be that the idea of an insurance company selling trips and tours and experiences would, during the innovation process, reveal itself to be bad because the insurance company's image is unchangeable and one of being boring. Should the project then be abandoned? All these questions need to be answered. Doing so requires creativity, which is the ability to ask untraditional questions, find new knowledge and combine it into solutions of the problem. Creativity at this stage is most efficient if it is a collective process within the organisational framework (Amabile, 1996). Providing a systematic formula for such a state of affairs is difficult; such innovation demands the effort of several individuals (employees, managers or other actors). Such effort is more likely to emerge if the innovation process is a social one in which many employees, managers and other actors are supposed to participate than if it is a top-down military-style process. Creativity-in-the-process is an advantage, because it may fine-tune the innovation or because it leads to radical changes in the process, thus an innovation that runs into difficulties may be replaced by another one.

12.4.2 Customer involvement

The other advantage of the traditional complex and labile innovation process in services is the involvement of the users of the service product, the customers. Since the service products (as all other contemporary products) must be sold on a market that is already fundamentally satisfied, it is very difficult to achieve market success via an innovation. It is very rarely the case that the service (or any other product) is of fundamental importance for the customers' life. They must see the service as an interesting offer that they will accept instead of using their money on other interesting (but not fundamentally necessary) products. Therefore, the more and the earlier the customer and their preferences can be involved in the innovation process, the better. There are only rare cases of innovation in which customers have been intensively involved from the beginning or where the customer has been allowed to carry out the innovation (as suggested by von Hippel, 2005). The relation to the customers usually goes via the employees and managers (Sundbo, 1998). They meet the customers and know – to varying degrees – the customers' lives, problems and wants. The customers' wants or needs cannot be directly observed; the customers are often not aware of them themselves. Therefore the innovating firm must rely on the best informants they can find, and that is in most cases their employees.

The employees are not only the ones who get ideas and, perhaps, are creative in the first phase of the innovation process, they also act as corporate entrepreneurs (Drucker, 1985; Kanter, 1983; Sundbo, 1998) (or intrapreneurs, Pinchot, 1985). Often they take ownership of the idea as an innovation. They communicate with other employees and managers and argue their case to convince them about the idea. Generally, several employees or managers are quickly involved in the innovation process as developers or supporters (champions, Pinchot, 1985). The process becomes a collective one, not individual entrepreneurship. The management often involves these intrapreneurs in the subsequent development and implementation of the innovation. Sometimes the intrapreneurs involve themselves.

The innovation process can be political (cf. Hage, 1980). One group or department may attempt to realise the innovative idea while others fight against it. These others may be afraid of losing their power or jobs. Sometimes the intrapreneurs even make alliances with external parties such as customers, experts or political administrators. The top management may often not have control over these political processes, and sometimes the top manager may not even know that an innovation process is going on, and neither therefore that there are organisational-political fights about it. The top manager may know about the innovation just before it is to be implemented, or they may not even know about it at all (particularly if it is a large company). This means that the innovations are not tested for their coherence with the firm's overall strategy and business policy, nor indeed its chances of being successful on the market. The driving force behind the innovation may be a group of employees' or a manager's desire rather than the overall interest of the company. This is problematic from a rational point of view as it leads to inefficient innovation processes. However, innovation is a risky business and it is difficult to pick the successes and the fiascos prior to their being placed on the market. However, these counter-intuitive innovations may nonetheless be market successes. Again, the value of the complex innovation system in services is difficult to assess. The undirected and anarchistic elements are generally a disadvantage, but can in some situations be an advantage.

Because of the deep involvement of the employees in innovation in the services, the firm gets an optimal customer orientation in the innovation process, at least if it is managed in the right way. The customer orientation can of course be left to a marketing department. It can contribute to the innovation process, but the employees in other departments encounter the customers on a daily basis and learn to know them in a more comprehensive way than the formal marketing departments can do. The employees often encounter the customers in services, because the service is co-produced by the service front personnel and the customer (cf. Eiglier and Langeard, 1988).

12.4.3 High success rate in unstable markets

These elements of the labile and complex service innovation process rarely lay the ground for radical innovations – those will probably be produced by a clear push-system based on new scientific results or where there is an attempt to realise one person's radical idea. However, in a world where radical innovation is very difficult, and therefore rare, these elements assure a higher success rate for the incremental innovations.

These characteristics of the service innovation process may not be advantages in stable markets, where the advantages of a steered, systematic innovation process (cost control, short development time, etc.) clearly outdo the creativity-in-the-process and customer-orientation advantages of the complex and labile service innovation process. This would also be traditional economic logic. However, if the market is turbulent and changing, the logic may be different. If the customers' preferences are uncertain and continuously changing, if the needs that products and services should fulfil are more or less redundant or at least a luxury (in relation to subsisting), demand does not necessarily follow the traditional economic laws. It might be better for the firm to invest in the more expensive and uncertain complex form of service innovation process.

The empirical research is not decisive as to which innovation system is the most efficient. We may cautiously conclude that the complex system has some advantages, but the lack of systematisation and steering also produces some disadvantages. The theoretical standpoint thus could be that a more systematic service innovation process would increase efficiency, but the advantages of the complex and labile process should not be abandoned. This leaves us with a combination or optimisation situation. Some aspects of systemisation should be added to the complex system if the service firms are to attempt to run their businesses optimally. In the next sections I will discuss what these aspects are, based on tendencies that can be seen from the case studies.

12.5 Standardisation or industrialisation and the use of technology

One might argue that service producers will attempt to standardise or industrialise services (standard products produced in a standardised way) (as done in Sundbo, 2002). The argument can be made from the point of view of economics. Industrialisation will reduce costs and thus produce a competitive advantage on the market. The case for industrialisation can even be argued from a production point of view because it implies more technology. New technology, not least information and communication technology (ICT), is developing and it is natural for service firms to include that in the service products and the production and marketing process. Traditionally, technology has led to industrialisation because it has the potential for systematic, less man-demanding production systems or even automation.

The tendency of service standardisation can lead to self-service: the customer produces the service themselves by using a technology (cf. Gershuny, 1978). Examples are the do-it-yourself repairing of houses, filling in your tax return by using web-based accountancy expert programmes, and fully automated hotels where the guest takes the key from a machine. In self-services there is no interaction between the service personnel and the customer. The customer may have some interaction with other customers, but that is – considered from the traditional market norms in our societies – outside the sphere of the producer. Thus, the producer avoids all the trouble with the complex social processes that are difficult to steer.

The attempt to standardise or industrialise services can also be found in the tendency to introduce objective, quantitative measures of service quality. In classical service marketing theory, service quality is considered to be a result of a social process: the customer is satisfied with the process. The satisfaction is caused by the technical solution of the problem, but also by the fact that the customer is satisfied with the outcome of the social interaction process (they won the power contest, got a pleasure experience, etc.). The classical service marketing theory (e.g. Lovelock, 1988; Grönross, 2000) suggests using this fact to please the customer and thus make them pay more and even become a loyal customer. This customer relations marketing (e.g. Gummesson, 1999) has been a core part of service theory. However, even if authors holding this position argue that this is the best practice, we can see that many service providers (public as well as private) introduce more standardised quantitative quality measures (cf. e.g. Deming, 1986) (for example in the academic world, the number of articles has become a measure of the quality of research, which also is a service; the subjective user satisfaction is not the most conspicuous quality measure).

This observation may lead to a viewpoint of services as just the production activities that we have not succeeded in standardising and technologising. Service is just a transition or a residual (e.g. Cohen and Zysman, 1987). But is this so? Should service production be standardised and industrialised as fast as possible and the customer's involvement abandoned as in manufacturing? Should the innovation process in services be systematised and removed from the anarchistic influence of customers and employees such as is the ideal in the R&D laboratory system of manufacturing?

This position is not acceptable in its extreme form. First, because it cannot explain or be useful to all service production (i.e. that which can not be industrialised). Second, because it will not, in the long term, be useful to either the individual service producer, or society. The industrial system only provides increased turnover and profit to a manufacturer for a short period of time; thereafter it has led to market saturation, price competition, bankruptcies and stagnation. It has not broken the law of the product life cycle (Vernon, 1966). For a society – the citizens – this industrial development has not necessarily led to greater happiness. Happiness is not only the satisfaction of all material and intellectual needs, it is also experiences of social processes (for example to win a fight over power or to get an experience flow; cf. Csikszentmihalyi, 2002 such as when one climbs a mountain). Thus, the social interaction aspects of the service and the service production have a social value in themselves and, therefore, also an economic value. Complete industrialisation of services and a complete scientification of the innovation process with laboratories, abstract R&D work and systematic push-processes is probably not the solution to the service sector's development problems. Thus, total industrialisation cannot explain the development of the innovation practice in service firms.

However, the service firms may benefit from introducing some elements from the industrial systematic, R&D-based innovation process into service innovations.

12.6 Service engineering as the salvation?

Thus the innovation process in services has traditionally been rather unsystematic, complex and unpredictable. This is a problem for economics and managers. For economics it is a problem because the service firms do not seem to strive for the most profitable outcome in their innovation processes, and that goes against traditional economic theory and wisdom. This issue will be discussed in this section. The managerial problems will be discussed in the next section.

Are there other tendencies (than the ones discussed above) suggesting that the innovation processes in services develop in a more organisational and rational-economic direction? Yes, at least in a direction that seems more organisationally rational. For a long time there has been a tradition of service operations (Voss et al., 1985; Heskett et al., 1990), which was inspired by the tradition of industrial engineering. This is an attempt to systematise service production and delivery; thus the procedures become planned and standardised. The discipline of service operations has inspired some practice-oriented researchers and practitioners to create systematic instruments for innovating new services. The service innovation or development process has been analysed, and structured models with different tools for design of new services have been provided (e.g. Edvardsson et al., 2000). This includes tools such as blueprinting (Shostack, 1987), new product development models (Edvardsson et al., 2000; Meyer and DeTore, 2001) and advanced quality securement systems (Edvardsson et al., 1994). Such instruments may be combined with more rational project management instruments which have been developed and used by engineers. By combining the development of new services with project management tools, the service firm gets more control over the expenses of the innovation process.

Lately the way of thinking inspired by industrial engineering that we know from manufacturing has been further developed into the idea of a service engineering field (particularly in the USA, and Germany which has a long manufacturing tradition) (e.g. Hefley and Murphy, 2008; Stauss et al., 2008). This is an attempt to standardise the service design process: thus certain methods and ways of thinking should be followed. The goal of the innovation process will be clearer and the costs will be under control. Innovation will be a craft instead of a free, and somewhat anarchistic, social process. The rational method can be taught in schools and universities, and thus competencies of a rational organisation of the service innovation process can be widespread. One might even think of particular practice-oriented academic research and education service-engineering institutions similar to the engineering schools or technical universities of the manufacturing period (cf. Hefley and Murphy, 2008). Service engineering is not yet fully developed as a discipline, thus we cannot say what it will look like in the future. However, it is a serious suggestion for a discipline that can make the service innovation process more rational.

Service engineering leads to industrialisation and more systematic and controlled innovation processes in services. As it is inspired by technical disciplines, it will also emphasise technology more than has previously been the case in services. Of course there is technology in services, but fundamentally a service is behaviour: a service is an act that the service operator carries out to solve a customer's problem. Technology has not been conspicuous in the service management and marketing tradition (e.g.

Grönroos, 2000). Although many service innovations, particularly contemporary ones, are based on technology, many are not technological, but involve the introduction of some form of new behaviour (e.g. European Commisssion, 2004a; SIC, 1999). However, technology, particularly ICT, is used in services. For example, all services provided via the Internet are based on ICT. The combination of this fact and service engineering attempts leads to a greater focus on technology. The behavioural elements will then recede into the background. The innovations will be technological and the new services will increasingly be technology-based.

A tendency towards developing self-service products may be connected to this development. Self-service means that the customer themselves will carry out the necessary behaviour to solve their problem by using self-service instruments which are technological (a physical tool). Such instruments are most obvious within knowledge services such as accountancy, information services, education and entertainment services. The Internet, mobile phones and other ICTs provide extensive new possibilities for developing self-services.

Other technologies (e.g. chemicals for cleaning, kitchen technology in restaurants etc.) could also be emphasised more. They could also be used for providing self-services for manual functions such as repairing things, cleaning and so forth. However, as earlier argued (Sundbo, 2002) this tends to remove the business for the service firm, or at least transform it to a manufacturer of the technology. Thus, as Gershuny (1978) has argued, manual self-service does not necessarily improve the service business. Further, self-service means that the service operation is left to the customers. This again means that to develop the right service for the potential customers, the service firm needs to involve the customers in the innovation process if the service is to fit their needs and lifestyle. This may also be relevant to manufacturers, but it is obviously relevant to service producers, even self-service producers.

When the customers are involved, or the service innovators should take the customers' future behaviour, wants and attitudes into consideration, a fluffy element is again introduced into the innovation. Rationalisation, systematisation and control become difficult and the innovation process becomes labile. Of course, the service provider can choose to rely only on the internal service engineering and thus have a fixed, systematic innovation process. If the service is a self-service, the service provider can squeeze the employees as parties because they will not be in direct contact with the future customers. However, the service provider will – at least according to the current trend of user-based innovation (e.g. Sundbo, 2008a; von Hippel, 2005) – have an interest in involving the customers or knowledge about the customers. If the service is a non-self-service one, even the employees are relevant parties for involving in the innovation process, because the new service will be created via their behaviour. Thus we are back to the labile innovation process.

Service engineering and industrialisation of services may solve some of the problems of the labile and unsystematic innovation process in services, for example maintaining the focus on the innovation process so that it is not forgotten. However, it will not solve all problems and if the innovation process is carried out following rules of service engineering only, the innovations might be less successful. Service is fundamentally behaviour or a way of thinking (if it is knowledge services). The behaviour and way of thinking of the customers and employees should be taken into consideration to ensure the success of the innovation attempts, and human behaviour and ways of thinking are not easy to systematise and put into an engineering formula.

12.7 Strategic reflexivity as managerial solution

Innovation processes in services are seldom deterministic systems where the process and outcome can be predicted (not even a dissipative chaotic system; cf. Fuglsang and Sundbo, 2005). However, more systematic and regular service design methods such as service engineering are emerging, and therefore the innovation process in the future may be more systemic and deterministic. As argued above, it might be dangerous for a service firm to move too far in that direction because the valuable customer and market focus then disappears. This complex situation creates a problem for the management of service firms. Innovation in services is a toilsome process which is difficult to maintain because it is not focused; several innovation processes may run simultaneously and managers in services normally have many tasks besides managing innovation. This complexity is a challenge for the manager as it demands a complex managerial approach. It is not sufficient to manage a clearly defined innovation project or an R&D department. Many actors, knowledge-providing acts, customer relations etc. must be taken into consideration. The management also needs to rely on intrapreneurship and decentralized management (the employees and employee–customer relations are managed by middle managers or even by the employees themselves).

Another managerial problem involves restricting the innovative ideas (cf. Sundbo, 1996). The anarchistic elements of the service innovation process entails that many new ideas are often presented as the employees' creativity and entrepreneurship is high. If more systematic engineering design of new services run parallel, a number of ideas developed from different logics are presented. The firm cannot use all these ideas, not even let the employees – or the R&D or engineering department – use resources on developing too many ideas as it could create a deficit, even if some successful innovations evolve. The service firm should also maintain its production and service delivery, and often it is the same people who carry out the production and the innovation activities. It may be difficult for the manager to assess the value of the many innovative ideas and to decide whether to continue or stop each innovation process. Further, managers have difficulties in multi-tasking and maintaining the focus of the single innovation project.

The management then has a toilsome task in managing the service innovation process. So what do they do? Managers generally resort to a general strategic management solution (Sundbo, 2001). To assess all the ideas, they create a general business policy or strategy for the service firm and let that be the guide for assessing the ideas. The strategy may even be a determinant or guideline for innovations because many innovation processes are planned by the top management (which, however, may easily go wrong, as with all other innovation processes). Today, almost all firms have a strategy – which can be more or less structured or plan-oriented (Mintzberg, 1994). The strategy does not need to be very rational and plan-oriented, it only serves as a guideline for the top management and thus they, at least, have some framework for managing the innovation process. Such strategies also have the advantage of being very market- and customer-oriented, at least according to some strategy theories (e.g. Kotler, 1997). If that is the case, a success factor for service innovation, namely the market's acceptance, is integrated into innovation management.

The strategic management of innovation not only has the purpose of regulating the stream of ideas and intrapreneurship from employees and managers. It also has the purpose of managing the more systematic and planned service design (Edvardsson et al., 2000) that the service engineering approach makes possible. Formal innovation departments must be established and their task defined. Such departments can be found in service firms (e.g. Sundbo, 1998). They are rarely research or even development departments. Service firms have had a weak tradition for doing research or basing their innovations on research. Although the service engineering approach has led to an emerging interest in defining and developing a specific service science (e.g. Maglio and Spohrer, 2008; Hefley and Murphy, 2008), the content and focus of this discipline are not yet clear. It contains elements of the old service marketing or management discipline (e.g. Grönroos, 2000), old service design methods (e.g. Edvardsson et al., 2000), and new engineering design elements particularly based on ICT. The innovation departments are thus a kind of strategic organisation. Their task is mostly to regulate, communicate and inspire others, to promote, and restrict, the types and number of innovations. These departments are often closely related to the strategy department, which is often a staff function closely related to the top manager. The departments may also be closely related to the marketing department since service innovations are considered to be very customer- or market-based (and not based on ideas developed in an internal service laboratory).

Since service innovation processes are so complex and employees as well as customers – and, sometimes, service design laboratories – must be included as actors in the process, innovation management in services becomes complex (cf. Tidd et al., 2001). The top manager may easily get confused and tired, even if they rely on the strategy. Case studies show that the employees often do not grasp the essence of the strategy and thus do as they like. From a managerial point of view, it is desirable that the employees and departments internalise a balanced and strategically determined creativeness and enterprise. This will lead to innovative activities being auto-acting – autopoietic (cf. Maturana et al., 1980 and Luhmann's, 1995, theory). Bessant (2003) has suggested such an organisational system based on a step-wise internalisation of strategic entrepreneurship in the organisation.

Unfortunately, the innovation task cannot be solved by inducing intrapreneurship and an understanding of the strategy into the organisation. The world is changing rapidly and the market is changing with it. Many innovative initiatives fail because the market rejects them. The strategy may even be outdated faster than the management thinks it will be. Therefore, the strategy and the innovation policy and guidelines are under continuous consideration in the most successful service firms. The service firm practises what has been termed strategic reflexivity (Sundbo and Fuglsang, 2002; Sundbo, 2008b). Often the employees are involved in this consideration because they meet the customers on a daily basis and thus know what is going on in the market. The innovation attempts are opportunities to test the market and the strategy, thus strategic reflexivity is connected to the innovation process. The autopoietic process will include strategic reflexivity, thus it will be an auto-acting system. It may be a utopia that the service organisation itself can carry out an efficient autopoiesis which would make it much less toilsome for the management. The strategic reflexive organisation will probably always require managerial regulation. However, as the managers gradually become aware of the autopoietic strategic reflexive principle and the actual strategy, the managerial task will be easier.

The strategic reflexive management of service innovation, which supports a complex innovation process, does not create the most efficient innovation system seen from the point of view of economic theory. Many resources are wasted and the innovation process is not as goal-oriented and rational as it could be. The process is not the most manageable and therefore not the most wanted by managers. The complex innovation process and strategic reflexivity is nevertheless the optimal innovation system in service firms because services must be socially accepted by potential future customers. Services should solve the customers' problems in a way that fits with the customers' lives in general. Innovations thus must fulfil the actual needs of the customers. The interaction between employees and customers in the service encounter (van Looy et al., 1998) provides more sustainable innovations in that they are surer of being accepted on the market. The user-oriented and less top-down planned innovations in services – what has been called pull-based innovation – combined with a certain element of push in the form of service engineered design is the optimal way of organising innovation in services. At the general level, nobody can say what the balance between the push and pull elements should be. Probably we will never establish a general law concerning this. Thus the management must decide the balance of the innovation system based on incomplete information. Management will always be necessary and management of service innovation will remain a toilsome task. However, the more awareness of the complexity of this task and the more practical experience is accumulated and communicated, the easier the task will be and the closer the innovation management will come to the ideal optimum.

One comfort for the management is that the risk of a service innovation project is on average much less than that of a project of innovating a manufacturing product. Service innovations are on average not as radical and profit-creating as in manufacturing, but the losses by a failure are on average less.

12.8 Conclusion

The innovation of services is a complex and toilsome task. There seem to be no easy solutions such as seeking a patent or establishing an R&D department. The management must find a balance between 'hard' engineering methods and 'soft' focusing on the human beings. Research and consultants can provide certain tools and contribute to maintaining the innovation process, but nobody can generally tell what the optimal balance between the 'hard' and the 'soft' methods is and how the innovation process should be organised.

It is difficult to maintain the focus of individual innovation projects and to balance the push and pull elements, or even to decide how much innovation and which innovative ideas should be promoted.

Innovation processes in services are generally unsystematic and not research-based. The emerging service engineering discipline provides some tools to get a better grip on the process and make it more systematic.

However, the success of the innovation also depends on social factors, i.e. that the new services fit into the life of the customer (whether business customer or consumer). Thus it is not optimal to engineer the service innovation process completely.

The solution for service firms is strategic reflexivity. This solution is not necessarily the most rational seen from the point of view of economic theory, but it is the most optimal when the social factors are taken into consideration. It makes innovation management hard and toilsome. However, more experience about the best ways to organise the innovation process will make their task easier. Service research can contribute to this by investigating the service innovation process and providing general models and tools.

References

Amabile, T. (1996), *Creativity in Context*, Boulder, Co: Westview.

Andersen, B., J. Howells, R. Hull, I. Miles and J. Roberts (eds) (2000), *Knowledge and Innovation in the Service Economy*, Cheltenham, UK and Northampton, MA, USA: Edward Elgar.

Bessant, J. (2003), *High-Involvement Innovation*, Chichester: Wiley.

Boden, M. and I. Miles (eds) (2000), *Services and the Knowledge-Based Economy*, London: Continuum.

Bourdieu, P. (1994), *Raisons pratiques. Sur la théorie de l'action*, Paris: Seuil.

Chesbrough, H. (2006), *Open Innovation*, Boston, MA: Harvard Business School Press.

Christensen, C.-M. (1997), *The Innovator's Dilemma: When New Technologies Cause Great Firms to Fail*, Boston, MA: Harvard Business School Press.

Cohen, S. and J. Zysman (1987), *Manufacturing Matters*, New York: Basic Books.

Cooper, R.G. and S.J. Edgett (1999), *Product Development for the Service Sector*, Cambridge, MA: Perseus Books.

Csikszentmihalyi, M. (2002), *Flow*, London: Rider.

Deming, W.E. (1986), *Out of Crisis*, Cambridge, MA: MIT Press.

Drucker, P. (1985), *Innovation and Entrepreneurship*, New York: Harper & Row.

Edvardsson, B., A. Gustafsson, M. Johnson and B. Sanden (2000), *New Service Development and Innovation in the Service Economy*, Lund: Studentlitteratur.

Edvardsson, B., B. Thomasson and J. Øvretveit (1994), *Quality of Service*, Maidenhead: McGraw-Hill.

Eiglier, P. and E. Langeard (1988), *Servuction*, Paris: McGraw-Hill.

European Commission (2004a), *Innovation in Europe*, Luxembourg (European Communities).

European Commission (2004b), Innovation in Services: Issues at Stake and Trends, INNO-Studies 2001: Lot 3 (ENTR-C/2001), Bruxelles (EU Commisssion).

Freeman, C. and L. Soete (1997), *The Economics of Industrial Innovation*, London: Pinter.

Fuglsang, L. and J. Sundbo (2005), 'The organizational innovation system: three modes', *Journal of Change Management*, **5** (3), 329–344.

Gallouj, F. (1994), *Economie de l'innovation dans les services*, Paris: L'Harmattan.

Gallouj, F. (2002), *Innovation in the Service Economy*, Cheltenham, UK and Northampton, MA, USA: Edward Elgar.

Gershuny, J. (1978), *After Industrial Society?* London: Macmillan.

Giddens, A. (1984), *The Constitution of Society*, Oxford: Policy.

Grönroos, C. (2000), *Service Management and Marketing*, Chichester: Wiley.

Gummesson, E. (1999), *Total Relationship Marketing*, Oxford: Butterworth-Heinemann.

Hage, J. (1980), *Theories of Organizations*, New York: Wiley.

Hefley, B. and W. Murphy (eds) (2008), *Service Science, Management and Engineering*, Berlin: Springer.
Heskett, J., W.E. Sasser and C. Hart (1990), *Service Breakthroughs*, New York: Free Press.
Kanter, R.M. (1983), *The Change Masters*, London: Unwin.
Kim, W.C. and R. Mauborgne (2005), *Blue Ocean Strategy*, Boston, MA: Harvard Businesss School Press.
Kotler, P. (1997), *Marketing Management*, Englewood Cliffs, NJ: Prentice Hall.
Lovelock, C. (1988), *Managing Services*, Englewood Cliffs, NJ: Prentice Hall.
Luhmann, N. (1995), *Social Systems*, Stanford, CA: Stanford University Press.
Maglio, P.P. and J. Spohrer (2008), 'Fundamentals of Service Science', *Journal of the Academy of Marketing Science*, **36** (1), 18–20.
Maturana, H.R., F.J. Varela and S. Beer (1980), *Autopoesis and Recognition: The Realization of the Living*, Dordrecht: D. Reidel.
Meyer, M. and A. DeTore (2001), 'Perspective: Creating a platform-based approach for developing new services', *The Journal of Product Innovation Management*, **18** (3), 188–204.
Mintzberg, H. (1994), *The Rise and Fall of Strategic Planning*, New York: Free Press.
Nelson, R. and S. Winter (1982), *An Evolutionary Theory of Economic Change*, Cambridge, MA: Belknap.
Pinchot, G. (1985), *Intrapreneuring*, New York: Harper & Row.
Pine, B.J. and J.H. Gilmore (1999), *The Experience Economy*, Boston, MA: Harvard Business School Press.
Schmookler, H. (1966), *Invention and Economic Growth*, Cambridge, MA: Harvard University Press.
Shostack, L. (1987), 'Service positioning through structural change', *Journal of Marketing*, **51** (1), 34–43.
SIC (Service Development, Internationalisation and Competence Development) (1999), 'Danish Service Firms' Innovation Activitites and Use of ICT', Report no. 2, Centre of Service Studies, Roskilde University, Denmark.
Stauss, B., K. Engelmann, A. Kremer and A. Luhn (eds) (2008), *Service Science*, Berlin: Springer.
Sundbo, J. (1996), 'Balancing empowerment', *Technovation*, **16** (8), 397–409.
Sundbo, J. (1997), 'Management of innovation in services', *Service Industries Journal*, **17** (3) 432–55.
Sundbo, J. (1998), *The Organisation of Innovation in Services*, Copenhagen: Roskilde University Press.
Sundbo, J. (2001), *The Strategic Management of Innovation*, Cheltenham, UK and Northampton, MA, USA: Edward Elgar.
Sundbo, J. (2002), 'The Service Economy', *The Service Industries Journal*, **22** (4), 93–116.
Sundbo, J. (2008a), 'Customer-based innovation of knowledge e-services: the importance of after-innovation', *International Journal of Service Technology and Management*, **9** (3–4), 218–233.
Sundbo, J. (2008b), 'Innovation and Involvement in Services', in L. Fuglsang (ed), *Innovation and the Creative Process*, Cheltenham, UK and Northampton, MA, USA: Edward Elgar, pp. 25–47.
Sundbo, J. and Fuglsang, L. (eds.) (2002), *Innovation as Strategic Reflexivity*, London: Routledge.
Sundbo, J. and F. Gallouj (2000), 'Innovation as a loosely coupled system in services', *International Journal of Service Technology and Management*, **1** (1), 15–36.
Tidd, J., J. Bessant and K. Pavitt (2001), *Managing Innovation*, Chichester: Wiley.
Tidd, J. and F. Hull (eds) (2005), *Service Innovation*, London: World Scientific Publications.
Toivonen, M., T. Touminen and S. Brax (2007), 'Innovation process linked with the process of service delivery: a management challenge in KIBS', *Economies et Sociétés*, **8** (3), 355–384.

van Looy, B., R. Dierdonck and P. Gemmel (eds) (1998), *Services Management*, Harlow: Prentice Hall.
Vernon, R. (1966), 'International investment and international trade in the product cycle', *Quarterly Journal of Economics*, **80** (2), 190–207.
von Hippel, E. (2005), *Democratizing Innovation*, Cambridge , MA: MIT Press.
Voss, C., C. Armistead, R. Johnston and B. Morris (1985), *Operations Management in Services and the Public Sector*, Chichester: Wiley.
Weeks, J. (2004), *Unpopular Culture: The Ritual of Complaint in a British Bank*, Chicago, IL: University of Chicago Press.
Weick, K. (1995), *Sensemaking in Organizations*, Thousand Oaks, CA: Sage.

13 Customer integration in service innovation

Bo Edvardsson, Anders Gustafsson, Per Kristensson and Lars Witell

13.1 Introduction

Services are one of the main bases for profitable businesses today. In a service-driven economy, companies try to increase their competitiveness through service innovations that create value for existing customers, attract new customers and at the same time produce shareholder value (e.g. Edvardsson et al., 2000; Gustafsson and Johnson, 2003). Service innovation can be divided into three categories: (1) type of innovation (service-level innovations); (2) the management of innovation (firm-level innovations); and (3) the innovation context (sector-level innovations). We will focus on the second category of managing firm-level service innovations, i.e. creating and managing innovative firm-level services by integrating customers in the innovation process.

By the innovation process, we refer to the various phases from idea generation to market launch with at least some acceptance of the new service in the market. By service innovation, we refer to new services for an organization and further development of existing services as well as services 'new to the world'. In other words, we do not distinguish between incremental innovations and radical innovations. The perspective we offer in this chapter is service innovation for, by and with customers. With customer integration in service innovation, we want to highlight the various roles that customers play in the service innovation process; a higher degree of customer integration means a change from service innovation for the customer to service innovation with the customer.

By customer involvement, we mean being proactive and 'getting close to customers' in order to learn from and with them beyond what traditional focus groups, observations, questionnaires and interviews can provide. Our point of departure is the significant business potential from integrating customers and learning with them throughout the innovation process. The customer as a co-innovator or co-creator of new service is a growing concept in service research, with relevance not only for service companies and public service providers but also for manufacturing companies in the transition from product orientation to service orientation.

In this chapter, we will share some of our insights from various projects carried out at the CTF-Service Research Center, Karlstad University, Sweden. The empirical foundation is a survey of Swedish service firms as well as a number of case studies and experiments carried out in cooperation with multinational companies such as Telia, Ericsson and Whirlpool. We begin by defining customer integration and explaining its importance for service innovation. We then discuss the customers' role in value creation through service and continue with different views on the possible roles of the customer in service innovation. We continue with examples from Swedish companies of customer integration in service innovation and finally, we present some guidelines about how to manage customer integration in service innovation.

13.2 Growth of services

Regardless of the statistics of development of gross domestic product (GDP) that we study, we find that the service sector is growing in both size and importance. The growth of services is a reflection of fundamental changes in both our cultures and economies. Consider the four forces illustrated in Figure 13.1. First, individuals have less time available to shop and do things on their own as the percentage of working women and single-parent households continues to grow. One result is more meals out, more in-home services and less time spent shopping for physical goods as a means to an end. Rather, people are more willing to trade time for money to buy services and experiences directly.

A second force is a growing availability of easy-to-use and affordable technology that allows customers to perform services themselves when and wherever possible. Paying bills, investing and shopping online and at home are increasingly common ways to cope with the growing complexity of our lives.

A third force contributing to our move towards services is a business climate in which organizations are striving to focus on core competencies and therefore outsourcing those services that they cannot cost-effectively provide.

Fourth, we believe that the move towards services is a fundamental reaction to the evolution of competition. This competition forces companies to evolve over time from competing on product value, to competing on service value, to competing on solution value or 'linked activities' and co-created service experiences. These linked activities are a constellation of related products and services that provide customers with an integrated solution to a particular need or problem. Whether selling cake mixes or mini-vans, one eventually finds intense competition in every niche of the market as it matures. Even the niches become host to similar competitors

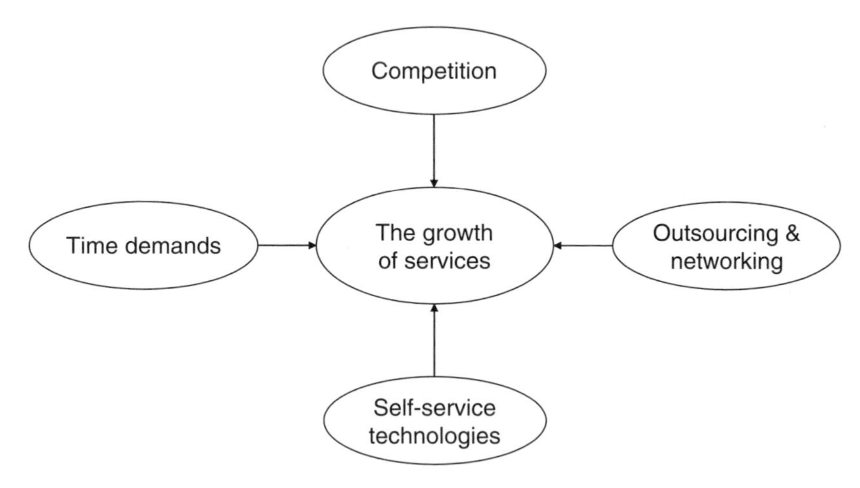

Figure 13.1 Factors driving the growth of services

where quality is a given and price is the main way to compete. This evolution has forced companies to look downstream, or down the value chain, to the possibility of competing with the services and experiences that surround their products. Rather than sell the cake mix, you bake, decorate and deliver the cake. As these new services continue to develop and link together in a value proposition, these linked activities or service constellations become a critical source of competitive advantage.

13.3 The changing role of the customer

The role of the customer has changed and employees no longer meet the customer on a daily basis in many businesses. The traditional face-to-face service interactions have been replaced by technology-based encounters. Many service innovations have the consequence of a changing role for the customer in service production. A paradox arises because the new technology creates a distance between the company and its customers, i.e., customers do not interact with employees – they meet technology. Technology not only increases the distance between customers and employees making it more difficult for employees to understand the customer, but also influences customers' ability to articulate what they need and want – they do not understand the possibilities and limitations that a complex technology may convey. The changing role of the customer in service production also has consequences for the role of the customer in the development process of products and services. Technology makes it possible to touch and learn from customers' actual behaviours over time and on an individual level. To understand the customer, it is no longer sufficient simply to conduct interviews or surveys; the customer must become an active participant and co-creator in the development process of new products and services. As a result, employees can better understand the customer and therefore the customers' potential as a source of new ideas can be better utilized. Consequently, customer integration has been suggested as one such new and important way of listening to the customer and translating customer information into value-creating offerings (Alam, 2002; Edvardsson et al., 2006). In the literature, customer integration generally denotes everything from a company using questionnaires, i.e., gathering of customer information, to an innovation developed by customers, i.e., lead user approach. There is limited knowledge about better means for understanding the customer. There is a need for future research, especially in the relationship between the implementation of such means and the financial performance of an organization. One more advanced mode of customer integration is customer involvement. By customer involvement, we mean being proactive and 'coming close to customers' in order to learn from and with them throughout the service innovation process and beyond what traditional focus groups, observations, questionnaires and interviews can provide.

Our goal is to be as close as possible to the customers' real life, not only when a new service is tested but also during the early phases of the service innovation process. A company needs to have access to the sticky information, i.e., information from the customers' everyday life that is difficult to transfer to a company because their actions may not be triggered by deliberate thought. The information may only emerge as a customer is using a service, but they may not think of it afterwards. We discuss this in more detail in the next section.

13.4 Rationales for customer integration in service innovation

The rationale for involving customers in the service innovation process is improved service performance and ultimately better business performance (Kristensson et al., 2004). A number of general studies conducted about why different product development processes succeed or fail support this rationale. Of these studies, Rothwell (1976) and Maidique and Zirger (1984) identified intensive communication with the customer as the main determinant of new product success. When investigating time-to-market, Gupta et al. (1986) found that poor definitions of product requirements were stated by 71 per cent of the respondents as the primary reason for product development delays. When asked how to improve and accelerate the new product development (NPD) process, respondents emphasized, among other things, the importance of early market testing and active customer integration.

In a largely refereed study, Rothwell (1976) shows that good design and successful new products (maximized fit to customer needs) are linked to early and close consultation with customers and an understanding of their 'real' and for them important needs. Von Hippel (1976, 1982) highlights the key role of the user in the innovation process with his findings that users have invented a majority of the innovations in his study. In addition, Rothwell and Gardiner (1983) emphasize that customers often play an active role as contributors of innovative ideas and relevant feedback in an iterative development process. Maidique and Zirger (1984, 1990) report close contact with leading customers as one factor that influences outcome positively. Finally, Alam (2006, 468) argues that: 'customer interaction in new service development has a positive impact on the performance of new services'.

There are at least three motives for customer integration in the service innovation process. First, what makes it difficult to understand customer needs is the 'sticky' and difficult-to-transfer information that the customers and their context possess (von Hippel, 2001). Customer involvement means an opportunity to learn from, with and about individual customers. Market-oriented companies have mainly focused on satisfying the expressed needs of the customer, typically by using verbal techniques such as focus groups and customer surveys to gain understanding of the use of current products and services (Slater and Narver, 1995). It has been claimed that the result has been minor improvements rather than innovative thinking and breakthrough products or services, but few empirical studies support this opinion. Focusing on hidden or latent needs, however, offers a high potential for differentiation, individualization and competitiveness. It is difficult for organizations to access, understand and meet the latent needs of the customers by using the traditional research methods. They 'impose fixed attribute descriptions and scaled response categories to obtain standardized and comparable responses from large samples' (Day, 2004b, 243–244). Furthermore, customers have trouble imagining and giving feedback about something they have not experienced (e.g., Leonard and Rayport, 1997; Ulwick, 2002; Veryzer, 1998; von Hippel, 1986).

Customer involvement facilitates organizational learning (Neale and Corkindale, 1998; Sinkula, 1994); as knowledge about the market increases, it becomes an intangible resource embedded within the firm, making it an important competitive advantage difficult to imitate (Day, 1994b). Organizational learning occurs as individuals acquire intelligence, share the intelligence throughout the organization, achieve a shared interpretation of the intelligence, alter organizational behaviours based on the shared interpretation and further develop and renew existing products over time (e.g. Day, 1994a; Sinkula, 1994; Slater and Narver, 1995). Organizational learning is valuable to a firm and its customers because it supports the understanding and satisfying of customers' expressed and latent needs through new products, services and ways of doing business (Day, 1994a; Sinkula, 1994). Organizational learning makes it possible not only to create products before competitors, but also to create them before the recognition of an explicit customer need (Hamel and Prahalad, 1991; Slater and Narver, 1995). Consequently, it is argued that a firm's long-term performance is dependent on both customer satisfaction with the present offering and the development of new products and services (Flint et al., 2002; Slater and Narver, 1995).

Customer involvement attends to the core of the marketing concept, i.e., assessing customer needs and increasing the organization's understanding of user value (Wikström, 1996; von Hippel, 1994, 1998; Nonaka and Takeuchi, 1995). Alam (2002) identified six objectives of user involvement: superior and differentiated services, reduced cycle time, opportunity for user education, rapid diffusion, improved public relations, and building and sustaining long-term relationships. Customer involvement may also facilitate idea generation (von Hippel, 1986; Shaw, 2007; Kaulio, 1998), reduce time-to-market (Alam, 2002; Gupta et al., 1986), and enhance user competence (Prahalad and Ramaswamy, 2000). In summary, customer involvement is important to product and service success; the question is, however, how and when best to involve customers in the innovation process. In addition, the extent to which customer involvement can lead to project success is still unknown.

13.5 The customers' role in value creation through the service-dominant logic

There are many trends in the academic field in terms of how companies should relate to their customers. Researchers suggest that companies should develop relationships with their customers (e.g. Grönroos, 2000), become market-oriented (e.g. Kohli and Jaworski, 1990; Narver and Slater, 1990), or become service-centred through adopting a service-dominant logic (SDL) (Vargo and Lusch, 2004a, 2004b, 2008). What distinguishes the management of these theories is the increasing importance of communication. Customer relationship building, market orientation and service-centred theories emphasize two-way communication through better listening to customers and interactivity. The idea is communication before, during and after transactions with customers. It also implies that organizations continually need to collaborate and learn with and from customers in order to respond to their current and future needs. The strong emphasis on customer communication and interactivity may be especially worth noticing since much of the past focus in marketing and innovation literature has been preoccupied with personal characteristics rather than situational characteristics (such as communication). Von Hippel's lead user theory (1986), for example, focuses on special individuals being of particular interest for firms, which also is the basic idea behind traditional models of market segmentation. The increasing focus on communication and interaction suggests that situational factors (such as communication) may be of greater importance than mere personal characteristics when it comes to understanding customers for purposes of innovation.

SDL suggests that customers create value through service experiences and relationships, especially in the co-creation and sharing of resources such as skills and knowledge with customers, partners and suppliers (Vargo and Lusch, 2004a; Lusch et al., 2007). Companies offer value propositions (promises) and marshal resources for customers. Customers use the offered resources together with their own resources to co-create services that render value. The goods-dominant logic (GDL) suggests that companies develop and offer products or services that have value, while SDL focuses on how value is co-created with and by customers. The difference between GDL and SDL has profound theoretical and managerial implications for service innovation.

SDL is inherently customer-oriented and relational. Operant resources used for the benefit of the customer place the customer in the centre of value creation, implying relationship dynamics and a new way of organizing value co-creation that includes the customer as an active player and resource in the value-creation system. Organizations exist to combine specialized competences into complex services that provide desired solutions and attract customer value-in-use. SDL suggests that service innovation is about developing value propositions and prerequisites for customers so that they can co-create value for themselves by providing resources with their knowledge and skills, resulting in attractive customer value and favourable customer experiences.

In summary, SDL emphasizes the key and active role of the customers in co-creating service. Their individual knowledge, skills experiences and values are important for understanding service and how the service is assessed on the basis of value-in-use in the customer's own context. Adopting SDL instead of GDL has major implications for the management of service innovations. Customer integration in the innovation process becomes natural in the design of value propositions, service concepts and service processes. Service developers must pay attention to the role of the customer to realize customer value – as defined and experienced by the customer – while in GDL. Service innovation pertains to the designed-in quality of value propositions, service concepts and service processes as defined by the provider.

13.6 Organizational views of the customer

The organizational view of the customer reflects a company's attitude towards its customers. There are several possible roles of the customer both in service production (Lengnick-Hall, 1996) and in service development (Alam, 2002). At one end of the continuum, the customer is viewed as a mere product with little to contribute and the organization is seen to have all the expertise. For certain services or during certain phases of a development process customers may actually have little to add in terms of knowledge. At the other end of the continuum, a customer is viewed as a resource that possesses knowledge critical for the development of future products and services. In this respect, there may be strongly held beliefs that customers are competent and an important ingredient in the production or development of service. In summary, due to the distance between these two views it seems important to note that the view of the customer may vary across different time-periods, services or parts of an organization.

13.6.1 The customer in service production

Gersuny and Rosengren (1973) provided one of the first models covering the different roles of the customer. They argued that the customer could have four different roles: resource, worker (co-producer), buyer or beneficiary (user). Building on this work, Lengnick-Hall (1996) concludes that the customer can be viewed as a resource, a co-producer, a buyer, a user or a product. The customer as a resource and the customer as a co-producer are input-based roles because they directly or indirectly influence the operations and outcomes of an enterprise. The other three roles are on the output side of the system.

As a resource, the role of the customer has mainly been to supply information and/or wealth. Customers are often the raw material of the production process and the more impersonal the service process, the more discretionary the number, intensity and continuity of the customer-resource contacts. If customers provide information that is incomplete or inaccurate, or if they make commitments of funds or time that they do not meet, their input reduces service performance.

The customer as a co-producer means that the customer is a co-creator of value when utilizing a service. The more customers who are co-producers, the more influence they have on the quality of the work processes. In these cases the production is largely dependent on the knowledge, motivation and experience of the customer. If customers know what they are intended to do and how they are supposed to do it, they are more likely to perform well.

Lengnick-Hall (1996) has three views of the customer as output: (1) the customer as a buyer; (2) the customer as a user; and (3) the customer as a product. These three views all focus on constructs related to service delivery such as expectations, customer satisfaction and intention to buy.

Customer participation has been defined as 'the degree to which the customer is involved in producing and delivering the service' (Dabholkar, 1990, 484). In relation to this definition, Meuter et al. (2000) distinguish among three types of customer participation: company production, co-production and customer production (wordings changed from Meuter et al., 2000). Company production implies a product made entirely by the firm and its employees, with no participation by the customer. Co-production implies both the customer and the firm's contact employees interacting and participating in the production. Customer production is a product made entirely by the customer, with no participation by the firm or its employees.

Some of these customer roles are specific to service production, while some of them are more general and applicable in the context of service development. In addition, there are consequences for the choice of role for the customer in the production process as to what and how the customer can contribute to the service innovation process.

13.6.2 The customer in service innovation

Within development work, additional models have been developed viewing the customer as buyer, subject of interest, provider of information, expert and co-developer (e.g. Finch, 1999; Nambisan, 2002). Alam (2002) suggests a model based on the degree of communication between the organization and the customers throughout the development process. In his model, the intensity of the communication ranges from passive acquisition of input, information and feedback on specific issues and extensive consultation with customers, to representation of customers in the development team. It is explained that organizations that treat their customers only as users will lose out to other firms that integrate their customers in a variety of roles that expand and deepen the relationship.

Voss (1985) suggests five categories of customer integration:

- User developed, not transferred: the user handles all stages of the innovation process and distributes the innovation commercially.
- User developed, transferred: the user handles most stages of the innovation process and then the supplier handles commercial diffusion.
- User innovation, supplier developed: the user recognizes the need and generates the idea for the solution and then the supplier handles the rest of the process.

- User initiated supplier innovation: the user expresses the need and the supplier handles the rest of the process.
- Supplier innovation: the supplier handles all four stages without user involvement, except of course later as a commercial customer.

It is noteworthy that the main source of idea generation in the first three categories is the user, while the supplier is the dominating party in the last two categories. As idea generation is likely to be the determining activity for the subsequent progress of the service innovation process, it seems likely that if the user takes on an active role in the early phases and generates ideas then this participation will be a defining occurrence in a service innovation project (e.g. Kristensson et al., 2004).

Voss (1985) takes the innovation as his point of origin, but we prefer to look at the organizational view of the customer. Based on the models previously mentioned, we suggest a classification system showing a gradually changing view of the customer ranging from the customer as buyer, the customer as a subject of interest, a customer as a provider of information, the customer as co-developer and finally the customer as developer.

When customers are seen as buyers, the company also sees them as passive recipients of a new service. The company may therefore have a technology-push belief, or they may believe that service innovation is driven by their own ideas or capabilities, created in the absence of any specific need that customers may have. In technology-push situations, innovations are created and then appropriate applications or user populations are sought that fit the innovation. Methods used are often forms of internal idea generation that rely on the know-how of the research and development (R&D) department as the source for new services.

When customers are viewed as subjects of interest, an organization uses passive information such as customer complaints or sales-force knowledge. In these cases the development is more driven by the things gone wrong than by future opportunities. The logic is that the company is not actively searching for information; instead, it passively waits for feedback. Often, a company is stuck in day-to-day activities and has difficulty finding time to create service innovations. Commonly used methods are customer comment cards, problem detection studies and critical incident studies.

When customers are seen as providers of information, traditional market research techniques are often used as a viable means for gaining knowledge about their needs. Common market research techniques include in-depth interviews, surveys or focus groups where a company questions a customer about present needs. Another common scenario involves the company testing almost-finished prototypes in order to understand what final steps need to be taken ahead of launch. Some companies would argue that using market research implies a form of co-creation with customers. This is seldom the case, however, as often the company has come a long way in its development work and the customer only speaks when spoken to (i.e., the customer takes a reactive role in the innovation process).

When customers are viewed as co-developers, there is a change in their role in the service innovation process from being reactive to proactive. Often the customer is involved earlier in the service innovation process and the involvement can be carried out over several phases of the development process. Companies and customers have joint roles in education, shaping expectations and co-creating market acceptance for products and services. Customers are part of the enhanced network; they co-create and extract business value. They are collaborators, co-developers and competitors at the same time. Successful innovations come from matching technical knowledge about a certain platform with knowledge about usage (where value occurs). Since needs arise from within the operating conditions that surround a user, it seems reasonable that users are competent enough to co-produce the service innovations of tomorrow. After all, the user represents a substantial part of the knowledge that is crucial for innovation.

From a company perspective, customers as developers seem to take over the responsibility of service innovation. It can be a rewarding strategy since the costs of service innovation are low or non-existent, but it is also a risky strategy. Taken too far, customers can become more knowledgeable than the company specialists and start to develop innovations without the company and start to market, distribute and sell the innovation themselves. Linux is a good example where the customers are developers. It started in 1991 with an e-mail from Linus Torwald, a young student in Finland. He asked for reactions and feedback on an idea for a new computer operative system, based on open source. Linux had a market share of 6.8 per cent in 1997, 24 per cent in 2003 and it is still growing. Companies such as IBM, HP, Intel, Volvo and Motorola are now using Linux. A network of customers forms a group of innovators and innovation takes place on a continual basis. The system is developed based on the users' needs, solutions to their problems and, particularly, the customers' expertise.

13.6.3 Relationships between the customer's role in service production and innovation

We argue that there should be a relationship between the role of the customer in production and the potential role of the customer in service innovation. When there is a change in the process of service production, for instance, the customer takes over a larger responsibility of the workload through technology, which ought to be followed by a change in the service innovation process. There are several reasons to pursue such a change. First, by changing from company production to co-production or customer production, the knowledge of the developers is reduced and the distance between the company and the customer increases. Second, the number of service encounters is decreasing, therefore limiting the number of face-to-face interactions. Fewer face-to-face interactions mean less opportunity to learn from the customer and possibly weaker customer relationships.

One example of a change in service production is IKEA and its creation of a kitchen planner. By using the kitchen planner, a customer can test-drive the kitchen before purchase and consumption. The customer can build different models of a kitchen suited to the measurements in their kitchen and test these models in a virtual environment. As such, a customer's interaction with a hyperreal service can create an experience which is more distinct and clearer than the reality we know, i.e., a hyperreality (Edvardsson et al., 2005). If this service is only used for the customers' benefit, then the change occurs only in the production process. If information is gathered in a systematic way, however, then letting customers test-drive service experiences can be used in service innovation (e.g., Edvardsson et al., 2005; see Figure 13.2).

In an example of microwave ovens, Whirlpool changed the customers' role in the innovation process while leaving the production process unchanged. To be able to get customers to generate new ideas about functions, services and features related to a microwave oven, customers were allowed access to a new microwave that was not available in stores. Customers received a bag with instructions about how to use the microwave, a camera, a diary, a bag of popcorn and a cake to bake in the microwave. Customers were told to use the new microwave oven for a week and during this time write a diary about how they used it. Each time a customer had an idea related to how to buy, use or dispose of a microwave oven, the customer was supposed to write this idea in a specific section of the diary. At the end of the week, the microwave ovens and the diaries were collected. During one week, 30 customers were enabled to use a new microwave oven. During this period, the customers generated 108 ideas related to microwave ovens (see Figure 13.2).

One illustration of a change in both dimensions is the introduction of the Lego Digital Designer in 2004. It is an Internet service where the customer can build Lego in a 3D CAD (three-dimensional computer-aided design) program on-line. The customer can create a customized model in the program, see the cost of it and order it online. As a result, the production process is changed from company production of Lego models to

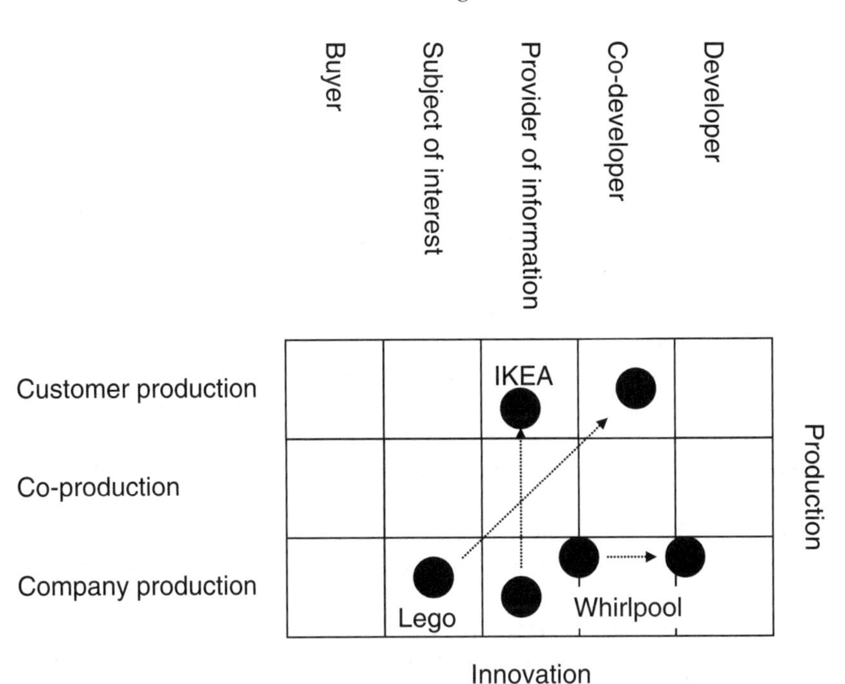

Figure 13.2 Changes of the role of the customer in service production and innovation

customer production. Customers create new, virtual models by interacting with one another online. The innovations are available on Lego's website, and other customers can add suggestions and input. The virtual models are further developed by other customers. Some models are considered innovative and well designed and can be mass produced and marketed by Lego. The customer then receives royalties. This service enables a change at Lego from '100 designers to 100,000 innovators' (Tidd and Bessant, 2009). The 100,000 innovators are new and existing customers. In this case, the change in service production was followed by a change in service innovation, where the role of the customer changed from subject of interest to co-developer (see Figure 13.2).

13.7 Guidelines on how to manage customer integration in service innovation

In this chapter, we have described the rationale for customer integration and we have shown how customers can be involved throughout the service innovation process. We have shown that learning from and with customers to capture and assess ideas, understand latent needs through sticky information and test-drive innovative service is fruitful and profitable – but not easy. There is still much to be learned about how best to integrate customers, which customers to select, what methods to use and when to involve customers in the innovation process. There is a clear trend in most industries and especially among the leading and successful companies to come closer to their new and existing customers. Companies can do this by using proactive methods to learn more about how to increase realized value-in-use in the customer's context, the customer's behaviour, drivers of favourable customer experiences, value creators and value destroyers as defined by customers and important customer values. Integrating various customers into the service innovation process will provide useful and critical information to make sure that the resulting new services will be 'wanted', 'chosen' and 'preferred'. Customer involvement in the 'service design shop', however, is no guarantee for commercial success but rather one important success factor in a service-driven economy. We will now conclude with five guidelines for customer integration in service innovation based on our research and the literature on service development and service innovation.

First, in the modern service-driven economy, customers trade time for money to buy services and experiences directly. The new services, often based on self-service technologies, must be renewed in a structured and planned way to result in value-in-use over time. The emphasis on renewal will be even more important in the future, especially in the market for business-to-business (B2B) services. Customers can be involved in the renewal with, e.g., ideas on what is most important for further development. Companies and other organizations must plan how to renew the service and service experiences over time to arrive at favourable customer experiences and stay competitive.

Second, the service-dominant logic emphasizes the key role of customers in the co-creation of service as well as service assessed on the basis of value-in-use in the customers' own context. The customers' knowledge, skills and experiences have a major impact on perceived customer value. Therefore, new service cannot require more of the customers than they are willing or able to do within their role as co-creators of the service. Since the customers often do not know how to use the full potential of a service, customer competence development or developing the customers' context should be seen as an important area in service innovation. This development will broaden the scope of how service innovation is framed and how new challenges and opportunities are introduced, making it even more natural to involve customers. Developing the customers' competence, resources and capabilities to use services to improve realized customer value should be emphasized more often.

Third, there are differences in organizational views about the customer in the innovation process. We have described five different roles: a buyer, a subject of interest, a provider of information, a co-developer and a developer. We suggest that companies and other organizations should be able to understand and manage these roles and that they complement one another. We claim that customers as providers of information are often not enough in the service innovation process, especially when the goal is to create new kinds of experiences for customers. Customers should also be involved as co-developers of personalized experiences. Customers are part of the enhanced network; they co-create and extract business values. They are collaborators, co-developers and competitors. Understanding various customer roles is crucial for long-term success in service innovation. Companies and other organizations must understand and manage various customer roles as they complement one another; close and in-depth integration of customers throughout the innovation process is important but at the same time challenging.

Fourth, when the basic requirements are met, favourable service-driven customer experiences become a powerful competitive weapon in service competition. Service innovation already focuses on new, individualized and attractive service experiences and will continue even more so in the near future. Test-drives or simulated service contexts allow customers to be involved in the creation of such services and to test-drive the service before purchase and consumption. In the case of IKEA, the stores are designed as a service landscape (servicescape) to host experience rooms, e.g., living rooms, kitchens or bedrooms. The furniture and other items are means to ends, resources and enablers for 'solutions to real life problems at home' (Edvardsson et al., 2005). Customers can experience the solutions in the store, in the catalogue and at the website. The customers can test-drive the solutions with their own furniture, size of kitchen, etc. using the simulation-tool kitchen planner provided by IKEA. The customer can also get help and advice from an architect or other expert in the store. Make it possible for customers to create their own, individualized service and to test-drive the service before purchase and consumption.

Fifth, technology makes many new opportunities possible for customer integration in service innovations. In this chapter, we have mentioned the creation of hyperrealities, simulations, service test-drives and open source. Technology also makes it possible to track and store data about customers' actual behaviours over time on websites, in various chat rooms and on blogs. These somewhat new sources are becoming more and more important in service innovation and customers can play an important role in interpreting and translating the information into action in service innovation. It is possible to capitalize on new sources of in-depth and fruitful customer information by integrating customers as interpreters and translators during various phases of the service innovation process.

References

Alam, I. (2002), 'An exploratory investigation of user involvement in new service development', *Journal of the Academy of Marketing Science*, **30** (3), 250–261.

Alam, I. (2006), 'Removing the fuzziness from the fuzzy front-end of service innovations through customer interactions', *Industrial Marketing Management*, **35**, 468–480.

Dabholkar, Pratibha (1990), 'How to improve perceived service quality by improving customer participation', in B.J. Dunlap (ed.), *Developments in Marketing Science*, Cullowhee, NC: Academy of Marketing Science, pp. 483–487.

Day, George S. (1994a), 'Continuous learning about markets', *California Management Review*, **36** (4), 9–31.

Day, George S. (1994b), 'The capabilities of market-driven organizations', *Journal of Marketing*, **58** (4), 37–52.

Edvardsson, B., B. Enquist and B. Johnston (2005), 'Co-creating customer value through hyperreality in the pre-purchase service experience', *Journal of Service Research*, **8** (2), 149–161.

Edvardsson, B., A. Gustafsson, M. Johnson and B. Sandén (2000), *New Service Development and Innovation in the New Economy*, Lund: Studentlitteratur.

Edvardsson, B., A. Gustafsson, P. Kristensson, P. Magnusson and J. Matthing (eds) (2006), *Involving Customers in New Service Development*, London: Imperial College Press.

Finch, B.J. (1999), 'Internet discussions as a source for consumer product customer involvement and quality information', *Journal of Operations Management*, **17**, 535–556.

Flint, D.J., R.B. Woodruff and G.S. Fisher (2002), 'Exploring the phenomenon of customers' desired value change in a business-to-business context', *Journal of Marketing*, **66** (4), 102–117.

Gersuny, C. and W.R. Rosengren (1973), *The Service Society*, Cambridge, MA: Schenkman.

Grönroos, C. (2000), *Service Management and Marketing*, New York: John Wiley & Sons.

Gupta, A.K., S.P. Raj and D. Wilemion (1986), 'Model for studying R&D–marketing interface in the product innovation process', *Journal of Marketing*, **50** (2), 7–17.

Gustafsson, A. and M.D. Johnson (2003), *Competing through Services*, San Francisco, CA: Jossey-Bass.

Hamel, G. and C.K. Prahalad (1991), 'Corporate imagination and expeditionary marketing', *Harvard Business Review*, **69** (4), 81–92.

Kaulio, M.A. (1998), 'Customer, consumer and user involvement in product development: a framework and a review of selected methods', *Total Quality Management*, **9** (1), 141–149.

Kohli, A.K. and B.J. Jaworski (1990), 'Market orientation: the construct, research propositions, and managerial implications', *Journal of Marketing*, **54** (2), 1–18.

Kristensson, P., A. Gustafsson and T. Archer (2004), 'Harnessing the creative potential among users', *Journal of Product Innovation Management*, **21** (1), 4–14.

Lengnick-Hall, L. (1996), 'Customer contributions to quality: a different view of the customer-oriented firm', *Academy of Management Review*, **21** (3), 791–824.

Leonard, D. and J.F. Rayport (1997), 'Spark innovation through empathic design', *Harvard Business Review*, **75** (6), 102–113.

Lusch, R.F., S.L. Vargo and M.O. O'Brien (2007), 'Competing through service: Insights from service-dominant logic', *Journal of Retailing*, **83** (1), 5–18.

Maidique, M.A. and B.J. Zirger (1984), 'A study of success and failure in the product innovation: the case of the US electronics industry', *IEEE Transactions on Engineering Management*, **31** (4), 192–200.

Meuter, A.L., R.I. Ostrom and M.J. Bitner (2000), 'Self-service technologies: understanding customer satisfaction with technology-based service encounters', *Journal of Marketing*, **64** (3), 50–64.

Nambisan, S. (2002). 'Designing virtual customer environments for new product development: toward a theory', *Academy of Management Review*, **27** (3), 392–413.
Narver, J.C. and S.F. Slater (1990), 'The effect of a market orientation on business profitability', *Journal of Marketing*, **54** (4), 20–35.
Neale, M.R. and D.R. Corkindale (1998), 'Co-developing products: involving customers earlier and more deeply', *Long Range Planning*, **31** (3), 418–425.
Nonaka, I. and H. Takeuchi (1995), *The Knowledge Creating Company*, Oxford: Oxford University Press.
Prahalad, C.K. and V. Ramaswamy (2000), 'Co-opting customer competence', *Harvard Business Review*, **78** (1), 79–87.
Rothwell, R. (1976), 'Marketing a success factor in industrial innovation', *Management Decision*, **14** (1), 43–53.
Rothwell, R. and P. Gardiner (1983), 'The role of design in product and process change', *Design Studies*, **4** (3), 166–170.
Shaw, C. (2007), *The DNA of Customer Experience*, Basingstoke: Palgrave Macmillan.
Sinkula, J. (1994), 'Market information processing and organizational learning', *Journal of Marketing*, **58** (1), 35–45.
Slater, Stanley F. and John C. Narver (1995), 'Market orientation and the learning organization', *Journal of Marketing*, **59** (3), 63–74.
Tidd, J. and J. Bessant (2009), *Marketing Innovation: Integrating Technological, Market and Organisational Change*, 4th edn, Wiley Europe.
Ulwick, A.W. (2002), 'Turn customer input into innovation', *Harvard Business Review*, **80** (1), 91–97.
Vargo, S.L. and R.F. Lusch (2004a), 'Evolving to a new dominant logic for marketing', *Journal of Marketing*, **68** (1), 1–17.
Vargo, S.L. and R.F. Lusch (2004b), 'The four service marketing myths – remnants of a goods-based, manufacturing model', *Journal of Service Research*, **6** (4), 324–335.
Vargo, S.L. and R.F. Lusch (2008), 'Service-dominant logic: continuing the evolution', *Journal of the Academy of Marketing Science*, **36** (1), 1–10.
Veryzer, R.W. (1998), 'Discontinuous innovation and the new product development process', *Journal of Product Innovation Management*, **15** (4), 304–21.
von Hippel, E. (1976), 'The dominant role of users in the scientific instrument innovation process', *Research Policy*, **5** (3), 212–239.
von Hippel, E. (1982), 'Get new product from customers', *Harvard Business Review*, **60** (3), 117–122.
von Hippel, E. (1986), 'Lead users: a source of novel product concepts', *Management Science*, **32** (7), 791–805.
von Hippel, E. (1994), 'Sticky information and the locus of problem solving: implications for innovation', *Management Science*, **40** (4), 429–439.
von Hippel, E. (1998), 'Economics of product development by users: the impact of sticky local information', *Management Science*, **44** (5), 629–644.
von Hippel, E. (2001), 'User toolkits for innovation', *Journal of Product Innovation Management*, **18** (3), 247–257.
Voss, C.A. (1985), 'The role of users in the development of applications software', *Journal of Product Innovation Management*, **2** (2), 113–121.
Wikström, S. (1996), 'The customer as co-producer', *European Journal of Marketing*, **3** (4), 6–19.
Zirger, B. J. and M.A. Maidique (1990), 'A model of new product development: an empirical test', *Management Science*, **36** (7), 867–883.

14 Collaborative innovation in services

Christiane Hipp

14.1 Introduction

According to Chandler (1990) it is assumed that companies with their various combinations of abilities and strategies can be regarded as key actors of innovation and technological change. This change is the result of knowledge generation and learning within organisations as well as between organisations based on collaborative arrangements. Often, new technologies and know-how are generated through the interaction of companies and their environment and are developed further internally (Dosi, 1988).

In particular, the flexibility and value creation potential of interfirm collaboration has accelerated a growing number of alliances during the past decade (for example Hansen and Nohria, 2004). Benefits in terms of innovation through collaboration have encouraged increasing research and development (R&D) partnerships in diverse sectors (Chakravorti, 2004). The literature has analysed collaboration between high-techs, start-ups and incumbents, between suppliers and clients, and more unusually between competitors (Miotti and Sachwald, 2003). Additionally, for the purpose of improved innovativeness, firms also ally with universities, public research institutes and private research institutes (for example Teece, 1989).

However, only a slowly growing number of studies have analysed innovation in service industries. The reasons reside in a strong tradition of innovation research that promotes new technologies and tangible artefacts whilst most services industries are rarely considered to undertake formal R&D processes or produce technically improved material objects (Miles, 2007). The gap in substantial findings escalated further when extending this underdeveloped research to studies in service-related collaborative or cooperative innovation activities.

From an evolutionary, neo-Schumpeterian perspective, Pavitt (1984) was able to summarise empirical results from a British innovation survey in four categories. Each category represents a model of technological change. Pavitt (1984) has classified services as 'supplier-dominated' users of technology without any ability to have their own internal R&D processes. In a later development, and inspired by the work of Barras (1986), Pavitt et al. (1989) added a new classification of 'information-intensive' companies to his model. This classification emphasises the importance of information and communication technologies in an innovation trajectory, and it is notable that this classification relates particularly to two service sectors – financial services and (large-scale) retailing – still assuming a supplier-dominated situation (Tether et al., 2001).

However, this new 'information-intensive' classification type does not concur with the importance of various service industries. In recent years, scholars have identified the importance of service innovations, highlighting the ability of service companies to have their own and specific knowledge creation processes (Soete and Miozzo, 1989; Evangelista, 2000). Therefore Gallouj and Gallouj (1997) suggest taking the service sector apart for a useful analysis of the innovation process including cooperative arrangements with external partners. Also Gallouj and Weinstein (1997) as well as Sundbo (1998) highlight differences between innovation in service and (technology-based) innovation in manufacturing.

Most recently, Sundbo and Gallouj (2000) as well as Potts and Mandeville (2007) predict a convergence of services and manufacturing innovation systems in which the often-described differences between both sectors will become blurred (Coombs and Miles, 2000). Also Tether (2002) can demonstrate that the old debates about industry, firm size, market structure and innovation are becoming outmoded, as the boundaries of the firm are becoming increasingly 'fuzzy'. This is also reflected in growing numbers of cooperative activities across formerly separate segments of society and the economy. Examples include business collaborations involving participants across and within industries, and collaborative arrangements between business and government such as public–private or community–business partnerships (Potts and Mandeville, 2007).

However, a complete convergence between manufacturing and services will not appear. Some characteristics will remain as 'service peculiarities', like the importance of person-to-person contacts, the customer orientation and, therefore, the service-oriented culture and special qualification requirements within the firm (Hipp, 2008).

This chapter focuses on the microeconomic level and aims to analyse the so-called service peculiarities and, in particular, the role and importance of collaborative service innovation. Based on empirical results from the German Community Innovation Survey (CIS) the choice of different partners and knowledge sources for enabling cooperative innovation activities are investigated. Additionally, the research explores different collaborative behaviour within different service industries and focuses on collaboration between service innovators and research institutes. The purpose is to show service-specific innovation collaboration behaviour and to prove the hypothesis that traditional R&D and research-oriented trajectories are non-applicable within service industries.

14.2 Theoretical analysis on collaborative systems

14.2.1 Traditional systems approach

In the field of innovation research the definition of the term 'system' is based on the considerations by the structural-functionalistic school, as mainly represented by Parsons (1976). Here a company, a region, an economy, a science, general regulative frame settings and the like are understood as systems. Carlsson et al. (2002, 234) define systems 'as a set of interrelated components working toward a common objective'. Both perspectives show common characteristics, like components, relationships and attributes. Components are seen as the operating parts of a system like individual actors or organisations; they can also be physical or technological artefacts or institutions, such as laws or social norms. Relationships are the links between the components, involving market and non-market links, feedback loops and the transfer of technological and non-technological knowledge. Attributes characterise the components, relationships and therefore the whole system. For example, the capabilities to generate, diffuse and utilise technologies, knowledge or competencies to identify and exploit business opportunities could be defined as attributes.

The aim, the specific achievement and the function of a system can be analysed. Interesting findings can be generated through the determination of the elements that enable specific processes, and the analysis of exchange relations within and between the systems, as well as the system behaviour and its position in a larger context. Smith (1997) points out that within the frame of knowledge generation different system approaches exist.

For example, the analyses of Lundvall (1988, 1992), Nelson (1993), Freeman (1995) and Perroux (1983) focus on different geographic levels (for example regional and national) and the institutional aspects and interactions between innovators that make the specific innovation and learning environment successful. A nation is, for example, a system, since it has clearly defined system borders deriving from geographical nearness, a common language, a common cultural basis, and homogeneous legal and economic political frame settings (Lundvall, 1988). In addition, an innovation system has grown out of the different actors and institutions and their relations to each other, which results in a stable, only slowly changing structure (Münt, 1996).

An alternative is the industrial cluster analysis or the sectoral system approach; components, relationships, and attributes are determined by specific sectoral or industrial conditions and located around key technologies or markets. Porter (1992) and Hirschmann (1979) concentrate here on environmental conditions and interactions between companies while

Malerba (2002) analyses sectoral systems of innovation based on a set of specific products and actors bringing them together.

In addition, there is a growing tendency to define innovations per se as systemic. Ropohl (1998) sees this development as an extension of the system horizon where, for example, innovators of the car may also consider the traffic system or the concept of mobility at the centre of their innovation activities. Similarly, Majer and Stahmer (1996) make a connection between service provision and systems marked not by material categories but by periods of time. This generates the chance for additional compensation for material goods with, for example, a focus on quality-of-life issues. The management of this growing complexity is important and adds further dimensions, such as the integration of time (for example the entire life cycle), more qualifications, interdisciplinary methods, collaborative organizational arrangements and new values within the innovation system (Hipp, 2000).

14.2.2 Institutional and loosely coupled systems

Sundbo and Gallouj (2000) follow a situational approach in identifying and combining driving forces that constitute a system. Their analysis results in a differentiation of two innovation subsystems: the 'institutional' and 'loosely coupled' systems. The distinction between the two systems is based on the level of clearly defined patterns. The institutional innovation system is a rational system with a limited series of pattern-based relationships between different actors through which knowledge and ideas for innovations are diffused. In an institutional innovation system these patterns are often long-term cooperations between actors and are often formalised through contracts. Companies using traditional innovation and R&D patterns and formal intellectual property mechanisms, or those that base their activities on long-term contracts, should be located within this system. For that reason, the strong traditional and institutionalised pattern approach is more common in manufacturing firms than for services, and also in those companies where radical innovations are less likely to appear.

Instead, a loosely coupled innovation system is not a fixed constellation between actors; in fact it can take numerous forms. Sundbo and Gallouj (2000) summarise that environments that are more competition-oriented can be characterised as loosely coupled systems. As a consequence, firms cooperate less, the integration of external knowledge is less formalised and institutionalised, and the diffusion process does not follow a straight line: 'This innovation system cannot be understood theoretically from a coherent explanatory model . . . because of the loose coupling of all elements and non-fixed behavioural patterns and traditions' (Sundbo and Gallouj, 2000,

61). Several weak patterns exist in parallel and although it is not possible to predict which pattern will be chosen, relationships can be explained and some rules can be formulated. Sundbo and Gallouj (2000) conclude that much of the service sector can be found within the loosely coupled system because of the lack of coherence in terms of technological and professional trajectories as well as the weak science-base. Hence, innovations are often quick, customer-oriented and practical in nature; it is hypothesised that: 'it would be difficult to find a route of imitation where different actors have a mutual relation and the diffusion of new ideas and concrete innovations could be followed through several links' (Sundbo and Gallouj, 2000, 62). Therefore, collaboration with different actors coming from different industries and which are changing over time are especially important for those service firms following the loosely coupled innovation approach.

14.3 Collaborative arrangements for innovation in the service industries

14.3.1 Increasing importance of collaborative arrangements for innovation

Interfirm collaboration covers a broad range of organisational combinations of various sizes, shapes and motivations in various levels of formal agreements (Borys and Jemison, 1989). Explanations as to why firms form interorganisational linkages centre either on the pattern of prior linkages that influences a firm's choice of future partners, or on incentives to collaborations (Ahuja, 2000). Findings by structural sociologists inspired the view that the ability of firms to form networks relies on their historical position or experiences within a network (Gulati and Gargiulo, 1999). A huge stream of literature from different viewpoints has investigated and explained firms' incentives to collaborate (Ahuja, 2000). Generally, incentives come from the numerous and flexible possibilities to merge tangible and intangible resources (Kogut and Zander, 1992).

The different authors agree that innovations in network structures help to cushion technological and market insecurities and dynamic change (Dodgson, 1996; Tether, 2002). A network organisation is the ideal form for coping more easily with increasing complexity and shorter product cycles (Carl and Kiesel, 1996; Rothwell, 1994). This form of organisation can be viewed as a mixture between market and hierarchy (Imai and Baba, 1991) as it allows companies the opportunity to focus on their core competencies and still remain flexible for new learning processes (Rammert, 1997; Nonaka and Takeuchi, 1995).

Also, according to Potts and Mandeville (2007), business collaboration has become a key segment of corporate strategy to cope with dynamic technological change and the risk of lock-in of scarce resources. From an evolutionary point of view, innovation collaborations are routines developed because of internal and external pressure. The generation of knowledge is based on a mixture of imitation and internal learning processes that change under the influence of a dynamic environment and changing internal characteristics (Niosi, 1996).

In particular, external partners with their complementary expert knowledge can give considerable innovative impulses (Hipp and Gassmann, 1999). Niosi (1996) regards the turbulent environment and the growing complexity caused by the new, technologically determined paradigm as the cause and driving force of new collective forms of learning. The developing theory of cooperative innovations is based on the assumption that technology is a quasi-public good and that learning mostly takes place locally. The new conditions force companies to cooperate; thus, to trade externalities (knowledge) and develop new forms of collective learning that develop in addition to the traditional learning processes. This process can be interpreted as a routine that companies integrate in their existing behavioural patterns.

Dahlander and Wallin (2006) summarise the recent developments along the open innovation paradigm proposed by Chesbrough (2003). They argue that firms need to rethink on traditional forms of knowledge ownership and to open themselves up to new forms of networked innovation projects. Now, innovation activities are much less hierarchically organised. The network of relationships between the firm and its external environment can play a dominant role in shaping innovative performance (Laursen and Salter, 2006).

14.3.2 The role of services in innovation network structures

The service sector as a whole is a very heterogeneous construction that can hardly be described by a single theory (Rubalcaba and Gago, 2003). Therefore, various service-specific criteria have to be considered to describe innovation behaviour in services adequately. One major characteristic of services discussed in the literature relates to the 'intangibility' of the service output (Hipp, 2008). Intangibility hinders the storing, transfer and transportation of the output (for example for trading), and creates difficulties in the demonstration of the output, in advance, to potential clients. In addition, intangibility hampers the employment of traditional instruments to protect innovations within the formal intellectual property system. With ongoing improvements of information and communication technologies, firms can – to a certain extent – relegate limitations of intangibility. However, the protection problem remains within the new information and communication regime, especially within the software development environment (Blind et al., 2001).

The increasing tradability of services and innovations in the field of communications and information technology promotes decentralisation, specialisation and thus the division of labour in service and industrial activities. In addition, there are new forms of flexibility and process parallelisation (for example in the automobile industry) that are based on demanding network structures. In those networks, services take over the functions of logistics, planning, controlling, coordinating and monitoring. The system's complexity and the many different network actors require new forms of governance structures, which essentially are carried out by these service companies.

In particular, knowledge-intensive service businesses play an important role as knowledge brokers in innovation activities. They, for example, absorb knowledge from their environment and place it at their partners' and customers' disposal for their innovation activities (for example engineering companies, consultancies). The service companies themselves are tied up in a knowledge and innovation network and play an important role in the learning processes of other companies (Hipp, 1999). Service-providing companies can also be active as network operators and establish, especially after the deregulation of the telecommunications market, new value-added services based on physical and mobile communication networks. Thus, they enable and support the networking and communication of diffused companies or parts of companies. Also Potts and Mandeville (2007) emphasise that today services can be better understood as networks rather than as 'producing' systems. In this respect, services can be seen as the increasing complexity of an evolving network rather than as the shifting-out of a company-internal function through, for example, outsourcing or offshoring.

Tether et al. (2000) – focusing on collaborative innovation projects – found that 26 per cent of the innovating service firms in Europe are engaged in cooperative arrangements to develop innovations, compared to 28 per cent in manufacturing. Therefore, service firms are collaborating at nearly the same intensity as manufacturing firms. However, Tether et al. (2000) can also show that there is a broad variety between different countries and industries. In a more recent article Tether and Tajar (2008) argue, based on empirical results from the Community Innovation Survey (CIS), that the most significant source of innovative ideas is still internal research and development, and that about a quarter of European innovators are engaged in cooperative arrangements. This proportion is higher in high-tech manufacturing, and lower for services. Main partner for innovators are customers and suppliers. Universities are less important; however, high-tech manufacturing companies in particular show some significant and positive relationship while low-technology-intensive services show a significantly negative relationship with universities (Tether, 2002).

These empirical findings contradict to a certain extent the more theoretical considerations towards a 'network' economy described above. The reason for this can be seen on the one hand in a possible dominance of loosely coupled network systems, especially for service innovators. These loosely coupled systems are difficult to analyse empirically. On the other hand barriers of collaborative innovation activities still exist, which hamper further development towards open or network innovations. The next section will go more into these details.

14.3.3 Problems of collaborative innovation in the service sector

Interfirm collaboration is populated with numerous uncertainties which help to explain the focus of many companies towards internal knowledge generation instead of collaborative innovation activities. Firstly, there is uncertainty in obtaining information regarding the competencies and requirements of the partners before the cooperative arrangement advances (Gulati and Gargiulo, 1999). Secondly, uncertainty emerges from incomplete information concerning the reliability of potential partners (Coase, 1937; Williamson, 1975).

This unpredictable behaviour creates uncertainties associated with moral hazard. Partners can either free-ride by limiting their contribution to the alliances or behave opportunistically by taking advantage of information (Gulati and Gargiulo, 1999). Uncontrolled and unwanted knowledge-spillovers can appear – especially in knowledge-intensive innovation networks or if common service innovations are based on intangibles which are hard to control and protect. Furthermore, unpredictable changes in the environment induce adjustments of the actors' behaviour that modify their contribution to collaboration and which cannot be regulated in advance. For that reason most collaborative contracts are imperfect (Williamson, 1989). Thus, service firms face the dilemma that on the one hand collaborative innovation offers a great potential to profit from specialised knowledge, technical and financial resources. But on the other hand companies can hardly predict exact efforts and expenses before the collaborative project itself has been finished.

Initial work by Kaiser (2000, 2002a, 2002b) tests game-theoretic models on research expenditures and research cooperation by means of a survey data of the German service sector. Kaiser concludes that an increase of horizontal knowledge-spillovers tends to reduce incentives to collaborate. He also illustrates that an increase in horizontal spillovers leads to a reduction of research efforts if services are complements and spillovers are large – and vice versa, if services are substitutes and spillovers are small.

The risk of opportunism will grow correspondingly to the extent to which the collaboration allows private benefits for individual R&D partners (Khanna et al., 1998). Private benefits offer outcomes from the collaboration, which any partner could use individually (Inkpen, 1998). The collaboration should instead focus on common benefits resulting from joint knowledge creation. Common benefits may emerge if partners learn to work together for a common purpose, capturing value from collective learning and installing joint procedures or assets that cannot be used without their partners (Khanna et al., 1998).

These barriers hamper especially service innovators, because intangibles and knowledge spillovers are often hard to control and are one reason for the empirically unexpected low proportion of collaborative service innovators. However, some methodological problems might also arise since most of the empirical studies ask for institutional collaborative arrangements. The importance of loosely coupled systems is hard to measure and has not been the focus of empirical research so far. The following analysis tries to close this gap to a certain extent by trying to show that loosely coupled activities are of main importance for service companies. In addition, the role of the science base as the classical indicator for a traditional, institutional and manufacturing-based innovation pattern will be included to illustrate the hypothesis of service peculiarities and convergence in service and manufacturing innovation activities.

14.4 Empirical analysis

14.4.1 Database

The loosely coupled innovation network system has been identified as appropriate for service innovators. However, the previous theoretical analysis as well as the first empirical results from the Community Innovation Survey (CIS) can show that there are some limits and barriers for collaboration in services. For the purpose of this analysis the CIS3 (2003) for Germany has been chosen because special attention has been paid to the analysis of different internal and external sources for innovation instead of just focusing on highly institutionalised innovation cooperation. According to critics of previous Community Innovation Surveys a number of additional questions have been included. Tether (2002) criticises that it is not possible to assess directly either how significant the collaborative relationships were to the innovation activities of the firm, or how successful the cooperations were. Moreover, we can also only indirectly understand the motives the firms had in establishing these relationships. Some interesting results can be derived from the German CIS3 questionnaire to get more insights on the way service innovators collaborate

– especially in loosely coupled arrangements. Also, an additional focus on cooperation with scientific partners has been chosen. The data can help to identify the importance of loosely coupled collaboration and give some insights into the previously neglected important role of universities within the innovation system of service industries.

The survey was carried out in 2003 for the German Ministry of Education and Research by the Centre for European Economic Research as well as the Institute for Applied Social Sciences, and according to the Community Innovation Survey, a common and highly standardised European survey on innovation activities. The survey was postal and voluntary. The sampling included firms with ten and more employees and was confined to 22 different industries from manufacturing and services, with sampling stratified by firm size, by sector and by location (East or West Germany). The service companies analysed come from nine different service industries including wholesale trade, retail trade, transportation, banking/insurance, telecommunication, technical services, business services, and other services. The sample is dominated by technical (20 per cent) and transport (18 per cent) services. Sixty-eight per cent of the service firms have less than 50 employees while 12 per cent of the sample have 250 or more employees (see Figure 14.1). Around 62 per cent of the service companies are located in West Germany while the other 38 per cent of the sample are located in East Germany.

A total of 3967 firms responded to the survey for manufacturing and services together – of which 2272 were classified as innovators. For the analysis of cooperative arrangements in the service industries, the analysis was restricted to the sample of 864 service firms that claimed to have been engaged in innovation activities. Question 7 of the questionnaire asked: 'Did your corporation have any cooperation arrangements on innovation activities with other enterprises or institutions between 2000 and 2002?' A definition of 'innovation cooperation' was also provided: 'Innovation cooperation means active participation in joint R&D and other innovation projects with other companies or non-profit organisations. It does not necessarily imply that both partners derive immediate commercial benefits from the venture. Pure contracting out of work, where there is no active participation is not regarded as cooperation.' A total of 770 of these service innovators answered the question on innovation collaboration. More than half of the innovative service firms have no cooperative arrangements to foster innovation activities while around 42 per cent of the innovative service firms provided details on cooperative arrangements – compared to around 40 per cent in manufacturing. Above average are technical services with 61 per cent, telecommunication and electronic data processing (EDP) with 53 per cent and business services with 44 per

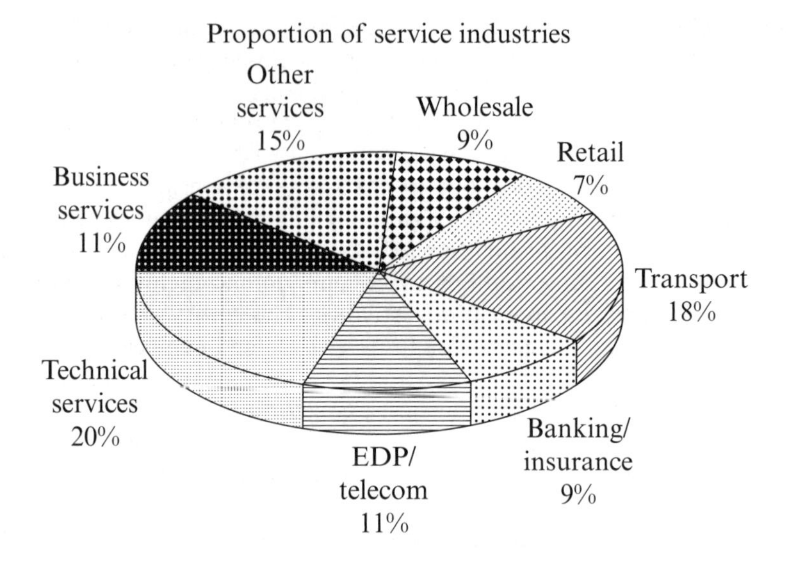

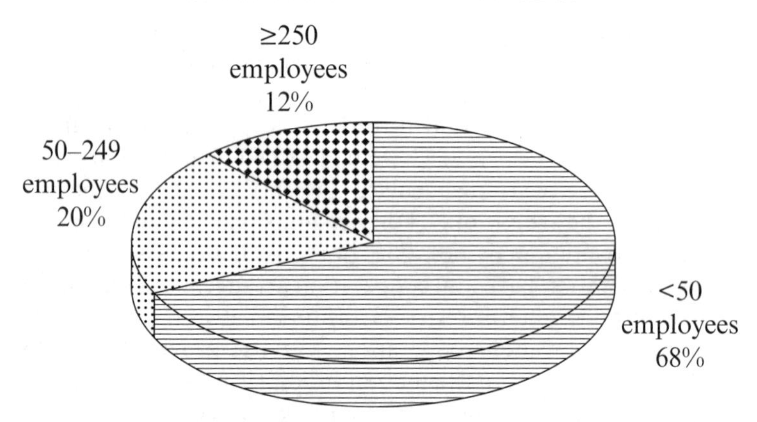

Source: Mannheim Innovation Panel – Community Innovation Survey, 2003.

Figure 14.1 Overview of the database

cent of collaborating firms. Retail trade has the lowest proportion of collaborative firms with 16 per cent. Compared to the previously described results offered by Tether (2002), German service and manufacturing innovators show much higher collaborative activities than companies from other European countries.

Table 14.1 Knowledge sources for innovation activities

	Result: product innovation (%)	Result: Process innovation (%)
Service industries		
Customers as knowledge sources for innovation	48	11
Competitors as knowledge sources for innovation	21	7
Science base as knowledge sources for innovation	17	4
Suppliers as knowledge sources for innovation	16	6
Law and regulation as knowledge sources for innovation	21	7
Manufacturing industries (without mining / food)		
Customers as knowledge sources for innovation	58	14
Competitors as knowledge sources for innovation	23	7
Science base as knowledge sources for innovation	13	5
Suppliers as knowledge sources for innovation	17	8
Law and regulation as knowledge sources for innovation	15	5

Source: Mannheim Innovation Panel – Community Innovation Survey, 2003.

14.4.2 Discussion of results

In the following analysis the empirical results are focused on innovators only – sometimes just for service industries, sometimes compared with manufacturing to identify different patterns of innovation of these two sectors. Question 9 of the questionnaire especially asked for loosely coupled collaboration activities in terms of the importance of external knowledge sources which had been indispensable for the innovation activities between 2000 and 2002. Additionally it was asked whether product and/or process innovations had been fostered by the loosely coupled external knowledge sources. The results are summarised in Table 14.1.

For services as well as manufacturing companies, customers are the main source for product and process innovations. However, customers are – surprisingly – much more important for manufacturing (58 per cent) than for the service sector in general (48 per cent). Here, the well-documented close relationship of service companies with their customers would have implied a different result. For manufacturing the second and third most important sources are competitors and suppliers, for both product and process innovations. These results are also surprising since competitors are not the preferred collaboration partners due to the danger of horizontal knowledge spillovers (Kaiser, 2000). However, within the loosely coupled system, knowledge can be transferred more easily but with less depth and danger of losing competitive advantages – for example just via collaborative benchmarking analyses. Another unexpected result is the relatively high percentage of companies getting ideas and knowledge from the science base. Here, 17 per cent of the service innovators have indicated this knowledge source as crucial for their product innovations.

Traditionally, a phase model and some division of labour between different actors are assumed for all kinds of innovation activities. Basic knowledge is firstly created by some research institutes (fundamental research). Companies as well as some applied research organisations use this knowledge, and transfer it into applicable products and technologies before these products and applications diffuse into the market or other companies. Altogether, innovation researchers come to the conclusion that a purely process-oriented approach does not suffice with respect to immediacy in describing the structures and sequences of the whole innovation process (Hipp, 2000). Criticised thus are the deterministic sequence of the individual phases, and the dominance of science and technology, as well as the insufficient consideration of the complex innovation-relevant environment in combination with the action strategies of individual actors. The traditional innovation concepts are applied by bigger companies in the manufacturing sector consciously planning their innovation processes and having a much higher affinity to science-based, technology-oriented research. For the service sector, which is dominated by smaller companies and characterised by other specific features, new concepts are needed (Hipp, 1999).

However, in many studies on knowledge- and technology-intensive service firms it is highlighted that especially high-tech services follow a more manufacturing innovation pattern, also integrating research institutes into their network structures. Table 14.2 summarises some descriptive results supporting the existing findings to a certain extent.

One main surprise is that the research system is linked with the service sector to a high extent. Informal contacts with research institutions are mostly important for manufacturing as well as for service companies. More than half of the service companies have some form of loosely coupled knowledge exchange with universities ('informal contacts'). However, more formal agreements are less important – but also for manufacturing

Table 14.2 Joint innovation activities with research institutions (universities)

	Joint research (%)	Contract research (%)	Joint theses (%)	Licensing (%)	Limited personnel exchange (%)	Qualification of personnel in research institutions (%)	Consulting through research institutions (%)	Informal contacts (%)
Manufacturing (without mining / food)	31	30	42	14	10	25	43	62
Services (all)	28	22	42	11	13	24	27	57
Wholesale	6	9	19	4	4	13	13	35
Retail	7	9	22	7	7	16	9	36
Transport	8	16	29	5	5	18	20	36
Banking/Insurance	11	12	37	7	7	20	11	42
Telecom / EDP	37	17	47	12	12	23	25	61
Technical services	50	40	59	19	22	34	42	79
Business services	29	20	52	6	13	25	26	66
Other services	16	16	25	8	10	25	34	42

Source: Mannheim Innovation Panel – Community Innovation Survey, 2003.

firms. Here, consulting through research institutions (43 per cent) and joint theses (42 per cent) come second while for service firms joint theses are ranked second (42 per cent) and then followed by joint research (28 per cent). Licensing and limited personnel exchange is of least importance, for the manufacturing sector as well as for service firms. These last two different forms of collaboration need a strong fit in terms of knowledge flow (licensing) and joint knowledge development (personnel exchange). Both forms also require some institutional arrangements like contracts (licensing) and non-disclosure agreements and agreements on joint developed exploitation of knowledge (personnel exchange), which might explain the barriers of application and the low percentage of joint use.

Going more into industry-specific network arrangements, Table 14.2 also shows that different network behaviour with research institutions can be observed within different service industries. The most intensive collaboration activities exist with technical services, business services, and telecommunication and information-technology (IT) services. These industries show an even higher percentage of collaborative arrangements in many respects than the average manufacturing firm.

Following the network approach further, Table 14.3 gives an overview of the propensity of interlinkages between the above-described loosely coupled knowledge sources supporting product innovations, and the more institutionalised forms of interaction with research institutes. In addition, the internal aspects of own knowledge creation capabilities are integrated into the models ('*Own R&D activities*') following the concepts of absorptive capacity (Cohen and Levinthal, 1990) and the strong interaction with internal and external knowledge generation and learning (Dosi, 1988). In addition the models are controlled for industry and size. One German peculiarity still is the difference between East and West German companies in terms of innovation performance and productivity which has been controlled for as well (see the variable '*East Germany*').

The results confirm that especially telecommunication and IT, as well as technical and business services, have a higher propensity to interact closer with research institutions. Joint theses and joint research are of main importance here and the difference is highly significant compared to the base service industry. Also important is the finding that own R&D activities significantly raise the propensity to work together with universities on almost all institutional levels – supporting the hypothesis that the absorptive ability to integrate and further develop knowledge offered and diffused by research institutions is of importance. Also the propensity to have closer contact with universities rises with firm size.

Interestingly, all kinds of knowledge sources significantly raise the ability to work together with universities, and all significant results are

Table 14.3 Relationship between collaboration activities with research institutes and other forms of knowledge sources

Dependent variable Model n	Joint research (1)	Contract research (2)	Joint theses (3)	Licensing (4)	Limited personnel exchange (5)	Qualification of personnel in research institutions (6)	Consulting through research institutions (7)	Informal contacts (8)
Customers as knowledge sources	0.1485 (.119)	−0.1270 (.131)	0.0953 (0.110)		−0.1753 (0.146)	0.0850 (0.118)	0.0614 (0.117)	0.1050 (0.113)
Competitors as knowledge sources	0.1466 (0.143)	0.3100** (0.147)	0.2540* (0.133)		0.2793* (0.170)	0.0943 (0.140)	0.1788 (0.139)	0.1790 (0.138)
Suppliers as knowledge sources for innovation	0.2705* (0.155)	0.1596 (0.162)	0.0476 (0.140)		0.3567** (0.170)	0.2618* (0.146)	0.4175*** (0.144)	−0.0298 (0.144)
Law and regulation as knowledge sources	0.1150 (0.134)	0.1352 (0.141)	0.2337* (0.124)		0.1954 (0.152)	0.1635 (0.131)	0.1154 (0.130)	0.3258** (0.132)
Own R&D activities	0.3995*** (0.080)	0.2816*** (0.082)	0.2374*** (0.078)		0.0899 (0.086)	0.1581** (0.076)	0.2333*** (0.078)	0.3726*** (0.083)

Table 14.3 (continued)

Dependent variable Model n	Joint research (1)	Contract research (2)	Joint theses (3)	Licensing (4)	Limited personnel exchange (5)	Qualification of personnel in research institutions (6)	Consulting through research institutions (7)	Informal contacts (8)
Wholesale	0.1079 (0.469)	0.562 (0.565)	0.3126 (0.388)		−0.0009 (0.518)	0.3079 (0.423)	−0.4756 (0.404)	0.0759 (0.335)
Retail	0.4017 (0.471)	0.7575 (0.563)	0.1717 (0.387)		0.2511 (0.484)	0.5420 (0.415)	−0.3930 (0.409)	0.2327 (0.344)
Transport	0.1059 (0.428)	0.7207 (0.522)	0.2457 (0.332)		−0.3962 (0.470)	0.2514 (0.372)	−0.3597 (0.331)	−0.1963 (0.300)
Banking / Insurance	0.3722 (0.415)	0.4224 (0.535)	0.2607 (0.333)		0.0378 (0.441)	0.4289 (0.372)	−0.5728 (0.350)	−0.2104 (0.303)
Telecom / EDP	1.0015*** (0.375)	0.7195 (0.496)	0.9088*** (0.310)		0.2242 (0.400)	0.3951** (0.344)	−0.1743 (0.301)	0.4792* (0.278)
Technical services	1.3660*** (0.370)	1.5472*** (0.488)	1.3058*** (0.306)		0.6800* (0.384)	0.8269* (0.336)	0.4274 (0.293)	10.08361*** (0.2731788)
Business services	0.8386** (0.394)	0.9225* (0.5121)	1.2222*** (0.326)		0.5061 (0.417)	0.6066* (0.362)	−0.0127 (0.322)	0.7974** (0.301)

Other services	0.6586 (0.404)	0.7770 (0.522)	0.2555 (0.343)		0.1486 (0.426)	0.5161 (0.362)	0.2444 (0.318)	0.0644 (0.301)
Real estate				BASE				
<50 empl.				BASE				
50–249 empl.	−0.2157 (0.142)	−0.0664 (0.148)	0.1834 (0.132)		0.0792394 (0.1667162)	0.0875591 (0.1378569)	0.0621 (0.139)	0.0948 (0.131)
>=250 empl.	−0.2311 (0.193)	0.3790* (0.193)	10.0948*** (0.174)		−0.1083 (0.240)	0.4158** (0.177)	0.2693 (0.178)	0.8958*** (0.177)
East Germany	0.1738 (0.116)	0.2887** (0.120)	0.1244 (0.111)		0.3583*** (0.130)	0.2860** (0.114)	0.3632*** (0.115)	0.3523** (0.114)
No of observations	703	692	700		693	696	692	706
Prob > chi^2	0.000	0.000	0.000	not significant	0.000	0.001	0.000	0.000
Pseudo R^2	0.172	0.123	0.139		0.086	0.053	0.102	0.1652

Note: Standard errors in parentheses * $p < 0.05$, ** $p < 0.01$, *** $p < 0.001$.

Source: Mannheim Innovation Panel – Community Innovation Survey, 2003.

positive. This means that existing and combined internal and external knowledge-sourcing strategies also open up the propensity to have loosely and close (institutional) contact with research institutions. The results of the probit analysis further deepen the descriptive results previously presented. Interactions with suppliers and competitors strengthen the propensity to integrate with research institutes via consulting contracts as well as via limited personnel exchange, qualification of personnel in universities and joint research. Also, the interaction with competitors increases the propensity to have limited personnel exchange with universities. In addition, contract research and joint theses are more likely when getting innovative ideas from competitors. Surprisingly, customers as knowledge sources might be so 'normal' that no significant different behaviour with respect to research institutes can be observed. Unexpected also is the result that a headquarters location in East Germany fosters the propensity to cooperate and collaborate with universities. The reason for that might be some knowledge catching-up processes in East Germany, locally concentrated publicly funded transfer projects bringing universities and companies together, and other East German policy instruments to support personnel exchange and joint qualifications.

14.5 Conclusions

Some main findings can be derived from the theoretical and empirical analysis and are summarised below.

- A high importance of collaborative arrangements in general can be analysed, supporting the predicted and described network innovation paradigm of societies. These collaborations are extremely important for product innovations. On the aggregated level no significant difference between manufacturing and services can be observed.
- Service innovators collaborate in loosely coupled network structures, however, the difference between services and manufacturing is lower than expected. Both sectors show similar loosely coupled innovation behaviour. The role of customers is mostly important for both sectors.
- Surprising is the importance of universities as knowledge sources and collaborative partners for service innovators. Collaborations with universities are highly underinvestigated. These findings contradict previous research results on innovation patterns of service companies. However, the reason for a possible increasing connectivity between services and universities needs to be analysed further.
- Different innovation patterns occur within service industries,

showing the heterogeneity of the service sector. In particular, telecommunication and IT, technical and business services are highly engaged in a more 'traditional' innovation pattern. Whether these innovation patterns really reflect a 'standard' manufacturing innovation pattern needs to be investigated in detail. The assumption is that within manufacturing different innovation patterns also exist, and that some service peculiarities remain which drive and influence the innovation pattern of the telecommunication and IT, technical and business services.

- The characteristics of the innovation activities of the other service industries like wholesale, retail, transport, banking and insurance, and other services is still underinvestigated. At least some informal contacts with universities exist; and these industries also work together with universities in a more institutionalised way. However, a specific pattern of innovation activities cannot be derived from the empirical findings. The CIS questionnaire uses a traditional understanding and working of innovation (like R&D investment). This might not be adequate to analyse and measure service innovation.
- In general, loosely coupled collaboration and external knowledge-sourcing strategies foster research collaborations with universities. It can be assumed that service innovators which are engaged in loosely coupled innovation activities have a strong focus on knowledge-building and learning capabilities which also require a strong internal knowledge generation process through own research and development capabilities.

It can be accepted as a result of the above findings that the boundaries of the firm are becoming increasingly 'fuzzy'. But a more detailed analysis is required to get insights into the network system with its components, relationships and attributes – which are also changing over time. Some of the actors and knowledge sources within innovation collaboration have been analysed in this chapter in more detail and some service peculiarities have been discussed. But for collaborative arrangements the differences between manufacturing and services are less significant than previously assumed. Loosely coupled and institutional network structures are important for both sectors, although some differences within service industries can be elaborated. While some service industries follow a more manufacturing-based innovation pattern, all other services industries show no specific innovation characteristics. The reason for this might be the very traditional measurement approach of the Community Innovation Survey.

Tether and Tajar (2008) hypothesise that the existing innovation measurement system provided by national agencies as well as by the

European Commission (definitions, questionnaires, surveys) has been historically focused on R&D-based and technological product and process innovation-based approaches. For that reason the collaborative practice – especially for service innovators – is much richer and more alive than is shown by the existing measurement concepts.

Measurement problems arise especially when analysing the relationships and attributes of network structures in detail. Much more effort has to be undertaken to get some insights into innovation patterns – especially for the non-technical service industries. And without opening this 'black box' no substantial arguments exist to support or refute the assumption of an increasing convergence or synthesis approach (Coombs and Miles, 2000) – with respect to collaboration. Here, new forms of open innovation, the role of the Internet, and arguments from different disciplines like transaction cost theory, organisational theory, systems theory and evolutionary theory need to be taken into consideration.

References

Ahuja, G. (2000), 'The duality of collaboration: inducements and opportunities in the formation of interfirm linkages', *Strategic Management Journal*, **21** (3), 317–344.

Barras, R. (1986), 'Towards a theory of innovation in services', *Research Policy*, **15** (4), 161–173.

Blind, K., J. Edler, R. Nack and J. Straus (2001), *Micro- and Macroeconomic Implications of Patentability of Software Innovations. Intellectual Property Rights in Information Technologies between Competition and Innovation – Study on behalf of German Federal Ministry of Economics and Technology*, Karlsruhe: Fraunhofer-Institute and Munich: Max-Planck-Institute.

Borys, B. and D. Jemison (1989), 'Hybrid arrangements as strategic alliances', *The Academy of Management Review*, **14** (2), 234–249.

Carl, N. and M. Kiesel (1996), *Unternehmensführung: Moderne Theorien, Methoden und Insturmente*, Landsberg/Lech: Verlag Moderne Industrie.

Carlsson, B., S. Jacobsson, M. Holmén and A. Rickne (2002), 'Innovation systems: Analytical and methodological issues', *Research Policy*, **31** (2), 233–245.

Chakravorti, B. (2004), 'The role of adoption networks in the success of innovations: A strategic perspective', *Technology in Society*, **26** (2–3), 469–482.

Chandler, A.D., Jr (1990), *Scale and Scope: The Dynamics of Industrial Capitalism*, Cambridge, MA: Belknap Press of Harvard University Press.

Chesbrough, H. (2003), *Open Innovation: The New Imperative for Creating and Profiting from Technology*, Boston, MA: Harvard Business School Press.

Coase, R.H. (1937), 'The nature of the firm', *Economica*, **4** (16), 386–405.

Cohen, W.M. and D.A. Levinthal (1990), 'Absorptive capacity: a new perspective on learning and innovation', *Administrative Science Quarterly*, **35** (1), 128–158.

Coombs, R. and I. Miles (2000), 'Innovation, measurement and services: the new problematique', in J.S. Metcalfe and I. Miles (eds), *Innovation Systems in the Service Economy*, Boston, MA: Kluwer, pp. 83–102.

Dahlander, L. and M.W. Wallin (2006), 'A man on the inside: unlocking communities as complementary assets', *Research Policy*, **35** (8), 1243–1259.

Dodgson, M. (1996), 'Learning, trust and inter-firm technological linkages: some theoretical associations', in R. Coombs, A. Richards, P. Saviotti and V. Walsh (eds), *Technological Collaboration: The Dynamics of Cooperation in Industrial Innovation*, Cheltenham, UK and Brookfield, VT, USA: Edward Elgar, pp. 54–75.

Dosi, G. (1988), 'The nature of the innovative process', in G. Dosi, C. Freeman, R. Nelson, G. Silverberg and L. Soete (eds), *Technical Change and Economic Theory*, London: Pinter, pp. 221–238.

Evangelista, R. (2000), 'Sectoral patterns of technological change in services', *Economics of Innovation and New Technology*, **9** (3), 183–221.

Freeman, C. (1995), 'The 'national system of innovation' in historical perspective', *Cambridge Journal of Economics*, **19** (1), 5–24.

Gallouj, C. and F. Gallouj (1997), 'EU-Project SI4S', Working Paper, Lille: IFRESI-CNRS.

Gallouj, F. and O. Weinstein (1997), 'Innovation in services', *Research Policy*, **26** (4–5), 537–556.

Gulati, R. and M. Gargiulo (1999), 'Where do interorganizational networks come from?' *American Journal of Sociology*, **104** (5), 1439–1493.

Hansen, M.T. and N. Nohria (2004), 'How to build collaborative advantage', *MIT Sloan Management Review*, **46** (1), 22–30.

Hipp, C. (1999), 'The role of knowledge-intensive business services in the new mode of knowledge production', *AI and Society*, **13** (1), 88–106.

Hipp, C. (2000), *Innovationsprozesse im Dienstleistungssektor*, Heidelberg: Physika.

Hipp, C. (2008), 'Service peculiarities and the specific role of technology in service innovation management', *International Journal of Services Technology and Management*, **9** (2), 154–173.

Hipp, C. and O. Gassmann (1999), 'Leveraged knowledge creation: the role of technical services as external sources of knowledge', *Management of Engineering and Technology, 1999. Technology and Innovation Management. PICMET '99. Portland International Conference on Management of Engineering and Technology*, Vol. 2, pp. 186–192.

Hirschmann, A.O. (1979), *Toward a New Strategy for Development*, New York, USA and Frankfurt, Germany: Pergamon Press.

Imai, K. and Y. Baba (1991), 'Systemic innovation and cross-border networks: Transcending markets and hierarchies to create a new techno-economic system', in OECD (eds), *Technology and Productivity*, Paris: OECD, pp. 389–405.

Inkpen, A. (1998), 'Learning, knowledge acquisition and strategic alliances', *European Management Journal*, **16** (2), 223–229.

Kaiser, U. (2000), 'R&D cooperation and R&D investment with endogenous spillovers: Theory and microeconometric evidence', ZEW discussion paper No. 00-25, Mannheim: ZEW.

Kaiser, U. (2002a), 'An empirical test of models explaining research expenditures and research cooperation: evidence for the German service sector', *International Journal of Industrial Organization*, **20** (6), 747–774.

Kaiser, U. (2002b), 'Measuring knowledge spillovers in manufacturing and services: an empirical assessment of alternative approaches', *Research Policy*, **31** (1), 125–144.

Khanna, T., R. Gulati and N. Nohria (1998), 'The dynamics of learning alliances: competition, cooperation, and relative scope', *Strategic Management Journal*, **19** (3), 193–210.

Kogut, B. and U. Zander (1992), 'Knowledge of the firm, combinative capabilities, and the replication of technology', *Organizational Science*, **3** (3), 382–397.

Laursen, K. and A. Salter (2006), 'Open for innovation: the role of openness in explaining innovation performance among U.K. manufacturing firms', *Strategic Management Journal*, **27** (2), 131–150.

Lundvall, B.-Å. (1988), 'Innovation as an interactive process: from user-producer interaction to the national system of innovation', in G. Dosi, C. Freeman, R. Nelson, G. Silverberg and L. Soete (eds), *Technical Change and Economic Theory*, London: Pinter, pp. 349–369.

Lundvall, B.-Å. (1992), *National Systems Of Innovation: Towards a Theory of Innovation and Interactive Learning*, London: Pinter.

Majer, H. and C. Stahmer (1996), 'Wie definiert, mißt und schließt man regionale Nachhaltigkeitslücken?', in U.-P. Reich, C. Stahmer and K. Voy (eds), *Kategorien der Volkswirtschaftlichen Gesamtrechnung*, Marburg: Metropolis, pp. 285–319.

Malerba, F. (2002), 'Sectoral systems of innovation and production', *Research Policy*, **31** (2), 247–264.
Miles, I. (2007), 'Research and development (R&D) beyond manufacturing: the strange case of services R&D', *R&D Management*, **37** (3), 249–267.
Miotti, L. and F. Sachwald (2003), 'Co-operative R&D: Why and with whom? An integrated framework of analysis', *Research Policy*, **32** (8), 1481-1499.
Münt, G. (1996), *Dynamik von Innovation und Außenhandel ⊠ Entwicklung technologischer und wirtschaftlicher Spezialisierungsmuster*, Heidelberg: Physica.
Nelson, R.R. (ed.) (1993), *National Innovation Systems: A Comparative Analysis*, New York and Oxford: Oxford University Press.
Niosi, J. (1996), 'Strategic technological collaboration in Canadian industry: Towards a theory of flexible or collective innovation', in R. Coombs, A. Richards, P.P. Saviotti and V. Walsh (eds), *Technological Collaboration: The Dynamics of Cooperation in Industrial Innovation*, Cheltenham, UK and Brookfield, VT, USA: Edward Elgar, pp. 98–118.
Nonaka, I. and H. Takeuchi (1995), *The Knowledge Creating Company: How Japanese Companies Create the Dynamics of Innovation*, Oxford: Oxford University Press.
Parsons, T. (1976), *Zur Theorie sozialer Systeme*, Opladen: Westdeutscher Verlag.
Pavitt, K. (1984), 'Sectoral patterns of technical change: towards a taxonomy and a theory', *Research Policy*, **13** (6), 343–373.
Pavitt, K., M. Robson, and J. Townsend (1989), 'Technological accumulation, diversification and organisation in UK companies, 1945–1983', *Management Science*, **35** (1), 81–99.
Perroux, F. (1983), *A New Concept of Development*, London: Croom Helm.
Porter, M.E. (1992), *Wettbewerbsstrategie: Methoden zur Analyse von Branchen und Konkurrenten*, Frankfurt/Main, Germany, adn New York, USA: Campus.
Potts, J. and T. Mandeville (2007), 'Toward an evolutionary theory of innovation and growth in the service economy', *Prometheus*, **25** (2), 147–159.
Rammert, W. (1997), 'Innovation im Netz – Neue Zeiten für technische Innovationen: heterogen verteilt und interaktiv vernetzt', *Soziale Welt*, **48** (4), 397–416.
Ropohl, G. (1998), *Wie die Technik zur Vernunft kommt*, Amsterdam: Fakultas.
Rothwell, R. (1994), 'Industrial innovation: success, strategy, trends', in R. Rothwell and M. Dodgson (eds), *The Handbook of Industrial Innovation*, Aldershot, UK and Brookfield, VT, USA: Edward Elgar, pp. 33–53.
Rubalcaba, L. and D. Gago (2003), 'Regional concentration of innovative business services: testing some explanatory factors at European regional level', *The Service Industries Journal*, **23** (1), 77–94.
Smith, K. (1997), 'Economic infrastructures and innovation systems', in C. Edquist (ed.), *Systems of Innovation: Technologies, Institutions and Organizations*, London: Pinter, pp. 86–106.
Soete, L. and M. Miozzo (1989), 'Trade and development in services: a technological perspective', Working Paper, Maastricht: Economic Research Institute on Innovation and Technology (MERIT).
Sundbo, J. (1998), *The Organisation of Innovation in Services*, Frederiksberg: Roskilde University Press.
Sundbo, J. and F. Gallouj (2000), 'Innovation as a loosely coupled system in services', in J.S. Metcalfe and I. Miles (eds), *Innovation Systems in the Service Economy*, Boston, MA: Kluwer, pp. 43–68.
Teece, D.J. (1989), 'Inter-organizational requirements of the innovation process', *Managerial and Decision Economics*, **10** (Special Issue), 35–42.
Tether, B.S. (2002), 'Who co-operates for innovation, and why: an empirical analysis', *Research Policy*, **31** (6), 947–967.
Tether, B.S., C. Hipp, and I. Miles (2001), 'Standardisation and particularisation in services: evidence from Germany', *Research Policy*, **30** (7), 1115–1138.
Tether, B., I. Miles, K. Blind, C. Hipp, N. de Liso and G. Cainelli (2000), 'Innovation in the service sector – analysis of data collected under the Community Innovation Survey (CIS-

2)', Report for the European Commission within the Innovation Program, Manchester: CRIC.
Tether, B. and A. Tajar (2008), 'The organisational-cooperation mode of innovation and its prominence amongst European service firms', *Research Policy*, **37** (4), 720–739.
Williamson, O. (1975), *Markets and Hierarchies: Analysis and Antitrust Implications*, New York: Free Press.
Williamson, O.E. (1989), 'Transaction cost economics', in R. Schmalensee, R.D. Willig, (eds), *Handbook of Industrial Organization*, Vol. 1, Amsterdam: North-Holland, pp. 135–182.

15 Knowledge regimes and intellectual property protection in services: a conceptual model and empirical testing

Knut Blind, Rinaldo Evangelista and Jeremy Howells[1]

15.1 Introduction

Traditionally services have been seen as being technological backward, passive, non-innovative sectors. This was enough to justify the little interest or involvement shown by service firms with the issue of appropriability and intellectual property rights (IPR), with the exception of artistic and text-based rights. Data from, for example, the Community Innovation Survey (CIS) in Europe seem to confirm such a view, highlighting the limited use by service firms of the most traditional IPR tool, namely patents. Thus, in the Second CIS (CIS2) only 7 per cent of innovating companies in services during the period 1994–96 applied for at least one patent compared with a quarter of all innovating manufacturing companies and this does not appear to have changed for the Third Community Innovation Survey (CIS3) (Blind et al., 2003).

These data are likely to provide a too narrow picture of the actual and potential relevance of the 'appropriability' issue in services. Recent studies show that patent activities are rapidly increasing in services. Patent applications at the European Patent Office by the top 50 European service companies grew from around 280 applications per annum in 1990 to almost 450 per annum in 1997 (Blind et al., 2003, 56). In any case patent statistics do not make justice of the increasing importance of the IPR issue outside the manufacturing sector. This is because of the highly heterogeneous nature of services, of the distinctive nature of 'knowledge assets' in this section of the economy and, consequently, of the presence of very different strategies through which service firms could protect such assets.

This is a difficult field to explore, both conceptually and empirically. On the one hand, we still know very little about the relevance and nature of innovation activities in the heterogeneous universe of services, and even less about how firms accumulate and protect their critical knowledge assets and competencies. On the other hand, the lack of reliable data and robust quantitative evidence of innovation hampers the empirical research in this area. Patents statistics represent inadequate metrics with which to measure the innovative output of service firms or, even less, the ability of firms to appropriate the returns of their investment in knowledge. This chapter, therefore, aims to contribute in filling this gap in our knowledge by exploring the relevance and nature of IPR-related behaviours in services. It starts by acknowledging that service sectors are very diverse from each other, and that firms located in different service sectors face very different knowledge, appropriation and intellectual property (IP) regimes (Andersen and Howells, 2000). Accordingly we first provide a conceptual framework of what is termed here 'knowledge regimes' within the service sector. After that, we develop a set of hypotheses regarding how these different regimes affect the specific strategies pursued by firms to appropriate and defend their knowledge assets. The evidence presented is based on a European survey on IPR strategies in services funded by the European Commission.[2]

The chapter is structured as follows. The next section discusses the key dimensions used here to distinguish between different 'knowledge regimes' in services. We will focus here on the codified or tacit nature of firms' knowledge assets as well as on their tangible or intangible nature. The following section operationalises such a conceptual model in more detail linking different 'knowledge regimes' to alternative IPR strategies and specific tools used by service firms ('IPR regimes'). The validity of this conceptual framework is then empirically tested against the data collected. The findings from this analysis are then reviewed before a short conclusion is presented.

15.2 Knowledge and intangibility

This study uses two key dimensions in discussing and conceptualising the issue of knowledge and appropriability in services, namely the level of codification or tacitness of knowledge and its degree of tangibility. Before the model itself is presented and discussed, how do we define both knowledge and tangibility? These two concepts will now be explored in turn.

15.2.1 Appropriability of codified and tacit knowledge

There have been various attempts to identify and classify different types of knowledge, but an early and seminal distinction has been made by Michael Polanyi (1958, 1961, 1962, 1967) who distinguished between explicit (or codified) knowledge and tacit knowledge. The difference between these two broad types of knowledge are linked to the degree of formalisation and the requirement of physical presence in knowledge formation. Explicit, or codified, knowledge involves know-how that is transmittable in formal, systematic language and does not require direct experience of the knowledge that is being acquired and it can be transferred in such formats as a blueprint or operating manual.[3] By contrast, tacit knowledge cannot be communicated in any direct or codified way. Tacit knowledge concerns direct experience that is not codifiable via artefacts. As such, it represents disembodied know-how that is acquired via the informal take-up of learned behaviour and procedures. Indeed, some tacit knowing is associated with learning without awareness; a process termed as 'subception' by Polanyi (1966). However, more formal learning mechanisms also play an important role here with 'learning by doing' (Arrow, 1962), 'learning by using' (Rosenberg, 1982) and 'learning to learn' (Ellis, 1965; Estes, 1970; Argyris and Schon, 1978; Stiglitz, 1987) seen as critical elements within tacit knowledge acquisition. Tacit knowledge can also be associated with scientific intuition (Ziman, 1978) and the development of craft knowledge within scientific disciplines (Delmont and Atkinson, 2001) and is 'sticky' and often difficult to transfer (von Hippel, 1994). Elsewhere, Polanyi sums up tacit knowing as an act of 'indwelling', the process of assimilating to ourselves things from outside (Polanyi, 1962). It also involves, though, more innate values, such as skills. In addition, it is generally accepted that tacit know-how cannot be directly or easily transmitted, as knowledge and task performance are individual and specific and involve the acquirer making changes to existing behaviour. However, the degree of tacitness does vary (Howells, 1996). Within the range of tacit knowledge itself, the less explicit and codified the know-how is, the more difficult it is for individuals and firms to assimilate it (Cohen and Levinthal, 1990). More specifically within the context of services, it is this difficulty of assimilating tacit knowledge which plays a prominent role in relationships with clients and suppliers and in the innovation process itself. This codified/tacit dimension of knowledge has important implications for the extent to which, and ways through which, the latter can be 'protected', i.e. the way economic returns are appropriated. The regime of appropriability represents a fundamental incentive to innovation, governing the innovator's ability to capture the economic returns from innovation. The level and ways in which innovation can be appropriated define the specific regime of protection characterising a given technology, innovation output or, more broadly, a knowledge asset (Teece, 1986).

It is not surprising that formal IPRs in general, and patents in particular, are much better at defining and protecting inventions and knowledge assets whose content can be expressed in a codified form. However, not all the stock of knowledge exists and can be transferred and used in a codified form. In any organisation, a variable portion of this knowledge is tacit in nature. Tacit knowledge, by definition, is in turn characterised by being more appropriable than codified 'knowledge' and thus reduces its 'public nature' (Ancori et al., 2000). It has been generally acknowledged that knowledge and competencies embedded in people, organisations or defined social contexts, are difficult to transfer and economically exploit elsewhere. Furthermore, in the case of tacit knowledge formal IPR tools are difficult to use and this is because of the difficulty of defining in a codified and unambiguous way the knowledge content of a given invention, technology or product. In most cases the tacit component of knowledge is dominant and there is little incentive to adopt any protection strategy which is anyway often expensive to establish and enforce. Furthermore, knowledge and the distinctive competencies of an organisation are often tacit insofar as they are embedded in the human resources. Accumulating human capital, and building competencies, are processes which take time, require substantial efforts and are affected by numerous contextual factors. In such cases informal IPR strategies, and in particular employment-centred mechanisms, might assume a crucial importance.

Before passing on to the next section it is important to make explicit that in the model proposed in this chapter we considerably simplify the complex and variegated nature of knowledge. This is justified by our attempt of linking knowledge regimes to different IPR task environments. The simplification we make is twofold. Firstly, the knowledge regime model that is presented below makes a simple distinction between codified (or explicit) knowledge and tacit knowledge. It is acknowledged that by so doing this is creating a crude bipolar dichotomy between codified and tacit knowledge. Polanyi was himself at pains to stress that explicit and tacit knowledge were not divided, but rather should be seen as a continuum between wholly explicit knowledge and a wholly tacit form of knowledge. Secondly, in the operationalisation of the model we use the word 'knowledge' not in its true form, outlined above of that which can only reside in and be held by an individual,[4] but rather use the term in a more generalised (and arguably subverted) sense, as a phenomenon that can essentially reside outside the individual. This prototype 'knowledge' is therefore more like a high-level form of information which can be more readily taken up in the form of codified knowledge (or not so readily absorbed as a less formal, articulated form of tacit knowledge). Both these aspects of the model are discussed in more detail later.

15.2.2 Intangibility and appropriability

The discussion so far has focused on the issue of knowledge, but associated with this is the issue of tangibility and intangibility. This is arguably a simpler and less problematic concept. Basically, intangibility is anything that cannot be seen or touched. However, the definition and classification of intangible activities, products and assets is generally applied in the negative sense in the first instance, i.e. it is anything that cannot be classified as tangible in nature. Indeed, the fact that service activities involved intangible outputs was an important factor in initial attempts to define and classify services (Greenfield, 1996). Most of the attention on intangibles has been on intangible investments rather than on more general issues associated with intangible activity within services, but whatever aspect of intangibility is considered, measurement problems remain (Den Hertog et al., 1997). More importantly here, although services are associated with intangible outputs and investments, they often involve very close relationships with tangible goods (i.e., artefacts and investments in the production and delivery of those services), such as transport or telecommunication services. The concept of tangibility which is relevant in our analysis is the one which refers to the technology and, more broadly, to the relevant knowledge assets of the firm (Evangelista, 1999, 2000). It is in adopting such a perspective that tangibility is linked to the issue of appropriability.

Tangible, embodied 'knowledge' has been seen as being more appropriable than intangible, disembodied 'knowledge' (Geroski, 1995). In the same vein, it has been generally acknowledged that it is harder to appropriate intangible innovations using formal IPR mechanisms. By contrast, IPRs in general, and patents in particular, are much better at defining and protecting tangible attributes. This may not hold in all cases though. Knowledge embodied in tangible artefacts may be very difficult actually to copy. This is for two reasons:

1. the potential copier may not know what is the key 'essence' (or knowledge attribute) that is of value and therefore worthy of copying; and
2. even once identified, the innovation may be difficult to copy because it involves a series of intangible distributed agents; assets which are difficult to bring together to reproduce that quality or essence.

Indeed this paradox – of the properties of tacitness and intangibility making it difficult to protect innovations by formal IPR mechanisms, but also the very same attributes making them difficult to copy – has been recognised by Geroski, who noted that knowledge transfers were rarely costless, making appropriability less easy than usually observed under these conditions (Geroski, 1995, 93).

15.3 Knowledge and intellectual property regimes in services: a model

15.3.1 The knowledge and intellectual property regimes of services

On the basis of the discussion made in the previous section, Table 15.1 seeks to provide a simple typology of intellectual property regimes in

Table 15.1 A stylisation of knowledge and appropriability regimes in services

Knowledge Type	**Level of Tangibility***	
	Tangible	Intangible
Codified	*1. Patents* Copyright Trademarks Contracts*	*2. Copyright* Trademarks Contracts*
Tacit	*4. Other informal IPR tools* Confidentiality clauses & labour contracts Trademarks	*3. Brands – Trademarks* Other informal IPR tools Contracts**

Notes:
* Referring to innovative output and technology and knowledge-based assets.
** These include confidentiality clauses and labour contracts that seek to restrict intellectual property within the firm.

accordance with the tacit or codified nature of knowledge and according to whether the knowledge assets of firms are tangible or intangible, producing a 2×2 matrix. With services associated with more codified forms of 'knowledge' and producing and working with tangible assets and goods, the intellectual property mechanisms to protect such property would be the patent system (Cell 1). This would cover services directly using and helping to run and construct capital goods and buildings (such as rail transport or logistical services, construction or technical engineering services). Here information associated with this knowledge regime is explicit, codifiable and (ostensibly) transmittable without bias between individuals. The tangible nature of the technological assets that such firms deal with means that the patent regime is ideally suited for protecting the intellectual property of the firm. The level of appropriability may indeed be assumed to be generally 'tight' under these conditions.

For firms working with services whose knowledge bases are codified in nature, but whose services and innovative outputs are intangible in form, it is hypothesised that copyright protection is the most applicable formal IPR regime. Copyright has long been associated with protecting traditional mediums, such as printed words and artistic creations, but copyrights have now been applied to software programs, and other new electronic media now is more important (Cell 2). However, this protection tool is also important for more client-intensive services, such as architectural services, where codified information is the dominant output; this includes drawings and technical specifications. As noted earlier, copyright protection is automatic on the creation of the work, although obviously its protection is retrospectively enforced by taking copyists to court.

As will be discussed below, however, although service firms may be operating under this type of intellectual property regime, they might not seek to use copyright, or indeed other formal IPR tools such as patents and trademarks, by activating it through legal proceedings. This is because innovation: (1) is not worth protecting because it is not novel or innovative enough; and/or, (2) it is too costly to protect because of the likely cost of the legal proceedings required to be deployed in gaining legal protection from copiers.

For services whose knowledge base is tacit and intangible in form, more generalised protection can only be obtained via brands (Cell 3) with recourse to protection by trademarks (although this is also available to services in Cells 1 and 2). Brands are essentially trade names which are used to define a specific product, or set of products. Companies operating under this regime, and not having recourse to patent or copyright, have to use branding via trademarks as the only default option for protection of intellectual property, apart from certain confidentiality, secrecy and labour contract rights (see below). Innovative companies in this knowledge and IPR regime therefore still generate new products which may be worth protecting, and because these new innovative products create new identities they can then be protected under new trademark applications. Thus, although brands may seem less associated with innovative activities, recent studies have shown that brands are more (not less) important for high-technology products than for traditional consumer goods (Ward et al., 1999).

For services involved with tacit knowledge but dealing with tangible goods and technologies, no direct protection measures are used since innovation activity and intellectual property generation are generally low (Cell 4). Trademarks are likely to be less applicable here. However, firms within this regime could still use other intellectual protection tools, such as 'secrecy' and 'lead-time advantages' (Andersen and Howells, 2000), but also more formal non-IPR mechanisms, such as 'confidentiality agreements', 'trade secrets' (Cheung, 1982) and labour contracts, to protect their knowledge bases. These other types of protection tools (which are also available to, and often used more frequently by, organisations in Cell 3 (as well as Cells 1 and 2)) can include a wide range of confidentiality and labour contract agreements associated with, for example:

- new employees required to sign confidentiality agreements restricting disclosure of the activities to third parties of what they are working on;

- employees on leaving the firm are not allowed to work for specified competitors for a specified time;
- restrictions on where employees undertake certain development tasks ('on-site restrictions');
- restrictions on what work employees can undertake away from the firm or what they physically take off-site ('off-site restrictions');
- incentives for long-term stayers, via enhanced pay rates and stock options (associated with such mechanisms as 'golden handcuffs'); and
- 'job design measures' (Liebeskind, 1996) which fragment key tasks so that no one individual can have complete knowledge about a key task, code or development activity within the firm.

These are therefore some of the employee-centred mechanisms that can be used to protect the knowledge base of the company without recourse to traditional IPR mechanisms. As noted earlier, there are other intellectual property protection tools, such as lead-time advantages and secrecy and others which parallel the list above. Thus 'fragmented job design' measures are paralleled by and associated with 'complex product designs' which are themselves decomposed into component parts (Sanchez, 2001). Decomposition is undertaken to overcome the problem of dealing with complexity all in one go, but it also has the advantage of making it much harder for competitors to duplicate the overall design task. The copying and relative loss of a particular component will mean that the competitor still does not have the overall capability to copy the whole product design. Strategies involving technical solutions using a complex product design and embodying intangibles in physical products characterise a final example of protection mechanisms.

In summary, on the basis of the knowledge regimes depicted by the four cells in Table 15.1, the following set of hypotheses can be drawn:

Hypothesis 1 With firms operating in services associated with more codified forms of knowledge, and working with tangible assets and goods, the intellectual property mechanisms to protect knowledge assets are most readily suited to using the patent system (Cell 1).

Hypothesis 2 Companies active in service sectors associated with a regime of codified knowledge and more intangible assets are more likely to use copyright protection in the first instance, but also use trademark protection (Cell 2).

Hypothesis 3 For firms operating in a knowledge regime characterised by a high degree of intangibility and tacit knowledge, the use of trademark registration is often seen as the most important protection mechanism (Cell 3).

Hypothesis 4 For companies in a regime where tangible assets are combined with tacit knowledge, formal IPR mechanisms are of little importance and instead informal strategies play a more relevant role (Cell 4).

15.3.2 An ex ante allocation of service sectors across the different knowledge and IPR regimes

It has already been pointed out that the service sector is far from being an homogeneous entity and this is also true as far as innovation is concerned. Several studies have in fact looked into the service sectors and highlighted the presence of very different (largely sector-specific) innovation patterns, and this should be enough to discourage any simple generalisation about innovation in services. How will the different service activities and sectoral patterns of innovation be distributed across the four knowledge regimes depicted in Table 15.1? We will seek to answer this question in the following section.

An *ex ante* allocation of sectors is provided based on previous studies relating to innovation in service, and more in particular on the few taxonomic exercises which have explicitly taken into account the dimensions of knowledge regimes considered in this study. Thus, Soete and Miozzo have identified different technological trajectories in services recalling Pavitt's, namely: 'supplier-dominated', 'scale-intensive' and 'specialized' and 'science-based' (Soete and Miozzo, 1989). Interestingly in light of the model presented here is the distinction made by Soete and Miozzo within the scale-intensive category between physical and tangible services (transportation and wholesale) and network-intensive services (banking, insurance and other financial services as well as large-scale communication services), which deal with more intangible, network-related technologies. Another relevant distinction is that between IT suppliers and IT users, which include business services providing information-intensive services.

Evangelista has proposed a sectoral taxonomy, which is coherent with the model presented here (Evangelista, 2000). In this taxonomy, service sectors are distinguished according to the overall innovative performance of firms, the nature of the innovation activities carried out, the different knowledge bases (tangible/intangible; codified/tacit) underlying the innovation processes, and the different patterns of interaction through which service firms innovate. Three main sectoral categories are identified: technology users heavily relying on the use of tangible technological assets (for example, transport, wholesale and waste disposal); science and technology-based and technical consultancy sectors, specialised in

Table 15.2 'A priori' allocation of service sectors across knowledge and appropriability regimes

Knowledge Type	**Level of Tangibility***	
	Tangible	Intangible
Codified	*1. Patents* Technology-based Manufacturing & Business Services Transport Services: 1 Communications & Media: 1	*2. Copyright* ICT & Software-based Services Communications & Media: 2
Tacit	*4. Other informal IPR tools* Transport Services: 2 Public Services Other Services	*3. Brands – Trademarks* Financial Services Wholesale & Retail Services

Notes:
* Referring to innovative output and technology and knowledge-based assets.
Transport Services: 1 – infrastructure-intensive.
Transport Services: 2 – customer-focused.
Communication & Media: 1 – infrastructure-intensive.
Communication & Media: 2 – content-based.

the provision of codified knowledge (such as research and development, engineering and technical services); interactive and IT based sectors (such as financial services, advertising and business services) whose distinctive feature consists of innovating through software and close interactions with customers and clients. In addition there are several case studies which have looked in more detail at the innovative patterns and IPR strategies characterising the different service sectors (Blind and J. Edler, 2003).

Bearing in mind the results of these studies the different service sectors can be deductively allocated in our conceptual matrix on knowledge and IPR regimes as shown in Table 15.2. In Cell 1 we expect to find service sectors which rely upon tangible technologies, i.e. sectors that 'deal' with technology that is codifiable and in most cases embodied in tangible infrastructure and artefacts (research and development services, transport and technology-based business services; however, see later). We in turn, therefore, expect that firms in these sectors can and will use formal IPR tools, especially patents. In Cell 2 are likely to be located knowledge-intensive sectors whose innovative output is codifiable and intangible (ICT and software-related sectors, communication and media sectors). The sectoral allocation in Cell 3 and Cell 4 is more difficult to be made on purely deductive grounds. The remaining service sectors are in fact less knowledge-intensive and innovative compared to those located in the other two cells, and the room for codification is in general rather limited. IPR strategies are likely to be less important, less formalised and based on the use of a mix of informal IPR mechanisms. Sectors which make large use of software and intangible technological infrastructure, such as financial services and wholesale and retail services, are likely to be located in Cell 3. In these sectors the key knowledge assets are tacit in nature, being embedded in the human resources employed. Very few sectors can be allocated to Cell 4. Knowledge regimes based on tangible technological assets and tacit competencies, while likely to characterise a few manufacturing industries, are rather difficult to find in services.

There are a couple of service sectors in which different knowledge and IPR regimes are, however, likely to coexist. This is the case for transport, media and communication services. In the case of transport services a first group consists of infrastructure- (and goods-) intensive, or network-based, services. The latter are associated with providing the basic infrastructure or 'backbone' facilities for other (or the same in the case of, for example, railway or basic terrestrial-based telephone services) service providers. These are more likely to be located in Cell 1. The second group is composed of more customer-focused ('front-end') transport services (for example, airlines rather than charter or leasing aircraft operators). As far as media and communication services are concerned, a distinction should be made again between companies providing the communication infrastructure and companies producing content, like news agencies or companies producing movies. As already mentioned the sectoral allocation presented in Table 15.2 is of course a crude one and is based on secondary empirical evidence and deductive reasoning. An empirical testing of our conceptual model will be undertaken in the next section.

15.4 Data, empirical findings and testing of the model

15.4.1 Relevance and nature of IPR strategies in services: data and overview

The empirical findings presented in this section are the research outcome of a much larger European project assessing the relevance and characteristics of IPR strategies in services (Blind et al., 2003). For this purpose a survey was undertaken in 2001 based on direct interviews of a selected sample of service firms across Europe. The objective of the project was the investigation of the basic characteristics of innovation in services and, in particular, of the strategies and specific means used by service firms to 'protect' their knowledge assets, their investments in innovation and the new services introduced into the market. The survey dealt with IPR issues in a broad sense, i.e. on the role played by patents, trademarks and copyrights, or any alternative strategy implemented by firms to appropriate the results of their innovation activities. More than 2000 service firms were contacted and sent a structured questionnaire. Direct interviews of a subsample of 65 firms were undertaken.[5]

The survey showed that each service firm had a very specific attitude both to innovation and to the IPR issue, depending on its business activities, its age and its competitive environment. However, in order to give a general descriptive overview, the distribution of the cases by country and service sector, and other selected characteristics, are displayed here. It should also be pointed out that the surveyed firms were obviously not meant to be fully representative of the universe of service firms in Europe, but instead sought to reflect those firms most likely to be involved and aware of IPR issues in the operation of their businesses. Nevertheless, overall the surveyed firms were reasonably evenly distributed across EU member states according to the relative size of the total firm population of their service sectors, although Scandinavian and Mediterranean countries (except Spain), and the United Kingdom and Germany, were slightly over-represented (Figure 15.1). The sectoral coverage of the survey was somewhat skewed because the study purposely sought to focus on innovative and technology-intensive services, like software development research and development (R&D) and other business services, which may have more of a need for intellectual property protection (Figure 15.2).

In terms of the overall distribution of employees in the different sectors, the wholesale and retail sector employs more than 40 per cent of the workforce in the whole service sector in Europe whereas it represented only 6 per cent of surveyed firms (Eurostat, 1999). This was because the sector is neither very innovative nor technology-intensive. By contrast, the innovative software and R&D service sectors together with the business services (consultancies) account for less than 20 per cent of the workforce, but represented 50 per cent of the surveyed firms. Lastly, in terms of size distribution of the surveyed firms, only 28 per cent of the firms employed less than 99 workers. The distribution is therefore not representative of the real size distribution in the service sector. However, several studies (Cohen et al., 2000) on the role of IPR and especially the use of patents in the manufacturing sector have shown that for small firms intellectual property protection is not only less important for their business success, but also less frequently used because of the high fixed costs to run, for example, a dedicated intellectual property department or to hire a specialist patent lawyer. One major interest of this study is to elucidate the motives of service companies to use IPR tools in their protection strategies. The focus was therefore on companies with some IPR activities and these are more

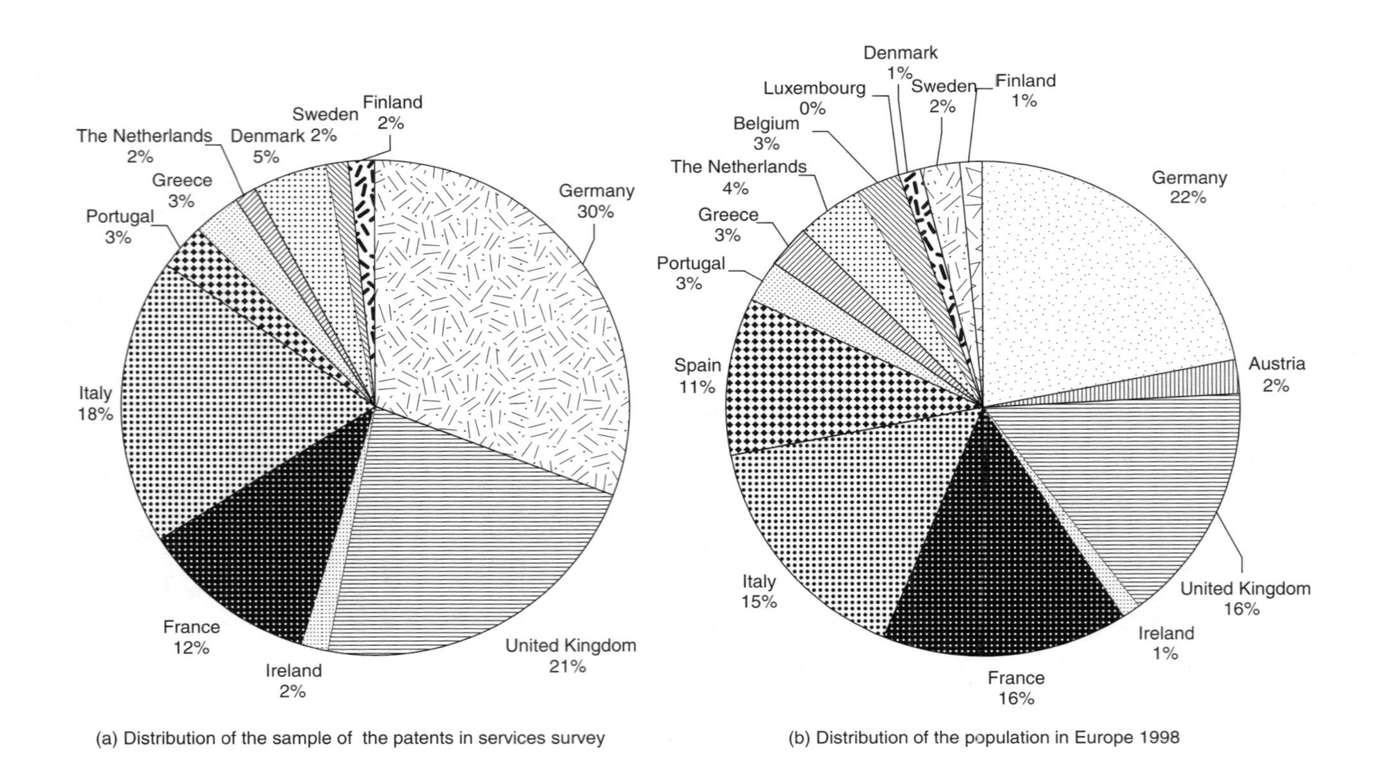

Source: (b) IWF, UNO.

Figure 15.1 Distribution of cases and total service business population by member states

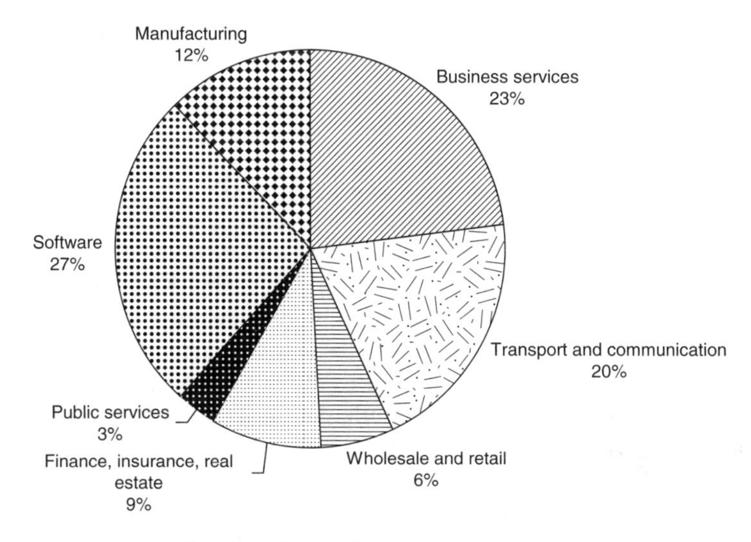

Figure 15.2 Distribution of cases by sector

likely to be large firms.[6] Nevertheless, we also included small companies (with and without patent activities) as a kind of control group.[7]

In relation to the relevance of the appropriability issue in services on the different IPR strategies, the following was ascertained from the survey respondents:

1. Appropriability was not felt as a critical issue by most of the firms although its importance is increasing and plays some role in some service sectors.
2. Patents were the least-used IPR tool (with the exception of very large firms and those firms who had a strong relation to hardware activities).
3. Copyright and trademarks were much more frequently used than patents even though in most cases they are seen as only ancillary IPR mechanisms.
4. A much more frequently used set of property protection mechanisms used by survey firms were more informal IPR strategies, such as employment contracts or secrecy agreements.

15.4.2 The empirical test of the model

In section 15.3 we have presented a conceptual framework linking the IPR strategies of firms to the specific knowledge regimes in which they operate. The model has been represented through a 2×2 matrix used to operationalise a possible correspondence between 'knowledge regimes' and 'IPR regimes'. Following the ways in which the conceptual framework has been developed, the empirical testing of the model will be articulated in two distinct parts, as detailed below.

(a) Verify the existence of distinct knowledge regimes along the tacit/codified and tangible/intangible dimensions This will be done by locating firms in the matrix reported in Table 15.1, using all information provided by firms regarding the tacit/codified and tangible/intangible nature of their knowledge assets, technologies used and innovative output. Firms will be then clustered in the matrix according to their similarity with respect to a set of variables reflecting the very nature of their knowledge assets. We will then assess the extent to which the different clusters have a clear sectoral connotation.

(b) Verify the correspondence between 'knowledge regimes' and 'IPR regimes' The second stage of the empirical testing will consist of assessing whether groups of firms belonging to the different cells differ in the way they protect their knowledge assets (and if such differences exist, whether they are consistent with those envisaged in the model as represented in Table 15.1). In other words, this will mean testing the set of Hypotheses 1–4 listed earlier in section 15.3.1.

Empirical testing of (a) The result of the empirical testing (a) is synthetically presented in Figure 15.3. The latter synthesises the results of a qualitative, but comprehensive analysis of the 65 individual cases. Due to the heterogeneity of the service sectors and individual cases covered, a quantitative multivariate analysis was not feasible. The clustering, therefore, was based on a comprehensive assessment of the companies' innovation activities as a whole. The figure shows that a large number of firms fall well within the four cells. In particular, at the two extremes of the diagonal axis between Cell 1 and Cell 3 we found several clusters of firms which belong to distinct 'knowledge regimes'. As expected, however (and also emerged from the *ex ante* allocation of sectors made in section 15.3), a few clusters are located on the borderline between the different cells, i.e., are characterised as mixed 'knowledge regimes'.

The different clusters (circles) have been labelled according to the most representative sector. It should be pointed out that while in some cases the different clusters have a clear sectoral connotation – like financial services, which also confirms the sectoral allocation outlined earlier – in other cases, such as technical services, sectoral affiliation is less clear-cut. However, overall, the picture confirms the *ex ante* sectoral allocation

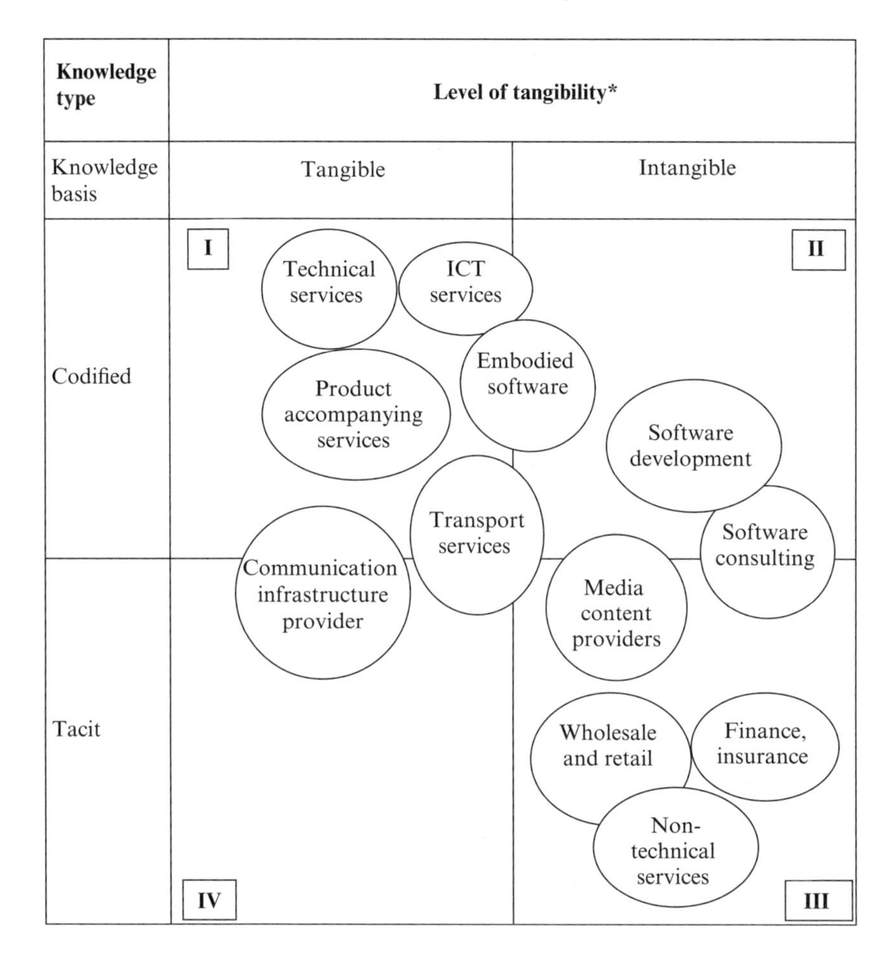

Notes: * Referring to innovative output and technology and knowledge-based assets.

Figure 15.3 Cluster of service sectors differentiated by knowledge regimes

presented in Table 15.2. Thus in quadrant one (Cluster 1), we find firms operating in services which deal mostly with tangible assets and delivering mostly services involving codifiable knowledge. This area covers services directly using and helping to run and/or construct physical goods. Knowledge within these framework conditions has an explicit nature, is codifiable and apparently transmittable. Technical engineering services fall under this category, since they rely or deal directly with tangible assets, like machines or other kinds of embodied technology. Furthermore, mostly codified knowledge in the form of technical documentation is involved. Closely related to technical services are ICT services, which deal with hardware components and are associated more closely with codifiable outputs, like manuals, instructions and switching diagrams. ICT services, and the development of software which is embodied in hardware components and certainly codified in its respective programming language, rely very strongly on codified, mostly intangible knowledge. Finally, product-accompanying services, like maintenance and repair, are closely linked to physical products. A group of companies which deal both with tangible and intangible assets are also included here, and these enterprises are involved in activities such as the development of software embedded in hardware, such as machinery, robots or cars.

The second quadrant (Cluster 2) classifies companies which deal mainly with intangible assets, like software or feature films, which themselves belong to the category of codified knowledge. Although codified forms of knowledge remain important in this regime, tacit knowledge is more involved here both as an input, associated with the problem solving capacity of software developers, and as an output, in the form of reputation or attention. The combination of a high degree of intangibility and tacit knowledge describes the service companies and sectors in the third quadrant (Cluster 3). Here, we find non-technical services in the widest sense including wholesale and retail companies and the financial sector. Customer relations and the skills of the staff, like those of brokers in banks, are decisive for the success of these companies.

The fourth quadrant (Cluster 4) is characterised by a high level of tangibility especially of the infrastructural inputs necessary for providing the services, combined with more or less tacit assets, like specific skills of the staff or individual customer relationships. Typical companies fulfilling these criteria are transport companies, such as freight forwarders, railway companies and airlines. In addition, telecommunication carriers also rely on large and complex technical infrastructures. Furthermore, the knowledge used in these sectors is also codified in the sense of software being used to run the digital telecommunication networks or the physical transport systems, such as airlines or railroads.

Empirical testing of (b) We now pass on to verify whether firms located in the different cells of Figure 15.4 pursue different IPR strategies, and whether these strategies are consistent with the expectation of our model as depicted in Table 15.1. In order to test the model empirically we need to simplify somewhat the picture provided by Figure 15.3. Accordingly the clusters with their focal point in Cell 1 have been aggregated and named 'Hardware-related services'. The clusters located both in Cell 1 and Cell 4 have also been aggregated and named 'Transport and communication

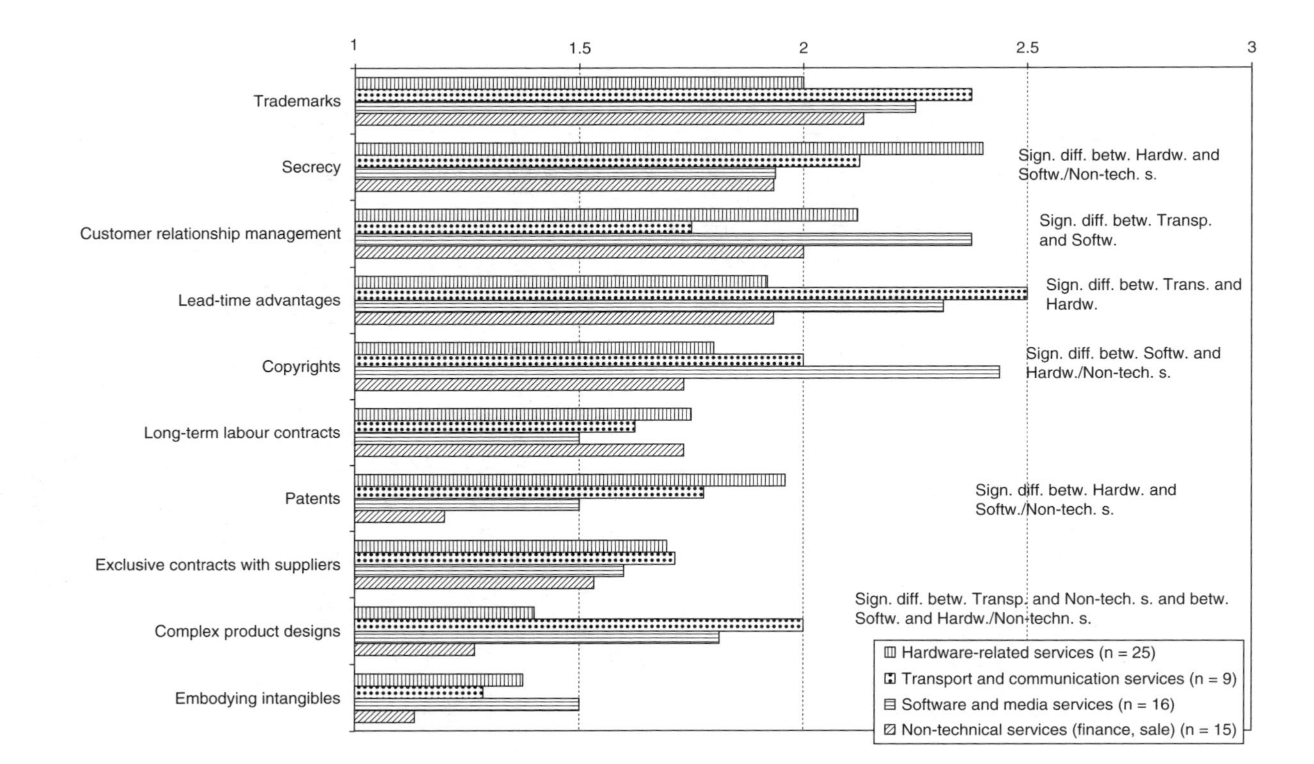

Notes: 1 = not very important to 3 = very important.

Figure 15.4 Importance of different protection instruments differentiated by clusters

sectors'. Similarly clusters in Cells 2 and 3 have been grouped together and labelled as 'Software and media services'. Finally, Cluster 4 is composed of 'Non-technical services', such as banks, insurance companies, wholesalers and non-technical consultancies.

Comparative t-tests were used to test the four hypotheses listed in section 15.3, linking different knowledge regimes to the importance attached by firms to the different IPR mechanism (Figure 15.4). In interpreting the results one has to take into account the small number of observations summing up to just nine in the case of 'Transport and communication services' and 25 in 'Hardware-related services'. These small numbers limit the statistical analysis that can be undertaken, but nevertheless some significant results[8] do emerge using t-tests. With regard to Hypothesis 1 and the use of patents the results are consistent with our expectations. 'Hardware-related services' assign a much higher importance to patents in comparison to 'Software and media services' and 'Non-technical services'. The differences to 'Transport and communication services' can be neglected, and this is in line with the location of this sector in-between Cells 1 and 4.

Furthermore, the second hypothesis emphasising the very high importance of copyrights especially by 'Software and media services' is also confirmed. Thus companies in this cluster assign a significantly higher importance to copyrights than do both 'Hardware-related services' and 'Non-technical services'. However, the difference between the 'Software and media services' and the 'Transport and communication services' is not so significant. This is also consistent with the assumption concerning the importance of copyright-protectable software for network-based transport and communication companies running large and complex hardware infrastructures.

The use of trademarks is not fully consistent with Hypothesis 3. In fact, all four clusters of service companies attach a similar level of importance to the registration of trademarks. This result suggests that trademarks complement more than substitute for the use of patents or copyrights. In relation to Hypothesis 4, concerning the importance of informal protection mechanisms, the picture emerging from Figure 15.4 is rather ambiguous. For the patent-emphasising 'Hardware-related services', secrecy is much more important than for both the software and the non-technical services. However, the hypothesis is supported in terms of the importance of lead-time advantage and complex product designs: 'Transport and communication services' evaluate them significantly higher than 'Hardware-related services' regarding the first informal strategy, and higher than 'Non-technical services' for the second informal strategy. This result confirms that lead-time advantages are crucial for infrastructure or network-based service industries. On the basis of the above results, therefore, the knowledge and appropriability regimes stylised in our model do reflect, reasonably well, the actual use of IPR and other informal protection strategies by the different clusters of service firms. Even if one takes into account the small number of observations involved, significant differences regarding the importance of protection strategies were detected, and these differences were generally consistent with the hypotheses derived from our conceptual framework.

15.5 Conclusions

On the basis of the above analysis, what can be concluded from this analysis and discussion? The study has proposed a model of knowledge property regimes which seeks to allocate and classify innovative service firms by sector according to the type of knowledge, level of the tangibility and intellectual property context that the firm or organisation is situated in. It has revealed that many, if not most, service firms are still not concerned with utilising formal intellectual property mechanisms, even when they do innovate. However this appears to be changing with growth in patenting and IPR activity by service firms across Europe (Blind et al., 2003). The model has successfully described and allocated service firms according to the four main types of knowledge property regimes presented in the chapter. The empirical testing has portrayed a picture which is more complex than that envisaged by crude bipolar distinctions of tangible/intangible, and tacit/codified knowledge, as presented in Tables 15.1 and 15.2. In some instances sectors actually cover more than a single regime group and there is an ongoing dynamic shift in the knowledge regime that different services inhabit. Nevertheless, the basic model that has been produced has been robust and does indicate the broad range of heterogeneous knowledge property regimes that firms operating in a variety of service sectors confront in the modern knowledge-based economy. The model is also useful for considering those firms that do not use formal IPR mechanisms by identifying and indicating in its typology non-formal protection mechanisms. In a dynamic perspective, and taking into account the increasing importance of innovation in services, it is likely that the recourse to IPRs and other protection tools will also increase in this sector.

Lastly, the model acknowledges that the service sector is a much more differentiated sector of the economy with regard to innovation, knowledge and intellectual property protection than the homogeneous and lumpen mass often attributed to it in more general innovation and business models. Service firms do appear to adopt different intellectual property protection strategies according to the different knowledge regimes they operate within. This model may imperfectly highlight such differences, but at the very least it acknowledges that such differences exist and makes an attempt at articulating and mapping out such differences and firm-level approaches.

Notes

1. This chapter arises out of research funded by the Commission of the European Communities (EC Contract ERPHPV2-CT-1999-06). Thanks go to other members of the Patents and Services team, namely Birgitte Andersen, Jacob Edler, Christiane Hipp, Ian Miles and Joanne Roberts and to the valuable comments of Bruce Tether on an earlier draft of the chapter.
2. Referred to hereafter as the 'Patents and Services' survey; see Blind et al. (2003).
3. What Polanyi (1958, 69–131) originally described as 'articulated' knowing; although articulation and explicitness are not directly equivalent.
4. Ancori et al. (2000, 257), though, would argue that this is taking an 'absolutist position on tacit knowledge'.
5. The most common reason for the non-cooperative attitude of the firms was that IPR issues were not important to the overall strategy of the firm. This was because of either too little internal innovative effort of the firms contacted, or because of the specific nature of innovation and knowledge assets in services, which meant that formal IPR mechanisms were not seen as important protection mechanisms. Some mainly large firms were also concerned about 'confidentiality' issues.
6. As already noted in the introduction, it should be acknowledged that less than 10 per cent of the innovative service companies questioned in CIS2, for example, applied for one or more patents.
7. For more details of the wide range in-depth qualitative and quantitative information collected as part of the survey see Blind et al. (2003).
8. Differences which reach a level of significance below 10 per cent are reported here.

References

Ancori, A., A. Bureth and P. Cohendet (2000), 'The economics of knowledge: the debate about codification and tacit knowledge', *Industrial and Corporate Change*, **9**, 255–287.

Andersen, B. and J. Howells (2000), 'Intellectual property rights shaping innovation in services' in B. Andersen, J. Howells, R. Hull, I. Miles and J. Roberts (eds), *Knowledge and Innovation in the New Service Economy*, Cheltenham, UK and Northampton, MA, USA: Edward Elgar, pp. 229–247.

Argyris, C. and D.A. Schon (1978), *Organisational Learning: A Theory of Action Perspective*, New York: Addison-Wesley.

Arrow, K. (1962), 'The economic implications of learning by doing', *Review of Economic Studies*, **29**, 155–173.

Blind, K. and J. Edler (2003), 'Idiosyncrasies of the software development process and their relation to software patents: theoretical considerations and empirical evidence', *Netnomics*, **5**, 71–96.

Blind, K., J. Edler, U. Schmoch, B. Andersen, J. Howells, I. Miles, J. Roberts, C. Hipp and R. Evangelista (2003), 'Patents in the service sector', Final Report to DG Research of the European Commission, Karlsruhe, 2003 (EC Contract ERPHPV2-CT-1999-06).

Cheung, S.N.S. (1982), 'Property rights in trade secrets', *Economic Inquiry*, **20**, 40–53.

Cohen, W.M. and D.A. Levinthal (1990), 'Absorptive capacity: a new perspective on learning and innovation', *Administrative Science Quarterly*, **35**, 128–152.

Cohen, W., R.R. Nelson and J. Walsh (2000), 'Appropriability conditions and why firms patent and why they do not', National Bureau of Economic Research, Working Paper 7552, February.

Delmont, S. and P. Atkinson (2001), 'Doctoring uncertainty: mastering craft knowledge', *Social Studies of Science*, **31**, 87–107.

den Hertog, P., R. Bilderbeek and S. Maltha (1997), 'Intangibles: the soft side of innovation', *Futures*, **29**, 33–45.
Ellis, H.C. (1965), *Transfer of Learning*, New York: Macmillan.
Estes, W.K. (1970), *Learning Theory and Mental Development*, New York: Academic Press.
Eurostat (1999), *Services in Europe: Data 1995–1997*, Eurostat Datashop Luxembourg.
Evangelista, R. (1999), *Knowledge and Investment: The Sources of Innovation in Industry*, Cheltenham, UK and Northampton, MA, USA: Edward Elgar.
Evangelista, R. (2000), 'Sectoral patterns of technological change in services', *Economics of Innovation and New Technology*, **9**, 183–221.
Geroski, G. (1995), 'Markets for technology: knowledge, innovation and appropriability', in P. Stoneman (ed.), *Handbook of the Economics of Innovation and Technological Change*, Oxford: Blackwell, 90–131.
Greenfield, H.I. (1996), *Manpower and the Growth of Producer Services*, New York: Columbia University Press.
Howells, J. (1996), 'Tacit knowledge, innovation and technology transfer', *Technology Analysis and Strategic Management*, **8**, 91–106.
Liebeskind, J.P. (1996), 'Knowledge, strategy, and the theory of the firm', *Strategic Management Journal*, **17** (Winter Special Issue), 93–107.
Polanyi, M. (1958), *Personal Knowledge: Towards a Post-Critical Philosophy*, London: Routledge & Kegan Paul.
Polanyi, M. (1961), 'Knowing and being', *Mind NS*, **70**, 458–470.
Polanyi, M. (1962), 'Tacit knowing', *Review of Modern Physics*, **34**, 601–616.
Polanyi, M. (1966), 'The logic of tacit inference', *Philosophy*, **41**, 1–18.
Polanyi, M. (1967), *The Tacit Dimension*, London: Routledge & Kegan Paul.
Rosenberg, N. (1982), *Inside the Black Box: Technology and Economics*, Cambridge: Cambridge University Press.
Sanchez, R. (2001), 'Product, process, and knowledge architectures in organizational competence', in R. Sanchez (ed.) *Knowledge Management and Organizational Competence,* Oxford: Oxford University Press, pp. 227–250.
Soete L. and M. Miozzo (1989), 'Trade and development in services: a technology perspective', Working Paper No. 89-031, MERIT, Maastricht.
Stiglitz, J.E. (1987), 'Learning to learn, localised learning and technological progress', in P. Dasupta and P. Stoneman (eds), *Economic Policy and Technological Performance*, Cambridge: Cambridge University Press, pp. 125–153.
Teece, D.J. (1986), 'Profiting from technological innovation: implications for integration, collaboration, licensing and public policy', *Research Policy*, **15**, 285–305.
von Hippel, E. (1994), '"Sticky information" and the locus of problem solving: implications for innovation', *Management Science*, **40**, 429–439.
Ward, S., L. Light and J. Goldstine (1999), 'What high-tech managers need to know about brands', *Harvard Business Review*, July–August, 85–95.
Ziman, J. (1978), *Reliable Knowledge: An Exploration of the Grounds for Belief in Science*, Cambridge: Cambridge University Press.

PART IV

INNOVATION IN SERVICES AND THROUGH SERVICES: IMPACT ANALYSES (GROWTH, PERFORMANCE, EMPLOYMENT AND SKILLS)

16 Innovation and employment in services

Rinaldo Evangelista and Maria Savona[1]

16.1 Introduction

This chapter addresses some of the key theoretical and empirical issues related to the role of innovation to explain the dynamics of employment in services. This is a very ambitious aim: the effect of innovation on employment is generally complex and this complexity increases in the case of the service sector. This is mainly because the theories, concepts, definitions and methodological tools available have been developed with (explicit or implicit) reference to the manufacturing industry, and rely on a narrow and technology-based concept of innovation.

It is well known that the effects exerted by technological change on employment are at the centre of the theoretical and empirical dispute in economic theory that goes back to the beginnings of this discipline and, after more than two centuries, is still unresolved. We start with the perhaps rhetorical, but not trivial question of why we should focus on the service sector to study the innovation–employment relationship. There are two answers to this question.

The first and most obvious response is that services, and particularly the analysis of the economic effects of innovation in services, have been rather overlooked in economics. Although services is the fastest-growing sector in advanced economies, our understanding of the specific role of innovation in fuelling the process of tertiarisation and employment growth remains rather vague and fragmented, based more on impressions and anecdotes than on robust and theoretically grounded empirical evidence.

The second justification for the focus on services is associated with the (so-called) peculiarities of the tertiary sector with regard to the variety of economic features of firms, activities and markets, whereby it is supposed to make the locus and nature of innovation activities different from those in the manufacturing sectors. This point has been much debated in the literature, and there are contrasting views and hypotheses but no definitive conclusion. Furthermore, this debate has focused on the supposed similarities and differences between manufacturing and service industries in the nature of innovation, giving much less attention to the implications of these particularities for the conceptualisation and analysis of the economic – and particularly employment – impact of innovation. Thus, we need to explore whether and to what extent the lessons and stylised facts applied to the manufacturing industry hold also for the service sector.

There is a fundamental methodological lesson to be learned from existing work (on the manufacturing sectors) which can be applied to services. This is the multilevel nature of the innovation–employment relationship. The employment effects of technological change can (and should) be analysed at different levels of aggregation (i.e., micro, sectoral, macro) as at each level of analysis different drivers of innovation exert an impact on employment. It is the multidimensional nature of this relationship that has produced the extreme heterogeneity in the hypotheses, approaches and empirical results in the literature.

In order to study the service sector, we need a broad perspective of the impact of innovation on employment that takes account of different sets of innovation-related 'drivers' of employment growth and combines them at the theoretical and empirical level. The first acts in the long run at the macro-structural level and nowadays is largely based on the paradigmatic change brought about by information and communication technologies (ICTs) in the models of production, trade and consumption. A second, micro-related, driver is associated with the increasing importance of innovation as a key competitive asset in firms' strategies. Between these two levels of the analysis (and connecting them) there is a third driver, which is related to the presence of very different (and largely sector-specific) innovation regimes and modes. It is the heterogeneity in innovation regimes that is responsible for most of the differences in productivity and growth performance among sectors and, therefore, is an essential component of the process of structural change in the composition of sectoral employment in economies.

This chapter aims to review the main hypotheses, theoretical approaches and empirical stylised facts on the relationship between innovation and employment in services, across the three levels of analysis identified above. Given the space constraints and the complexity of the issue, this review does not pretend to be exhaustive; rather, we want to provide a way to systematise the literature to enable the identification of still unresolved and debatable issues, and to set priorities for future research agendas.

The chapter is structured as follows. Section 16.2 reviews the literature on the structural changes in the sectoral composition of the economy which have led to the growth in service employment over recent decades. We provide a macro-structural reading of these changes, emphasising and contrasting the role of demand- versus supply-related drivers of the process of tertiarisation. Section 16.3 focuses on the 'meso' level of analysis and the sectoral heterogeneity in innovation regimes and its diverse effects on employment, and section 16.4 discusses the micro-dynamics of innovation and its effect on employment. The concluding section summarises the main stylised facts that emerge from our review of the conceptual and empirical literature and identifies future challenges in this field of research.

16.2 Innovation and employment at the macro level of analysis: old and new views on structural change and tertiarisation

The advanced economies – and increasingly some of the fast-growing developing economies such as India (Dasgupta and Singh, 2005) – are experiencing processes of structural change that are producing profound modifications to the sectoral structure of employment towards services. This process of tertiarisation has been ongoing for several decades in Europe and advanced Organisation for Economic Co-operation and Development (OECD) countries, resulting in an increasing number of attempts to identify, conceptually and empirically, its determinants and impact on aggregate economic growth (see Peneder, 2003; Peneder et al., 2003; Parrinello, 2004; Savona and Lorentz, 2005; Pugno, 2006; Schettkat and Yocarini, 2006; Kox and Rubalcaba, 2007, among the most recent studies). With some variation across countries, there is a general pattern of a monotonically increasing share of service employment and a reduction in the manufacturing employment base (OECD, 2008a; Kox and Rubalcaba, 2007). Employment growth, at least among the most advanced economies, is increasingly depending on the expansion of the service sector and, particularly, on the capacity of services to generate new jobs (or 'absorb' redundant manufacturing jobs).

Identifying the relationship among technical change, innovation and employment growth in services at the macro level of analysis implies a theoretical and empirical assessment of how technological progress and innovation affect structural change in the sectoral composition of employment in economies. The determinants and effects of tertiarisation of the labour force have been a concern for many scholars – now considered the 'classics' in the economics of services.

The traditional explanation of the process of tertiarisation is linked to the cost-disease argument and the 'technological stagnancy' in services, which negatively affects labour productivity growth. This explanation has been superseded by an account that emphasises the propulsive effect on productivity gains of ICTs in the 'New Economy'. These arguments – and the streams of literature that they have produced – provide opposing visions of the impact of technical change on employment growth in services. Below we provide a selective review of the main arguments in the theoretical and empirical literature explaining the macro-structural patterns behind employment growth in services.

16.2.1 Technological stagnancy and Baumol's cost disease

Traditional explanations of employment growth in services that refer to the (lack of) technical change stem from classic writings in the economics of services, culminating with Baumol's cost-disease argument (Baumol, 1967). Are such arguments valid today? Does the most recent empirical evidence support them?

The earliest speculations on the shift in employment from manufacturing to services date back to the 1930s (Clark, 1940; Fisher, 1935; Fourastié, 1949). Many of these contributions were ancestors of Baumol's cost-disease argument, which is based on the differentials in labour productivity growth in services with respect to the rest of the economy. The second generation of debates on the nature and consequences of services growth developed in the 1960s. The cost-disease concept has continued to be implacably associated with the poor productivity performances of service industries, since Baumol and Bowen (1966) first referred to the 'cost disease of a string-quartet performance'. The concept was extended by Baumol (1967) in a very influential paper, and in later contributions by Baumol and colleagues (Baumol et al., 1985, 1989; Baumol, 2001, 2002).

The productivity gap between services and the rest of the economy is argued to be due to the lack of opportunities for economies of scale and low capital intensity.[2] The concepts of technical progress and productivity gains are tightly linked to capital intensity, which is a characteristic of the manufacturing sector. The positive employment performance in services is therefore the result of low capital intensity and productivity gains.

Baumol assumes that the demand for services and goods, measured at constant prices, does not depend on income levels and that, at the aggregate level, the share of services in total output is constant over time and across countries. However, as the increase in (labour) productivity levels is lower in services than in manufacturing, and combines with low productivity growth over time, high-income countries will experience higher shares of employment in services. The cost-disease argument has long been evoked to explain the employment performance of services from the gloomy perspective of technological stagnancy, which negatively affects productivity performance and, therefore, favours employment growth in services.

Baumol's original argument was 'revised' by Baumol himself in a more recent contribution (see Chapter 4 by Baumol in this volume).

The latest strand of the literature on structural change and the tertiarisation of employment attempts to embrace and account for the role of technical change and its impact at the macro level of analysis. While the old view of structural change in employment towards services in the

Baumolian and post-Baumolian literature was concerned more with productivity differentials and productivity slowdown over time, Baumol and Fuchs's (see Fuchs, 1968, 1977) idea of a 'structural burden' linked to the tertiarisation of employment is being challenged empirically, more and more frequently (Peneder, 2003; Peneder et al., 2003; Savona and Lorentz, 2005; Kox and Rubalcaba, 2007; see also Van Ark et al., 2003a, 2003b, 2007; Djellal and Gallouj, 2008).

Some of the most recent critiques of Baumol's 'constancy of services' hypothesis show empirically that up to the mid-1970s this argument held. Also, as ten Raa and Schettkat (2001) and Appelbaum and Schettkat (2001) argue, the main assumption underlying the cost disease is linked to an assumption of zero price elasticity in demand for services or to a mechanism of perfect compensation between negative price elasticity and positive income elasticity of demand for services, which is rarely (practically never) the case.

Ten Raa and Schettkat (2001) point out that there continues to be a 'service paradox' at the centre of the (macro)economic debate on the shift in employment towards services. This paradox lies in the empirical fact that advanced economies are continuing to experience sustained growth in real output and employment in services, despite the trend towards increasing input costs and prices. These issues have been reprised in an edited collection by Gregory et al. (2007), and the role played by patterns of final consumption is analysed empirically in the case of Europe and the US. Data on consumption by private households, particularly in the case of US compared to Europe, also confirm the existence of a paradox: the patterns of final consumption explain most of the service growth in the US, despite increasing prices (Gregory et al., 2007).

In an attempt to explain this paradox, ten Raa and Schettkat (2001) call for a general 'change in demand conditions' not related exclusively to the changing patterns of final consumption which are linked to pure prices and income effects.

The change in demand conditions has been tested in some formalised contributions. Lorentz and Savona (2008) propose a number of scenarios related to the process of tertiarisation on the basis of a micro-based growth model, calibrated using actual data from OECD input–output tables. Lorentz and Savona (2008) test the validity of Baumol's cost disease versus a more Schumpeterian-like scenario and find that some of the most recent employment trends can be explained by innovation and changing patterns of intermediate demand.

The effects of the joint role of changes in production and consumption patterns on structural change and growth were also simulated by Ciarli et al. (2009). They find that the impact of technical change on output and employment structural changes is crucially linked to the pattern of changes in final consumption triggered by the introduction of new products. This confirms that the idea of technological stagnancy has been superseded by the notion that the role of change in patterns of consumption and intermediate demand triggered by innovation favour sectoral structural changes in employment towards services.

As in the old debate on structural change and tertiarisation in which advocates of the post-industrial era and economic progress[3] were challenged by the 'pessimist' views of a 'productivity slowdown', sustaining the cost-disease argument, the most recent literature encompasses hyper-optimistic views which are tempered by more balanced views of what is meant by the knowledge economy. We explore these issues in more depth below.

16.2.2 ICTs, knowledge-intensive business services and related myths

The 'changes in demand conditions' evoked by ten Raa and Schettkat (2001) are likely to be linked to changes in the division of labour across sectors and in the composition of intermediate demand. Since the mid-1990s, growth in intermediate services has been receiving renewed and increased attention (among the most recent, see Savona and Lorentz, 2005; Guerrieri and Meliciani, 2005; Kox and Rubalcaba, 2007). These contributions highlight technical change and the role of innovation – traditionally confined to the manufacturing sector – as crucial to the service domain. Technology and innovation are becoming ever more important in sustaining and shaping long-term growth in services. ICTs are playing a pivotal role in changing the ways in which most 'traditional' services are produced, traded and delivered, and offering opportunities for the generation of new activities in many service industries (Petit, 1995; Petit and Soete, 2001; Van Ark et al., 2003a, 2003b, 2007). Within services, a subset of high-value-added industries, the so-called knowledge-intensive business services (KIBS) are supposed to play a fundamental role in the ongoing paradigmatic change towards a knowledge-based economy (Boden and Miles, 2000).

More particularly, the literature on ICTs and KIBS emphasises two fundamental aspects as being responsible for the emergence of the knowledge-based and 'post-industrial' economy. One is related to the changing patterns of final consumption towards luxury and intangible products; the other is related to the changing composition of intermediate demand towards knowledge-based inputs and processes (see Savona and Lorentz, 2005; and Montresor and Vittucci Marzetti, 2007; for recent reassessments). The latter implies the emergence of new firms, markets and sectors and the changing organisation and size of existing ones, based on increased codification and commoditisation of knowledge inputs and outputs.

The overall impact on employment of these processes is still being debated. To our knowledge, the role of labour productivity change due to ICTs in the net impact on service employment is not supported by solid empirical evidence, and many questions remain:

- Is the new ICT paradigm responsible for the general increases in labour productivity in – mostly intermediate – services across the advanced countries?
- If so, are these labour productivity gains the results of higher output growth in these sectors rather than employment cuts, such that they translate into a net employment impact at the aggregate level?
- If not, do countries experience net positive employment effects at the aggregate level – demonstrating that a sort of cost-disease scenario still holds?
- Are the two ICT-related growth patterns of services – due to final and intermediate demand – self-reinforcing or is there a dominant intermediate demand-led mechanism?

Empirically supported answers to these questions are not unanimous and thus do not confirm the panacea of the knowledge-based economy for countries' employment performance, especially in the tertiary sector. Many empirical contributions point to the strong growth of KIBS by focusing on the determinants of structural change on aggregate growth, trade and competitiveness (Guerrieri and Meliciani, 2005; Savona and Lorentz, 2005; Kox and Rubalcaba, 2007; Gregory et al., 2007). The relative role of intermediate demand versus other components of aggregate demand, which sustains the importance of changes in the intersectoral division of labour and higher sectoral specialisation, is confirmed by empirical work based on input–output structural decomposition analysis (see Savona and Lorentz, 2005; reprised in Kox and Rubalcaba, 2007).

Most of these contributions confirm that the role of intermediate demand accompanied by the use of ICTs has determined the positive employment performance in KIBS. Guerrieri and Meliciani (2005), for instance, argue that national specialisation in manufacturing sectors which are intensive users of KIBS determines the ability to specialise in internationally competitive business services.

The majority of contributions dealing with the impact of ICTs on services, however, focus on KIBS; very few contributions attempt to extend the analysis of the impact of ICT use to other service sectors (exceptions include Djellal, 2000, 2002; and Djellal and Gallouj, 2005, on the cleaning and health services). The crucial issue within a macro-structural perspective is to assess the overall impact of ICTs on the service sector as a whole, including poor innovators.

Few scholars have challenged the knowledge-based economy rhetoric or pointed to the potential drawbacks of digitalisation on employment, particularly low-skilled jobs which comprise the majority of tertiary employment (see, e.g., Daveri, 2002; Crespi et al., 2006; Marrano et al., 2007, for interesting assessments of the UK case). Despite the tendency for the literature to stress the positive economic effects of ICTs, the impact on employment is less clear-cut as services are among the heaviest users and generators of ICT and these technologies are often introduced to rationalise production and delivery structures and to lower costs.

It should be stressed that the preoccupation of authoritative institutions, such as the OECD (2008a), has shifted towards other issues related to specific segments of the labour market, such as youth or the disabled, or issues such as gender, or the informal labour market. Although all these aspects are very important, they refer to specific segments of the labour market. The question of whether and the extent to which technical change has an overall negative impact on aggregate employment, depending on the different effects at sectoral level and the relative sectoral weight of different services, seems to be losing its place on institutions' and policymakers' agendas.

In this context, a careful empirical assessment of the overall impact of technological change on employment in services is even more important. It implies taking account empirically of the joint presence of positive and negative direct effects, and the existence of the complex set of compensating mechanisms operating within the service sector, between the service and the manufacturing sectors, and increasingly at the international level.

16.2.3 Services and the compensation theory

The debate on the actual functioning and strength of the compensation mechanisms (see Spiezia and Vivarelli, 2002; Vivarelli and Pianta, 2000, for a review of the traditional compensation mechanisms) is still open, crucially depending on the existence of perfect competitive markets, complete input substitutability and the validity of the so-called Say's Law which guarantees that changes in 'supply conditions' (i.e. productivity growth and supply of new products) always generate corresponding (market-clearing) changes in demand (Katsoulacos, 1986). On this controversial issue, the empirical literature is far from having provided a conclusive answer. However, on both theoretical and empirical grounds there seems to be some agreement that the compensation mechanisms exert most of their beneficial effects on employment in services rather than in the manufacturing industry (Petit and Soete, 2001; Spiezia and Vivarelli, 2002; Pianta, 2005). The presence of a positive link between technological change and employment in services is explicitly or implicitly based on the following assumptions.

The first one is that the employment–output growth elasticity in services is, and will continue to be, positive and in any case much higher than that characterising the manufacturing sector. This is in turn due to the intrinsic labour-intensive nature and the high 'qualitative requirements' of many service functions and outputs. These factors reduce the opportunity of adopting labour-saving technologies and conversely tend to favour innovation strategies aiming at maintaining or even enhancing the qualitative content of the services provided (Wolff, 2002).

Secondly, as already mentioned, some services branches, specifically those dealing with the generation and dissemination of information and other types of knowledge assets, are probably the most dynamic players in the current shift towards the so-called knowledge-based economy. As mentioned above, the traditional view, which portrayed services as 'sheltered' sectors, characterised by low productivity and poor technological performance, is now superseded by one emphasising the high technological performance of sectors such as ICT-related services, telecommunications, or high-value-added business services (Miles et al., 1995; Miozzo and Miles, 2002; Kox and Rubalcaba, 2007). In other words an important section of services could benefit from the compensation mechanism 'via new machines', which nowadays takes the form of ICTs and other tradable knowledge assets.

Thirdly, the fact that ICTs might also be used to cut down costs, rationalise production and delivery processes, and save jobs and skills cannot be overlooked. However, it is assumed that such effects are confined to a limited section of services, and/or that they could be offset by compensation mechanisms operating via the reduction of prices, especially in markets characterised by a high demand–price elasticity.

Fourthly, and closely connected to the previous points, it might be argued that it is in the service sector where the new technological paradigm provides the highest opportunities for the introduction of entirely new or improved services.

Fifthly, technological imperative might be less stringent in (most) services than in the manufacturing sector, allowing for some degree of input (capital–labour) substitutability in accordance with changing conditions in the labour market.

Last but not least, the benefits of compensation mechanisms operating through decreased prices and increased incomes might get concentrated in services rather than in manufacturing. It is in fact plausible to assume that the increased purchasing power and profits generated outside the tertiary sector end up being mostly spent in services (according to Engel's Law) (see also Gregory et al., 2007).

Each of these assumptions is plausible but, unfortunately, most of them are not supported yet by systematic empirical evidence.[4] Furthermore the highly heterogeneous nature of services makes any generalisation of such arguments risky and likely misleading (see Gallouj and Savona, 2009 for a recent reassessment).

16.3 The heterogeneous nature of innovation in services: implications for employment

We have shown that an empirical exploration of the complex links between technological change and employment in services is very difficult. Part of this difficulty is due to the highly diverse nature of the service sectors and the locus and nature of the innovation activities taking place in this heterogeneous universe of firms, activities and markets. The multiform nature of innovative activity and its sectoral specificity are underlined in the huge literature on this subject, although almost exclusively with reference to the manufacturing industry. In addition to activities that generate new technological knowledge, such as research and development (R&D), attention is being paid to less-formalised technological activities linked to design, and to the processes of technology adoption and diffusion (Evangelista, 1999; OECD, 2005). This wider perspective on the nature and locus of innovation activities is even more necessary for services where formalised R&D and 'hard' technological activities can be expected to play only a marginal role (Evangelista, 2000; Abreu et al., 2008; Djellal and Gallouj, 2007; Gallouj and Savona, 2009). In-depth analyses of Community Innovation Survey (CIS) data show that the service industries differ in the amount of resources devoted to innovation and ways in which new knowledge is generated or adopted, and that interindustry variance in firms' innovation strategies and performances is greater in the service branch than in the manufacturing sector (Evangelista, 2006). This heterogeneity is related to the type of knowledge bases characterising the different service industries, the type of innovation inputs used, and the technological and knowledge outputs that they deliver to other sectors (Evangelista, 2000; Evangelista and Savona, 2003; Cainelli et al., 2006). This implies that the economic effects of innovation on economic performance and employment are likely also to be rather heterogeneous.

Several taxonomic exercises have been conducted to explore the heterogeneous universe of services and identify the main typologies of services and sector specfic innovation patterns. Soete and Miozzo (2001) identify different technological trajectories in services, recalling Pavitt's

(1984) 'supplier-dominated', 'scale-intensive', 'specialized supplier' and 'science-based' categories. Within the scale-intensive category, Soete and Miozzo distinguish between physical or tangible services (transportation and wholesale) and network-intensive services (banking, insurance, other financial services and large-scale communication services), which deal with more intangible, network-related technologies. They also distinguish between information technology (IT) suppliers and IT users, which include business services providing information-intensive services. Evangelista (2000) proposes a sectoral taxonomy comprising three main sectoral categories: technology users, which rely heavily on the use of tangible technological assets (e.g., transport, wholesale, waste disposal); science- and technology-based and technical consultancy sectors, specialised in the provision of codified knowledge (such as R&D, engineering and technical services); and interactive and IT-based sectors (such as financial services, advertising, business services), which are distinctive in innovating through software and maintaining close relations with customers and clients.

Castellacci (2006, 2007) proposes a taxonomy that integrates manufacturing and services, and stresses the increasing importance of vertical linkages and intersectoral knowledge exchange between these interrelated branches of the economy. Sectors are grouped according to their function in the economic system as providers and/or recipients of advanced products, services and knowledge, and the dominant innovative mode characterising their technological activities. The idea is based on the emergence of the ICT paradigm and the nodal role assumed by a new set of general-purpose technologies, which has changed the structure of the sectoral linkages that fuel the process of technological accumulation and diffusion, and economic growth. This taxonomy is applied to the innovative activities and economic performance of the manufacturing and service industries in Europe, based on information from the fourth CIS.

Crespi and Pianta (2008) and Bogliacino and Pianta (forthcoming) provide substantial evidence of the sector-specific nature of technological regimes, based on a unique sectoral dataset matching a wide set of CIS-based indicators. Both studies emphasise the distinction between cost (process-oriented) competitiveness (CC) and technological (product-oriented) competitiveness (TC), strategies that are shown to play a crucial role in differentiating the technological regimes of sectors, including services.

Although these taxonomies were not designed with the purpose of highlighting the effects of innovation on employment, all of them – to varying degrees – are informative. Soete and Miozzo (2001) and Castellacci (2007) emphasise the role of a particular subset of service sectors (ICT-related and KIBS) as the new drivers of the process of structural change. The high and increasing demand for these new knowledge-intensive services is an important factor explaining the rapid employment growth in these industries.

As already mentioned, ICTs play a key role in the service sector although the employment impact of these technologies may differ depending on the specific role (and set of technological opportunities) of ICTs in the strategies of firms and innovation regimes. The Soete and Miozzo (2001) and Evangelista (2000) taxonomies make a clear distinction between ICT user and producer services. The relevance of this distinction for the innovation–employment relationship is addressed empirically in a contribution by Evangelista and Savona (2003), which is one of the first of very few attempts to investigate the diversified effects of innovation on employment, and take explicit account of the presence of very diverse and sector-specific innovation regimes and of the different roles played by ICTs in these strategies.

What is common across all service industries is that high-skilled jobs replace blue-collar employment. The net employment outcome, however, varies widely across sectors. In the most innovative and knowledge-intensive sectors, there is generally a positive impact, while in the case of financial-related sectors, the typical capital-intensive service industries (transport-related services) and some of the most traditional branches of services (trade and waste disposal), there is a negative effect. Compared with much of the current literature emphasising the positive impact of ICTs on employment, the evidence provided in Evangelista and Savona's (2003) article draws a more varied picture, emphasising the diversified impact of ICTs. A crucial distinction is made in this contribution between sectors producing, adopting and disseminating new ICT services (and knowledge-based inputs more broadly), and sectors that use these technologies as process innovations to cut costs and rationalise production and distribution structures. While in the first group of sectors (science- and technology-based) innovation activities are associated with the introduction of new services and the emergence of new markets, in the second group of sectors (technology and ICT users) the labour-saving nature of ICTs usually prevails.

Bogliacino and Pianta (forthcoming) present a similar argument and provide a certain amount of evidence that the distinction between CC and TC strategies is also applicable to services and represents a powerful conceptual tool to explain the different long-term dynamics of productivity and employment across the manufacturing and service industries.

16.4 The micro-dynamics of innovation, economic performance and employment outcomes

Since the end of the 1990s there has been a sort of microeconomic shift in the empirical literature investigating the economic impact of innovation, due to: (1) the increasing dissatisfaction with aggregate analyses that do not grasp the heterogeneity of firms' innovative behaviour or the different technological sources of firms' competitiveness;[5] (2) the widely perceived need to strengthen empirically the micro-foundation of the innovation–economic performance relationship grounded in the evolutionary literature; (3) the growing availability of large firm-level datasets which allow the use of more sophisticated econometric techniques, dealing more effectively with endogeneity problems and controlling for a wide set of fixed effects. The most recent contributions in this field attempt to go beyond the linear and R&D-centred vision of the innovation process and its effects on performance, through increased use of comprehensive measures of firms' innovation inputs and outputs. CIS data are being used increasingly in this new body of empirical research, although the potential of this statistical source is far from being fully exploited.[6] Most studies followed the so-called CDM model proposed in Crepon et al.'s (1998) seminal contribution.

Within this new micro-oriented stream of empirical research, services are an underinvestigated area, the majority of analyses being focused on the manufacturing sector.[7] However, there are a few empirical works focusing on the effects of innovation on the economic performance of both manufacturing and service firms. Although these pioneering studies do not explicitly address the innovation–employment relationship, or deal in detail with the specificities of services, they have some implications for these areas in services and show that, similar to the situation in manufacturing, innovation plays a significant role in productivity in services (OECD, 2008b) and in other dimensions of firms' economic performance (Cainelli et al., 2006; Lööf and Heshmati, 2006; Van Leeuwen and Klomp, 2006; Abreu et al., 2008, 2010; Castellacci and Zheng, 2008).

The extent to which increased firm-level performance generates new jobs is an aspect that has been overlooked by both conceptual and empirical studies.[8] The way that innovation affects the different measures of economic performances and the existence of possible trade-offs between economic and productivity growth is a fundamental aspect that requires investigation, and one that is clearly linked to the ultimate impact on employment of firms' innovation strategies. Thus, it is crucial to distinguish conceptually (and disentangle empirically) the direct and indirect employment effects of innovation (see Figure 16.1). While the direct effects are similar to those discussed in section 16.2, the indirect firm-level effects are those produced as the result of competition among firms. Direct effects are produced (in the short run) by the introduction of labour-saving (process-oriented) innovations and consequently are expected to be negative; indirect (positive) effects on employment are expected (although with

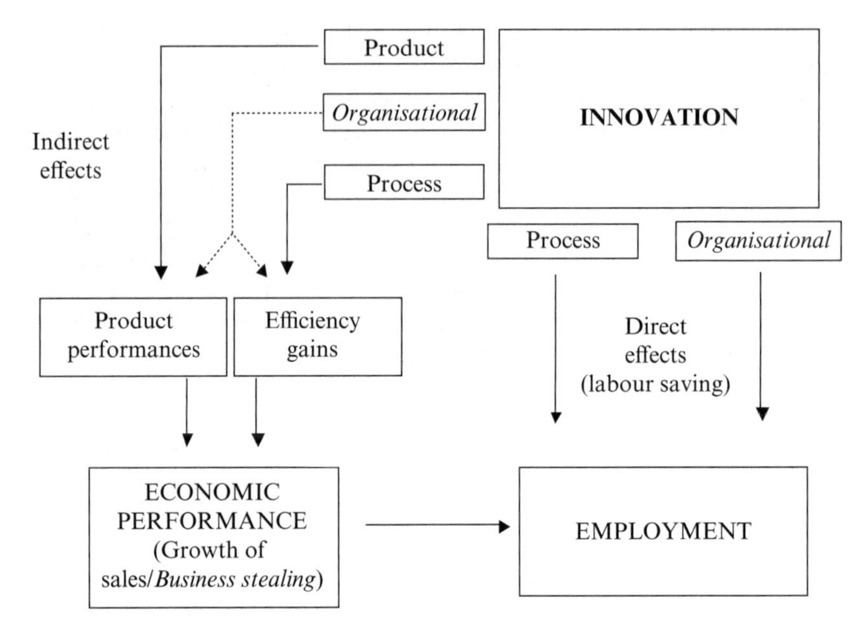

Source: Evangelista and Vezzani (2008).

Figure 16.1 The employment impact of innovation at a firm level

some time lag) if (and depending on the extent to which) the introduction of an innovation provides a competitive advantage compared to non-innovators. This applies also to process innovations which might have positive effects on employment if the achieved efficiency gains provide a competitive premium in terms of (above-average) growth rates, and market share gains, that compensate for the initial labour-displacing effects of these innovations. The strength and size of these firm-level direct and indirect effects of innovation on employment (and consequently their net employment outcome) is also influenced by a set of other factors, such as the specific innovation strategy underlying the introduction of product and/or process innovation, the organisational structure of the firm (mainly firm size), the level of competition, the innovation regime and the demand conditions characterising the specific sector or market in which the firm operates.[9]

This conceptual framework and the scheme in Figure 16.1, based on the contrasting effects of product and process innovation, clearly draw on our understanding of innovation in the manufacturing industry.[10] When this type of framework is applied to services, two obvious complications and question marks emerge: the first is related to the relevance and effectiveness, in the case of service industries, of the traditional distinction between product and process innovation; the second emerges when the concept and domain of innovation are extended beyond the technological boundaries.

Whether and to what extent the traditional conceptual toolbox used to analyse innovation in the manufacturing sectors can be applied to the service sector is the subject of lively debate (Gallouj, 2002; Drejer, 2004; Evangelista, 2006; Tether, 2005; Tether and Tajar, 2008; Gallouj and Savona, 2009). Scholars of the so-called 'demarcation approach' argue that the co-terminality of production and consumption that characterises most service activities makes the distinction between product and process innovation in services less clear-cut, or even meaningless, while the 'assimilation approach' tends to minimise the differences between the manufacturing and service industries in terms of their fundamental knowledge assets and types of firm innovation activities. Since the mid-2000s, this polarisation has been overtaken somewhat by a more balanced perspective that argues that:

> services and manufactures do not follow entirely different approaches to innovation, but that studies of services and their innovation activities (such as those undertaken in the demarcation tradition) bring to the fore neglected aspects of the innovation process, which although most prominent in services, are (increasingly) widely distributed throughout the economy. (Tether, 2005, 156)

In this synthetic perspective, the answer to the question about whether the product–process distinction applies to services is probably yes, although less so than in the manufacturing sector, and to different extents among different service industries. Some of the sectoral-level studies and taxonomies referred to in the previous section support this view and show clearly that both manufacturing and service industries differ in the product or process orientation of their technological activities (Evangelista, 2000; Hollenstein, 2003; Evangelista and Savona, 2003; Bogliacino and Pianta, forthcoming).

There are also some more direct firm-level studies that provide empirical support for the presence in both manufacturing and services of different effects of product and process-related strategies on employment. Evangelista and Savona (2003), on the basis of CIS data for Italy, confirm that the impact (both quantitative and qualitative) of innovation on employment is clearly affected by the strategies adopted by firms.[11] These authors show that a positive impact is more likely in the case of innovative strategies focused on the introduction of new services, internal generation of knowledge (R&D and design), absorption of competencies from external sources (acquisition of know-how), and marketing. Process-oriented innovation activities, measured by the acquisition of new technological equipment, have been shown to have labour-displacing effects and some firm-level studies using CIS data confirm the positive impact of product innovation on employment. For process innovation, the evidence is less clear-cut due to the difficulty involved in disentangling its direct and indirect effects (Harrison et al., 2008; Evangelista and Vezzani, 2008)

Another major problem emerges if we include non-technological forms of innovation (especially organizational change), in the framework depicted in Figure 16.1 (the dotted boxes and lines). Despite the need to broaden the technological boundaries of innovation in the case of both the manufacturing and service industries, it is the latter sector where progress is vital. One of the peculiarities of innovation in services that is heavily underlined is related to the dominant role played by non-technological types of knowledge assets, competencies and strategies. This has clear implications for the analysis and empirical investigation of the employment impact of innovation. In terms of the scheme presented in Figure 16.1, and confining our discussion to the role and impact of organizational innovation, several questions emerge. Do these types of innovation represent a third autonomous innovation mode or are they complementary to product or process innovation (and related strategies)? What are their (individual or combined) effects on firms' economic performance and employment? There is a lack of empirical studies on these areas; some of the difficulties involved in exploring these phenomena are related to the multiform dimension of firms' 'organisations' and the high degree of firm- and industry-level heterogeneity in their organisational strategies and assets (Brynjolfsson and Hitt, 2000; Bresnahan et al., 2002).

Existing studies based on different approaches and data all seem to indicate that technological and non-technological innovations are complements rather than substitutes. This is supported by the results of both CIS3 and CIS4 (Eurostat, 2004; 2008; OECD, 2008b). However, CIS data also show that a non-negligible proportion of European firms innovate exclusively or mainly by introducing non-technological innovations or by changing their organisational structures and routines, and that these types of innovation are more widespread within services than manufacturing (OECD, 2008b; Abreu et al., 2008, 2009; and Tether and Tajar, 2008, for the UK case). It should be noted, however, that surveys such as the CIS (focused on a technological approach to innovation) are not the most appropriate tool to grasp the complex and composite nature of organisational changes and, especially in the case of services, are likely to underestimate the relevance of this type of innovation (Tether and Tajar, 2008).

The literature generally points to the co-existence, in both the manufacturing and service industries, of different modes of innovation, characterised by different mixes of product, process and non-technological activities (Hollenstein, 2003; OECD, 2008b; Tether and Tajar, 2008; Abreu et al., 2008), and that high economic performances are usually associated with more complex or systemic types of strategies. Brynjolfsson and Hitt (2000), Bresnahanet et al. (2002) and Gu and Gera (2004) provide firm-level evidence showing that productivity is positively associated with the introduction of product and process innovation and the extent to which these are combined with the adoption of IT and organisational change. In line with the arguments proposed by evolutionary economists, these authors suggest that this interfirm variety is related to the best and worst managerial practices, to different routines and strategic choices undertaken in a context of bounded rationality, and to firm-level differences in the adjustment costs and speeds associated in particular with changes in work organisation and human capital.

The employment impact of these different innovation modes remains to be investigated – at least with reference to the service sector. Most empirical work is confined to manufacturing and focuses on the effects of technological and organisational innovations on the skill composition of employment.[12] An exception is a study by Evangelista and Vezzani (2008) which (in a manufacturing–services comparative framework) explores the relationship between technological and non-technological innovation in firms' innovation strategies, and the employment impact of these different types of innovation. Using firm-level data from the fourth Italian CIS, Evangelista and Vezzani identify four distinct innovation strategies: two based on the introduction respectively of process and product innovations; a third based on organisational innovation (particularly relevant in services), and a more complex mix of product, process and organisational innovation. Although Evangelista and Vezzani's econometric analysis does not identify any direct negative effects of process innovation (in either macro sector)[13] the size and statistical significance of the positive indirect effects of innovation on employment (i.e., through improved performance) are found to vary across different innovation modes. More particularly, product-related strategies (especially when complemented with process and organisational innovation) are found to show the strongest employment impact in both the manufacturing and service industries, while the employment impact of process innovations is shown to be positive only in the manufacturing sector. The introduction of stand-alone organisational changes is shown to boost the economic and employment performance of firms and, rather surprisingly, is shown to be stronger in the manufacturing than in the service industries.

The empirical evidence from this set of studies, which extend the analysis of innovation (and its impact) beyond the technological domain, should be considered preliminary and influenced by the crude definitions and indicators of organisational innovation adopted (especially in the case of CIS data). More conceptual and empirical work is needed to validate, consolidate (or refute) these types of findings. Nonetheless, they seem to provide a general methodological indication in favour of the adoption of a synthesis approach to both the study of innovation and its economic effects. This represents a challenge for research, since it requires to bridge between two completely separate bodies of literature and empirical research, namely those dealing with technological innovation and with organisational innovation. The difficulty of this task is increased by the asymmetric state of the art in the literature on these two broad domains of empirical and theoretical investigation. While the concepts, definitions and datasets related to firms' technological activities are fairly established, this is not the case for organisational changes. Future advances in a synthesis perspective will depend crucially on the possibility for advancing and integrating robust and meaningful conceptual frameworks and data, in both domains.

16.5 Concluding remarks

This chapter has addressed some of the theoretical and empirical issues around the impact of innovation on employment in services. This topic is of crucial importance because the role of services in creating jobs is controversial, and also the impact of innovation and technical progress on economic performance in services has changed dramatically over time. As pointed out in section 16.2, the structural change leading to the growth of services in most of the advanced countries over the last few decades has often been seen as the result of sluggish labour productivity, large labour productivity gaps with respect to the manufacturing sector, and overall technological stagnancy. The positive role of services as an antidote to unemployment – especially in relation to deindustrialisation – has been counterbalanced by technological stagnancy, described in the Baumolian and post-Baumolian literature. However, this view has been superseded by one in which technological progress and innovation in services are increasingly being recognised.

This has led (a few) scholars to explore whether innovation has an impact on employment in services, to investigate its main determinants and its dominant sign. In this chapter we have tried to systematise these attempts according to the level of analysis adopted, based on a belief that assessing the impact of innovation in services and its effect on employment, and the mechanisms by which it occurs, is crucial.

We started from the macro level of analysis by challenging Baumol's cost-disease argument related to employment growth based on labour productivity gaps, with a more recent wave of contributions. The latter rather overreact to the pessimism of the cost-disease argument and see the knowledge-based economy, sustained by the ICT paradigm and relying on the propulsive role of KIBS, as a very positive development. Several empirical contributions find evidence of dramatic increases in the share of employment and productivity performance of KIBS, based on profound changes in intersectoral linkages and increases in intermediate demand for ICT-specific intermediate inputs from the whole economy.

Sections 16.2 and 16.3 have highlighted a number of caveats to the aggregate net impact of ICTs on service employment and on the mechanisms through which the ICTs exert an effect on employment. First, the relative weight of KIBS in most of the advanced economies might not be sufficient to drive positive net employment performance at the aggregate level. Although it may be possible to find empirical evidence that ICTs have a positive impact on employment in KIBS, this may not necessarily apply to the less innovative service sectors, which represent the highest share of employment and, with a few exceptions (see Djellal, 2000, 2002; Djellal and Gallouj, 2005) are generally ignored in the literature.

Second, a crucial aspect that has not received the attention it deserves is the net employment effect of the complex intersectoral compensation mechanisms that operate when the whole set of the linkages between the service sectors and the rest of the economy are taken into account.

Third, a proper assessment of the net impact of ICTs on employment should be made in order to capture the range of mechanisms underlying it, at the sectoral and micro levels of analysis. In this chapter we have stressed that firm behaviour is an important determinant of the employment outcome of innovation, but even more important are the specific industries, markets and technological regimes in which firms operate. The sectoral and micro-level literature reviewed in sections 16.3 and 16.4 shows that the employment opportunities afforded by ICTs vary widely across the service industries, which has important implications for employment. What is important is the firm's underlying strategy in terms of how these technologies are used in a process-oriented way, or in order to create new services and open up new markets.

In a 'synthesising' perspective on innovation in services (Gallouj and Weinstein, 1997; Gallouj and Savona, 2009; and Chapter 1 in this volume), broadening the analysis of innovation–employment linkages to go beyond the technological domain is vital to take account of the organisational and non-technological aspects of innovation. This effort represents perhaps the most urgent and challenging issue for the service innovation research agenda, and should consider the problems related to conceptualising and measuring non-technological innovation and its effects on employment.

Notes

1. This contribution is based on research funded by the Italian Ministry of Education, University and Research (PRIN project No 2007CC2A2T). Although this was joint research, the sections of the chapters can be attributed as follows: Evangelista (sections 16.1, 16.3 and 16.4); Savona (sections 16.2 and 16.5).
2. It is argued that the presence of a Kaldor–Verdoorn (Verdoorn, 1949; Kaldor, 1966, 1975) mechanism linking productivity and output growth to a positive impact on employment is valid only for manufacturing activities.
3. Bell (1973) is a seminal contribution in this stream of the literature.
4. See again Spiezia and Vivarelli (2002) and Pianta (2005) for a review of the existing empirical evidence on these points.
5. Compared to aggregate studies, firm-level analysis, especially when combined with an appropriate set of technological indicators, is also more effective at capturing or incorporating the effects of qualitative changes in firms' output while sectoral or macro-level analyses implicitly assume homogeneous inputs and outputs. This is crucial in the tertiary sector given the customised nature of many service activities.
6. The effort to exploit CIS micro-data within a cross-country framework was led by the OECD, which ran a project on the determinants of productivity that involved teams of researchers with access to national data conducting the empirical tests using a common econometric regression model, the CDM (see OECD, 2008b, for a synthesis of this project).
7. In studies using CIS data, this is due to persisting data constraints (for many services sectors micro data are not available) and the fact that, according to many scholars, this type of survey cannot grasp the specific nature of innovation activities in services or, consequently, its economic impact (Tether and Tajar, 2008).
8. Among the few exceptions see Evangelista and Savona (2003), Peters (2005), Harrison et al. (2008), Evangelista and Vezzani (2008) and Nahlinder (2005) for a specific focus on KIBS.
9. On purely conceptual grounds, it should be noted that firm-level analysis and econometric estimations, though interesting and illuminating in many respects, cannot provide indications of the overall aggregate impact of innovation on employment. In fact, the aggregated employment impact of the sum of all firms' innovation activities is the outcome of both the functioning and strength of the compensation mechanisms discussed in section 16.2, and the competitive game entered into by firms within each industry and across different industries. Leaving aside the effects of the various compensation mechanisms, it is important here to stress that the existence of a positive relationship between innovation and employment at the firm level (as found e.g., in most firm-level studies in the manufacturing sector) may be compatible with an overall loss of jobs at a more aggregate level. This is likely to occur in the case of weak (declining) aggregate demand, which makes the competition between firms particularly severe. In such cases, the introduction of new services and, especially, new processes by the most innovative firms might displace production capacity and jobs in other (less innovative) firms (Evangelista, 1995; Pianta, 2001)
10. See Spiezia and Vivarelli (2002) and Pianta (2005) for a review of the micro-level studies on the innovation–employment relationship. Firm-level studies covering the manufacturing industry confirm the positive impact of product innovation on employment, while as far as process innovation is concerned the evidence is less clear-cut, based in part on the difficulty involved in disentangling the direct and indirect effects of process innovations (Harrison et al., 2008; Evangelista and Vezzani, 2008).
11. In the Italian CIS2 and CIS3 questionnaires there is a question aimed at acquiring direct information about the impact of the innovation on employment. Both manufacturing and service firms are asked whether the introduction of innovation has led to: (a) an increase; (b) a decrease; or (c) no effect on the use of labour. Firms are also to distinguish between the impact of innovation on total employment, and on high-, medium- and low-skilled personnel. This information is used to estimate the employment effects

of different types of innovation strategies and firm-specific characteristics (Evangelista and Savona, 2003).

12. See, among others, Berman et al. (1998), Bresnahan et al. (2002), Bauer and Bender (2004), Tether et al. (2005), Piva and Vivarelli (2002) and Piva et al. (2006).
13. This finding confirms the results of previous studies on the manufacturing sector, which find no significant direct effects of process innovation (Harrison et al., 2008).

References

Abreu, M., V. Grinevich, M. Kitson and M. Savona (2008), 'Understanding hidden innovation: services in the UK. Indicators, empirical evidence and policy implications', NESTA, National Endowment of Science, Technology and Arts, London, May 2008.

Abreu, M., V. Grinevich, M. Kitson and M. Savona (2009), 'Policies to enhance the "hidden" innovation in services: evidences and lessons from the UK', *Service Industries Journal*, ifirst article, pp. 1–23 (advanced access online: DOI: 10.1080/0264206080223616).

Appelbaum, E. and R. Schettkat (2001), 'Are prices unimportant? The changing structure of the industrialised economies', in T. ten Raa and R. Schettkat (eds), *The Growth of Service Industries: The Paradox of Exploding Costs and Persistent Demand*, Cheltenham, UK and Northampton, MA, USA: Edward Elgar, pp. 121–131.

Bauer, T. and S. Bender (2004), 'Technological change, organizational change, and job turnover', *Labour Economics*, **11** (3), pp. 265–291.

Baumol, W.J. (1967), 'Macroeconomics of unbalanced growth: the anatomy of an urban crisis', *American Economic Review*, **57**, 415–426.

Baumol, W.J. (2001), 'Paradox of the services: exploding costs, persistent demand', in T. ten Raa and R. Schettkat (eds), *The Growth of Service Industries: The Paradox of Exploding Costs and Persistent Demand*, Cheltenham, UK and Northampton, MA, USA: Edward Elgar, pp. 3–28.

Baumol, W.J. (2002), 'Services as leaders and the leader of the services' in J. Gadrey and F. Gallouj (eds), *Productivity, Innovation and Knowledge in Services*, Cheltenham, UK and Northampton, MA, USA: Edward Elgar, pp. 147–163.

Baumol, W.J., S. Blackman and E. Wolff (1985), 'Unbalanced growth revisited: asymptotic stagnancy and new evidence', *American Economic Review*, **75**, 806–816.

Baumol, W.J., S. Blackman and E. Wolff (1989), *Productivity and American Leadership*, Cambridge, MA: MIT Press.

Baumol, W.J. and W. Bowen (1966), *Performing Arts: The Economic Dilemma*, New York: Twentieth Century Fund.

Bell, D. (1973), *The Coming of Post-Industrial Society: A Venture in Social Forecasting*, New York: Basic Books.

Berman, Eli, John Bound and Stephen Machin (1998), 'Implications of skill-based technological change: international evidence', *Quarterly Journal of Economics*, **113**, 1245–1279.

Boden, M. and I. Miles (eds) (2000), *Services and the Knowledge Based Economy*, London: Continuum.

Bogliacino, F. and M. Pianta (forthcoming), 'Breaking the innovation-employment spell: an investigation on Revised Pavitt classes', *Research Policy*.

Bresnahan, T.F., E. Brynjolfsson and L.M. Hitt (2002), 'Information technology, workplace organization, and the demand for skilled labour: firm-level evidence', *Quarterly Journal of Economics*, **117** (1), 339–376.

Brynjolfsson, E. and L.M. Hitt (2000), 'Beyond computation: information technology, organizational transformation and business performance', *The Journal of Economic Perspectives*, **14** (4), 23–48.

Cainelli, G., R. Evangelista and M. Savona (2006), 'Innovation and economic performance in services: a firm level analysis', *Cambridge Journal of Economics*, **30**, 435–458.

Castellacci, F. (2006), 'Innovation and international competitiveness of manufacturing and service industries: a survey', DIME WP 2006.05.

Castellacci, F. (2007), 'Technological paradigms, regimes and trajectories: manufacturing

and service industries in a new taxonomy of sectoral patterns of innovation', Paper presented at the DRUID Conference 2007.
Castellacci, F. and J. Zheng (2008), Technological regimes, Schumpeterian patterns of innovation and firm level productivity growth', NUPI Working Paper, No. 730/42.
Ciarli, T., A. Lorentz, M. Savona and M. Valente (2009), 'The effects of consumption and production structure on growth and distribution: a micro to macro model', *Metroeconomica*, pp. 1-89 (advanced access online: DOI: 101111/j1467-999X.2009.04069.x).
Clark, C. (1940), *The Conditions of Economic Progress*, London: Macmillan.
Crepon, B., E. Duguet and J. Mairesse (1998), 'Research, innovation, and productivity: an econometric analysis at the firm level', *Economics of innovation and new technology*, **7** (2), 115–158.
Crespi, G., C. Criscuolo, J. Haskel and D. Hawkes (2006), 'Measuring and understanding productivity in UK market services', *Oxford Review of Economic Policy*, 22 (4), 560–572.
Crespi, F. and M. Pianta (2008), 'Demand and innovation in productivity growth', *International Review of Applied Economics*, **22** (5), 655–672.
Daveri, F. (2002), 'The new economy in Europe, 1992–2001', *Oxford Review of Economic Policy*, **18** (3), 345–362.
Dasgupta, S. and A. Singh (2005), Will services be the new engine of economic growth in India? Discussion paper, Working Paper Series, no. 310, Centre for Business Research, University of Cambridge.
Djellal, F. (2000), 'The rise of information technologies in non informational services', *Vierteljahrshefte zur wirtschaftsforschung*, **69** (4), 646–656.
Djellal, F. (2002), 'Innovation trajectories in the cleaning industry', *New Technology and Employment*, **17** (2), 119–131.
Djellal, F. and F. Gallouj (2005), 'Mapping innovation dynamics in hospitals', *Research Policy*, **34**, 817–835.
Djellal, F. and F. Gallouj (2007), 'Innovation and employment effects in services: a review of the literature and an agenda for research', *Service Industries Journal*, **27** (3–4), 193–213.
Djellal, F. and F. Gallouj (2008), *Measuring and Improving Productivity in Services: Issues, Strategies and Challenges*, Cheltenham, UK and Northampton, MA, USA: Edward Elgar.
Drejer, I. (2004), 'Identifying innovation in surveys of services. a Schumpeterian perspective', *Research Policy*, **33**, 551–562.
Eurostat (2004), 'Innovation in Europe. Results for the EU, Iceland and Norway. Data 1998–2001', Luxembourg: European Commission.
Eurostat (2008), 'Science, technology and innovation in Europe – 2008 edition', Luxembourg: Eurostat.
Evangelista, F. (1995), 'Innovations e occupazione nell'industria italiana: un 'analisi per imprese e settori', *L' Industria*, **1**, 113–137.
Evangelista, R. (1999), *Knowledge and Investment: The Sources of Innovation in Industry*, Cheltenham, UK and Northampton, MA, USA: Edward Elgar.
Evangelista, R. (2000), 'Sectoral patterns of technological change in services', *Economics of Innovation and New Technology*, **9**, 183–221.
Evangelista, R. (2006), 'Innovation in the European service industries', *Science and Public Policy*, **3** (9), 653–668.
Evangelista, R. and M. Savona (2003), 'Innovation, employment and skills in services: firm and sectoral evidence', *Structural Change and Economic Dynamics*, 14, 449–474.
Evangelista, R. and A. Vezzani (2008), 'The employment impact of technological and organizational innovations: firm level evidence', paper presented to the International Workshop on Non-technical Innovation: Definition, Measurement & Policy Implications, Karlsruhe, 16–17 October.
Fisher, A.G.B. (1935), *The Clash of Progress and Security*, London: Macmillan.
Fourastié, J. (1949), *Le grand espoir du XX siècle*, Paris: Presse Universitaire de France.
Fuchs, V. (1968), *The Service Economy*, New York: National Bureau of Economic Research.

Fuchs, V. (1977), 'The service industries and US economic growth since World War II', NBER Working Paper Series 211 (November).
Gallouj, F. (2002), *Innovation in the service economy: The New Wealth of Nations*, Cheltenham, UK and Northampton, MA, USA: Edward Elgar.
Gallouj, F. and M. Savona (2009), 'Innovation in services: a review of the debate and a research agenda', *Journal of Evolutionary Economics*, **19** (2), 149–172.
Gallouj, F. and O. Weinstein (1997), 'Innovation in services', *Research Policy*, **26** (4–5), 537–556.
Gregory, M., W. Salverda and R. Schettkat (ed.) (2007), *Services and Employment: Explaining the US–European Gap*, Princeton, NJ, USA and Oxford, UK: Princeton University Press.
Gu, W. and S. Gera (2004), 'The effect of organizational innovation and information technology on firm performance', *International Productivity Monitor*, **9** (Fall), 37–51.
Guerrieri, P. and V. Meliciani (2005), 'Technology and international competitiveness: the interdependence between manufacturing and producer services', *Structural Change and Economic Dynamics*, **16**, 489–502.
Harrison, R., J. Jaumandreu, J. Mairesse and B. Peters (2008), 'Does innovation stimulate employment? A firm-level analysis using comparable micro data of four European countries', NBER Working Paper No. 14216.
Hollenstein, H. (2003), 'Innovation modes in the Swiss service sector: a cluster analysis based on firm-level data', *Research Policy*, **32** (5), 845–863.
Kaldor, N. (1966), *Causes of the Slow Rate of Growth in the United Kingdom*, Cambridge: Cambridge University Press.
Kaldor, N. (1975), 'Economic growth and the Verdoorn Law: a comment on Mr. Rowthorn's article', *Economic Journal*, **85** (340), 891–896.
Katsoulacos, Y. (1986), *The Employment Effects of Technical Change: A Theoretical Study of New Technology and the Labour Market*, London: Wheatsheaf.
Kox, H.L.M. and L. Rubalcaba (2007), 'Analysing the contribution of business services to European economic growth', BEER Bruges European Economic Research Paper No. 9, February.
Lööf, H. and A. Heshmati (2006), 'On the relationship between innovation and performance: a sensitivity analysis', *Economics of Innovation and New Technology*, **15** (4–5), 289–299.
Lorentz, A. and M. Savona (2008), 'Evolutionary micro-dynamics and changes in the economic structure', *Journal of Evolutionary Economics*, **18** (3–4), 389–412.
Marrano, M.G., J. Haskel and G. Wallis (2007), 'What happened to the knowledge economy? ICT, intangible investment and Britain's productivity record revisited', Working Paper no. 603, June, Department of Economics, Queen Mary University of London.
Miles, I., N. Kastrinos, P. Bilderbeek and P. den Hertog (1995), 'Knowledge intensive business services: their role as users, carriers and sources of innovation', Report to DG13 SPRINT-EIMS, PREST, Manchester, March.
Miozzo, M. and I. Miles (eds) (2002), *Internationalisation, Technology and Services*, Cheltenham, UK and Northampton, MA, USA: Edward Elgar.
Montresor, S. and G. Vittucci Marzetti (2007), 'The deindustrialization/tertiarisation hypothesis reconsidered: a sub-system application to the OECD', University of Bologna, Mimeo.
Nahlinder, J. (2005), 'Innovation and employment in services: the case of KIBS in Sweden', Department of Technology and Social Change, Linkoping University, Linkoping, Sweden.
OECD (2005), *Oslo Manual: Proposed Guidelines for Collecting and Interpreting Technological Innovation Data*, 3rd edn, Paris: OECD.
OECD (2008a), *Employment Outlook 2008*, Paris: OECD.
OECD (2008b), *Science, Technology and Industry Outlook*, Paris: OECD.
Parrinello, S. (2004), 'The service economy revisited', *Structural Change and Economic Dynamics*, **15**, 381–400.

Pavitt, K. (1984), 'Sectoral patterns of technical change: towards a taxonomy and a theory', *Research Policy*, **13**, 343–374.
Peneder, M. (2003), 'Industrial structure and aggregate growth', *Structural Change and Economic Dynamics*, **14**, 427–448.
Peneder, M., S. Kaniovsky and B. Dachs (2003), 'What follows tertiarisation? Structural change and the role of knowledge-based services', *Service Industry Journal*, **23**, 47–66.
Peters, B. (2005), 'Employment effects of different innovation activities: microeconomic evidence', ZEW Discussion Paper No. 04-73.
Petit, P. (1995), 'Employment and Technological Change', in P. Stoneman (ed.), *Handbook of the Economics of Innovation and Technological Change*, Amsterdam: North Holland Publishing Company, pp. 366–408.
Petit, P. and L. Soete (2001) (eds), *Technology and the Future of European Employment* Cheltenham, UK and Northampton, MA, USA: Edward Elgar.
Pianta, M. (2001), 'Innovation, Demand and Employment', in P. Petit and L. Soete (eds), *Technology and the Future of European Employment*, Cheltenham, UK and Northampton, MA, USA: Edward Elgar, pp. 142–165.
Pianta, M. (2005), 'Innovation and employment', in J. Fagerberg, D. Mowery and R. Nelson (eds), *The Oxford Handbook of Innovation*, Oxford: Oxford University Press, pp. 568–598.
Piva, M. and M. Vivarelli (2002), 'The skill bias: comparative evidence and an econometric test', *International Review of Applied Economics*, **16** (3), 347–358.
Piva, M., E. Santarelli and M. Vivarelli (2006), 'The skill bias effect of technological and organisational change: evidence and policy implications', *Research Policy*, **34** (2), 141–157.
Pugno, M. (2006), 'The service paradox and endogenous economic growth', *Structural Change and Economic Dynamics*, **17**, 99–115.
Savona, M. and A. Lorentz (2005), 'Demand and technological contribution to structural change and tertiarisation: an input-output structural decomposition analysis', LEM (Laboratory of Economic and Management, School of Advanced Studies S. Anna, Pisa) Working Paper Series, 2005/25, December.
Schettkat, R. and L. Yocarini (2006), 'The shift to services employment: a review of the literature', *Structural Change and Economic Dynamics*, **17**, 127–147.
Soete, L. and M. Miozzo (2001), 'Internationalisation of services: a technological perspective', *Technological Forecasting and Social Change*, **67**, 159–185.
Spiezia, V. and M. Vivarelli (2002), 'Technical Change and Employment: a critical survey', in N. Greenan, Y. L'Horty and J. Mairesse (eds), *Productivity, Inequality and the Digital Economy: A Transatlantic Perspective*, Cambridge, MA: MIT Press, pp. 101–31.
ten Raa, T. and R. Schettkat (eds) (2001), *The Growth of Service Industries: The Paradox of Exploding Costs and Persistent Demand*, Cheltenham, UK and Northampton, MA, USA: Edward Elgar.
Tether, B. (2005), 'Do services innovate (differently)? Insights from the European Innobarometer Survey', *Industry and Innovation*, **12**, 153–184.
Tether, B., A. Mina, D. Consoli and D. Gagliardi (2005), 'A literature review on skills and innovation: how does successful innovation impact on the demand for skills and how do skills drive innovation?' Report by the Centre for Research on Innovation and Competition (CRIC).
Tether, B. and A. Tajar (2008), 'The organisational-cooperation mode of innovation and its prominence amongst European service firms', *Research Policy*, **37** (4), 720–739.
Van Ark, B., L. Broersma and P. den Hertog (2003a), *Services Innovation, Performance and Policy: A Review.* Synthesis Report in the Framework of the Structural Information Provision on Innovation in Services (SIID). June 2003.
Van Ark, B., R. Inklaar and R. McGuckin (2003b), 'Changing gear – productivity, ICT and service industries in Europe and the United States', in F. Christensen and P. Maskell (eds), *The Industrial Dynamics of the New Digital Economy*, Cheltenham, UK and Northampton, MA, USA: Edward Elgar, pp. 56–99.

Van Ark, B., M. O'Mahony and G. Ypma (eds) (2007), 'The EU KLEIMS Productivity Report: an overview of results from the EU KLEIMS Growth and Productivity Accounts for the European Union, EU member states and major other countries in the world', EU KLEIMS Report, 1, March.

Van Leeuwen, G. and L. Klomp (2006), 'On the contribution of innovation to multifactor productivity growth', *Economics of Innovation and New Technology*, **15** (4–5), 367–390.

Verdoorn, P.J. (1949), 'Fattori che Regolano lo Sviluppo della Produttivitá del Lavoro', *L'industria*, **1**, 45–53.

Vivarelli, M. and M. Pianta (eds) (2000), *The Employment Impact of Innovation: Evidence and Policy*, London: Routledge.

Wolff, E.N. (2002), 'How stagnant are services?', in J. Gadrey and F. Gallouj (eds), *Productivity, Innovation and Knowledge in Services*, Cheltenham, UK and Northampton, MA, USA: Edward Elgar, pp. 3–25.

17 Innovation and services: on biases and beyond

Pascal Petit

17.1 Introduction: services – a potential of development that needs to be redefined

In all developed economies the service sector is the only one where employment has kept on growing in the 1980s, 1990s and 2000s, turning them into fully fledged tertiary economies. The decrease in the shares of manufacturing and agriculture in total employment in these economies has been accompanied by relatively high productivity gains in these two sectors. By contrast productivity in services has appeared rather stagnant, displaying on average low gains.

This broad sector which by now represents over two-thirds of employment was thus mainly identified as not innovative, neither in terms of process nor in terms of product. It led to a tertiary economy being featured as one of unbalanced growth between a broadening stagnant sector and a shrinking dynamic one. This model was well coined by Baumol (1967), some 30 years ago. In effect in a two-sector closed economy, where the demands of the two products remain in a constant ratio, employment is bound to concentrate more and more in the sector with the lower productivity gains and overall economic growth is also going to be determined by this slowly growing sector: a really inhibiting cost disease (see also Chapter 4 in this volume).

Still, this should not imply that tertiary economies are condemned to stagnation. This representation relies much on how one assesses services in real terms, e.g. in volumes. This measurement is based on conventions which will be reviewed later on in this chapter. Revisions of the conventions on which national accounts are based are on the agenda after the successive crises that occurred at the beginning and at the end of the 2000s. The investigation of this potential for redesigning growth patterns remains, though, highly tentative.

In the first place I shall stress that this service sector is in fact composed of a large variety of activities, a diversity which has increased in the last decades with the rapid expansion of business services, e.g. of services used by other firms, of which many are manufacturing ones.

Indeed in this diversity some activities appear to be only asymptotically stagnant, creating a third category with an effective potential of productivity gains (see Baumol et al., 1985). But my contention goes further. First, the stagnation of productivity directly depends upon the conventions we retain to measure it. Changing such conventions requires legitimisation. New assumptions to assess the quality of services (and therefore the volumes) have to be backed by the preferences of would-be majorities of users. This constitutes a risky bet. The only support comes, on the one hand, from the rising questioning of a growing divergence between measures of GDP and of well-being (Gadrey and Jany-Catrice, 2007), and on the other hand, from industry surveys that show that there exist as many innovative steps taken in services as in manufacturing (Djellal and Gallouj, 2008). Second, the actual development of innovations in services tends to be strongly dependent on existing environments. This sometimes leaves little room for learning processes that would open the way towards more 'ambitious' transformations. I shall investigate some of these biases, as taking them into account may offer possibilities to redesign growth patterns. Still, we are here on shaky ground and one can explore this frontier only with a maximum of caution.

Indeed in most cases the conventions retained to measure prices and volumes of production of services in order to assess the productivity gains are very crude. In many cases volumes of inputs played a central role in evaluating the volumes of output. By the mid-1990s an expert in the field like Zvi Griliches (1994) could still consider that some 50 per cent of the measures were highly disputable. But going from such criticisms to the assessment of a new growth potential is another matter.

I shall contend here that the conventions of measurement are crucial, and can vary with time, following some learning processes whereby consumers and producers come to value some characteristics more and others less, therefore changing their views and strategies on the quality of services.

The main stimuli for such learning processes to take place will rely on what are considered in economics as sources of endogenous growth such as changes in education, technology and institutions. Still, the trajectories of these learning processes may follow rather bumpy roads, with lock-ins, pushes and pulls, synergies and inertia. The internalisation of the many externalities brought by the above changes in the conditions of production and uses of services, which are at the core of the dynamics of endogenous growth, are bound to raise many conflicting issues.

It may not seem to be the case with services to persons. We tend to gain our first view of services primarily from menial jobs, such as those of servants, which have for long, even unfairly, been considered as low-ranking activity, of little value in the economic sphere, and maybe too close to the domestic sphere.[1]

But such schemes of interpersonal services refer to contexts where services are rendered in a simplified context where externalities are minimal and the skills involved are basic. Were the skills more exceptional, then the perception one might have of these 'menial' services would change. In our societies the high skills of some service providers can lead them to acquire a very high status, be they a cook, a hairdresser or a musician.

But my topic is not so much about the wide range of personal services that one can see in our modern society. I want to focus more on general changes of contexts in the provision and in the use of services, developing cooperative effects and interdependencies in ways that could help to modify our perception of growth processes.

Many structural changes are implied in this transformation, namely: (1) increased levels of education of the labour force; (2) development and diffusion of a technological system centred around information and communication technologies, modifying the time and space of interpersonal relations, and therefore the organisation of activities; (3) deregulations of product markets facilitating the circulation of production factors and products between places, along with tighter regulation of norms and quality standards of products, which implies greater know-how from both producers and consumers.

Paying attention to all these changes as well as to the tertiary nature of our developed economies helps us to take another view of growth processes. It should help us to see the potential for some redefinition of development patterns that would be more in accord with some of the aspirations expressed by all those in search of more sustainable growth trajectories both in terms of environment and social justice.

In order to move in this direction I shall first recall the variety of activities concerned which are the common and distinctive features of these distributions among countries where different modes of coordination prevail. Services are in essence loci of strong idiosyncrasies, and looking at them through the filter of the diversity of capitalism can help (section 17.2). In a second step I shall check how these diverse services activities are faring in terms of productivity gains (section 17.3). I shall pay specific attention to the last two decades (from the 1990s onwards) when a new growth regime, characterised by dense internationalisation and dominance of finance, has tended to diffuse across the major developed economies.

During this period the obvious importance of the diffusion of a new technological system centred around communication and information technologies has led its role in the apparent stagnancy of productivity to be questioned. Section 17.3 will take stock of the controversy over this paradox. But other factors could similarly be questioned, namely the rise in education (section 17.4) and the changes in the institutional contexts that represent both the deregulation of entrepreneurial behaviours and the regulation of products by means of norms and standards (section 17.5).

I shall identify different types of bias in section 17.6 and try to see how they affect the dynamics of services. I shall then be in a position to outline some of the implications of this shifting context for the dynamics of innovation in services. Dry assessment will not pretend to be exhaustive, but simply to outline some of the potential for innovation that learning processes can offer. It will also point to some drawbacks and help, in the conclusion in section 17.7, to present some policy recommendations.

17.2 On the diversity of service activities

Services have always been a composite set of activities, even if one considers the paradigmatic personal services, the figures of the exceptional artist of or the inventive doctor along with the figures of the domestic servant or of the assistant shopkeeper. The increasing number of civil servants and of all kinds of professionals brought new figures at the turn of the twentieth century into this picture of services.

One can give a rough account of this diversity using four categories of services, namely: the classic personal services; the social services including the large network of services such as health and education; the intermediate services for distribution, communication, transport and banking; and the producer services, grouping activities that contribute to the productive tasks of firms (either manufacturing or services).[2]

The big rise in the employment share of services has followed a clearly cut pattern. In the golden age of capitalism, e.g. the 30 years that followed the Second World War, 1945–75, one can observe across all countries a steady rise in the share of producer services (starting from rather low levels) as well as in the share of social services. By contrast the share of employment in personal services and distribution services remained rather steady, though at levels which varied from one country to another. In the transition period of ten years or so that followed the crisis of the mid-1970s the same trends could be observed, with a quite noticeable acceleration in the rise of social services in all countries.

In the 1990s this rise in social services clearly petered out under the pressure of deregulation and restrictive budget policies (as can be seen in Table 17.1), to leave the employment share of producer services as the only category still on the rise.

The steady rise in the employment share of business services followed a flow of externalisations of tasks from the manufacturing sectors. Indeed such internal reorganisation explained more than two-thirds of the decline in the employment share of manufacturing (see Table 17.2). It contributed to blur further the frontier between services and industries (see Gadrey and Gallouj, 1998; Gallouj, 2002).

Table 17.1 Service employment by subsector and type of capitalism (% of total employment in 1998)

Type of capitalism	Producer services	Distributive services	Personal services	Social services	Total services
Market-based					
United States	15.8	21.2	12.1	24.8	73.8
United Kingdom	14.7	21.8	9.2	25.7	71.4
Canada	16.5	19.4	11.7	22.3	69.9
Australia	14.7	24.6	11.8	22.2	73.3
Social-democratic					
Denmark	11.4	21.1	5.8	31.2	69.5
Finland	11.3	18.8	6.2	28.0	64.2
Sweden	12.2	19.4	5.9	33.4	70.9
Meso-corporatist					
Japan	22.6	26.8	–	–	59.4
Korea	9.3	24.9	–	–	59.7
Continental Northern Europe					
France	11.9	19.9	8.3	29.2	69.2
Germany	10.9	19.9	7.1	24.8	62.6
The Netherlands	14.3	22.0	6.2	27.6	70.2
Belgium	11.7	21.8	6.8	29.8	70.2
Mediterranean					
Italy	9.3	21.6	8.0	22.0	60.8
Greece	7.4	23.3	10.4	17.7	58.8
Spain	9.0	22.4	11.8	18.5	61.7
Portugal	5.5	17.7	10.7	16.2	50.2
Average OECD	11.4	21.3	9.2	24.0	63.5
(yearly growth rates between 1984 and 1998)	(1.3)	(-1.3)	(0.4)	(0.7)	

Source: OECD Employment Outlook, June 2000, Annex 3.C. Typology of capitalism drawn from Amable and Petit (2001).

Only in social services do we see a similar continuous rise until the mid-90s. As I have said, it ended with the pressure to limit public expenditures. In broad terms policies to support rises in employment rates fuelled increases in services, mainly in producer services, in social services and personal services. An oversupply of labour does stimulate employment by means of what Kaldor (1980) called sponge effects. But other factors

Table 17.2 Explaining deindustrialisation 1992–2002

	Changes in % share of manufacturing employment	Due to internal changes	Due to external trade	Unexplained residual
United States	−4.0	−3.4	−1.8	1.5
European Union	−3.7	−2.8	−1.0	0.0
Japan	−5.1	−2.9	−1.3	−0.9
Korea	−6.5	−4.2	−1.4	−0.8
Taiwan	−3.2	−2.7	−3.3	3.7

Source: Extracted from Table 2 in Rowthorn and Coutts (2004).

contributed to this expansion of service employment: namely, a better coverage by public means of some social needs for an expansion of social services in fields like education and health.

The type of jobs concerned may vary widely. Services jobs may be of high quality when they accompany the spread of new innovative activities requiring information and communication technologies and/or a qualified labour force.[3] Low-qualified services jobs tend to be in greater proportion when excess labour supply is a major factor driving expansion. Surveys on the quality of jobs stress this duality. It shows also in the measures of the productivity gains.

There is a strong specific effect of the type of capitalism in these overall dynamics of services. Clearly social services are quite widespread in Scandinavian countries or in continental Europe as opposed to Anglo-Saxon countries or Mediterranean countries. Conversely, personal services as well as producer services are important in Anglo-Saxon countries. In Mediterranean countries, if personal services are relatively important (as they are in Anglo-Saxon countries), business services seem less developed.

By and large these broad figures seem to correspond with the various levels of development, but also with various modes of organisation of activities (practices of externalisation) and mode of welfare organisation (on a public or private basis[4]). These three dimensions are effectively at the root of the differentiation of the various types of capitalism retained in Table 17.1. Part of these differences are bound to be reduced, if only with the rise in what we call the levels of development, understood here as levels of gross domestic product (GDP). Part will remain as a lasting characteristic of national preferences.

To progress in assessing the changes that can occur in the structure and modes of organisation of services one can now look at the productivity gains, even if crudely measured, that have been associated with the recent expansion of services in this post-Fordist period.

First, note that the shares of distributive services (namely distribution, transport, communication and finance) in employment are rather similar across all types of capitalist countries displayed in Table 17.1, with a few exceptions such as Portugal and Japan. Such has been the case more or less all along since after the Second World War (with some catching up in the immediate post-war period in countries like France[5]).

This relative stability should not convey the idea of any stagnancy. On the contrary, interestingly enough these activities underwent major technological and organisational changes over the whole period under review. Checking their achievements in terms of productivity would then be all the more revealing of the direction of their transformation.

17.3 Productivity gains in the post-Fordist period

The period identified as post-Fordist started at the end of the 1970s with the demise of the old growth regime, mainly driven by a search for economies of scale and Taylorian division of labour. The 1980s were largely, in most developed economies a period of transition with a lot of technological and organisational transformations. These did not translate into productivity gains. In the early 1990s this productivity slowdown had become a major puzzle[6] while changes in technology with the diffusion of the new system of ICTs (information and communication technologies) were so obvious.

We can take stock of this debate looking at the structure of productivity changes by sectors in Europe and in the US over the decade 1995–2005, a period when the diffusion of Internet uses came to complete the productive system centred around ICTs.

The attention paid to the effects of this technological diffusion led many studies to distinguish industries directly involved in the production of ICT technologies from activities that only used them. Tables 17.3 and 17.4 thus show how activities respectively in the US and in Europe involved in manufacturing ICT equipment enjoyed, in both subperiods, relatively sustained productivity gains. These gains were higher at the peak of the period of ICT diffusion (e.g. 1995–2004),[7] and also noticeably higher in the US than in the EU-15.[8]

As for the other activities, whether services or non-ICT manufacturing, the productivity gains appear relatively low, both in the transition period as well as in the core period of technological change, 1995–2004, for the EU-15 sample and for the US.

There is an important caveat to this general assessment of a widespread productivity slowdown in all activities which are not directly involved

Table 17.3 Growth and productivity in some of Europe-15 (by sector, 1980–2004)

EU-15 1980–95 Less GR, IR, LU, Pl, SW	VA	L	VA/L	H	K	Kit	Mfp
Market economy (82.2)	1.9	0.0	1.9	−0.3	1.1	0.4	0.3
Electrical machinery, post and communication (4.1)	3.9	−0.7	3.2	−0.8	1.6	0.9	2.9
Other manufacturing (21.6)	1.2	−1.3	2.5	−1.5	0.8	0.2	1.7
Other goods producing industries (20.7)	−0.2	−1.2	1.0	−1.4	0.9	0.2	0.2
Distribution services (19.5)	2.6	0.4	2.2	0.0	0.8	0.3	1.4
Finance and business services (8.1)	3.6	2.2	1.4	1.9	1.9	0.8	−0.7
Personal and social services (8.1)	1.8	1.8	0.0	1.5	1.0	0.3	−1.1
EU-15 1995–2004 (Less GR, IR, LU, Pl, SW)	**VA**	**L**	**VA/L**	**H**	**K**	**Kit**	**Mfp**
Market economy (79)	2.2	0.7	1.5	0.4	1.2	0.6	0.3
Electrical machinery, post and communication (3.4)	6.0	−0.4	6.4	−0.6	1.7	1.2	4.7
Other manufacturing (16.4)	1.0	−0.3	1.3	−0.6	0.7	0.3	0.6
Other goods producing industries (14.5)	1.2	0.0	1.2	−0.2	0.7	0.1	0.5
Distribution services (20.3)	2.3	0.7	1.6	0.6	1.2	0.5	0.4
Finance and business services (13.5)	3.5	2.1	1.4	1.9	2.3	1.3	−1.3
Personal and social services (10.8)	1.7	1.5	0.2	1.4	0.9	0.3	−0.9

Notes: Average growth rates of VA: gross value added, L: labour input, VA/L: productivity gains, H: hours worked, K: capital input, Kit: ICT capital, Mfp: multifactor productivity.
In parenthesis under the name of the sectors is the average share in the total hours worked.

Source: Van Ark et al. (2007).

Table 17.4 Growth and productivity in the US (by sector, 1980–2004)

US 1980–95	VA	L	VA/L	H	K	Kit	Mfp
Market economy	3.0	1.2	1.8	1.0	1.1	0.5	0.7
Electrical machinery, post and communication	6.6	0.1	6.7	−0.3	1.9	1.0	4.6
Other manufacturing	1.7	0.1	1.6	−0.2	0.6	0.3	0.9
Other goods producing industries	0.7	0.7	0.0	0.4	0.7	0.2	−0.7
Distribution services	3.9	1.3	2.6	1.2	1.2	0.6	1.3
Finance and business services	4.4	2.9	1.5	2.7	1.8	1.0	−0.3
Personal and social services	2.8	2.5	0.3	2.5	0.5	0.2	−0.2
US 1995–2004	**VA**	**L**	**VA/L**	**H**	**K**	**Kit**	**Mfp**
Market economy	3.7	0.7	3.0	0.3	1.4	0.8	1.6
Electrical machinery, post and communication	8.9	−0.3	9.2	−0.9	2.5	1.5	6.8
Other manufacturing	0.7	−1.1	1.8	−1.5	0.7	0.4	1.1
Other goods producing industries	1.6	1.0	0.6	0.9	0.9	0.2	0.3
Distribution services	4.7	0.5	4.2	0.2	1.4	1.0	2.8
Finance and business services	4.9	2.0	2.9	1.6	2.0	1.2	0.9
Personal and social services	2.6	1.7	0.9	1.4	1.0	0.4	0.0

Notes: Average growth rates of VA: gross value added, L: labour input, VA/L: productivity gains, H: hours worked, K: capital input, Kit: ICT capital, Mfp: multifactor productivity.

Source: Van Ark et al. (2007).

in the production of ICTs: somehow productivity gains seem to have regained some momentum in the 1995–2004 period in the US in distribution services and to a lesser extent in financial and business services over the same period.

The column 'Kit' (ICT capital) in Tables 17.3 and 17.4 gives an idea of the effort in ICT equipment in all sectors. It shows a rising effort in finance and business services, leading to moderate productivity gains in the US, still bigger than those observed in the European Union (EU). The contrast between Europe and the US is more striking in distribution services where the diffusion of ICTs increased relatively in the US, while it remained relatively lower in Europe, which seems to explain the productivity boost over the period 1995–2004 in the US. The Wal-Mart 'revolution' played a role in this event. The period saw a diffusion across the US of Wal-Mart establishments with their specific business model, installing big stores in mid-size towns that outcompeted local shops and constituted local monopolies afterwards. This competition was also supported by an efficient network-based management of supplies.

Tables 17.3 and 17.4 also show rather low productivity gains in personal services accompanying low levels of investment in ICTs. This is in accord with the traditional view on services recalled above. However, at a time when personal computers and the Internet allow more extensive and frequent social networking one could have expected some quality improvement and therefore some productivity enhancement to show up. Overall the conclusion one draws from this growth accounting exercise is that despite a sizeable diffusion of ICTs their impact on productivity remains modest, especially in services.

The relative extra growth in productivity in the US in distribution and financial services remains to be explained. The dominant role of the US financial sector, at an international level, is certainly part of the explanation of this better result. If in the US one observes in distribution services a modernization of the structure of the industry that had been achieved much earlier in Europe, in countries like France and the UK, then one cannot expect similar 'extra' productivity gains to be achieved in Europe in the near future in the these industries. Moreover one should recall that the two services under view, namely distribution and financial services, are also those where the measurement issue is more problematic.

Overall, looking at the standard statistics of productivity (real value added per person employed) the diffusion of ICTs seems, everything else being equal, to have had only a rather marginal effect until the mid-2000s.

However, such a statement is highly conditional. It may be the case that ICT equipment requires some specific combination of factors to be productive, a complementarity condition that is precisely claimed in the debate on the skill bias nature of ICTs. I will investigate in the next section to what extent this requirement explains the 'productivity paradox' we are looking at.

17.4 Another important structural change: a more educated labour force

An increase in the quality of the labour force should have a similar boosting effect on productivity that we expected from ICTs. In the 1980s, 1990s and 2000s developed countries have experienced a continuous rise in schooling rates. The upward turn was marked in tertiary education from the mid-1970s onwards when unemployment rose and education appeared as a premium to enter tighter labour markets. This upskilling of the labour force could have come from a demand for more qualified workers by the firms or from some screening effects on the labour market whereby more educated people are given priority in the access to jobs. This increased demand could have been effectively stimulated by slacking labour markets, which induce longer stays in initial formal education designed to delay entry and to improve the chances of finding a proper job. Such an effect has been underlined by Howell and Wolff (1992) for the US, and by Goux and Maurin (1995) for France.

Reports praising the emergence of so-called 'knowledge-based economies' have certainly encouraged this trend. The Lisbon Agenda in 2000 constituted in that respect a major consecration of such presentation of the future of developed economies (unfortunately just before the crash of the dot.com bubble and the downturn of the stock markets).[9]

The rise in schooling rates appears as a steady trend across all developed economies, little affected by the cyclical crises of these so-called knowledge-based economies. Table 17.5 shows how universal this trend was. In the early 2000s over 40 per cent of the 20–24 year-olds of most of the developed economies were enrolled in post-secondary education (see columns 2 and 3 of Table 17.5). One can have an idea of the yearly improvement of the labour force qualifications brought by these entries when comparing with the education of the stock of the labour force (given in column 1 in Table 17.5). Such an increase in quality should have contributed to boosting the productivity of all benefiting activities in all so-called knowledge-based economies.

Still looking at the assessment of productivity gains (given in Tables 17.3 and 17.4), this rise in the education levels of the young entering the labour markets does not seem to have had a widespread impact. It could account for some of the sectoral differences I had underlined. Thus, entries of highly educated young people in financial and business services (see Table 17.6) could have contributed to the moderate but slightly bigger productivity rise one observes in these activities. It would in this case be combined with the effect of the diffusion of ICTs.

Again, this 'other' productivity paradox suggests that the expected positive impact of an increase in the quality of the labour force is conditional on the presence of some other factors. In this case the effectiveness of ICTs could be linked with the presence of skilled labour.

Indeed during the 1980s many observers contended that investment in the new ICTs was accompanied by a marked change in the structure of employment in favour of skilled labour. The shift was even more marked in manufacturing where low-skill employment was actually decreasing (see Table 17.6). But this called attention to the fact that this bias could result from the concomitant rise in the competition of low-wage countries in trade. The whole issue gave way to an interesting debate on the skill-bias nature of technical change (SBTC hereafter).

The shift observed in the distribution of skilled labour suggested a

Table 17.5 Structural change indicators: education and diffusion of ICT at the end of the 1990s

	Education			Diffusion of ICTs			
	1	2	3	4	5	6	7
Australia	43			8.85	30.9	75.0	469
Austria	26			4.82	7.2	57.6	257
Belgium	43	40	45	5.88	7.9	39.7	315
Canada	21			8.52	30.4	127.2	361
Denmark	20	50	59	6.94	26.0	72.5	414
Finland	28	42	53	5.88	68.1	159.1	360
France	38	42	47	5.96	5.3	19.2	222
Germany	19	41	43	5.27	10.3	31.7	297
Ireland	49			6.48	13.0	31.1	405
Italy	56	30	34	4.72	3.7	32.6	192
Greece	50	30	31	5.51	2.8	13.0	60
Japan	19			7.06	8.4	32.5	287
Korea	34			4.42	2.1	10.8	182
Netherlands	35	52	48	7.13	21.9	81.6	360
Norway	15			6.93	40.9	116.5	447
Portugal	79	32	42	5.31	3.1	13.4	93
Spain	65	40	51	4.03	4.0	15.7	119
Sweden	23	43	52	9.28	35.0	106.3	451
Switzerland	18			7.48	20.7	63.5	462
UK	18	41	42	9.35	15.7	52.5	303
European Union	39	40	43		10.2	37.4	
USA	13			8.87	56.5	234.2	511

Notes and Sources:
Column 1: Share of population aged 25–64 with less than senior matriculation, 1996, OECD (2001, 173).
Column 2: Rate of post-secondary education enrolment for men aged 20–24, 1999/2000 European Union, Eurostat.
Column 3: Rate of post-secondary education enrolment for women aged 20–24, 1999/2000 European Union, Eurostat.
Column 4: ICT expenditures as a percentage of GDP, 1999, UNESCO *World Development Indicators,* 2001.
Column 5: Internet hosts per 1000 inhabitants, July 1997, OECD (2001, p. 181).
Column 6: Internet hosts per 1000 inhabitants, October 2000, OECD (2001, p. 181).
Column 7: Number of personal computers per 1000 inhabitants, 1999, UNESCO *World Development Indicators,* 2001.

new complementary relationship between capital and skills (both being a substitute for unskilled labour), which contrasts with the fact that during the 'Fordist' era capital was more often a substitute for skilled labour. Griliches (1969) explains this complementarity in terms of a decline in the

Table 17.6 Employment growth breakdown by skill level in manufacturing and services (annual % growth rates, 1980s)

	Blue-collar low-skilled		Blue-collar high-skilled		White-collar Low-skilled		White-collar high-skilled	
	Manuf. Ind.	Market Services	Manuf. ind.	Market services	Manuf. ind.	Market Services	Manuf. ind.	Market services
United States 1983–93	−0.2	0.2	−0.1	0.2	−0.1	1.4	0.2	0.9
Canada 1981–91	−0.7	0.1	−0.2	–	0.1	0.8	0.4	1
Japan 1980–90	0.1	0.3	−0.2	−0.1	0.5	1.3	0.4	0.9
Germany 1980–90	−0.8	−0.2	0.2	0.3	0.2	1	0.3	0.6
France 1982–90	−1.4	0.2	−0.3	–	−0.2	0.5	0.4	1.2
Italy 1981–91	−0.5	0.6	−1.0	0.1	−0.1	1.2	0.2	1.0

Source: OECD (STI/EAS Division).

relative price of capital,[10] whereas Denny and Fuss (1983) for instance attributed the phenomenon to specific effects of technical change.

Berman et al. (1994), Machin and Van Reenen (1998) and Berman et al. (1998) all stress the importance of this skill bias in manufacturing[11] and the difficulty of distinguishing this skill bias effect from a standard trade effect induced by the competition of low-wage countries.

Looking at Table 17.6, changes in white-collar employment in services tend to show, in contrast with what we have in manufacturing, that the growth rates of low-skilled jobs overtook those of high-skilled ones. Again this is observed for a majority of the countries in the sample, but not for all, France and Canada being the exceptions. This is in accordance with the stress I put earlier on the variety of capitalism. According to country, the dynamics of low-skill services are expected to differ. But in that case one would have expected a more rapid growth of low-paid jobs in market-oriented countries (as is the case for the US, but then raising questions for Germany) and less rapid in countries with high welfare (as is the case with France, but raising questions for Canada). One should also consider that the data concern a period where reorganisations were just starting, and one should look over a longer period to check the consistency of the typology of capitalism used earlier.

This issue of skill-biased technological change (SBTC) is of interest for my own issue of innovation in services. If it applies mainly to manufacturing, what then is conditioning the innovative potential of ICTs in services? It could be the case that the complementary factors required are ill-defined when referring to a rough dichotomy between skilled and unskilled labour. It could also be the case that learning how to use ICTs takes more time in services. In effect, both issues have been raised in the literature.

First, using a slightly more sophisticated distinction led to more informative results. The notion of skill used in many of the studies referred to above made only a crude distinction between production and non-production workers.[12] By considering three major dimensions of skill – cognitive (analytic and synthetic reasoning, numerical and verbal facilities), motor (physical abilities, coordination, dexterity) and interpersonal skills (supervisory skills, leadership) – Howell and Wolff (1992) and Wolff (1996) constructed three skill scales on which basis they graded occupations by sector, and documented changes in these grades during the 1980s. These studies confirmed the general decline in the demand for unskilled labour, but associated it with a drop in the demand for motor skills. The shift towards service activities turned out to be a major cause of the decrease in demand for motor skills. Meanwhile, the demand for higher cognitive and interactive skills was shown to increase. These studies also suggested that the effects of investment on the skill structure could differ depending on whether investment was in general equipment or ICTs. As the latter represent only a small fraction of total investment (Oliner and Sichel, 2000), the specific effects of ICTs on skills may sometimes be obscured by conventional capital–labour substitution effects.

Another alternative to the dichotomic specification between skilled and unskilled is to consider organisational forms directly. Most of the studies discussed so far analyse the skill-bias of new technology as if it was an individual issue, disregarding the fact that the primary concern of a firm is with work organisation as a whole. These organisational issues are certainly difficult to follow as firms initially react differently in times of structural changes (Attewell, 1994) before best practices are identified and diffused.

Aoki's (1988) distinction between the information structures embodied in the different forms of work organisation in Japan (horizontal structure of decentralised decision-making processes) and the US (vertical structure of centralised decision-processes) has been one major way to address this issue.[13] Hence Piore (1988) has suggested that the US model has used the new technological context to change towards more horizontal channels of communication, with less hierarchy and reduced specialisation (thus moving closer to Aoki's Japanese model of organisation). But this is only a broad assessment of organisational change that transfers in rather hybrid ways and is therefore difficult to track down at firms level.

Considering organisational issues as possible preconditions for a successful innovative use of ICTs by firms, either in manufacturing or in services, leads one to question in turn the conditions under which firms operate to implement such organisational changes. If these changes are of a certain magnitude they are bound to be conditioned by the regulatory framework in which firms operate. Therefore one is led to consider the changes in regulations that do constitute another major structural change of the 1980s and 1990s.

17.5 Deregulation as a prerequisite for innovative reorganisations

The Thatcher and Reagan governments launched in their respective countries a wave of deregulation in the early 1980s that then diffused to most Organisation for Economic Co-operation and Development (OECD) countries. Such a process, often step-wise to ease its implementation, takes time. It went on throughout the 1990s. The OECD tried to assess the magnitude of this deregulation, asking experts to rank between 0 and 6 the relative level of regulation for its member countries. The exercise, beyond its arbitrariness, showed differences in degrees of regulation (see Table 17.7), as expected, as well as a trend of deregulation across all member countries during the 1980s and 1990s (see Figure 17.2). I have used in Table 17.7 the same typology of capitalisms as in Table 17.1.

The assessment of the degrees of regulation in 1998 in most OECD countries displayed in Table 17.7 deserves some comment. It concerns various types of regulation. Table 17.7 shows how the overall degree of regulation (column 6) is the combination of various types of regulation, namely of trade, of general administrative regulation and of the conditions of competition. First this inventory leaves out all the regulations applying to the products themselves (in terms of norms and standards); it concerns mainly the conditions of production. Figure 17.1 summarises the relative importance of these components, showing that barriers to trade are very limited[14] and that degrees of regulation are very much tied to internal conditions of production.

Figure 17.2 shows in turn that the trend in deregulation has been quite general even if it somehow conserved the relative positions of countries over the whole of the two decades under review, 1978–98. Countries which have moved the least are on the left-hand side of the figure.

One may think that most of this deregulation trend concerned labour markets following the frequent call in the 1980s for more flexibility. Indeed labour markets were deregulated over the 1980s and 1990s, but in a very

Table 17.7 Degrees of business regulation in 1998 by type of capitalist countries

Type of capitalism	International[1]	State regulatory control		State discretionary control		Overall regulatory environment[6]
		Domestic[2]	Enterprise[3]	Public Sector[4]	State Control[5]	
Market-based						
Australia	0.4	1.2	1.1	0.8	1.8	0.9
Canada	2.2	1.0	0.8	1.2	1.4	1.5
USA	0.9	1.1	1.3	0.8	0.9	1.0
GB	0.4	0.5	0.5	0.	1.2	0.5
Nordic						
Sweden	0.8	1.7	1.8	2.3	0.6	1.4
Finland	0.6	2.3	1.9	3.3	1.9	1.7
Denmark	0.5	1.9	1.3	2.3	2.7	1.4
Mesocorporatist						
Japan	1.0	1.8	2.3	0.7	2.1	1.5
Korea	1.7	2.7	3.1	2.5	2.2	2.4
European continental						
Germany	0.5	2.7	2.1	1.2	2.5	1.4
France	1.0	2.7	2.7	2.3	3.0	2.1
Belgium	0.6	2.7	2.6	2.0	3.8	1.9
Ireland	0.4	1.1	1.2	1.3	0.5	0.8
Netherlands	0.5	1.8	1.4	2.6	1.9	1.4
Mediterranean						
Spain	0.7	2.2	1.8	2.0	3.4	1.6
Portugal	1.1	2.1	1.5	2.7	3.0	1.7
Italy	0.5	3.3	2.7	4.4	3.3	2.3
Greece	1.3	2.7	1.7	3.4	4.5	2.2

Notes: Column 1: degree of regulation of international business transactions, barriers to trade and investment; Column 2: degree of regulation of domestic business transactions; Column 3: degree of small business regulations, barriers to entrepreneurship; Column 4: size of the public sector; Column 5: degree of control of economic activities by the State; Column 6: general degree of regulation of economic activities.

Constructed using experts' scores from 0 (extremely liberal) to 6 (extremely strict regulation) for the various fields. The indicator in Column 2 is the average of the indicators in columns 3, 4, and 5.

Source: Nicoletti et al. (2000).

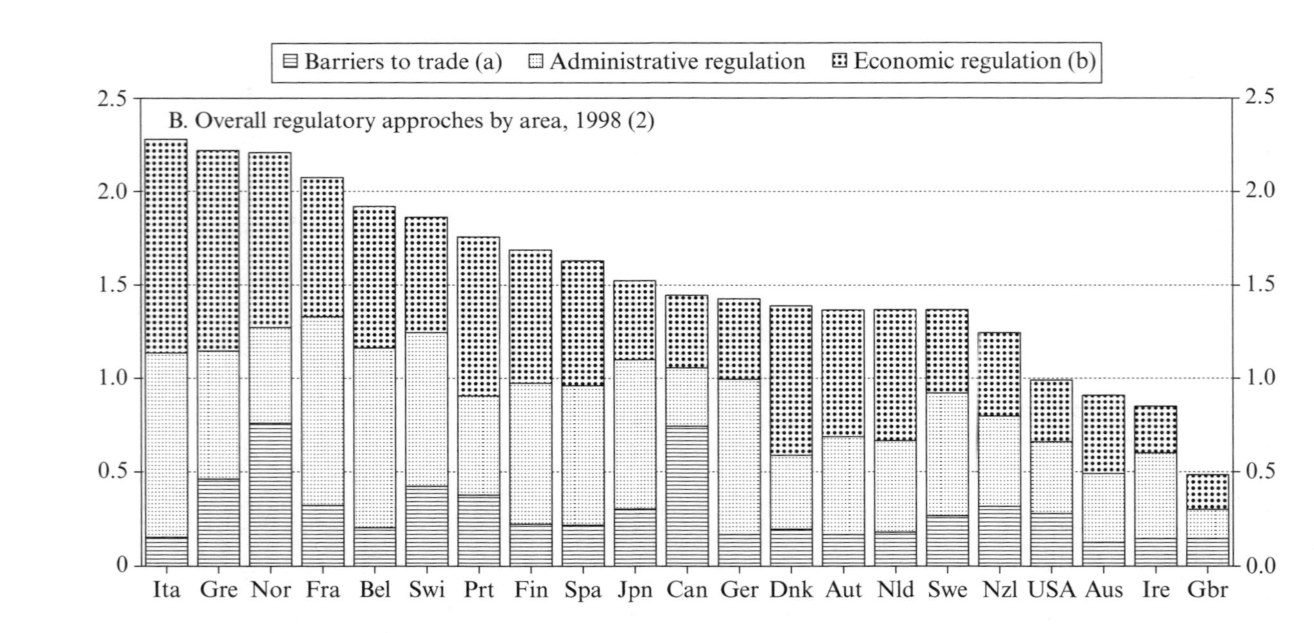

Notes:

1. Reports changes in the regulatory stance in seven non-manufacturing industries (gas, electricity, post, telecommunications, passenger air transport, railways and road freight) between 1978 and 1998. The regulatory stance is measured by a synthetic indicator ranging from 0 (least restrictive) to 6 (most restrictive).
2. a) Includes trade and FDI restrictions b) includes barriers to competition and state control.

Source: Nicoletti et al. (2000).

Figure 17.1 Relative importance of the various types of regulation

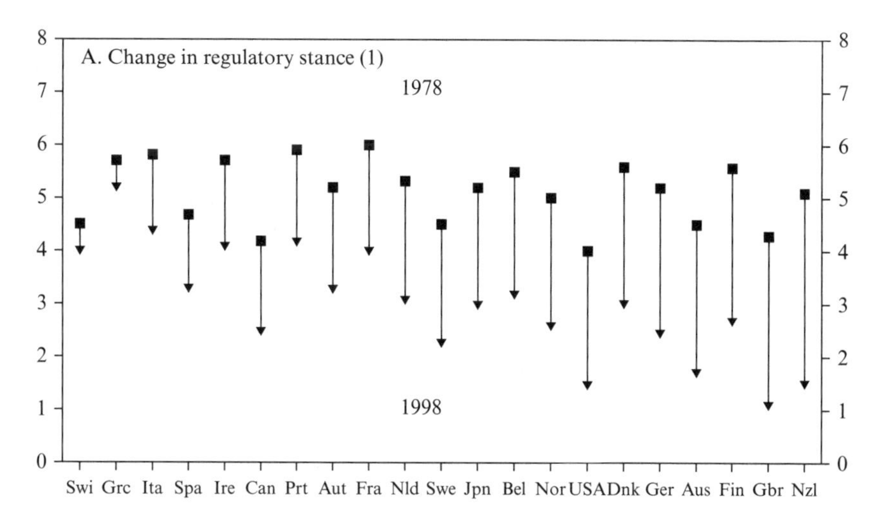

Source: De Serres (2003).

Figure 17.2 Changes in degrees of regulation between 1978 and 1998

specific manner whereby most of the deregulation concerned temporary work in countries with relatively high strictness of employment legislation, while the conditions of regular employment remained much the same (as shown in Figure 17.3).

The fact that most of the deregulation of labour markets in the 1990s related to temporary employment is of interest as these jobs are quite numerous in service industries. This increased flexibility in the use of temporary employment may well have influenced the mode of organisation of services.

Finally, an important factor constituting these degrees of regulation could well be tied to specificities of some networked service sectors which have long been activities under public tutelage. The picture we get in Table 17.8 is more composite. Indeed services like retail trade, railways and electricity have on average rather stricter regulatory conditions. But one observes a wide diversity of cases whereby countries with overall low degrees of regulation may have conserved some strict regulations in one or two service industries. In the US one finds for instance that the electricity sector remained rather regulated in 1998; in the UK retail trade remained also quite regulated to pick up the case of the overall more deregulated OECD countries. This underlines, first, that arrangements in these large networked services have often been country-specific; and second, that these organisations present some inertia.

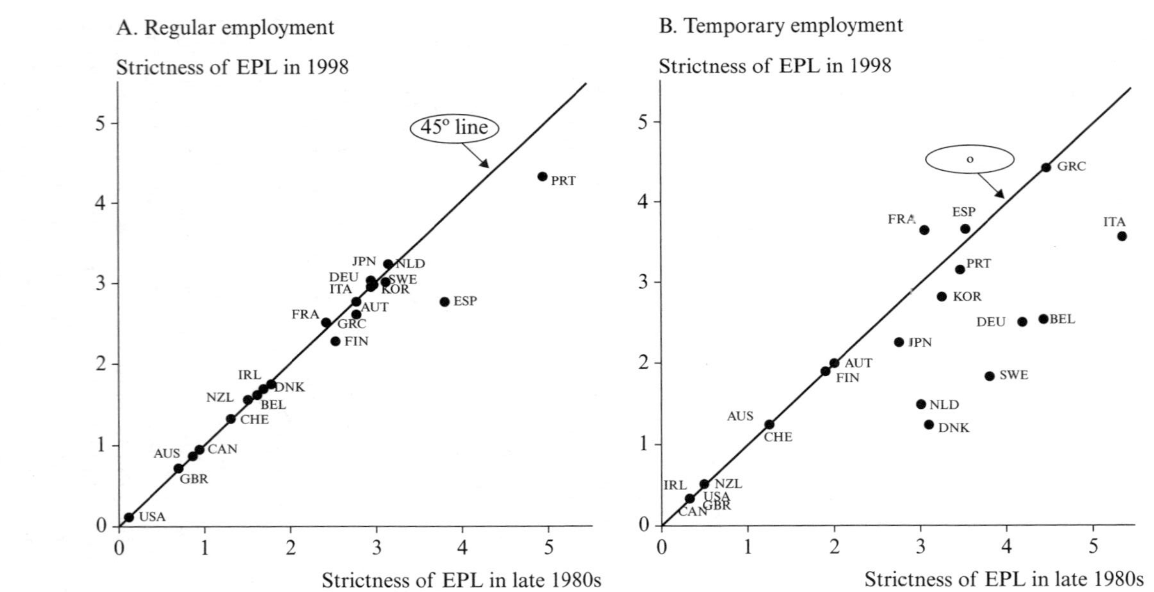

Notes:
1 For definition, see Source.
A higher indicator value implies a more restrictive policy stance.

Source: OECD (1999).

Figure 17.3 Strictness of employment protection and legislation, 1988–98 (synthetic OECD restrictiveness indicators)

Table 17.8 Degrees of regulation in some large networked service industries, 1998

Type of capitalism	Retail trade	Truck transport	Mobile phone	Fix phone	Air passenger transport	Electricity	Rail
Market-based							
Australia	0	0	2	2	2	0	0
Canada	2	2	–	2	4	6	2
USA	0	0	2	2	0	4	0
GB	4	0	2	2	2	0	2
Nordic							
Sweden	2	2	2	2	2	0	2
Finland	4	2	2	2	2	0	6
Denmark	4	–	2	2	2	4	6
Mesocorporatist							
Japan	6	4	–	–	4	4	–
Korea	2	0	2	2	4	–	6
European continental							
Germany	2	4	2	2	2	2	0
France	6	–	–	–	4	6	6
Belgium	4	4	2	2	2	4	6
Ireland	2	–	2	2	2	4	6
Netherlands	2	4	2	2	0	4	0
Mediterranean							
Spain	4	4	6	2	2	4	6
Portugal	4	2	2	6	6	4	–
Italy	4	6	6	2	2	6	6
Greece	6	6	2	6	6	6	–

Source: Nicoletti et al. (2000).

What can we draw from this rapid survey of the main features of the deregulation policies implemented in the 1980s and 1990s? The assessment proposed by the OECD studies, based on expert advice, does confirm the universal trend to liberalise activities, but it also shows that this deregulation occurred in specific ways which can be summarised as follows:

- It concerned labour markets mainly affecting the flexibility of the status of temporary workers.
- Trade barriers have been significantly reduced across all OECD countries.
- Services overall have been transformed unevenly, whatever the type of capitalism, hinting at the specific and lasting nature of some national organisational arrangements.

It should also be noticed that the regulations under view in these OECD studies mainly concerned the conditions of production and of access to markets. This assessment did not take into account the regulations, in terms of norms and standards applying to the products themselves. If such had been the case one would have observed, parallel to the decrease in the strictness of the regulation of the conditions of production, a reverse trend whereby products tended to be more and more submitted to regulations, and more specifically so in manufacturing.

These last comments may help us to understand why despite the diffusion of ICTs, and despite the deregulation trend, productivity gains in services activities remained relatively low (even by their own standards), and why we tend rapidly to conclude that innovations were few and their potential weak.

17.6 Implications for the potential for innovation in services

So far I have mainly stressed to what extent such important structural changes as the diffusion of new technologies, the increase in the level of education of the labour force or the changes in regulatory framework have had little impact on the growth of productivity in services. One way to answer this generalised paradox may be to consider that it takes time for all these changes to lead to innovations of processes and products. The learning processes at work may require some specific coordination, the implementation of which is not straightforward. This has led some authors studying SBTC to distinguish a time of adoption of ICTs and a time of use (Chun, 2003; Basu et al., 2003),[15] implying that SBTC could be a reaction in an early phase, followed by a more mature adjustment where SBTC effects could disappear. In such a perspective SBTC would be a rather temporary effect. Similar comments can be made regarding the other two structural changes. In effect, difficulties setting up learning processes could directly follow from the conditions in time and space in which these external structural changes occurred.

Being simultaneous and global these structural changes presented firms with opportunities they had to seize for their specific use, without realising the long-run implications of these choices, leading to changes in production processes more perceived as first reactions to new conditions of production than as well-monitored innovations processes in relation to their own product markets. The networking dimension of the structural changes under view may explain such behaviours.

This led in many cases to what Barras (1986) called a reverse cycle of innovations whereby innovations in production processes pre-exist and precondition product innovations. In the Fordist period (1945–75) manufacturing had shown the opposite sequence where product innovations, designed by research and development (R&D) departments of large firms, led in turn to process innovations to reduce the price of the new goods and to remain competitive, once the rent on product innovation started to disappear. Barras noticed, at the beginning of the diffusion of ICTs, namely of mainframe computer networks, that a 'reverse cycle' was emerging in large intermediary services (banking, distribution, telecommunications, transport). Deregulation, the Internet and internationalisation seem to have extended this possibility of reverse innovation cycles to manufacturing industries.

The diffusion of ICTs, the rise in education and the changes in regulatory environment all looked like external challenges for firms to remain competitive and be more reactive to changes in demand. These structural changes (at least the first two of them) were not neutral. ICTs obviously favoured some decentralisation of tasks and a more educated labour force helped in that direction, allowing externalisation of menial tasks and delocalisation of parts of the manufacturing production processes. Changes in the regulatory environment also had an impact of their own. The deregulation of finance was clearly a precondition for its globalisation but it also opened the way to a drastic move in the governance of large firms, whereby shareholder value mirrored by stock markets and globalised financial investors became the prevailing criteria. A consequence of this shift was also for the firms to favour short-term objectives, all of which impacted in turn upon their innovation policies. These policies became more diverse, mixing traditional strategies relying on in-house R&D with 'open innovation' strategies trying to take advantage of inventions from diverse sources (foreign affiliates, partners, competitors, public research bodies).[16] Free entry and exit of foreign direct investments, mergers and acquisitions, franchising and all kinds of international accords thus facilitated the development of international network production processes.

Large network services which were the first to seize these opportunities produced emblematic examples of this modernity from Benetton to Wal-Mart, not to mention yield management pricing in transport industries and the globalisation of financial operations.

The specific change in labour laws that I have underlined, whereby temporary employment became more flexible in all countries, had a specific effect: the rapid organisational adjustment of large network services took advantage – directly or indirectly when using local subcontractors – of an unskilled temporary labour force. This occurred chiefly in distribution and transport but also affected the organisation of social and personal services. As a consequence technical change appeared overall in services as without skill (once local externalisation of menial or peripheral jobs had been accounted for). It follows that in these first phases (concerning innovations in processes) of the generalised inverted innovation cycles, à la Barras, manufacturing and services tended to be affected in opposite directions. The question is then to know how this impacted upon the second phase of the cycle, namely the product innovation policies induced by the new organisations of production.

I have already mentioned how product innovation policies in manufacturing industries diversified and relied more on 'open innovation' strategies. Let us now consider some of the impacts this had on innovation in large network services. Clearly the new flexibility in the use of temporary employment led to the favouring of some neo-Taylorian work organisations, which in turn conditioned the opportunities for product innovations. The first reactions of large service firms confronted with the three supply shocks I detailed above induced cases of path-dependency[17] in organisational patterns, meaning that adjustments quickly done in a given environment were bound to condition strongly the future organisational choices.

This is particularly the case with large firms in services which can somehow impose standards and norms of products. While in manufacturing, standards and norms of products are the combined output of a broad and open competition and of specific product regulations, this is not exactly the case with service products, which tend to be much more dependent on local conditions of provision and are sometimes less regulated or still subjected to country-specific regulations. As mentioned above, the assessment of the deregulation trend is much less focused on product regulations than on the regulatory framework of the general conditions of production. The deregulation of the entrepreneurial conditions of production may well be matched by increases in product regulations, where norms and standards from public and private origins have been developing steadily, especially in manufacturing. By contrast some service products seem only lightly constrained by external product regulations. Financial products, emerging from the wave of deregulation of the 1980s, have largely escaped such product supervision, fuelling the 2008 financial crisis. Other large network services, responding to the challenges raised by structural changes, have also set up organisations which have impacted strongly upon ways of life, leaving little ground for alternatives trajectories to take place.

One of the best examples is Wal-Mart, a firm that has grown over the 1990s and 2000s to the first rank of the Fortune 500 on a basic business model which took advantage of:

1. a location (suburb of a middle-sized town, where no competitor could then enter, while local downtown trade was heavily hurt and while access was facilitated to a flexible, low-paid, part-time labour force[18]);
2. ICT's use to speed up efficiently financial as well as merchandise flows;
3. the reduction of trade barriers, to import massively from low-wage countries.

In the end the strategy of such a firm has had a strongly structuring impact on the way of life of millions of Americans. Such a major transformation has, though, never been discussed by the communities concerned, even if there have been waves of protestation. Beyond the extreme of such a case as Wal-Mart, one can see the asymmetry that existed in the development of large network services in the three decades under review. Large firms can adopt the changes rapidly and profitably. Customers (as citizens) will not be in a position, during this adoption phase, to discuss the opportunity of changes in their way of life.[19] I have already stressed (Petit, 2002) that this asymmetry in the timing of the learning processes, on the supply and on the demand side respectively, led to suboptimal organisation of the society in the sense that it was not only detrimental for citizens and customers, locked in a too rapidly fixed organisation with many drawbacks, but also for the firms and the society as a whole, deprived of the advantages that could have brought learning processes on the demand side. Considering that in such a process customers have not been in a position to develop their customer skills, what we observe is again a case of skill-biased technological change (SBTC). But while in manufacturing SBTC occurred at the time of adoption and could eventually be adjusted in the use period, I characterise the asymmetry in learning processes between the supply and demand sides in services (stressed above) as a case of SBTC at the time of use, when product innovations are taking place. This may have lasting consequences, reducing strongly the long-term innovation potential of large network services.

In a knowledge-based economy a society has learning processes taking place on both demand and supply sides. The new industrial economics tends to take a very static view of the positions of agents, in taking no account of the different timing of the learning processes. The deregulation policies that they inspired gave little attention to these phenomena. In large networked services with an important infrastructure such as transport industries, the decoupling between natural monopolies in charge of the infrastructure, and competing service providers exploiting it, did not really cope with this drawback. It led to the development of two rather damaging logics: first, chronic underinvestment on the side of the local monopolies in charge of running the infrastructures profitably (with as main illustrations the railways infrastructure in the UK, and the electricity infrastructure in the US); and second, conservative yield management pricing from the firms competing to reap the short-term profits of their use of the infrastructure on the other side. Very little remained of the debates of the early twentieth century around the access to some universal services as basic elements of citizenship. In the same vein the development of automated services has been mainly driven by the will to reduce labour costs, and little attention has been paid to discussing user needs, to enlarge the capacities of citizens to make the best of these large network services.

Pressure to set norms based on the view that consumer surplus could only stem from short-term reductions in consumer prices has been growing over the 1980s, 1990s and 2000s. The case of Japanese retail trade and electricity is rather telling in that respect (Jones and Yoon, 2008). Japan was in the 1980s a country with a relatively high degree of regulation. This was especially noticeable in the retail trade, and the fact that large retail stores and foreign affiliates were few seemed a direct consequence of this restrictive regulatory framework. To comply with the international claim for liberalisation from its trade partners, Japan tried to align its regulations with general OECD practices. It succeeded, and Japan was ranked within the OECD average in terms of regulatory pressure by the end of the 1990s. Still, the number of large stores and of foreign affiliates decreased.[20] In effect, beyond the will of the central government to reduce visible and invisible barriers to free establishment, there is a strong local concern to check that all the impacts are identified and acceptable. Such a precautionary approach is quite understandable. But under the general pressure for fully fledged liberalisation these concerns tend to be discarded by the regulator. The reasons given for the success of a Japanese innovation in retailing like the convenience store chain Seven Eleven suggest that local community issues should be taken into account. Nonaka et al. (2008) explain the success of Seven Eleven Japan (SEJ), by now the biggest convenience store chain in Japan, by its capacity over 30 years constantly to anticipate and satisfy the changing needs of consumers, utilising 'multilayered "*ba*" (shared context of interaction) and networked knowledge to introduce products and services faster than its main competitors', creating a balance between supportive information systems (objectivity) and human insight (subjectivity). At a time when climate change is a real threat, an even more explicit debate on adjustments in ways of life is objectively on the agenda. The quality and usefulness of the SEJ interface with consumers will then be fully obvious. The assessment that the small number of large-scale outlets is responsible for the low level of productivity in the retail sector in Japan (Aoki et al., 2000) does then look like a tautology, following a rather narrow (and common) definition of real output in retailing.

A similar type of issue is raised with the electricity industries in Japan, where prices appear to be relatively high, especially for households in the mid-2000s compared with other OECD countries.[21] In 2003 the degree of regulation (as measured by the OECD) was just above average (largely due to a strong vertical integration) after a noticeable reduction from the 1995 level. OECD experts are still advocating more deregulation. Again this focus on short-term price reductions may miss essential points. It may be the case that incumbent firms are exploiting monopolistic positions for their own profit. But it may also be the case that in a country liable to earthquakes and lacking national resources of energy you need to have special quality of infrastructures, resistant and saving imported raw resources. Such requirements (and the list is open) have to be taken into account before any policy recommendation.

One could go on surveying studies on other service activities in Japan or elsewhere; most of them would confirm that a one-sided view, based on short-term price reduction, is taken, stressing the need for a more comprehensive approach of needs and the quality of services. To revive learning processes so that citizens could have their say in the long-term structuring of their daily way of life is largely a political issue. The economic crisis begun in 2008 may help such reformatting of the innovation potential of services.

17.7 Redefining innovation in services

If one wants to take stock of various observations to assess the potential of innovation in services, it would be misleading to rely on standard productivity measures as indicators. In many cases these measures embed, from the start, assumptions on productivity, especially when output measures are based on input quantities.

Strategies of development mainly concur, then, to cost-cutting strategies, all of which lead to further reductions in employment and in wages. Such cost-cutting strategies may have a detrimental effect on the very production of services. Deregulation as it has been practised, that is, mostly concerned with loosening conditions of production, tends to back such cost-cutting strategies. The uses made of structural changes, such as the diffusion of new technologies, are then mainly aimed to rapidly cut per-unit costs of production. This hampers the development of a longer and broadly based learning process that would incorporate full phases of adaptation on the user side. In industries where users are often involved in the provision of services, trajectories that do not allow for the full development of user learning processes are strongly reducing the potential for innovations.

To get out of the deadlock generated by this asymmetry between learning processes on the demand and supply sides is a major political objective. It may be channelled by forging new product regulations, new norms and standards, helping to gauge the utility of these various services. Goods safety conditions as well as ecological rules have played such roles. It is much less the case with services. Beyond services that remain under important public tutelage, like health and education, product regulations in services have generally lost ground with the wave of deregulation and the retreat of the state. The debates of the early twentieth century on universal services in all large network services (from banking to transport, telecommunication, distribution and electricity) as being parts of citizenship are bygone. The credo that private operators ensure efficient management has nearly eliminated in the 1990s and 2000s all debates on the positive externalities linked to the concerted definition and satisfaction of collective needs. Financial services have given, with the subprime crisis, a tragic illustration of the chaos that an uncontrolled development of products can provoke.

Unfortunately while our societies are facing various major challenges, from global warming to ageing, notwithstanding increasing exposures to poverty or pandemics, the developments of many large networked services are shortsighted and centred on costs reductions. Clearly costs have to be monitored, but the cost disease, that was stressed in the 1960s, should not be transformed into dysfunctional trends in large services systems. The debates on the need to revise our measurement of growth and welfare (see Gadrey and Jany-Catrice, 2007) tend to point to many such dysfunctions. A very telling example is given by the health sector in the US where rising expenditures did not prevent a net downturn of indicators of the health situation of populations.[22] This example is all the more striking in that health expenditures in the US, as a share of GDP, are roughly twice as big as the OECD average, while life expectancy is close to that of a developing country. A general rise in inequalities within developed countries can 'explain' this dysfunction. Indicators on education will lead to rather similar results if one takes into account both the quality of training received and the ability to find a job in accordance with it.[23]

Respective balance sheets of the impacts of our transport systems and of our distribution systems (retailing) in terms of their contribution to global warming would show similar 'destructive' trends, while the concern has been on the agenda well before, let us say, the widespread diffusion of Wal-Mart centres across the US. The list would be long showing how little our large service systems have so far helped to manage our environment in a direction compatible with any sustainable development.[24]

It is clearly in such directions that developed economies have to look for sizeable service innovations. Their systemic aspects imply the involvement of communities and give them a political dimension. For this reason it is hard to believe that such involvement could emerge from private management, obviously too focused on short-term cost reductions and private profit maximisation, when the transformations that would be radically innovating have a long-term horizon and generate a wide range of social benefits. Clearly the redefinition of innovation in services goes together with the development of a more collective approach of long-term objectives of sustainable development, embracing its three dimensions: environmental, social and economic. How such a turn in policies can be achieved when for so long efficiency has been looked for only by means of 'liberalisation' remains a question, even in the middle of a major crisis. Countries have not all evolved in the same way regarding effective levels of liberalisation and degrees of rejection of public intervention. The crisis will also affect them differently, their exposure to hazardous financing having been different as well as their part in the international division of labour. New forms of collective actions will have to be invented in support of the major transformations of large service systems that are required. Whether market economies, which have rejected most kinds of public intervention, will succeed more rapidly than economies still relying on non-market coordination to invent these new forms of collective actions, remains to be seen. In a very open world the international coordination of these transformations will be a real challenge. Some countries will react quickly, thinking that moves towards more public intervention are temporary. Clearly the new social conventions that we need to get out of the present tri-dimensional crisis will have to be based on sound sets of indicators to avoid a return to the hazards of the unsustainable paths of the 1990s and 2000s.

Notes

1. The relationship involved in the provision of such menial personal services is perceived according to one's culture. D'Iribarne (1989) thus stresses the difficulty posed by this relation in a country like France, where it tends to be connoted with some hierarchical order, in contrast with Anglo-Saxon experiences where it remains more factual, precisely limited to certain tasks, without implying any subordinate status.
2. See Petit (2008). The definition of these items often varies from one study to another,

with finance activities often being joined with producer services, for instance in the FIRE item of most OECD categorisation of service activities. Indeed comparisons are sometimes made difficult, with some activities being highly country-specific.
3. On professional services in knowledge-based economies, see Gallouj (2002).
4. Esping-Andersen (1990) underlines the importance of these various modes of welfare provision, stressing the role retained by the family in the Mediterranean countries.
5. See Elfring (1992) for the distribution of services by subsector between 1960 and 1987, reproduced in Petit (2008).
6. This issue has been called the Solow paradox to echo a remark made by Robert Solow as he received his Nobel Prize: 'we see a lot of seemingly important (modernizing) changes . . . with little impact on the statistics of productivity gains. We see the transformations . . . but not so much the results'.
7. These gains have been especially outstanding in the manufacturing of ICT products, driven by the regular process of miniaturisation of chips (a reduction doubling the capacity of microprocessors every 18 months, as predicted and induced by Moore's self-fulfilling prophesy; Moore was a former chief executive of the firm Intel in the 1960s).
8. In the tables presented here the EU-15 is reduced with Greece, Portugal, Luxembourg, Ireland and Sweden being excluded from the sample.
9. According to this agenda the European Union was bound to become the first knowledge-based economy in the world within a decade.
10. The sharp reduction in the price of microprocessors is an important feature of ICTs.
11. An observation confirmed by Caroli and Van Reenen (2001) at the level of the establishment.
12. This distinction is correlated with a number of alternative distinctions based on broad characteristics of educational attainment or of occupational category, as Berman et al. (1994) have shown for the US.
13. These structures display different capacities for facing and organising technological change, with the Japanese horizontal structure of information (in its work organisation) being more fit to diffuse and implement incremental innovations in products or processes; whereas US firms, with their more vertical structures of information (as characterised by their hierarchical work organisation) would be better at facing radical technical changes.
14. In Table 17.8 trade barriers remain relatively important in Canada and in Norway, chiefly because of the importance of primary industries with specific regulations, like quarrying and mining.
15. O'Mahony et al. (2008b) showed that such lags between adoption and use could also exist among countries, with the US showing a more rapid learning process.
16. See Chesbrough (2003).
17. Path-dependency and lock-in phenomena as exposed in Arthur (1989), David (1991) and North (1990).
18. Wal-Mart has been investing in the neighbourhood of its shopping centres, leading the firm paradoxically to score high when measuring what firms are spending under the heading of corporate social responsibility.
19. I have already stressed this asymmetry in the timing of the learning processes on the supply and on the demand side.
20. It went down from 21 per cent of the sales in 1997 to 16 per cent in 2004 (or from 13.1 per cent of the shops to 11.7 per cent).
21. In 2005 for the household sector a kilowatt per hour was priced $0.18 in Japan, versus $0.1 in the US (2008), $0.14 in France (2008) but quite the same in the UK (2008) and $0.21 in Germany (2005) and $0.23 in Italy (2008).
22. The indicator used by Miringoff and Miringoff (1999) combines 16 variables: four age groups, each with three or four criteria varying with the age group. At the turn of the 1970s the trajectory of the health indicator started decreasing while GDP in real terms went on steadily increasing.
23. The PISA OECD survey on the performances of the trainees displayed large differences

of results among countries, and hinted at growing dysfunctions in terms of both capacities acquired and employment found.
24. Indeed indicators of sustainable development (where costs of environmental degradation are included) display a similar deviation from GDP trend in the 1970s (see Cobb and Cobb, 1994).

References

Acemoglu, D. (1998), 'Why do new technologies complement skills? Directed technical change and wage inequality', *Quarterly Journal of Economics*, **113** (November), 1055–1089.
Amable, B. and P. Petit (2001), 'The diversity of social systems of innovation and production during the 1990s', document de travail CEPREMAP 2001–15, http://ideas.repec.org/p/cpm/cepmap/0115.html.
Aoki, M. (1988), *Information, Incentives and Bargaining in the Japanese Economy*, Cambridge: Cambridge University Press.
Aoki, M., A. Garber and P. Romer (2000), 'Why the Japanese economy is not growing: micro barriers to productivity growth', McKinsey Global Institute, Washington, DC.
Arthur, B.W. (1989), 'Competing technologies, increasing returns and locks-in by historical events', *Economic Journal*, **99**, 116–131.
Attewell, P. (1994), 'Information Technology and the Productivity Paradox', in D. Harris (ed.), *Organizational Linkages Understanding the Productivity Paradox*, National Research Council, Washington, DC: National Academy Press.
Barras, R. (1986), 'Towards a theory of innovation in services', *Research Policy*, **19**, 215–237.
Basu S., J.G. Fernald, N. Oulton and S. Srinavasan (2003), 'The case of the missing productivity growth: or does information technology explain why productivity accelerated in the US but not in the UK', NBER Working Paper 10010.
Baumol, W.J. (1967), 'Macroeconomics of unbalanced growth: the anatomy of urban crisis', *American Economic Review*, **57** (June), 415–426.
Baumol, W.J., S.A. Blackman and E.W. Wolff (1985), 'Unbalanced growth revisited: asymptotic stagnancy and new evidence', *American Economic Review*, **4**, 807–817.
Bell, B.D. (1996), 'Skill biased technical change and wages: evidence from a longitudinal data set', Institute of Economics and Statistics, University of Oxford, mimeo.
Berman, E., J. Bound and Z. Griliches (1994), 'Changes in the demand for skilled labor within US manufacturing: evidence from the annual survey of manufactures', *Quarterly Journal of Economics*, May, **109** (2), 367–397.
Berman, E., J. Bound and S. Machin (1998), 'Implications of skill-biased technological change: international evidence', *Quarterly Journal of Economics*, **113** (November), 1245–1280.
Berman, E., J. Bound and S. Machin (1995), 'Implications of skill biased technological change: international evidence', presented at Expert Workshop on Technology, Productivity and Employment: Macroeconomic and Sectoral Evidences', Paris: OECD.
Bryson, J.R. and P.W. Daniels (eds) (2008), *The Handbook of Services Industries*, Cheltenham, UK and Northampton, MA, USA: Edward Elgar.
Caroli, E. and V. van Reenen (2001), 'Skill-based organisational change: Evidence from a panel of British and French establishments', *The Quarterly Journal of Economics*, **116** (4), 1449–1492.
Chesbrough, H. (2003), *Open Innovation: The New Imperative for Creating and Profiting from Technology*, Harvard, MA, Harvard Business School Press.
Chun, H. (2003), 'Information technology and the demand for educated workers: disentangling the impact of adoption versus use', *Review of Economics and Statistics*, **85** (1), 1–8.
Cobb, C. and J. Cobb (1994), *The Green National Product: A Proposed Index of Sustainable Economic Welfare*, New York: University Press of America Press.
D'Iribarne, P. (1989), *La logique de l'honneur*, Paris: Seuil.
David, P.A. (1991), 'Computer and dynamo: the modern productivity paradox in a not too distant mirror', OECD Technology and Productivity, Paris.
Denny, M. and M. Fuss (1983), 'The effects of factor prices and technological change on the

occupational demand for labour: evidence from Canadian telecommunications', *Journal of Human Resources*, **17** (2), 161–176.
De Serres, A. (2003), 'Structural Policies and Growth – A Non-Technical Overview', Economics Department Working Paper no 355, Paris: OECD.
Djellal, F. and F. Gallouj (2008), *Measuring and Improving Productivity in Services*, Cheltenham, UK and Northampton, MA, USA: Edward Elgar.
Elfring, T. (1992), 'An international comparison of service sector employment growth', in United Nations Commission for Europe, 'Personal and collective services: an international perspective', Discussion paper, 2 (1), 1–13.
Esping-Andersen, G. (1990), *The Three Worlds of Welfare Capitalism*, Princeton, NJ: Princeton University Press.
Freeman, C. (1987), 'Information technology and the change in techno-economic Paradigm', in C. Freeman and L. Soete (eds), *Technical Change and Full Employment*, Oxford: Basil Blackwell, pp. 49–69.
Gadrey, J. and F. Gallouj (1998), 'The provider customer interface in business and professional services', *Service Industries Journal*, **18**, 1–15.
Gadrey, J. and F. Gallouj (eds) (2002), *Productivity, Innovation and Knowledge in Services*, Cheltenham, UK and Northampton, MA, USA: Edward Elgar.
Gadrey, J. and F. Jany-Catrice (2007), *Les nouveaux indicateurs de croissance*, Collection nepéres, Paris: La Découverte.
Gallouj, C. (1997), 'Asymmetry of information and the service relationship: selection and evaluation of service provider', *International Journal of Service Industry Management*, **8** (1), 42–64.
Gallouj, F. (2002), *Innovation in the Service Economy: The New Wealth of Nations*, Cheltenham, UK and Northampton, MA, USA: Edward Elgar.
Goodman, P., J. Lerch and T. Mukhopakhyay (1994), 'Individual and organizational productivity: linkages and processes', in D. Harris (ed.), *Organizational Linkages Understanding the Productivity Paradox*, National Research Council, Washington, DC: National Academy Press, pp. 54–80.
Goux, D. and E. Maurin (1995), 'Les transformations de la demande de travail en France. Une étude sur la période 1970–1993', Document de travail INSEE, April, mimeo; English version: 'Changes in the demand for labour in France: a study for the period 1970–93', presented at Expert Workshop on Technology, Productivity and Employment: Macroeconomic and Sectoral Evidences, Paris: OECD, 19–20 June.
Griliches, Z. (1969), 'Capital–skill complementary', *Review of Economics and Statistics*, **51**, 465–468.
Griliches, Z. (1994), 'Productivity, R&D, and the data constraint', *American Economic Review*, **84** (March), 1–23.
Harris, D. (1994), *Organizational Linkages Understanding the Productivity Paradox*, National Research Council, Washington, DC: National Academy Press.
Hausman, J. (1994), 'Valuation of new goods under perfect and imperfect competition', NBER Working Paper no. 4970, December.
Howell, D. and E. Wolff (1992), 'Technical change and the demand for skills by US industries', *Cambridge Journal of Economics*, **16**, 128–146.
Jones, R.S. and T. Yoon (2008), 'Enhancing the productivity of the service sector in Japan', OECD Working paper no. 651, December.
Kahneman, D. (1998), 'Assessments of individual well-being: a bottom up approach', in D. Kahneman, E. Diener and N. Schwarz (eds), *Understanding Well Being: Scientific Perspectives on Enjoyment and Suffering*, New York: Russell Sage Foundation.
Kaldor, N. (1972), 'The irrelevance of economic equilibrium', *Economic Journal*, December, **82**, 1237–1255.
Kaldor, N. (1980), *Reports on Taxation I: the Economics of the Selective Employment Tax*, London: Duckworth (2III; 6I).
Landauer, T.K. (1995), *The Trouble with Computers: Usefulness, Usability and Productivity*, Cambridge, MA: MIT Press.

Lucas, R. (1988), 'On the mechanics of economic development', *Journal of Monetary Economics*, **22** (1), July, 3–42.
Machin, S. (1996), 'Changes in the relative demand for skill in the UK', in A. Booth and D. Snower (eds), *Acquiring Skills*, Cambridge: Cambridge University Press, pp. 129–146.
Machin, S. and J. Van Reenen (1998), 'Technology and changes in skill structure: evidence from seven OECD countries', *Quarterly Journal of Economics*, **113** (4), 1215–1244.
Miringoff, M. and M.L. Miringoff (1999), *The Social Health of the Nation: How America is Really Doing*, Oxford: Oxford University Press.
Nelson, R. and S.G. Winter (1982), *An Evolutionary Theory of Economic Change*, Cambridge, MA: Harvard University Press.
Nicoletti, G., S. Scarpetta and O. Baylaund (2000), 'Summary indicators of product market regulation with an extension to employment protection regulation', OECD Economics Department Working Paper no. 226.
Nonaka, I., V. Peltokorpi and D. Senoo (2008), 'Knowledge creation in a Japanese convenience store chain: the case of Seven Eleven Japan', in J.R. Bryson and P.W. Daniels (eds), *The Handbook of Services Industries*, Cheltenham, UK and Northampton, MA, USA: Edward Elgar.
North, D.C. (1990), *Institutions, Institutional Change, and Economic Performance*, New York, NY: Cambridge University Press.
OECD (1999), 'Employment protection and labour market performance', *Employment Outlook*, Chapter 2, June.
OECD (2001), *Science, Technology and Industry Score Board: Towards a Knowledge Based Economy*, Paris: OECD.
O'Mahony, M., M. Peng and F. Peng (2008a), 'Skill bias, age and organisational change', EU KLEMS working paper no. 36.
O'Mahony, M., C. Robinson and M. Vecchi (2008b), 'The impact of ICT on the demand for skilled labor: a cross country comparison', *Labour Economics*, **15** (6), 1435–1450.
Packard, V. (1957), *The Hidden Persuaders*, London: Longman.
Petit, P. (2002), 'Growth and productivity in a knowledge based economy', in J. Gadrey and F. Gallouj (eds), *Productivity, Innovation and Knowledge in Services*, Cheltenham, UK and Northampton, MA, USA: Edward Elgar, pp. 102–123.
Petit, P. (2008), 'The political economy of services in tertiary economies', in J.R. Bryson and P.W. Daniels (eds), *The Handbook of Services Industries*, Cheltenham, UK and Northampton, MA, USA: Edward Elgar.
Piore, M. (1988), 'Corporate reform in American manufacturing and the challenge to economic theory', MIT Working Paper no. 533.
Pratt, J., A. Leyshon and N. Thrift (1996), 'Financial exclusion in the 1990s: the changing geography of UK retail financial services', Working Paper no. 34, Department of Geography, University of Bristol.
Rowthorn, R. and R. Coutts (2004), 'De-industrialisation and balance of payments in advanced economies', *Cambridge Journal of Economics*, **28** (5), 767–790.
Van Ark, B., R. Inklaar and R. McGuckin (2002), 'Changing gear: productivity, ICT and services: Europe and the United States', Research Memorandum GD60, Growth and Development Center, University of Groningen.
Van Ark, B., M. O'Mahony and M.P. Timmer (2007), 'EU KLEMS growth and productivity accounting: an overview', *International Productivity Monitor*, Spring.
Wolff, E.N. (1996), 'Technology and the demand for skills', *OECD Science, Technology and Industry Review*, **18**, 96–123.
Wolff, E.N. (2002), 'How stagnant are services?' in J. Gadrey and F. Gallouj (eds), *Productivity, Innovation and Knowledge in Services*, Cheltenham, UK ad Northampton, MA, USA: Edward Elgar, pp. 3–25.
Wölfl, A. (2003), 'Productivity growth in service industries: an assessment of recent patterns and the role of measurement', OECD STI Working Paper, 2003/7.

18 How important are knowledge-intensive services for their client industries? An assessment of their impact on productivity and innovation

José A. Camacho and Mercedes Rodriguez

18.1 Introduction

Services began to be taken into consideration in innovation studies during the 1980s with the pioneering works of Richard Barras, Jonathan Gershuny and Ian Miles, after decades of having been virtually ignored. But it was not until the 1990s that services captured the attention they deserved. Among the different types of service activities analysed, a group of industries stood out because of their 'special' characteristics – those called knowledge-intensive services (KIS). In a widely cited paper Miles et al. (1995) summarised these characteristics in terms of the three features they have in common, that is, they rely greatly on professional knowledge, are themselves sources of knowledge and are of competitive importance for their clients.

In addition to the relatively recent recognition of the key role of services and more concretely of KIS in innovation, we have been witnessing a great increase in international trade in services. Far from the traditional belief that services are, by nature, non-tradable, the fact is that the provision of services is becoming more and more global, with the result that foreign direct investment (FDI) in services is nowadays more relevant than FDI in manufacturing. Many factors can be highlighted as drivers of the upsurge in the internationalisation of service activities (changes in business strategies, higher competition), but one stands out above all: the impact that information technologies (IT) have had on services. Services are not only the main users of IT but they are also, thanks to advances in technology, more divisible than ever, in the sense that many service functions can be separated and as a result traded, such as in the case of software development. Consequently, within this global context it becomes essential to adopt an 'international perspective'.

Starting from these premises, in this chapter we try to analyse the impact that the increasing role of KIS in production processes, in combination with the growing internationalisation of this type of activities, has on the performance of their client industries. In particular, we examine two aspects of the use of KIS: their impact on productivity and their effects on the diffusion of research and development (R&D). In the case of the effects on productivity, we analyse whether the use of KIS exerts a positive impact on the productivity of their clients. Given their knowledge intensity, KIS are more than intermediate inputs for their clients: they do not only provide services, but also contribute to improve the performance of their clients, in terms of both productivity and innovation. In relation to innovation, we centre on the diffusion of product-embodied R&D across national borders. To do so, we estimate the product-embodied R&D diffusion that takes place in 11 European countries by means of the use of some KIS as intermediate consumptions. In this diffusion we differentiate between domestic flows and flows coming from imports. Given the limited amount of statistical information available, we centre on a group of KIS called by Eurostat 'knowledge-intensive high-tech services' or 'high-tech services', composed of the three following industries of the ISIC revision 3: 64 – post and telecommunications; 72 – computer and related activities; and 73 – research and development.

The structure of the chapter is the following. In section 18.2 we briefly review recent theories about the role of services in productivity, innovation and in international trade. In particular, we show how considerably the interest in innovation in services has risen during recent years. An important consequence of this growing attention is that KIS are regarded nowadays as essential elements in the innovation processes of their client firms. Different perspectives can be identified in the analyses that support this general vision. For example, some studies, based on knowledge creation theories, such as that elaborated by Nonaka and Takeuchi (1995), adopt an 'organisational' perspective (Strambach, 2001; Glückler, 1999). Others take the national innovation system concept to characterise the functions of KIS (Fischer, 2001).

Furthermore, the fact that KIS are ever more easily provided from long distances (Lommelen and Matthyssens, 2005) makes their 'innovation supporting' role more global. The papers on the internationalisation of business service firms highlight the following of clients abroad as the major incentive for going international (Bryson, 2001; Fernandez-Fernandez, 2001; Leo and Philippe, 2001; O'Farrell and Wood, 1998; O'Farrell et al., 1996). This fact reinforces the above-mentioned hypotheses regarding the positive impact of the use of KIS on knowledge diffusion and ultimately on innovation and productivity (Miozzo and Soete, 2001).

In section 18.3 we examine the increasing importance of KIS as intermediate consumptions within the different production systems, paying special attention to growth in the imports of intermediate consumptions of these services. We also highlight the increasing R&D efforts carried out by KIS.

Next, in section 18.4, we evaluate the impact of the use of KIS on the productivity of their client industries by starting from a modified production function, in which we assume that output is produced by means of labour that acts on two types of inputs: 'material inputs' and 'inmaterial inputs' or KIS.

We study the diffuser role of KIS within the production systems in section 18.5. In particular we analyse whether the diffusion of product-embodied R&D carried out by KIS is mainly developed through the use of domestic or imported intermediate consumptions. The methodology employed (Rodriguez, 2003; Camacho and Rodriguez, 2006) is a modified version of the input–output model introduced by Papaconstantinou et al. (1998). The novel element is the attempt to estimate, for the first time, the product-embodied R&D diffused by imported intermediate consumptions of KIS. In this way it is possible to 'go beyond' traditional analyses which commonly centre on domestic economies by demonstrating that the functions of KIS surpass national borders.

The empirical analysis is developed for 11 of the EU-15 countries: Belgium, Denmark, Finland, France, Germany, Ireland, Italy, the Netherlands, Spain, Sweden and the United Kingdom, in the year 2000 in the case of the diffusion of product-embodied R&D. In the case of the productivity equations the US, Canada and Japan are also included. Four databases are employed: the OECD Input–Output Database 2006, the ANBERD database, the STAN database and the Eurostat International Trade in Services Database. The main results of the analysis and future lines of research are summarised in the conclusions, section 18.6.

18.2 Innovation, productivity and KIS: recent trends

The traditional vision of services as non-innovative and low-productivity activities has radically changed in recent years. In relation to the non-innovative character of services, the long-time predominance of manufacturing within the economies established a technologically biased vision of innovation. Given the supplier-dominated nature of services (Pavitt, 1984) and their scant contribution to technological change, services were not included in innovation studies. One of the first attempts to break with this trend was the work of Gershuny and Miles. In their book *The New Service Economy: The Transformation of Employment in Industrial Societies* (1983), they recognised the potential impact of information technologies (IT) on services. This laid the foundation for the elaboration of the first model of innovation in services by Barras in 1986. Starting from the famous 'product cycle' described by Abernathy and Utterback (1978),

Barras observed that, despite information technologies being implemented in order to improve efficiency, the use of IT brought about learning processes, which led, in a first phase, to improvements in quality and other aspects, and in a second phase, to the emergence of new services (innovation). Nonetheless, it was not until the latter half of the 1990s that in-depth works on services and innovation came to light, in which new theories were developed (Gadrey et al., 1995; Gallouj and Weinstein, 1997; Miles, 1996) and empirical analyses were carried out (Hipp et al. 1996; Sundbo, 1998).

Despite this belated interest in innovation activities of services, some studies that called attention to the potential role of services as knowledge diffusers appeared earlier. During the 1960s, economists such as Machlup (1962) or Greenfield (1966) described the role of certain services, specifically business services, as creators and diffusers of knowledge.

Since the second half of the 1990s, we have witnessed a dramatic upsurge of interest in these functions of services in the innovation domain, and specifically of those called knowledge-intensive services (KIS). KIS are 'special' due to their close relationship with knowledge as well as being innovative in their own right. One of their main capacities is the provision of knowledge to other industries. Antonelli (2000, 171) states that KIS function as: 'holders of proprietary "quasi-generic" knowledge, from interactions with customers and the scientific community, and operate as an interface between such knowledge and its tacit counterpart, located within the daily practices of the firm'. In other words, they can act as 'bridges' for knowledge (Czarnitzki and Spielkamp, 2000) or, to quote Den Hertog and Bilderbeek (2000), 'as a second knowledge infrastructure', even substituting for functions traditionally ascribed to the public sector.

Among the various analyses on the functions of KIS, some studies adopt what can be described as a 'management' or 'organisational' perspective to describe the activities of KIS in transmitting knowledge to their clients. This is the case of the papers by Strambach (2001), Glückler (1999) and Schulz (2000), where the emphasis is placed on the generation, diffusion and creation of knowledge by the interactions between KIS and their clients.

Another approach consists in starting from the concept of a national innovation system. In this case KIS are considered to carry out three major functions: they are facilitators, carriers and sources of innovation for their client firms (Fischer, 2001; den Hertog and Bilderbeek, 2000; Hipp, 2000). Concretely, they act as facilitators of innovation when they collaborate in the innovation process of their client firms, but they do not directly generate innovations or transfer them from other firms. They function as carriers of innovation when they directly participate in the innovations developed by their client firms. Finally, KIS are sources of innovation when they generate innovations for their client firms.

More recent papers centre on the behaviour of KIS firms in order to clarify how they innovate and collaborate in innovation (Freel, 2006; Leiponen 2005, 2006; Wong and He, 2005).

Concerning the effects of KIS on the productivity of their client industries, we can consider that the knowledge-diffuser role described below translates into more efficient production processes and, ultimately, into higher productivity. Since the beginning of the study of service activities, it was commonly accepted that services show low productivity growth rates. However, if we take recent data, the picture obtained is not so clear. For instance, in a recent study carried out by Pilat (2007), he distinguishes very different patterns in the evolution of productivity in business services in the European countries during the decade of the 1990s. Thus, whereas Germany and France report negative productivity growth, the Netherlands, Denmark, the United Kingdom, Italy and Spain show slightly positive growth rates of productivity in business services. Part of these variations can be explained by the difficulties still existing in measuring productivity in services. Nevertheless, on a general basis, evidence provided by studies at the macro level (such as those described in the previous section) and also at the micro level (like the papers by Abramovsky and Griffith, 2005, or Mann, 2003) confirm that the use of business services can enhance productivity.

From all the above we can conclude that the close relationships that tie KIS with their clients influence both innovation and productivity. KIS collaborate with innovation in other firms and besides, thanks to IT, they can easily be provided at long distances. This poses various questions about the different effects brought about by the increasing international sourcing of KIS. For instance, how the international provision of KIS contributes to knowledge diffusion,

Given the scant availability of internationally comparable data on knowledge, in this chapter we take as a proxy R&D expenditures.[1] In particular, the objective is to estimate the impact of the use of KIS on productivity and to evaluate the product-embodied R&D diffuser role of certain KIS (the high-tech services) from a double perspective: domestic and imported.

18.3 An analysis of the role of KIS within the production systems using input–output tables

As has been mentioned, in this chapter we focus on a specific group of KIS, the 'high-tech services',[2] namely, post and telecommunications, computer and related activities, and research and development. Before entering into the estimation of their product-embodied R&D diffuser role, in this section we examine two aspects.

Firstly, we evaluate whether KIS are acquiring greater importance as intermediate consumptions within the production systems. In line with what theories described above point out, KIS are becoming key sources of value added and, as a result, they account for a growing share of the cost of products. Therefore, it is reasonable to suppose that their participation in intermediate consumptions is growing as well.

Secondly, we analyse whether economies domestically produce the KIS they use as intermediate consumptions, or whether they are increasingly acquiring them from abroad. That is to say, we differentiate between domestic and imported intermediate consumptions of KIS. In order to examine the relevance of KIS within the production systems, the shares of KIS in intermediate consumptions in the 11 European countries previously listed as well as the US, Canada and Japan, in the years 1995 and 2000, are shown in Table 18.1. The annual average growth rate of intermediate consumptions of KIS and the percentages of imported intermediate consumptions are also reported for each country.

As can be observed, the share of KIS in total intermediate consumptions has increased considerably during the period studied. Thus, whereas in 1995 there were noticeable differences between the EU countries and Japan and the US, these differentials decreased significantly in only five years. In 1995 KIS accounted for more than 6 per cent of total intermediate consumptions in Japan and the US, while only one European country, France, showed a similar participation. In contrast to this situation, in the year 2000 the participation of KIS in intermediate consumptions was above 7 per cent not only in Japan and the US, but also in four European countries: the United Kingdom, France, Sweden and Denmark.

However, we have to highlight substantial differences among the various EU countries: e.g., in the group of four countries mentioned above the share of KIS in intermediate consumptions in 2000 was more than double the participation in other EU countries, such as Ireland or Spain.

In any case, a trend towards reduction of these differences is observed. If we take the annual average growth rates of KIS – excepting the US, France and Japan – all the countries analysed have experienced annual average growth rates in the intermediate consumption of KIS above 10 per cent. The growth in four European countries – Belgium, Finland, Denmark and Spain – deserves special attention, since their annual average rates were higher than 15 per cent.

In the two last columns of Table 18.1 we show the percentage of intermediate consumptions of KIS from abroad, in order to study the origin

Table 18.1 Evolution of the participation of KIS in intermediate consumptions, 1995–2000

	Share in total intermediate consumptions		Annual average growth rate	Percentage of intermediate consumptions imported	
	1995	2000	1995–2000	1995	2000
Belgium	2.53	4.23	18.34	21.07	21.62
Denmark	4.16	7.02	16.68	9.11	12.26
Finland	3.19	4.61	16.94	9.12	6.66
France	6.44	7.43	8.93	2.05	2.49
Germany	3.54	4.83	10.90	11.31	15.37
Ireland	n.a.	3.04	n.a.	n.a.	7.38
Italy	3.04	3.79	10.16	4.43	9.60
Netherlands	4.13	5.61	13.11	16.61	18.87
Spain	2.33	3.32	15.45	5.29	6.54
Sweden	5.40	7.18	11.98	7.78	14.39
United Kingdom	5.56	8.04	14.00	8.60	5.10
US	6.00	7.12	9.77	0.16	0.16
Canada	n.a.	4.09	n.a.	n.a.	7.10
Japan	6.43	7.49	4.33	1.10	1.50

Note: In the case of Ireland the most recent year available is 1998 and information on the origin of intermediate consumptions is available only for post and telecommunications.

Source: OECD Input–Output Database 2006.

(domestic or imported) of the intermediate consumptions of KIS. We notice that the imported intermediate consumptions of these services grew during the period analysed in all countries except Finland and the United Kingdom. The countries that imported the greatest volume of KIS in 2000 were respectively Belgium, the Netherlands, Germany, Sweden and Denmark, where more than 10 per cent of the KIS employed as intermediate consumptions were imported. The percentages are especially remarkable in Belgium (21.62 per cent) and the Netherlands (18.87 per cent). At the opposite end of the scale we find such countries as the US, Japan and, within Europe, France and to a lesser extent the United Kingdom that domestically produce the major part of the KIS employed as intermediate consumptions.

As was mentioned in the previous section, KIS are highly innovative activities. So as to evaluate the innovation efforts carried out by KIS, in Table 18.2 we report their R&D intensities in 2000 (defined as business

Table 18.2 Evolution of R&D expenditures in KIS, 1995–2000

	R&D intensity			Annual average growth rate		
	Post	Computer	Research	Post	Computer	Research
Belgium	0.81	2.20	0.55	32.97	14.10	−1.55
Denmark	n.a	6.39	3.60	n.a	20.53	40.46
Finland	3.33	3.79	n.a.	33.15	17.17	n.a
France	n.a.	1.04	n.a.	n.a	4.17	n.a
Germany	n.a.	1.99	5.74	n.a	46.37	38.15
Ireland	2.25	4.45	34.17	16.04	46.60	55.06
Italy	0.01	0.65	12.94	−41.62	18.41	18.88
Netherlands	0.47	2.10	3.45	−1.12	52.12	32.34
Spain	0.98	3.07	7.75	31.91	33.34	44.67
Sweden	1.83	2.97	9.82	10.64	36.51	4.86
United Kingdom	1.37	1.55	6.99	9.91	−2.26	11.29
US	0.43	4.18	16.53	−12.11	13.60	25.05
Canada	0.17	4.02	n.a.	−21.66	12.34	5.90
Japan	n.a.	1.37	n.a.	n.a	n.a	n.a

Notes:
In Denmark the R&D intensities are for 2001 and the annual average growth rate refers to the period 1995–99.
In Germany and Japan the intensities for computer services are for 2001.
In the US the intensity for post and telecommunications refers to 1998 and the growth is for the period 1996–98.
In Spain the intensity for research and development is for 1999.

Source: OECD Input-Output Database 2006, STAN and ANBERD.

expenditures on R&D divided by production) and the annual average growth rates of their business expenditures on R&D during the period 1995–2000. Both the intensities and the growth rates are calculated for each of the three industries included within KIS.

With respect to the industry of post and telecommunications we observe quite varied evolutions. One the one hand, we find those countries with the highest R&D intensities and where the R&D expenditures carried out in this industry grew at a fast pace: Finland and Ireland. On the other hand, we note those countries with the lowest R&D intensities that, in addition, have experienced a considerable reduction in R&D expenditures: Italy, Canada and the US. In the rest of the countries the annual average growth rates of the R&D expenditures were relatively high, except in the Netherlands, where the expenditures decreased slightly.

In the case of computer and related activities the picture is clearer: the

R&D expenditures carried out in this industry grew significantly in all countries with the sole exception of the United Kingdom, where there was a minor decline. Denmark is the country with the highest R&D intensity in computer services, followed by Ireland, the US, Canada and Finland. Again Italy is the country that shows the lowest R&D intensity.

The situation in the industry of research and development is very similar to that of computer services. The R&D expenditures carried out in this industry increased a lot in all the countries analysed except in Belgium, where the expenditures diminished a little. The countries with higher R&D intensities are Ireland, the US and Italy, whereas Belgium is the country with the lowest R&D intensity.

In brief, we can confirm that KIS are, in fact, services that carry out important efforts in R&D. Thus, they not only show high R&D intensities, but they also spend more and more on R&D. Within this general trend we can only exclude the industry of post and telecommunications, because of the disparate evolution presented in the countries analysed. Having in mind these trends, in the following section we try to estimate the effects of the use of KIS on productivity.

18.4 Use of KIS and productivity: is there a link?

The trends pointed out in the previous section support the hypothesis of a positive impact of the use of KIS on productivity. The major part of the empirical analyses aimed at studying the relationship between the use of services of a 'knowledge-intensive' nature and productivity take a wider definition of KIS than us, and include the whole group of business services. Moreover, although the analyses that evaluate the effects of the use of business services start from a production function, we can distinguish two approaches. The first (Katsoulacos and Tsounis, 2000; Antonelli, 1998, 2000; Drejer, 2002) introduces business services as an additional input in a traditional production function. The second one, which is slightly different (Windrum and Tomlinson, 1999; Tomlinson, 2000a, 2000b), assumes the hypothesis that output is produced by means of labour that acts on material inputs and/or business services. Therefore, a distinction is made between 'immaterial' inputs (B) and 'material' inputs (the remainder of intermediate inputs). In this section, we employ this latter approach.[3]

Thus, Q (gross output) is considered a function of M, B and L, where M is the amount of material inputs, B is the amount of KIS and L is the employment:

$$Q = A(ML)^{a}(BL)^{b} \qquad (18.1)$$

Taking logarithms, we obtain the following equation to estimate:

$$\log Q = \log A + a \log M + b \log B + (a + b) \log L \qquad (18.2)$$

where A is a constant and a and b are the parameters to estimate.

Starting from equation (18.1) we can also obtain an equation for productivity if we divide each term by the amount of labour employed (L):

$$Q/L = A(M/L)^a (B/L)^b \qquad (18.3)$$

taking logarithms:

$$\log Q/L = \log A + a \log M/L + b \log B/L \qquad (18.4)$$

Table 18.3 shows the results of the estimation of the productivity equation for our fourteen countries under analysis in 1995 and 2000.[4] As can be noticed, the estimated values differ considerably among countries.

For one part, Germany and the Netherlands are the countries with a higher impact of the use of KIS on productivity, along with the US and Canada. Special attention is deserved by the case of Finland, where the effect of KIS has considerably increased during the period.

At the opposite end of the scale we find countries like the United Kingdom, where the impact of what are defined as KIS has turned negative and non-significant. No relationship between higher use of KIS and higher productivity is detected: the United Kingdom, France, Sweden and Denmark, the countries with greater participations of KIS in intermediate consumptions, show non-significant estimated values in 2000, although a disparate evolution is observed. Thus, whereas in the United Kingdom and Denmark the estimations were considerably high in 1995, in France and Sweden they were non-significant.

On the contrary, a potential correlation between the use of imported intermediate consumptions and higher productivity could exist (excepting the US): the Netherlands and Germany, the countries with higher estimated values, are among the top importers of intermediate consumptions of KIS. We must note, anyway, that all the significant estimated values obtained for KIS are positive and considerably high, ranging from 0.08 in Belgium to 0.20 in the US in 1995 and from 0.12 in Finland to 0.18 in Germany in 2000.

Concerning the results of previous analyses, we can differentiate between those that estimate a traditional production function (Antonelli, 2000; Katsoulacos and Tsounis, 2000; Drejer, 2002) and those that employ the same methodology as here (Tomlinson, 2000a, 2000b; Windrum and

Table 18.3 Estimation of the effects of the use of KIS on productivity, 1995–2000

	1995					2000				
	c	m/l	b/l	Adj. R	F-stat	c	m/l	b/l	Adj. R	F-stat
Belgium	1.23**	0.66**	0.08*	0.81	81.63	1.09**	0.67**	0.05	0.85	109.27
	(0.17)	(0.05)	(0.05)			(0.15)	(0.05)	(0.05)		
Denmark	1.46**	0.66**	0.14**	0.71	55.81	1.35**	0.70**	0.11	0.61	35.53
	(0.19)	(0.07)	(0.06)			(0.23)	(0.09)	(0.08)		
Finland	1.16**	0.62**	0.05	0.83	98.54	1.30**	0.61**	0.12**	0.85	112.95
	(0.16)	(0.05)	(0.05)			(0.15)	(0.05)	(0.05)		
France	1.19**	0.57**	0.06	0.73	54.97	1.16**	0.60**	0.05	0.80	82.88
	(0.13)	(0.06)	(0.05)			(0.12)	(0.05)	(0.04)		
Germany	1.53**	0.64**	0.19**	0.78	73.92	1.48**	0.64**	0.18**	0.79	75.39
	(0.17)	(0.06)	(0.05)			(0.15)	(0.05)	(0.05)		
Ireland	1.16**	0.68**	0.05	0.81	82.51					
	(0.16)	(0.05)	(0.04)							
Italy	1.27**	0.66**	0.06	0.75	62.30	1.24**	0.69**	0.06	0.78	71.15
	(0.19)	(0.06)	(0.06)			(0.20)	(0.06)	(0.06)		
Netherlands	1.53**	0.64**	0.18**	0.81	87.85	1.44**	0.64**	0.16**	0.83	96.03
	(0.15)	(0.05)	(0.05)			(0.14)	(0.05)	(0.05)		

Spain	1.41**	0.60**	0.12*	0.79	77.90	2.11**	0.61**	0.15	0.47	18.96
	(0.23)	(0.05)	(0.07)			(0.30)	(0.11)	(0.10)		
Sweden	1.01**	0.70**	0.04	0.83	95.64	0.94**	0.70**	0.01	0.87	137.86
	(0.15)	(0.06)	(0.05)			(0.13)	(0.05)	(0.05)		
United Kingdom	1.26**	0.60**	0.11*	0.73	53.98	1.16**	0.65**	−0.11	0.32	10.51
	(0.14)	(0.06)	(0.05)			(0.23)	(0.15)	(0.09)		
US	1.49**	0.74**	0.20**	0.95	406.47	1.30**	0.68**	0.14**	0.88	169.97
	(0.10)	(0.03)	(0.03)			(0.13)	(0.04)	(0.05)		
Canada	n.a.	n.a.	n.a.			1.43**	0.63**	0.13**	0.85	121.29
						(0.15)	(0.04)	(0.05)		
Japan	1.18**	0.60**	0.05	0.73	63.18	1.15**	0.62**	0.04	0.78	82.69
	(0.15)	(0.07)	(0.06)			(0.15)	(0.07)	(0.06)		

Notes:
** Significant at 5%. * Significant at 10%.
Standard errors in brackets.

Source: OECD Input–Output Database 2006.

Tomlinson, 1999). Although in all of the cases the definition of business services is wider than the one employed in this chapter, we can compare the sign of the effects and their dimension.

We begin with the studies that estimate a traditional production function. Antonelli (2000) and Katsoulacos and Tsounis (2000) analyse two aspects: the co-evolution of communication services and business services and the effects of the use of these two types of services on productivity in several countries – Antonelli for four countries, France, Germany, Italy and the United Kingdom; and Katsoulacos and Tsounis for Greece. The impact on productivity is examined using data for 1990, except in Italy and Greece where the input–output tables employed are for 1988. In both cases, the effects on productivity of what they define as KIS (communication services and business services) are positive and significant. The paper by Drejer (2002) tries to extend the analysis by introducing a sectoral dimension and examining a longer period of time. Thus, she estimates a production function using data for 52 industries during the period 1970–95. The results show the existence of considerable sectoral differences in the importance of the use of business services.

In the case of those studies that estimate the same equations, we can compare our estimations for the impact of KIS on productivity with those obtained by Windrum and Tomlinson (1999) and Tomlinson (2000a).[5] In Windrum and Tomlinson (1999) KIS are defined as communications and most of the industries included under 'business services'.[6] They estimate equations for four countries for the years indicated in parenthesis: Germany (1993), Japan (1990), the Netherlands (1994) and the United Kingdom (1990). The values they obtain for the effects of KIS on production and productivity are higher than those we estimated. Tomlinson (2000a) compares the situation of Japan and the United Kingdom in the years 1970 and 1990. He defines KIS as communications and business services. We obtain a lower estimated value in 1995 than he does (0.11 in comparison with 0.18).

Therefore, we can conclude that, in line with the conclusions reached by previous studies, the use of KIS as intermediate inputs has, in general, a positive impact on productivity.

18.5 Domestic versus imported R&D diffusion by KIS: an evaluation

In section 18.3 we corroborated the key role that KIS play as intermediate consumptions within the European production systems as well as the general growth in their R&D efforts. The objective of this fifth section is to evaluate the product-embodied R&D diffuser activity carried out by KIS, both domestically and by means of imports.

We define the R&D intensity in industry i as the Business Expenditure on R&D (R_i) divided by value added (W_i):[7]

$$r_i = \frac{R_i}{W_i} \; (i = 1, 2, \ldots, n) \tag{18.5}$$

The methodology employed (Rodriguez, 2003) starts from the model elaborated by Papaconstantinou et al. (1998) and applied to service activities by Amable and Palombarini (1998). Nonetheless our objective is substantially different from that of these previous studies. Whereas they centred on the incorporation of product-embodied R&D, our aim is to approximate the diffusion of product-embodied R&D that takes place by means of the acquisition of intermediate consumptions.

Using the output inverse matrix introduced by Ghosh (1958), the equilibrium in the domestic supply model can be expressed as follows:

$$X = W(I - B)^{-1} \tag{18.6}$$

where X is the vector of domestic gross outputs, W is the vector of value added and $(I - B)^{-1}$ is the domestic output inverse (Ghosh) matrix. We can define the matrix of domestic embodied R&D diffusion, D, by introducing a diagonalised matrix of R&D intensities in equation (18.6) as follows:

$$D = W\hat{r}(I - B)^{-1} \tag{18.7}$$

where r-hat ($\hat{r}$) indicates a diagonal matrix whose elements are those of vector r.

Equation (18.7) relates domestic product-embodied R&D diffusion to the value-added components (compensation of employees and gross operating surplus).

Thus, the domestic product-embodied R&D diffusion per unit of value added of industry i, DUD_i, can be obtained from the sum of the ith row of matrix $\hat{r}(I - B)^{-1}$:

$$DUD_i = \sum_{j=1}^{n} r_i q_{ij} \; (i = 1, 2, \ldots, n) \tag{18.8}$$

where r_i is the R&D intensity for industry i and q_{ij} are the elements of the Ghosh inverse.

Since the ith row of the Ghosh inverse measures the impact on domestic production when the utilisation of primary inputs (valued added) of the ith industry varies by one unit, equation (18.8) provides the amount of domestic product-embodied R&D diffused per unit of value added of industry i.

We can obtain the total domestic product-embodied R&D diffused through the domestic intermediate consumptions, *DTD*, by pre-multiplying equation (18.8) by the value added:

$$DTD_i = w_i \sum_{j=1}^{n} r_i q_{ij} \tag{18.9}$$

where w_i is the value added of industry i.

Similarly, the imported product-embodied R&D diffused per unit of value added, *IUD*, can be defined as follows:

$$IUD_i = \sum_{j=1}^{n} \sum_{k=1}^{l} \alpha_k r_{ik} q_{ij} \; (i = 1, 2, \ldots, n) \tag{18.10}$$

where α_k is the import share of country k, r_{ik} is the R&D intensity for industry i in country k and q_{ij} are the elements of the Ghosh inverse.

Again, the total product-embodied R&D diffused through the imported intermediate consumptions, *ITD*, can be obtained by pre-multiplying equation (18.10) by the value added:

$$ITD_i = w_i \sum_{j=1}^{n} \sum_{k=1}^{l} \alpha_k r_{ik} q_{ij} \; (i = 1, 2, \ldots, n) \tag{18.11}$$

The calculation of the domestic flows is a way of approximating the relevance of KIS as transmitters of their own R&D efforts within the production systems. The comparison of the diffusion carried out by domestic and imported intermediate consumptions allows us to evaluate to what extent the countries benefit from the R&D efforts carried out by KIS in other countries.

In Table 18.4 the product-embodied R&D diffused per unit of value added through domestic and imported intermediate consumptions of KIS is shown.[8] We report independently the flows for each industry.

As can be expected, on a general basis, the industry that diffuses more product-embodied R&D in most of the countries is research and development, and the industry that diffuses less product-embodied R&D is post and telecommunications. The high domestic diffusion that this latter industry carries out in Finland, probably due to the great importance of the telecommunications industry within its production system, can be noted.

Concerning the industry of computer services we observe a high correspondence between the R&D intensity (Table 18.2) and the product-embodied R&D diffused per unit of value added. Thus, Denmark, Finland and Sweden, which were the countries with higher R&D intensities, are also the countries where the diffusion of both domestic and imported product-embodied R&D is greater. On the contrary, Italy and France, the countries with lower R&D intensities, are the ones where less product-embodied R&D per unit of value added is diffused.

Table 18.4 Product-embodied R&D per unit of value added diffused through domestic and imported intermediate consumptions of KIS, 2000

	Domestic			Imported	
	Post	Computer	Research	Computer	Research
Belgium	0.034	0.113	0.015	0.053	0.380
Denmark	n.a.	0.280	0.096	0.059	0.209
Finland	0.116	0.129	n.a.	0.059	n.a.
France	n.a.	0.047	n.a.	0.045	n.a.
Germany	n.a.	0.052	0.167	0.053	0.274
Ireland	0.065	0.091	1.788	0.051	0.174
Italy	0.0004	0.026	0.331	0.042	0.187
Netherlands	0.017	0.070	0.117	0.047	0.326
Spain	0.037	0.061	0.277	0.040	0.374
Sweden	0.077	0.127	0.317	0.062	0.268
United Kingdom	0.060	0.073	0.255	0.053	0.217

Note: In Ireland the data refer to 1998.

Source: OECD Input–Output Database 2006, STAN and ANBERD.

The exception to be highlighted is Ireland, since it occupies a modest position in terms of product-embodied R&D diffusion per unit of value added, despite being the country with the second-highest R&D intensity in computer services. In a midway position, the industry of computer services in the United Kingdom and the Netherlands carries out medium R&D efforts, the diffusion of product-embodied R&D being moderate too.

Two cases should be pointed out: Germany and Spain. Germany can be classified as the country that benefits the most from imports (in terms of product-embodied R&D). While its R&D intensity is not very high (and in consequence the domestic product-embodied R&D diffused per unit of value added is not very high either), it is the fourth-highest ranking country that incorporates more product-embodied R&D through imports of intermediate consumptions. The opposite phenomenon is found in Spain. Although its R&D intensity is quite high, it is the country that diffuses the least product-embodied R&D via imported intermediate consumptions of computer services.

If we turn to the industry of research and development, the panorama changes greatly. Ireland and Italy, the countries with higher R&D intensities, are also the countries that diffuse more domestic product-embodied R&D. Yet they are the countries that diffuse less R&D through imported intermediate consumptions.[9] The country that benefits the most from the product-embodied R&D diffused by imported intermediate consumptions is Belgium: it shows the lowest R&D intensity, but is the country where more product-embodied R&D per unit of value added is diffused via imports. To a lesser extent, Spain and the Netherlands also benefit from the R&D incorporated in their imported intermediate consumptions of research and development services.

The rest of the countries – that is, Sweden, Germany, the United Kingdom and Denmark – show unitary flows of product-embodied R&D in line with their own R&D efforts. We have to note, however, that the product-embodied R&D diffused per unit of value added is higher for imported intermediate consumptions than for domestic intermediate consumptions in Denmark and Germany.

As was pointed out earlier, we can easily obtain the total product-embodied R&D diffused by multiplying the unitary flows by the value added of each industry. In Table 18.5 we report the total product-embodied R&D diffused, expressed as a percentage of the total business expenditure on R&D.[10]

By taking into account the extent of these industries within the production system of each country, we can note the considerable increase of the

Table 18.5 *Total product-embodied R&D diffused through domestic and imported intermediate consumptions of KIS, 2000 (% of total BERD)*

	Domestic			Imported	
	Post	Computer	Research	Computer	Research
Belgium	5.29	7.86	0.17	3.69	4.38
Denmark	n.a.	20.64	1.64	4.35	3.57
Finland	13.36	7.27	n.a.	3.31	n.a.
France	n.a.	6.11	n.a.	5.87	n.a.
Germany	n.a.	4.64	3.45	4.78	5.69
Ireland	19.48	9.79	4.95	5.45	0.48
Italy	0.17	6.18	20.01	9.75	11.32
Netherlands	3.54	10.71	3.97	7.20	11.04
Spain	16.82	10.19	1.78	6.63	2.41
Sweden	5.72	8.52	6.81	4.17	5.77
United Kingdom	13.58	13.06	7.93	9.51	6.73

Note: In Ireland the data refer to 1998.

Source: OECD Input–Output Database 2006, STAN and ANBERD.

role played by post and telecommunications, its diffusion being especially noticeable in Ireland and Spain.

In the case of computer services, Denmark, the United Kingdom, the Netherlands and Spain are the countries where the diffusion of product-embodied R&D has greater importance. The case of Italy draws our attention. Here the major part of the product-embodied R&D is obtained from imports because of the low R&D effort of the industry. Again, and in line with previous comments with reference to the unitary flows, Germany is, along with Italy, the only country where the product-embodied R&D diffused by imported intermediate consumptions is superior to the domestic product-embodied R&D diffused.

With reference to the industry of research and development, Italy, the Netherlands, the United Kingdom and Sweden are the countries where the diffusion carried out has higher relevance.

As in the case of computer services in Italy, in Belgium because of the low R&D intensity of the industry, the product-embodied R&D diffused through imported intermediate consumptions is higher than the product-embodied R&D diffused domestically. The same phenomenon occurs in Germany, the Netherlands and Spain, where more product-embodied R&D is diffused via imported intermediate consumptions than through domestic intermediate consumptions.

In section 18.3 a general upsurge in imported intermediate consumptions of KIS was shown. In light of the diffuser role carried out both by domestic and imported intermediate consumptions of KIS, a question emerges: do imported intermediate consumptions of KIS contribute to raising the volume of product-embodied R&D diffused in countries with low R&D intensities in these services? To try to answer this question, in Table 18.6 we compare the shares of imported intermediate consumptions with the domestic R&D intensities and with the imported product-embodied R&D flows per unit of value added.

As can be observed, the imports of intermediate consumptions perform an important compensatory role in some countries where the domestic R&D efforts are not very high. The clearest examples are Germany, in the case of computer services, and Belgium, in the case of research and development. Germany is the third-highest ranking country in its share of intermediate consumptions of computer services imported, and at the same time, shows a relatively low R&D intensity. Nevertheless, it is the fourth-highest ranking country in terms of product-embodied R&D diffused per unit of value added through imported intermediate consumptions. Belgium is the country where the participation of imported intermediate consumptions of research and development is the highest: more than 80 per cent. This high dependence on imports compensates for its low

Table 18.6 Imported intermediate consumptions, R&D intensities and product-embodied R&D diffused through imports, 2000

	Computer services			Research and development		
	Share	Intensity	Diffusion	Share	Intensity	Diffusion
Belgium	28.47	5.69	0.053	81.87	1.20	0.380
Denmark	15.34	15.03	0.059	28.98	6.31	0.209
Finland	4.50	6.75	0.059	n.a.	n.a.	n.a.
France	1.54	1.97	0.045	n.a.	n.a.	n.a.
Germany	13.90	2.67	0.053	49.17	11.92	0.274
Italy	5.16	1.05	0.042	15.93	20.92	0.187
Netherlands	10.66	3.53	0.047	57.99	7.27	0.326
Spain	7.91	4.51	0.040	51.19	11.60	0.374
Sweden	8.36	6.92	0.062	49.65	19.89	0.268
United Kingdom	4.70	2.96	0.053	14.52	11.98	0.217

Note: As there are no data on the origin of intermediate consumptions in Ireland, it is not included in this table.

Source: OECD Input–Output Database 2006, STAN and ANBERD.

R&D intensity: Belgium is the country where more product-embodied R&D per unit of value added is diffused through imported intermediate consumptions of research and development services.

18.6 Conclusions

The essential functions that KIS perform in the innovation domain have been highlighted by many scholars. Currently, as our economies are ever more knowledge-based, KIS are acquiring greater relevance in the production processes of the different industries. In addition to their growing 'domestic' importance, we note the significant rise in internationalisation of KIS, mainly due to the changes in transportability introduced by IT.

In this chapter we have tried to show that the contribution of KIS to knowledge diffusion processes expands beyond national borders. To do so, we have calculated the effects of the use of KIS on productivity and the product-embodied R&D diffused both by domestic and by imported intermediate consumptions of KIS.

Using data from the input–output tables, we have confirmed the growing relevance of KIS as intermediate consumptions within production processes. In only five years the differentials between the European countries and the US and Japan have lessened considerably, thanks to the fast pace of growth experienced by these services. Intermediate consumptions of KIS grew at an annual average rate higher than 10 per cent in all the European countries analysed except France (which was the only European country with a share of KIS in intermediate consumptions above 6 per cent in 1995).

In addition to the rise in the use of these services, the percentage of intermediate consumptions of KIS imported has also substantially increased. Finland and the United Kingdom are the only two countries where this did not occur. The share of imported intermediate consumptions is specially striking in some cases. In research and development 82 per cent of the intermediate consumptions were imported in Belgium and 58 per cent in the Netherlands in the year 2000. The share was around 50 per cent in Spain, Sweden and Germany. In relation to computer services the share of intermediate consumptions imported was 28 per cent in Belgium and more than 10 per cent in Denmark, Germany and the Netherlands.

In terms of R&D intensities, the trend is quite similar. R&D expenditures in KIS rose in almost all the European countries examined during the period 1995–2000. This growth was especially noteworthy in computer services and research and development.

The estimation of productivity equations has confirmed the existence of a positive relationship between the use of KIS as intermediate consumptions and the productivity of their client industries (all the significant estimated values were positive and quite high). Two countries deserve special attention: Germany and the Netherlands, that along with the US showed the highest significant estimated values.

The combination of a rising use of KIS as intermediate consumptions and their increasing R&D intensity translates into what we observe as a key role for product-embodied R&D diffusion. Given their rising internationalisation, it is of significant interest to differentiate between the diffusion carried out by domestic and by imported intermediate consumptions of KIS.

By comparing domestic and imported product-embodied R&D diffused through intermediate consumptions of KIS, we have corroborated that imported intermediate consumptions are very relevant diffusers. In some cases, these diffusions, per unit of value added, are higher than those carried out by domestic intermediate consumptions. This is the case of Germany and Italy in computer services and Belgium, the Netherlands, Germany, Denmark and Spain in research and development. Among those countries that diffuse more product-embodied R&D by means of imports two cases stand out: Germany and Belgium. In these countries the relatively low domestic R&D intensity is 'compensated' by imported intermediate consumptions. Thus, Germany is the third-highest ranking country in product-embodied R&D diffused by imported intermediate consumptions of computer services, and Belgium is first in terms of product-embodied R&D diffused per unit of value added via imported intermediate consumptions of research and development services.

In conclusion, we feel that this study is a small step in the understanding of knowledge flows generated through international provision of KIS. We have obtained evidence concerning a potential compensation of low domestic R&D intensities via imports that could also translate into productivity gains. A next stage in this line of analysis could be an in-depth examination of the specialisation patterns of each economy better to link specialisation, innovation and international flows of knowledge. This analysis, in combination with statistical information for various years, could shed more light on the international diffusion of knowledge through KIS – a diffusion that seems to be acquiring growing importance within our ever more global 'knowledge-based economies'.

Notes

1. KIS are considered quite similar to innovative manufacturing industries in terms of their R&D efforts. In this line, the study by Wong and He (2005) highlights that there are no differences in the propensity to carry out R&D activities between KIS and manufacturing firms.
2. Since the establishment of the pioneering distinction between 'traditional professional services' or p-KIS and 'new technology-based services' or t-KIS by Miles et al. in 1995, the number of studies that seek a correspondence between 'theoretical' and 'statistical' classifications of KIS has increased considerably (an excellent revision is provided by Nählinder, 2002). In our case, the combination of four different databases makes it difficult to employ an 'accurate' classification of KIS. We have preferred to analyse a smaller number of KIS industries in order to facilitate the interpretation of the results obtained.
3. One of the main reasons for the selection of this second approach is the absence of detailed capital series at the desegregation level employed in the input–output tables. Moreover, the option of estimating a traditional production function that includes capital inputs and distinguishes between business services and other business services has additional problems such as the potential multicollinearity that arise (especially in some countries) from the high correlation between the use of business services and the use of other services.
4. Non-autocorrelation tests were carried out for the different countries. The Durbin–Watson test accepts the null hypothesis of non-autocorrelation in Denmark, France, Italy, Portugal and the United Kingdom. In Belgium, Finland, Spain and Sweden the test is not conclusive. This is why we calculated the Breusch–Godfrey statistic, which accepts the null hypothesis of non-autocorrelation at the 5 per cent level in all of the cases.
5. Tomlinson (2000b) again examined the case of the United Kingdom starting from the 1990 input–output table using the same definition of KIS as Windrum and Tomlinson (1999). The novelty is the estimation of separate equations for each industry included within the group of KIS.
6. The group of KIS comprises the following industries. In Germany: post and telecoms, banking and financial services, insurance, real estate agencies, science and culture and publishing business services and other business services. In Japan: communication services, financial and insurance services, real estate agencies and business services. In

the Netherlands: post and telecoms, banking, financial services, insurance, real estate agencies, letting services, legal, accounting and business services, computing services, advertising agencies, general business services and employment agencies and recruitment services. In the United Kingdom: postal services, telecoms, banking and finance, auxiliary finance, insurance, real estate agents, legal services, accounting, other professional services, advertising, computer services and other business services.

7. We choose to compute R&D intensities with respect to output instead of employment because, as Amable and Palombarini (1998) point out in their paper, the use of employment entails two problems when comparing the growth of R&D intensities among countries: firstly, differences can hide variations in capital and labour ratios; and, secondly, increases in R&D intensities can be explained by labour productivity gains. In reference to the output indicator, we consider value added a better measure of output than production or turnover. Since production includes intermediates, any output of intermediate goods consumed within the same sector is also recorded as output. As a result, the impact of such intra-sector flows depends on the coverage of the sector. Turnover, for its part, refers to the actual sales in the year and can be greater than production in a given year if all products are sold together with stock from previous years. Consequently, the turnover can be higher or lower in an industry, depending on how perishable the stock is.
8. The product-embodied R&D diffused by imported intermediate consumptions cannot be calculated in the case of the industry of post and telecommunications because of the scant statistical information available.
9. This fact can be explained by the scarce international trade in services between these two countries. Thus, the reciprocal share of imports between Ireland and Italy is less than 4 per cent of the total imports in other services from the EU-15.
10. We report the total flows as percentages of total business expenditure on R&D in order to control for the size of the different countries when comparing the total amount of product-embodied R&D diffused.

References

Abernathy, W. and J. Utterback (1978), 'Patterns of innovation in technology', *Technology Review*, **80** (7), 40–47.

Abramovsky, L. and R. Griffith (2005), 'Outsourcing and offshoring of business services: how important is ICT?' Institute for Fiscal Studies, Working Paper WP05/22.

Amable, B. and S. Palombarini (1998), 'Technical change and incorporated R&D in the service sector', *Research Policy*, **27** (7), 655–675.

Antonelli, C. (1998), 'Localized technological change, new information technology and the knowledge-based economy: the European evidence', *Journal of Evolutionary Economics*, **8** (2), 177–198.

Antonelli, C. (2000), 'New information technology and localized technological change in the knowledge-based economy', in M. Boden and I. Miles (eds), *Services and the Knowledge-Based Economy*, London: Continuum, pp. 69–82.

Barras, R. (1986), 'Towards a theory of innovation in services', *Research Policy*, **15**, 161–173.

Bryson, J.R. (2001), 'Services and Internationalisation: annual report on the progress of research into service activities in Europe in 1998', *Service Industries Journal*, **1**, 227–240.

Camacho, J.A. and M. Rodriguez (2006), 'Knowledge intensive services and R&D diffusion: an input–output approach', in J.W. Harrington and P.W. Daniels (eds), *Knowledge-Based Services and Internationalization*, Aldershot: Ashgate, pp. 41–59.

Czarnitzki, D. and A. Spielkamp (2000), 'Business services in Germany: bridges for innovation', Discussion paper No. 00-52, Mannheim: ZEW.

den Hertog, P. and R. Bilderbeek (2000), 'The role of technology-based knowledge-intensive business services', in M. Boden and I. Miles (eds), *Services and the Knowledge-Based Economy*, London: Continuum, pp. 222–246.

Drejer, I. (2002), 'A Schumpeterian perspective on service innovation', DRUID Working

Paper No. 02-09, DRUID, Copenhagen Business School, Department of Industrial Economics and strategy / Aalborg University, Department of Business Studies.
Fernández-Fernández, M.T. (2001), 'Performance of business services multinationals in host countries: contrasting different patterns of behaviour between foreign affiliates and national enterprises', *Service Industries Journal*, **21** (1), 5–18.
Fischer, M.M. (2001), 'Innovation, knowledge creation and systems of innovations', *Annals of Regional Science*, **35** (2), 199–216.
Freel, M. (2006), 'Patterns of technological innovation in knowledge-intensive business services', *Industry and Innovation*, **13**, (3), 335–338.
Gadrey, J., F. Gallouj and O. Weinstein (1995), 'New modes of innovation: how services benefit industry', *International Journal of Service Industry Management*, **6** (3), 4–16.
Gallouj, F. and O. Weinstein (1997), 'Innovation in services', *Research Policy*, **26** (4–5), 537–556.
Gershuny, J. and I. Miles (1983), *The New Service Economy: The Transformation of Employment in Industrial Societies*, London: Pinter.
Ghosh, A. (1958), 'Input–output approach to an allocative system', *Economica*, **25**, 58–64.
Glückler, J. (1999), 'Management consulting – structure and growth of a knowledge intensive business service market in Europe', Institut für Wirtschafts- und Sozialgeographie, Department of Economic and Social Geography, Johann Wolfgang Goethe-Universität Frankfurt, IWSG Working Papers 12-1999.
Greenfield, H.I. (1966), *Manpower and the Growth of Producer Services*, New York: Columbia University Press.
Hipp, C. (2000), 'Information flows and knowledge creation in knowledge-intensive business services: scheme for a conceptualization', in J.S. Metcalfe and I. Miles (eds), *Innovation Systems in the Services Economy: Measurement and Case Study Analysis*, Boston, MA: Kluwer Academic Publishers, pp. 149–167.
Hipp, C., M. Kukuk, G. Licht and G. Muent (1996), 'Innovation in services: results of an innovation survey in the German service industries', paper presented at the Conference on the new S&T indicators for the knowledge-based economy, OECD.
Katsoulacos, Y. and N. Tsoinis (2000), 'Knowledge-intensive business services and productivity growth: the Greek evidence', in M. Boden and I. Miles (eds), *Services and the Knowledge-Based Economy*, London: Continuum, pp. 192–208.
Leiponen, A. (2005), 'Organization of knowledge and innovation: the case of Finnish business services', *Industry and Innovation*, **12** (2), 185-203.
Leiponen, A. (2006), 'Managing knowledge for innovation: the case of business to business services', *Journal of Product Innovation Management*, **23** (3), 238–258.
Léo, P.Y. and J. Philippe (2001), 'Internationalisation of service activities in Haute-Garonne', *Service Industries Journal*, **21** (1), 63–80.
Lommelen, T. and P. Matthyssens (2005), 'The internationalization process of service providers: a literature review', *Advances in International Marketing*, **15**, 95–117.
Machlup, F. (1962), *The Production and Distribution of Knowledge in the United States*, Princeton, NJ: Princeton University.
Mann, C.L. (2003), 'Globalisation of IT services and white collar jobs: the next wave of productivity growth', Institute for International Economics Policy Brief No. PB03-11, December.
Miles, I. (1996), *Innovation in Services: Services in Innovation*, Manchester Statistical Society.
Miles, I., N. Kastrinos, K. Flanagan, R. Bilderbeek, P. den Hertog, W. Huitink and M. Bouman (1995), 'Knowledge intensive business services: their role as users, carriers and sources of innovation', EIMS Publication No. 15, Innovation Programme, DGXIII, Luxembourg.
Miozzo, M. and L. Soete (2001), 'Internationalization of services: a technological perspective', *Technological Forecasting and Social Change*, **67** (2), 159–185.
Nählinder, J. (2002) 'Innovation in knowledge intensive business services: state of the art and conceptualisations', Arbetsnotat, No. 244, Linköping University, Sweden.

Nonaka, I. and H. Takeuchi (1995), *The Knowledge-Creating Company*. New York: Oxford University Press.
O'Farrell, P.N. and P.A. Wood (1998), 'Internationalisation by business service firms: towards a new regionally based conceptual framework', *Environment and Planning*, **30** (1), 109–128.
O'Farrell, P.N., P.A. Wood and J. Zheng (1996), 'Internationalization of business services: an inter-regional analysis', *Regional Studies*, **30** (2), 101–118.
Papaconstantinou, G., N. Sakurai and E. Ioannidis (1998), 'Domestic and international product-embodied R&D diffusion', *Research Policy*, **27** (3), 301–314.
Pavitt, K. (1984), 'Sectoral patterns of technical change: towards a taxonomy and a theory', *Research Policy*, **13** (6), 343–373.
Pilat, D. (2007), 'Productivity in business services', in L. Rubalcaba and H. Kox (eds), *Business Services in European Economic Growth*, Cheltenham, UK and Northampton, MA, USA: Edward Elgar, pp. 62–73.
Rodriguez, M. (2003), 'Services and innovation: towards the knowledge-based economy', PhD thesis, University of Granada.
Schulz, C. (2000), 'Environmental service-providers, knowledge transfer, and the greening of industry', presented at IGU Commission on the Organization of Industrial Space Annual Residential Conference: Industry, Knowledge and Environment, Dongguan, China, 8–12 August.
Strambach, S. (2001), 'Innovation process and the role of knowledge-intensive business services', in K. Koschatzky, M. Kulicke and A. Zenker (eds), *Innovation Networks: Concepts and Challenges in the European Perspective*, Heidelberg, Germany and New York, USA: Physica-Verlag, pp. 53–68.
Sundbo, J. (1998), 'Standardisation vs. customisation in service innovations', SI4S Topical paper No. 3, STEP, Oslo.
Tomlinson, M. (2000a), 'Information and technology flows from the service sector: a UK–Japan comparison', in M. Boden and I. Miles (eds), *Services and the Knowledge-Based Economy*, London: Continuum, pp. 209–221.
Tomlinson, M. (2000b), 'The contribution of knowledge-intensive services to the manufacturing industry', in B. Anderson, J. Howells, R. Hell, I. Miles and J. Roberts (eds), *Knowledge and Innovation in the New Service Economy*, Cheltenham, UK and Northampton, MA, USA: Edward Elgar, pp. 36–48.
Windrum, P. and M. Tomlinson (1999), 'Knowledge-intensive services and international competitiveness: a four country comparison', *Technology Analysis and Strategic Management*, **11**, 391–408.
Wong, P.K. and Z.L. He (2005), 'A comparative study of innovative behaviour in Singapore's KIS and manufacturing firms', *Service Industries Journal*, **25** (1), 23–42.

PART V

INNOVATION IN SERVICES AND NATIONAL AND INTERNATIONAL SPACES

19 Services innovation in a globalized economy

Peter Daniels

19.1 Introduction

Only very recently has work on services innovation moved beyond 'Cinderella status . . . being neglected and marginal' (Miles, 2000, 371; Tether, 2003). Research has tended to focus on conceptualization of service innovation together with case studies of selected service activities (see for example Boden and Miles, 2000; Gallouj, 2002; Gallouj and Weinstein, 1997; Metcalfe and Miles, 2000). The need to understand services innovation better has increased as services have become increasingly globalized with the emergence of new economies, notably India and, to a lesser extent, China. It seems that:

> the challenge to business is to find ways to gain productivity and economy of scale in services, while still maintaining a differentiation among competing services that creates loyal customers and economic profits. We can begin this process by gaining a better understanding of how services . . . innovation can be embedded in information technology and systems in order for the innovator to be able to capture the economic rewards of their innovative efforts and investments. (Goldhar et al., 2007, 10)

A services sector that is productive and innovative is the key to overall national economic performance as well as the welfare of citizens (OECD, 2005a, 2005b). The capacity of national economies to adjust to economic globalization in services and their growing share of employment and growth will rely heavily on innovation. This applies both to service producers and to the policies that will be required to facilitate and deepen innovation capacity. While India and to a lesser extent China have become leading offshore locations for undertaking software development and all types of information technology (IT)-enabled business services, the US and Canada and many of the EU countries (except those that have joined since the early 1990s) need to consider how to enable the growth of high-wage service jobs and businesses in response to a more global environment (see for example, Baker et al., 2008). There are of course many components of a strategy designed to release the growth opportunities associated with services such as the impact of barriers to trade, information and communications technology (ICT), or labour force skills; in this chapter the focus is on the role of innovation in services in a globalized economy. The importance of innovation in services has been reinforced by globalization and by advances in ICT. Both of these have stimulated new forms of competition and opened new markets for the creation and delivery of innovative services.

Service-producing firms are, in general, less likely than manufacturing firms to be innovative, but those that are information- and knowledge-intensive, such as financial intermediation and business services, do show above-average levels of innovation (see for example Lucking, 2004; Tether, 2003). However, it may be unwise to focus purely on the innovative capacity of service businesses because the manufacturing sector also now relies more than ever on computer, business and telecommunications services, for example, to stimulate improvements in productivity or as part of a competitive strategy that is aligned towards product–service packages or the supply of solutions rather than tangible products in the traditional sense (Australian Expert Group in Industry Studies, 2002; van Ark et al., 2003; Goldhar et al., 2007; Baker et al., 2008a). IBM illustrates this shift very well; in 1998 some 38 per cent of its revenues came from services (42 per cent from software) but in 2006 services accounted for 53 per cent of revenues and software just 25 per cent. Computer hardware accounted for 16–20 per cent of revenues throughout the period (Droogenbroeck, 2007). This servicization of manufacturing means that some 25 per cent of innovations in manufacturing involve a combination of goods and services while, conversely, approximately 40 per cent of the innovators in retail and distribution rate themselves as largely goods innovators (Howells, 2004). Thus, it is not necessarily helpful, even if it is methodologically more convenient, to concentrate on innovation in services or in manufacturing. While acknowledging this problem, in the interests of simplicity, the focus here will remain on the services sector.

The survey that takes place every four years in the EU countries to investigate levels of innovation in business (the latest Community Innovation Survey as of 2009 covered the period 2002–04) defines innovation as 'a new or significantly improved product (goods or service) introduced to the market or the introduction within the enterprise of a new or significantly improved process' (Eurostat, 2007; Arundel et al., 2007).[1] Put another way, services innovation 'is the successful exploitation of new ideas' (DTI, 2007), while a more comprehensive definition is provided by van Ark et al. (2003, 14) as:

> a new or considerably changed service concept, client interaction channel, service delivery system or technological concept that individually, but most likely in combination leads to one or more (re)new(ed) service functions that

are new to the firm and do change the service/good offered on the market and do require structurally new technological, human or organisational capabilities of the service organisation.

There may also be sector-specific ways of defining service innovation. For example Tesco, a major UK retailer that is also engaged in an internationalization strategy, has described innovation in its sector in terms of: 'focus on the consumer enables us also to develop the hundreds of innovations, both small and large, that keep us ahead of our rivals – be they in product specification, distribution, design and refit of stores, staff training, or operating processes in store' (Tesco, 2006; cited in DTI, 2007). While agreement about typologies of services innovation is elusive, the Tesco definition is an example of innovation that is partly 'client-led' combined with in-house innovation by Tesco. As an alternative to these types of innovation, there may be a reliance on 'supplier-dominated' (or external) innovation or innovation through services (service firms support other firms to innovate). In the case of innovation through services, some of the most important players are knowledge-intensive business services (KIBS) that act as facilitators, sources and carriers of innovation to client firms (Bettencourt et al., 2002).

Service product innovation occurs when a product is either new or significantly improved; this can arise from changes to its fundamental characteristics, to its technical specifications, to any software incorporated within it, to other immaterial components, to its intended uses or to its user-friendliness. Service process innovation occurs where the technology used is completely new or improved, when the methods of supplying a service are improved, or where the delivery is in some way enhanced significantly. Product and process innovations by service enterprises should impact upon levels of output, the quality of their products, the costs of production and of distribution. It is relatively easier to define types or sources of service innovation than it is to measure it; in part this is because the production of a service involves interaction with the consumer. This may be obvious with respect to a haircut or dental surgery, but it is also true for a marketing project or the receipt and use of a management consultant's report by the client.

Notwithstanding the earlier comment about the breakdown of the distinction between innovation in services and in manufacturing, there are some aspects of innovation by the former that distinguish them from the latter (Howells, 2001; Tether and Howells, 2007). Innovation in manufacturing tends to be led by investment in research and development (R&D), while in services it is led by the acquisition of knowledge, most of which is accessed via sources external to the enterprise or via continuous professional development or specialist training courses used by sources internal to an enterprise. It follows that if knowledge is the principal source for innovation in services, the development of human resources is absolutely vital to success, and in particular a reliance on highly skilled and highly educated workers to lead the innovation process. Because innovation in manufacturing requires investment in R&D facilities (laboratories, testing facilities etc.) it is much less likely to be dependent on the formation of new businesses than is the case for services, where innovation activity is often the prerogative of newly established firms (perhaps spin-offs from larger, established firms); entrepreneurship is therefore another key driver of service innovation. The structure of innovation expenditure by sector for UK firms illustrates some of the differences between services and manufacturing, as well as some of the commonalities (Figure 19.1) (see also Pilat and Wölfl, 2005).

The quality and the quantity of highly skilled labour in the economy is therefore crucial to services innovation (OECD, 2007). This reliance is not just upon scientific and technical skills linked to the utilization of ICT, but also upon the so-called 'soft' skills associated with interpersonal communications, adaptability, social networking skills and entrepreneurial ability. It has been suggested that: 'it is ICT, in combination with innovative business concepts, client interfaces and organisational innovations, which have transformed important parts of service activities [and] . . . Hence technological and non-technological innovation are so strongly connected that it is hard to treat them separately' (van Ark et al., 2003, 12). This, in turn, places a premium on national education systems that have shifted towards transmitting and transferring skills that enable workers to function efficiently in a real world that is complex and requires the ability to structure problems, to adopt continuous learning strategies, and to manage sophisticated concepts that are relevant, for example, to service design or delivery (see for example, Gadrey and Gallouj, 1998).

There is a presumption here that services innovation actually means something and leads to economic benefits to firms and/or the economy. While there is a general assumption that this is the case, some commentators are not convinced that the economic rewards equate with those associated with innovation in traditional physical product or traditional factory businesses (Goldhar et al., 2007). While not suggesting that there will be no rewards for services from innovation, Goldhar et al. suggest that gains do not arise from the direct ownership of intellectual property or 'trade secrets' by service providers, rather, the gains are achieved indirectly by using intellectual property embodied in physical products supplied by others, such as point-of-sales cash machines or ATM, that are used to develop strategies, organizational structures or marketing strategies that

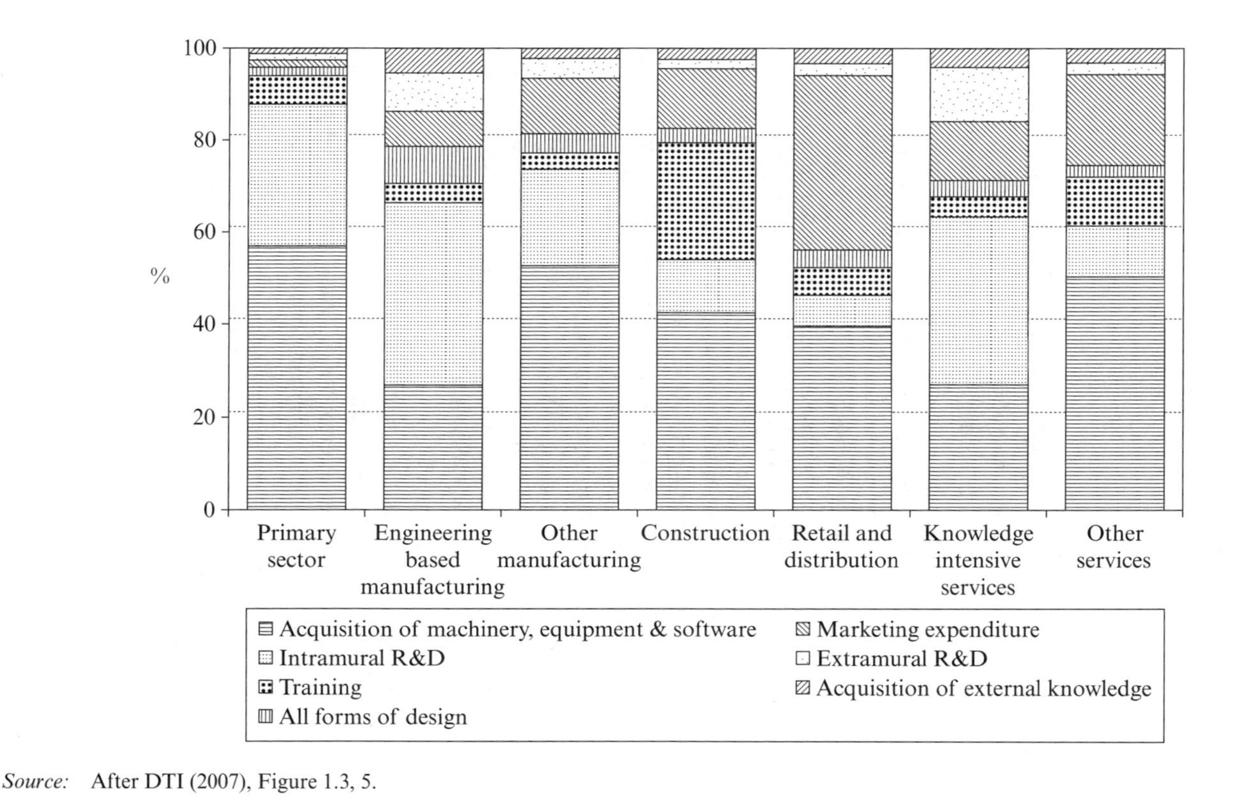

Source: After DTI (2007), Figure 1.3, 5.

Figure 19.1 Shares of innovation expenditure by sector, UK

will achieve some service advantages over competitors. They are not, in the majority of circumstances, reliant on their own intellectual property in patents to compete in the marketplace. Innovative service providers such as Tesco, DHL or Starbucks use patentable physical products made available by others but carefully use pricing, branding and execution of their service offer to 'protect' their service 'monopoly' (Goldhar et al., 2007; Berg, 2006).

19.2 Global services innovation

The requirement for services to be more innovative with respect to processes or products has been magnified by the extension of markets associated with globalization. In order to protect and expand market share, some services (although by no means all, see for example, Baker et al., 2008b) that are tradable find it necessary to extend their reach or to improve product competitiveness in the face of external competition, both within their home markets because of the access to them by outside providers, and in external markets. These effects are underpinned by the commitment towards greater transparency and openness in international services (and goods) trade at a global level (such as the role of the World Trade Organization in the rules of services trade between nations) as well as within trade blocs such as the North American Free Trade Association (NAFTA), the Southern Common Market (MERCOSUR) centred on Latin America, Asia-Pacific Economic Cooperation (APEC), or the European Union (EU). While progress on the liberalization of trade in services via the General Agreement on Trade in Services (GATS) has been almost glacial during the last few years, tradable service activities have recognized the opportunities presented by globalization to achieve operating efficiencies and to access new markets. But in order to capitalize on these opportunities they must innovate.

A useful basis for understanding the potential for global service innovation is the distinction between product and process innovation. The former is largely about creating superior value, usually through a product rather than a service, and is often accompanied by increased operational complexity. The lead time to market is lengthy but acceptance of the innovation by the market is rapid. Process innovation is about delivering and capturing value; actions can be implemented relatively quickly but the benefits for a firm's bottom line may take some time to become apparent. Simplification of operation complexity is the key to successful process innovation by, for example, finding new ways to integrate an innovative product or service with those that already exist, or engaging in customization involving reduced complexity in assembling the product or in delivery. It will be apparent that process and product innovation are not mutually exclusive, but addressing problems of process complexity is perhaps the main driver of services innovation and globalization. Typically, a services process incorporates the creation of large, often routine, datasets that require validation before they can be used or analysed. The latter requires knowledge, expertise, interpretation and judgement and underlines the role and availability of suitably skilled human resources when identifying locations that can contribute to global service innovation.

Innovation in the global markets for services is also dictated to a significant degree by ICT, alongside trade liberalization, the growing mobility of skilled human resources (Mode 4 in the GATS), or policies to attract services foreign direct investment (FDI). ICT may have a pervasive influence but it is not available to an equally high standard everywhere (see for example ITU, 2003). A frequently cited exemplar is the way in which ICT has enabled service companies to consider relocation of some of their activities offshore (European Foundation for the Improvement of Living and Working Conditions, 2004; Merino and Rodriguez, 2007). This may occur alongside outsourcing to external providers who may be located locally, or if they use suppliers located in other parts of the world they will be engaged in offshore outsourcing (UNCTAD, 2004). Another form of offshore outsourcing occurs when a multinational company elects to centralize, say, its corporate accounting services at one location and then decides to outsource that function to a specialist provider at some other geographical location, possibly as part of a partnership arrangement (Kedia and Lahiri, 2007).

When service firms make a decision to trade in the global market, or that it is more cost-effective to outsource all their functions apart from core activities, it is unlikely that they will possess in-house all the necessary expertise, especially relating to ICT systems and innovative ways to deploy them. They may well then turn to specialized global business service providers such as Siemens IT Solutions and Services which offers 'global delivery with onshore excellence' (Siemens, 2008, 5). Siemens has established a series of Global Production Centers that use its long-term presence in global markets to ensure that client service firms can access mature skill sets and embedded expertise that can be combined on a very flexible 'mix and match' basis to address the full range of corporate needs, from application management to development and any point along the value chain in global delivery. The innovation incorporated into the services provided by global business solution providers like Siemens takes the form of tailored solutions based on detailed knowledge of client firms' business processes and goals to create a 'landscape [that] must function in an integral, synchronized way irrespective of borders . . . and turned into business advantage' (Siemens, 2008, 6). Western European clients may, for example, specify that some work should be 'near-shored' to countries in Eastern Europe, that only certain functions should be outsourced to locations where English is not the principal language, or that only highly skilled, specialist staff should be used even if this is at the expense of higher wage costs. The ability of a globally organized service provider to pull together relevant information about wage and non-wage costs, skill sets, language skills, time zone, culture and the geopolitical circumstances of global regions then becomes a significant advantage.

Nissi Outsourcing Services, based in Coimbatore, South India, is one of several major Indian business process outsourcing (BPO) services supporting companies worldwide. It is typical of service providers that enable companies to improve their customer service management, allowing them to concentrate on their core areas, while outsourcing their non-core areas of business. The company works closely with its clients, recognizing that each of them has varying business needs and expectations, to provide reliable and scalable solutions that are quality assured to international standards by an in-house team. In addition to providing back-office services to clients in the UK, US, Israel, Malaysia and Singapore, Nissi (which is a relative newcomer to BPO) offers data conversion and transcription, online data research and data mining, international call centre service, website maintenance, engineering outsourcing, geographical outsourcing, project outsourcing, project consultancy, accounting services, e-publishing, e-accounting, legal services, medical billing, web design outsourcing, web development outsourcing and employee outsourcing.

Global relocation of manufacturing production activities is long established and, as co-production is increasingly the key to successful innovation, it has required service providers to follow their clients offshore. This is especially the case for large service providers offering services that gradually move up the value chain, starting with delineated and standardized tasks, and gradually including more activities, until a full business process is provided (European Foundation for the Improvement of Living and Working Conditions, 2005). When a full business process is provided the service firm becomes a strategic partner for the outsourcing company. In some cases it can lead to 'reverse offshoring'; large offshoring firms like India's Tata Consultancy Service, Infosys and Wipro are IT service firms that are re-employing US workers. Wipro, for instance, has a presence in a number of US cities such as Atlanta, Phoenix or Seattle that possess substantial pools of skilled software and related workers and reasonable salary costs. It is also one of a number of global firms such as Dell and General Motors that promote the idea of offshore (or global) innovation networks that incorporate significant linkages across firm boundaries. Others refer to those networks as global collaboration (MacCormack and Forbath,

2008) that extend beyond outsourcing to enhancement of capabilities by investing in experiments to learn which processes work best, which partners are best at cooperating with others, or which skills or organizational arrangements need to be improved in order to make process innovation through collaboration better. MacCormack and Forbath (2008) show that leading global firms (such as Boeing) make strategic investments in collaboration targeted at four areas: human resources (people), processes, platforms (infrastructure for sharing data), and programmes.

19.3 Innovation and global service firms: case studies

Innovation and the development of service firms in a globalizing economy are inextricable. This is most readily demonstrated using case examples of the kind used by Eröcal (2005) for a study of successful large international firms in the service sector. Innovation, whether related to a process or a product, is one of three factors identified by Eröcal (2005) as common to successful multinational service firms (the others are the opening up of markets and a motivational work environment). While public policies relating to innovation or to ICT contribute to service firm success, their effects are very limited unless they occur within open and contested markets.

Federal Express (FedEx)

Following a very modest but pioneering start, when Federal Express Corporation was founded in 1971 in Little Rock, Arkansas as the precursor of the modern air/ground express delivery industry, FDX Corporation was set up in 1998 and became FedEx Corporation in January 2000. FedEx has its headquarters in Memphis, Tennessee, to which it moved in 1973, and employs more than 141,000 worldwide (approximately half of the total for the FedEx Corporation), servicing 220 countries and handling some 3.5 million packages daily through 375 airports using a fleet of 671 aircraft and more than 44,000 motorized ground vehicles. The motivation for establishing FedEx was the realization on the part of its founder, Frederick W. Smith, that the advantageous use of aircraft to transport time-sensitive freight such as computer parts, pharmaceuticals or electronic equipment had hardly been realized, and there was likely to be significant latent demand provided that a reliable process for handling, despatching and delivering such goods could be organized. For such an arrangement to be effective it was necessary to be able to use airports allowing 24-hour aircraft movements and to be able to use the largest aircraft available. Memphis saw the potential for an innovative freight service in 1973 and deregulation of the air freight carrier sector was put in place in 1977.

The initial problem was how to establish a system that was robust and comprehensive for the needs of clients within the United States who would expect package handling from initial pick-up though to delivery at a designated address by a predefined time. This guaranteed delivery time is the key offering and the innovations made by FedEx have been introduced to ensure that these times are as reliable as possible and the outcome of a package handling process that optimizes time efficiency with minimum risk of error (Williams and Frolick, 2002; Farhoomand et al., 2003). The reliability problem was solved by the introduction of a hub system (large warehouses located in or near airports and served by a comprehensive network of ground transport services) for receiving, sorting and resending packages; FedEx was the first company to develop a system that has since been emulated by its major competitors (DHL and United Parcel Service, UPS) as well as state-owned postal systems such as the Royal Mail (UK). FedEx pioneered the use of a centralized computer system known as COSMOS, introduced in 1979, that enabled the company to manage vehicles, employees, packages routes and weather scenarios on a real-time basis. The extent to which guaranteed delivery times are sustained and, preferably, improved is one of the major sources of competitive advantage, underpinned by innovations in client services such as the use of the Internet to allow customers to track their packages (based on the use of an innovative hand-held barcode scanner) through each step from origin to destination. By 1986, FedEx was so confident in its package handling and tracking system that it offered full refunds to customers whose delivery schedules were not fulfilled.

Having constructed a successful express parcel business serving all corners of the United States, FedEx was well positioned to take the next logical step at a time when globalization of business was driving up demand for international freight services. The ability to use larger, long-haul jet aircraft from 1978 onwards was clearly advantageous and in 1984 FedEx began to provide express parcel services from and to Europe and, shortly after, Asia. The first foreign FedEx hub was located in Brussels in 1985. It now has 737 stations (collection points) outside the United States and four air express hubs at Subic Bay, Philippines (Asia Pacific), Toronto (Canada), Paris (Europe/Middle East/Latin America) and Miami (Latin America/Caribbean). With one recent study (Li et al., 2006) showing that there was very little to choose between FedEx and a close competitor such as UPS with respect to customer satisfaction, there continues to be much scope for technological innovations that provide some service advantage at the margin. R&D investment targets questions such as: how to maintain constant temperatures during transit of medicines or blood products whose properties are critically dependent on temperature; how to fulfil the requirement of a big company that all packages from wherever source, of whatever size, and not necessarily all transported by the same shipper, are delivered at the same time every day (seven days a week); or how to manage packages for which the customs rules relating to the contents are different at the origin of the consignment from those in place at the destination.

Other examples involving continuous innovation are refining official weather forecasts even more so as to optimize aircraft or ground vehicle scheduling and improving safety, or routing delivery vehicles on the ground in ways that reduce vehicle-miles and therefore the amount of fuel used. UPS, for example, has been able to reduce annual fuel consumption in the US by several million gallons by minimizing the number of left turns that its delivery vehicles are required to make (Deutsch, 2007). Likewise, aircraft radar systems developed in association with national organizations such as the Federal Aviation Authority (FAA) in the US allow freight aircraft that are always working to very tight schedules to slot into very tight aircraft queues as they approach destinations or to descend more steeply nearer their destinations, thus saving on time and fuel. Finally, constant innovation is also required in package identification and the incorporation of customer requirements (such as a delivery after a certain time) into barcodes that can then be used to remind delivery staff en route as to special requirements (such as the need for a customer signature), whether they have overlooked a delivery, or even a last-minute change in the details of the delivery location. Most of the innovations in the service provided are associated with the use of information technology which is about minimizing any delays in the package handling process since, ultimately, even a one-minute delay that affects several thousand packages on any one day can cost FedEx if it means that its guaranteed delivery time is not fulfilled.

Starbucks

FedEx is an example of a firm that has developed a market-creating service innovation based on a flexible solution. This is one of four types of market-creating service innovation suggested by Berry et al. (2006) using a matrix based on type of service and type of benefit (Figure 19.2). These service innovations are distinguishable by whether they are a new core benefit or a new way of delivering a core benefit, and the extent to which the supplier and the user (or beneficiary) are separable. In the case of FedEx the client and the supplier do not need to be in the same place for the service to be provided, while Virgin Atlantic Upper Class involves delivery of a journey experience that starts with booking, pick-up from home, express check-in, use of a lavishly equipped Virgin Atlantic Clubhouse, continues with

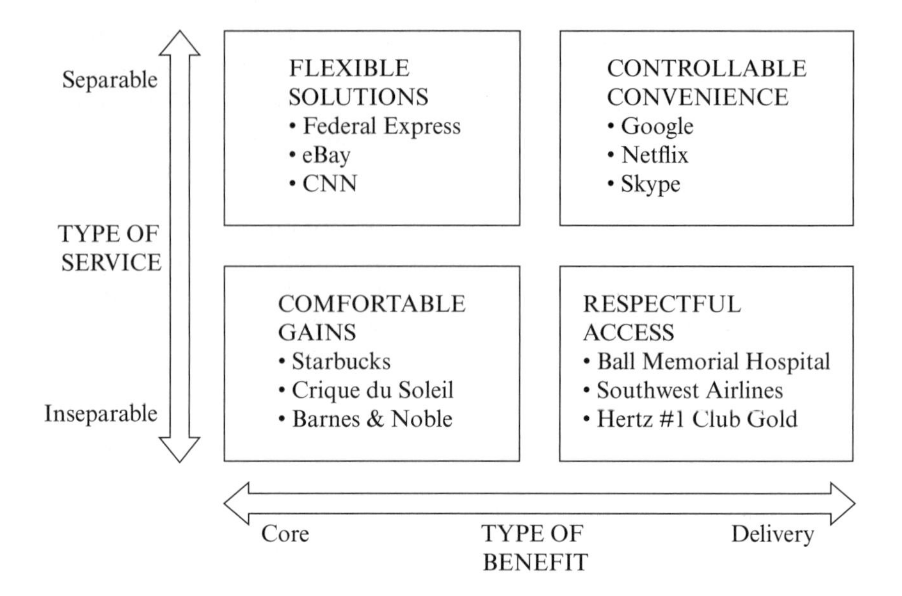

Source: After Berry et al. (2006, 59).

Figure 19.2 Market-creating service innovations

flat-bed seats and a cocktail bar area on the flight, and ends with transport home from the airport; the client and the supplier are not separable.

In the 25 years since Starbucks opened its first store in Pike Place Market (Seattle) it has focused on service innovations that customers see as a core benefit consumed at the time and place of production (or consumption). This benefit comprises a new experience that contributes to the emotional or physical comfort of customers (for a full discussion of innovation in experiential services see Verhoef et al., 2004; Voss and Zomerdijk, 2007). In its most recent Annual Report (Starbucks Corporation, 2007) it is stressed that Starbucks is a coffee company that celebrates the interaction with its customers via the 'Starbucks Experience' which more than 170,000 employees (partners) bring to life every day in over 15,000 stores and in 43 countries. It claims to be an enduring global company that has been built in a different way, i.e. on trust that is founded upon the way its partners work together, its relationships with its customers, the respectful way it treats coffee farmers and the contributions of Starbucks Corporation to the communities it serves. Starbucks has relied heavily on employee-friendly policies for its success. The group prizes respect for its partners. Internal communications are seen as vital and all workers employed for more than 20 hours per week receive health insurance

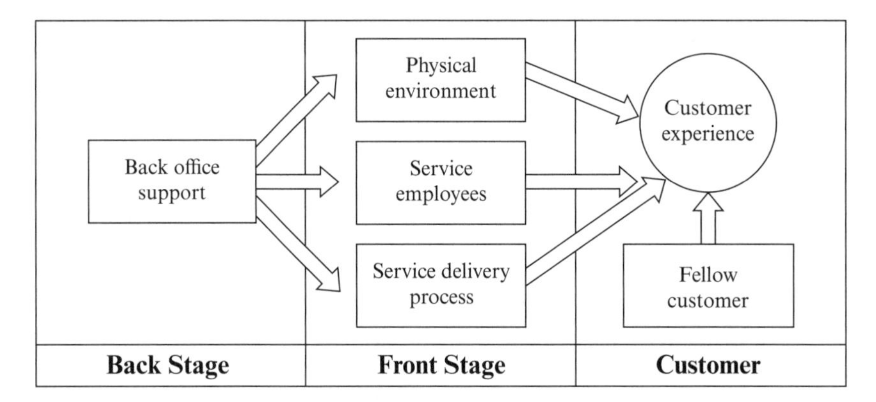

Source: Voss and Zomerdijk (2007, 105).

Figure 19.3 Elements of experiential service design

and stock options (for US employees in particular the former is a major employer innovation). Employees are remunerated at rates considerably better than the minimum wage (US$6 to US$8 per hour) and receive training from the outset. Low levels of employee turnover (65 per cent annually compared with 150 to 350 per cent for the fast food industry more generally) reflects loyalty, commitment to the 'Starbucks Experience', and lowers costs. The innovations are therefore 'people-centred' and experiential (Figure 19.3); the quality of the coffee and the variety of flavours is highly consistent from day to day and from one location to the next, to the extent that consumers are prepared to pay premium prices.

For example, its newest blend 'is ground fresh. Brewed fresh. Every 30 minutes all day. Every day' (http://www.starbucks.com/retail/beverages.asp, accessed 8 April 2008). The ambience of the 'spaces of coffee consumption' are also carefully designed to be relaxing, relatively low density so that the privacy of conversations or the use of PDAs or laptops is not easily compromised, and round tables are preferred to square or rectangular versions because they evidently do not make customers who are on their own feel isolated or uncomfortable. This is encapsulated in the way that the company envisages its stores as a 'third place' (apart from home and work) to spend time. Many of the stores therefore provide free electricity for customers as well wireless Internet access. This is not in itself innovative but is promoting innovations in the way, for example, that people work (Fletcher, 2008). The media in the US and the UK have recently highlighted the growing number of workers, sometimes labelled as the 'new Bedouin' who, equipped with mobile phones, laptops and wi-fi

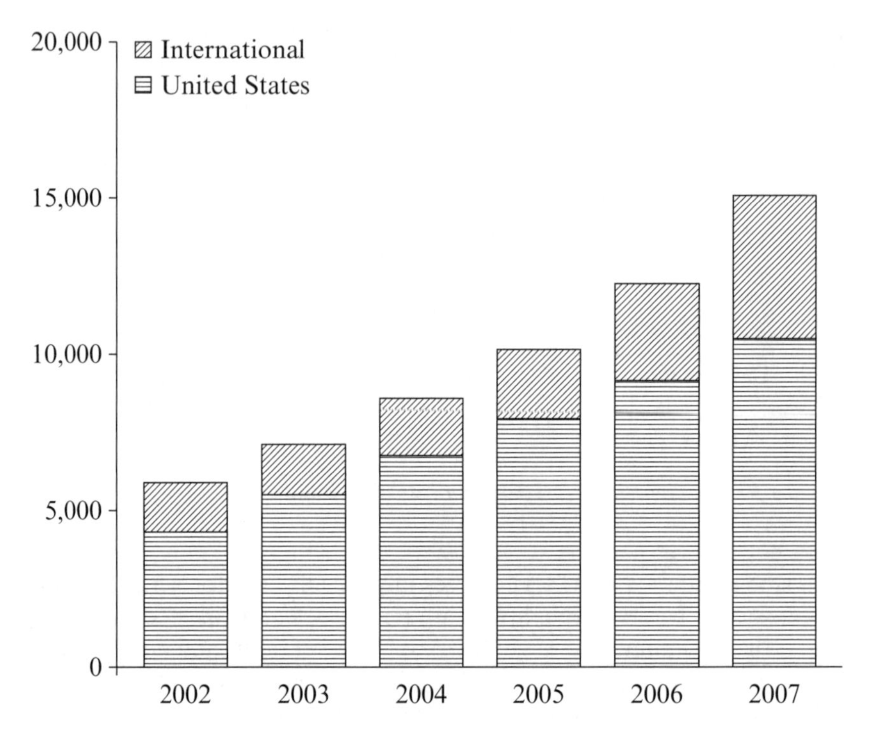

Source: Starbucks Corporation (2007).

Figure 19.4 Growth of Starbucks stores, 2002–07: United States and international

access are setting up office in coffee shops like Starbucks. Even though the wi-fi access is charged for and etiquette requires the purchase of a beverage at least once an hour, it is argued that it is still cheaper than renting conventional office space.

Starbucks Corporation had 17 coffee bars in North America at the end of 1987, including two in Chicago and Vancouver. It began to spread its brand name the following year by offering mail-order coffee throughout the United States. The innovations that were integral to its business plan were clearly successful; by 1992 there were 165 Starbucks cafe-bars across the United States. It was then large enough to go public and to raise capital from the stock market that would enable it to expand internationally (Figure 19.4).

In relative terms the growth of US outlets slowed while international expansion has increased, starting with a joint venture in Tokyo in 1996 that has ultimately made Starbucks the world leader in coffee distribution, retailing hot and cold drinks, food items and coffee-making equipment and other related accessories. Stores outside the US and Canada now account for some one-third of the total. In late 2007 Starbucks owned 8505 outlets worldwide, of which 1712 were outside the US. There were also 6506 joint-venture and licensed outlets with 2615 in other countries and US territories.

19.4 Conclusion

Globalization of markets combined with the opportunities created by advanced ICT have enabled new service delivery mechanisms and have reduced the time required to develop and introduce new services which will also, as a result, have a shorter shelf life requiring constant reappraisal of the service offer and ways of reaching clients or customers. The place of ICT in services innovation is also secured by its relatively low capital costs and usage charges. But while ICT has opened up global business opportunities for services, it has also mediated strong links with the knowledge economy and as a consequence has spurred demand by service firms for workers with more specialized skills. These may be accessed from within or, increasingly, from outside individual firms, and locally or internationally, and are the sources of creativity and flexibility that are a necessary part of services innovation. Because of the reliance on external sources of specialist expertise, there is a parallel tendency for global clusters of expertise to emerge which can be accessed from a wide cross-section of service activities.

Process rather than product innovation is the main driver of global service innovations. Offshore outsourcing is one of the most visible outcomes with scope for almost all tasks that are not core business to be outsourced. The tasks regarded as 'core' will vary according to the type of business but it can be expected that commercially sensitive, innovative and creative functions such as design and development will remain in-house; the risks of losing core competencies to competitors are more important than the fiscal and other benefits generated by outsourcing. This still leaves scope for standardized core tasks that do not require face-to-face contact or may not require specific or legal data protection, but over which a company may still want control of the process, to be handed over to business process providers, along with any processes outwith a company's core activities.

Few countries have incorporated service innovation into their programmes for stimulating economic development; industrial policies invariably attract involvement by manufacturers rather than service providers. There is no doubt that awareness of innovation programmes or policies amongst service providers is generally poor, while the commitment, at all levels from local to national, to the design of innovation support initiatives targeted at services is also well below requirements. Nevertheless, there are some minimum requirements for the attraction and retention of innovation-driven business process providers. These include national education and training policies that ensure a supply of well-trained and flexible labour that is also adaptable to change, as global service firms constantly assess and evaluate process innovations and the best locations to access them.

Finally, one other global trend that will have an important impact on innovation in services is demographic shift (Forfás, 2006). With some exceptions, the world population is ageing, with consequences for certain segments of the global service economy such as health, lifelong learning, social or tourism services. There are therefore opportunities for both new services geared to the needs and expectation of a 'greying' population, especially in the advanced market economies, and the identification of new ways of delivering those services.

Note

1. CIS4 (2002–04) produced a broad set of indicators on innovation activities, innovation expenditure, effects of innovation, public funding, innovation cooperation, sources of information for innovation, main obstacles on innovation activity and protection methods of intellectual property rights.

References

Arundel, A., M. Kanerva, A. Cruysen van and H. Hollanders (2007), *Innovation Statistics for the European Service Sector*, Maastricht: UNU-MERIT.

Australian Expert Group in Industry Studies (2002), *Selling Solutions: Emerging Patterns of Product-Service Linkage in the Australian Economy*, Sydney: Australian Expert Group in Industry Studies.

Baker, P., I. Miles, L. Rubalcaba, N. Plaisier, S. Tamminen and I. de Voldere (2008a), *Study on Industrial Policy and Services: Final Report – Part I*, Rotterdam: Ecorys Consulting, for European DG Enterprise and Industry.

Baker, P., N. Plaisier, S. Tamminen and I. de Voldere (2008b), *Study on Industrial Policy and Services: Final Report – Part II*, Rotterdam: Ecorys Consulting, for European DG Enterprise and Industry.

Berg, D. (2006), 'Analysis of the service sector', *International Journal of Information Technology and Decision Making*, **5** (4), 699–70.

Berry, L.L., B. Shankar, J.T. Parish, S. Cadwallader and T. Dotzel (2006), 'Creating new markets through service innovation', *Sloan Management Review*, **47** (2), 56–63.

Bettencourt, L.A., A.L. Ostrom, S.W. Brown and R.I. Rowntree (2002), 'Client co-production in knowledge-intensive business services', *California Management Review*, **44** (4), 100–105.

Boden, M. and I. Miles (eds) (2000), *Service and the Knowledge-Based Economy*, London: Continuum.

Deutsch, C.H. (2007), 'UPS embraces high-tech delivery methods', *The New York Times*, 12 July.

Droogenbroeck, P.V. (2007), 'IBM overview', Presentation to a meeting on Services Science, Luxembourg, 13 November.

DTI (2007), 'Innovation in services', DTI Occasional Paper No. 9, June, London: DTI.
Eröcal, D. (2005), 'Case studies of successful companies in the services sector and lessons for public policy', STI Working Paper 2005/7, Paris: OECD.
European Foundation for the Improvement of Living and Working Conditions (2004), *Outsourcing of ICT and related services in the EU*, Luxembourg, Office for the Official Publications of the European Communities.
European Foundation for the Improvement of Living and Working Conditions (2005), 'Proceedings of a seminar on Offshore Outsourcing of Business Services: Threat or Opportunity, Tallinn, 17–18 March', accessed at http://www.eurofound.europa.eu/emcc/content/source/tn05001a.htm, 1 December 2008.
Eurostat (2007), *Fourth Community Innovation Survey*. Luxembourg: EUROSTAT.
Farhoomand, A.F., P.S.P. Ng and W.L. Conley (2003), 'Building a successful e-business: the FedEx story', *Communications of the ACM*, **46** (4), 84–89.
Fletcher, H. (2008), 'Coffee shops beat the office grind', *Sunday Times*, 23 February.
Forfás (2006), *Services Innovation in Ireland: Options for Innovation Policy*, Dublin: Forfás.
Gadrey, J., and F. Gallouj (1998), 'The provider–customer interface in business and professional services', *Service Industries Journal*, **18** (2), 1–15.
Gallouj, F. (2002), *Innovation in the Service Economy: the New Wealth of Nations*, Cheltenham, UK and Northampton, MA, USA: Edward Elgar.
Gallouj, F. and O. Weinstein (1997), 'Innovation in services', *Research Policy*, **26** (4–5), 537–556.
Goldhar, J., Y. Braunstein and D. Berg (2007), 'Services innovation in the 21st century: it all begins with defining services vs. products and factory vs. service operations', paper presented at UC Berkeley–Tekes Service Innovation Conference, Berkeley, CA, 26–28 April.
Howells, J. (2001), 'The nature of innovation in services', in *Innovation and Productivity in Services*, Sydney, Paris: OECD, pp. 57–82.
Howells, J. (2004), 'Innovation, consumption and services: encapsulation and the combinatorial role of services', *Service Industries Journal*, **24** (1), 19–36.
ITU (2003), 'ITU Digital Access Index: world's first global ICT ranking', http://www.itu.int/newsroom/press_releases/2003/30.html, accessed 1 December 2008.
Kedia, B. and S. Lahiri (2007), 'International outsourcing of services: a partnership model', *Journal of International Management*, **123**, 22–37.
Li, B., M.W. Riley, B. Lin and E. Qi (2006), 'A comparison study of customer satisfaction between the UPS and FedEx: an empirical study among university customers', *Industrial Management and Data Systems*, **106** (2), 182–199.
Lucking, B. (2004), *International Comparisons of the Third Community Innovation Survey (CIS3)*, London: Department of Trade and Industry.
MacCormack, A. and T. Forbath (2008), 'Learning the fine art of global collaboration', *Harvard Business Review*, January, 10–11.
Merino, F. and D.R. Rodriguez (2007), 'Business services outsourcing by manufacturing firms', *Industrial and Corporate Change*, **16** (6), 1147–1173.
Metcalfe, S. and I. Miles (2000), *Innovation Systems in the Service Economy: Measurement and Case Study Analysis*, Dordrecht: Klewer.
Miles, I. (2000), 'Science innovation: coming of age in the knowledge based economy', *International Journal of Innovation Management*, **4** (4), 371–389.
OECD (2005a), *Enhancing the Performance of the Service Sector*, Paris: OECD.
OECD (2005b), 'Promoting innovation in services', Working Paper on Innovation and Technology Policy, Paris: OECD.
OECD (2007), *Innovation and Growth: Rationale for an Innovation Strategy*, Paris: OECD.
Pilat, D. and A. Wölfl (2005), 'Measuring the interaction between manufacturing and services', STI Working Papers, 2005(5), Paris: OECD.
Siemens (2008), *Strategy Realized: Mastering SAP® Solutions and Services*, Munich: Siemens AG, www.siemens.com/it-solutions.
Starbucks Corporation (2007), *Fiscal 2007 Annual Report*, Seattle: Starbucks Corporation.

Tesco (2006), 'Tesco Main Submission to the Competition Commission Inquiry into the UK Grocery Market, 2006', cited in DTI (2007), 'Innovation in services', DTI Occasional Paper No. 9, June, London: DTI, p. 10.

Tether, B. (2003), 'The sources and aims of innovation in services: variety between and within sectors', *Economics of Innovation and New Technology*, **12** (6), 481–505.

Tether, B. and J. Howells (2007), *Changing Understanding of Innovation in Services: From Technological Adoption to Complex Complementary Changes to Technologies, Skills and Organisation*, Report for the DTI, CRIC and Manchester Business School, Manchester: University of Manchester.

UNCTAD (2004), *Service Offshoring Takes off in Europe: In Search of Improved Competitiveness*, Paris: UNCTAD.

van Ark, B., L. Broersma and P. den Hertog (2003), *Service Innovation, Performance and Policy: A Review*, The Hague: Ministry of Economic Affairs.

Verhoef, P.C., G. Antonides and A.N. De Hoog (2004), 'Service encounters as a sequence of events: the importance of peak experiences', *Journal of Service Research*, **7** (1), 53–64.

Voss, C. and L. Zomerdijk (2007), 'Innovation in experiential services: an empirical view', in 'Innovation in services', DTI Occasional Paper No 9, London: DTI, pp. 97–134.

Williams, M.L. and M.N. Frolick (2002), 'The evolution of EDI for competitive advantage: the FedEx case', in J.B. Ayers (ed.) *Making Supply Chain Management Work: Design, Implementation, Partnership, Technology and Profits*, New York: Auerbach, pp. 185–194.

20 Outsourcing and offshoring of knowledge-intensive business services: implications for innovation

Silvia Massini and Marcela Miozzo

20.1 Introduction

Outsourcing and offshoring of services has grown from basic software coding and call centre work to a range of back-office functions such as payroll and accounting, financial and legal research, and even tightly regulated work such as drug development and interpretation of radiology images. International trade of commercial services increased from under $400 billion in the early 1980s to more than $2.1 trillion in 2004 (UNCTAD, 2004). International trade of services includes both international outsourcing and in-sourcing. Outsourcing refers to the decision to buy what was previously made internally. It relates to the fundamental question of why firms exist, whether and what a firm should make or buy, and it has been studied using transaction cost economics (Williamson, 1975), core competences (Prahalad and Hamel, 1990), the evolutionary and resource-based view of the firm (Penrose, 1959; Nelson and Winter, 1982) and dynamic capabilities (Teece et al., 1997). Offshoring refers to the transfer of activities outside the domestic boundaries, whether as in-house operations (captive or fully owned subsidiaries) or outsourced to third-party providers.

This chapter focuses on the outsourcing and offshoring of a range of high-skill services, which include more creative and knowledge-based services, such as research and development (R&D), design services, new software development, medical testing or analysis, and architecture drawings (Evangelista, 2000; Miles, 2001; Miozzo and Soete, 2001). These services, which are fundamental components of innovation processes, are also called knowledge-intensive business services (KIBS) and are defined by Miles (2001) as all those business services founded upon either technical knowledge and/or professional knowledge. This broad definition captures both the social and institutional knowledge involved in many of the traditional professional services (such as management consultancy and legal services) and the continuously evolving technological and technical knowledge involved in high-tech services (services such as R&D, product design, engineering services, and software design).

The outsourcing and offshoring of KIBS raise new issues and questions for innovation. In contrast to outsourcing and offshoring of manufacturing, outsourcing and offshoring of KIBS involve concerns about the development of new organisational and technological abilities by firms to relocate tasks based on intangible assets, and coordinate an interfirm and geographically dispersed network of activities (Levy, 2005). Outsourcing and offshoring of KIBS reveal key features of the international division of innovative labour, including the growing interdependence between firms and between developed and less-developed countries. Although there is a great amount of business cover stories and policy reports on the increasing outsourcing and offshoring of high-value-adding knowledge-intensive activities, there is little work on the consequences of the relocation of KIBS for innovation, especially at the level of the theory of the firm, and international division of labour in relation to innovation processes.

This chapter considers a number of emergent issues with implications for innovation processes. First, we consider implications for the theory (and boundaries) of the firm. Decentralisation of corporate activities (in the form of shared services, outsourced and offshored services that are corporate functions) calls into question traditional models of the multidivisional firm (Sako, 2005). Also, as innovation processes become increasingly distributed across firms, suppliers, customers, research organisations and other institutions (Chesbrough, 2003; Coombs et al., 2003; Rothwell, 1992), the changing boundaries of the firm and the outsourcing and offshoring of KIBS bring new challenges to organisations, such as governance issues, and the coordination and integration of decentralised knowledge-creating activities.

Second, we explore implications regarding the globalisation of innovation. Accompanying outsourcing and offshoring, we are witnessing the development of both large service suppliers which start to compete among themselves globally, and small independent entrepreneurial ventures in developed and less developed countries. This brings in new players in the global market for the production of innovation (captive multinational enterprises – MNEs; specialist MNEs; independent firms; and host-country captives) (Dossani and Kenney, 2004), while host locations are experiencing transformations, including wage inflation, overcrowded infrastructure in hot-spot locations and high attrition rates in service providers offshore. These processes raise issues of evolving market structure and competition, the emergence of technical and knowledge clusters where companies compete for activities and skills across a range of sectors, and raise questions on the extent to which these processes result in an erosion of the knowledge-based comparative advantage of developed countries.

The chapter is organised as follows. Section 20.2 outlines the growth of outsourcing and offshoring of KIBS. Section 20.3 examines attempts to analyse the implications for innovation of domestic outsourcing, international in-sourcing and the combined challenges of offshore outsourcing. The following section examines the emergent issues for innovation studies from offshore outsourcing. A final section discusses the implications of outsourcing and offshoring of KIBS for the knowledge-based competitive advantage of developed countries.

20.2 How are outsourcing and offshoring growing?

Offshoring goes back to the period of classic multinationalisation (1950s–1970s) when foreign direct investment (FDI) was guided by the characteristics of host economies or 'locational advantages' (abundant natural resources, lower labour costs, available skills, market protection) (Hymer, 1976; Vernon, 1966). During this period, repetitive jobs in manufacturing, first in traditional industries (shoes, textiles, toys, mature and standardised electronics), then in high-tech manufacturing (electronic components, electronic goods assembly), moved to less-developed countries in Latin America and Asia.

Outsourcing as a business strategy is a more recent phenomenon, linked to a series of organisational changes, with vertical disintegration increasing in the 1980s in part as a response to the rise of 'flexible specialisation' (Piore and Sabel, 1984; Womack et al., 1990) with the development of long-term supplier relations. In the 1980s and 1990s, leading firms in capital-intensive sectors such as automobiles and electronics set up international production networks not only to assemble their finished goods but also to develop a supply base for intermediate products and subassemblies. Outsourcing developed further in the 1990s together with other organisational changes and restructuring related to downsizing, delayering, de-scaling and de-scoping (e.g. Pettigrew and Massini, 2003).

A further wave of these developments is the movement of white-collar occupations to less-developed countries, especially to China and India. In recent years, outsourcing and offshoring have expanded to information technology (IT) applications, finance and accounting, engineering and R&D, human resources and contact or call centres, and is emerging as a widely popular business practice (e.g. Lewin and Peeters, 2006). This wave includes increasingly innovative activities and R&D. This, again, is not new, and carries on from the trend of global product development started by the establishment of corporate labs adapting products from developed economies to the new markets in Asia and Latin America for cost-saving reasons. Niosi (1999) provides a very useful summary of the state of the art of the literature and theoretical trends on internationalisation of R&D in the 1990s, and its contrast with the previous wave, up to the 1980s. The various contributions of the special issue edited by Niosi present models to explain the new empirical findings on large multinational enterprises (MNEs), which discuss the types of international management of innovation projects, motives and extent (these are discussed at more length later). However this literature focuses on the strategies and missions of large multinationals, which are by definition operating over a number of countries. What is new about the current offshoring wave is the phenomenon that also less internationalised companies and small and medium-sized firms are going offshore, and not for expanding their existing activities to new markets, but for cost-saving opportunities. Offshoring has been defined as the relocation of existing activities (or functions) from the home country to low-cost countries (not relocation of activities within the Triad, which is not new). Also, these activities are not designed to serve the local market of the host country, but instead to serve activities based in the home country or other global operations (Kenney et al., 2009).

The 2004 UNCTAD report presents a useful description of export-oriented FDI projects which are offshored (Table 20.1). The first three categories describe projects which companies may offshore in the various categories: call centres, shared services and IT services – which include what we call KIBS. However, regional headquarters may simply have a support role for overseas operations, rather than being offshoring projects themselves.

Doh et al. (2009) categorise further the above types of services outsourced or offshored according to the relative extent of interactivity, repetition and innovativeness, with call and contact centres (technical support and advice, claims, customer relations management) typically requiring a relatively high level of interactivity between employee and customer, compared to those in shared services centres and IT, and little or no innovative element, as most of the activities here are highly standardised and routinised (Figure 20.1). Some call centre activities can also have some degree of repetitiveness, but not as much as activities in shared services centres, which on the other hand have very little interaction with customers and require less innovative ability, since also in this case the work involved is highly repetitive and standardised, remains fairly the same over time, and does not involve any novelty. Finally, IT services (and other KIBS) tend to require innovativeness (novelty, creativity and change) in the development of customised and specific applications, processes and solutions. According to Doh et al. (2009), both interactivity with customers and repetition are small for many IT services, although it could be argued that some element of interactivity with clients may be relevant and crucial at some point in the development of the project.

Table 20.1 Export-oriented FDI projects offshored

Call/contact centre services	Shared services centres	IT services centres	Regional headquarters
Help desk	Claims processing	Software development	Headquarters coordination centre
Technical support/advice After-sales	Account processing Transaction processing	Application testing Content development	
Employee inquiries	Query management processing	Engineering and design	
Customer support/advice Market research	Customer administration processing	Product optimisation	
Answering services	HR/payroll processing		
Prospecting	Data processing		
Information services	IT sourcing		
Customer relationship management	Logistics processing Quality assurance		
	Supplier invoices		

Source: UNCTAD (2004), p. 159.

The main problem regarding research on outsourcing and offshoring of KIBS is the lack of data (Sturgeon et al., 2006), especially fine-grained data at the micro level of the projects such as those reported in the UNCTAD report. Available statistics are mainly about international trade and they tend to be highly aggregate, richer for developed countries, but very limited for destination countries, therefore allowing only very crude analysis or only on specific sectors, like IT and computing and business services (e.g. Amiti and Wei, 2005) and normally static analyses. Few

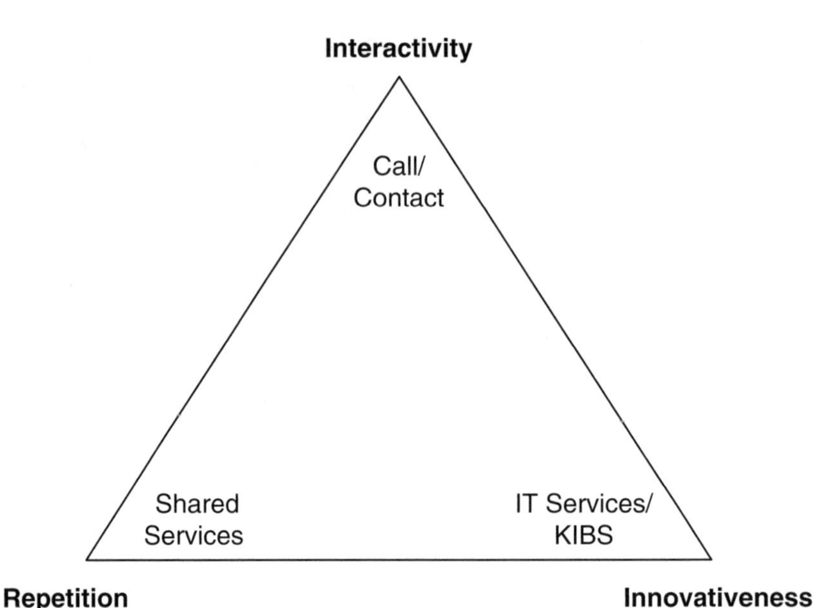

Source: Adapted from Doh et al. (2009).

Figure 20.1 Characteristics of services outsourced/offshored

studies are now collecting firm-level and project-level data (e.g. the ORN project, see Lewin and Peeters, 2006; Lewin et al., 2009; Manning et al., 2008; or the offshoring survey in Denmark, Maskell et al., 2007; Pedersen and Orberg, 2007).

The need for better data on outsourced and offshored services becomes clearer when we look at recent evidence on offshoring projects by different business functions (Figure 20.2). Data from the ORN research project on offshoring of technical and administrative work show that offshoring is not limited to standardised IT or business processes, but has increasingly involved product development activities, such as engineering, R&D and product design, which started as early as IT in the early 1990s and has become the second most frequently offshored business function after IT; and administrative functions such as finance and accounting and human resources (HR) (e.g. payroll); while call centres, which attract the media attention, have been lagging behind more advanced functions (Manning et al., 2008). It is those services based on emerging technological and technical knowledge (Miles, 2001), especially IT and R&D (although IT can be argued to have a more 'operational' nature and R&D a more 'strategic' nature), that are more amenable to offshoring.

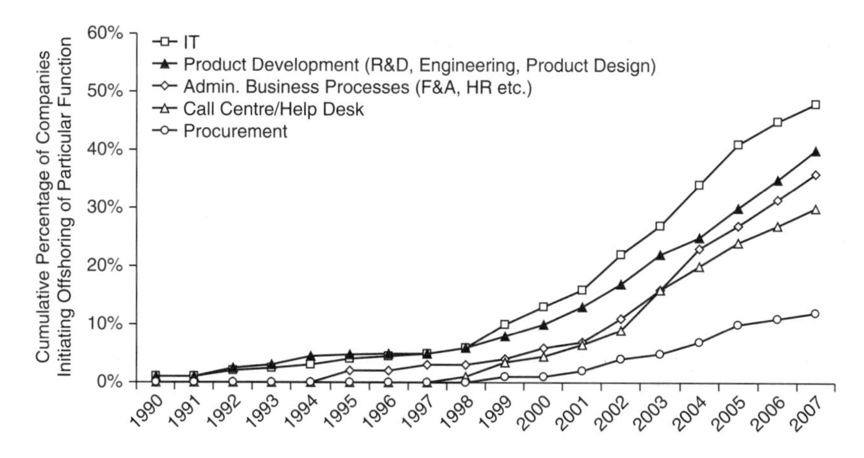

Source: Manning et al. (2008).

Figure 20.2 Offshored projects by different business functions

20.3 What are outsourcing and offshoring?

Outsourcing and offshoring are often used as interchangeable terms, although they indicate distinctive processes which relate respectively to firm and country boundaries and may occur independently or jointly. Outsourcing refers to a company buying products or services from another domestic or offshore company, while offshoring refers to a domestic company obtaining services from a foreign-based company, be that a subsidiary (captive or international in-sourcing) or an independent service provider (offshored outsourcing).

The four alternatives are represented in the matrix in Table 20.2. This matrix shows how economic activities are organised across firms and country boundaries. The top left quadrant corresponds to the ideal-type notion of a domestic firm, with in-house tasks, where no outsourcing or offshoring are undertaken. The top right quadrant relates to outsourcing, that is, when companies no longer undertake some tasks in-house and these are carried out by domestic providers. The bottom left quadrant represents the case when a company moves or expands some of its functions or administrative and technical tasks offshore, as fully owned or captive operations. Finally, in the bottom right quadrant both outsourcing and offshoring take place, that is, when a company's tasks are outsourced offshore to a local or international service provider. We outline below the issues and challenges that have been addressed by the literature on innovation for domestic outsourcing (quadrant 2) and international in-sourcing

Table 20.2 Offshoring/outsourcing matrix

		Outsourcing	
		NO	YES
Offshoring	NO	1) Internal domestic provision	2) Domestic outsourcing
	YES	3) Captive/foreign subsidiary (international in-sourcing)	4) Offshore outsourcing

(quadrant 3), and argue that these issues and challenges will be combined for offshore outsourcing (quadrant 4).

20.3.1 Domestic outsourcing

For the top right quadrant (2), much of the literature on innovation has focused on the outsourcing of R&D, which has increased significantly since the early 1990s, but is not a new phenomenon. As Howells (1999) illustrates, at the turn of the twentieth century in-house R&D laboratories were novel and most firms that needed to undertake research would contract it out to universities or independent research scientists when necessary. Even in sectors with a relatively long scientific tradition, such as the pharmaceutical industry, this was still the most usual method of conducting research up until the First World War. The development of in-house R&D first occurred in the US and in Germany. But effectively only after the interwar period (for the US and Germany) and especially after the Second World War (for the UK and for most other leading European countries) did the large, centralised R&D laboratories emerge. However, the difference between the external R&D of the late nineteenth and early twentieth centuries and the current outsourcing is that in many cases companies are continuing to develop in-house research and technical capacity, which leads to other hybrid forms of collaboration and R&D-related interfirm and interorganisational relations, to continue to grow research and technical competences.

The studies on specific factors that have contributed to the growing use of external research and technical resources by firms, and why firms should seek to collaborate or contract out parts of their R&D, design and engineering activities, tend to be based on the transaction cost model developed by Williamson (1975) which explained why there may be a shift from an in-house vertical and hierarchical provision of such research needs towards a more market-mediated approach. Alternative, or complementary, frameworks are found in the notions of 'strategic assets' (Winter, 1987), 'core competences' (Prahalad and Hamel, 1990) and the 'resource-based view' (Barney, 1991). These approaches emphasise that firms have key assets or competences that have resulted from previous investments and from learning-by-doing. These core competences can be seen as 'resources' as well as capabilities and knowledge sets which are accumulated over the long term, which firms seek to both develop and deploy to gain competitive advantage. Because of the cumulative nature of technological change and learning processes (Dosi et al., 1990), Coombs (1996) suggests that firms may be over-reaching themselves in their desire to decentralise and outsource their R&D portfolios, ultimately weakening their core technological competences.

Howells (1999) distinguishes between 'push' and 'pull' factors for the increasing outsourcing of R&D. Among the push factors are the increasing complexity and fast-changing nature of the research process and the cost and risks of R&D. Products are becoming more sophisticated, creating new, more complex problems and solutions, while at the same time demanding consumers increase pressure for more innovative products. Technologies have become increasingly interdisciplinary and many companies do not have all the necessary scientific resources and competences to develop new products fully in-house. The increasing complexity of the R&D process implies that the costs of research tend to exceed internal financial resources and projects may not be carried out without partners to share cost and risk. The result is an emerging collaborating mode of innovation between organisations with strengths in different fields. Among the 'pull' factors there are the relative gap between internal and external expertise and resources, which firms may not possess or only very inadequately, especially in the field of emerging technologies or new products and processes; when broadening the scope and testing of in-house scientific and technical activities; as a general scanning mechanism of technological opportunities existing outside the company and as a means to network with other organisations and to enable in-house staff to be part of a wider 'invisible college' within the specific research community, is vital for longer-term competitiveness (Howells, 1999). As a result, the boundaries of firms engaged in distributed innovation processes (Coombs et al., 2003) are becoming increasingly undefined and fuzzy.

The literature on innovation in complex products and systems has addressed the consequences for firm innovation of outsourcing (Davies and Brady, 2000; Gann and Salter, 2000; Hobday, 2000; Prencipe, 2000). Prencipe (1997) rejects the simple notion of core competences that recommend the outsourcing of production and, more importantly, the outsourcing of the development of components and subsystems, arguing that decisions based on economic factors alone may compromise the future technological competences of the firm. In his analysis of the technological competences of Rolls-Royce, Prencipe (1997) shows that this firm is vertically integrated regarding the components, the performance of which influences the entire engine performance, such as the jet engine's inner core, and retains design and manufacturing capability and research. Although it contracts out the outer core of the engine, it maintains in-house understanding of the contracted technology to integrate it into the system. Brusoni and Prencipe (2001) argue that outsourcing requires an intense effort of knowledge and organisational coordination. This effort of knowledge and organisational coordination in 'loosely coupled' network structures is played by 'systems integrator' firms, which 'know more than they do' and which may outsource detailed design and manufacturing to suppliers but keep in-house concept design and the ability to coordinate R&D, design and manufacturing by suppliers (Brusoni et al., 2001).

Much of the literature on outsourcing focuses on the externalisation of IT, which has shown remarkable growth since the early 1990s and has been the engine of growth for the software and computer services sector. While the growth of IT outsourcing and expansion of the supply base (with a growing number of specialist IT suppliers) may be easily interpreted as evidence of the separability of IT from internal production activities, numerous studies suggest this is not the case: IT plays a critical role in strategy formulation and implementation (Venkatraman, 1991); IT is important for the coordination of the firm and it is increasingly difficult to distinguish between information and production technologies (Jonscher, 1994); and IT capabilities can be used to transform business structures and processes (Applegate, 1994; Henderson and Venkatraman, 1994), as well as to provide opportunities for increased connectivity, enabling new forms of inter-organisational relations and enhanced network productivity (Scott Morton, 1991). The inseparability of IT from internal production activities means that, even in situations of total outsourcing, a minimum set of capabilities are often retained in-house by the client firm (the so-called 'residual IT organisation') (Willcocks and Fitzgerald, 1994) to maintain the linkages between IT provision and their business prerequisites.

Recognition of the peculiarities of IT as an activity which is integral to the coordination of the firm means that the development of IT outsourcing poses questions regarding the boundaries, coordination and control of the modern firm. A detailed study of IT outsourcing in the UK and Germany (Miozzo and Grimshaw, 2005; Grimshaw and Miozzo, 2006) shows that IT outsourcing, as an example of design modularity of KIBS, can enable economies of scale, production efficiencies and the introduction of new technologies. However, recourse to external KIBS involves not just a simple substitution of internal services but instead a rather more complex process of knowledge transfer that requires reciprocal learning and interaction (Miles, 2003). In most of the IT outsourcing cases examined by Miozzo and Grimshaw (2005), inseparability of production and information technologies meant that IT outsourcing was accompanied by wider transformation in the client's business. As such, the focus of change was towards improved measurement and monitoring of a range of areas of business performance, in line with the metrics developed in the course of developing and running the IT outsourcing contracts. Indeed, management of IT outsourcing relations means that the client is therefore concerned not only with relations with the external supplier, but also with internal strategies of knowledge and organisational coordination and control. These involved two particular practices: the transfer of employees with industry- and firm-specific expertise from the client firm to the supplier; and the creation of an in-house 'retained organisation' to coordinate the link between the IT and business strategies of the client. The client has an incentive to retain in-house capabilities to ensure coordination between IT and business strategy. Moreover, the intangibility of services exacerbates the tensions between client and supplier objectives, bringing into question the sustainability of innovation (Miozzo and Grimshaw, 2005).

The use of outsourcing of R&D and other KIBS such as IT is clearly a means for rationalising limited internal resources, to be able to deploy existing ones to core projects, products and technologies, and at the same time being able to utilise the latest technologies and knowledge that suppliers and partners may have developed or own. These practices, however, demand important efforts of knowledge and organisational coordination by the outsourcing firm.

20.3.2 Captive or foreign subsidiary (international in-sourcing)

The bottom left quadrant (3) relates to the transfer of activities from the home base to a foreign location in the form of a fully owned subsidiary (captive or international in-sourcing). This model is well established in manufacturing (Hymer, 1976; Vernon, 1966; Dunning, 1998) and although it has grown also for R&D-related FDI especially since the 1980s, some scholars posit that it was not all new even then and that back in the 1930s, the largest European and US firms carried out about 7 per cent of their total R&D at locations abroad (Cantwell, 1995), and that since the 1960s this figure has been steadily rising, particularly in technologically intensive industries (Kuemmerle, 1999b). However, in the 1990s, international R&D and patenting activities were still mainly conducted in the home country (Edler et al., 2002; Meyer-Krahmer and Reger, 1999; Patel and Pavitt, 1991), and FDI in R&D occurred primarily between a small number of highly industrialised countries (Florida, 1997; Kuemmerle, 1999a), mainly concentrated within the Triad nations (Archibugi and Iammarino, 2002).

Kuemmerle (1999a) identifies distinct waves of FDI in R&D by country of origin. US companies were pioneer investors in R&D facilities abroad and invested first in Europe, then in Japan, and then in the rest of the world (primarily Canada, Australia and a small number of Asian countries). European companies invested first in other European countries, then in the US and then in Japan, but only to a very limited degree in the rest of the world, whereas Japanese FDI started simultaneously in the US, Europe and in the rest of the world in the early 1980s and rose strongly only in the late 1980s and 1990s, to the US and Europe. Overall, the US was the most attractive location for FDI in R&D, attracting 30 per cent of all R&D sites established abroad.

The international business literature argues that FDI occurs when firms seek to exploit firm-specific capabilities in foreign environments (Dunning, 1998; Hakanson, 1990; Hymer, 1976; Vernon, 1966) and suggests that a high level of local R&D is carried out primarily to adapt products to local markets (Hakanson and Nobel, 1993; Howells, 1990). Traditionally, most FDI into manufacturing and marketing units have fallen in this category. In the case of R&D, these are often called asset-exploiting R&D (Dunning and Narula, 1995) or home-base exploiting (HBE) R&D (Kuemmerle, 1999b). Home-base exploiting R&D is mainly concerned with adapting home-base R&D to local requirements and is likely to be closely connected to and located in proximity of foreign manufacturing and marketing facilities. Several researchers have described the importance of FDI in R&D for exploiting firm-specific capabilities in foreign environments (e.g. Bartlett and Ghoshal, 1990; Hakanson, 1990; Vernon, 1966) and argue that as local demand grows, increasingly sophisticated, local R&D facilities are crucial to adapting existing products to local needs. As firms establish manufacturing facilities abroad and assign increasingly complex products to them, locating R&D sites in close proximity to factories becomes a requisite feature. These sites facilitate the transfer of knowledge and prototypes from the firm's home location to actual manufacturing.

A second driver for FDI in R&D has been identified as the need to augment a firm's knowledge base (Cantwell, 1991; Dunning, 1998; Florida, 1997; Howells, 1990). These efforts are often called asset-augmenting R&D (Dunning and Narula, 1995) or home-base augmenting (HBA) R&D (Kuemmerle, 1999b). Home-base augmenting R&D requires the development of links with host-country R&D organisations and systems to enhance the knowledge base at home and to connect more closely to the foreign R&D environment and access local knowledge (Florida, 1997). Specific nations or regions might be particularly attractive locations for R&D facilities because of potential knowledge spillovers from existing and productive local R&D organisations, such as research universities, publicly funded research institutes and innovative competitors.

Gammeltoft (2006) reviews the literature on R&D internationalisation and identifies six types of motives: (1) market-driven to customise products for the specific market; (2) production-driven, i.e. R&D close to manufacturing facilities (which is similar to Kuemmerle's HBE R&D); (3) technology-driven (pull), i.e. to access and monitor knowledge bases in foreign nations (similar to Kuemmerle's HBA R&D); (4) innovation-driven (push), i.e. to generate new ideas from the foreign environment; (5) cost-driven, namely to access less-expensive R&D resources; and (6) policy-driven, i.e. to satisfy foreign governments that demand local R&D in return for market access (see also Murtha, 1991). This classification summarises the main drivers for internationalisation of R&D; however, the new wave of offshoring of innovative activities and other KIBS seems to be driven by multiple reasons, such as innovation and technology, as well as cost-driven. It would combine categories 3, 4 and 5 in the Gammeltoft's categorisation. Moreover, a dynamic view should be maintained when studying firms' offshoring strategies, as it is plausible that the motives for offshoring evolve over time. For example companies may start offshoring low-skill and routinised work for cost reasons, but then expand their offshore activities to include more advanced and complex activities which relate to technology and innovation (Lewin et al., 2009; Maskell et al., 2007). This may also be the case because of the decreasing supply of domestic scientists and engineers, and the discovery of good pools of skilled and educated workers in emerging economies (Lewin et al., 2009; Manning et al., 2008).

The rapid advances in ICT have greatly enabled the disintermediation and externalisation of innovation processes through outsourcing and remote relocation of R&D groups and laboratories overseas (Howells, 1990, 1995). Moreover, companies seem increasingly to choose offshore locations independently of geographical distance and have located their IT or business process outsourcing and other functions in less-developed countries. Recalling the earlier definition of offshoring, it appears that offshoring strategies are evolving from home-base augmenting to what have been defined as home-base replacing (HBR) innovation capabilities (Lewin et al., 2009). Lewin et al. (2009) argue that this seems to be the case especially for larger multinational enterprises (MNEs), whose strategies have been extensively discussed in the international business literature, whereas small and medium-sized enterprises (SMEs), in general neglected by the mainstream international business literature, seem to be adopting innovation offshoring strategies in order to augment their limited innovation capabilities (HBA).

From the discussion above, three features seem to be central to differentiate the past wave of internationalisation of R&D and the current offshoring of KIBS. First, this phase involves the participation of SMEs and less-internationalised companies previously unable to compete with larger MNEs. Second, offshoring now involves the relocation of activities to countries (rather than the Triad). Third, activities offshored are not aimed at serving local less developed markets but are part of domestic operations.

20.3.3 Offshore outsourcing

The bottom right quadrant (4) in the table refers to offshore outsourcing. This process has not received attention from the innovation literature. We suggest that this practice combines the benefits and problems of both pure outsourcing (the make or buy decision) and fully owned activities offshore (managing and coordinating activities across national boundaries), and faces new ones which are particularly relevant in the case of KIBS (knowledge transfer, outsourcing of non-core projects and functions, while maintaining enough internal knowledge and absorptive capacity). Lower value-adding tasks within the innovation process can be outsourced to third-party providers offshore, in low-cost countries, with clear cost and, sometimes, time advantages. Tasks can be discussed and commissioned during working hours in Europe and the US, be worked on during night time in Europe and the US, which is daytime in East Asia and India, and be ready for the next working day in Europe and the US. However for knowledge-intensive tasks there may be the need for more frequent face-to-face interactions to facilitate tacit knowledge flows and creation of new knowledge resulting from the combination and recombination of existing knowledge held by people residing in remote locations (e.g. Kogut and Zander, 1993; Minbaeva et al., 2003). This brings some limits to the extent to which an innovation process can be effectively modularised and divided into sub-problems and tasks, and may indicate that companies may find it easier to outsource offshore entire research processes for non-core new products than develop them in-house, domestically.

Other challenges resulting from (re)locating knowledge-intensive activities offshore concern the knowledge transfer, coordination and motivation issues as well as practices used to resolve them. Successful offshore activities would need to replicate the business process knowledge developed and deployed domestically. This is a clear challenge for captive operations but it is even greater in the case of outsourcing offshore, when the vendor is required to replicate the business process knowledge of the client and provide a seamless integration between stages and functions in the client operations. Thus the ability of the vendor to absorb knowledge needed to execute the client processes, and the ability of the client to manage the vendor remotely, are two central capabilities that define offshoring relationships. Success therefore depends upon the joint competence of vendors and clients rather than on client firms alone – and the process and capability in the selection of the right vendors.

One practice increasingly used in outsourcing offshore is the development of master service agreements (see also Miozzo and Grimshaw, 2008). Contract design is of course central to transaction cost economics, and it is argued that all contracts are inherently incomplete (Grossman and Hart, 1986; Williamson, 1996), due to incomplete and asymmetric information, and difficulties in foreseeing all the possible future contingencies under which a contractual hazard can emerge because of bounded rationality characterising economic actors (Simon, 1957). However, as companies increase the scale and scope of operations, they develop expertise in contract design and learning, and contracts might serve as repositories for knowledge about how to govern collaborations (Mayer and Argyres, 2004); while offshorers that repeatedly outsource to service providers can develop learning and contract design capabilities on how much and what kinds of detail to include in a contract (Argyres and Mayer, 2007).

At the same time service providers can also develop capabilities to relate to clients as they repeatedly interact over time and on a variety of projects, to improve cost-efficiency and project execution (Ethiraj et al., 2004). Client-specific and other infrastructure-building investments can also enhance vendors' capabilities which in turn contribute to build reputation, and to mature project management capabilities to compete with larger international service providers. The next section outlines the emergent issues for innovation studies of offshore outsourcing.

20.4 Emergent issues for innovation studies of the new phase of outsourcing and offshoring

This section explores the challenges of innovation posed by the new phase of outsourcing and offshoring, especially the challenges to the theory of the firm, the implications for integration of decentralised knowledge, the role of ICTs, the opportunities for small and large firms, and the challenges to trade and investment theories. Table 20.3 summarises the differences in the decisions, nature of investment, suppliers, clusters and locations of the previous wave of (mainly manufacturing) ourtsourcing and offshoring, and the new wave of KIBS outsourcing and offshoring, that are developed in this section.

20.4.1 Challenges to the theory (and boundaries) of the firm

The recent phase of outsourcing and offshoring is part of the general process of corporate restructuring and vertical disintegration that has

Table 20.3 Features of previous and present wave of outsourcing/offshoring

	Previous wave of (mainly) manufacturing outsourcing/offshoring	Present wave of KIBS outsourcing/offshoring
Outsourcing decision	Operational decision, made at the factory or divisional level Typically made by large MNEs	Board decision, related to firm restructuring Made by large MNEs and SMEs
Nature of suppliers	Monopsonistic power of clients Supplier operates in the same sector as client Emergence of new intermediaries, fuzzy boundaries	Suppliers do not face competition from clients Know-how that can be applied to clients in a broad range of sectors Development of large global suppliers (IBM, EDS, Accenture) and evolution of small local providers into large offshoring service providers (TCS, Wipro)
Role of ICTs	Enabling role	Enabling role Organisational function that can be outsourced/offshored Coordination tool
Nature of investment	To exploit domestic capabilities and serve local markets Co-location of manufacturing and marketing To augment firm's innovation capabilities	To serve home market or global operations of the firm Coordination of globally and interfirm dispersed activities and knowledge
Clusters	Industry based Competition in the supply chain	Technology/knowledge/function based Hybrid organisations and complex networks
Location	Triad and sequential internationalisation to Asia and Latin America in the 1960s	Rapid relocation of existing of activities to China and India in the 1990s and in East Asia and Eastern Europe in the 2000s Important set of activities remain rooted in advanced economies

received attention in the theory of the firm (and its boundaries). There are different views on the consequences of vertical disintegration for the typical Chandlerian firm. On the one hand, Langlois (2003) argues that vertical disintegration challenges the extension of Chandler's (1977, 1990) managerial revolution into the late twentieth century, when vertical disintegration began replacing the multidivisional firm. For Langlois, vertical disintegration ('the vanishing hand') is a further continuation of the Smithian process of division of labour in which Chandler's managerial revolution, with its organisational capabilities to manage scale and scope, was a way-station (Langlois, 2003). Langlois (2003) interprets the changes since the 1980s as a result of changes in coordination technology and the extent of markets, which led to the specialisation of function with generalisation of capabilities, and the hiving off not only of unrelated divisions but also vertically related divisions. These arguments, however, refer mostly to outsourcing of manufacturing tasks.

On the other hand, Pavitt (2003) argues that the recent trend for disintegration of product design and manufacture, made possible by advances in modularity (Baldwin and Clark, 2000; Sturgeon, 2002) and ICTs are, instead, a prolongation of the Chandlerian firm, that is able to exploit economies of scale, speed and scope through 'systems integration'. This division of labour will be incomplete, as large specialised firms will need to maintain and develop technological competences beyond what they make themselves (see Brusoni et al., 2001, above), and associated manufacture will remain an important resource. Pavitt (2003) argues that as large firms outsource manufacture they cannot be regarded as 'services' firms (and we are not moving to a 'post-industrial' stage); instead, these firms are focusing on the knowledge-intensive element of industrial activity.

These two contributions, however, do not examine explicitly the outsourcing and offshoring of KIBS. Sako and Tierney (2005) were among the first to stress the difference between outsourcing of manufacturing and outsourcing of KIBS. They draw a distinction between traditional vertical disintegration (vertical disintegration of the production of inputs that go into a firm's final products or services) and the unbundling of corporate functions and recentralisation of these services in an outside supplier of services. They argue that while the former is an operational decision made at the factory or divisional level, the latter is associated with the rise of shareholder value, as decisions are made by corporate executives and chief finance officers at the corporate headquarters (as bonuses are linked to cost saving and return on assets, and large outsourcing deals include the sale of a shared service centre; see Adler, 2003). The latter occurs when corporations are undergoing restructuring of corporate functions. There are different implications from these two different types of vertical disintegration. In traditional vertical disintegration, suppliers expand through the exploitation of economies of scale across a broad customer base; this expansion is based on developing new capabilities to take on higher-value-added processes in the supply chain, entering into competition with the client firm's business (which remain powerful due to monopsonistic power vis-à-vis suppliers). In outsourcing involving corporate function unbundling, instead, corporate functions in human resources, finance and accounting, procurement and logistics may be downsized and sold to service suppliers that have developed specific know-how that can be applied to clients in a broad range of industrial sectors. Suppliers expand through the exploitation of economies of scale and cross-sectoral learning without facing direct competition with clients in the same market (which cannot exercise monopsonistic power over them).

The outsourcing of KIBS is regarded by Sako and Tierney (2005) not as vertical disintegration in its conventional sense, but as the rebundling of an administrative function or process, to be carried out by a third party. As argued above, it is a decision taken by the company board, affecting the administrative support structure of professional managers, rather than the previous wave of outsourcing that affected mainly blue-collar workers. Whatever the extent of dissection of the value chain for outsourcing and offshoring, one thing is clear: that, as we will see below, this disintegration of activities that are no longer within the firm but still part of the overall production process requires efforts to coordinate and integrate knowledge.

20.4.2 Implications for coordination and integration of decentralised knowledge: the challenge of managing science and engineering capabilities worldwide

There is an emerging belief that there have been a series of organisational innovations since the 1980s, bringing a change from the central R&D lab which was key to the development of large high-tech firms after the Second World War, to an increasingly distributed mode (Coombs and Richards, 1993; Christensen, 2002), or fifth generation model of innovation (Rothwell, 1992), or open innovation (Chesborough, 2003). This includes the downsizing of central labs and the delegation of responsibility for technical innovations to product divisions and subsidiaries and, in many cases, the internationalisation of R&D (Kuemmerle, 1999a; Gerybadze and Reger, 1999). Innovation processes are becoming more open and increasingly distributed across firms, suppliers, customers, research organisations and other institutions, by disintegrating and externalising the innovation process through outsourcing and remotely dispersed R&D groups (Quinn, 2000; Howells, 1995), recombining knowledge (Kogut and Zander, 1992), and utilising knowledge management practices, knowledge and boundary spanners, and reverse knowledge transfer (Lewin et al., 2009). All together these trends indicate that we are not just revisiting the external R&D labs typical of the pre-war periods, but we are witnessing a new trend of hybrid organisations and complex networks emerging and developing.

The international dimension in the new wave of outsourcing and offshoring of parts of the value chain and business processes to highly specialised external suppliers creates new challenges to develop strategies and policies for managing knowledge interfaces and transferring and reconnecting knowledge across spatial and organisational boundaries (Lewin et al., 2009). Increased emphasis needs to be placed on the management of knowledge, both to identify sources of external knowledge and to link that knowledge with internal knowledge. Firms still need to develop core competences that are unique, complex and difficult to imitate. Christensen (2006) suggests that large firms have expanded the diversification of their technological profiles, placing greater emphasis on the development of 'background competencies' or absorptive capacity (Cohen and Levinthal, 1990) to explore new opportunities emerging from scientific and technological breakthroughs outside the firm. Also, as large firms take the role of coordinators of increasingly distributed innovation chains, they need to develop system integration competences involving experience-based and firm-specific architectural knowledge. An important aspect of this process is that firms outsource and offshore tasks and functions without eroding underlying knowledge (Brusoni et al., 2001) and weakening their core competences (Coombs, 1996).

Manning et al. (2008) argue that the evolution of offshoring should be seen as a co-evolutionary process (e.g. Volberda and Lewin, 2003) where demand and supply or push and pull factors interplay and affect one another to result in emerging dynamics. One outcome of this co-evolutionary process is the emergence of technology and knowledge clusters for firms in different sectors. Clusters of service providers are developing in less developed countries. Local and international service providers tend to be located in concentrated areas, in a similar way to the industrial clusters in Western economies, such as Silicon Valley for IT companies or Southern Denmark for biotechnology companies. However, these new geographic clusters tend to develop around particular functions or upstream services rather than industries, as knowledge-based clusters that provide services to clients across industries (Manning et al., 2008), and fuel diffusion of knowledge and practices perhaps originated initially as industry-specific to cross different industrial sectors.

The interest in the geography of international business activities and especially in the role of clusters has grown remarkably since the 1990s, although the importance of industrial clusters, or districts, was first noted by Alfred Marshall (1920) who identified three fundamental elements: clusters of subcontractors, readily available skilled talent, and a knowledge base shared by a local community of firms and people. These elements are also present in 'hot spots' and emerging second- and third-tier cities in India, China and other low-cost countries which are destinations for offshoring of services. Clustering provides synergistic agglomeration effects where traded and untraded dependencies can assist in economic growth and industrial upgrading (Porter, 1990; Storper, 1997). There are many cities in India and other nations that have attracted operations and developed cluster dynamics, such as Bangalore, New Delhi and Mumbai. Bangalore is the exceptional and well-known case, as it is the centre of the Indian IT services industry, and experienced a rapid expansion of not only multinationals' R&D operations, but also Indian providers and entrepreneurial start-ups (Athreye, 2005; Zaheer et al., 2009). In terms of financial services offshoring, Mumbai has become an important destination, though Grote and Taube (2006) concluded that much of the relocated work has been routinised and not, in general, higher-value-added research activities.

20.4.3 ICTs as facilitators of outsourcing and offshoring

The developments in ICTs are playing a crucial role in the evolution of offshoring. This is not only as enablers, in the form of more efficient and cheaper communication. ICTs facilitate outsourcing and offshoring of services as they enable automation of some tasks requiring rules-based logic, e.g. through interactive voice recognition; they create standardised work processes and machine-paced operations through automated call distribution systems, and routinisation of work through the use of scripts that reduce operational risks and enable electronic monitoring. These tasks can then be more easily and cheaply transported between producers and consumers located remotely (see Bardhan and Kroll, 2003; Levy and Murnane, 2004; Batt et al., 2006).

Also, however, the new phase of outsourcing and offshoring is facilitated by ICTs because they are functions and activities which are themselves offshored, and because the information systems and other web-based collaborative technologies which are developed in offshore locations (e.g. system analysis and program development – SAP; business services management – BSM) come to help cope with the managerial challenges of coordinating globally dispersed high-value activities. Lehrer (2006) shows how ERP depends on standardised and codified processes. For example, web-based collaborative technologies, such as Electronic Notebook Systems (e.g. at BMS), have been revealed as efficient and powerful tools for tracking on a daily basis the work product of scientists at remote locations. These have been adopted very effectively by only a few organisations, but they could soon be implemented on a wide scale not only for R&D but also for project and administrative applications.

20.4.4 Opportunities for large firms or small entrepreneurial ventures?
Outsourcing and offshoring have created new opportunities for 'global supplier' firms. This trend is more visible in manufacturing, with the emergence of large global contract manufacturers in electronics and in the auto industry (see Sturgeon, 2002). In manufacturing, the consolidation and geographic expansion of global suppliers providing key intermediate products and subassemblies has been dramatic, with the top contract manufacturers such as Solectron, Flextronics, Sanmina/SCI, Celestica and Jabil Circuit expanding largely as the result of acquisition of outsourced manufacturing plants from electronic firms such as IBM, Hewlett-Packard and Lucent. Similarly, global first-tier suppliers of auto-parts, such as Bosch, Johnson Controls and Lear, have consolidated their international presence. This generates both opportunities and challenges for local producers. On the one hand, this can lead to the upgrading of local producers, such as the co-evolution of electronics contract manufacturing in Taiwan and the US, through which leading firms in the computer industry, such as Hewlett-Packard, Dell and Apple have relied on Taiwanese contract manufactures, leading to the development and increased reputation and brand value of the Taiwanese firm Acer. On the other hand, there is the fear that large suppliers tend to concentrate good jobs in relatively few locations and may crowd out local producers (Gereffi, 2005).

A number of MNEs (e.g. General Electric, American Express and Citigroup) have pioneered the relocation of back-office operations to countries such as India. As MNEs shift a variety of back-office service functions to offshore locations, either by establishing their own offshore affiliates or by outsourcing to foreign contractors and local contractors with foreign operations, we observe changes in the international division of labour. Operating an MNE captive in a different environment requires management talent and organisational coordination (Dossani and Kenney, 2004).

The Indian operations of MNEs are also climbing the value ladder. For example, Hewlett-Packard GlobalSoft, which is a software development and IT services division, is headquartered in Bangalore. Bangalore has profit/loss responsibility for global operations, including Eastern Europe and Mexico. HP Global eBusiness Operations, the business process outsourcing division providing financial and other services, with approximately 6000 employees worldwide, also is headquartered in India. SAP Labs India employs over 3000 persons and is the largest SAP laboratory outside of Germany, with a leadership role for the development of certain software products and services. Adobe India has global responsibility for PageMaker and Framemaker software upgrades. Cisco has recently established Cisco Center East in Bangalore, India, as the new Cisco corporate headquarters for Cisco-wide innovation, under the leadership of the first Chief Globalisation Officer of the company (first ever for a US company). Its former San Jose US headquarters is now referred to as Cisco Center West, reflecting its new role in the corporation. These anecdotes indicate that at certain MNEs, their Indian operations have matured sufficiently to receive global mandates, indicating their enhanced and developing capabilities. Created as cost-saving operations, they are now acquiring strategic roles.

Even as MNEs dramatically expand their operations in India, the number and size of Indian firms exporting services is expanding rapidly. In the early 1990s, Indian firms were largely confined to low-level coding and programming and other simple processes (Dossani, 2006). This is evolving as the largest Indian system integrators (SIs) are growing fast and increasingly competing for larger and more sophisticated projects. Among the three largest Indian SIs, in 2000, the largest Indian service provider, TCS, had 17,000 employees and Infosys and Wipro had approximately 10,000 each; six years later, in 2006, TCS employed 78,000 persons globally, Infosys had grown to 66,000 and Wipro had 61,000. Though the Indian SIs are still smaller than IBM with its global employment of approximately 330,000 or Accenture with 140,000 employees, they are expanding not only in India, but also in advanced countries (Western Europe and the US) and other low-cost countries in Eastern Europe, and becoming serious competitors to services providers from these countries. These Indian firms, however, are also facing the same challenges as Western offshorers as they move up the value chain, including high attrition rates.

Similarly to earlier outsourcing and offshoring of manufacturing, we are witnessing the growth and consolidation of large MNE suppliers in services outsourcing and offshoring. Miozzo and Grimshaw (2008) describe the capabilities developed by large IT services suppliers IBM and EDS to create and appropriate value in the IT outsourcing market to serve a number of large clients internationally. These firms have expanded through the development of unique organisational capabilities involving three key features: investment in complementary assets; the development of organisational routines in the form of company-wide processes; and staff transfer from client firms as a means to acquire skills in the IT outsourcing market. The challenge for these firms is to combine corporate processes with tacit and codified knowledge specific to the client firm's business. An important strategic response is the implementation of a phased organisational learning model that brings increased productivity and efficiency in the provision of an increasing number of projects (Miozzo and Grimshaw, 2008.

Large MNE outsourcers with an international presence have also developed in the area of payroll and accounting, such as ADP; in call centre and customer relationship management, such as Convergys, Sitel and Sykes; and large consultancy firms such as Accenture. These large MNE outsourcers and offshorers are in competition with large global outsourcing firms from less-developed countries, such as HCL, Infosys, Satyam, TCS and Wipro, and other local firms in the IT services sector.

It is said that, in manufacturing, SMEs have disadvantages due to their small scale and limited resources for innovation, but advantages coming from lower hierarchical structures and higher flexibility in making decisions. In the case of outsourcing and offshoring of KIBS, flexibility may result in an advantage in deciding to offshore. In the case of outsourcing, probably the scale of operations may be too small to reach substantial cost savings; but offshoring may provide opportunities for start-ups to locate some activities and functions in low-cost countries from the beginning (Manning et al., 2008). More recently, small entrepreneurial firms are increasingly offshoring new product development and other knowledge-intensive activities because these allow them to grow more and faster, increase speed to market, or simply as a last chance for survival (Dossani and Kenney, 2007) especially in knowledge-driven industries (Murtha, 2004). Lewin et al. (2009) find that smaller firms have higher probability of offshoring innovation projects, indicating that offshoring enables smaller and more agile companies to augment their innovation capabilities (HBA) in contrast to larger more resourceful companies that are also using offshoring strategies to replace innovation capabilities (HBR). Asia in particular is playing a central role in the growing global innovation networks, as indicated, e.g., by the growth in US patents granted to companies in Asia (South Korea, Singapore, China, Taiwan and India) between 1986 and 2003 (Ernst, 2002, 2006). However, Hirshfeld and Schmid (2005) argue that, although firms in the US and Europe are increasingly attracted to and are exploring new science and engineering clusters in emerging countries, advanced economies are likely to remain at the forefront of innovation activities, at least in the foreseeable future (Manning et al., 2008).

20.4.5 Challenges to the relations between trade/investment theories and technology

As manifested by the reaction to Mankiw's controversial claim that offshoring is no more than an extension of typical trade (Blinder, 2006), the opinions of economists and other social scientists regarding the 'threat' of offshoring are divided. Some argue that offshoring, like other forms of trade, creates value for firms and economies. For example, Farrell (2005) from McKinsey Global Institute argues that:

> A call centre in Bangalore, for instance, might be filled with Dell computers, Siemens telephones, HP printers, and Microsoft software. We estimate that for every dollar of US corporate spending that moves to India, US exports to India increase by an additional five cents. This partly explains why exports from the US to India grew from $3.7 billion in 2000 to $5 billion in 2003. (Farrell, 2005, 677)

Others, however, argue that offshoring is altogether different. Levy (2005) argues that although there is evidence of income growth in some areas of China and India, offshoring shifts the relations of power rather than bringing efficiencies. Offshoring leads to wealth creation for shareholders, but not necessarily for countries or employees, as it brings a growth in the corporate capacity to manage dispersed networks, with the implication that the core of these clusters will become less 'sticky' and increasingly devoid of workers (decoupling value creation and geographic location). This is also argued by Gereffi (2005) who posits that global value chains have created labour markets that respond to the demands for jobs in production, design, marketing, logistics and finance that now cut across industries. And, indeed, because service occupations are more widely distributed through the economy, negative effects could be more broadly based (in developed countries) than the effects of the offshoring manufacturing (Bardhan and Kroll, 2003).

Scholars working on international trade attempt to make progress in the understanding of the outsourcing and offshoring of services using existing theories, especially the eclectic theory (Dunning, 1998), which relies on the ownership, location and internalisation (OLI) paradigm. Markusen (2006) suggests outsourcing is a 'mode' choice (in transaction cost theory), related to the internalisation or make-or-buy decision about the boundaries of the firm, and offshoring is a 'location' choice, claiming that we can understand offshoring of white-collar work at the theory level from the existing portfolio of models. Other scholars are more doubtful about the usefulness of existing theories. Doh (2005) argues that offshoring both reaffirms and challenges the OLI framework, with location being an important variable but ownership and internalisation advantages being less relevant.

Recent developments in outsourcing and offshoring challenge established views on the relation of technology to market structure, including firm entry, product differentiation and standardisation involved in the product and industry cycle (Vernon, 1966; Klepper, 1997). According to these theories (developed to explain the internationalisation of manufacturing), as sectors reach maturity and products become standardised, concern over production costs begins to replace concern over product characteristics, and the location of production shifts to less-developed countries, which would then export to advanced countries. New sectors would develop in advanced countries, based on their domestic innovation. However, the sectors that are relocated in the new wave of outsourcing and offshoring are not mature (as expected in the product and industry cycle), but instead include a variety of sectors at different stage of industry evolution. Also, the destruction of industries and jobs in advanced countries is faster than the creation of new ones. At the early stages of the development of high-tech sectors, some of the more labour-intensive and less-skilled jobs are moving to less-developed countries. Soon afterwards, product development and R&D move offshore. Contrary to slow, sequential internationalisation of manufacturing (Levy, 2005), the low capital intensity and purely electronic form of services delivery means that services offshoring can grow faster than has been the case with manufacturing (Dossani and Kenney, 2004). Vernon (1966) emphasised the importance of local demand as a catalyst for export abroad, but offshoring is to serve home rather than host markets. It is argued that offshoring, if unrelated to domestic demand, may exacerbate the reliance of less-developed countries on the capital and resources of industrialised countries, and may make them more vulnerable to the vagaries of MNEs that may choose to shift production from developed to less-developed countries and from one less-developed country to another (Doh, 2005; Miozzo and Grimshaw, 2008).

20.5 Some conclusions: is outsourcing and offshoring of KIBS the next step in the evolution of the globalisation of innovation?

The developments in the outsourcing and offshoring of KIBS discussed in the present chapter raise the question as to whether the offshoring of KIBS is an important step towards the globalisation of innovation and the end of the knowledge-based competitive advantage of developed countries. It is important to clarify that globalisation of R&D is different from the new trend of outsourcing and offshoring R&D and innovation processes. As discussed earlier, past waves of internationalisation and globalisation of R&D seemed to be motivated by exploiting domestic capabilities in the foreign market, and were carried out in support of local manufacturing and marketing activities, or for accessing foreign resources, laboratories, connection with local universities and networks, to enhance existing internal and domestic knowledge and capabilities, often with captive operations and possibly to serve local or regional markets. The more recent trend of offshoring R&D, and KIBS in general, is more varied: it includes captive and outsourced operations; it tends to involve peripheral, routinised activities of the innovation process currently undertaken domestically, replacing existing domestic activities often originated by cost-saving motives, although companies are discovering pools of skilled workers in such less developed countries and are increasing the scale and scope of their offshoring operations.

Outsourcing and offshoring of KIBS is simpler than manufacturing outsourcing in terms of resources, space and equipment requirements and may therefore proceed much more quickly. Services outsourcing affects overwhelmingly white-collar middle-class occupations and jobs, unlike manufacturing outsourcing, which impacted primarily upon blue-collar workers. Also, a different set of countries are in contention for these jobs. Services offshoring has attracted the attention of less-developed countries' policy-makers and analysts. On the one hand, they are interested as host countries to MNEs that are looking for new locations to lower costs, skilled human resources, diversification of risks and possibilities to export services. On the other hand, domestic firms in less-developed countries are looking for opportunities to offshore to industrialised countries. While some claim that offshoring is beneficial for less-developed countries, others argue that it leads to 'immiseration growth' with less-developed countries competing with one another to offer lower operating costs (Kaplinsky, 2000).

Despite the emergence of China and India as important hubs of activity, outsourcing and offshoring has not meant a wholesale transfer of economic activity out of developed countries and into less-developed countries. It is argued that an important set of activities remain rooted in advanced economies, even as they have become tightly interlinked to activities located elsewhere. Firms do not just export products and services, but participate in complex cross-border networks of partners, customers and suppliers (Gereffi and Kaplinsky, 2001; Gereffi et al., 2005). Although most of the concerns about offshoring have been directed to the developed countries' jobs and wages, as argued above, when offshoring is unrelated to domestic demand, it may exacerbate the reliance of less-developed countries on the capital and resources of industrialised countries, and may make them more vulnerable to the vagaries of multinationals (Doh, 2005).

The transfer of product design to less-developed countries requires training in skills. Some selected emerging economies have begun to implement national policies and tax incentives designed to 'reverse' the brain drain, and evolve their infrastructures and institutions partly based on, partly deviating from advanced countries' models, in order to continue to attract an ever-increasing number of foreign operations, resulting in virtuous cycles which will make the destinations even more attractive.

At the same time, however, it appears that research and innovation policies in the home countries of offshoring companies have not kept up with the latest global developments and seem to be struggling to counteract the relocation of high-end innovation activities. The interrelation between developed and less-developed countries is becoming tighter and the interdependencies of education, business and innovation systems ever closer. Policy-makers in both developed and less-developed economies need to become more aware of these interdependencies, not least in order to anticipate better the effects and consequences of their policy decisions. Moreover managers in developed countries need to become more involved in the discussions and formulation of national policies affecting technology policies and other policy discussions and interventions at the international levels, which may affect the outcome of their offshoring strategies and plans.

References

Adler, P. (2003), 'Making the HR outsourcing decision', *Sloan Management Review*, **45** (1), 53–60.

Amiti, M. and S.-J. Wei (2005), 'Fear of service outsourcing: is it justified?', *Economic Policy*, **20**, 308–347.

Applegate, L. (1994), 'Managing in an information age: transforming the organisation for the 1990s', in R. Baskerville, S. Smithson, O. Ngwenyama and J. DeGross (eds), *Transforming Organisations with Information Technology*, Amsterdam: North-Holland, pp. 15–94.

Archibugi, D. and S. Iammarino (2002), 'The globalisation of technological innovation: definition and evidence', *Review of International Political Economy*, **9** (1), 98–122.

Argyres, N. and K.J. Mayer (2007), 'Contract design as a firm capability: an integration of learning and transaction cost perspectives', *Academy of Management Review*, **32** (4), 1060–1077.

Athreye, S. (2005), 'The Indian software industry and its evolving service capability', *Industrial and Corporate Change*, **14** (3), 393–418.

Baldwin, C.Y. and K.B. Clark (2000), *Design Rules: The Power of Modularity*, Cambridge, MA: MIT Press.

Bardhan, A. and C. Kroll (2003), 'The new wave of outsourcing', Research Report Fisher Center for Real Estate and Urban Economics Paper 1103, Berkeley, CA: University of California.

Barney, J.B. (1991), 'Firm resources and sustained competitive advantage', *Journal of Management*, **17** (1), 99–120.

Bartlett, C. A. and S. Ghoshal (1990), 'Managing innovation in the transnational corporation', in C.A. Bartlett, Y. Dox and G. Hedlund (eds), *Managing the Global Firm*, New York, USA and London, UK: Routledge, pp. 215–255.

Batt, R., V. Doellgast and H. Kwon (2006), 'Service management and employment systems in US and Indian call centers', in S.M. Collins and L. Brainard (eds), *Brookings Trade Forum 2005: Offshoring White-Collar Work*, Washington, DC: Brookings Institution, pp. 335–372.

Blinder, A.S. (2006), 'Offshoring: the next industrial revolution?' *Foreign Affairs*, **85** (2), 113–128.

Brusoni, S. and A. Prencipe (2001), 'Unpacking the black box of modularity', *Industrial and Corporate Change*, **10** (1), 179–205.

Brusoni, S., A. Prencipe and K. Pavitt (2001), 'Knowledge specialisation, organisational coupling, and the boundaries of the firm: why firms know more than they make?' *Administrative Science Quarterly*, **46**, 597–621.

Cantwell, J. (1991), 'The international agglomeration of R&D', in M. Casson (ed.), *Global Research Strategy and International Competitiveness*, Oxford: Blackwell, pp. 104–132.
Cantwell, J. (1995), 'The globalisation of technology: what remains of the product cycle model', *Cambridge Journal of Economics*, **19** (1), 155–174.
Chandler, A.D. Jr (1977), *The Visible Hand*, Cambridge, MA: Harvard University Press.
Chandler, A.D. Jr (1990), *Scale and Scope: Dynamics of Industrial Capitalism*, Cambridge, MA: Harvard University Press.
Chesbrough, H. (2003), *Open Innovation: The New Imperative for Creating and Profiting from Technology*, Boston, MA: Harvard Business School Press.
Christensen, J.F. (2002), 'Corporate strategy and the management of innovation and technology', *Industrial and Corporate Change*, **11** (2), 262–288.
Christensen, C. M. (2006), 'The ongoing process of building a theory of disruption', *Journal of Product Innovation Management*, **23** (1), 39–55.
Cohen, W.M. and D.A. Levinthal (1990), 'Absorptive capacity: a new perspective on learning and innovation', *Administrative Science Quarterly*, **35** (1), 28–152.
Coombs, R. (1996), 'Core competences and the strategic management of R&D', *R&D Management*, **26** (4), 345–355.
Coombs, R., M. Harvey and B. Tether (2003), 'Analysing distributed processes of provision and innovation', *Industrial and Corporate Change*, **12** (6), 1125–1155.
Coombs, R. and A. Richards (1993), 'Strategic control of technology in diversified companies with decentralised R&D', *Technology Analysis and Strategic Management*, **5** (4), 385–396.
Davies, A. and T. Brady (2000), 'Organisational capabilities and learning in complex product systems: towards repeatable solutions', *Research Policy*, **29** (7–8), 931–953.
Doh, J. (2005), 'Offshore outsourcing: implications for international business and strategic management theory and practice', *Journal of Management Studies*, **42** (3), 695–704.
Doh, J.P., K. Bunyaratavej and E.D. Hahn (2009), 'Separable but not equal: the location determinants of discrete services offshoring activities', *Journal of International Busienss Studies*, **40** (6), 926–943.
Dosi, G., K. Pavitt and L. Soete (1990), *The Economics of Technological Change and International Trade*, Hemel Hempstead: Harvester Wheatsheaf.
Dossani, R. (2006), 'Globalization and the offshoring of services: the case of India', in S. Collins and L. Brainard (eds), *Offshoring White-Collar Work*, Washington, DC: Brookings Institution, pp. 241–267.
Dossani, R. and M. Kenney (2004), 'The next wave of globalization? exploring the relocation of service provision to India', Berkeley Roundtable on the International Economy Working Paper 156, September, Berkeley, CA: University of California.
Dossani, R. and M. Kenney (2007), 'The next wave of globalization: relocating service provision to India', *World Development*, **35** (5), 772-791.
Dunning, J.H. (1998), *Multinational Enterprises and the Global Economy*, 3rd edn, Wokingham: Addison-Wesley.
Dunning, J.H. and R. Narula (1995), 'The R&D activities of foreign firms in the United States', *International Studies of Management and Organization*, **25** (25), 39–73.
Edler, J., F. Meyer-Krahmer and G. Reger (2002), 'Changes in the strategic management of technology: results of a global benchmarking study', *R&D Management*, **32** (2), 149–164.
Ernst, D. (2002), 'Global production networks and the changing geography of innovation systems: implications for developing countries', *Economics of Innovation and New Technology*, **11** (6), 497–523.
Ernst, D. (2006), 'Complexity and internationalization of innovation: why is chip design moving to Asia?', *International Journal of Innovation Management*, **9** (1), 47–73.
Ethiraj, S K., K. Prashant, M.S. Krishnan and J.V. Singh (2004), 'Where do capabilities come from and how do they matter? A study in the software services industry', *Strategic Management Journal*, **26** (1), 25–45.
Evangelista, R. (2000), 'Sectoral patterns of technological change in services', *Economics of Innovation and New Technology*, **9**, 183–221.

Farrell, D. (2005), 'Offshoring: value creation through economic change', *Journal of Management Studies*, **42** (3), 675–683.
Florida, R. (1997), 'The globalization of R&D: results of a survey of foreign-affiliated R&D laboratories in the USA', *Research Policy*, **26**, 85–103.
Gammeltoft, P. (2006), 'Internationalisation of R&D: trends, drivers and managerial challenges', *International Journal of Technology and Globalization*, **2** (1–2), 177–199.
Gann, D.M. and A.J. Salter (2000), 'Innovation in project-based, service-enhanced firms: the construction of complex products and systems', *Research Policy*, **29**, 955–972.
Gereffi, G. (2005), 'The new offshoring of jobs and global development: an overview of the contemporary global labour market', ILO 7th Nobel Peace Prize Social Policy Lectures, Kingston, Jamaica, December 5-7.
Gereffi, G. and R. Kaplinsky (2001), 'The value of value chains: spreading the gains from globalisation', Special Issue of the *IDS Bulletin*, **32** (3).
Gereffi, G., J. Humphrey and T. Sturgeon (2005), 'The governance of global value chains', *Review of International Political Economy*, **12** (1), 78–104.
Gerybadze, A. and G. Reger (1999), 'Globalization of R&D: recent changes in the management of innovation in transnational corporations', *Research Policy*, **28** (2), 252–274.
Grimshaw, D. and M. Miozzo (2006), 'Institutional effects on the market for IT outsourcing: analysing clients, suppliers and staff transfer in Germany and the UK', *Organisation Studies*, **27** (9), 1229–1259.
Grossman, S.J. and O.D. Hart (1986), 'The costs and benefits of ownership: a theory of vertical and lateral integration', *Journal of Political Economy*, **94** (4), 691–719.
Grote, M.H. and F.A. Taube (2006), 'Offshoring the financial services industry: implications for the evolution of Indian IT clusters', *Environment and Planning A*, **38**, 1287–1305.
Hakanson, L. (1990), 'International decentralization of R&D: the organizational challenges', in C.A. Bartlett, Y. Doz and G. Hedlund (eds), *Managing the Global Firm*, London: Routledge, pp. 256–278.
Hakanson, L. and R. Nobel (1993), 'Foreign research and development by Swedish multinationals', *Research Policy*, **22**, 373–396.
Henderson, J.H. and N. Venkatraman (1994), 'Strategic alignment: a model for organisational transformation via information technology', in T.J. Allen and M.S. Scott Morton (eds), *Information Technology and the Corporation of the 1990s*, Oxford: Oxford University Press, pp. 202–220.
Hirshfeld, S. and G. Schmid (2005), 'Globalisation of R&D', *Technology Review* 184/2005, Helsinki TEKES.
Hobday, M. (2000), 'The project-based organisation: an ideal form for managing complex products and systems', *Research Policy*, **29** (7–8), 895–911.
Howells, J. (1990), 'The location and organisation of research and development: new horizons', *Research Policy*, **19** (2), 133–146.
Howells, J. (1995), 'Going global: the use of ICT networks in research and development', *Research Policy*, **24** (2), 169–184.
Howells, J. (1999), 'Regional systems of innovation?', in D. Archibugi, J. Michie and J. Howells (eds), *Innovation Systems in a Global Economy*, Cambridge: Cambridge University Press, pp. 67–93.
Hymer, S. (1976), *The International Operations of National Firms*, Cambridge, MA: MIT Press.
Jonscher, C. (1994), 'An economic study of the information technology revolution', in T.J. Allen and M.S. Scott Morton (eds), *Information Technology and the Corporation of the 1990s*, Oxford: Oxford University Press, pp. 5–42.
Kaplinsky, R. (2000), 'Spreading the gains from globalization: what can be learned from value chain analysis', Working Paper 110, Institute of Development Studies, University of Brighton.
Kenney, M., S. Massini and T. Murtha (2009), 'Offshoring administrative and technical

work: new fields for understanding the global enterprise', *Journal of International Business Studies*, **40** (6), 887–900.
Klepper, S. (1997), 'Industry life cycles', *Industrial and Corporate Change*, **6** (1), 145–181.
Kogut, B. and U. Zander (1992), 'Knowledge of the firm, combinative capabilities and the replication of technology', *Organization Science*, **3**, 383–397.
Kogut, B. and U. Zander (1993), 'Knowledge of the firm and the evolutionary theory of the multinational enterprise', *Journal of International Business Studies*, **24** (4), 625–645.
Kuemmerle, W. (1999a), 'Foreign direct investment in industrial research in the pharmaceutical and electronics industries: results from a survey of multinational firms', *Research Policy*, **28** (2–3), 179–193.
Kuemmerle, W. (1999b), 'The drivers of foreign direct investment into research and development: an empirical investigation', *Journal of International Business Studies*, **30** (1), 1–24.
Langlois, R. N. (2003), 'The vanishing hand: the changing dynamics of industrial capitalism', *Industrial and Corporate Change*, **12** (2), 351–385.
Lehrer, M. (2006), 'Two types of organisational modularity: SAP. ERP product architecture and the German tipping point in the make/buy decision for IT services', in M. Miozzo and D. Grimshaw (eds), *Knowledge Intensive Business Services: Organizational Forms and National Institutions*, Cheltenham, UK and Northampton, MA, USA: Edward Elgar, pp. 187–204.
Levy, D. (2005), 'Offshoring in the new global political economy', *Journal of Management Studies*, **42** (3), 685–693.
Levy, F. and R.J. Murnane (2004), *The New Division of Labour: How Computers are Creating the Next Job Market*, Princeton, NJ: Princeton University Press.
Lewin, A.Y., S. Massini and C. Peeters (2009), 'Why are companies offshoring innovation? The emerging global race for talent', *Journal of International Business Studies*, **40** (6), 901–925.
Lewin, A.Y. and C. Peeters (2006), 'Offshoring work: business hype or the onset of fundamental transformation?', *Long Range Planning*, **39** (3), 221–239.
Manning, S., A.Y. Lewin and S. Massini (2008), 'The globalization of innovation: a dynamic perspective on offshoring', *Academy of Management Perspectives*, **22** (3), 35–54.
Markusen, J.R. (2006), 'Modeling the offshoring of white-collar services: from comparative advantage to the new theories of trade and foreign direct investment', in L. Brainard and S. Collins (eds), *Brookings Trade Forum: 2005 Offshoring White-Collar Work*, Washington, DC: Brookings Institution Press, pp. 1–34.
Marshall, A. (1920), *The Principles of Economics*, 8th edn, London: Macmillan.
Maskell, P., T. Pedersen, B. Petersen and J. Dick-Nielsen (2007), 'Learning paths to offshore outsourcing: from cost reduction to knowledge seeking', *Industry and Innovation*, **14** (3), 239–257.
Mayer, K.L. and N. Argyres (2004), 'Learning to contract: evidence from the personal computer industry', *Organization Science*, **15** (4), 394–410.
Meyer-Krahmer, F. and G. Reger (1999), 'New perspectives on the innovation strategies of multinational enterprises: lessons for technology policy in Europe', *Research Policy*, **28** (7), 751–776.
Miles, I. (2001), 'Knowledge-intensive business services and the new economy', paper presented at the Evolutionary Economics Unit, Max Planck Institute for Research into Economic Systems, September.
Miles, I. (2003), 'Business services and their contribution to their clients' performance: a review', contribution to Ecorys/CRIC project, Manchester: CRIC, University of Manchester.
Minbaeva, D., T. Pedersen, I. Bjorkman, C.F. Fey and H.J. Park (2003), 'MNC knowledge transfer, subsidiary absorptive capacity, and HRM', *Journal of International Business Studies*, **34** (6), 586–599.
Miozzo, M. and D. Grimshaw (2005), 'Modularity and innovation in knowledge-intensive

business services: IT outsourcing in Germany and the UK', *Research Policy* **34**, 1419–1439.
Miozzo, M. and D. Grimshaw (2008), 'Capabilities of large knowledge-intensive business services suppliers: the case of EDS and IBM', paper presented at the 2008 International Schumpeter Society Conference, 2–5 July, Rio de Janeiro, Brazil.
Miozzo, M. and L. Soete (2001), 'Internationalisation of services: a technological perspective', *Technological Forecasting and Social Change*, **67** (2), 159–185.
Murtha, T.P. (1991), 'Surviving industrial targeting: state credibility and public policy contingencies in multinational subcontracting', *Journal of Law, Economics, and Organization*, **7** (1), 117–143.
Murtha, T.P. (2004), 'The metanational firm in context: competition in knowledge-driven industries', *Advances in International Management*, **16**, 101–136.
Nelson, R. and S. Winter (1982), *An Evolutionary Theory of Economic Change*, Cambridge, MA: Harvard University Press.
Niosi, J. (1999), 'The internationalisation of industrial R&D: from technological transfer to the learning organisation', *Research Policy*, **28**, 107–117.
Patel, P. and K. Pavitt (1991), 'Large firms in the production of the world's technology: an important case of "non-globalization"', *Journal of International Business Studies*, **22** (1), 1–21.
Pavitt, K. (2003), 'Specialisation and systems integration: where manufacture and services still meet', in A. Prencipe, A. Davies and M. Hobday (eds), *The Business of Systems Integration*, Oxford: Oxford University Press, pp. 75–91.
Pedersen, T. and P.J. Orberg (2007), 'Offshoring of advanced activities', paper presented at EURAM Annual Conference, Paris, 16–19 May.
Penrose, E.T. (1959), *The Theory of the Growth of the Firm*, New York: John Wiley.
Pettigrew, A. and S. Massini (2003), 'Innovative forms of organizing: trends in Europe, Japan and the USA in the 1990s', in A.M. Pettigrew, R. Whittington, L. Melin, C. Sanchez-Runde, F. van den Bosch, W. Ruigrok and T. Numagami (eds), *Innovative Forms of Organizing International Perspectives*, London: SAGE, pp. 1–32.
Piore, M. and C. Sabel (1984), *The Second Industrial Divide: Possibilities for Prosperity*, New York: Basic Books.
Porter, M. (1990), *The Competitive Advantage of Nations*, New York: Basic Books.
Prahalad, C.K. and G. Hamel (1990), 'The core competences of the corporation', *Harvard Business Review*, May–June, 79–91.
Prencipe, A. (1997), 'Technological competencies and product's evolutionary dynamics: a case study from the aero-engine industry', *Research Policy*, **25**, 1261–1276.
Prencipe, A. (2000), 'Breadth and depth of technological capabilities in CoPS: the case of the aircraft engine control system', *Research Policy*, **29** (7–8), 895–911.
Quinn, J.B. (2000), 'Outsourcing innovation: the new engine of growth', *Sloan Management Review*, **41** (4), 13–28.
Rothwell, R. (1992), 'Developments towards the fifth generation model of innovation', *Technology Analysis and Strategic Management*, **4** (1), 73–75.
Sako, M. (2005), 'Outsourcing and offshoring: key trends and issues', Background Paper prepared for the Emerging Markets Forum, November.
Sako, M. and A. Tierney (2005), 'Sustainability of business service outsourcing: the case of human resource outsourcing', prepared for the Stanford Conference on the Globalisation of Services at Stanford University, 17 June.
Scott Morton, M.S. (1991), *The Corporation of the 1990s: Information Technology and Organisational Transformation*, Oxford: Oxford University Press.
Simon, H. (1957), *Models of Man*, New York: Wiley.
Storper, M. (1997), *The Regional World: Territorial Development in a Global Economy*, London: Guilford Press.
Sturgeon, T.J. (2002), 'Modular production networks: a new American model of industrial organization', *Industrial and Corporate Change*, **11** (3), 451–496.
Sturgeon, T. J., F. Levy, C. Brown, J. B. Jensen and D. Wel (2006), 'Why we can't measure

the economic effects of services offshoring: the data gaps and how to fill them', Services Offshoring Working Group, Final report, Industrial Performance Center Massachusetts Institute of Technology.
Teece, D.J., G. Pisano and A. Shuen (1997), 'Dynamic capabilities and strategic management', *Strategic Management Journal*, **18**, 509–533.
United Nations Conference on Trade and Development (UNCTAD) (2004), *World Investment Report 2004: The Shift Towards Services*, New York: United Nations.
Venkatraman, N. (1991), 'IT-induced business reconfiguration', in M.S. Scott Morton (ed.), *The Corporation of the 1990s: Information Technology and Organisational Transformation*, Oxford: Oxford University Press, pp. 122–158.
Vernon, R. (1966), 'International investment and international trade in the product cycle', *Quarterly Journal of Economics*, **80**, 190–207.
Volberda, H.W. and A.Y. Lewin (2003), 'Co-evolutionary dynamics within and between firms: from evolution to co-evolution', *Journal of Management Studies*, **40** (8), 2111–2136.
Willcocks, L. and G. Fitzgerald (1994), 'Toward the residual IS organisation? Research on IT outsourcing experiences in the United Kingdom', in R. Baskerville, S. Smithson, O. Ngwenyama and J. DeGross (eds), *Transforming Organisations with Information Technology*, Amsterdam: North-Holland, pp. 129–152.
Williamson, O.E. (1975), *Markets and Hierarchies: Analysis and Antitrust Implications*, New York: Free Press.
Williamson, O.E. (1996), *The Mechanisms of Governance*, Oxford: Oxford University Press.
Winter, S. (1987), 'Knowledge and competence as strategic assets', in D. Teece (ed.), *The Competitive Challenge: Strategies for Industrial Innovation and Renewal*, Cambridge, MA: Cambridge University Press, pp. 159–184.
Womack, J.P., D. Jones and D. Roos (1990), *The Machine that Changed the World*, New York: Macmillan.
Zaheer, S., A. Lamin and M. Subramani (2007), 'Cluster capabilities or ethnic ties? Location choice by foreign and domestic entrants in the services offshoring industry in India', *Journal of International Business Studies*, **40** (6), 944–968.

21 Innovation and internationalization: a dynamic coupling for business-to-business services

Jean Philippe and Pierre-Yves Léo

21.1 Introduction

This chapter proposes to study the connections between innovation in services and the internationalization of service firms. These two axes of development have always been considered separately. Bringing them together should enable us to understand better, from the conceptual and managerial point of view, the relationships which these two dynamics share. Service innovation obviously has consequences for the international development of service firms. But internationalization is also a powerful driver for service innovations and notably so in business-to-business services.

Research on innovation in services has shown up the role of technical systems as a source of innovation, but also the importance of innovations in methods and organization (Gallouj, 2002, 2003). From a methodological perspective, these research works had to overcome analytical difficulties specific to services due to the immaterial nature of their production, to their interactivity and to the contribution of clients to the final outcome of the service delivery. These works identified several types of innovation: radical, improvement, incremental, ad hoc, recombination and formalization.

Also faced with a great variety of forms, research on service internationalization used a quite similar approach. The first step was to describe the different types of internationalization: role and type of international networks; solutions adopted to master the difficulty to maintain high-quality standards in different cultural environments; use of information and communication technologies (ICTs) replacing or simply supporting traditional networks; have all been highlighted, and so have the firms' strategies concerning what services they offer abroad, but also how they organize their relations with clients.

The connection between innovation and internationalization becomes evident when observing the business world: globalization is the rule, whether it is commanded by markets or competition. Business-to-business services must satisfy increasingly worldwide clients by tackling international competition stirred by the generalization of deregulation policies. In such a context, the quest for lower costs is often supported by innovation in methods but is reinforced by growing internationalization, which may well entail some scale economies (Ghemawat, 2007). The search to standardize offers is a very similar process. The quest for reducing costs is a tendency which remains fundamental for firms, but changes in the market can provoke more basic innovations. New countries are emerging on the world's economic scene, and expectations for better-adapted, more localized offers can be observed. A new dialectic may therefore exist, inducing more radical innovations to be implemented.

Our chapter intends to analyse the relationships between these drivers of the transformation of firms and offers, to distinguish clear profiles and specific contexts and draw up the managerial implications from them. Our research programme on the links between international development and innovation in services is still at its beginnings and our approach is therefore exploratory. Our field is that of business services, for this sector employs all the ways and means of international trade and it makes an intensive use of information and communication technologies (ICTs). We will use the results of research on innovation and studies concerning the internationalization of business services. In section 21.2 we will make a brief comparison of different approaches to innovation and to the internationalization of services, and then we will give, in section 21.3, indications on the surveys upon which we support our research. Section 21.4 will present a primary analysis of the behaviour of service firms which innovate and export. Internationalization is a fantastic accelerator for service innovation, just as innovation is an essential factor of international competitiveness: we will show how the coupling of innovation to internationalization operates, firstly concerning research and development (R&D) activities linked to export, then concerning the service offering through rationalization innovations, and lastly concerning the international service delivery through relational innovations. Section 21.5 concludes.

21.2 Two ways to transform a service firm

Within the development strategies of firms, internationalization and innovation occupy the front line: sometimes in a complementary manner, when used to describe the product life cycle (as in Vernon's model); or quite the opposite, when the issue is to measure the risks generated by the development strategy. In the first case, internationalization follows in the steps of the domestic market to maintain the rate of the sales growth after saturation of the product's initial market. In the second, the analysis of risks brings to light the difficulty of proposing new products on markets where the firm is not established. This reminder shows that the two strategies have identical elements in common: creativity, growth, but also risk. Below, we propose to present the two approaches in parallel, showing three themes: the basic model, dynamics at work and the organization models adopted by firms.

21.2.1 Basic models of innovation and international organization

The works of Faïz Gallouj (2003, 1994) and Camal and Faïz Gallouj (1996) are taken as the base for our information on innovation, especially their analysis of Barras's (1986) inversed product life cycle model and of Soete and Miozzo's (1990) technological trajectories. Innovation in services is strongly linked to technology and in particular to information technology (IT). Barras's model describes a sequence of transformation for services which is the opposite of the traditional cycle of innovation for production in the manufacturing industry. The cycle starts in the back office with a phase of incremental innovations on methods (computerization of procedures, for example), continues in the front office with a phase of radical methodological innovations (for example, installation of automatic cash machines by banks) and ends up with 'product' innovation, when the firm is able to propose new services to its clients. This analysis shows that service innovations are not situated in the technical systems themselves, but in the changes which are made possible by them. The main criticism expressed of this model underlines that it tends to represent a diffusion model for technical innovation originating from industry in services, more than a real theory on service innovation. In particular, it does not take into account non-technological service innovations, for example organizational ones.

This model is completed by the taxonomy of service sector technological trajectories which isolate three kinds of activities. First, low-innovation activities which are dominated by technology suppliers; they can be found in services to persons sectors and in public or social services. Second, network services whose overriding strategy of lowering networks' running costs imposes specifications adapted to their organization upon technology suppliers. The third category concerns knowledge-intensive services with an offer founded on science; in this case, firms may be highly innovative and they may well demand that their suppliers meet their very specific needs. This taxonomy underlines the heterogeneous behaviour of services properly but it presents some weaknesses pointed out by the analysts of service innovation: the concept of trajectories seems to be too deterministic and univocal, excluding the possibility for a firm to belong simultaneously to two of them. Otherwise, as with Barras's model, non-technological trajectories are clearly omitted.

Bartlett and Ghoshal's (2002) strategic matrix provides a fundamental analytical framework, synthesizing two forces which all multinational firms must take into account to define their strategy: global integration and local adaptation. These two criteria define four types of international positioning for firms which can be international, transnational, 'multidomestic' or global. The positioning in this matrix can be seen as the result of a sequential development process, schematized by Doz et al. (2001) by which the firm reaches the multinational stage (third level) once its competitive advantage has been secured at home (first level) and it has then developed internationally according to the product life cycle (second level). This model was formulated from the analysis of large manufacturing firms and it can be generalized to very large service firms; nevertheless it does not fit the possibilities available to most service firms, whether because of their smaller size, or because of the specificity of their service delivery. Lovelock and Yip (1996) consider that the global strategy represents 'the development horizon for international service firms'. The elements of the service offering must be appraised according to their impact on the firm's globalization. But it must be noted that this global strategy merely concerns some ten or so very large firms. Most international firms adopt other strategies, in particular the international strategy which, although it appears to be the least comprehensive, however rallies the largest number.

The comparison of the basic models and the actual behaviour of service firms show great similarity between innovation and internationalization: innovative service firms or those which develop at international level are privileging a progressive approach to change. Great behavioural heterogeneity prevails within the trajectories, although the quest for cost-cutting leads to a very significant demand for IT to support them.

21.2.2 Innovation and internationalization dynamics

In his research on the innovation theory, Gallouj (2002) proposes a representation of service as the simultaneous mobilization of technical characteristics (material and immaterial), and internal (the provider) and external (the client) skills, to produce a service with clearly defined characteristics. Then he defines innovation as being any change affecting one or several terms of one or several vectors of the characteristics (technical, service) or skills. The forms of change are manifold: evolution or variation, extinction, emergence, association, dissociation, formatting. Innovation is not considered as a result, but as a process leading to 'modes' which may be described by the particular dynamics of characteristics. Radical innovation can be observed when a new set of characteristics is created; improvement innovation, when the 'weight' of certain characteristics is increased; incremental innovation operates by adding or suppressing characteristics, but maintaining the general structure; ad hoc innovation consists in offering an original solution for a problem, modifying the skills and technical characteristics; a 're-combining' innovation keeps the characteristics and skills unchanged but associates and/or dissociates some of them differently; and lastly, formalization innovation aims at better quality control but it also reduces the flexibility of the organization by formatting and standardizing the characteristics.

This representation of services as vectors of skills and characteristics can be applied to all the major questions concerning service dynamics and notably to internationalization, which is going to affect certain characteristics of the offer. In a sense, according to Gallouj's definition, internationalization could be considered as a specific case of innovation. But such a definition would be far too reducing because internationalization is not only an innovation for the firm: it has direct and relatively rapid consequences in terms of turnover, operations, costs and benefits (Farrell, 2004; Gottfredson et al., 2005). The main difference with innovation lies in the extent of changes, probably more limited in the case of internationalization because of the quest to cut costs and the risk associated with new markets. According to a tradition which is well anchored from the very beginnings of research work on service management (Shostack, 1977), growth strategies in service activities are often based on the service concept itself. Välinkangas and Lethinen (1994) identified three main strategic axes for international development: standardization, specialization and customization. Standardization can be considered as similar to the global strategy: it resides in the pursuit of scale economies, aiming the service at the widest-ranging market possible, if needs be, via the use of franchise agreements, in order to improve the rate of development of the network. Specialization differs from the previous strategy, because it does not propose the search for scale economies as a priority concern: the firm seeks to take advantage of one single and unique service, totally differentiated from those of competitors. Therefore, the costs preoccupation ranks behind the quest for excellence. Customization is akin to adaptation. The service is customized to fit closely the consumer's needs.

Whatever, sooner or later, internationalized small firms have to reconsider their service delivery, either by adapting it to the local market or by enabling it to avoid this obligation. Service standardization is often described as an advantageous way of developing on the international scale, but its practicability may depend on the type of service activity and on factors which are specific to each firm (Quester and Conduit, 1996). Furthermore the opportunities offered by this method seem far more limited in the field of business services. The theories elaborated from cases of multinational service firms put the accent on the importance of the standardization of services and the formalization of procedures in gaining access to markets all over the planet. For the smaller firms, the stakes are probably not worldwide, but there is no doubt that the way the services are adapted entails specific future growth opportunities.

The question of the services offered, of the combination of the core service (or services) with peripheral services, is essential for the management and marketing of the firm. This combination conditions the firm's positioning and operational management. For just this reason, it is to be found at the very centre of numerous analyses (Heskett, 1995; Eiglier and Langeard, 1987). This question must be examined too when dealing with the delivery of services on foreign markets. The quest for competitive advantage, just like the quest for growth, can lead the firm either to diversify or to specialize. We were able to observe (Léo et al., 2006) how firms arbitrate between their distinctive competencies and the need to transfer them within the context of a different market: the services they effectively offer in foreign markets do not always correspond to the main services they deliver in the domestic market. However our surveys show that on the whole, the services offered abroad remain mostly in the same field of service expertise (Léo and Philippe, 2008).

The analysis of the dynamics of internationalization shows it as being a particular case of the dynamics of innovation, with a whole set of changes brought about by the necessity to adapt to the market and/or reduce costs.

21.2.3 Organization models

From the very beginning, strategic analysis has insisted on the fact that any strategy requires the setting up of organizational means, and dedicated organizational structures are to be found heading the list of operational structures. This issue takes on a specific prominence in the context of services which often adopt flexible modes of organization. Even for innovative service firms, innovation or R&D departments are the exception, but it does not mean that innovation remains unorganized: Barcet et al. (1987) proposed two organization models, completed by two further models from Sundbo and Gallouj (2000). The professional partnership model is typical of intellectual services firms, often average in size and delivering skills and expertise-based services. Such organization allows firms to react rapidly to the clients' demands but the process of innovation may not be carried out to term, or not sufficiently formalized. The managerial model characterizes firms with no innovation department, but which are capable of getting organized to develop a new service project. The industrial model, with a well-defined R&D department is to be found in very large service firms. Gallouj (2003) underlines their scarcity in the field of services and identifies a neo-industrial model, where innovation is the result of the interaction between several functional departments of companies, mobilized on a specific project. The entrepreneurial model describes firms created on the basis of a radical innovation. Lastly, the small-scale or craft industry model corresponds to small operational services firms with no R&D department but which nevertheless constantly try to improve the services they deliver. Statistically speaking, the professional, managerial and neo-industrial models are the most common amongst service firms, and this illustrates the difficulty to be found in structuring the innovation process, and at the same time maintaining a flexible organization.

Relating to internationalization, we observe that the original objectives are not devoid of consequences on the way the firm gets organized to manage its foreign clients. The needs and constraints for client contact are not identical for all types of services. In business-to-business services in particular, the firm often travels to meet its clients, whereas in most services to the public, the clients travel to consume the service. The impact of ICTs will also be different, because the possibilities of substituting to displacement will not take on the same importance in different activities. Lastly, the types of establishments abroad are more flexible for business services, because many do not require heavy set-ups.

ICTs convey new solutions to solve the critical difficulties of international development: access to clients and the international organization of management. These new solutions open up a range of possible strategies, but also oblige firms to make choices in the function of their objectives and the specific constraints of their service deliveries. Alongside the strategic global integration model by which very large service firms can master the 'cultural transfer' of their concept, forms of international development are emerging which are more limited and allow smaller firms simply to exploit the advantages they have built in their original profession. These developments call intensely upon IT and communications. The organization of the international service firm is therefore strongly marked by the dilemma of service tradability. The World Trade Organization has made this question the reference in the commitments of states concerning the liberalization of service business.

Four large modes of international business organizations are identified. The first concerns purely transborder trade, requiring no displacement on behalf of either the service firm, or the client. This first kind of organization is made up either of industrialized services, incorporated in manufactured goods, or of informational distance services or 'teleservices' using the vector of telecommunications to deliver service. The second mode concerns service providers who move to the client's country to deliver the service; they also operate without a set-up abroad. Management, legal and engineering consultancy firms are mainly concerned, but also some training services which are delivered abroad. The transport may be connected to this category. In the third mode the client has to travel to the service firm's country to obtain the service. Three services are very concerned: tourism, health service and training. This third mode is developing due to the relative drop in the price of air transport for passengers. In this configuration, most of the time internationalization does not create a need for specific organization, and the international client is treated exactly the same as the national one. The last mode appears when a firm develops internationally through a foreign set-up. Very complex types of organization are to be found here, because the evolution of ICT facilitates new management models which complete, or supersede the traditional foreign subsidiary.

Vandermerwe and Chadwick (1989) establish an explicit relationship between internationalization modes and ICTs, by proposing a classification of the modes and the internationalization strategies originated from two criteria: the degree of interaction between the producer and the service consumer, on the one hand, and on the other hand, the degree of tangibility of services which opposes 'pure' services (with low tangibility) and services associated with a product or delivered through a product and, lastly, to services fully incorporated in a product. The contemporary development of the telecommunication technology modifies this panorama by enabling new organizational models (offshore back-office, better control and tracking of distant front-office operations) and/or creating a new mode of internationalization for certain services which can now be exported without any kind of temporal or permanent presence in the foreign country: direct banking and all online financial activities, the sale of insurance policies in airports, access to foreign data banks and online computer technology services are all evidence of this phenomenon. In any case, technological developments in services encourage integrated international development strategies as well as the complementarities between transborder trade and local set-ups more than it replaces traditional modes of trade (Zimny and Mallampally, 2002).

21.3 Surveys and samples of internationalized service firms

Here our objective was first of all to revisit a survey which had already been carried out, and pertaining to the means of competitiveness used by French exporters, amongst which a significant number of service activities could be isolated. Among the quantity of questions it was possible to make use of those concerning international performance, those regarding the technological dimension of the innovation and its process upstream of R&D, and lastly those in relation to the hierarchy of assets upon which firms leaned to be competitive on external markets.

This first survey was carried out in 1999 in cooperation with Telexport, an organization grouping together French Chambers of Commerce and dedicated to foreign trade; it organizes a yearly survey concerning all registered exporters. An inset dealing specifically with competitiveness means was annexed and sent to all the contacted firms, whatever the sector of activities (more than 30,000). If we exclude the responses from the commercial or agricultural companies, 3370 exploitable responses were obtained, amongst which we were able to identify 93 firms exporting only services: they are related to professional services (26 suppliers in computer services, 24 in technical studies or engineering, 7 management consultants, 19 logistics services and 5 operational services) or to industrial activities with a purely tertiary international activity (12 cases): technical assistance, maintenance, training services etc. The very low proportion of tertiary responses (93 out of more than 3000, that is, 2.8 per cent) comes from the difficulty, so often mentioned, of identifying tertiary companies operating in foreign markets and from the historical orientation of the survey database, particularly focused on the exportation of tangible goods. The rest of the database concerns manufacturing and extractive sectors and will be used to bring to light what is specific to service activities.

This survey supplies us with a sample of a suitable size for us to be able to observe the relations between the international development of services and their innovatory profiles. Evidently it is hard to say to what degree such a sample is representative, but we may compare it to the characteristics of subsequent surveys (Léo et al., 2006) to mark off service firms from internationalized firms. These two sources converge concerning the most implicated activities: engineering, computer software services and logistics. Indeed, these three sectors make up the main poles of international activity for these firms. In our other surveys, however, business or management consultancy and operational services are proportionally double those found in this sample.

The sample also presents the following characteristics: 35 per cent of the answers were collected in the Paris area; the rest of France is represented very unevenly: 5 per cent from Bordeaux, the same from Toulouse, whereas Marseilles and Lyons are barely represented. A total of 83 per cent of the firms answering the inset are compact according to location (no other establishment), and 74 per cent of the answers come from independent companies. We find a majority of small firms (76 per cent employ less than 50 persons), but also a few larger-sized companies (up to 550 employees). Apart from the geographical pattern, this sample profile tallies with our other surveys, but here there are far fewer very small-size firms (31 per cent employ less than ten people, versus 57 per cent in our second set of surveys in Marseilles and Toulouse).

The information requested in the inset concerned the assets explaining the company's performance in external markets and is completed by information collected from the normal Telexport survey (turnover, exportations, employment, organization, export countries etc.). The inset asked specific questions on the type of product (or service) exported, patents held by the firm, degree of novelty of innovations incorporated in exports, the R&D specialized department (does it exist, how many persons does it employ, importance of the dedicated budget), the export experience (how long the exportation has been running), the number of market countries and the recent evolution of general and international turnovers. A list of assets, possible strong features of the firm, was then proposed and responders were asked to tick those which were strong points and those which could eventually be a handicap in international markets. A second question invited them to sort their strongest points (five maximum).

A second set of surveys were carried out in collaboration with the Chambers of Commerce and Industry (CCI) of Toulouse and Marseille-Provence. A two-step method of survey was used. A first very brief questionnaire was sent to all members of the CCI in the business service sector, in the simple aim of identifying which firms had international activities (7000 e-mails, faxes or letters). This operation enabled the identification of a target of 980 companies, raised to 1140 after adding the firms already acknowledged as such by the exporters' files (Telexport). The second, far more detailed questionnaire was sent through the post and only to the targeted firms. A total of 274 answers were received and could be used as a database: only a very few had omitted to answer a very small number of questions.

A rapid overview of the characteristics of this sample shows that the answers mostly (84 per cent) come from French regional firms having their headquarters located inside the investigated region, 82 per cent of these being compact firms (with only one establishment in France). Of the sample, 74 per cent are independent firms, in a great majority small, or very small companies (57 per cent count less than ten employees). Seven large firms (with more than 500 employees) also answered our questionnaire. Answering firms belong mainly to four sectors: technical consulting and computer services provide 41 per cent of the responses, logistics and transportation services come next (26 per cent), followed by management consulting (21 per cent). Lastly, a certain number (12 per cent) of operational services (cleaning, janitorial, translation, secretarial services, opinion polls, factoring, etc.) are also to be found.

Beyond the economy of means offered by this method, a cursory glance shows the importance of sweeping so wide: small-sized companies strongly involved in international markets appear to be more numerous than could be expected. If global or giant firms have frequently been studied, smaller ones are little known and this is probably one of the specific contributions of our surveys.

21.4 Coupling innovation and internationalization

Being present in foreign markets assumes that the firm profits from competitive advantage presenting some specific characteristics. According to Barney (1991), competitive advantages have to be distinctive and durable: they must be of value or rare, and they must be difficult to imitate or replace. Most research which has analysed export performance has studied industrial goods (Lefèbvre et al., 1998; Wakelin, 1998) and highlighted the intensity of R&D as a favourable factor for exporting and gaining durable competitive advantage. Beyond the existence of R&D in the firm, the concept of capability of innovation which was proposed describes an aptitude to introduce new products and to adopt new processes. Christensen (1995) retains four dimensions for this concept; Guan and Ma (2003) expand the number of dimensions to seven: learning capability, R&D capability, manufacturing capability, marketing capability, organizational capability, capability to exploit resources and strategic capability. Drawing inspiration from this research, in the Telexport database we have analysed and compared the assets of manufacturing and service firms explaining their competitiveness. Next, founding our research on the inset dedicated to service firms, we show the types of innovation provoked by internationalization in the service offering, then in the international client relationship modes.

21.4.1 Means of competitiveness and exporting performances

At first glance, the answers from service firms can hardly be differentiated from those of manufacturing firms. Both types of activities are faced with similar difficulties on foreign markets. The exportation of services does not appear significantly less dynamic and the degree of commitment to exportation of the two kinds of firms are alike (28 and 27 per cent export rate on average, 17 and 18 per cent for the median value). Service exportations are however perceptibly less 'ancient' (11 years, versus 14 years for the median value) and less widespread geographically (in five countries instead of eight; two geographical zones instead of three for the median values). The number of geographical market zones is the main significant difference between the two samples ($F = 12.15$; $p > F = 0.0005$).

Concerning innovation, 48 per cent of the 93 service firms in the sample declare having a R&D department. This is an unexpected and fairly high proportion which could be interpreted as a consequence of the sample being limited to internationalized firms. However this percentage is considerably less than in the industrial sample (55 per cent) and this number must be taken in all relativity because in numerous cases, the very activity of the firm is to function as an externalized R&D department for client firms. Obviously, a process of innovation exists, but it does not, strictly speaking, concern the renewal of the service concept which is offered to the client nor the technology used to deliver quality services. This result poses the methodological problem of using identical items for questioning manufacturers and service providers when a given formulation does not have the same meaning for the two populations.

Thirty-four per cent of the sample declared incorporating technical innovations into exported services, which is very close to what can be observed in the manufacturing sector (31 per cent). What is more, when looking at the type of innovation, answers from tertiary firms make up a slightly larger part of the truly new innovations (at world, European or French levels), totting up 55.5 per cent of exported innovations (versus 52 per cent for manufactured goods). Such a difference indicates that it is more complex for services to export commonplace 'products', and this highlights the importance of innovation for a service firm wishing to develop abroad. It would seem, however, that it is far more difficult to protect technical innovations in the services field, for only 29 per cent of the firms in our sample declare holding a patent or licence, whereas 38 per cent of exporting manufacturers have at least one.

Analysis of the assets supporting the companies' competitiveness reveals the specificities of the tertiary sector better. As shown by Figure 21.1, service exporters do not base their competitiveness upon the same means as manufacturers. Search for quality is certainly among their first preoccupations. This does not oppose them to manufacturers, but it is worth pointing out (and this does not show in this statistic) that the means to reach quality are not identical for both. The second important lever is manpower skills, which are far from being placed so high by the manufacturing companies. It is evident that the impact of a service firm depends first and foremost on this factor associated to its workforce, but also to the technologies they are able to use. The third strong point quoted by exporting firms was the specificity of their products or service concepts, and this fits in with the industrials' opinion. It is certainly more difficult to make a place on foreign markets for a product (goods or service) if it is commonplace than if it is well designed and specified. The capacity to adapt comes next, and has already been underlined as a prerequisite when a firm wishes to operate in a differentiated market context, as is the case for exportation. The succession of the assets and the importance awarded to each of them also demonstrates that service firms depend far less on the range of the offer or on the brand and its image.

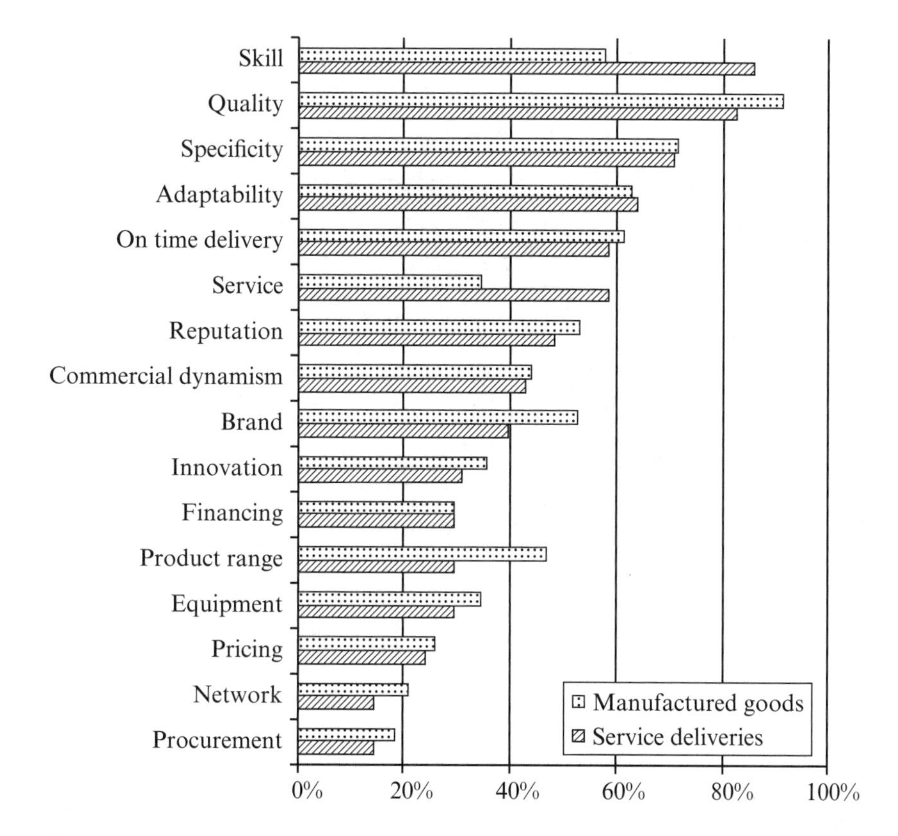

Source: Data processed by the CERGAM-GREFI from the Telexport survey 1999.

Figure 21.1 Assets explaining competitiveness of French firms according to whether they export services or manufactured goods (frequency of being selected within the 5 main strong points)

From the point of view of innovation, we must insist on the fact that few of the firms are basing their international development on the novelty of their products or services. Moreover the role played by equipment is not put forward as a trump card, neither by industrials nor by service suppliers. On the contrary, if we consider as innovative only the firms holding a patent and those whose exported innovation is a technological premier (at world, European or French levels) the competitiveness profile is considerably modified, especially for service exporters, as demonstrated in Figure 21.2.

Innovative exporters therefore present a specific competitiveness profile which can be distinguished from non-innovative exporters: the novelty of

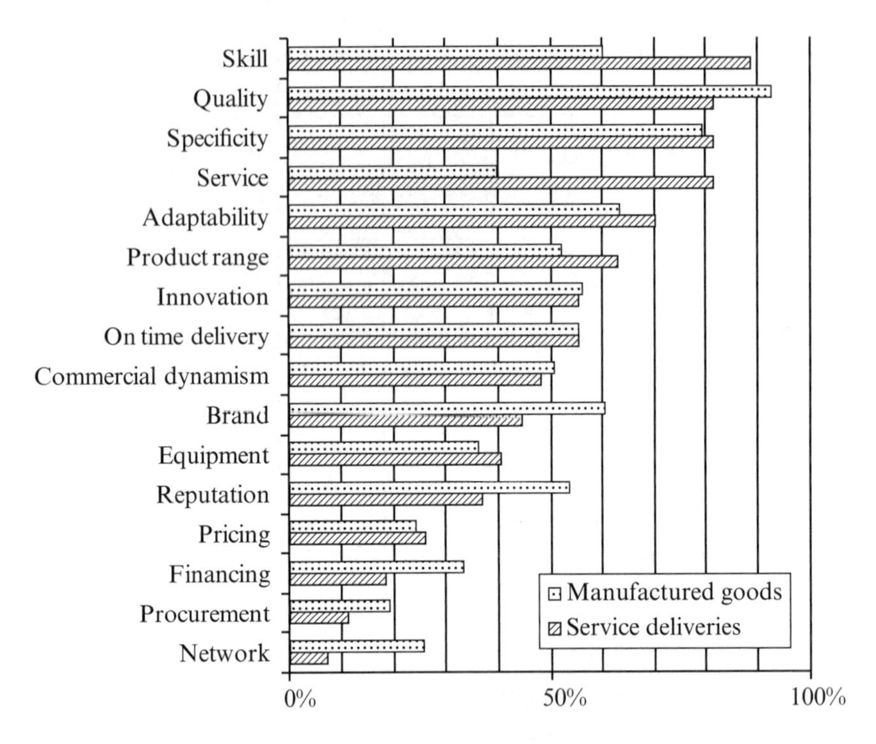

Source: Data processed by the CERGAM-GREFI from the Telexport survey 1999.

Figure 21.2 Assets explaining competitiveness of innovative French firms according to whether they export services or manufactured goods (frequency of being selected as a strong point)

the products (goods or services) of course plays a more significant role, but so does product range, equipment, adaptability, service spirit, commercial dynamism, specificity of the 'product', manpower's skills and even brand image. The relatively low ranking of the 'service spirit' amongst industrial goods exporters should not be interpreted as natural: French manufacturers may be lagging behind in the service 'revolution' which is so well described by Karmarkar (2004). Tertiary innovators diverge from industrials by their weakness in terms of brand, reputation, financial means and network. They gamble more heavily on their staff's know-how, service spirit and adaptability.

21.4.2 Rationalization of the international offer

The content of the service offering is evidently a strategic question for each service firm and the choices which are made in this field exert deterministic influence on the firm's future performance. Internationalization, just like innovation, modifies the strategic positioning of the firm and affects the various dimensions of the service offer. Two strategic objectives are in action here and they can appear as antagonists in some cases: the search for growth and the need to master costs. These objectives guide the choices in the service contents offered abroad: focusing the offer on the core service limits the risks and the costs entailed by international growth; on the other hand a diversified offer can meet the clients' true needs and facilitate the access to new markets better. Standardizing the service allows better cost control when adapting it gives the firm a better relation with its new foreign clients. This debate can also take place on a third dimension: is it worth customizing the service in foreign markets and would such a strategy be workable?

Specialization versus diversification In his analysis of commercial innovation Camal Gallouj (2007) presents a centre–periphery approach to commercial innovation which – founded on the distinction between core and peripheral services (Eiglier, 2004) – identifies radical innovations affecting the core service and incremental innovations concerning the peripheral ones. Gallouj notes that this approach gives rise to certain criticism because some service providers modify or add services their customers do not really need and without assessing the pertinence of such modifications in their offer. Nevertheless when they are well specified these service 'flowers', as named by Lovelock (1992), prove to be very useful for the marketing of the firm because they are able to enhance the clients' loyalty. However such combined service offers have to be questioned when the firm expands to foreign markets because they have direct consequences on the global costs the firm has to face up to. For firms developing internationally the pressure of cost and competition often entails a rationalization of the service offer which has to be streamlined better. Concerning the core services, a return to the basics is often necessary. The need and the utility of each peripheral service have to be reappraised in this new context.

As stated before, internationalization can lead the firm either to diversify or to specialize, according to its main strategic objectives. Diversification concerns some two-thirds of the firms in our sample; on the contrary, firms in the transportation and logistics sector seem to have chosen specialization. Are these choices fairly free or determined by the kind of activity performed? What is best: to focus the offer on the core service, which personifies most effectively the distinctive competencies of each firm, or to surround it with peripheral, secondary services, to make it more attractive?

On a list of some 27 types of services,[1] nearly a third of the responding firms (34 per cent) declared only delivering services of one type abroad, either directly or through the intermediary of subsidiaries or local partnerships. Seemingly, a majority of companies present a diversified offer on export markets and this might be a major source of expanding costs in their international development. Are these firms offering really diversified services on foreign markets? Most of them (76 per cent) deliver no more than five types of service. A principal components analysis has helped us identify what kind of services the firms actually do associate in their international offer. The categories brought to light by this analysis quite clearly reveal the specialization logic within the same large service expertise field. A similar tendency has already been observed in a different domain, concerning services associated with products exported by French manufacturers (Léo and Philippe, 1999).

The principal components analysis (PCA) identifies six axes which assemble an important part of the total variance. Each of the axes clearly corresponds to a well-differentiated field of service professions: the first axis expresses the affinity of different technical services peripheral to the exportation of goods, including the training of the final user. The second axis shows that various consulting services in management, human resources and accounts auditing are often associated by firms offering them abroad. The third axis groups together services with a commercial or marketing orientation: commercial intelligence, commercial agent services, advertising and legal advice. The fourth axis clearly opposes the domain of technical engineering and computer technologies to logistics and transportation services, associated with financial service (insurance, factoring). The fifth axis emerges as being structured by rental and hiring professions. The last axis illustrates that operational services (janitorial, secretarial, translations and telecommunications) are often associated by firms offering them abroad. The logical coherence of these different axes shows the organizing power of service professions which play a leading role when the firm projects its offer onto the international scene.

The results of the PCA enable us to categorize the service offerings according to seven fields of service activity (six axes, one being bipolar). A total of 142 firms of the sample are offering services in one unique field at international level; 98 in two and 23 in three fields. In more than three fields we only find 11 firms. This plainly shows that the majority of diversified service strategies remain concentrated on one or two service domains. Two fields of activity – engineering with 152 firms and logistics with 82 – focus the majority of the firms. It is within these two fields that the most numerous offers limited to one unique field can be observed (43 and 62 per cent respectively). The other fields of service are rarely mobilized as a unique axis of international development (between 18 and 8 per cent of the cases, depending on the field) and therefore appear within far more diffuse offers.

Standardization versus adaptation On foreign markets, firms do not only have to choose whether they lean on a specialized offer or on a diversified one; they have to decide about another question: to adapt, or not, the service they deliver. Our sample is divided in two quasi-equal parts by the answer to this question: 49 per cent are selling a service abroad identical in every way to the one sold in France, the 51 per cent left adapting this service more or less profoundly.

The first option is the one favoured by high-technical-content services: computer technology services and logistics activities where the service delivery appears to remain strictly unchanged 70 per cent of the time. The second option is observed for service offerings with a more cultural or 'legally controlled' content. Firms proposing advertising services abroad do adapt their service in 80 per cent of the cases. Similar proportions can be observed for accountancy or financial services (78 per cent) and commercial intelligence (77 per cent). The differences between countries concerning regulations and cultural background therefore influence the possibility for some services to be traded internationally to a high degree.

Among the firms having adapted their services abroad, nearly one-third had to redefine a new service concept in order to be able to supply international markets. Such an upheaval is quite a problem because they have to build a new competitive base from zero. This behaviour is not therefore that of the majority and our survey actually reveals that other firms predominantly put into operation marginal adaptations. In a majority of firms (55 per cent), the adaptations introduced are limited to one or two aspects. More often than not, it is the way in which the service is delivered (55 per cent) which is reconsidered. Pricing is the second point to be modified (37 per cent of the firms adapting their service delivery), either in order to take into account (and pass on) additional costs or, in some cases, to respond to competitors by offering special payment conditions (26 per cent). The third option (30 per cent) is to modify the type of clientele targeted. Lastly, advertising communication is the part which is the least alluded to (16 per cent), probably because business-to-business services call for little of this sort of promotion.

Firms were asked to summarize what objectives they pursued when introducing adaptations to their services abroad. Two major ambitions stand out: firstly, to follow up and control the quality of the service offered at a distance (mean quote 3.6 on a scale ranging from 1: not important at all, to 5: very important); and secondly, to make this service acceptable by making it conform to cultural (3.6) or legal (3.5) contexts or other aspects of local specificity (3.4). The quest for competitiveness, faced with local competitors, does not seem to weigh too much in the balance (3.1).

A relationship can be observed between adaptation and diversification of the offer: firms which adapt their services offer significantly more services than those carrying out no adaptation ($F = 15.15$; $p > F = 0.0001$) and these offerings concern a wide variety of service fields ($F = 18.68$; $p > F = 0.0001$). Symmetrically, adaptation, measured by an indicator cumulating the intensity of the adaptation and the number of aspects upon which it is carried out, appears more intensive and widespread for firms which have diversified their offer in several fields ($F = 9.83$; $p > F = 0.0020$). Standardization brings about simplification of the offer and its concentration in a single domain. The latitude left to contact personnel is thereby reduced and this leads to a severe curb on the possibilities of adaptation of the services.

Service customization and formalization of procedures A majority of the firms investigated have transformed their service offerings since the early stages of their internationalization. These firms show a characteristic profile, since we find specifically among this group[2] companies which have already developed a service delivery network; companies which are also highly involved in international markets as indicated by the importance of their exportations; companies which adapt their service; and companies offering diversified services. The changes made to the services reinforce this last aspect, since they are oriented, above all, toward a more diversified international offer: 57 per cent have diversified their core services as well as their peripheral ones, others (26 per cent) only their core services. On the contrary, a small number have reduced their scope of main services and/or complementary services abroad. In most cases this evolution is identical to that operated in the domestic market, but nevertheless firms which have diversified more rapidly abroad than in France (18 per cent) are more frequent than the opposite case (8 per cent).

The procedures used, when they have evolved, have done so toward more formal forms in foreign markets 84 per cent of the time. The evolution toward standardization is more uncommon on the content of services supplied: only 30 per cent of the firms having modified their service content increased standardization. The others (70 per cent) declare having evolved towards a more customized service delivery than the original offer. We can observe here an essentially flexible attitude which develops under the double pack of new possibilities offered by ICTs and the demands of clients who integrate these changes into their service expectations.

We can therefore say that the internationalization of these firms can be translated by a quest for internal efficiency, by the improvement of their procedures and often by quality certification (43 per cent of the companies display a 'label' or are 'quality certified', 55 per cent of which conform to ISO standards), whilst trying to approach their clients to the best of their ability, by offering customized services to each kind of circumstance, which leads them to widen their range of offers. The frequency with which we come across this sort of strategy in our sample authorizes us to sweep aside the idea that they could just be a coincidence. They constitute without a doubt an original line of attack on internationalization for small service firms, in particular in the domain of consulting.

21.4.3 Relational innovations

The modes of relationships with clients are of strategic importance in the success of international development. To settle the difficulties related to developing foreign markets, firms must not only innovate to present an offer adapted to the client, but also to contact the client, present the offer and follow it up on a regular basis. It appears that business service internationalization strategies are far from being homogeneous and that on analysis they show a multiple and varied panorama of radical and incremental innovations.

Offering services internationally obliges firms to master space in order to deliver their services to their clients and to develop or maintain relationships with them. The organizational modes of this client relation are to be found as follows in our sample: 71 per cent move their personnel to meet their client, 37 per cent depend on foreign clients coming to obtain the service in France, 48 per cent have branches abroad which carry out the service delivery and 12 per cent carry out 'pure' cross-border transactions. The choice of a delivery network, if not obligatory (RESER, 1995), really seems to be an efficient mode once international transactions are established on a long-term footing: 60 per cent of the most experienced firms (with ten years or more of practice in international markets) have developed at least one local agency to ensure their service delivery.

Apart from 'pure' cross-border transactions (as defined in our survey), these modes of relationships are not restricted one to another, and two modes (sometimes three) coexist within one and the same firm in 49 per cent of the sample. Most often, when one unique mode of relationship is used, it concerns the movement of home-based service personnel to foreign clients (55 per cent) or the delivery by a local establishment (35 per cent). Other delivery modes are rarely used alone by the firms in our sample: 12 firms rely only on non-equity foreign partners to provide their services abroad; nine others service foreign clients only in France.

These relationship modes are backed up in 85 per cent of the cases by the use of means of communication, whether they are traditional or not.

Phone or fax is used by 72 per cent of the respondents; traditional mail by 57 per cent. New means of telecommunications (Internet, Electronic Data Interchange – EDI, Data Banks) are used by 60 per cent of the respondents, which clearly shows the high-speed breakthrough of these technologies and their adaptation to the needs of this type of firm. The weight of new technologies in telecommunications can well be assessed by the high proportions of firms in our sample who have an electronic address (79 per cent) or developed their own, personal Internet site (47 per cent). The relatively few firms of our sample opting for 'pure' cross-border transactions thoroughly use these new relationship modes: 68 per cent take advantage of new telecommunication modes, 65 per cent use traditional telecommunication systems.

Table 21.1 Relationship modes with foreign clients according to service sector

Relationship mode	Technical consulting		Management consulting		Logistics		Operational services	
	Number	%	Number	%	Number	%	Number	%
'Pure' cross border	12	10.7	5	9.1	11	15.9	3	10.3
The staff moves	86	76.8	41	74.5	41	59.4	19	65.5
The client moves	45	40.2	21	38.2	19	27.5	12	41.4
Delivery by local agency	50	44.6	22	40.0	45	65.2	11	37.9
Together	112	100.0	55	100.0	69	100.0	29	100.0

Note: Column sum may exceed 100% due to the possibility of multiple answers. Sample used 265 firms (9 answers missing).

From a sector-related point of view, relationships with foreign clients differ from logistical activities to the two consulting sectors and operational services: as shown in Table 21.1, 'pure' cross-border service delivery and services ensured by local branches are more frequent when considering logistical services. Simultaneously, the moving of the client is far more atypical in this sector which includes many activities pertaining to the organization of chains of transportation using IT know-how and a network.

Having a network or not, and its type, seems to play a very discriminating role for many variables. As a rule, due to the fact that a subsidiary network remains out of the reach of small or medium-sized firms, companies with such a network present a very specific profile. Among the firms with this type of network, we only list 36 per cent of small independent firms, which represented in our sample 78 per cent of the partnership networks, 68 per cent of the commercial networks and 86 per cent of the firms with no network at all. Size indicators (manpower, turnover, capital, number of establishments) are all very marked by this characteristic. Analyses seemingly show, nevertheless, a logical path which leads firms step by step from international positioning to the gateway to multidomestic or global strategies.

A long-lasting international experience, like the maturity of the company, is significantly related with a service delivery carried out through a subsidiary network. Taking into consideration the diversity of the market countries the network appears to be an efficient solution: firms which have developed a partnership network, and to a greater extent those relying on subsidiaries, are the most present on non-European markets. International dynamism is significantly higher ($F = 7.30$; $p > F = 0.0001$) when a network has been set up, whatever its form.

The classification tree presented in Figure 21.3 has been established using the C&RT method[3] as developed by Breiman et al. (1984). The purpose is to explain the variations of a dynamism indicator ('Dyn') according to the characteristics of the network, the extent of the use of ICTs and the type of firm (small or medium-sized – SME – or not).

The indicator of dynamism which is used here proposes an exploratory means of approach to the evolution of the firm's international involvement. The evolution of exportations can be accepted as a measure of international vitality only when the company has no network aboard. As soon as a network is in place and delivers services to clients, the measure of international involvement, and consequently of its dynamic performance, becomes practically impossible: beyond the actual export turnover, sales made by the subsidiaries, but also the sales made by partners (joint ventures, agents, franchises and other) should also be taken into account, but probably not completely, since these partner firms may already have an activity in their country, independent of the activity which could be credited to the network. Each firm sets up its own assessment and control policies and no figures can be systematically requested from these nebulous organizations. On the other hand, it is possible to ask the manager how he assesses the development of this global figure. This is what we have done. Answers were quoted on a scale ranging from 1 (severe decrease) to 7 (very rapid growth). The quality of the responses is improved, without a doubt, if we associate this question to other questions asking for domestic sales evolution and for specifically exported turnover, which are grasped

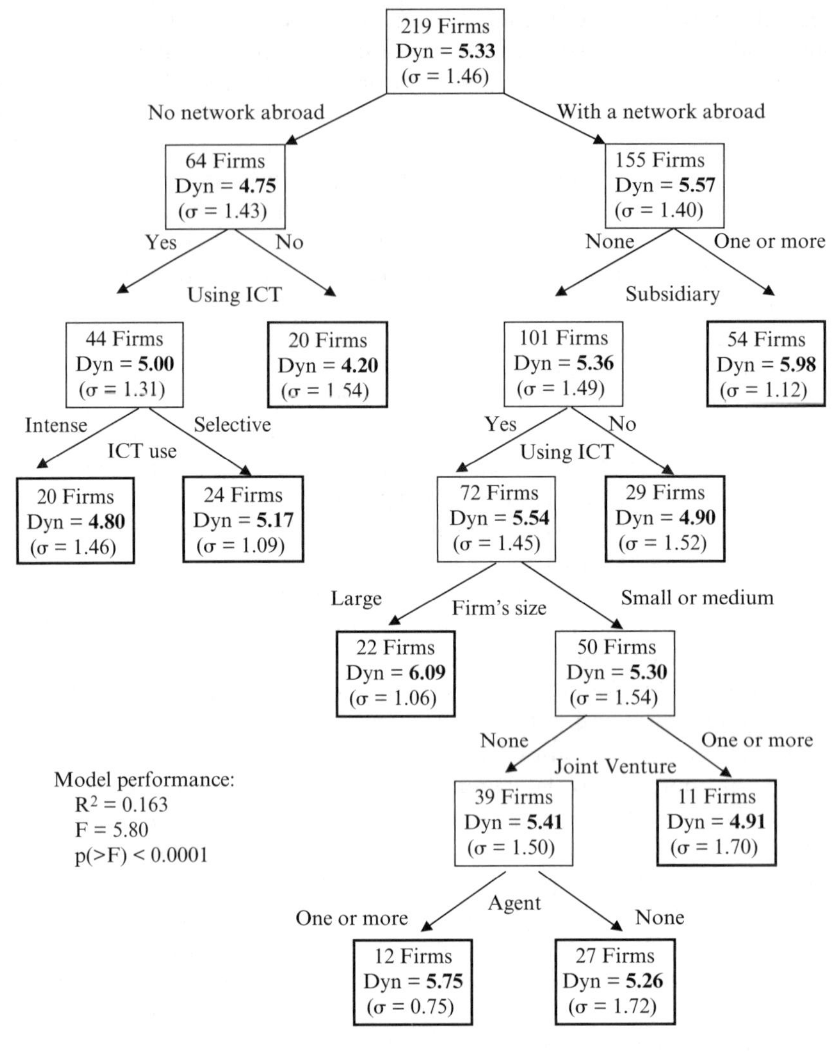

Figure 21.3 Global international dynamism according to network, ICT and firm's size

far easier by the manager. We must make a remark here: this indicator presents no significant difference according to the four sectors covered by this research (logistics, engineering, management consulting and operational services). Now, either these four sectors experience parallel development internationally or, and this seems more probable, the managers have more or less keyed their estimation upon that of their field of activity.

The step by step dichotomous analysis enabled us to identify groups of service firms experiencing significantly different international dynamism. The reliability of the 'diagnosis' decreases as we move down the classification tree which, a priori, can be more or less developed according to size range of the selected groups. Taking the size of our sample into account we programmed our software not to divide groups of less than 30 firms and not to keep groups of less than ten. The explicative variables entered into the program were: the fact that the unit was, or was not, a small or medium-sized independent firm; that it had at least one branch abroad (delivering services or simply a commercial one); the structure of the network (subsidiaries, agents or other partnerships); and lastly an indicator of the use of ICTs by the firm: none, just a simple Internet site, a selective use (limited to prospecting or to service delivery), an extensive use (associating prospecting and service delivery). At each stage, the program ran comparisons between the different possible combinations and in the case of the variables presenting more than two modes (ordinal variables are more suitable in this case) retains the most efficient dichotomy regarding the targeted variable (international dynamism here).

Many results can be drawn from this analysis. The first step opposes firms with no network of any kind abroad to the others. At this stage, the type of branch does not seem significantly to influence dynamism. The lack of a network considerably hinders development potential, but it would seem that firms can partially compensate for this absence by using ICTs: those doing so appear to be less handicapped but do not reach high levels of dynamism. It would also appear that a personalized Internet site is not really sufficient to make a company stand out from one which does not use ICTs at all. Paradoxically, it seems that a selective use of ICTs would prove to be more efficient than the systematic use of them. We can legitimately suppose here that ICTs are still in a fairly experimental phase and that this will tend to be inverted in the future.

Firms which have invested abroad through subsidiaries are amongst the most dynamic, even if they are independent small firms. This well and truly confirms that this sort of network is the most achieved one and the most adapted to international service development. The size of firms is certainly also a factor of international dynamism, since the most dynamic category is made up of elements from large groups which do not possess subsidiaries in their own name, but who make liberal use of ICTs. On the other end of the scale, two groups show signs of counter-performances which approach them to firms with no branches whatever: that is, those absolutely never using ICTs (not even a simple Internet site) and small firms which have developed their network using joint ventures. This last result is worth being verified on larger samples, since it merely concerns 11 small firms in our sample. This kind of network is frequently recommended for small businesses because it reduces investment but it seems that it would be less appropriate for service firms.

These results clearly demonstrate that relational innovation pays whether it is carried out with the technological support of ICT or by the creation of foreign agencies. On the contrary, firms which do not integrate new technologies, or do not develop a foreign network, appear to have a significantly less dynamic international development.

21.5 Conclusion

Global competition is growing for business-to-business services. Foreign competitors are entering domestic markets in exactly the same way as local firms are penetrating foreign markets. The robotizing of procedures is furthermore gaining ground; and software in the back and front office can now do tasks previously carried out by the personnel. If they want to continue on existing and maintain their competitiveness, firms have to innovate, adopt very active behaviour and provoke changes in their organization, methods and relation modes with clients. Internationalization, which up to the present has only marginally concerned service firms, now represents a menace, but also offers opportunities which innovative firms should be able to seize.

This analysis of business services brings to light the diversity of offers and the types of interaction that these activities have with their clients. Client–provider interactions can be dedicated to the commercial deal, the definition of the service and, if necessary, its co-production; they may be punctual or sustained, according to the type of services, concerning contact personnel as much as other categories of employees. All these elements are brought together under the term of 'service relationship' (De Bandt, 1994). International market operations are strongly constrained by the importance of the conditioning of the service relationship by social-cultural factors: language, scales of values, attitudes, behaviours. The way firms lift these constraints by innovating in their organization and in their relations with clients defines their internationalization profile: does the firm maintain identical offers to those offered on the home market or does it yield more or less to adaptations that the change of market context implies? Does the firm try to be physically near to its clients, or does it make do with temporary and 'ad hoc' proximity, far less expensive as long as markets are gained one after the other?

The first results we have shown in this chapter show that innovative exporters have trumps in hand regarding service offering and commercialization, which the simply exporting firms or only innovative ones do not have. This also brings to light that the great difference with manufacturing firms concerns the importance of the skills of staff to increase international competitiveness. This places the internationalization of services in an intercultural management context which is not the easiest to deal with. Lastly, these first results confirm that services are lagging behind compared to industrial companies in the creation of intangible assets – brand, reputation or network – which are nevertheless necessary to support international development.

Notes

1. Technical consulting & surveys; Computer software engineering; Staff training; Technical assistance; Maintenance & Repairing; Equipment setting & fitting, After sale service; Transportation & Logistics; Insurance & Financial services; Management consulting; Auditing & Accountancy; Human resources; Legal or fiscal advice & consulting; Commercial intelligence; Commercial agent services; Advertising; Audio-visual services & shows; Manpower hiring; Equipment rental; Trade fairs, Congresses & Travel organizations, Patent related services; Telecommunications; Translating; Secretarial work; Janitorial; Cleaning of premises; Catering & cafeteria.
2. This is true even when taking into account only the firms which have more than two years of international experience.
3. The classification and regression tree (C&RT) method is available within the SPSS 'Answer Tree' software.

References

Barcet, A., J. Bonamy and A. Mayère (1987), 'Modernisation et innovation dans les services aux enterprises', rapport pour le Commissariat Général du Plan, Lyon: Economie at Humanisme.

Barney, J.B. (1991), 'Firm resources and sustainable competitive advantage', *Journal of Management*, **17**, 99–120.

Barras, R. (1986), 'Towards a theory of innovation in services', *Research Policy*, **15**, 161–173.

Bartlett, C.A. and S. Ghoshal (2002), *Managing across Borders: the Transnational Solution*, Boston, MA: Harvard Business School Press.

Breiman, L.J., H. Friedman, R.A., Olshen and C.J. Stone (1984), *Classification and regression trees*, Belmont, CA: Wadsworth.

Christensen, J.F. (1995), 'Asset profiles for technological innovation', *Research Policy*, **24**, 727–745.

De Bandt, J. (1994), 'La notion de marché est-elle transposable dans le domaine des services informationnels aux entreprises', in J. De Bandt J. and J. Gadrey (eds), *Relations de service, marchés de service*, Paris: CNRS Editions, pp. 217–240.

Doz, Y., J. Santos and P. Williamson (2001), *From Global to Metanational: How Companies win in the Knowledge Economy*, Boston, MA: Harvard Business School Press.

Eiglier, P. (2004), *Marketing et stratégies des services*, Paris: Economica.

Eiglier, P. and E. Langeard (1987), *Servuction, le marketing des services*, Paris: McGraw Hill.

Farrell, D. (2004), 'Beyond offshoring: assess your company's global potential', *Harvard Business Review*, **82** (12), 82–90.

Gallouj, C. (2007), *Innover dans la grande distribution*, Brussels: De Boeck.

Gallouj, C. and F. Gallouj (1996), *L'innovation dans les services*, Paris: Economica.

Gallouj, F. (1994), *Economie de l'innovation dans les services*, Paris: L'Harmattan.

Gallouj, F. (2002), *Innovation in the Service Economy: The New Wealth of Nations*, Cheltenham, UK and Northampton, MA, USA: Edward Elgar.

Gallouj, F. (2003), 'Innovation dans une économie de service', in P. Mustar and H. Penan (eds), *Encyclopédie de l'innovation*, Paris: Economica, pp. 109–128.
Ghemawat, P. (2007), 'Managing differences: the central challenge of global strategy', *Harvard Business Review*, **85** (3), 59–68.
Gottfredson, M., R. Puryear and S. Phillips (2005), 'Strategic sourcing: from periphery to the core', *Harvard Business Review*, **83** (2), 132–139.
Guan, J. and N. Ma, (2003), 'Innovative capability and export performance of Chinese firms', *Technovation*, **23**, 737–747.
Heskett, L.J. (1995), 'Strategic services management: examining and understanding it', in J.W. Glynn and G.J. Barnes (eds), *Understanding Services Management*, Chichester: Wiley, pp. 449–473.
Karmarkar, U. (2004), 'Will you survive the Service Revolution?', *Harvard Business Review*, **82** (6), 101–107.
Lefebvre, E., L.A. Lefebvre and M. Bourgault (1998), 'R&D related capabilities as determinants of export performance', *Small Business Economics*, **10**, 365–377.
Léo, P.-Y., J.-L. Moulins and J. Philippe (2006), 'Profils d'internationalisation pour les services', *Revue Française de Gestion*, **167**, 15–31.
Léo, P.-Y. and J. Philippe (1999), 'Stratégies tertiaires des exportateurs industriels', *Economie et Sociétés, série Economie et Gestion des Services*, **1** (May), 17–43.
Léo, P.-Y. and J. Philippe (2008), 'Internationalisation et stratégie d'offre de prestations pour les services aux entreprises', *Economies et sociétés, série Economie et Gestion des Services*, **9** (March), 437–458.
Lovelock, C.H. (1992), 'Cultivating the flower of services: new ways of looking at core and supplementary services', in P. Eiglier and E. Langeard (eds), *Marketing, Operations and Human Resources: Insights into Services*, Aix en Provence: IAE, pp. 296–316.
Lovelock, C.H. and G.S. Yip (1996), 'Developing global strategies for service businesses', *California Management Review*, **38** (2), 64–86.
Quester, P.G. and J. Conduit (1996), 'Standardization, centralization and marketing in multinational companies', *International Business Review*, **5** (4), 395–421.
RESER, (ed.) (1995), 'Réseaux et sociétés de conseil en Europe', Aix en Provence: Serdeco.
Shostack, G.L. (1977), 'Breaking free from product marketing', *Journal of Marketing*, April, 73–80.
Soete, L. and M. Miozzo (1990), *Trade and Development in Service: A Technological Perspective*, Netherlands: MERIT.
Sundbo, J. and F. Gallouj (2000), 'Innovation in services as a loosely-coupled system', in J.S. Metcalfe and I. Miles (eds), *Innovation Systems in the Service Economy*, Dordrecht, The Netherlands: Kluwer Academic Publishers, pp. 43–68.
Välinkangas, L. and U. Lethinen (1994), 'Strategic types of services and international marketing', *International Journal of Service Industry Management*, **5** (2), 72–84.
Vandermerwe, S. and M. Chadwick (1989), 'The internationalization of services', *Service Industry Journal*, **9** (1), 79–93.
Wakelin, K. (1998), 'Innovation and export behavior at the firm level', *Research Policy*, **26**, 829–841.
Zimny, Z. and P. Mallampally (2002), 'Internationalization of services: are the modes changing?', in M. Miozzo and I. Miles (eds), *Internationalization, Technology and Services*, Cheltenham, UK and Northampton, MA, USA: Edward Elgar, pp. 87–114.

22 The role of standards for trade in services: hypotheses and first insights

Knut Blind

22.1 Introduction

After addressing the reduction of non-tariff barriers to trade in goods markets since signing the Maastricht treaty in 1992 in order to achieve the Single Market within the European Community, and the Tokyo round of the GATT (General Agreement on Tariffs and Trade) negotiations in 1979, the focus in the discussion about the liberalisation of trade has shifted towards the trade in services. In 1995, the GATS (General Agreement on Trade in Services) was published, and in 2006, the European Union published the European Directive on the trade of services, which is based de facto on the country of destination principle, but relies heavily in its implementation on service standards to be produced by the European standardisation bodies.

Despite the high and still increasing relevance of trade in services, trade statistics covering the service sectors – in contrast to trade in manufacturing sectors or of products – are still not well developed. Nevertheless, first studies based on simulation studies (Copenhagen Economics, 2005) show that the liberalisation of trade in services will have positive impacts on trade volumes, and also on growth and employment. However, in contrast to the regulatory non-tariff barriers to trade, the issue of standards published by national, European or international standardisation bodies, as well as standardisation consortia, has not yet been addressed, either in the theoretical or in the empirical trade literature. Swann et al. (1996) followed by Blind and Jungmittag (Blind and Jungmittag, 2005a, 2005b) analysed the role of technical standards in the goods market. They found some ambivalent effects, although international technical standards especially are in general trade-enhancing. Nevertheless, assessing the impacts of standards in trade with services is more complicated, since besides the simple cross-border trade equivalent to the trade in products, other types of service trade – i.e. cross-border consumption, commercial presence or temporary movement of natural persons, which are even more relevant – must be taken into account, especially in the future.[1] These other types of service trade require an in-depth analysis of the relevance and impacts of standards in trading services. The objective of this chapter is to present first hypotheses about the influence of the different modes of trade on the requirements for different types of standards. Since empirical investigations addressing this complex relationship do not exist, I present results of a survey among service companies of their requirements for service standards and their role in companies' success. The chapter concludes with an outlook on future empirical research to address this complex relationship.

22.2 Modes of trade in services

One key characteristic of many services is their intangibility and this means they are not tradable in the conventional sense.[2] Services cannot usually be put on boats, trucks or planes. The distinction between the intangibility of services and the tangibility of goods is less a sharp demarcation than a continuum along which outputs contain varying proportions of services and goods or tangible and intangible elements. We can observe an increasing trend that more goods embody some accompanying service and most services embody some intermediate goods. Consequently, trade in some services may be achieved through trade in the goods in which they may be embodied. For other services, trade may require accompanying goods to be traded in order for a transaction to take place. Further services which are or come closer to being pure services may require little or no accompanying trade in goods. However, if they are based on information and communication technology (ICT) for international delivery, these are necessarily goods required as infrastructures and peripherals.

From the perspective of international transaction in services, and especially the statistics dealing with such transactions, the service component embodied in many goods cannot be separately identified from the trade in goods themselves. Thus service outputs that are readily embodied in goods are excluded from the definition of services in trade statistics. Computer software is one example of this case. Essentially, computer software provides a service to the user rather than a tangible good, yet where this service is supplied in a physical good such as a diskette or CD-ROM, it is counted as trade in goods in trade statistics.

The technical possibility of embodying services in goods is not the only determinant of the tradability of services. A closely related feature of services, which affects their tradability, is the need for interaction or proximity between producer and consumer. In the extreme case, this interaction requires that both the supplier and the consumer be present for a transaction to take place, so-called co-terminality. Cross-border supply alone is only an option where the required level of supplier–consumer interaction is low, even though this does not necessarily mean that the frequency of interaction is low. Nonetheless, cross-border trade is important for some service sectors. There are various attempts to characterise international service transactions (Sampson and Snape, 1985; Sapir and Winter, 1994; Vandermerwe and Chadwick, 1989). A basic distinction is drawn between services that may be traded in a conventional sense by cross-border supply and those that require factor movements, i.e. the movement of the supplier either temporarily or permanently to the location of the client.

However, the General Agreement on Trade in Services (GATS) provides a useful way to conceptualise the various forms of services trade and to capture different kinds of linkages across these forms of trade, through a mode-wise classification. The four GATS modes of services trade include (see, e.g., the description in Chanda, 2006):

- Cross-border supply (mode 1), when the service is delivered within the territory of the consumer from the territory of the service supplier and actually crosses national borders. Cross-border supply may entail delivery by mail, telecommunication or satellite-based data or information flows (e.g. distance education, telemedicine, financial transfers done electronically) from one country to another.
- Consumption abroad (mode 2), which refers to trade in services where the consumer of the service moves to consume a service in another country (e.g. tourism, persons seeking medical treatment abroad, students pursuing higher education in other countries).
- Commercial presence (mode 3), which refers to the establishment of any type of business or professional enterprise in the foreign market for purposes of supplying a service. This mode involves granting of rights to a foreign interest to establish an investment in the host country. Commercial presence includes foreign equity participation, establishment of corporate subsidiaries, joint ventures, partnerships, sole proprietorships, associations, representative offices, and branches or acquisition of such entities (e.g. financial, telecom, distribution services).
- Movement of natural persons (mode 4), which refers to the temporary cross-border movement of individuals, either in an independent capacity or as part of a commercial establishment (e.g. software, health, and construction services). One of the distinguishing characteristics of mode 4 is that it is temporary and that it is distinct from economic migration, which involves entry into the permanent labour market and citizenship.

While this taxonomy is a conceptual framework for understanding and analysing trade in services, in reality it is often difficult to distinguish between the different modes of trade in services. Multiple modes may be simultaneously involved in the course of services trade and one mode may be subsumed within another. Often a particular mode may be the primary form of commercial exchange between two countries, but exchange through the latter mode may only be possible when supported by other modes. For instance, a service may be primarily traded through information and data flows over the Internet and via satellite links, but the latter may only be possible through some supporting temporary movement of professional service suppliers between the two countries to execute the offshore project. While in the purely commercial sense the primary form of trade and foreign exchange earnings here is through the offshore component, there is related reliance on temporary cross-border movement of service providers to supervise and coordinate the work. From the perspective of a firm that is engaged in services trade, the two modes would be seen as an integrated package and as contributing together to the foreign exchange earned by executing this work. Thus, in reality it is difficult to map the GATS modes directly as to how services are actually traded.

Karsenty (1999) tries to give an indication of the relative importance of the diffusion modes of supply. More than 40 per cent of total services are provided by cross-border supply, e.g. financial services transmitted via telecommunication. Another 20 per cent is taken by consumption abroad in the way that a service consumer moves into another country's territory to obtain a service, e.g. tourism and travel. A third mode of service trade requires the commercial presence of service suppliers in another country to provide the services there, e.g. hotel or fast food chains. This third mode accounts for more than a third of service trade. Although the movement of personnel to supply a service abroad, e.g. in consulting or education services, is responsible for less than 1 per cent of total trade, there are trends, which will lead to increasing shares of this mode of service trade.

22.3 Hypotheses on the role of standards for the modes of trade in services

According to Swann et al. (1996) empirically supported by Blind and Jungmittag (2005a, 2005b), standards in general have a positive impact on companies' and countries' trade performance by allowing products to be provided at lower costs and higher quality. The trade impact for all actors together is higher if common international standards are published, whereas idiosyncratic national standards may create frictions in the exchange of goods across national borders. Furthermore, the additional administrative burden of applying national standards may diminish the competitiveness of the national suppliers. However, the sole existence of standards is trade-enhancing because of their positive cost-decreasing effect and the reduction of information asymmetries between the supply and the demand side, especially in the case of cross-border transactions.

These positive impacts of standards in the product area are also relevant for the trade in services, although the different modes of trade in services must be taken into account. Furthermore, Swann et al. (1996) only distinguish between general cost-saving, quality-ensuring and variety-reducing standards following David's taxonomy of standards according to their economic function (David 1987). However, the role of standards for services or the service industries is not the same and requires further differentiation. Despite the fact that no commonly agreed taxonomy of standards exists, I rely on previous attempts by De Vries (2001) and recently Blind (2006a) to provide typologies or taxonomies of service standards summarised in tables 22.1 and 22.2.

In order to determine the requirements to develop standards for the different modes of trade in services, I rely on the general concept of De Vries (2001), who defines standards as coordination instruments for entities or for relations between entities. The four different modes of trade in services influence either the distance between existing entities involved in the service-providing process or the creation of new entities. The assessment is always measured in comparison to the baseline of the domestic provision of services. Under distance I understand both the physical distance in kilometres and cultural differences between the service provider and the customer, which correlate highly with each other.

The distance to the customer will increase in all modes of service trade with the exception of the consumer visiting the service provider.[3] The distance from the home-base to the personnel actually producing or delivering the service is especially high in the case of commercial presence abroad and in moving personnel abroad. However, if we consider the real distance between the service provider and the customer, the cross-border supply is characterised by the largest distance, whereas it is only somewhat higher than in the domestic service provision process.[4] The different modes of trade in services also have an influence on the heterogeneity of input factors, another kind of entity in the service process. The commercial presence abroad, but also the temporary movement of personnel abroad, require reliance on input factors, either equipment or even supporting staff, which is likely to be different from the inputs used for the domestic service-providing process. Similar to the inputs, the service process itself is likely to be different, if the companies establish a commercial presence abroad due to different external factors and framework conditions.

Besides different input factors and processes, the four modes of service trade also require adaptations in the service companies themselves, e.g. by the creation of affiliations abroad or by establishing a department in the home-base responsible for the cross-border supply. If the driving factors for standards in the context of service trade are summarised, the

Table 22.1 Services standards typology

Standards for: entities or relations between entities	. . . may concern:
Service organisation	Quality management, environmental management, occupational health and safety management Solvency and other financial aspects Crew, e.g. minimum number of staff and their educational level
Service employee	Knowledge Skills Attitude Ethical code (e.g. confidentiality)
Service delivery	Specification of activities Trustworthiness Privacy aspects Safety aspects Code of conduct
Service result	Result specification Trustworthiness
Physical objects supporting service delivery	E.g. technical requirements for trains in public transport services
Workroom	E.g. requirements for daylight access in offices
Precautions	Emergency measures Complaints handling Guarantee
Additional elements to the core service – delivery	E.g. waiting facilities
Additional elements to the core service – results	
Communication between customer and service organisation (before, during and after providing the service)	Semantics (e.g. data elements to be used) Syntax (e.g. forms layout, syntax rules for electronic messages)

Table 22.1 (continued)

Standards for: entities or relations between entities	. . . may concern:
	Specification of Information and Communication Technology to be used Protocols Code of conduct Approachability (e.g. hours of accessibility per telephone and average waiting time)
Communication within the service organisation or between this organisation and its suppliers	Semantics Syntax Specification of ICT to be used Protocols Code of conduct Approachability

Source: De Vries (2001).

commercial presence abroad will generate the largest and broadest needs. However, the movement of natural persons requires some additional efforts in the area of standardisation and even the simple cross-border supply of standards causes some organisational changes, which can more easily be solved by adequate standards. This assessment of general driving forces for standardisation related to the trade of services allows me, in a next step, to derive more specific needs of standards differentiated by the four modes of service trade (Table 22.3).

The combination of the insights of Table 22.3 with modifications of the two typologies or taxonomies of service standards in Table 22.1 and Table 22.2 leads us to the following mapping of standards needs illustrated in Table 22.4.

This mapping of additional demand for services in Table 22.4 reflects the general assessment of the additional needs for standards in the four modes of service trade. However, the categorisation also reveals that cross-border supply and consumption abroad have similar requirements, on the one hand, for service standards, and on the other hand for commercial presence and the movement of natural persons. Furthermore, there are still very specific needs for standards in the four different modes of trade in service. For example, standards for accessibility in cross-border trade mean a rather remote access to the service providers, e.g. via a telephone or e-mail hotline, whereas accessibility standards for consumption abroad

*Table 22.2 Standards-related factors of services**

Service management	Service employee	Service delivery	Customer interaction	Data flows and security
Quality management	Qualifications and skills	Classification of services	Evaluation of services by customers	Data flows formats (customer interaction)
Environmental management	Further education	Service description	Code of conduct (customer contact)	Data security (customer interaction)
Health and safety management	Ergonomics	Equipment supporting service delivery	Customer and consumer information	Data flows formats (internal interaction)
		Service process	Accessibility	Data security (internal interaction)
		Customer satisfaction	Customer satisfaction	
			Code of conduct (internal interaction)	
			Organisation models	
			Information systems	

Note: * I have not included 'Terminology', since it applies to all companies and consequently factors in the same intensity.

require that foreigners coming to the home-base of the service company have easy physical access to the offices of the service providers.

Depending on the dominant mode of service, trade service sectors require different types of standards. However, the service-specific characteristics dominate the mode of service trade distribution, because according to Chanda (2006), sectors are mostly characterised by a mixture of trade modes.

In a final step, it has to be discussed whether the various types of standards mapped in Table 22.4 have to be published as formal standards published by national or, better by international, accredited standardisation

Table 22.3 Criteria for determining the relevance of standards depending on the modes of service trade

Modes of service trade	Distance to customer from home-base	Distance to service provider from home-base	Distance between service provider and customer	Heterogeneity of inputs (labour, capital) compared to home-base	Heterogeneity of process compared to home-base	Needs for organisational change	General needs for standards
Cross-border supply (mode 1)	high	low	high	low	medium	medium	low-medium
Consumption abroad (mode 2)	low	low	low	low	low	low	low
Commercial presence (mode 3)	high	high	low	high	high	high	high
Movement of natural persons (mode 4)	high	high	low	medium	medium	medium	medium-high

*Table 22.4 Additional needs for standards depending on the modes of service trade**

Mode	Service management	Service employee	Service delivery	Customer interaction	Data flows and security	Organisation
Cross-border supply	Quality, safety, environmental management		**Service description** Service process	User feedback Code of conduct Customer information Accessibility	Data flow formats Data security	Organisation models Information system
Consumption abroad	Quality management	Qualifications and skills	Service description	User feedback Code of conduct Accessibility	Data flow formats Data security	
Commercial presence	**Quality, safety, environmental management**	**Qualifications and skills**	**Service description** **Equipment supporting service delivery** **Service process**	**User feedback** **Code of conduct** **Customer information** Accessibility	**Data flow formats** **Data security**	**Organisation models** **Information systems**
Movement of natural persons	**Quality, safety, environmental management**	**Qualifications and skills**	**Service description** **Equipment supporting service delivery** **Service process**	**User feedback** **Code of conduct** **Customer information** **Accessibility**	**Data flows formats** **Data security**	Organisation models Information systems

Note: * The presentation of a standard type in bold indicates a 'high' relevance.

bodies or whether firm-specific solutions are sufficient or more adequate. Again, the conceptual approach to perceive standards as instruments to coordinate relations between entities or within entities is helpful to set priorities, since, in general, formal standards are more beneficial from the perspective of the whole economy or society. However, this approach neglects the costs of producing all these standards, and the incentives of companies to develop and implement company-specific solutions in order to achieve a competitive advantage.[5]

At first, standards which organise the relationship between service companies and other entities involved in the service-providing process should be developed within formal standardisation bodies and should be published as formal standards, because they address market failures. If standards reduce negative externalities generated for the environment during the service-providing process, then formal standards or even governmental regulations are more adequate from the efficiency point of view than firm-specific solutions. Furthermore, the reduction of information asymmetries or deficits of customers related to the characteristics – e.g. quality – of the provided service should basically be achieved by formal standards and not by company-specific solutions. This also covers classifications of services. Since these two types of standards are closely linked to regulatory issues, they should even be provided by European or international standardisation bodies in order to avoid distortions or even races to the bottom (Sinn, 1997) in the market of service providers.

Furthermore, the trade of services often leads to the creation of new entities, e.g. affiliations abroad. Depending on their legal status, formal standards may facilitate their interaction, especially in case of independence, e.g. the foreign affiliate does not only provide supporting services abroad for one company, but for several companies located in their different home markets. Trade in services often requires a further differentiation of the value chain by further outsourcing parts of the whole service package to newly established niche providers. The development of a set of formal standards organising these new relationships will foster the development of new service markets more efficiently than company-specific or bilateral solutions.

Finally, the relationship between service companies and their employees, often also independent entrepreneurs, becomes more complicated if companies set up commercial presences abroad or even send persons into other countries temporarily to fulfil services. In general, their relationship is defined by the domestic labour market regulations complemented by company-specific labour contracts. However, since an international labour market regulation is missing, international formal standards could contribute towards harmonisation, especially in the case of setting up a commercial presence abroad or moving staff temporarily into other countries.

Since the trade of services is often accompanied by significant flows of data between the service providers and their customers, broadly accepted and implemented standards are required to allow and facilitate these data flows. Closely related to data flows is the issue of data security. Both aspects are already dealt with in international standardisation organisations, like ISO.

In contrast to the aspects discussed above, which require formal standards and in some circumstances even governmental regulation, all company-internal standardisation issues, like the internal organisation of the service-providing process and even codes of conduct, are of secondary importance. However, the establishment of successful, but also transferable internal solutions would be a good starting point for launching formal standardisation processes in order to release general models, which would be beneficial for specific service sectors or the whole service industry.

22.4 First empirical investigations on the role of standards for service trade

22.4.1 The sample

A survey performed in 2002 (see Blind, 2003 for details) covered all member states and the sample was distributed according to the size of service companies and to their sectors. The addressees were approached via e-mail and a link to an Internet-based questionnaire. Due to the general lack of systematic knowledge about the use and preferences of service companies for service-related standards, it is crucial for the interpretation of the results to describe the organisation of the survey and the target group approached and the responding companies. A total of 364 service companies answered the questionnaire adequately, which is just below 2 per cent of the total sample of more than 17,000 addressed companies that were randomly chosen. This response rate is rather low compared to other non-mandatory business surveys in Europe, for two reasons. Firstly, the issue of standards within service companies is not yet broadly acknowledged or taken into account. Secondly, the response to European-wide Internet surveys is significantly lower than to paper-based surveys.

Although there is a strong bias towards Germany, the United Kingdom and the Scandinavian and Benelux countries are well represented. The Southern European countries are slightly underrepresented. For the further analyses, I built clusters of the Scandinavian, Benelux, Southern European and the German-speaking countries.

Since I assume that the differences between countries are not so relevant regarding the use and the preferences for standards, I have to consider the sector distribution more closely. One-third of the service companies deliver business-related services, such as computer and software services or research and development services. Educational and social services are provided by 22 per cent of the companies, financial and real estate services by 13 per cent. Finally, 9 per cent of the sample are active as wholesalers and resellers, and 7 per cent in the transport and communication business. In addition, 13 per cent of the sample is composed of manufacturers, who also provide services. This grouping of the very heterogeneous service sector allows for a significant number of companies with similar services in each class.

Regarding the focus of this analysis, the identification of specific needs of companies exporting services, the database does not allow a distinction into the four modes of trade of services, but only a division into those companies exporting services at all, and those without any export activities. From the answers of the 364 companies, I was able to identify 151 with export activities and 166 without exports. The remaining 47 could not be classified due to missing data. Data restrictions also do not allow expanding the analysis of export activity down to the sectoral level, which could at least give an indirect indication of the interdependence between modes of service trade and role of standards.

22.4.2 Empirical results

In presenting the survey results, I focus on the export-related aspects, which unfortunately cannot differentiate between the modes of service trade as presented above. Figure 22.1 indicates clearly that service companies actively involved in export activities are much more likely to participate actively in formal standardisation processes, which has already been observed for companies in the manufacturing sector (Blind, 2006b). This likelihood increases if we look at standardisation processes at the European or even international level.[6]

Whereas it is obvious that the actively exporting companies are trying to further their interests within national, but even more within supranational standardisation bodies, the overview of the additional requirements to standardise for trade in services leads to the conclusion that service companies active in standardisation should place greater importance on several types of standards. However, the qualitative assessment of the relevance of standards for different service-related aspects only shows significant differences in very few dimensions between companies involved in exports and those not involved (Figure 22.2). In several dimensions, the companies not involved in export activities even perceive a significantly higher relevance for standards.

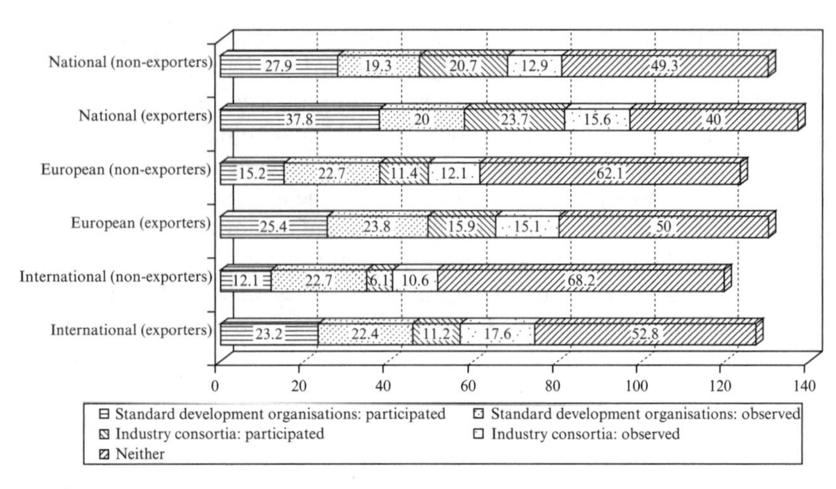

Figure 22.1 Activities in standardisation

However, the general assessment of the relevance of standards for specific service-related dimensions may be appropriate for identifying priorities for future standardisation activities, but not for actual practice. Therefore, I present the average number of standard types used, differentiating between company-specific, industrial, formal national, European and international standards respectively (Figure 22.3). Here, it becomes obvious that companies involved in standardisation implement both more company-specific, but also a larger number of formal standards released by national or supranational bodies. However, the differences are not significant.

In a last analytical step, I applied Chi-Square tests in order to identify significant differences between exporting and non-exporting companies in each separate standardisation category, differentiated by the type of standard implemented. In half of the categories, this analysis reveals a significantly higher implementation of standards by exporting companies. In the categories of health and safety management, further education, ergonomics and general terminology they indicate significantly less often that they do not rely on standards at all. Company-specific standards are more often used in structuring the code of conduct in relation to customers, but also internally, and in internal issues of data security. Industry standards are only more often used by exporting companies in organising the internal code of conduct. Formal standards released by national bodies are more likely to be used by exporting companies regarding the classification of services and related to further education of employees. Finally, European and international standards are more frequently implemented

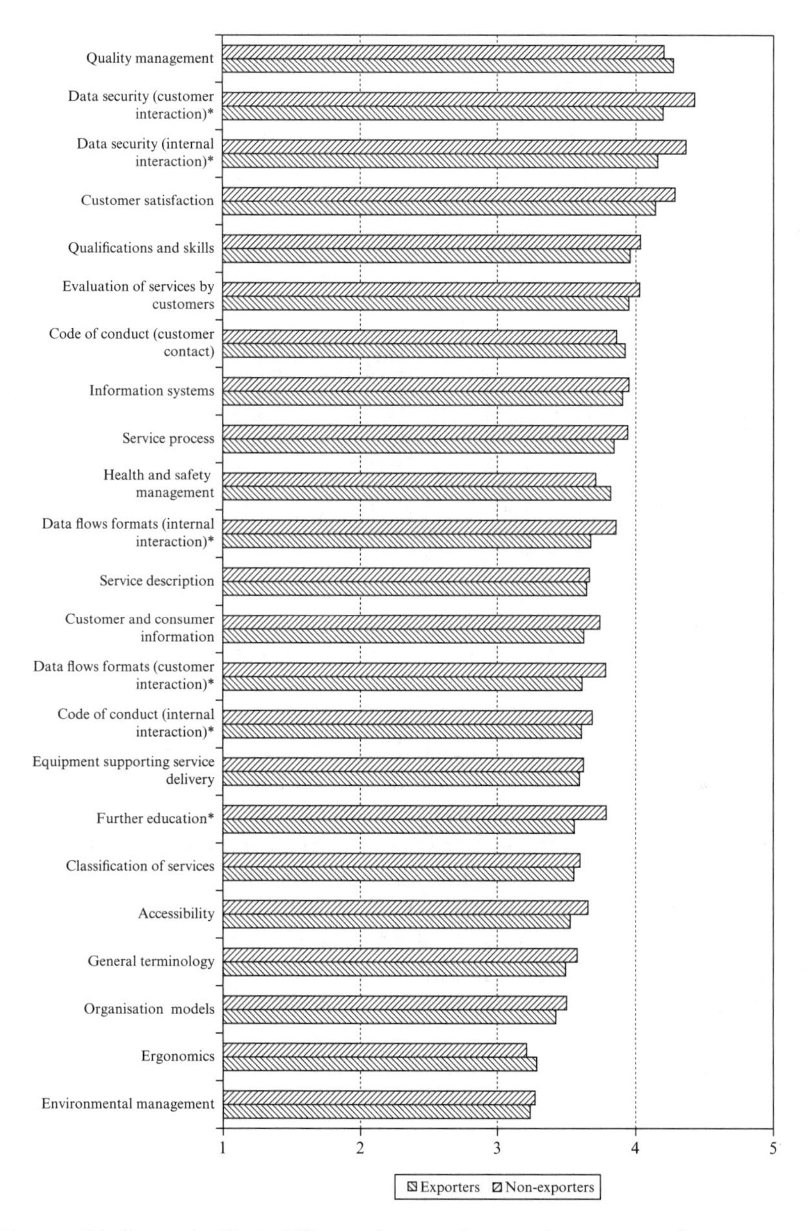

Note: a * indicates significant differences between the two subgroups according to a t-test.

Figure 22.2 Relevance of standards for different service-related aspects[a] *(1 = low relevance to 5 = high relevance)*

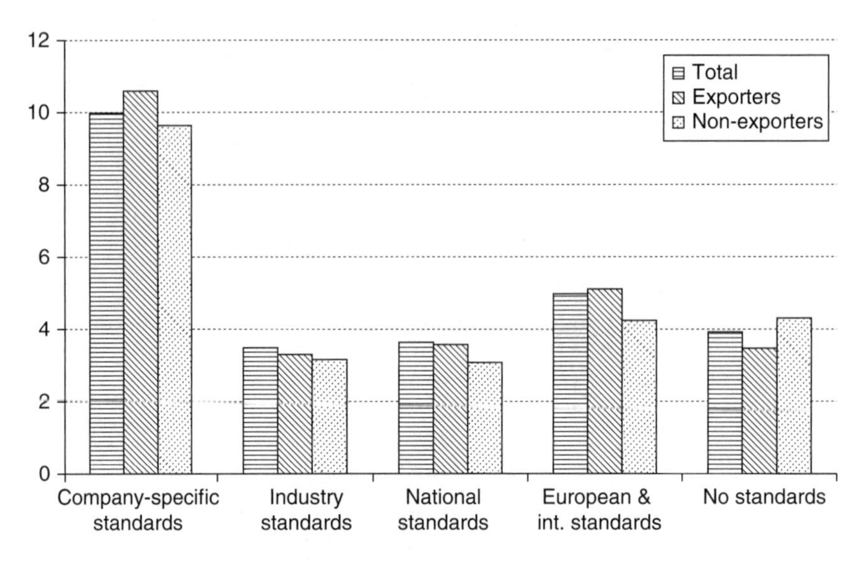

Figure 22.3 Average use of standards types

in the quality management of services, e.g. ISO 9000, but also to describe services and to coordinate the evaluation of services by customers.

In particular, the analysis of the implementation of specific standards supported my general hypothesis that companies actively involved in service trade have a higher demand for standards. It was found that supranational standards, especially for quality management and service description, but also for user feedback (i.e. the evaluation of services by customers), are used with significantly greater frequency by exporting companies, which confirms my hypotheses on the higher needs for specific standards by companies involved in service trade (see Table 22.4). However, the database does not allow differentiation of the requirements for standards according to the different modes of service trade.

22.5 Outlook for further research

The previous analysis of the dataset revealed that the hypotheses on the requirements for standards in service industries in relation to the different modes of service trade cannot be tested. However, the empirical investigations confirmed in general the greater needs of service companies involved in service trade to implement various types of standards. Future research, which will be performed within the German project Standard:IS, service standards in successful internationalization strategies, (http://www.standard-is.de/index.php?lang=en&na_id=standardis), will, on the one hand, address the differentiation of different types of service trade. On the other hand, some further research is still required to establish a typology of service standards, because they are much more difficult to define compared with standards describing nuts and bolts. Finally, the discussion about liberalising the cross-border trade of services has not yet taken the role of service standards into account. For example, at the European level, service standards can be considered to play a major role in the harmonisation of the various service markets in the 'New Approach' context, i.e. implementing the Service Directive by European service standards. Further research will also be required into the interplay between service regulation and service standardisation.

Notes

1. Meanwhile the traditional trade of goods is often accompanied by these different forms of service trade, e.g. by sending engineers abroad temporarily to install, repair or maintain the machines or even open an affiliation responsible for maintenance and repairing services.
2. The following section follows partly Baker et al. (2002).
3. Here the difference in the cultures between the service provider and the customer does not coincide with the physical distance. However, it can be assumed that customers going abroad to receive a specific service are rather tolerant to the conditions in the host country.
4. Blind (2004) already argued that providing services via e-business increases the distance to the customers, which generates specific needs for standards organising and stabilising the relationship to the customers.
5. The survey by Blind (2003) provides an assessment by companies as to whether they prefer company-specific solutions or formal standards. This reveals that companies often favour both formal standards and company-specific solutions. The latter allows them to generate competitive advantages.
6. The differences between exporters and non-exporting companies are statistically significant based on Chi-square tests.

References

Baker, P., M. Miozzo and I. Miles (2002), 'The internationalization of European services: what can data on international services transactions tell us?', in M. Miozzo and I. Miles (eds), *Internationalization, Technology and Services*, Cheltenham, UK and Northampton, MA, USA: Edward Elgar, pp. 59–86.

Blind, K. (2003), 'Standards in the service sectors: an explorative study', final report for DG Enterprise of the European Commission, Karlsruhe: ISI.

Blind, K. (2004), 'A taxonomy of service standards and a modification for e-business', in P. Cunningham and M. Cunningham (eds), *Adoption and the Knowledge Economy: Issues, Applications, Case Studies*, Amsterdam: IOS Press, pp. 264–270.

Blind, K. (2006a), 'A taxonomy of standards in the service sector: theoretical discussion and empirical test', *Service Industries Journal*, **26** (4), 397–420.

Blind, K. (2006b), 'Explanatory Factors for Participation in Formal Standardisation Processes: Empirical Evidence at Firm Level', *Economics of Innovation and New Technology*, **15** (2), 157-170.

Blind, K. and A. Jungmittag (2005a), 'The impacts of innovations and standards on German–French trade flows', *Économie Appliquée*, **58** (2), 99–125.

Blind, K. and A. Jungmittag (2005b), 'Trade and the impact of innovations and standards: the case of Germany and the UK', *Applied Economics*, **37** (12), 1385–1398.

Chanda, R. (2006), 'Inter-modal linkages in services trade', OECD Trade Policy Working Paper, Paris: OECD.
Copenhagen Economics (2005), 'Economic assessment of barriers to the internal market for services', Brussels: DG Enterprise.
David, P.A. (1987), 'Some new standards for the economics of standardisation in the information age', in P. Dasgupta and P. Stoneman (eds), *Economic Policy and Technological Performance*, Cambridge: Cambridge University Press, pp. 206–239.
De Vries, H.J. (2001), 'Systematic services standardization from consumer's point of view', contribution to the ISO Workshop in Oslo, May.
Karsenty, G. (1999), *Just How Big are the Stakes? An Assessment of Trade in Services by Mode of Supply*, World Trade Organization.
Sampson, G.P. and R.H. Snape (1985), 'Identifying the issues in trade in services', *World Economy*, **8**, 171–181.
Sapir, A. and C. Winter (1994), 'Services trade', in D. Greenaway and L.A. Winters (eds), *Surveys in International Trade*, Oxford: Blackwell, pp. 273–302.
Sinn, H.-W. (1997), 'The selection principle and market failure in systems competition', *Journal of Public Economics*, **66**, 247–274.
Swann, G.M.P., P. Temple and M. Shurmer (1996), 'Standards and trade performance: the UK experience', *Economic Journal*, **106**, 1297–1313.
Vandermerwe, S. and M. Chadwick (1989), 'The internationalization of services', *Services Industry Journal*, **9**, 79–83.

23 Global and national cooperation in service innovation

Xavier Vence and Alexandre Trigo

23.1 Introduction

The overdue attention to innovation in services could be attributed to the traditional disregard for services in general and the widespread assumption about their potential lack of innovative capacity, being considered just users of the innovation produced by the manufacturing activities (Pavitt, 1984; Barras, 1986; Evangelista, 2000). In short, services did not receive due credit for being innovative, to a certain extent, because of both their low research and development (R&D) intensity and patent application (Miles and Boden, 2000; Salter and Tether, 2006). Maybe the situation is also attributable to the traditional reluctance of economics against the analysis of nonprofit innovation, just when the public sector was responsible for most important service activities. However, when examples of innovation in services such as telecommunications, computer activities and logistics became better known, researchers recognized some services as having innovative characteristics. Maybe the wave of public service privatization around the world has contributed to attracting the attention of economists, providing conditions for an easier application of a 'market conceptual toolkit' of economic analysis.

Nowadays, services have become a topic under the focus of the community of scholars dealing with innovation issues (Gallouj, 2002a, 2002b, 2007; Miles, 2001, 2005; Salter and Tether, 2006). But an understanding of innovation in most of the service activities has required much greater focus and study, as well as a reconsideration of previously held concepts about innovation (Miles and Tether, 2003; Foray, 2004).

The attempt to recognize the innovative character of the service activities has demanded a new perspective beyond the traditional view of innovation, allowing an easier identification of such characteristics. Aspects such as non-technological innovation or client–producer interaction permit us to understand better the real scope of innovation in services and so the innovation concept. The assumed 'specificities of innovation' in the service activities (Sundbo and Gallouj, 1998; Sundbo, 2000) led researchers to a reconsideration of the concepts of industrial innovation in order to be applied to the services. This change has meant a significant advancement in the economics of innovation. However, a new step forward is required because innovation in services does not follow a uniform pattern. Significant differences among service innovation processes have been found due to the diversity of activities. The scope and the nuance of that diversity and differentiation have been discussed critically elsewhere (Vence and Trigo, forthcoming).

Although many studies still attempt to highlight the inferiority of the innovative character of the service sector as compared to the manufacturing sector, innovation in specific service activities may actually be considered higher than the manufacturing average (Miles, 2001; Vence and Trigo, forthcoming). Service firms are more prone to collaborate with customers and suppliers, whereas manufacturing firms are more prone to collaborate through in-house R&D (Tether, 2005). Newer analyses have shown services to be a source of innovation, in contrast to the traditional view (Barras, 1986; Tidd et al., 2001). However, innovation in some service activities lags behind the manufacturing innovation average (similarly to less-innovative manufacturing). That evidence shows that innovation in services may not be analysed using a one-size-fits-all model, and that these activities have different innovative goals. While many firms find their innovation activities constrained due to weak managerial and workforce skills, 'some services are amongst the most sophisticated businesses in the world' (Salter and Tether, 2006, 2).

Looking ahead, beyond a simple recognition of the aggregate similarities and differences between industrial and service innovation, we need to understand the internal diversity of services and overcome the simplistic vision of the 'specificities of the innovation' in that sector. Analysing the innovation patterns of this sector and attempting to create a common model which describes all service activities is a difficult (and perhaps inappropriate) task, given their heterogeneity and complexity. Therefore, the aim of this chapter is both to understand better the differences in innovation patterns within the service sector and to draw a taxonomy of innovation patterns in services, paying special attention to the differences in the geography of cooperative innovation links. This taxonomy will be made from the intra-industrial analysis of a set of innovation indicators provided by the third and fourth Community Innovation Survey (European Commission, 2004). Of course, in order to do so a fine disaggregation of services activities is needed. In this sense, our empirical analysis is constrained by the CIS3 classification of service activities into four broad subsectors: wholesale trade and commission trade; transport and communication; financial intermediation; and knowledge-intensive business services (KIBS).

23.2 Overview on innovation patterns in services

A review of the recent studies on service innovation shows that there are not only certain specificities compared to the innovation process developed by the manufacturing industry, but also that there exist considerable differences among services themselves (cf. den Hertog and Bilderbeek, 1999; Sundbo and Gallouj, 2000; Miles, 2001; Vence and González, 2002; Tether and Metcalfe, 2004; Chapter 1 in this volume). However, empirical analyses demonstrate that many of these particularities that are typically service-related (e.g. the close interaction between service providers and customers, the customization of service products, the cooperation with other agents in order to innovate) can also be found in other industries. In short, services do not innovate as differently as expected. Hence, considering sectors as a whole, innovation patterns in both manufacturing and services more and more present comparable features. Intra-industrial analyses, conversely, point out considerable differences among activities.

By the 1980s the continuous rise of service activities had attracted the attention of scholars and policy-makers. Until then the service sector had been disregarded by innovation researchers. The first attempts to understand better the innovation in services took into account the current innovation research and so the conceptual tools developed in a manufacturing sphere of innovation (Barras, 1986). That approach, called the 'assimilation perspective', consists of including services into the current innovation research. Researchers, investigating traditional manufacturing innovation, tend to study service innovation in a manner similar to that in which they study the manufacturing industry (Coombs and Miles, 2000; Miles, 2001; Howells, 2000, Chapter 1 in this volume). Barras considered, fundamentally, the impact of ICT – information and communication technology – on finance services. In brief, the assimilation perspective perceives service innovation in the same way as manufacturing innovation, but nowadays it is well known that the technological dependence is not equal for all activities.

In order to understand better the innovation patterns in services, Soete and Miozzo (1989), Miozzo and Soete (2001) and Bell and Pavitt (1993) created a taxonomy for service innovation founded on the former Pavitt taxonomy (Pavitt, 1984). Pavitt, in the first version of his well-known taxonomy, classifies services, in particular private ones, as supplier-dominated, but in fact, in the 1993 version he includes a new pattern, information-intensive, represented by some service activities such as finance, retailing and publishing (Bell and Pavitt, 1993). In the same vein Soete and Miozzo identify different innovation patterns in services, apart from the 'supplier-dominated' one mentioned before. They were also pioneers in describing interactive networks enterprises from an innovation perspective. Hipp and

Grupp (2005) remembered that network-based innovation types were not included in the Pavitt taxonomy. The reason is related to the distributive nature of these activities, which cannot be found in the manufacturing sector. However, Gallouj and Gallouj (2000) remarked that the network character should constitute a transversal characteristic in services rather than one described in the taxonomy. They observed that even 'supplier-dominated' services had such features. Another criticism regarding Soete and Miozzo's contribution concerns the idea of services as simple users of technologies (Gallouj and Gallouj, 2000).

A different perspective of research on innovation in services became eminent in the middle of the 1990s. According to this approach, named the 'demarcation perspective', service innovation is highly distinctive. Subsequently, the use and even the adaptation of conceptual and empirical tools commonly utilised in manufacturing innovation analysis are not suitable (Salter and Tether, 2006). On the contrary, this service innovation approach ascertains the need to use instruments of measure capable of embracing services' particularities such as their intangibility and high levels of interaction. Whereas the assimilation perspective focuses on the technological trajectory of the innovation process, the demarcation approach concentrates attention on organizational innovation and innovation in knowledge-based business services (Salter and Tether, 2006; Chapter 1 in this volume).

Den Hertog and Bilderbeek (1999) proposed a service taxonomy composed of seven types of innovation patterns. They highlighted both services' capacity to innovate by themselves and the non-technological character of the innovation. All innovation patterns indicated consist of a combination of connections between suppliers of inputs, innovative service companies and clients. However, they observed that more innovation patterns could be identified by using different variables such as the government and its role as an innovation promoter, as well as customers' interaction during the service production.

The service taxonomy suggested by Sundbo and Gallouj (2000) consists of an important representation of this research line. They observe that enterprises may present different patterns for different innovation, although certain patterns are more common than others. They observe, also, that despite the existing specificities of services innovation, there is a growing convergence between services and manufacturing innovation (Miles, 2001; Castellacci, 2008).

A third approach to research on innovation in services, the 'synthesis perspective', suggested that service innovation made evident neglected aspects of the innovation process (Coombs and Miles, 2000; Chapter 1 in this volume). Moreover, this approach moved the focus of research from a self-determining sector analysis towards a network of a particular market, in which services and manufacturing were interrelated activities. In that way, the gap between the innovation studies in both services and manufacturing is becoming more and more narrow.

Recently, Hipp and Grupp (2005) empirically evaluated innovation patterns previously ascribed to services activities. Their results show that innovation patterns in services depend to a lesser extent on the sector classification found in each of the studied service industries. This evidence strengthens, thus, the conclusion of Sundbo and Gallouj previously alluded to. Moreover, Hipp and Grupp affirm that the present service innovation taxonomy is fit for services that hold a classical innovation structure. They add that, due to the existence of other innovative service enterprises, new patterns have still to be recognized from both alternative concepts and alternative measure instruments. With a view to upcoming works they, as well as other authors, support the idea that manufacturing and services should be studied together, taking into account 'the products' instead of the sectors in which such outputs were produced.

All these contributions have made substantial progress towards a better understanding of innovation in services. One important consideration is that not all services are 'supplier-dominated'; on the contrary, they can also be sources of new technologies. In fact, some of these technologies are outcomes from co-production between users and producers (Tether, 2002). In that way, services 'are not merely passive recipients of others' innovations' (den Hertog and Bilderbeek, 1999, 4). However, most services possess the 'supplier-dominated' character as to the companies' innovation activities. Another important point concerns the role of organizational innovation in services, highly neglected, particularly, by the assimilation perspective. Lastly comes the recognition of a complex and multidimensional character of services and manufacturing that claims new perspectives regarding industrial innovation analyses. It is thus clear that there is diversity of theoretical approaches to innovation in services and, consequently, different routes to comprehending service innovation.

23.3 Empirical evidence of innovation and cooperation patterns in services

The aim of this section is to identify the main differences among innovation and cooperation patterns across the service sector by using the available statistical data. In order to do so we use the recent data of the third and fourth Community Innovation Survey for European Countries provided by Eurostat. Since our purpose is to establish universal innovation and cooperation patterns for services instead of a specific breakdown for a particular country, we used the EU-15 member states average plus Iceland

and Norway. In order to avoid possible national bias we are using mean values at the European level for each service industry.

A very detailed empirical analysis is constrained by the Eurostat classification of the service activities used basically in the CIS3: wholesale trade and commission trade (NACE 51), transport and communication (NACE I), financial intermediation (NACE J) and business services (computer activities; R&D; engineering activities and consultancy; technical testing and analysis) (NACE 72, 73, 74.2, 74.3).[1]

According to the results of the CIS3 and CIS4, the differences in cooperative and innovative patterns within services are substantial (European Commission, 2004, 2005). The empirical results are summarized in Table 23.1. Here we display main innovation indicators for the service activities. Although the diversity is not clear-cut in certain indicators, the differences in the relative relevance of these attributes may indicate important dissimilarities among subsectors.

23.3.1 *Unlike innovation intensities in service activities*

The proportion of innovative enterprises by subsectors shows that there are substantial differences among the various groups of activities. Transport and communication seem to lag behind the average while business services have the highest percentage (Table 23.1). These activities, clearly innovative, exceed both the manufacturing average and the service average. Wholesale trade and commission trade innovation is closer to the service average. Thus, the extent of the innovative character seems to show considerable heterogeneity within services.

Another indicator, the innovative intensity, used to measure the innovation effort of enterprises (total innovation expenditure/total turnover), confirms the assumption of diversity. However, the comparison between these indicators points out a relevant impact of both the nature of the productive activity and the market structure on the innovation process. In other words, the financial intermediation subsector exhibits the lowest rate of innovation effort, despite the high proportion of innovative enterprises present there (in a sector dominated by the presence of very large firms). Thus, these two indicators will be very important and complementary in the attempt to define innovation patterns.

23.3.2 *Diversity of R&D intensity in service activities*

R&D activities are an important way of internal knowledge creation; not only for manufacturing industries but also for service ones, particularly for some of them. This is more evident after widening the R&D concept in order to include as R&D some creative works devoted to generating new knowledge and innovation. Actually, intra-sectoral analyses reveal that

Table 23.1 Synthesis of the main aspects of the innovation in services, EU-15, Iceland and Norway, 1998–2000

	Main aspects of the innovation	Services		Wholesale trade and commission trade		Transport and communication		Financial intermediation		Business services	
General aspects	*Proportion of enterprises with innovation activity*		40%		35%		28%		58%		64%
Innovation activities	*Most frequent innovation expenditure**	Intramural R&D	34.8%	Intramural R&D	26.3%	Acquisition of machinery and equipment	47.4%	Acquisition of machinery and equipment	47.9%	Intramural R&D	57.0%
		Acquisition of machinery and equipment	30.7%	Training and market introduction expenditure	25.8%	Intramural R&D	20.5%	Training and market introduction expenditure	19.6%	Acquisition of machinery and equipment	13.7%
	*Total Innovation expenditure / Total turnover**		1.6%		1.1%		1.8%		0.8%		8.3%
R&D	*Proportion of enterprises with intramural R&D **		42%		29%		34%		44%		69%
	*R&D expenditure / Total turnover**		0.7%		0.3%		0.3%		0.1%		4.8%
	Personnel engaged in R&D / total employees		4%		3%		1%		2%		11%

Table 23.1 (continued)

	Main aspects of the innovation	Services	Wholesale trade and commission trade	Transport and communication	Financial intermediation	Business services
	Personnel engaged in R&D / total employees with higher education	20%	23%	55%	16%	27%
Cooperation	*Proportion of enterprises with innovation activity involved in innovation co-operation*	22%	16%	15%	21%	34%
	Geographic location of the agent with whom they cooperate more	National 19%	National 13%	National 13%	National 18%	National 31%
	*Main agents in the nationwide extent**	Suppliers 10.5% Clients 10%	Clients 7.9% Suppliers 7.7%	Suppliers 11.2% Clients 10.5%	Consultants 11.7% Suppliers 11.6%	Clients 15.7% Universities 15.7%

Notes:
* The data for these indicators correspond to the year 2000. The calculation has been made from the average of the available countries.

Source: *Own elaboration based on Eurostat data*

the intensity of R&D activities is higher in some service industries than in some manufacturing ones.

As far as the personnel engaged in R&D is concerned, the CIS3 data show that its ratio in relation to total employees is relatively high in services, but with large differences among subsectors. This ratio is relatively higher in the business services, where knowledge is a key factor in the R&D process, compared to other service activities (Table 23.1). Other recent investigations confirm the supremacy of business services indicating also a high share of employees with higher education in many service activities like scientific libraries and universities, publishing houses, hospitals, news offices, architectural practices and others (Hipp and Grupp, 2005).

23.3.3 Types of innovation expenditures in services

Examining the distribution of expenditure on innovation is a good starting point for identifying the diversity of innovation patterns in services. The results of the CIS3 reveal that the distribution of innovation expenditures is not very uniform across different service activities (Figure 23.1). To a certain extent, such expenditure depends upon the technological level of innovation, the stage of the product life cycle, and the intensity of the human capital used in the innovation process.

Clear differences regarding the distribution of expenditure on innovation in each subsector are evident. Although business services earmark, on average, 57 per cent of their expenditures on 'innovation in intra-mural R&D', the financial intermediation subsector spends, on average, less than 10 per cent of total innovative investment for the same purpose. Just as there are stark differences across subsector spending, so there are also some similarities; for instance, the financial intermediation subsector and the transport and communication sector spend proportionally similar amounts on the acquisition of machinery and equipment. Additionally, the wholesale trade and commission trade subsectors may be characterized as the most similar ones with respect to the distribution of their innovation expenditures. Thus, these first results highlight a diversity of the innovation patterns in services.[2] Concerning the propensity to engage in R&D activities, similar empirical analysis has confirmed a significant variation across services subsectors, highlighting computer and related services, telecommunication and R&D services as the three most R&D-intensive sub-sectors (den Hertog et al., 2006).

Notable differences in values across service subsectors, in addition to attributes previously mentioned, contribute to the heterogeneity which exists in the service sector. The nature of activities pursued by service subsectors, as well as the configuration of the present elements of the innovation process, are other key factors in this heterogeneity.

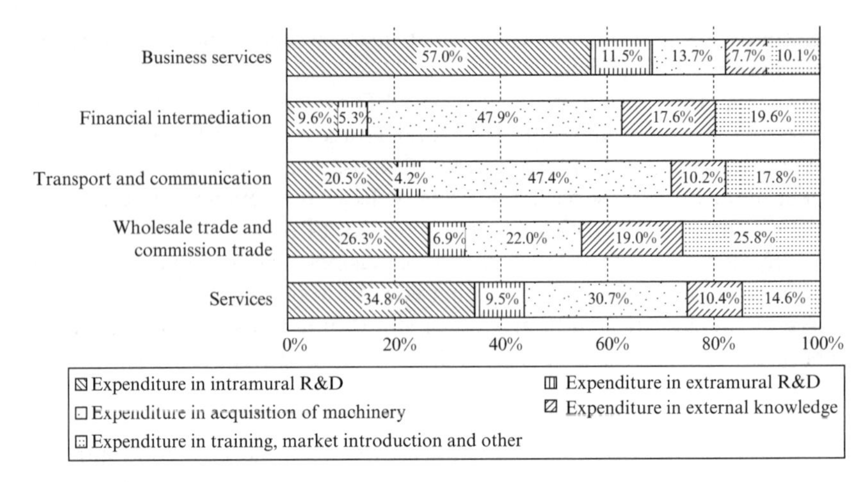

Source: Based on Eurostat data.

Figure 23.1 Distribution of the innovation expenditure, by type of enterprise, EU-15, 2000 (%)

23.3.4 Diversity of the scope of innovation cooperation in services

Innovation, from a systemic perspective, is a dynamic and interactive process of creating new products and services that involves the firm as well as other agents which constitute the firm's business environment. The increasing complexity of the innovation process – characterized by the large and diverse amount of knowledge that must be gathered in order to innovate, coupled with an exponentially increasing amount of available knowledge – has substantially increased the number of specialized agents in diverse functions and tasks (governmental regulations, standards, marketing, engineering and financing, for example). Another important aspect for understanding the relationship between service enterprises and other agents is the cooperation in the innovation process and the understanding of this cooperation as a formal collaboration. Moreover, formal collaboration provides the opportunity to exchange complementary information in the innovation process due to common effort and goals.

After checking the empirical results, it is possible to conclude that business services are much more inclined to cooperate than other enterprises involved in service activities (Table 23.1). This cooperation rate is quite high when compared with the corresponding data for the other service activities.[3] The data thus reveal an important trend: service enterprises, in general, have a higher propensity to cooperate in the innovation process than do manufacturing enterprises – this is especially true for the business and financial intermediation service subsectors.

Differences of innovation cooperation intensity in services The increased number of innovation cooperation companies in the service subsectors, however, significantly differ in their innovation performances (Figure 23.2). While R&D activities are at the forefront of innovation and cooperation in joint innovation projects, the subsector targeted at exercising financial intermediation is seen to be as innovative as or even less so than air and water transport activities, and other supporting and auxiliary transport activities (including activities of travel agencies). In this respect it can be observed that financial intermediation together with R&D companies share high levels of innovation cooperation in the service sectors, but the former possess quite different propensities to innovate.

The empirical analyses have assumed the diversity of innovation patterns in the service sectors, as has been indicated throughout this chapter. Nonetheless, it can be clearly spotted that this diversity stretches out much farther than a simple subsector division. In other words, the variety of innovation patterns exists in specific activities, such as KIBS. Both innovation performance and innovation cooperation propensity vary considerably according to the activity in question. We can also confirm that KIBS are very heterogenous when it comes to innovation and cooperation.

Wholesale and commission trade as well as transport and communication are two groups that, among other things, are characterized by a small number of innovative firms and also by low participation in innovation cooperation. However, post and telecommunication activities, apart from being more innovative than the sector average, also exhibit as high innovative propensity as that of the financial intermediation sector (Figure 23.2). It would be even more evident when the telecommunication industry is analysed separately.

In accordance with Figure 23.2, services can be classified into three groups according to the intensity of their innovation character and cooperation: (1) activities with a small number of innovative firms and a medium intensity of cooperation; (2) activities with few innovative firms and a high intensity of cooperation; and (3) activities with a high innovation intensity and a considerable heterogeneity in terms of their cooperation intensity (Figure 23.2).

Multiplicity of innovation cooperation linkages in services By briefly analysing the partners with which service enterprises cooperate, we believe the suppliers of equipment, materials or software prove to be top choices for service companies (Figure 23.3). This fact reinforces even more the

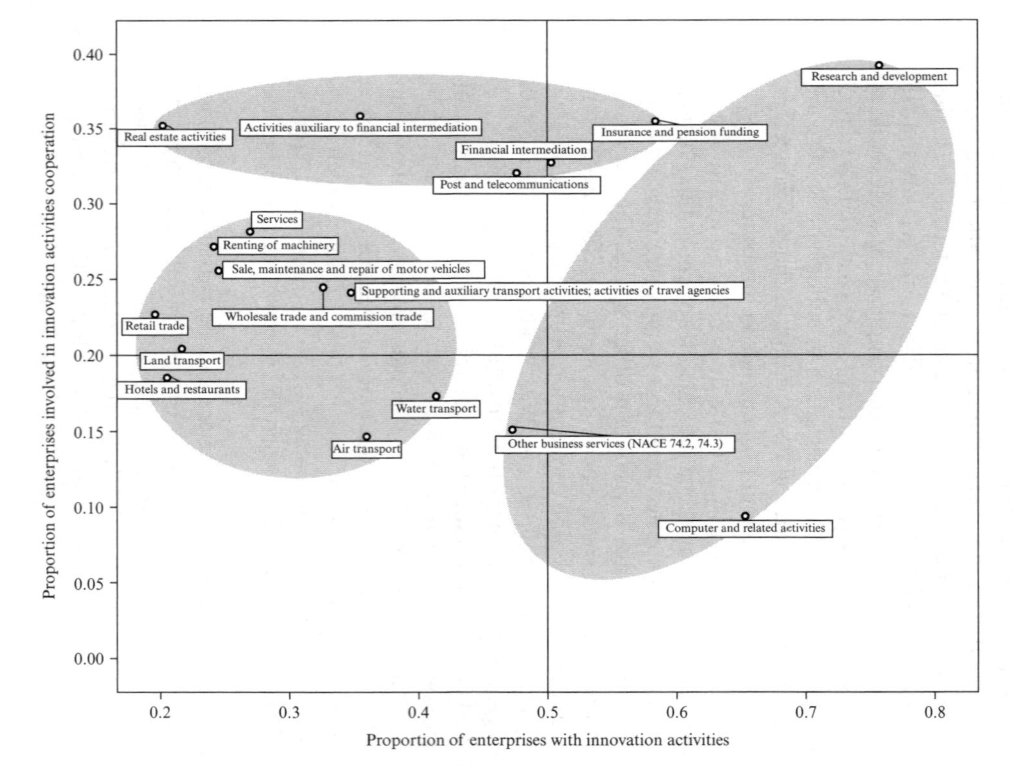

Note: Geographical coverage: 27 EU Member States, Iceland and Norway. Exceptions: Ireland, Slovenia and Malta.

Source: Based on Eurostat data.

Figure 23.2 Proportion (per unit) of enterprises involved in innovation cooperation and enterprises with innovation activities, EU-27, CIS-4

initial belief that services as such follow a supplier-dominant pattern of innovation (Pavitt, 1984), although this is not the case at an intra-sectoral level. Suffice it to say that R&D industry possesses a very high propensity to cooperate especially when accompanied by scientifically established partners, such as universities and other higher education institutions.

The financial intermediation sub-sector differs from other service sub-sectors not only in the extent to which this cooperation is undertaken, but also in its preference as to which types of agents to cooperate with (Figure 23.3). Whereas the main agents chosen by the business services are clients and universities, financial intermediation enterprises prefer to innovate with consulting firms and suppliers. The third-most-utilized innovation partner for financial intermediation enterprises is another enterprise within the enterprise group, which reveals that these large enterprises have strategies to create specialized firms for covering specific needs, especially technological needs.

The transport and communication sub-sector tends to have weak cooperation intensity with all kind of agents; in any case, the highest values are observed in the cooperation with suppliers, clients and competitors. Nonetheless, it is worthwhile to study the case of telecommunication companies. The most detailed data provided by CIS4 allow us to see with great precision the true extent of innovation and innovation cooperation those activities have reached.

The wholesale trade and commission trade subsector has the lowest ratio of enterprises with cooperative innovation activities as opposed to the other subsectors; in this case also, the agents preferred for cooperation are also suppliers, clients and competitors.

Geography of innovation cooperation in services: reduced and uneven international networking Many authors recognize the importance of proximity in innovation issues for service activities, to a greater extent than for manufacturing ones (Miles, 1994; Vence and González, 2008; Vinding and Drejer, 2006). This could be explained by the different types of collaborations and linkages. A large and diverse number of service activities contribute to the innovativeness of their clients by means of acquiring, producing, assembling, storing, monitoring, processing and analysing data, information and knowledge. The fact that not all of these interactions may be made through ICTs provides incentive for these services to locate in areas geographically near potential clients, particularly in metropolitan areas (Coffey and Polèse, 1987; Heinrich, 2001; Keeble and Nachum, 2001). Moreover, collaboration with distant partners in the innovation process will therefore often be of a more explorative nature than interaction with local partners, which is more associated

Suppliers of equipment, materials, components or software

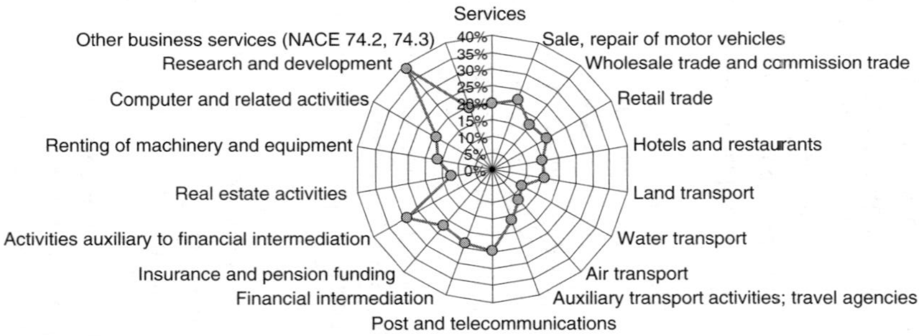

Services
Sale, repair of motor vehicles
Wholesale trade and commission trade
Retail trade
Hotels and restaurants
Land transport
Water transport
Air transport
Auxiliary transport activities; travel agencies
Post and telecommunications
Financial intermediation
Insurance and pension funding
Activities auxiliary to financial intermediation
Real estate activities
Renting of machinery and equipment
Computer and related activities
Research and development
Other business services (NACE 74.2, 74.3)
0%
5%
10%
15%
20%
25%
30%
35%
40%

Clients or customers

Services
Sale, repair of motor vehicles
Wholesale trade and commission trade
Retail trade
Hotels and restaurants
Land transport
Water transport
Air transport
Auxiliary transport activities; travel agencies
Post and telecommunications
Financial intermediation
Insurance and pension funding
Activities auxiliary to financial intermediation
Real estate activities
Renting of machinery and equipment
Computer and related activities
Research and development
Other business services (NACE 74.2, 74.3)
0%
5%
10%
15%
20%
25%
30%
35%
40%
45%

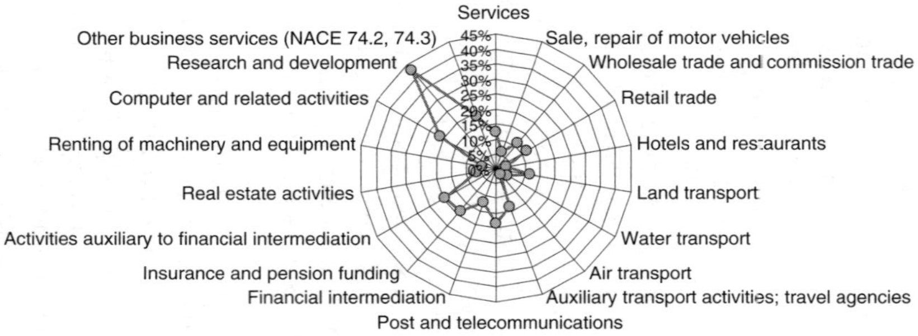

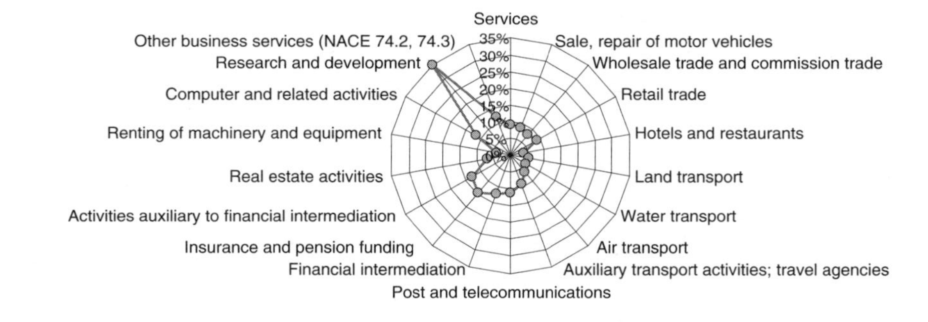

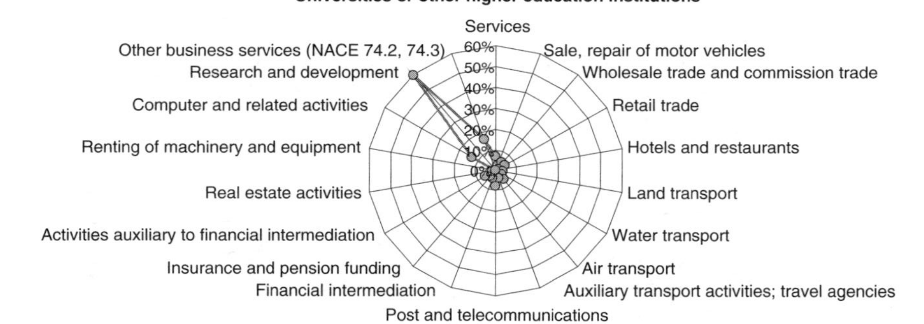

Note: Geographical coverage: 27 EU Member States, Iceland and Norway. Exception: Ireland, Slovenia and Malta.

Source: Based on Eurostat data.

Figure 23.3 Proportion of enterprises involved in innovation cooperation, by partner, national level, 2002–04, EU-27, CIS-4 (%)

with the exploitation of knowledge and hence incremental innovations (Nooteboom et al., 2007).

The agglomerative effects of high-tech services and all knowledge-intensive services also tend to spread to other sectors (depending on the intensity of links among activities). Thus, these services interactively contribute to all products and processes of innovation, acting as channels transferring knowledge across companies and sectors. The wider a client's network, the greater capacity there is to create, capture and accumulate tacit knowledge. And the outcome of this increasing (shared) knowledge among clients and companies leads to a reinforced innovation capability which benefits mainly the local productive system, generating a feedback process of cumulative innovation.

The empirical data lead to the conclusion that there are a fairly relevant number of international innovation networks in most service sectors. Cooperation in innovation remains an activity mainly performed at a national or regional level. Results show that networking and cooperation in innovation tend to be nation-bound. Only a small part of cooperation involves international partners. Among all services, it is R&D enterprises that cooperate most actively on a global scale (approximately 15 per cent, whereas 36 per cent cooperate domestically) (Figure 23.4).

Anyway, an emergent role can be identified in some specific knowledge-intensive services. R&D activities, business and financial services are much more inclined to cooperate at international level than other service activities.

However, an important difference in the propensity to cooperate on a global level can be observed when European countries are analysed separately (Figure 23.5). Service enterprises in Denmark, Lithuania, Luxembourg and Slovakia show a much higher propensity to cooperate internationally with other European countries, though not to the same extent with the United States and other countries (Figure 23.5).

23.4 Global and national cooperation and the taxonomy of innovation patterns in services

The different patterns of global and national cooperation in innovation activities reveal the existence of different patterns of the overall innovation process across service activities.

The analysis of the innovation attributes of services reflected by the third and fourth Community Innovation Survey allows us to build up a taxonomy of service innovation and cooperation patterns. The result leads us to a service classification that considers the innovative character of the sector, the innovative intensity of the enterprises and the type of innovation activities developed, as well as the propensity to cooperate in order to innovate, both at national and international level.

The data used in the analysis include all services activities (NACE sections G, H, I, J and K). Total number of enterprises in the population in the CIS3 is approximately 178,000 (with at least ten employees) for the EU-15 plus Iceland and Norway, among which 71,000 are enterprises with innovation activities (40 per cent).

The service taxonomy identified here consists of three broad service types according to their innovation process attributes: (1) low innovation-intensive sectors (represented by 'wholesale and commission trade' and 'transport and communication'); (2) technology-intensive and moderately innovation-intensive sectors (represented by 'financial intermediation'); (3) knowledge and innovation-intensive sectors (represented by 'Knowledge-intensive business services – KIBS'). In Table 23.2 we present these groups and their respective characteristics.

Obviously, the differentiating borders among the various types are not very clear-cut and there is some overlapping among the different activities studied. One of the reasons is the inevitable internal heterogeneity of the groups considering the broad classification used here. Thus, this taxonomy should be considered as a first step towards a better understanding of innovation and cooperation in a highly heterogeneous sector and should lead us to reject excessively simple and generic conclusions about the service sector. Furthermore, we agree that enterprises belonging to the same subsector may exhibit different patterns (cf. Sundbo and Gallouj, 2000).

Anyway, the three types of innovation and cooperation patterns in services identified show remarkable differences. The analysis shows clearly that not all services are as non-innovative as previously supposed. Nor does innovation proceed in a 'one-size-fits-all' way. Actually, service activities possess different innovative characters, intensity and levels of formalization of innovation. Also, they differ considerably in the innovation inputs and the intensity of innovation cooperation. The other variables studied indicate diversity too.

Additionally, underlying productive patterns can be identified in this taxonomy. They surely determine their innovative performances and thus the innovation patterns in services. This verification confirms the belief that productive nature is narrowly linked to innovative performances.

The first type identified, *Low innovation intensive sectors*, is composed of trade enterprises, including goods transport and storage. The Organisation for Economic Cooperation and Development (OECD) classification (OECD, 2001), calls them 'distributive services'. The sectors assigned to this type are wholesale and commission trade, repair of motor

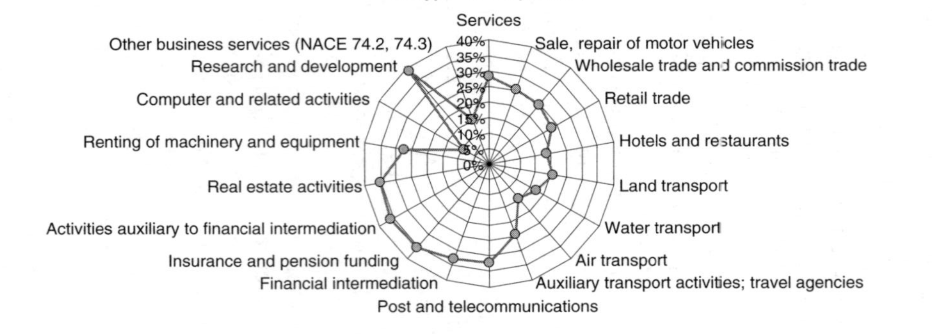
All types of cooperation
Services
40%
35%
30%
25%
20%
15%
10%
5%
0%
Sale, repair of motor vehicles
Wholesale trade and commission trade
Retail trade
Hotels and restaurants
Land transport
Water transport
Air transport
Auxiliary transport activities; travel agencies
Post and telecommunications
Financial intermediation
Insurance and pension funding
Activities auxiliary to financial intermediation
Real estate activities
Renting of machinery and equipment
Computer and related activities
Research and development
Other business services (NACE 74.2, 74.3)
Enterprise engaged in any type of innovation cooperation, national
Services
40%
35%
30%
25%
20%
15%
10%
5%
0%
Sale, repair of motor vehicles
Wholesale trade and commission trade
Retail trade
Hotels and restaurants
Land transport
Water transport
Air transport
Auxiliary transport activities; travel agencies
Post and telecommunications
Financial intermediation
Insurance and pension funding
Activities auxiliary to financial intermediation
Real estate activities
Renting of machinery and equipment
Computer and related activities
Research and development
Other business services (NACE 74.2, 74.3)

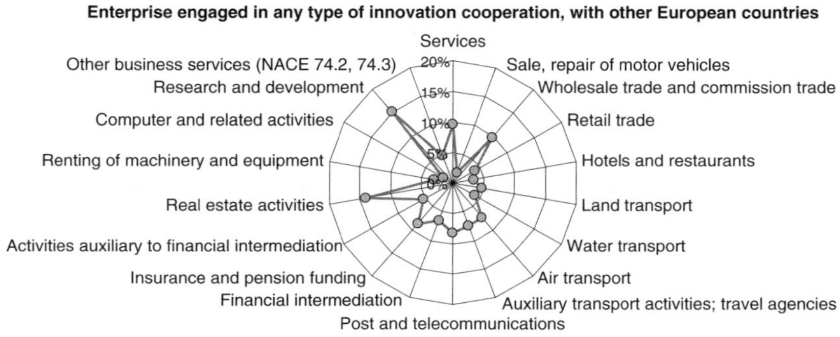

Enterprise engaged in any type of innovation cooperation, with other European countries

Services
Sale, repair of motor vehicles
Wholesale trade and commission trade
Retail trade
Hotels and restaurants
Land transport
Water transport
Air transport
Auxiliary transport activities; travel agencies
Post and telecommunications
Financial intermediation
Insurance and pension funding
Activities auxiliary to financial intermediation
Real estate activities
Renting of machinery and equipment
Computer and related activities
Research and development
Other business services (NACE 74.2, 74.3)
20%
15%
10%
5%
0%

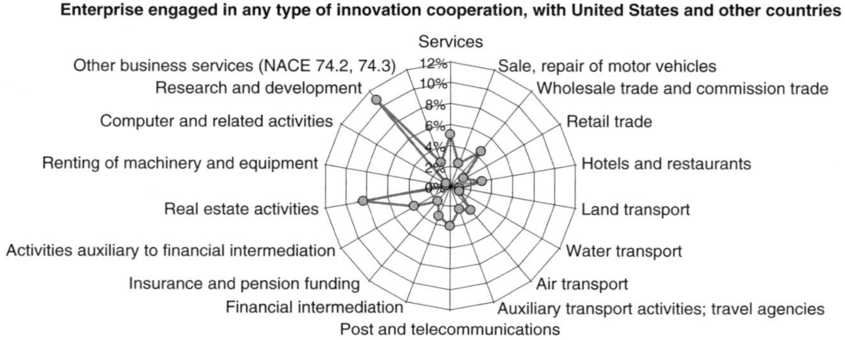

Enterprise engaged in any type of innovation cooperation, with United States and other countries

Services
Sale, repair of motor vehicles
Wholesale trade and commission trade
Retail trade
Hotels and restaurants
Land transport
Water transport
Air transport
Auxiliary transport activities; travel agencies
Post and telecommunications
Financial intermediation
Insurance and pension funding
Activities auxiliary to financial intermediation
Real estate activities
Renting of machinery and equipment
Computer and related activities
Research and development
Other business services (NACE 74.2, 74.3)
12%
10%
8%
6%
4%
2%
0%

Note: Geographical coverage: 27 EU Member States, Iceland and Norway. Exception: Ireland, Slovenia and Malta.

Source: Based on Eurostat data.

Figure 23.4 Proportion of enterprises involved in innovation cooperation, by geographic scope, 2002–04, EU-27, CIS-4 (%)

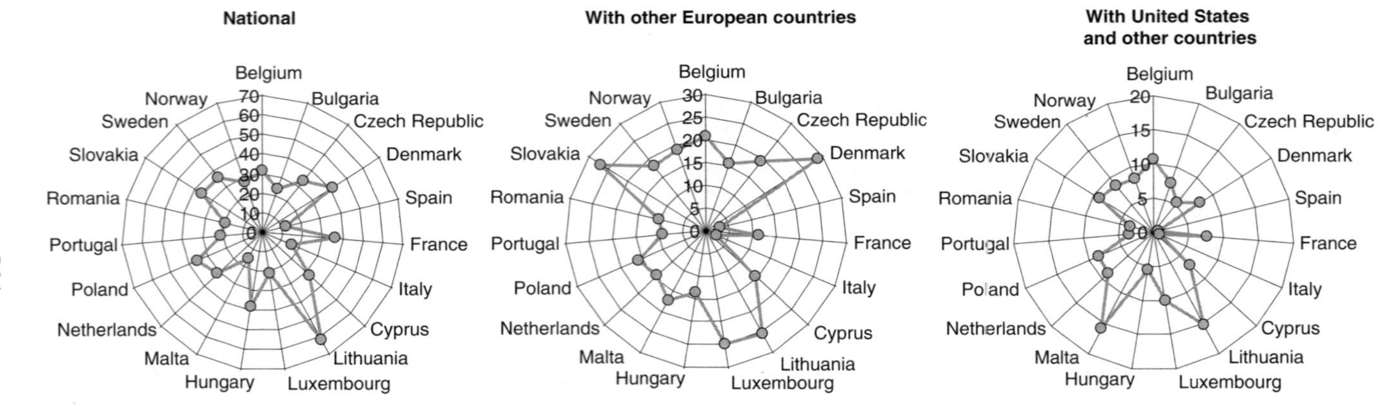

Source: Based on Eurostat data.

Figure 23.5 Proportion of service enterprises involved in innovation cooperation, by geographic scope and European countries, 2002–04, CIS-4 (%)

Table 23.2 Services taxonomy based on the characteristics of the innovation intensity and the innovation cooperation spreading

	Low-innovation-intensive sectors	Technology-intensive and moderately innovation-intensive sectors	Knowledge and innovation-intensive sectors
Innovative character of the sector	Low	Medium	High
Innovative intensity of the enterprises	Low	Low	High
Formalization of the innovation intensity in R&D	Low	Medium-low	High
Main source of the innovation (innovation expenditure)	Mix: Acquisition of machinery and equipment; Intramural R&D; Training; Market introduction expenditure	Acquisition of machinery and equipment	Intramural R&D
Intensity in innovation cooperation	Low	Medium	High
Main agents chosen to cooperate	Clients; suppliers	Consultancies; specialized suppliers	Clients; universities
Location of the partners	Mostly national and little international cooperation	Mostly national and relative international cooperation	Mostly national and relative international cooperation
Nature of the production	Goods-related services	Information-intensive networks	Knowledge-intensive
Sectors assigned	Wholesale and commission trade Repair of motor vehicles Transport Communication*	Bank Insurance and pension funding Activities auxiliary to financial intermediation	Computer and related activities R&D Architectural and engineering activities and related technical consultancy Technical testing and analysis

Note: * Telecommunications is included because the aggregation level of the available data.

vehicles and transport.[4] The performance of these activities is weaker than in the rest of the subsectors as far as most of the innovation indicators are concerned. The innovation cooperation intensity is very low and basically nationally bounded.

Hollenstein (2003) identifies a similar cluster in his analysis of the Swiss case. He called 'low-profile innovators with hardly any external links' what is composed essentially of personal services, real estate, hotels and restaurants, retail trade and transport. This author pinpoints as important characteristics of these activities the weak demand prospects, strong price competition, low appropriability and innovation opportunities, and the relatively poor human capital endowment. Most of these activities have been classified by several authors as supplier-dominated.

The second type, *Technology-intensive and moderately innovation-intensive sectors*, contains productive processes based principally on information. The enterprises belonging to this type are also sensitive to the innovation coming particularly from suppliers of new technologies and ICT, though non-technological innovations are also relevant to these activities. In any case, the ICT evolution could best explain what is behind the innovation developed by those enterprises (COTEC, 2006). The innovation cooperation intensity is very high though the propensity to innovate is quite diverse depending on the activity. The sectors assigned are fundamentally banking, insurance and other financial services.

Many authors have classified those activities as network-based (cf. Hipp and Grupp, 2005; Sundbo and Gallouj, 2000), due to their high level of interaction with clients and the information flows of these relations. In fact, the innovation related to the contact channel with clients has probably been the most important innovation type for financial intermediation in the last decade. That kind of innovation is often supposed to be product and process innovation (COTEC, 2006).

The most common tool for obtaining direct information from external clients is focus groups,[5] as well as surveys that measure the quality of the product. Another important tool whose use is more and more frequent in those activities is data mining (COTEC, 2006). These examples reveal, among other things, a growing participation of clients into the innovation process in the financial sector. In fact, this sector has become used to developing innovation jointly with lead users.[6] Some authors, in fact, classify the innovation developed by financial services as client-led innovation (den Hertog and Bilderbeek, 1999). Nowadays, the links between financial entities and certain organizations considered traditional sources of knowledge such as universities, technology institutes (TIs) and public research institutes, for instance, are practically non-existent.

Those activities normally incorporate new technologies, basically ICT, from technology suppliers, though some technologies are developed by the activities themselves. For that reason, some innovation pattern typologies consider innovation in financial services as IT-oriented network-integrated developers (Hollenstein, 2003).

The last type identified in our analysis, *Knowledge and innovation-intensive sectors*, is composed of activities situated at the forefront of innovation (Miles, 1994). They are known to use knowledge, mainly tacit knowledge, largely for developing their services. However, knowledge could be considered not only an important tool, but also the core of their final product. They are 'responsible for the combination of knowledge from different sources and for the distribution of knowledge itself' (Hipp and Grupp, 2005, 518).

In parallel, these service activities are commonly the ones that exercise innovation the most. The significant use of knowledge for both producing services and creating new outputs, in that way, allow these activities to innovate more frequently than any other. The so-called way to accumulate knowledge, learning-by-doing, permits these activities to develop innovation while providing their services. Additionally, the intense level of interaction, the ad hoc mode of innovation and the high level of interface with clients in business services contribute positively to the innovation process.

Innovation in business services could be perceived as the result of co-production with the client; in fact, the innovation frequently takes place within the client firm. In this manner, den Hertog and Bilderbeek (1999) named this innovation pattern 'innovation through services'. A great part of these activities play the role of source, carriers and facilitators, providing knowledge to the innovation process at the client firm. Sundbo and Gallouj (2000, 54) observed that these enterprises 'do not really sell product-services, but competencies, abilities to solve problems in different expertise areas'. Similarly, Miles (1994) classified them as 'information service producers'.

Computers and other service-related practices, R&D, architecture and engineering, and technical consultancy as well as technical testing and analysis, are examples of activities belonging to this cluster.

As far as innovation cooperation is concerned, relevant differences can be highlighted among those sectors. The higher the R&D and technology intensity the higher the propensity to innovation cooperation. This result suggests that the internal capacity of firms, measured for instance by R&D effort or the proportion of personnel engaged in R&D, increases the capability to interact with external partners and to absorb and value external knowledge.[7] These differences also apply for international innovation cooperation. International innovation cooperation usually requires an important capacity to create new knowledge, good reputation and the capacity to manage alliances and to identify when the knowledge of the partner complements the internal knowledge. In this sense, Vinding and Drejer (2006) show that 'the main innovation partner of knowledge intensive services firms with a high level of absorptive capacity is likely to be located further away (abroad) than the main innovation partner of firms with a low level of absorptive capacity'.

23.5 Conclusion

The intra-sectoral study of innovation in services leads to a recognition that services increase innovation activities but they do it at a different pace, a different intensity and following a different pattern.

This chapter has comparatively analysed four different service subsectors using a wide range of innovation attributes. It is important to stress that the service categorization used was determined by the availability of data and the previous classification established by the source (Eurostat – CIS3 and CIS4). The low level of disaggregation could not account for certain differences between the subsectors. A more thorough breakdown would likely produce more nuances, and could lead to the identification of new innovation patterns. Nonetheless, it was possible to identify important behavioural differences for certain attributes and to identify possible patterns of cooperation and innovation. Additionally, the activities belonging to each identified type possess a similar productive nature, which leads to the conclusion that innovation and cooperation performances will likely depend on the features of production of the service activities. These patterns could be summed up in three broad types: Low-innovation-intensive sectors (LIIS), Technology-intensive and moderately innovation-intensive sectors (TIMIIS), and Knowledge and innovation-intensive sectors (KIIS). The later profile is widely reported in the literature on innovation under the label of KIBS (Knowledge-intensive business services).

KIIS may be considered the leading service sector with regard to cooperation and innovation; conversely, other activities might be characterized by low intensity in both cooperation and innovation. Several reasons could explain this disparity. KIIS intrinsically use and transfer knowledge; they develop their activities in direct contact with clients and therefore have a more intense level of interaction than do enterprises in the other service subsectors. The ad hoc mode of innovation and the high level of interface with clients in KIIS lead to the development of customised products, whereas LIIS offer mainly standardized services.

An analysis of the innovation inputs in each subsector also reveals the existence of specific patterns. TIMIIS exhibit an innovative effort – the rate between innovation expenditure and turnover – well below the service sector average, whereas the KIIS rate is actually higher than that of some industrial activities. The other types exhibit rates around the service sector average. Almost half of all innovation expenditure in TIMIIS and in some LIIS activities is concentrated on the acquisition of machinery and equipment. In the latter case, such expenditure reflects the high percentage of ongoing innovation compared with the service sector average. The KIIS clearly consider R&D activities as a main source of knowledge for innovation. Additionally, the commonly held notion that R&D activity is focused only on KIIS and on some activities belonging to TIMIIS loses merit. Empirical findings show that the other types also use R&D as a means of increasing knowledge for innovation.

In general, a higher proportion of innovative enterprises cooperate in the service sector than in the manufacturing sector. However, an intrasectoral analysis reveals, as anticipated, that there are differences regarding the tendency to cooperate among enterprises in the different service subsectors. For nearly every type studied, the supplier is still the main agent with whom enterprises choose to cooperate, with the exception of KIIS in which the client plays an important role throughout the innovation process. All types except KIIS consider clients the second main source of information – or rather, their main source of external information.

Finally, the empirical data lead us to conclude on a modest relevance of international networks for innovation in most service activities. Cooperation in innovation remains as an activity mainly performed at national and regional level. The empirical results lead us to conclude that networking and cooperation in innovation tends to be national-bounded. The explaining mechanisms behind that could be social interactions and informal face-to-face interactions; regional labour markets and mobility; as well as region-specific institutions for knowledge accumulation and diffusion. Only a small part of cooperation involves international partners. Anyway, an emergent role can be identified in some specific knowledge-intensive services. R&D activities, business and financial services are much more inclined to cooperate at international level than other service activities. That behaviour could be associated to the firm's absorptive capacity, that is, formal qualifications of its employees and its R&D activities that influence the ability to evaluate and utilize external knowledge and to interact with different types of partners, both at a national and a global extent. To sum up, the trends in international cooperation for innovation also show relevant differences across sectors and countries.

Notes

1. The CIS4 data allow us to overcome certain difficulties in grouping activities in CIS3, as was the case with telecommunications. CIS3 grouped together communication and

transport activities in such a way that the former were placed among low-innovation–low-cooperation services. The CIS4 industry distribution, on the other hand, made it possible to place telecommunications in a separate group, which in the upcoming sections of this chapter will be called technology-intensive and moderately innovation-intensive sectors.
2. Similar analyses were conducted for some EU-15 countries and, in general, in all cases the distribution structures were alike, except in the case of Finland.
3. It is also interesting to note that the service sector cooperation average is much higher than the manufacturing sector average, in which the proportion of cooperative enterprises is 17 per cent (European Commission, 2004).
4. Due to the high level of aggregate data available we have been forced to classify 'communication activities' into this type, despite knowing the large extent of innovation in certain activities such as telecommunication.
5. The 'focus group' is a marketing concept for a type of qualitative research in which a group of people are asked questions concerning their attitudes towards products, services, concepts, advertisements, ideas or packaging. Focus groups are an important tool for acquiring feedback regarding new products, as well as various topics (Marshall and Rossman, 1999).
6. 'Lead users' is a term created by von Hippel (1986) to qualify certain costumers. In his own words: 'lead users are users whose present strong needs will become general in a marketplace months or years in the future' (von Hippel, 1986, 6).
7. The notion of absorbtion capacity induces us to the assumption that innovation and cooperation embrace a two-way relationship. Thus, it is not only cooperation that fosters the firm's innovative capacity, but also the innovative effort spent by enterprise confers the necessary knowledge and capacity to interact with others. This second statement is even more apparent in medium and high-tech subsectors.

References

Barras, R. (1986), 'Towards a theory of innovation in services', *Research Policy*, **15** (4), 161–173.

Bell, M. and K. Pavitt (1993), 'Technological accumulation and industrial growth: contrasts between developed and developing countries', *Industrial and Corporate Change*, **2** (2), 157–211.

Castellacci, F. (2008), 'Technological paradigms, regimes and trajectories: manufacturing and service industries in a new taxonomy of sectoral patterns of innovation', *Research Policy*, **37**, 978–994.

Coffey, W. and M. Polèse (1987), 'Trade and location of producer services: a Canadian perspective', *Environment and Planning*, **19**, 597–611.

Coombs, R. and I. Miles (2000), 'Innovation measurement and services: the new problematique', in J.S. Metcalfe and I. Miles (eds), *Innovation Systems in the Service Economy: Measurements and Case Study Analysis*, Boston, MA: Kluwer, pp. 85–103.

COTEC (2006), *Innovación en los Servicios Financieros*, Madrid: Fundación Cotec para la Innovacción Tecnológica.

den Hertog, P., H. Bouwman, J. Gallego, L. Green, J. Howells, T. Merrer, I. Miles, I. Moerschel, A. Narbona, L. Rubalcaha, J. Segers and B. Tether (2006), 'Research and development needs of business related service firms', RENESER Project, European Commission.

den Hertog, P. and R. Bilderbeek (1999), 'Conceptualising service innovation and service innovation patterns', mimeo, DIALOGIC, Utrecht.

European Commission (2004), *Innovation in Europe: Results for the EU, Iceland and Norway – Data 1998–2001*, Luxembourg: Office for official Publications of the European Communities.

European Commission (2005), 'European innovation scoreboard 2005: comparative analysis of innovation performance', Report under the European Trend Chart on Innovation Initiative, European Commission.

Evangelista, R. (2000), 'Sectorial patterns of technological change in services', *Economics of Innovation and New Technology*, **9**, 183–221.

Foray, D. (2004), *Economics of knowledge*, Cambridge, MA: MIT Press.

Gallouj, F. (2002a), *Innovation in the Service Economy: The New Wealth of Nations*, Cheltenham, UK and Northampton, MA, USA: Edward Elgar Publishing.

Gallouj, F. (2002b), 'Innovation in services and the attendants old and new myths', *Journal of Socio Economics*, **31**, 137–154.

Gallouj, F. (2007), 'Economia da Inovação: um balanço dos debates recentes', in R. Bernades and T. Andreassi (eds), *Inovação em Serviços Intensivos em Conhecimento*, São Paulo: Editora Saraiva, pp. 3–28.

Gallouj, C. and F. Gallouj (2000), 'Neo-Schumpeterian perspective on innovation in services', M. Boden and I. Miles (eds), *Services and the Knowledge-Based Economy*, London: Continuum Books, pp. 21–37.

Heinrich, C. (2001), 'Metropolitan information producers and services and their relevance to innovation theory', conference on The Future of Innovation Studies, Eindhoven University.

Hipp, C. and H. Grupp (2005), 'Innovation in the service sector: the demand for service-specific innovation measurement concepts and typologies' *Research Policy*, **34**, 517–535.

Hollenstein, H. (2003), 'Innovation modes in the Swiss service sector: a cluster analysis based on firms-level data', *Research Policy*, **32**, 845–863.

Howells, J. (2000), 'Innovation and services: new conceptual frameworks', CRIC Discussion Paper No. 38.

Keeble, D. and L. Nachum (2001), 'Why do business service firms cluster? Small consultancies, clustering and decentralization in London and Southern England', WP 194, ESRC Centre for Business Research, University of Cambridge.

Marshall, C. and G.B. Rossman (1999), *Designing Qualitative Research*, 3rd edn, London: Sage Publications.

Miles, I. (1994), 'Innovation in services', in M. Dodgson and R. Rothwell (eds), *The Handbook of Industrial Innovation*, Aldershot, UK and Brookfield, VT, USA: Edward Elgar, pp. 243–256.

Miles, I. (2001), 'Services innovation: a reconfiguration of innovation studies', PREST Discussion Paper 01-05, Manchester.

Miles, I. (2005), 'Innovation in services', in J. Fagerberg, D. Mowery and R.R. Nelson (eds), *The Oxford Handbook of Innovation*, Oxford University Press, pp. 433–458.

Miles, I. and M. Boden (2000), 'Introduction: are services special?', in Mark Boden and Ian Miles (eds), *Services and the Knowledge-Based Economy*, London: Continuum Books, pp. 1–21.

Miles, I. and B. Tether (2003), 'Innovation in the service economy', IPTS Report no. 71 February, pp. 45–51.

Miozzo, M. and L. Soete (2001), 'Internationalization of services: a technological perspective', *Technological Forecasting and Social Change*, **67**, 159–185.

Nooteboom, B., W. Vanhaverbeke, G. Duysters, V. Gilsing and H. Oord (2007), 'Optimal cognitive distance and absorptive capacity', *Research Policy*, **36**, 1016–1034.

OECD (2001), *OECD Employment Outlook*, Paris: OECD.

Pavitt, K. (1984), 'Sectoral patterns of technical change: towards a taxonomy and a theory', *Research Policy* **13**, 343–373.

Salter, A. and B. Tether (2006), 'Innovation in services: through the looking glass of innovation studies', Background paper for AIM Grand Challenge on Services, Oxford University.

Soete, L. and M. Miozzo (1989), 'Trade and development in services: a technological perspective', Working Paper No. 89-031, MERIT, Maastricht.

Sundbo, J. (2000), 'Organization and innovation strategy', in M. Boden and I. Miles (eds), *Services and the Knowledge-Based Economy*, London: Continuum Books, pp. 109–128.

Sundbo, J. and F. Gallouj (1998), *Innovation in Services*, SI14S Project Synthesis, WP 3–4, IFRESI – University of Lille.
Sundbo, J. and F. Gallouj (2000), 'Innovation as a loosely coupled system in services', in J.S. Metcalfe and I. Miles (eds), *Innovation Systems in the Service Economy: Measurements and Case Study Analysis*, Boston, MA, Kluwer, pp. 43–68.
Tether, B. (2002), 'The sources and aims of innovation in services: variety between and within sectors', CRIC Discussion Paper no. 55.
Tether, B. (2005), 'Do services innovate (differently)? Insights from the European Innobarometer Survey, *Industry and Innovation*, **12** (2), 153–184.
Tether, B. and J.S. Metcalfe (2004), 'Services and "systems of innovation"', in F. Malerba (ed.), *Sectoral Systems of Innovation*, Cambridge: Cambridge University Press, pp. 287–322.
Tidd, J. et al. (2001), *Managing Innovation: Integrating Technological, Market and Organizational Change*, 2nd edn, Chichester: John Wiley & Sons.
Vence, X. and M. González (2002), 'Los servicios y la innovación. La nueva frontera regional en Europa', *Economía Industrial*, 347, 41–66.
Vence, X. and M. González (2008), 'Regional concentration of the knowledge-based economy in the EU: towards a renewed oligocentric model?' *European Planning Studies*, **16** (4), 557–575.
Vence, X. and A. Trigo (forthcoming), 'Sobre el alcance real de las particularidades de la innovación en los servicios: un chequeo a la evidencia empírica', *Economía Industrial.*
Vinding, A.L. and I. Drejer (2006), 'The further the better? Knowledge intensive service firms' collaboration on innovation', DRUID Working Paper No. 06-31.
von Hippel, E. (1986), 'Lead users: source of novel product concepts', *Management Science*, **32** (7), 791–805.

24 Entrepreneurship and service innovation: a challenge for local development

Marie-Christine Monnoyer-Longé[1]

24.1 Introduction

French labour market statistics show that business services are offsetting job losses due to deindustrialization. Macroeconomic indicators and employment figures for the European Union confirm the relatively dynamic performance of business services. However, up to the 1980s, regional development policy was focused mainly on the industrial sector with business services to manufacturing being virtually ignored (Gallouj, 2006). Policy changes have had two main objectives: to create jobs locally, and to avoid any discouragement for investment in new industry based on the shortage of services in the metropolitan areas (Soldatos, 1991). Outside of the regional capitals, unstable demand for services prevents them from truly benefiting from any start-up support incentives (Léo et al., 1991). After 1985, there is evidence of growth in some cities (Moyard, 2000; Léo and Philippe, 2007), which is consistent with economic base theory (Planque, 1993; Léo and Philippe, 1998) and has led to better-balanced policy that recognizes services as being an active component of the local economy.

Research into innovation in services has supported this policy trend. The results of various studies provide breakdowns of innovation by type and domain and help to further our understanding of its potential impact on users and partners (Shostack, 1984; Miles, 1993; Gadrey et al., 1995; Gallouj, 1994, 2002; Sundbo, 1996). These studies dispel the notion of services as a cottage industry that is not involved in export; suggest a range of tools that can be used to analyse service innovation, in all its various forms; and discuss its evolution and incentives for service providers.

Realizing the economic potential of business service start-ups, some chambers of commerce introduced support programmes because many new companies were folding within two years and not creating new jobs. Business performance is affected by a range of factors: (1) differences between business model and business plan (Gartner, 1988; Fonrouge, 2002); (2) legitimacy of an innovative enterprise (Tornikovski et al., 2007); (3) the operating environment (Aldrich, 1999). Chambers of commerce and government should facilitate the transformation of an idea into a competitive business that will contribute to the local economy.

In 2007, the Chamber of Commerce and Industry (CCI) in Toulouse, which operates Reliantis, a programme for innovative start-ups, observed that among 86 start-ups qualified as innovative, there were 34 industrial services suppliers, 19 software developers and 9 'innovative other' versus 24 manufacturers. The managers of the Reliantis programme were impressed by these figures and decided that more knowledge was needed about the nature of innovative service firms and the kind of help they needed. The programme set out to compare the selection and progress of individual projects against current theories related to local development, innovative services and start-up support. It was necessary to identify information and indicators based on: (1) analysis of the links between new business services and local development; (2) service innovation; and (3) analyses of the start-up support process.

This chapter details a three-step approach to establishing a working relationship between start-up support actors and researchers. The first step involves situating business services within the context of regional development; the second identifies the various forms of innovation and the actors involved; the third identifies ways of improving the support provided by Reliantis for innovative new service firms. This last is achieved by comparing input from entrepreneurship theory with the results of a small, exploratory survey of innovative start-ups that received support from Reliantis (31 responses from 43 start-ups contacted).

24.2 Local development and services to innovative enterprises

Opinion CCMI/035 of the European Economic and Social Committee states that: 'The importance of business services for the growth, competitiveness and employment levels of European manufacturing and service industries needs to be recognized'. It further 'stresses the interdependence between manufacturing and the service sector'. This perhaps marks the first European affirmation that development of services and the development of manufacturing are equally important, and indeed may involve a symbiotic relationship. How is this relationship to be understood? Can we detect it at the local level? In order to answer these questions, we looked to Jean Gadrey's (1996) research into services productivity and the concepts of direct result (or 'product') and indirect (or 'mediated') result (see also Djellal and Gallouj, 2008) and examine the issue of services innovation and local development.

24.2.1 Direct and indirect impact of services

Business services account for an ever-growing share in European national economies (Rubalcaba and Kox, 2007). Industry statistics for the dynamic region considered here (i.e. the Toulouse region), indicate that

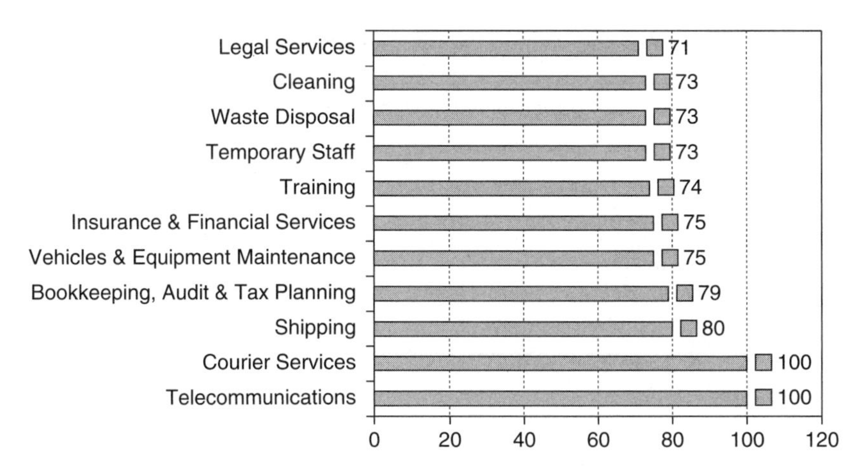

Source: SESSI (2007).

Figure 24.1 Percentage share of companies purchasing business services

manufacturing sectors spent 8.7 per cent of turnover on outsourced services in 2005 (SESSI, 2007).[2] The immediate impact of services is demonstrated by an increased volume of consumption. Gadrey (2002) notes that intermediate consumption of services to business and government agencies climbed by 50 per cent between 1980 and 1990. These figures (see Figure 24.1) show that services are procured on a regular basis and that regularity translates into an immediate impact on the manufacturing processes. The analysis of outsourced services shows that the larger the firm is, the more extended are its outsourced services (see Table 24.1).

Seventy-five per cent of outsourced services are from regular suppliers, mostly local to the user (Monnoyer and Zuliani, 2007; Spath and Ganz, 2008). Outsourcing of services has become inevitable and is built into the products and services with innovative content impacts. They are now an integral part of the manufacturer's products. Gadrey (1996, 98) terms this 'mediated impact' in describing the various effects of services on users across time and levels of manifestation.

Manufacturers reluctant to outsource services (CPCI, 2007) should note the findings in Léo and Philippe (2007) that competitive tendering for services sometimes holds the key to greater competitiveness, a maxim that should be borne in mind by those interested in local development (Gallouj, 2006).

24.2.2 Economic potential of services innovation

The service innovation literature (see Chapter 1 in this volume for a survey) tracks services innovation by subsectors, for example, hospitals,

Table 24.1 Business services to manufacturing (%)

Service	Independent SME	Corporate-owned SME	Firms 250+ employees
IT services	35.1	41.7	65.4
Architecture, engineering, testing and quality control	17.4	22.9	47.0
Temporary staff	64.1	80.3	93.1
R&D	8.5	15.8	35.8

Source: SESSI (2007).

supermarkets, cleaning and consultancy (Djellal, 2002; Djellal et al., 2004). The findings cover the potential benefits of outsourced services to users, and the synergies they generate (Djellal and Gallouj, 2007b). Business services make a powerful contribution to economic development and development planning because of the inextricable relationships they create among families of disciplines (see Table 24.2). Their interactions with users and other partners generate what Wood (2008) calls 'hidden innovation'.

Table 24.2 Selected benefits of innovative services to industry

Service	Benefit
Management & management consultancy	Improved organizational efficiency
Legal & accounting	Implementation of internal auditing procedures and interdepartmental coordination
Hiring, training and temporary staffing	Improved hiring, training and use of manpower
	Improvements to specific tasks and operations

Source: Rubalcaba (1999).

The benefit of innovative services to users stems not only from joint development of the service, but also from the frequency of interaction between the service supplier and the in-house user and producers of the service (Sundbo, 1996): one-third of businesses perform research and development (R&D) in-house and also buy in R&D (SESSI, 2007; Spath and Ganz, 2008).

For example, one of the firms in the survey is an architectural and construction consultancy specializing in piping, lighting and insulation. Its clients want to cut building costs. One of the services it provides is to model heat consumption for each volume of a structure and its intended function. The trade-offs from the savings on heating cost can be used by architects and construction companies in designing their products.

For businesses with a good awareness of the increased competition afforded by outsourcing, innovative new services will have a head start. We need also to examine their effects in the context of economic base theory and its ramifications.

24.2.3 Clustering and the local economy

Economic base theory has been validated through its application for more than 50 years to the analysis of regional imbalances in the growth of new jobs. North (1955) held that some services should be considered basic because, although local, the majority of users were not households. Subsequently, geographers drew attention to the fact that business services were concentrated in major cities and other large urban areas, which prompted them to investigate the link between the proximity of user head offices to their service provider, and the potential for growth in services (Harrington and Daniels, 2006; Jakobsen and Aslesen, 2004). It was found that the viability of the service provider is more closely correlated to the quality of its services and the likelihood of demand for those services, than to relative location. Léo and Philippe (2007) saw business services as being basic to the conurbations and searched for a link between total employment and an enlarged base in urban areas with over 200,000 inhabitants. Although it was found that such a link existed, it was stronger at the regional level than at the urban one. In their quest for other factors that skewed the patterns of job growth among cities, they identified two variables to explain the differences in jobs growth between the urban areas: the high share of qualified employees in urban areas and the high share of owners-managers to employees in business services.

This confirms that business services stimulate the creation of new jobs, as well as the value to local development of having service enterprises with a high share of management staff, which is the case for innovative business services.

From this point, the provision of start-up support stands out as an initiative that exerts a positive upstream effect on the creation of new jobs in a given urban area. This impact is enhanced through the provision of support for start-ups, which have to cope, for example, with demands for more staff very early in their lifecycles. Finally, even very small start-ups are more dependent than larger firms on the local economy, for both clients and suppliers (Mallard, 2007; Léo et al., 1991). It now becomes necessary to investigate how innovation happens in the services industry.

24.3 Services innovation

Research on services innovation was focused on firms that operated technologies borrowed from manufacturing R&D, until 1990 when studies began to include non-technological innovation (Gallouj, 2002). The start-ups in our analysis include service providers that encompass innovative services with or without technology. This is informative about both innovator thinking and the adjustment potential of the start-ups to their market and the added value they can contribute to the economic environment. I review innovation by type and key actors.

24.3.1 Types of innovation

'Product' does not codify neatly in services. What distinguishes it and makes it unique is the way it interacts with the client. We need to examine this interaction to see how a particular service responds to a particular need in a particular context. The first observation is that the design and development of a new service is sometimes a joint, interactive effort involving both client and the workers and management of the service supplier (see Table 24.3). As Barcet and Bonamy (2000) note: 'Services innovation starts with a concept which becomes a set of procedures, not mechanical effects, that exerts impact on the client'.

The breakdown presented in Table 24.3 excludes the technological dimension in order to focus on the user–supplier relationship, which is the crux of the innovation that Eiglier and Langeard (1987) call 'servuction' (from service and production), which is the interactive user–supplier

Table 24.3 Types of innovation

Type	Definition
Custom	New service tailored to a client's specifically expressed need or demand
Ad hoc	Existing service largely co-modified with a client to meet her specification
Recombination	New service resulting from segmentation of an existing off-the-shelf service or bundling of such services
Incremental	New service resulting from innovative or modified add-ons to existing services
Experiential	Existing service delivered or experienced in a new way

Source: Hauknes (2002, 127) based on Sundbo and Gallouj (1999).

process in which the supplier's employees recognize shortcomings in a service and begin to tweak it; innovation often occurs in this way. This leads to incremental improvements, and/or a recombination of the components of a service, although it can also be a new concept, evolved to meet a specific user need. Some innovation is intangible: while most research ranks custom-built, add-on and recombinant variants, ad hoc and experiential innovation also deserve attention. All five aspects are important although start-up support has often been unavailable for the last two.

In their 2002 survey of French business services, Djellal and Gallouj (2002, 148) found that 72 per cent of innovations were more or less non-technological even when technology was a non-negligible direct or indirect component, but that 65 per cent of innovative services were technology-based. Djellal and Gallouj (2002) highlight the importance of market forces and point to the importance of start-up support going beyond support for technology, and facilitating the mapping of an entrepreneur's approach to the design and development of their innovation. I next review the various origins and paths to services innovation.

24.3.2 The start-up support process

Sundbo and Gallouj (1999) identified the players involved in the start-up support process (see Figure 24.2). Innovation involves new business models, technological upgrades to an existing process, application of an R&D platform idea, and a product developed through user–supplier interaction.

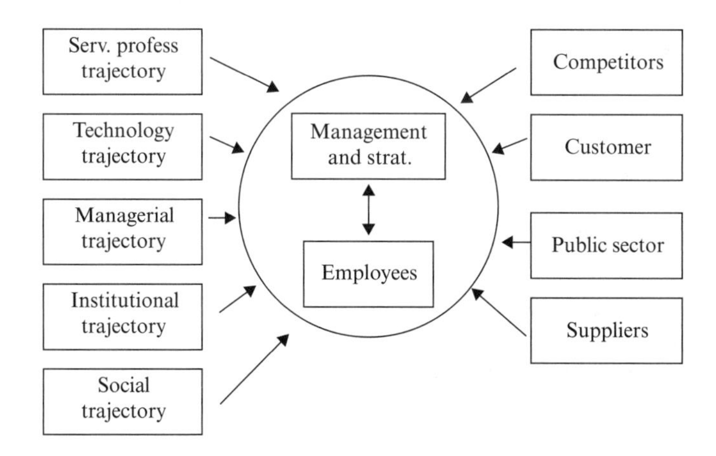

Source: Sundbo and Gallouj (1999).

Figure 24.2 Pathways of innovation

For a service innovation to occur, suppliers need high degrees of openness to the technological, social and industrial developments in their environments, and trust in relation to exchanges related to industrial or commercial issues. Openness is difficult for entrepreneurs: lack of time, small experience and lack of personal contacts often constrain the ability of start-ups to become embedded in industry, economic sectors or particular locations.

To identify the route from detection of a need to an innovation, Gallouj (1999) followed Gadrey (1991) to identify the following functions that combine to constitute a given service:

1. Raw materials processing and logistics.
2. Information processing and logistics.
3. Experiential processes.
4. Knowledge processing (for example operating procedures, etc.).

These require individuals with specific skills or technologies. Figure 24.3 depicts the know-how inbuilt in each function and the potential usefulness of the service to users.

<table>
<tr><th>1. Input</th><th colspan="2">2. Content</th><th>3. Output</th></tr>
<tr><td rowspan="3">Technological
know-how
(T)</td><td>Logistical operations
+ raw material processing +
supporting technology (M)</td><td rowspan="3">Experiential
operations
(E)</td><td rowspan="3">Features
of the actual service
(S)</td></tr>
<tr><td>Logistical operations
+ information processing +
supporting technology (I)</td></tr>
<tr><td>Knowledge processing
+ supporting technology (K)</td></tr>
</table>

Note: M, I, K and E are the functional components between T and S.

Source: Adapted from Gallouj (1999).

Figure 24.3 Functional phases of services

Thus, service innovations come in a variety of forms, all dependent on the quality of the needs analysis, which may be difficult for backers and other support actors to evaluate.

24.4 From innovation to actual start-up

Business service suppliers have for a long time been undercapitalized, which is doubtless why their R&D efforts have been limited to applied research and testing of new industrial products and processes. Research reported by

Barras in 1986 virtually ignores services innovation, as falling outside the realm of industrial technology. However, numerous service suppliers have overturned 'their subordination to industry' in terms of technological innovation, with 'systems of their own, which they develop themselves or do so from a strong bargaining position' (Djellal and Gallouj, 2007b).

Thus, the design by key local actors of a policy to support innovative start-ups of all types has been important in jump-starting the fabric of their economies. Start-ups upgrade technology, develop new products and re-energize areas hit by international competition and obsolescence. This is the idea behind the provision of start-up support in Greater Toulouse, whose potential benefits and limitations are investigated in this chapter.

24.4.1 Overview of Reliantis

If, as Gartner (1988) and Julien (2005) note, the enterprise is the entrepreneur and his or her actions, it is the operating environment that nourishes the enterprise with resources and generates the need for services. Market intelligence plays an important role in the establishment of new business (Kirzner, 1979) and it is this intelligence that the entrepreneur transforms into a business opportunity (Vaghely and Julien, 2010). However, accessing intelligence and knowledge is a long, difficult process (Sammut, 2003) and entrepreneurs are wary of risk and of missing opportunities (Dickson and Giglierano, 1986). It is understandable, therefore, that policy will tend to emphasize the availability of information.

Borges et al. (2008) highlight that more research is needed on how the start-up process happens and on its impact on firm performance because this will affect the firm's future and its impact on local economic development. Borges et al. (2008) draw on research conducted by Delmar and Shane (2002) to validate the critical role of support agencies to understand markets, secure finance and resolve sundry problems. I analyse this under-investigated area in the context of innovative service start-ups, paying particular attention to the evaluation of innovation and the contribution of business support actors.

The Toulouse CCI set up Reliantis, which is a complex but flexible mechanism designed to help innovators transform their ideas into viable firms. Entrepreneurs are selected through a four-step approach:

1. A screening committee vets applications.
2. Successful applicants are given help with their feasibility studies and business plans.
3. They are given a start-up kit which enables access to Reliantis legal, financial and other advisory services.
4. Follow-up services and coaching are provided to the start-up.

These steps are punctuated by working meetings in which entrepreneurs report on the status of their business to a committee comprised of all stakeholders in the new business. These meetings help to avoid delays that can arise if financers, lawyers or other interested players are left out of the loop and do not have the information required to move forward. These meetings promote interaction among the participants who quickly become part of the entrepreneur's socio-cognitive network (Rothaermel, 2000).

In order to focus its resources on start-ups whose services would increase the competitiveness of clients, Reliantis defines innovation according to the third edition of the *Oslo Manual* (OECD, 2005) as: 'the implementation of a new or significantly improved product (good or service), or process, a new marketing method, or a new organisational method in business practices, workplace organization or external relations'. OSEO,[3] the French support agency for small and medium-sized enterprises, adds to this definition: 'Any process that creates a competitive advantage for products, processes, services or working methods based on technology or know-how new to business.'

24.4.2 Methodology and results

One difficulty in responding to the Reliantis request for indicators is methodological: 43 start-ups is a very small sample for a standard quantitative survey. Can I submit the results of this theoretical analysis without reference to the operating environments of these start-ups? Or 'Are we just [the] impotent authors and speakers at courses and conferences?' described by E. Gummerson in his address to the Seventh International Research Seminar in Service Management (29 May 2002).

Survey methodology Because of the good quality of personal relations between the survey frame and Reliantis and the small size of the sample, I opted with the agreement of the CCI for a postal survey that would enable me to compare the initial planning phase (characteristics and setup) against the status of a firm at the time of the survey (size, age, R&D setup and revenue growth) in order to understand what role support measures could play. We contacted 43 business services and a further nine qualified as 'other innovative'. We received 31 usable responses, which we backed up by interviews with six managing directors. Each of them were conducted by two interviewers.

We found that with support from Reliantis, 84 per cent of successful applicants to the scheme resulted in the establishment of new innovative service enterprises and that innovation initially was 'important' or 'vital' to 77 per cent of business operations and continued to be so for 70 per cent. Some 75 per cent of firms were satisfied with the market response to their services,[4] with 68 per cent reporting 'strong' increased demand for their services and 38 per cent having more than ten employees as of January 2007, up from 16 per cent at the start of the business operation. Ninety-three per cent were optimistic about sales within a three-year time horizon and 77 per cent expected to hire more staff within that time frame.

These figures confirm the importance of innovation to the establishment and expansion of small and medium-sized enterprise (SME) service start-ups. However, innovation in services evolves in ways that CCI start-up support partners often find complex and hard to track. I try to identify these issues in order to simplify support through the definition of indicators and the roles of certain players.

Implementing innovation Services innovation should not be seen in isolation. It is market-driven and is 'shaped by what innovators perceive to be the opportunities and challenges' out there (Hauknes, 2002, 118). Thus, innovation is a response to a client need, sometimes in the absence of a pre-existing R&D or other platform.

We tested for, but did not find, a link between the availability of an R&D facility and the establishment of innovative business services. This is not surprising in the context of start-ups, and especially service firms where the individual duties of the founders tend to overlap.

However, the responses to our survey did show a link between availability of R&D facilities and the innovative services business model (see Table 24.4). The importance of this link is intensified by the link between the importance of innovation to the original and final business model. Expansion depends on how the original innovation evolves over time.

Table 24.4 R&D platform

R&D Platform	Innovative product/service	Non-innovative product/service	Total
Pre-existing	8	0	8
None	6	17	23
Total	14	17	31

Note: Pearson's $\chi^2 = 13.093$; 2-tailed asymp. Sig.

Thus, an R&D platform is not a prerequisite for the initial phase of service innovation although it may well contribute to expansion subsequently.

I tried to correlate current vs original importance of innovation to start-ups in order to assess the market response to business expansion. The results are significant (see Table 24.5) and show that adjustment to market demand promotes expansion.

Table 24.5 Value of innovation to the original business model and current business operation

		Current value			Total
		Slight or moderate	Important	Vital	
Original Value:	Slight or moderate	7	0	0	7
	Important	2	7	0	9
	Vital	0	2	13	15
Total start-ups		9	9	13	31

Note: Pearson's $\chi^2 = 54.0011$; df = 12 and 2-tailed asymp. sig. = 0.000.

Although an innovative service is a mix of technology and response to market needs, the typical innovator is someone able to spot emerging needs and markets rather than a pure technology fanatic. This explains why a pre-existing R&D platform is not a necessity, but also points to the isolation of innovation. This contrasts with research that stresses the value of teamwork and the creative cognitive dynamics it engenders for entrepreneurs in the early phases of firm set-up (Schoonhoven and Romanelli, 2001; Prax, 2007). In examining the difficulties involved in the phase they call 'preparation', Borges et al. (2008) confirm the contribution of start-up support – both tangible and intangible – to the performance of 175 enterprises, in the form of marketing plans and negotiating, mutual encouragement and market knowledge.

Ideally, support actors will avoid imposing templates and nurture new ventures with relevant feedback and appropriate help with problematic areas (Sammut, 2003). Our focus on the innovative aspects of start-ups steered us towards investigation of the role of start-up support agents in filing patent applications.

Second, from innovation to patenting service innovations often involve intangibles that cannot be protected by patents. However, Djellal and Gallouj (2002, 138) note that some services can be packaged into patentable products for sale as units. From our sample of firms we found a handful of patented services: all patent application filings had been prompted and supported by start-up policies. Start-up support should scan for patentable products/services and provide adequate support for application filings.

24.5 Conclusion

Despite the small sample size of our survey I can confirm the results in the literature. Start-up support enhances the performance of new enterprises; already confirmed for the industrial sector, this statement also applies to business services. Regardless of the economic sector, start-up support needs to be operational, informative and interpersonal. Because innovative service suppliers are exposed to particular risks, the need for knowledge relating to the operating environment and other information is great, but this is not to imply that support teams require an R&D platform from the outset. However, some sort of R&D facility will be necessary if a start-up is to maintain its innovative lead.

The literature on innovative services extends our knowledge on entrepreneurship and provides some guidelines for start-up support agencies in the phases of initial vetting and feasibility assessment.

1. Even services that are innovative to only a small degree generate added value for the user if they fit smoothly into a critical operation. However, the innovator should evolve the service according to the template in Figure 24.3. Start-up support players should concentrate on how well the service adheres to this template rather than on the intrinsic originality of the innovation.
2. The functional breakdown serves as a basis for analysing how entrepreneurs handle each component of the Gallouj (1999) model. Careful handling helps the service evolve, while cursory treatment may lead to dissatisfied users.
3. Regardless of its technological content, services innovation enriches the fabric of the local economy through the interaction with users and partners and the range of contributions it makes to their business activities. A diagram of innovation pathways can be used to map the entrepreneur's sources of innovation and the quality of their relationship with market demand.

In terms of economic base theory, I can highlight links between the economic development of regional towns and cities and business services, especially the most innovative. As Djellal and Gallouj (2007a) note: 'many service suppliers have shed their subordination to industry when it comes to technological innovation. In other words, they build their own hardware systems, or (at least) operate from a position of strength as they reinforce synergy with their clients and partners.'

The structure of the firm in the start-up phase affects the fabric of the local economy. The number of staff increases rapidly and very small start-ups are much more dependent on local clients and suppliers than are major corporations (Mallard, 2007; Léo et al., 1991).

Business developers and local economic actors who want to jump-start their regional economies should design policy to support innovative start-ups that encompasses more than industrial innovation. Innovative service enterprises provide new technologies and upgrades as well as new services that act to rejuvenate both local industry and the wider economic fabric as these services migrate beyond regional borders. This is the thinking behind the design of the start-up support in Greater Toulouse, whose potential is demonstrated by the research described in this chapter.

Notes

1. The author thanks Wael Touzi, a doctoral researcher in the IMS Laboratory, LAPS Department, University of Bordeaux I, for his help in conducting the survey and processing the data.
2. The head offices of the aircraft maker Airbus, and also Fabre and Sanofi pharmaceuticals are in Greater Toulouse.
3. http://www.oseo.fr/a_la_une/dossiers/development_durable_pme_innovation_financement.
4. Data are in percentages for easier reading.

References

Aldrich, H. (1999), *Organizations Evolving*, London: Sage.

Barcet, A. and J. Bonamy (2000), 'L'innovation de service, conditions macro-économiques', *Actes de la conférence internationale Economie et socio-économie des services*, Lille, pp. 3–19.

Barras, R., (1986), 'Towards a theory of innovation in services', *Research Policy*, **15**, 161–173.

Borges, C., L.J. Fillion and G. Simard (2008), 'Processus de création de nouvelles entreprises: temps, difficultés, changements et performance', 9° congrès international francophone en entrepreneuriat et PME, Louvain, CDRom.

Comité économique et social européen (2006), 'Services et industrie manufacturière européenne: les interactions entre ces secteurs et l'impact de celles-ci sur l'emploi, la compétitivité et la productivité', CMCI/035.

CPCI (2007), 'Rapport annual sur l'industrie française en 2006–2007', INSEE Paris.

Delmar, F. and S. Shane (2002), 'What firm founders do: a longitudinal study of the start up process', in W.D Bygrave et al. (eds), *Frontiers of Entrepreneurship Research*, Wellesley, BA: Babson College, pp. 632–645.

Dickson, P.R. and J.J. Giglierano (1986), 'Missing the boat and sinking the boat: a conceptual model of entrepreneurial risk', *Journal of Marketing*, **50** (3), 58–70.

Djellal, F. (2002), 'Le secteur du nettoyage face aux nouvelles technologies', *Formation emploi*, **77**, 37–49.

Djellal, F. and F. Gallouj (2002), 'A propos de la nature de l'innovation dans les services: les enseignements d'une enquête postale', in F. Djellal and F. Gallouj (eds), *Nouvelle économie des services et innovation*, Paris: L'harmattan, pp. 135–163.

Djellal, F. and C. Gallouj (2007a), *Introduction à l'économie des services*, Grenoble: PUG.
Djellal, F. and F. Gallouj (2007b), 'La relation innovation-emploi dans les services: un bilan et un agenda de recherche', in M. Monnoyer and P. Ternaux (eds), *Mondialisation des services*, Paris: L'harmattan, pp. 23–43.
Djellal, F. and F. Gallouj (2007c), 'Productivity strategies in services: assimilation, particularism and integration', Reser conference, http://reser.net/2007-RESER-Conference papers_a292.html?PHPSESSID=bd0fdd1ff51a96c1bdbd40755aa71dc2.
Djellal, F. and F. Gallouj (2008), *Measuring and Improving Productivity in Services: Issues, Strategies and Challenges*, Cheltenham, UK and Northampton, MA, USA: Edward Elgar.
Djellal, F., C. Gallouj, Gallouj, F. and K. Gallouj (2004), *L'hôpital innovateur: de l'innovation médicale à l'innovation de service*, Paris: Masson.
Edvardsson, B., M. Holmlund and T. Strandvik (2008), 'Initiation of business relationships in service-dominant settings', *Industrial Marketing Management*, **37** (3), 339–350.
Eiglier, P. and E. Langeard (1987), *Servuction, le marketing des services*, Paris: Macgraw Hill.
Fonrouge, C. (2002), 'L'entrepreneur et son entreprise: une relation dialogique', *Revue française de gestion*, **138**, 145–158.
Gadrey, J. (1991), 'Le service n'est pas un produit: quelques implications paor l'analyse économique et la gestion', *Politiques et management public*, **9** (1), 1–24.
Gadrey, J. (2002), 'Les mutations du système productif', *Cahiers Français,* n° 311.
Gadrey, J. (1996), *Services: la productivité en question*, Paris: Desclée de Brouwer.
Gadrey, J., F Gallouj. and O. Weinstein (1995), 'New modes of innovation: how services benefit industry' *International Journal of Service Management*, **6** (3), 4–16.
Gallouj, C. (2006), 'Les politiques locales et régionales de développement des services: le cas des chambres consulaires françaises', 16th Reser conference, Lisbon, http://www.reser.net/2006-RESER-Conference-papers_a287.html.
Gallouj, F. (1994), *Economie de l'innovation dans les services*, Paris: L'Harmattan.
Gallouj, F. (1999), 'Les trajectories de l'innovation dans les services: vers un enrichissement des taxonamies évolutionistes', *Economies et sociétiés, série Economie et Gestion des Services*, **1** (5), 143–169.
Gallouj, F. (2002), *Innovation in the Service Economy*, Cheltenham, UK and Northampton, MA, USA: Edward Elgar.
Gartner, W.B. (1988), 'A conceptual framework for describing the phenomenon of new venture creation', *Academy of Management Journal*, **19**(4), 696–706.
Granovetter, M. (1985), 'Economic action and social structure: the problem of embeddedness', *American Journal of Sociology*, **91**, 481–510.
Harrington, P.W. and P.W. Daniels (2006), *Knowledge Based Services: Internationalization, Regional Development*, Aldershot: Ashgate.
Hauknes, J, (2002), 'L'innovation sous l'angle des services: faut-il créer de nouveaux concepts?' in F. Djellal and F. Gallouj (eds), *Nouvelle économie des services et innovation,* Paris: L'harmattan, pp. 103–134.
Jakobsen, S.E and H.W. Aslesen (2004), 'Location and knowledge interaction between head office and KIBS in city areas', Reser conference Toulouse, http://reser.net/index.php?action=telechargement&startdownload=23_S2_AS.PDF&startid=3231&classeur=&PHPSESSID=b101e9a3ff7bb80fe73ab9a053875302
Julien, P.A. (2005), *Entrepreneuriat régional et économie de la connaissance*, Montréal: Presses de l'université du Québec.
Kirzner, I.M. (1979), *Perception, Opportunity and Profit*, Chicago, IL: University of Chicago Press.
Léo, P.-Y., M.C. Monnoyer-Longé and J. Philippe (1991), *Métropoles régionales et PME, l'enjeu international,* Aix en Provence : Serdeco.
Léo, P-Y and J. Philippe (1998), 'La transformation des métropoles françaises', in RESER (ed.), *Services et métropoles, formes urbaines et changement économique*, Paris: L'Harmattan, pp. 25–45.

Léo, P.-Y. and J. Philippe (2007), 'Executives and business services: key factors of French metropolitan growth', *Service Industries Journal*, **27** (3–4), 215–233.
Madrid, C. (2003), 'Importance of service for manufacturers: conclusions of an empirical study', *Service Industries Journal*, **23** (1), 167–194.
Mallard, A. (2007), 'La pluralité des rapports au marché dans les très petites entreprises, une approche typologique', *Economie et statistiques*, **407**, 51–71.
Miles, I. (1993), 'Services in the new industrial economy', *Futures*, **25** (July), 653–672.
Monnoyer, M.-C. and J.M. Zuliani (2007),'The decentralisation of Airbus production and services', *Service Industries Journal*, **27** (3–4), 251–262.
Moyard, L. (2000), 'La diffusion spatiale des services aux entreprises et ses dynamiques: application aux régions urbaines belges', *Economies et sociétés, Série Economie et gestion des services*, **6**, 131–153.
North, D.C. (1955), 'Location theory and regional economic growth', *Journal of Political Economy*, **63** (June), 243–258.
OECD (2005), *Oslo Manual: Guidelines for Collecting and Interpreting Innovation Data*, 3rd Edition.
Planque, B. (1993), 'La distribution spatiale des fonctions et des qualifications', in M. Savy and P. Veltz (eds), *Les nouveaux espaces de l'entreprise*, DATAR, La Tour d'Aigues, France: Editions de l'Aube.
Prax, J.Y. (2007), *Le manuel du knowledge management*, Paris: Dunod.
Rothaermel, F.T. (2000), 'Technological discontinuities and the nature of the competition', *Technological Analysis and Strategic Management*, **12** (2), 149–160.
Rubalcaba, L. (1999), 'Business services and the European industry: growth, employment and competitiveness', Brussels: Commission of the European Union.
Rubalcaba, L. and H. Kox (eds) (2007), *Business Services in European Economic Growth*, London: Palgrave MacMillan.
Sammut, S. (2003), 'L'accompagnement de la jeune entreprise', *Revue française de gestion*, **144**, 152–164.
Schoonhoven, C.B. and E. Romanelli (2001), 'Emergent themes and the next wave of entrepreneurship research', in C.B. Schoonhoven and E. Romanelli (eds), *The Entrepreneurship Dynamic: Origins of Entrepreneurship and the Evolution of Industries*, Palo Alto, CA: Stanford Business Books, pp. 383–408.
SESSI (2007), *Enquête recours aux services par les entreprises industrielles en 2005*, Paris: SESSI.
Shostack, G.L. (1984), 'Designing services that deliver', *Harvard Business Review*, **62** (1), 133–139.
Soldatos, P. (1991), *Les nouvelles villes internationales: profil et planification stratégique*, Aix en Provence: Serdeco.
Spath, D. and W. Ganz (2008), *The Future of Services*, Munich: Hanser.
Sundbo, J. (1996), 'Service development: from quality assurement to innovation', Proceedings 4th international research seminar, La Londe les Maures, France.
Sundbo, J. and F. Gallouj (1999), 'Innovation as a loosely coupled system in services', *International Journal of Services Technology and Management*, **1** (1), 15–36.
Tornikovski, E.T. and S.L. Newbert (2007), 'Exploring the determinants of organizational emergence a legitimacy perspective', *Journal of Business Venturing*, **22** (3), 311–335.
Vaghely, I.P. and P.A. Julien (2010), 'Are opportunities recognized or constructed?', *Journal of Business Venturing*, **25** (1), 73–86.
Wood, P. (2008),'Policies for promoting knowledge intensive services in cities: reflections on British experience', Reser conference, http://reser.net/2008-RESER-Conference papers_a353.html?PHPSESSID=bd0fdd1ff51a96c1bdbd40755aa71dc2.

25 A dominant node of service innovation: London's financial, professional and consultancy services

Peter Wood and Dariusz Wójcik

25.1 The financial services: a paradigm for knowledge-intensive business service innovation?

The innovativeness of the London economy, and its basis in financial and other business service functions, has been a well-established feature of the UK economy since at least the early 1990s (Wood, 2008). It has thus presented a paradox for conventional innovation studies. In 2004, according to the Community Innovation Survey (CIS), the city's labour productivity was the highest of any UK region and its businesses showed the highest percentage turnover of products new to the firm or significantly improved. Nevertheless, they also spent the lowest regional share of workplace gross value added (GVA) on research and development (R&D): 0.5 per cent compared with 3.3 per cent in the Eastern region and 2.2 per cent in the South East. London's innovativeness has therefore depended, not on the creation of new technologies, but on the flexibility with which its service labour skills are adapted to new demands, often facilitated by new or established technologies (Wilson, 2007, 13–14). Such innovation is marked by successful change in international markets arising from interactions between the city's specialist financial, professional, business and creative expertise and that of its clients.[1]

The financial services have attracted most attention, mainly because of their historic role in managing UK capital resources and, in modern times, supporting international trade and arbitrage. In spite of their diversity, they also display strong commercial, cultural and political cohesion, with international and national institutions concentrated in and around the 'City' of London (Thrift, 1987, 1994).[2] Commercial innovation by individual firms has also traditionally been framed within a national regulatory framework designed to sustain trust in the wider financial system and protect retail investors. From the mid-1990s, however, regulation was eased, specifically to encourage innovation and support London as the dominant European base for international financial exchange. One consequence was to weaken the transparency and scrutiny of market processes and outcomes, as competition extended innovative forms of deal-making across international banking, insurance, equities, bonds, securities and commodity exchange.

The many non-financial forms of the 'knowledge-intensive business services' (KIBS) are less easy to characterize.[3] They encompass both formally constituted professions, including commercial law, accountancy and real estate management, as well as many forms of commercial business consultancy. Most markets are dominated by large multifunctional firms, but there is a myriad of medium-small businesses in the professions, and in consultancy specialisms such as computer systems and database management, management and human resource consulting, engineering, construction and architecture, graphic and industrial design, and advertising and market research. Non-financial KIBS are more diffuse than the financial services, functionally, institutionally and, within central London, geographically. Local clusters are evident only in disparate areas such as the legal quarter west of the City, around Holborn; media and advertising in Soho–Fitzrovia; and design and other creative services in Clerkenwell (Walker and Taylor, 2003; Nachum and Keeble, 2003; Keeble and Nachum, 2002; Hutton, 2006; Faulconbridge, 2007). It is also often assumed, especially in The City, that their primary role is to support the financial services and augment London's financial competitiveness. In fact, London's professional and business KIBS serve clients in every other national sector, and many overseas. Their independent influence on national and UK regional innovation and competitiveness, however, is generally neglected (Wood, 2008). This is partly because in London they are assumed to shadow the financial services, but also, as we shall argue, because there are real difficulties in separately identifying and measuring this influence.

Conventional studies of innovation, of course, focus on the introduction of new technologies and, for services, on how these have transformed service supply (Barras, 1986, 1990; Miles, 2002, 169–71). In the wholesale financial services, however, while technological developments have undoubtedly transformed market operations, they have not been the main driver of London's success. On the contrary, in many activities widespread access to such technologies elsewhere has often challenged the City's domination. London has nevertheless thrived because of the flexibility with which its financial expertise has been realigned to new market opportunities as they arise from national and international business exchange (Taylor et al., 2003; Pain, 2007). Similar forms of expertise-based responsiveness underpin the innovativeness of the non-financial KIBS. Until recently, however, their experience was commonly compared with manufacturing, and associated with the application of new product or process technologies (Coombs and Miles, 2000; Nijssen et al., 2006). The evidence for low formal R&D, reflected in London's performance, thus inevitably placed them in a poor light. Other commentators, however, emphasizing the distinctive qualities of service exchange, have focused more on the outcomes of interactive service processes between providers and clients, and the conditions supporting such service co-production (Wood, 1991, 2002b; Gallouj and Weinstein, 1997; Gallouj, 2002a, 2002b; Miles, 2002; Drejer, 2004; den Hertog, 2002). These processes, combining non-technological and technological expertise from both KIBS and their clients, certainly offer a more appropriate basis for examining the innovativeness of central London's KIBS.

Innovation in the financial services has received considerable attention since 2003 primarily to explain their extraordinary global expansion up to 2007–08. In this chapter, we examine this experience in central London, and consider its relevance for other forms of KIBS innovation in the city. The comparison assumes no primacy of financial over other KIBS, however, and important differences between them become evident, most obviously the dependence of financial services on returns on capital assets rather than the operational value to clients of their expertise alone. The events of 2007–08 have also highlighted the critical role of regulation, which takes quite different forms among non-financial KIBS. The financial services are nevertheless still essentially 'knowledge-intensive' services, whose innovativeness, as for all other KIBS, is expressed through service relations with clients. The variable opportunities and constraints affecting these relations are the key to understanding innovativeness among all the types of KIBS active in central London.

25.2 The scale of central London's financial, professional and business services

The comparative importance of financial and non-financial KIBS for employment in central London up to 2006 is demonstrated in Table 25.1.[4] This shows estimated data from the Office of National Statistics Annual Business Inquiry for 2000, 2003 and 2006, distinguishing between financial, professional and business services (National On-line Manpower Information Services, 2009). Financial services here include banking, insurance, and the so-called 'auxiliary' financial activities, which are particularly significant in London, including exchange administration, securities broking and specialist fund management. For the professional services, whose core functions relate to their support for statutory provisions based on legal, accountancy and property surveying expertise, their commercial work now commonly extends into wider business consultancy. Like the financial services, however, a significant share of their market is among private consumers. The business services, on the other

Table 25.1 *Estimated employment data, Central London, City of London, Tower Hamlets and West End, 2000–2003–2006*

ALL CENTRAL LONDON SIC activity	2003 employment	Change 2000–03	% Change 2000–03	2006 employment	Change 2003–06	% Change 2003–06
65: Banking, etc	138,390	−15,959	−10.3	140,802	2,412	1.7
66: Insurance	25,300	1,164	4.8	17,775	−7,523	−29.7
67: Auxilliary to banking, etc	88,810	3,162	3.7	97,467	8,657	9.7
All finance services	252,500	−11,633	−4.4	256,044	3,546	1.4
7011/32: Real estate	22,325	−1,025	−4.4	22,566	242	1.1
7411: Legal	68,115	−1,040	−1.5	73,892	5,777	8.5
7412: Accountancy, etc.	41,551	−8,526	−17.0	46,472	4,921	11.8
All professional services	131,991	−10,591	−7.4	142,930	10,940	8.3
72: Computer and database	33,688	−1,019	−2.9	45,772	12,084	35.9
742/3;73: Architecture/ engineering/testing/R&D	39,806	−2,405	−5.7	44,734	4,931	12.4
7413/744: Advertising and market research	29,639	−5,922	−16.7	40,140	10,501	35.4
7414: Business consultancy	43,396	792	1.9	58,055	14,659	33.8
All business services	146,529	−8,554	−5.5	188,701	42,175	28.8
TOTAL KIBS:	531,020	−30,778	−5.5	587,675	56,661	10.7

CITY OF LONDON						
65: Banking, etc	76,456	−17,151	−18.3	67,079	−9,376	−12.3
66: Insurance	18,672	1,522	8.9	14,603	−4,068	−21.8
67: Auxilliary to banking, etc	49,693	2,367	5.0	48,823	−869	−1.7
All finance services	144,821	−13,262	−8.4	130,505	−14,313	−9.9
7011/32: Real estate	1,894	−19	−1.0	1,766	−127	−6.8
7411: Legal	38,875	1,617	4.3	39,647	773	2.0
7412: Accountancy, etc.	13,255	−1,861	−12.3	18,819	5,564	42.0
All professional services	54,024	−263	−0.5	60,232	6,210	11.5
72: Computer and database	8,869	1,258	16.5	9,550	681	7.7
742/3;73: Architecture/ engineering/testing/R&D	4,074	189	4.9	4,615	541	13.3
7413/744: Advertising and market research	1,112	−526	−32.1	1,216	105	9.5
7414: Business consultancy	7,759	2,031	35.5	7,778	19	0.2
All business services	21,814	2,952	15.7	23,159	1,346	6.2
TOTAL KIBS:	220,659	−11,943	−5.1	213,896	−6,757	−3.1

Table 25.1 (continued)

TOWER HAMLETS *SIC activity*	2003 employment	Change 2000–03	% Change 2000–03	2006 employment	Change 2003–06	% Change 2003–06
65: Banking, etc	18,645	5,366	40.4	42,447	23,802	127.7
66: Insurance	957	−792	−45.3	957	0	0
67: Auxilliary to banking, etc	18,012	5,579	44.9	22,682	4,670	25.9
All finance services	37,614	10,153	37.0	66,086	28,472	75.7
7011/32: Real estate	1,988	145	7,9	1,851	−136	−6.9
7411: Legal	1539	−81	−5.0	5,020	3,481	226.2
7412: Accountancy, etc.	1,100	361	48.8	977	−122	−11.1
All professional services	4,627	425	10.1	7,848	3,223	69.7
72: Computer consultancy	4,287	384	9.8	5,391	1,104	25.8
742/3;73: Architecture/ engineering/testing/R&D	1,522	372	32.2	1,476	−44	−2.9
7413/744: Advertising and market research	814	208	20.4	999	186	22.9
7414: Business/mngment consult.	2,059	238	13.1	4,563	2,504	121.6
All business services	8,682	1,202	12.2	12,429	3,750	43.2
TOTAL KIBS:	50,923	11,780	30.1	86,363	35,445	69.6

65: Banking, etc	43,289	−4,174	−8.8	31,276	−12,014	−27.8
66: Insurance	5,671	434	8.3	2,215	−3,455	−60.9
67: Auxilliary to banking, etc	21,105	−4,784	−18.5	25,962	4,856	23.0
All finance services	70,065	−8,524	−10.8	59,453	−10,613	−15.1
7011/32: Real estate	18,443	−1,151	−5.9	18,949	505	2.7
7411: Legal	27,701	−2,576	−8.5	29,225	1,523	5.5
7412: Accountancy, etc.	27,196	−7,026	−20.5	26,676	−521	−1.9
All professional services	73,340	−10,753	−12.8	74,850	1,507	2.1
72: Computer consultancy	20,532	−2,661	−11.5	30,831	10,299	50.2
742/3;73: Architecture/ engineering/testing/R&D	34,210	−2,966	−8	38,643	4,434	13.0
7413/744: Advertising and market research	27,713	−5,604	−16.8	37,925	10,210	36.8
7414: Business/mngment consult.	33,578	−1,477	−4.2	45,714	12,136	36.1
All business services	116,033	−12,708	−9.9	153,113	37,079	32.0
TOTAL KIBS:	259,438	−30,615	−10.6	287,416	27,973	10.8

Source: NOMIS (2009).

hand, more unambiguously serve the operational requirements of business firms and public agencies, through processes of commercial consultancy and outsourcing.

Non-financial KIBS, and especially the business services, appear to be at least as important to the success of the central London economy as the financial services. The area supported 0.25 million financial service jobs in 2006, and more than 0.33 million in the professional and business services. Taking account of the unknown share of consumer markets for the financial and professional services, therefore, business-directed innovation in central London appeared at least as likely to come from the non-financial business services as from financial services. Further, the combined business services and business-related 'auxiliary' financial services employed a total of 280,000, almost as many as in mainstream banking, insurance and professional services, with their significant share of consumer business. The former also grew faster between 2000 and 2006.

Two significant trends were apparent during 2000–2006, including the international recession of 2000–03 and subsequent recovery. The first was a shifting geography of financial services. The City lost almost 28,000 financial jobs, with little sign of recovery after 2003 (Table 25.1). Central London to its west (for short, the 'West End') also lost almost 20,000.[5] These losses, however, were largely balanced by the emergence of Docklands, east of the City, especially around Canary Wharf, reflected in the data for the Borough of Tower Hamlets. Here, financial services expanded by 40,000 jobs, to reach 66,000 by 2006. Figure 25.1 illustrates the significance of these developments. It also demonstrates the second component of recent trends: the continuing vitality of commercial business services in the West End, despite a sharp setback during the recession, which especially affected advertising and market research. By 2006, these losses had been more than recovered, with 37,000 jobs gained after 2003. The business-oriented 'auxiliary' financial services also recovered around 4,000 jobs in both the West End and Tower Hamlets, while City numbers continued to decline (Table 25.1).

This evidence confirms that the central London economy depends as much on professional and business services in the West End as on financial activities in the City and Canary Wharf.[6] Even their different geographies challenge any suggestion that non-financial KIBS innovation depends predominantly on the financial sector. Total KIBS employment in central London, including Tower Hamlets, fell by only 30,000 between 2000 and 2003, and regained these and a further 27,000 by 2006, to reach a total of over 580,000. Financial service employment fell slightly, while the professions declined and then returned to about their 2000 level. Net growth was largely confined to the business services, which increased by over 20,000

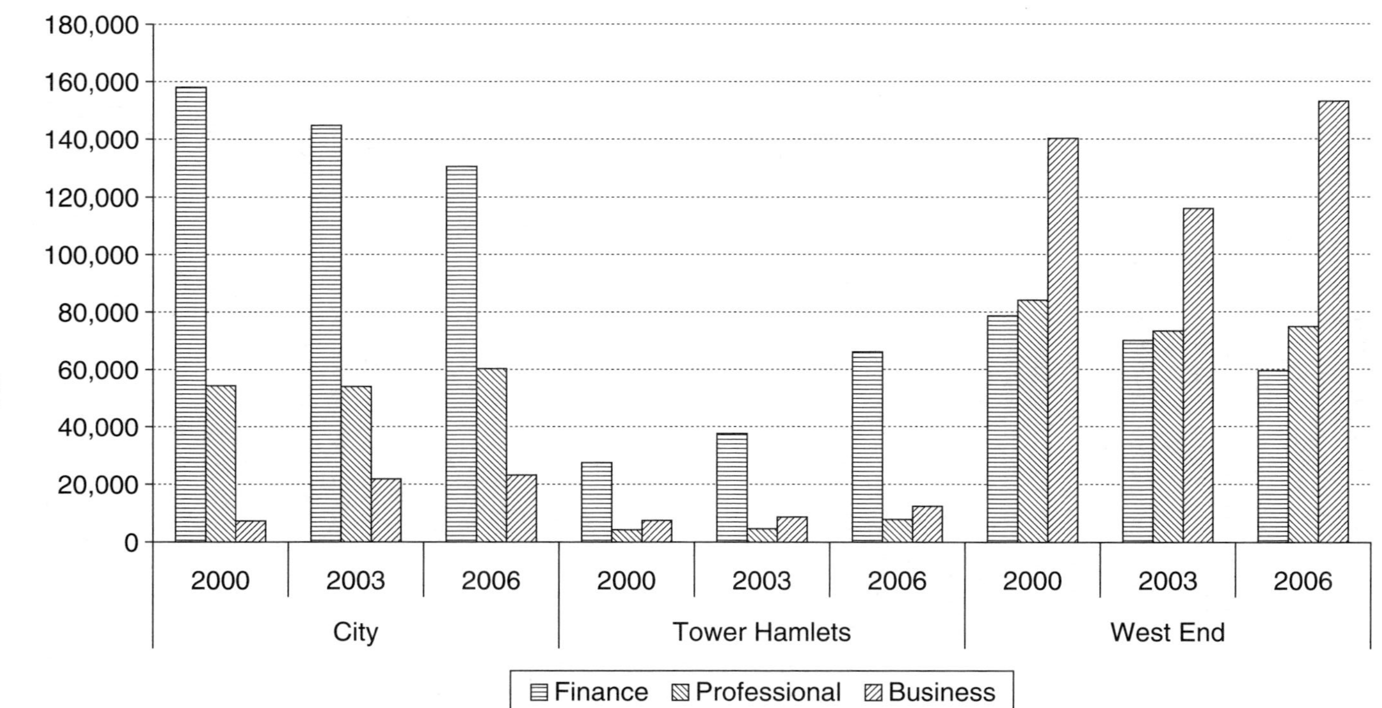

Figure 25.1 *Central London areas: total KIBS employment, 2000, 2003, 2006*

jobs, led by computer, advertising, and business and management consultancy. The innovative potential and growth of the non-financial business services in London may therefore have been as significant as that of the financial services. As the importance of financial innovation was more recognized during this period, therefore, what might this experience for London tell us about other, arguably equally significant, KIBS activities?

25.3 Processes of change and the innovative response of London's financial services

In recent decades financial sector innovation has included electronic stock exchanges, automated trading techniques, the introduction of credit cards on a mass scale, and the securitization of mortgages, hedging, swaps, options and other derivatives. This pace of innovation was a response primarily to three interlinked worldwide trends. Firstly, the deregulation of financial markets in the post-Bretton Woods era, which led to international financial flows of unprecedented volumes and high volatility, creating demands for new ways of managing them and the associated risks (Leyshon and Thrift, 1997). Secondly, a growing number of individual and institutional customers, both within and outside the financial sector, had accumulated sizeable financial assets seeking profitable investment opportunities, and required ever more sophisticated investment and risk management methods (French and Leyshon, 2004; Clark and Wójcik, 2007). Thirdly, in responding to this booming demand financial firms were able to take advantage of technological progress, particularly in the field of information and computer technology (ICT) (O'Brien, 1992; Cox and Jones, 2007).

In the context of these well-known drivers of worldwide financial innovation, research has emerged only since 2000 which directly investigates the factors making particular financial centres more or less innovative (Frame and White, 2002). How did location benefit institutions in creating new value for international clients, often in competition with other cities? What innovative opportunities were offered in changing market and regulatory conditions to exploit their specialist human resources and institutional arrangements? How did the interaction between such 'local' capacities and 'global' market opportunities apparently create new value? Which of these processes were most distinctive to certain centres, such as London, compared with others, including its major competitor, New York, or newly emerging centres such as Dublin? The evidence suggests that four place-specific factors provide the answers to these questions: labour quality; institutional flexibility, including interfirm clustering; financial infrastructure as it affects institutional behaviour; and, perhaps most crucial for the financial services, the outcomes of formal regulation, including market transparency.

25.3.1 Labour

It is widely accepted that financial centre innovativeness depends primarily on a large and diverse pool of skilled, creative and entrepreneurial labour (Clark, 2002; McKinsey, 2007). In a survey of 1500 financial firms in London, a pool of talented labour was found to be the most important factor helping these firms to innovate (Taylor et al., 2003, 45–6). A study for the Corporation of London characterized the dominant business culture: 'people constantly have to invent better mousetraps to stay ahead in a market where inventions cannot be routinely patented' (CSFI, 2003, 31). In the decade up to 2007, innovative employees were therefore financially rewarded and promoted, and their success stories advertised to create a circuit in which successful innovation attracted further innovators to particular companies and the City in general. London was helped in this respect by a relatively liberal UK immigration regime, particularly in comparison with the USA after 9/11, and specifically the ability to draw on the pool of graduates and young finance professionals from the rest of the EU. Anecdotal evidence suggested that the most talented young French- and German-speaking finance professionals were found not in Paris and Frankfurt, respectively, but in the City of London (CSFI, 2003; Cox and Jones, 2007, 59–60).

25.3.2 Institutional flexibility

Institutional flexibility was needed to support the responsiveness of this large and diverse pool of expertise to changing market conditions. Of crucial importance was access to sophisticated customers, as well as to other financial, professional and business service firms: 'The localized nature of relationships between skilled labour, customers and suppliers is a critical factor which helps firms achieve innovative solutions, develop new markets and attain more efficient ways to deliver services and products to clients' (Taylor et al., 2003, 3). In operating across financial markets, the City of London also combined the benefits of various forms of cluster relationship: (1) A 'hub-and-spoke' cluster based around the dominant market position of the largest, mostly foreign-owned investment banks (Clark, 2002); (2) a 'Marshallian' concentration of small, highly specialized financial firms, including corporate finance boutiques (Hall, 2007) and hedge funds; and (3) elements of a 'state-anchored' cluster, based on institutions such as the Bank of England and the Financial Services Authority (FSA) (Pandit and Cook, 2003). This hybrid and complex character underscored the multiple sources of London financial innovativeness (see also Amin and Thrift, 1992).

In each of these forms of relationship, proximity contributed to the exchange of tacit knowledge, creating an information-rich environment, which spurred both competition and innovation (Pryke and Lee, 1995; HM Treasury, 2006):

> For banks especially, location in either the City of London or Canary Wharf is rated as being a vital contributor to their ability to compete through product innovation. Moreover, the research shows that banks, and in particular investment banks, are at the hub of the cluster, and their inter-relationships with other investment banks provide the impetus for innovation and product differentiation as they search for new market share (Taylor et al., 2003, 9)

Nevertheless, the role of small and highly specialized financial firms in innovation should not be underestimated. According to a study by Lerner (2004), based on the data from the USA, financial innovation was characterized by a balanced participation of firms across the size spectrum, or even a disproportionate representation of smaller firms. The continuing innovative role of small financial firms was illustrated after 2000 by the growth of the London hedge fund industry and various forms of corporate financial boutique (Hall, 2007).

25.3.3 Financial infrastructure

In recent years, the implementation of interinstitutional action increasingly depended on the benefits of shared technological infrastructure, including an efficient electronic payments system. From the 1970s, new technologies moved from screen-based support for floor- and telephone-based exchanges, through electronic deal execution, to the emergence of standard protocols allowing direct investor access to markets and machine-based trading, and then of increasingly 'virtual' markets for many exchanges. As capital markets expanded, these developments supported a virtuous circle of increased speed, reduced unit costs and widening participation in equities, foreign exchange and derivatives markets. The pooled liquidity of multiple exchanges also allowed new trading venues to emerge in competition with traditional market centres (Cox and Jones, 2007).

In theory, participants can now access markets from anywhere in the world. Technical infrastructure no longer provides a significant competitive advantage for one centre over any other. Such developments therefore created new competitors, supporting shifts of functions between centres. London transacted a declining share of the more transparent and lower-value-added international exchanges, as these became outsourced to lower-cost centres such as Dublin (Sokal, 2007). On the other hand, its global institutional investors attracted clients from a wider international hinterland, especially in Europe. These sought to benefit from the presence in London of many financial specialists and support services, as well as a diversity of financial market infrastructure. Even more significantly,

London remained the major European base from which the senior managers of international financial institutions might deal with market change, and higher-value, customized and higher-risk forms of business. The common location of senior personnel in London was also thought to insure against unexpected events, while more scattered agencies, under varying regulatory rules, might find crisis resolution more complicated and time-consuming (Cox and Jones, 2007, 46).

In practice, therefore, rapidly improved ICT capabilities promoted the concentration of increasingly specialized and high-risk business into London, where it was assumed that it could be best managed by the City's institutions and regulatory arrangements. By 2007, however, it appears that the volume, speed and global reach of transactions driven by these technologies had also reduced their ability to carry out this task effectively.

25.3.4 Regulation

The competitiveness of all business is generally subject to regulations and laws such as those governing business registration, taxation, customs or employment. Regulation is fundamental, however, to the operations of the financial services, directly impinging on both their competitiveness and their innovativeness. The primary purpose of regulation is to protect the financial system as a whole from the adverse consequences of action by individual institutions, especially by requiring banks to maintain sufficient capital reserves to cover their likely debt liabilities. To this end, legislators and regulators decide which, and how, financial products are regulated. These rules were regularly redrawn following the ownership deregulations of the 1980s, especially during the 1990s. Regulation is generally based on a 'regulatory dialectic', whereby innovation responds to regulatory constraints, which in turn become adjusted in response to these innovations (Tufano, 2003). Financial products which fall under particular regulations are either made compliant or modified to avoid sanctions. Process innovation is also directly affected since, in principle, financial firms are expected to maintain adequate control systems and avoid excessive risks that might spill over to the rest of the financial system. Their clients often also seek ad hoc advice about the investment, taxation and other financial regulations affecting them. In some situations innovation involves the formalization of products, for example offering standardized support for managing clients' financial resources. Such was the 'aggressive tax planning' developed by major accountancy firms in the 1990s, assisting corporations and wealthy individuals to reduce their tax exposure. Innovation lay both in making the products available to multiple, rather than individual clients, and in their adaptability to changing national and international tax regimes (NESTA, 2006, 30).

The objective of financial regulation is to strike a balance between over- and under-regulation (Pryke and Lee, 1995). Over-regulation may lower risks in the financial system but hamper innovation and financial system efficiency. Under-regulation may encourage innovation but involve excessive risk of systemic crisis. By the mid-2000s, the interaction between the regulators and regulated firms, as well as their clients, favoured 'light touch' regulation. The aim, in principle, was to be responsive to market changes, and consistent, while being clear in the application of rules, weighing the benefits and costs of compliance, and inducing confidence among the regulated firms and their customers. In retrospect it now seems that these relationships had become too close, and the assessment of risk too lax. More risky but potentially highly profitable practices, employing complex derivates and market hedging strategies, which were developed for specialized clients and therefore often under-regulated, also began to attract significant investment from mainstream investment banking, insurance, mortgage and pension institutions.

As recently as in 2007 a McKinsey report praised the virtues of the UK Financial Services Authority (FSA). It is the sole regulator for the UK financial services industry, offering a single institution to which regulated parties are held accountable. Since the FSA was created in 1997, many other countries had similarly consolidated financial regulation. The FSA implemented principles-based regulation in two tiers:

> First, in an effort to provide greater clarity and predictability to regulated entities, the FSA has issued a set of eleven high-level principles that embody the essence of what is expected of regulated firms. This . . . includes, among others, the requirement that firms conduct themselves with integrity, and that they maintain adequate financial resources . . . The second set of principles relates to the FSA itself . . . This . . . provides the market with greater certainty about the regulator's future course of action and ensures that all new regulations will be subject to a rigorous analysis weighing the costs and benefits to the market. (McKinsey, 2007, 82)

One of the FSA's principles was not to discourage the launch of new financial products and services. Another was to consider the international mobility of financial businesses and, perhaps most significantly, avoid damaging the UK's competitiveness. Following the events of 2008, it became difficult to deny that these approaches were under-regulating the UK's financial services. High street banks were shown to have excessive exposure to opaque and risky derivatives such as Collateralized Debt Obligations and Credit Default Swaps.[7] As the weakness of the US subprime mortgage market, disguised by such instruments, spread throughout UK financial markets, banks such as Northern Rock, HBOS and Bradford & Bingley had to be recapitalized from public funds. Regulators had not effectively addressed weaknesses in risk and financial transactions reporting, the underdevelopment of risk management departments in banks and, most importantly, the effects of remuneration structures for traders that encouraged short-term opportunistic behaviour (ACCA, 2008). Under-regulation since the 1990s had therefore allowed experimentation and innovation, but these were focused on short-term pay-offs for individuals, with shareholders and, ultimately, the public bearing the risks.

25.3.5 London vs New York

The innovativeness of the City of London during this period, and the positions adopted by the FSA, were dominated by its relationship to its main rival, New York City. They remained in a class of their own as global financial centres (Z/Yen, 2007). London led according to many measures of international financial activity, including the volumes of foreign exchange trading, 'over-the-counter' (OTC) derivatives trading,[8] international bank lending, international asset management, marine insurance and other forms of specialized trading. New York, however, was traditionally associated with financial innovation. In 1999, Sassen had stated: 'What London lacks is Wall Street's brilliant financial engineering' (Sassen, 1999, 83). Securitization, hedge funds, non-investment grade bonds, option-pricing model and many other financial innovations all originated in the USA. While the world's largest investment banks remained headquartered in the USA, however, their London offices started to exercise considerable autonomy (Jones, 2002). In the mid-2000s, the European revenues of US banks such as JP Morgan, Morgan Stanley and Citigroup grew faster than US revenues, and a considerable part of their decision-making and innovative activity also shifted to London (McKinsey, 2005, 2007).

It appeared, however, that the innovative roles of London and New York were different, and even complementary. In London, financial innovation was mainly occurring through the development of new markets rather than new products (Taylor et al., 2003). While product innovations therefore still came predominantly from the USA, their international adoption, with many variations and adaptations, was often directed through London, predominantly by US banks (see also CSFI, 2003). This relationship appeared to be specifically encouraged by London's regulatory environment (McKinsey, 2007). There is a long history of competition and cooperation in financial innovation between New York and London, based especially on differential regulation. In the 1960s high levels of domestic regulation led US banks to create the Eurodollar and Eurobond markets centred on London (Kim and Stulz, 1988). More recently, the Sarbanes–Oxley Act of 2002 diverted firms wishing to cross-list their shares on foreign markets from US stock exchanges to London, and particularly to the lightly regulated Alternative Investment Market (Oxera, 2006; Piotroski and Srinivasan, 2007). Nevertheless, as the crisis of 2008 developed, it became evident that the pressure for innovation in the US had drawn many UK institutions into poorly regulated investments, including trade in derivatives based on subprime mortgages.[9] One of the key questions for the future, therefore, will be how the regulatory relationship between London, New York and other global financial centres is reshaped in an era when the leading US investment banks will have either been dissolved or absorbed into mainstream banking, and play a much less prominent part in global finance.[10]

25.3.6 Hedge funds

London also began to challenge New York's global dominance in the more specialized and risky hedge fund sector, with its very different structure and innovative focus. As with clustered high-technology industries, hedge fund firms are small and flexible, typically employing fewer than ten people. Many are created and close every year, often as spin-offs from investment banks, as individual traders and other banking professionals set up their own businesses (Gieve, 2006). Hedge fund managers are typically younger than conventional investment managers (Boyson, 2006). More conventional institutional investors generally tend to invest funds passively across a spectrum of assets, assuming the market to be efficient, with individuals able to outperform average market returns only by chance. Hedge funds, on the other hand, employ active investment techniques aimed at beating average market returns to achieve high absolute investment returns.

While the total assets managed by hedge funds are still many times smaller than those managed by pension, mutual or insurance funds, most hedge funds heavily leverage their managed assets with borrowed money, and trade at incomparably higher frequencies. This exacerbates their influence on international financial markets (Tsatsaronis, 2000). Management incentives are especially crucial to their success: 'Hedge funds remunerate managers with fees that are highly geared to performance. Hedge fund managers typically charge a 1–2% basic fee plus a 15–25% performance fee' (IFSL, 2009, 4). The quest for high absolute returns also implies that 'hedge funds are often specialized and operate within an industry or specialty that requires a particular expertise' (IFSL, 2009, 4). Managers have therefore tended to experiment with innovative investment and trading techniques, such as merger and risk arbitrage, in which they take positions betting on whether a merger deal will be completed. Overall, therefore, they innovate through the intensified exploitation of London's characteristic assets: highly motivated expertise and institutional flexibility, directed to serving customized client requirements.

About 80 per cent of the world's hedge funds are managed in the USA or the European Union (EU), and London handles about 90 per cent of EU hedge fund assets (IFSL, 2009). Their value more than tripled between 2002 and 2005 to $213 billion (HM Treasury, 2006, 19). UK hedge fund assets grew by 63 per cent per year between 2002 and 2005, compared with only 13 per cent in the US. In 2002, 28 of the world's 50 largest hedge funds were based in New York, with only three in London. By the end of 2005, 18 were in New York and 12 in London (McKinsey, 2007). This growth partly reflected the underused hedge fund potential of the EU, as UK hedge fund managers took advantage of the EU Investment Services Directive allowing services to be offered across other European Economic Area countries. Expansion also marked hedge fund investment as a growing component of the portfolios of mainstream financial institutions. Hedge fund development was also supported by other London services, including legal services, accounting, consultancy, administration and, in particular, the prime brokerage services of large investment banks. Once more, however, their innovativeness was most encouraged by London's light regulatory environment. Without this they would not have such a wide range of investment options, and could not produce such high returns (McKinsey, 2007; FSA, 2005).

25.4 Service relations supporting financial and non-financial services innovation: similarities and differences

25.4.1 Some similarities

London's financial services innovation was therefore primarily stimulated by international demand, through which competition from other centres, notably New York, created new opportunities to exploit the City's expert labour and institutional resources. These increasingly depended on attracting, developing and rewarding international financial talent. A similar pattern of international tradability also developed among London-based non-financial KIBS as they extended to new sector and geographical markets. Globally the commercial status of trading cities within the global urban hierarchy has come to depend on the presence of, and flows between, these international KIBS offices, with London as one of the two 'mega command centres' of the world city network (Taylor, 2004, 89; Z/Yen, 2007).

Like the financial services, therefore, the rise of London's non-financial KIBS after the 1980s reflected burgeoning national and international demands for specialist professional, technological, management and marketing expertise. More recently, they also began to manage cost-based outsourcing of non-core functions by clients, including reorganized public services.[11] In spite of their supposedly low technological innovativeness, UK KIBS are major adopters of new information and communication technologies (Miles, 2002, 177). More significantly for their innovative role, however, computer and other business consultancies support the installation of customized ICT software and systems for clients across all other sectors, while also helping them to manage the consequent changes in wider business procedures. Similar demand trends have supported consultancy growth in financial management; strategic planning; advertising and market research; human resource management; engineering, architecture and construction; and many branches of design, multimedia production, and operational and logistics support. In central London, much of this work takes place for international clients, while national clients also seek international standards of service quality and innovation (O'Farrell et al., 1996; Jones, 2002, 2005; Miozzo and Miles, 2002).

London's financial and non-financial KIBS also share operational characteristics that foster service innovation (Taylor et al., 2003). These include the ability to assemble and deploy novel arrays of knowledge and expertise from international sources of specialized labour (Gallouj, 2000; Drejer, 2004). Once more, such competencies, supporting client management of change, are associated with institutional flexibility, through which both financial and non-financial KIBS firms work closely with clients, while also competing, and sometimes collaborating, with each other and public agencies.[12] As among the financial services, there are 'Marshallian' concentrations of specialized firms, for example in advertising, design and media production, and also 'hub-and-spoke' relationships between major consultancies and smaller specialists. Some professional services even show signs of 'state-anchored' clusters, for example in the proximity of corporate law to the higher courts. In all cases, however, the dominant institutional relationships are with clients.

Flexibility is commonly reflected in the formation of multifunctional, project-based teams, able to cross formal institutional boundaries (Tidd and Hull, 2006). The major consultancies often employ matrix forms of organization, matching their internal resources of expertise and experience to changing global patterns of clients, projects and places (Jones, 2005). While new staff are recruited internationally and trained primarily through project experience, more established personnel adapt or move on, so sustaining human resources circulation within and between firms (for example Grabher, 2002a, 2002b). Hundreds of specialist medium and small firms also thrive, many spinning off the large companies, while offering niche services to both large and medium-small clients (Wood et al., 1993).

The 2004 Community Innovation Survey evidence cited at the beginning of this chapter confirms that the commitment of UK KIBS to innovation, especially in its 'non-technological' and market-driven forms, is at least as great as in much innovative manufacturing. The financial and non-financial services also hold in common a concentration of such national innovativeness into London, where the key market relationships required to stimulate it, and the operational conditions necessary to support it, are both focused (Wilson, 2007; Miles and Green, 2008; Wood, 2008).

25.4.2 Some significant differences

There are, however, substantial differences between financial and other KIBS, associated with the characteristics of capital markets compared to those for professional and business expertise. These are most apparent through: (1) the significance of the ICT infrastructure; (2) the critically different form taken by regulation, and especially its implications for market transparency; and, in consequence, (3) the variable autonomy of KIBS firms in relation to clients. The last has significant implications for the assessment of the innovative contribution of the non-financial KIBS.

ICT infrastructure As we have seen, ICT has become central to the interaction of financial firms with their clients, particularly retail clients, and their relationships with each other. This has exerted a significant influence on the international distribution of financial functions. Among other KIBS, however, there are only limited parallels with Internet banking and insurance, or with the ICT-led growth of financial transactions and the trading of financial instruments, mainly because there is little communal demand for operational information. In fact, the opposite is often the case, with KIBS-customer confidentiality dominating many fields. Of course, virtual communications are widely used, but mainly for low-value external exchanges and intra-firm communications (Pain, 2007). This has increased the overall speed and complexity of interchange supporting projects, however, raising the required intensity of strategic scrutiny, for example when developing overseas markets or during merger and acquisition negotiations. If anything, therefore, the uses of ICT have increased reliance on such confidential, tacit forms of knowledge exchange based on direct interaction between key senior staff (Taylor et al., 2003). Similar processes explain Jones's conclusions about the international operations of consultancy firms, which contrast the ideal of virtual electronic international integration with the reality of increasing senior business mobility (Jones, 2002, 2008).

At the same time, ICT developments have made possible the formalization of many business services processes (Drejer, 2004, 558). Since the late 1990s this has promoted financial and business process outsourcing, including to 'offshore' locations, in which KIBS firms manage routine, but administratively and technically complex, functions on behalf of clients. These include data processing, customer relations, accountancy and financial administration, human resource management, stock control, logistical monitoring and design software. New markets have thus been created for consultancy firms, in which they advise service and manufacturing clients on outsourcing strategies and often implement them on their behalf. Many consultancies also increasingly outsource their own cost-sensitive processes on an international scale (Abramovsky et al., 2004; Sako, 2006). The uses of ICT to support processes of service formalization perhaps most clearly demonstrate how KIBS innovativeness takes place, not only through the installation of new technologies but, more crucially, in the design of wider business models enabling clients to reap their benefits. While contributing to KIBS growth, and intensifying their wider connectivity with clients, however, these developments have not generally challenged the need for regular face-to-face interaction to deliver core consultancy services on a global scale (Jones, 2007).

Regulation The significance of the regulatory environment for financial service innovation was vividly demonstrated by the events of 2007–08. Such institutional oversight governing provider–client service relationships is absent for many other KIBS and, where it does operate, takes different forms. The traditional professions, for example, with client relations nationally regulated, compare to some extent with the financial services. The very existence of commercial law and business accountancy depends on statutory business regulation, and professional KIBS are themselves subject to the external supervision of training, qualifications and codes of practice. Although not usually regarded as vanguards of innovation, like financial services such professions may support client innovation by advising on the avoidance of, or adaptation to, regulatory requirements. Over time, processes of 'regulatory dialectic' may similarly modify legal and fiscal regulations. Nevertheless, the primary purpose of commercial regulation is to ensure market stability, thus supporting the commercial market imperatives that are the dominant drivers of KIBS innovation. London-based corporate lawyers and corporate taxation accountants thrive on such work (Faulconbridge, 2007; Faulconbridge et al., 2008).

Many aspects of commercial consultancy, however, are not formally regulated. In principle, outside the legal provisions of any business contract, sophisticated clients should not require regulatory protection. Neither does individual project failure or default generally create systemic problems for other firms. As with bilateral OTC trading in derivatives, therefore, there is apparently no need for a consultancy equivalent of the FSA, the Law Society or the Institute of Chartered Accountants. Nevertheless, the needs to satisfy often demanding clients, and to achieve significant market impacts, impose strong quasi-regulatory pressures on service–client relationships. After all, the critical outcome for KIBS firms in London lies in sustaining their reputation, affecting their ability to gain repeat projects from clients and also recommendations from them to others (Wood, 1996; Glückler, 2007).

Some KIBS in the UK have established institutions to promote their professional status through voluntary codes of conduct designed to reassure clients and certify practitioners, including management and ICT consultancy, advertising, market research, logistics and human resource management. Consultancies are also required to address internal conflicts arising, for example, from the liabilities of individuals in partner-owned and managed companies, and the need for 'chinese walls' of confidentiality between their work for competing clients, or when auditing and consulting for the same client (see Donaldson and Dunfee, 2002). The intangibility of the inputs required, especially for innovative services, make such 'self-regulatory' standards particularly important (Lazega, 2002; Tidd and Hull, 2006). Some take an 'impersonal' form, embodied in documentation, codes of practice, software and cost control procedures. More significant, however, are 'interpersonal' standards, based on trust, networking and reputation, commonly built through the co-location of teams (Glückler, 2005).

Within UK jurisdiction, therefore, different forms of KIBS expertise are subject to different processes of accreditation in their relations with clients. When institutional market regulation applies, as in the financial and professional services, this may sometimes itself form the basis for innovative advice to clients. In practice, however, most consultancy projects combine requirements for advice on financial or legal regulations with commercial management, technical, marketing and human resource expertise. Large consultancies are invariably multifunctional, employing the whole range of such expertise. Smaller KIBS firms, on the other hand, collaborate with other specialists, or depend on clients for regulatory advice. In the great majority of business consultancy projects, however, the net innovative outcome is dominated by commercial markets, rather than by specific regulatory conditions.

Regulation-related innovation may be possible, however, in international business projects, when consultancies advise clients about legal, cultural, linguistic and political barriers to trade and establishment (Jones, 2002, 2005). The example of the operation of the Sarbanes–Oxley regulations in the USA has already been noted, which created a boom in international demand for consultancy advice. Many barriers to international service operation remain, even within the European Union, although the 2006 Service Directive is designed to reduce these from the end of 2009 (Commission of the European Communities, 2007; BERR, 2009). Moves towards the alignment of international regulations themselves also offer scope for KIBS market growth. For example, as the International Financial Reporting Standards (IFRS) became compulsory for listed companies in the European Union in 2005, accountancy firms, led by the 'Big Four', were able to support clients in the transition from domestic standards to the IFRS. Corporate scandals in the USA and Europe (for example Ahold and Parmalat) also boosted consultancy demand, including specialized rating agencies, to advise on corporate governance standards (Clark et al., 2006). The growth of services specializing in international corporate social and environmental responsibility has also been spurred by climate change and labour exploitation legislation (Hebb and Wójcik, 2005). Significantly, some of these new fields of KIBS now span the boundaries between financial and non-financial services.[13]

KIBS autonomy Largely as a consequence of these different market and regulatory environments, assessment of the innovative performance of financial, professional and business KIBS activities in London is confronted by major differences in the comparative autonomy allowed to firms in their relationships with clients. In many financial dealings, the scale and technical complexity of international capital markets and exchange procedures mean that clients tend to award financial service firms considerable autonomy in their management of resources. This is why effective regulation depends on high levels of market transparency, so that the performances of financial agents may be identified and monitored. There is also comparative transparency in the outcomes of much specialist legal, accounting or other professional KIBS advice. In contrast, the operations of commercial KIBS in London are often based on the customization of services to individual clients under conditions of strict commercial confidentiality. The very nature of the service process thus obscures the KIBS contribution to the co-production inherent in service innovation.

This raises a further intractable problem in such cases. How can routine consultancy activities, or those that might lead to innovation, be distinguished from those that are actually innovative (Drejer, 2004)? At the very least, KIBS use may improve knowledge exchange for the client, and reduce costs through formalization. Such changes might appear to be innovative, especially when new technologies are employed. In Schumpeterian terms, however, innovation occurs only when new products, processes or organizational structures have asserted a distinctive market impact (Drejer, 2004). If the innovativeness of consultancy processes depends on client behaviour and market success, therefore, their distinct contribution cannot be identified, except subjectively by clients, in retrospect (Wood, 2005). It is certainly not necessarily related to the labour productivity of consultancy firms themselves, or their investment in new techniques and methods, as measured by conventional surveys such as the Community Innovation Survey. These throw no light on the client-directed processes that determine the actual significance for innovation of much commercial consultancy.

Of course, a similar conundrum arises from some of the recently expanding elements of London's financial KIBS. These include customized OTC dealings and hedge fund management, based on confidential, and therefore opaque, exchanges. But at least the strictly financial outcomes of such transactions for individuals become evident within agreed time scales. On the other hand, as we have seen, major non-financial consultancies have developed comparatively formalized, usually ICT-based proprietary services, including those supporting outsourced contracts. These supplement intensive, high-cost consultancy with more regular, contract-based sources of income, thus becoming tied into closer and longer-term collaboration with clients. Like the financial services, although through operational improvements, these activities are designed to deliver better value from client investment. The outcomes are therefore comparatively transparent, and subject to client management and shareholder scrutiny based on formal performance benchmarks.

25.5 Conclusions: functional distinctions and overlapping roles in London's knowledge-intensive business services

The central London economy has become dominated since the 1980s by a complex social division of specialist financial, professional and business service labour, in which many expert functions may be combined to respond to international commercial and regulatory change. Its innovative capacities depend as much on these overlapping and interdependent characteristics as on the expertise of particular functions. The various modes of KIBS innovativeness correspond only loosely to the differences between the financial, professional and commercial services. More fundamentally significant is the degree of formalization or customization of the service process.[14] This reflects the application of ICT, the balance of institutional or individualized market regulation, and the consequent degree of autonomy and transparency of KIBS activities.

While there has been a tendency for the experience of successful customization to be developed into more formalized services, central London's advantage, across the financial, professional and commercial KIBS, continues to lie in higher-level, knowledge-intensive processes. These are the basis of the most innovative KIBS, but the project-based co-production upon which they depend renders it inherently difficult to identify their contribution, even when the client's market performance has been demonstrably enhanced.

Modern technologies have expanded the range and speed of information exchange within and between organizations, but this has intensified the need for customized, face-to-face interaction between key personnel in places such as London, especially where innovation is required. For many KIBS, if markets are to operate successfully, some form of regulation to reassure clients is inherent in their intangible nature. Many financial and professional services depend on formal institutional regulation, requiring them to operate in a comparatively transparent manner. It also often allows firms to operate relatively autonomously on behalf of clients, sometimes to develop forms of regulation-derived innovation. The breakdown of these regulated arrangements lies behind the crisis of 2007–08.

Much business consultancy work, on the other hand, including that undertaken by some financial and professional and most business KIBS, depends on confidential liaison with clients, and is not externally regulated. This is more likely to be innovative than more formalized exchanges, and most characteristic of London. Instead, self-regulation prevails, based on internal and interfirm working standards. Client scrutiny determines the vital outcomes for consultancy firms, in the form of reputation, repeat contracts and referrals to new clients. While such circumstances offer little scope for national regulation-based innovation, international KIBS often encounter regulatory diversity. While posing operational problems, such diversity offers opportunities for innovative approaches on behalf of clients, especially by professional and business services.

The ultimate arbiter of KIBS innovativeness, however, is the market – the sectoral and even global impacts of the success they may induce in clients. Improving KIBS' own labour productivity, or investing in new technologies, is not enough. Many firms may pursue such policies, for example to reduce their costs, but are not economically innovative in the true sense. London nevertheless possesses a high share of those that are, simply because of the interaction its location sustains with innovative clients in international markets. Many KIBS firms are also global players in their own right, and actively promote client process and product innovation. Any formal verification of such claims, however, requires intensive evaluations of successes and failures from the client perspective. In practice, the most common metrics come from client satisfaction surveys and rates of client repeat work and referrals. KIBS innovation appears to be greatest, however, when their specialized financial, professional, technological or business expertise, including international experience, closely complements that of the client (Wood, 2002a, 78–85).

A comparison of the processes of financial services innovation with that in other KIBS in central London therefore offers limited, although revealing, insights. The distinctive impact of differential regulatory conditions on service relationships with clients seems to be most significant, along a spectrum from highly regulated to completely customized exchange. This has implications for the manner in which different KIBS can operate, and even on our ability to analyse their innovative role. In these circumstances, the market success of clients which, for many in the past, has commonly been associated with location in London, needs to be the starting point of enquiries. Such an approach is certainly more relevant than conventional technology-directed comparisons with manufacturing. Even though key processes remain elusive, it is clear that more direct assessment of KIBS' innovative impact on national and international economic systems is urgently needed, despite these daunting methodological challenges, and notwithstanding the crisis that faced London after 2008.

Notes

1. Gallouj argues that service innovation cannot be defined by the competencies of service firms themselves, but by designated innovative outputs from their activities. These may either be projects whose innovative quality are clearly identified a priori, or changes that are generally recognized to be innovative *a posteriori* (Gallouj, 2002b, 279). Here, it is assumed that the outcomes of innovativeness by financial, professional and business service firms, together forming the 'knowledge-intensive business services' (KIBS), must be associated with improvements in the market position of clients.
2. 'The City' is the area still administered by the historic Corporation of London, also known as the 'Square Mile', broadly defined by the lines of the medieval walls. This chapter focuses on the much larger area of central London, composed of seven inner London boroughs (see note 4).
3. Before the 1980s 'producer services' were invariably defined to include the financial services, as well as accountancy, advertising, legal services, real estate, research and development and, where they could be separately identified, technical, computer and management consultancy. Since then, specialist studies of financial services have emphasized their distinctiveness from other KIBS, although analyses of KIBS trends still usually include them (see Wood, 2009, Table 2). This chapter adopts the latter approach.
4. Central London is defined to include the cities of London and Westminster, and the boroughs of Camden and Islington to the north, Kensington and Chelsea to the west, Lambeth and Southwark to the south, and Tower Hamlets to the east.
5. The 'West End' is defined here to include all central London boroughs except the City of London and Tower Hamlets. KIBS are nevertheless concentrated largely into the City of Westminster and the boroughs of Camden and Kensington and Chelsea.
6. There are, of course, other substantial sources of service innovation in central London, including the creative and the tourist and consumer sectors.
7. A Collateralized Debt Obligation (CDO) is an investment security made up of tranches of various types of bonds, loans and other assets, including mortgages. Each tranche is associated with different maturities and risks. The higher the risk, the more the CDO undertakes to pay the investor. There are also specialized Collateralized Mortgage

Obligations (CMO) and Collateralized Bond Obligations (CBO). The purchaser of a Credit Default Swap (CDS) receives a pay-off from the seller if a financial instrument defaults. In effect, this spreads the purchaser's risk, which may be further offset to other purchasers who may not own the underlying security or even suffer a loss from its default.

8. Unregulated over-the-counter (OTC) trading in financial instruments involves high risks, and takes place between specialist issuers and experienced buyers outside formally organized exchanges.
9. December 2008 saw the collapse of a fraudulent New York hedge fund worth an estimated $50 billion run by Bernard Madoff, a former chair of the Nasdaq stock market. Some European banks made significant losses, including Santander, HSBC, RBS and BNP Paribas.
10. In September 2008, the last two specialist investment banks in the US, Morgan Stanley and Goldman Sachs, ceased trading independently, as reliance on trading in securities and offering clients financial advice ceased to be viable as a business model. They applied to the Federal Reserve to become regulated lending banks, intending to build a deposit base, mainly through mergers (Harper and Torres, 2008). This followed a summer in which Lehman Brothers had gone into bankruptcy and Merrill Lynch was absorbed into the Bank of America. Each of these independent investment banks had a substantial presence in London, and their demise had major employment impacts on the City.
11. For UK national KIBS demand trends for 1993–2003, see Wood (2006, 342–346).
12. Drejer (2004), drawing on Djellal and Gallouj, defines the ability to respond to new labour needs 'through a procedure of accumulating knowledge and expertise within services' as 'expertise field innovation'. Institutional flexibility, 'the establishment by a firm of particular relationships with partners', is termed 'external relationship innovation'.
13. Similarly 'niche' innovations in financial services have included the Alternative Investment Market (AIM), established in 1995 as a submarket of the London Stock Exchange to allow smaller companies access to the public stock market (McKinsey, 2007, 52). Since 2000, London has also become the leading Western financial centre for Islamic (Sharia-compliant) banking (Pollard and Samers, 2007; Financial Services Authority, 2006, HM Treasury, 2006, 19). The UK Carbon Emissions Trading Scheme was also launched in August 2001, as the world's first economy-wide greenhouse gas trading scheme (Tapley et al., 2008). It remains to be seen how far these initiatives are affected by the crisis of credit and confidence associated with the 2008 meltdown in mainstream banking.
14. This distinction echoes the first four of Miles's service innovation patterns (Miles, 2002, 175–176). These include formalized services, based on classic R&D, by large firms, for example in data processing or telecommunications. There are also various types of innovation based on customization; by service professionals, mainly medium-sized consultancies engaged in collective problem-solving with clients; organized strategic innovation by large firms often through ad hoc project teams; and entrepreneurial patterns, based on small firms created around a radical new service.

References

Note: all websites accessed December 2008.

Abramovsky, L., R. Griffith and M. Sak (2004), 'Offshoring of business services and its impact on the UK economy', London Advanced Institute of Management Research (AIM), November, http://www.aimresearch.org/uploads/pdf/Academic%20Publications/offshoring%20doc%20_1.pdf.

Amin, A. and N. Thrift (1992), 'Neo-Marshallian nodes in global networks', *International Journal of Urban and Regional Research*, **22**, 571–587.

Association of Chartered Certified Accountants (ACCA) (2008), 'Climbing out of the Credit Crunch', Policy Paper, www.accaglobal.com/pdfs/credit_crunch.pdf.
Barras, R. (1986), 'Towards a theory of innovation in services', *Research Policy*, **15**, 161–3.
Barras, R (1990), 'Interactive innovations in financial and business services', *Research Policy*, **19**, 215–237.
BERR (2009), *The European Services Directives: Guidance to Competent Authorities*, 1st edn, London: Department of Business, Enterprise and Regulatory Reform, June, http://www.berr.gov.uk/files/files1666.doc.
Boyson, N.M. (2006), 'Another look at career concerns: a study of hedge fund managers', Northeastern University, College of Business Administration, http://www.ssrn.com/.
Centre for the Study of Financial Innovation (CSFI) (2003), 'Sizing up the City: London's ranking as a financial center', London: Corporation of London, June, http://www.cityoflondon.gov.uk/Corporation/business_city/research_statistics/research_publications.htm.
Clark, G.L. (2002), 'London in the European financial services industry: locational advantage and product complementarities', *Journal of Economic Geography*, **2**, 433–53.
Clark, G.L. and D. Wójcik (2007), *Geography of Finance: Corporate Governance in the Global Marketplace*, Oxford: Oxford University Press.
Clark, G.L., D. Wójcik and R. Bauer (2006), 'Geographically dispersed ownership and inter-market stock price arbitrage: Ahold's crisis of corporate governance and its implications for global standards', *Journal of Economic Geography*, **6**, 303–322.
Commission of the European Communities (2007), *Handbook on Implementation of the Services Directive*, Luxembourg: Internal Market and Services DG,: Office for Official Publications of the European Communities, http://ec.europa.eu/internal_market/services/docs/services-dir/guides/handbook_en.pdf.
Coombes, R. and I. Miles (2000), 'Innovation, measurement and services: the new problematic', in J.S. Metcalfe and I. Miles (eds), *Innovation Systems in the Service Economy: Measurement and Case Study Analysis*, Boston, MA: Kluwer Academic, pp. 85–103.
Cox, P. and L. Jones (2007), 'The competitive impact of London's financial market infrastructure', London: City of London Corporation, April, http://www.cityoflondon.gov.uk/Corporation/business_city/research_statistics/research_publications.htm.
den Hertog, P. (2002), 'Co-producers of innovation: the role of knowledge intensive producer services', in J. Gadrey and F. Gallouj (eds), *Productivity, Innovation and Knowledge in Services*, Cheltenham, UK and Northampton, MA, USA: Edward Elgar, pp. 223–255.
Donaldson, T. and T.W. Dunfee (2002), 'Ties that bind in business ethics: social contracts and why they matter', *Journal of Banking and Finance*, **26**, 1853–65.
Drejer, I. (2004), 'Identifying innovation in surveys of services: a Schumpeterian perspective', *Research Policy*, **33**, 551–562.
Faulconbridge, J.R. (2007), 'London and New York's advertising and law clusters and their networks of learning', *Urban Studies*, **44**, 1635–1656.
Faulconbridge, J.R., J.V. Beaverstock, D. Muzio and P.J. Taylor (2008), 'Global law firms: globalization and organizational spaces of cross-border legal work', *GaWC Research Bulletin*, **256**, http://www.lboro.ac.uk/gawc/rb/rb256.html.
Financial Services Authority (FSA) (2005), *Hedge Funds: A Discussion of Risk and Regulatory Engagement*, London: FSA, www.fsa.gov.uk.
Financial Services Authority (FSA) (2006), 'Islamic banking in the UK', Briefing Note BN016/06, London, 9 March, www.fsa.gov.uk/pages/About/Media/notes/bn016.shtml.
Frame, W.S., and L.J. White (2002), 'Empirical studies of financial innovation: lots of talk, little action?', Working Paper 12, Federal Reserve of Atlanta, www.ssrn.com.
French, S. and A. Leyshon (2004), 'The new, new financial system? Towards a conceptualisation of financial reintermediation', *Review of International Political Economy*, **11**, 263–288.
Gallouj, F. (2000), 'Beyond technological innovation: trajectories and varieties of service innovation', in M. Boden and I. Miles (eds), *Services and the Knowledge-Based Economy*, London: Continuum, pp. 129–145.

Gallouj, F. (2002a), *Innovation in the Service Economy: The New Wealth of Nations*, Cheltenham, UK and Northampton, MA, USA: Edward Elgar.
Gallouj, F. (2002b), 'Knowledge-intensive business services: processing knowledge and producing innovation', in J. Gadrey and F. Gallouj (eds), *Productivity, Innovation and Knowledge in Services*, Cheltenham, UK and Northampton, MA, USA: Edward Elgar, pp. 256–284.
Gallouj, F. and O. Weinstein (1997), 'Innovation in services', *Research Policy*, **26**, 537–556.
Gieve, J. (2006), 'Hedge funds and financial stability', speech at the Hedge 2006 Conference, London, 17 October, www.bankofengland.co.uk/publications/speeches/2006/speech285.pdf.
Glückler, J. (2005), 'Making embeddedness work: social practice institutions in foreign consulting markets', *Regional Studies*, **41**, 949–961.
Glückler, J. (2007), 'Geography of reputation: the city as the locus of business opportunity', *Environment and Planning A*, **37**, 1727–1750.
Grabher, G. (2002a), 'Cool projects, boring institutions: temporary collaborations in social context', *Regional Studies*, **36**, 205–214
Grabher, G. (2002b), 'The product ecology of advertising: tasks, talents and teams', *Regional Studies*, **36**, 245–262.
Hall, S. (2007), '"Relational marketplaces" and the rise of boutiques in London's corporate finance industry', *Environment and Planning A*, **39**, 1838–1854.
Harper, C. and C. Torres (2008), 'Goldman, Morgan Stanley bring down curtain on an era' *Bloomberg.com*, http://www.bloomberg.com/apps/news?pid=20601068&sid=aSfyFs2LTxYs&refer=home.
Hebb, T. and D. Wójcik (2005), 'Global standards and emerging markets: the institutional-investment value chain and the CalPERS investment strategy', *Environment and Planning, A*, **37**, 1955–1974.
HM Treasury (2006), 'Financial services in London: global opportunities and challenges', London: HM Treasury, ww.hm-treasury.gov.uk/media/1E0/E6/bud06_cityoflondon_262.pdf.
Hutton, T.A. (2006), 'Spatiality, built form and creative industry development in the inner city', *Environment and Planning A*, **38**, 1819–1841.
International Financial Services London (IFSL) (2009), *Hedge Funds*, City Business Series, April, www.ifsl.org.uk/output/ReportItem.aspx?NewsID=73.
Jones, A. (2002), 'The "global city" misconceived: the myth of 'global management' in transnational service industries', *Geoforum*, **33**, 335–350.
Jones, A. (2005), 'Truly global corporations? Theorizing "organizational globalization" in advanced business services', *Journal of Economic Geography*, **5**, 177–200.
Jones, A. (2007), 'More than "managing across borders"? The complex role of face to face interaction in globalizing law firms', *Journal of Economic Geography*, **7**, 223–246.
Jones, A. (2008), 'The rise of global work', *Transactions, Institute of British Geographers*, **33**, 12–26.
Keeble, D. and L. Nachum (2002), 'Why do business service firms cluster? Small consultancies, clustering and decentralization in London and southern England', *Transactions, Institute of British Geographers*, **NS 27**, 67–90.
Kim, Y. and R. Stulz (1988), 'The Eurobond market and corporate financial policy: a test of the clientele hypothesis', *Journal of Financial Economics*, **22**, 189–205.
Lazega, E. (2002), 'Networks, distributed knowledge and economic performance: evidence from quality control in corporate legal services', in J. Gadrey and F. Gallouj (eds), *Productivity, Innovation and Knowledge in Services*, Cheltenham, UK and Northampton, MA, USA: Edward Elgar, pp. 124–143.
Lerner, J. (2004), 'The new new financial thing: the sources of innovation before and after State Street', NBER Working Paper 10223, Cambridge, MA: National Bureau of Economic Research.
Leyshon, A. and N. Thrift (1997), *Money/Space: Geographies of Monetary Transformation*, London: Routledge.

McKinsey (2005), '$118 trillion and counting: taking stock of the world's capital markets', Washington, DC: McKinsey Global Institute Report, February, http://www.mckinsey.com/mgi/publications/index.asp?RT=All&IS=All&Rg=All&srtby=Date&pg=5#brtop.
McKinsey (2007), 'Sustaining New York's and the US' global financial services leadership', report to Senators Michael R. Bloomberg and Charles E. Schumer, Washington, DC, www.senate.gov/~schumer.
Miles, I. (2002), 'Services innovation: towards a tertiarization of innovation studies', in J. Gadrey and F. Gallouj (eds), *Productivity, Innovation and Knowledge in Services*, Cheltenham, UK and Northampton, MA, USA: Edward Elgar, 164–196.
Miles, I. and L. Green (2008), 'Hidden innovation in the creative industries', NESTA, July, http://www.nesta.org.uk/hidden-innovation-in-the-creative-industries-report-pubs/.
Miozzo, M. and I. Miles (2002), *Internationalization, Technology and Services*, Cheltenham, UK and Northampton, MA, USA: Edward Elgar.
Nachum, L and D. Keeble (2003), 'MNE linkages and localised clusters: foreign and indigenous firms in the media cluster of central London', *Journal of International Management*, **9**, 171–193.
National On-line Manpower Information Services (NOMIS) (2009), *Annual Business Enquiry*, Office of National Statistics, http://www.nomisweb.co.uk/Default.asp.
NESTA (2006), 'The Innovation Gap: why policy needs to reflect the reality of innovation in the UK', London: National Endowment for Science, Technology and the Arts, Research Report, October, http://www.nesta.org.uk/informing/policy_and_research/highlights/innovation_gap_report.aspx.
Nijssen, E.J., J. Edwin, B. Hillebrand, P.A.M. Vermeulen and R.G.M. Kemp (2006), 'Exploring product and service innovation similarities and differences', *International Journal of Research in Marketing*, **23**, 241–251.
O'Brien, R. (1992), *Global Financial Integration: The End of Geography*, London: Pinter.
O'Farrell, P., P.A. Wood and J. Zheng (1996), 'Internationalization in business services: an inter-regional analysis', *Regional Studies*, **30**, 101–118.
Oxera (2006), 'The cost of capital: an international comparison', London: report for City of London Corporation, ww.cityoflondon.gov.uk/Corporation/media_centre/research+publications.htm.
Pain, K. (2007), 'City of London – global village: understanding the square mile in a post-industrial world economy', in S. Barber (ed.) *The Geo-Politics of the City*, London: Forumpress, pp. 19–38.
Pandit, N.R. and G.A.S. Cook (2003), 'The benefits of industrial clustering: insights from the British financial services industry at three locations', *Journal of Financial Services Marketing*, **7** (3), 230–245.
Piotroski, J.D. and S. Srinivasan (2007), 'The Sarbanes Oxley Act and the flow of international listings', University of Chicago, Graduate School of Business, Working Paper, http://papers.ssrn.com/sol3/papers.cfm?abstruct_id=956987.
Pollard, J.S. and M. Samers (2007), 'Islamic banking and finance: postcolonial political economy and the decentring of economic geography', *Transactions, Institute of British Geographers*, **32**, 313–330.
Pryke, M. and R. Lee (1995), 'Place your bets: towards an understanding of globalization, socio-financial engineering and competition within a financial centre', *Urban Studies*, **32**, 329–344.
Sako, M. (2006), 'Outsourcing and offshoring: implications for productivity of business services', *Oxford Review of Economic Policy*, **22**, 499–512.
Sassen, S. (1999), 'Global financial centers', *Foreign Affairs*, **78**, 75–87.
Sokal, M. (2007), 'Space of flows, uneven regional development, and the geography of financial services in Ireland', *Growth and Change*, **38**, 224–259.
Tapley, B., P. Settes and R. Brooke (2008), 'Sustainability assessment of global financial centres', London: Corporation of London, March, http://www.cityoflondon.gov.uk/economicresearch.
Taylor, P.J. (2004), *World City Network: A Global Urban Analysis*, London: Routledge.

Taylor, P.J., J.V. Beaverstock, G. Cook, N. Pandit, K. Pain and H. Greenwood (2003), *Financial Services Clustering and its Significance for London.* London: City of London Corporation, http://lboro.ac.uk/gawc/pubcol1.html.
Thrift, N. (1987), 'The fixers: the urban geography of international finance capital', in J. Henderson and M. Castells (eds), *Global Restructuring and Territorial Development*, Beverly Hills, CA: Sage.
Thrift, N. (1994), 'On the social and cultural determinants of international financial centres: the case of the City of London', in S. Corbridge, R. Martin and N. Thrift (eds), *Money, Power and Space*, Oxford: Blackwell, pp. 327–355.
Tidd, J. and F.M. Hull (2006), 'Managing service innovation: the need for selectivity rather than "best practice"', *New Technology, Work and Employment*, **21**, 139–161.
Tsatsaronis, K. (2000), 'Special feature: hedge funds', *BIS Quarterly Review*, Special Feature November, Basel, Bank of International Settlements.
Tufano, P. (2003), 'Financial innovation', in G. Constantinides, M. Harris and R. Stulz (eds), *The Handbook of the Economics of Finance*, Volume 1A, North Holland: Elsevier, pp. 307–336.
Walker, D.R.F. and P.J. Taylor (2003), *Atlas of Economic Clusters in London*, Global and World Cities Network (GaWC), University of Loughborough http://www.lboro.ac.uk/gawc/visual/lonatlas.html.
Wilson, R. (2007), 'Innovation in London', Greater London Authority, Economics Working Paper 19, http://www.london.gov.uk/mayor/economic_unit/docs/wp_19_innovation_in_london.pdf.
Wood, P. (1991), 'Flexible accumulation and the rise of business services', *Transactions, Institute of British Geographers*, **16**, 160–172.
Wood, P. (1996), 'Business services, the management of change, and regional development in the UK: a corporate client perspective', *Transactions, Institute of British Geographers*, **NS 21**, 649–65.
Wood, P. (ed.) (2002a), *Consultancy and Innovation: The Business Service Revolution in Europe*, London: Routledge.
Wood, P. (2002b), 'Knowledge-intensive services and urban innovativeness', *Urban Studies*, **39**, 993–1002.
Wood, P. (2005), 'A service-informed approach to regional innovation – or adaptation?', *Service Industries Journal*, **25**, 429–445.
Wood, P. (2006), 'Urban development and knowledge-intensive business services: too many unanswered questions?', *Growth and Change*, **37**, 335–361.
Wood, P. (2008), 'Service competitiveness and urban innovation policies in the UK: the implications of the "London paradox"', *Regional Studies*, http://www/informaworld.com/smpp/content~db=all~content=a902407063.
Wood, P. (2009), 'Knowledge-intensive business services' in R. Kitchin and N. Thrift (eds), *International Encyclopedia of Human Geography*, Oxford: Elsevier, 6, 37–44.
Wood, P., J. Bryson and D. Keeble (1993), 'Regional patterns of small firm development in the business services: evidence from the UK', *Environment and Planning A*, **25**, 677–700.
Z/Yen (2007), *The Global Financial Centres Index 3*, London: City of London Corporation, www.cityoflondon.gov.uk/Corporation/media_centre/research+publications.ht.

PART VI

INNOVATION IN SERVICES AND PUBLIC POLICY

26 Policy frameworks for service innovation: a menu-approach

Pim den Hertog and Luis Rubalcaba

26.1 Introduction[1]

The development towards a service-driven economy is a process that has been under way for some decades now. It is increasingly acknowledged that in advanced economies most room for productivity growth stems from services and service innovation. In both our private and professional lives new (combinations of) service concepts and experiences, new type of service organisations and occupations and associated – mostly information and communication technology (ICT)-based – service technologies are ubiquitous. Developing and exploiting these innovative services requires technological as well as non-technological innovation, knowledge and capabilities.

The concomitance between the role of services in modern economies and the role of innovation in economic growth has increased the interest in service innovation from different angles, from academic research to statistical developments, from new theories and typologies to a wider management and specific business plans for services innovation, from the inclusion of services in existing research and development (R&D) programmes to the development of new policy interventions.[2] An increasing number of firms are managing service innovation more explicitly. In some countries innovation policy-makers have started to explore new, more services-innovation-friendly R&D and innovation policy frameworks, and in a few even service innovation schemes (den Hertog, 2010).

The overall goal of this chapter is to discuss policy frameworks for furthering service R&D and innovation and to offer some policy options for furthering service innovation. By way of introduction we will in section 26.2 present three well-known approaches to services R&D and innovation (assimilation, demarcation and systemic approach; see also Chapter 1 in this volume). Here it is argued that the current debate on service innovation and service innovation policies still suffers from too much assimilation and demarcation thinking and lacks a vision as to how services are part and parcel of wider innovation systems. In section 26.3 we continue by reviewing possible rationales for having service innovation policies in the first place. We start off by reviewing some of the more practical macroeconomic argumentation when considering furthering service innovation. Subsequently, we will deal extensively with the market failure argumentation which is still dominant in most administrations. We will argue that some service industries do invest in technological R&D and innovation, and here market failure argumentation is as relevant as it is in manufacturing industries. Additionally, we will argue that when investments in non-technological R&D and innovation are made – which are strongly associated (but not exclusively, we would say) with service industries – the different categories of market failure might apply as well to all service industries and indeed manufacturing (although not in a similar fashion). Finally, we argue to look beyond market failure argumentation and take into consideration systemic failure.

In section 26.4 we analyse how service firms – within the limitations of statistical datasets available – benefit from standing European Union (EU) R&D and innovation policies. The empirical evidence suggests that services receive relatively less public support to their innovation, but results vary depending on the type of service and country considered. It will be argued that – although 'fair shares' cannot be the starting point of any R&D and innovation scheme – participation in 'industry-neutral' standing schemes seems to point at discrimination against some service industries and, more widely, service innovation. Section 26.5 then discusses some emerging policy frameworks and initiatives for furthering service innovation. Based on the proposition to consider services innovation as a systemic dimension of innovation systems, we offer a simple menu-approach with some policy options for furthering service innovation. Here we also argue that there is no fixed recipe available, as for every individual innovation system a (temporary) mix of instruments and policies will have to be developed to suit that particular innovation system. In section 26.6 we summarise and make a plea to develop the systemic approach towards service innovation (policies) further to counterbalance the current myopic view on services R&D and innovation (policies).

26.2 Three perspectives on service R&D and innovation[3]

Quite a number of innovation scholars, statisticians, policy analysts and policy-makers have analysed, measured, reported and discussed services R&D and innovation. This has not yet resulted in a coherent vision as to how to conceptualise, categorise, measure and facilitate services R&D and innovation. Innovation policy-makers seem to be increasingly aware of the key role service R&D and innovation play in driving economic growth. Some of them even recognise that there is a mismatch between the sheer size and economic role played by services and service innovations, and the sort of policy initiatives taken to further services R&D and innovation. However, in practice and with a few exceptions (Finland, Germany, Denmark, the European Commission), only a few countries have managed to define a policy framework for furthering service innovation and to design concrete programmes and schemes aimed at furthering service R&D and innovation. In our view this is to an important degree due to principally opposing views on the role of services R&D and innovation in innovation systems and the best way – if any – to facilitate these. Earlier, the many ideas and views on service R&D and innovation have been reduced to basically three approaches (see Boden and Miles, 2000; Coombs and Miles, 2000; and earlier, Gallouj, 1994; see also Chapter 1 in this volume).[4] These approaches – leaving out the approach of complete ignorance, as we presume this approach is no longer seriously supported[5] – are introduced briefly below.

26.2.1 Assimilation approach

The assimilation approach mainly starts from the idea that R&D and innovation is still about technological R&D and technological innovation,[6] that some services industries may play a role here, and that existing policies mostly need to be made somewhat more accessible for those service firms that perform technological R&D and innovation as there are no theoretical reasons justifying the exclusion of services. The central idea is that services and innovation in services can be analysed and supported using or adapting the concepts and tools developed for manufacturing and innovation in manufacturing (Tether, 2005). In policy terms this means unavoidable biases for technological support as most services innovation is less technology-driven (Howells and Tether, 2004), even if there are indeed indications that technological R&D and innovation in service firms are under-reported (Miles, 2005; RENESER, 2006). Most EU and national innovation policies can be characterised as belonging to the assimilation approach: services are ignored or are considered together with manufacturing under generic innovation policies. The horizontality is more theoretical than real since the bias towards technological R&D and innovation is more useful to goods industries than to service activities.

26.2.2 Demarcation approach

The demarcation approach is quite popular among services innovation researchers. They pointed at and reported on the peculiarities of services, the predominance of types of innovation other than pure technological innovation, and the different innovation processes or innovation styles of services R&D and innovation. As services R&D and innovation are conceptualised as rather different from the 'regular and well-known' technological R&D and innovation, a plea is made for new ways of measuring it and developing specific policy initiatives suited to services R&D and innovation. This would mean a development towards more specific or vertical R&D and innovation policies. There are also some views in favour of a specific focus on services. Heterogeneity of individual (service) industries and the fact that not all services are similar in structure, innovation behaviour and barriers to innovation encountered, would lead to the need for specific measures in particular markets. It does for not, example, seem so easy to argue that a horizontal policy – beyond supporting business R&D – is enough to fulfil the interests of such uneven sectors as telecommunications, tourism, retailing or professional services. At the same time it can be argued that a demarcation approach would require a quite detailed understanding of individual industries by policy-makers.

Further, the increased intertwining of services and manufacturing industries, the rise of service innovations deriving from manufacturing industries, and the fact that most innovations in both manufacturing and services are about combinations of technology and new services concepts and about how these are translated into sound business propositions, makes the demarcation approach a form of overshoot which in the end might not be that productive. Examples of typical demarcation-type service innovation policies are specific innovation programmes for health services or the promotion of innovation in the tourism sector.

26.2.3 Systemic approach

The synthesis or systemic approach refers to a more integrated view of the service economy and starts from the idea that technology and non-technological innovation, manufacturing and services innovation typically support and need each other. Every firm is made up of a (different) mix of service- and manufacturing-like functions. Most innovative firms realise that they also have to differentiate themselves from the competition by adding extra service functionality to both manufactured goods and services. The systemic approach also starts from the observation that the differences within industry and within services are in many cases more marked than between the two main categories.[7] It further acknowledges the decisive role that intangible elements play in innovation and growth, and hence their relevance as objects of innovation policy-making. Seen from a systemic or evolutionary innovation perspective, a two-way approach is in our view extremely relevant.

Firstly, is the established innovation system (especially in terms of the wider institutional setup, built in incentives) designed and modelled well enough towards facilitating services R&D and innovation? Service innovation is understood as a horizontal dimension or intrinsic quality of an innovation system, and the lack of service innovation, as a failure of the system.

Secondly, do the sets of firms dealing with or playing a role in service innovations contribute well enough to the overall quality of the NIS? Service innovation capability is a system characteristic that may help in improving the overall functioning of an innovation system. The systemic approach in this context points to the need to improve the relationships between 'goods' companies and 'service companies'. Especially, the role of knowledge-intensive (business) services is key here as these may provide necessary intermediate input to improve the competitive and innovative capacity of any manufacturing or service company (Rubalcaba, 1999; Antonelli, 2000; Wood, 2001), their connections with new technologies and especially their consolidation as part of the innovation system (Antonelli, 1999; Miles, 1999; Boden and Miles, 2000; Metcalfe and Miles, 2000; Muller, 2001; Zenker, 2001; Hipp and Grupp, 2005).

Given existing evidence we consider that innovative services (in both manufacturing and services) improve the dynamics of the overall innovation system or sectoral innovation systems (Miles et al., 1995; den Hertog, 2000; OECD, 2006; Rubalcaba and Kox, 2007). We think that especially business-related services – and probably most importantly knowledge-intensive business services (KIBS) – can play multiple roles in making innovation systems more dynamic and better-performing. Increasingly KIBS are seen as being part and parcel of an effective entrepreneurial or innovation climate, and as a basic element necessary for developing wide and deep clusters and networks. In almost all competitive clusters and networks you will find highly specialised service firms that help other firms to become more innovative and competitive. Therefore, in our view, adopting a systemic perspective when looking at services R&D and innovation explains not only how the innovation impacts upon the service firm, but also how a well-built-in service sector or 'service functionality' in the innovation system can increase the overall performance of this innovation system considerably. This requires a widening in terms of the rationale used for (service) innovation policies. This critical issue will be discussed in the next section.

26.3 Arguments for a service innovation policy

A logical question when discussing a policy framework for service innovation policies is whether or not there is a rationale for having these in the first place. In our view we should differentiate here between contextual and macroeconomic argumentation, and the more microeconomic argumentation to have service innovation policies. In the latter category we make a distinction between the well-known market failure argumentation and evolutionary systemic failure argumentation, which in some cases in some industries may both apply to services innovators.

26.3.1 Macroeconomic and contextual rationale

There are policy-makers and scholars that advocate not having any specific service innovation policy at all. Typical macroeconomic assumptions include in the first place the idea that high economic and productive growth rates are localised in countries where manufacturing industry is performing well. This is due to their relative costs (China, India) or because of the higher manufacturing performance associated with a high technological capability (e.g. some industrial medium- and high-tech sectors in Europe and the United States). This assumption forgets that many high-performance countries and sectors around the world are based on services and a high service performance. Moreover, service innovation may be at least as important as innovation in goods in explaining performance differences among countries.

A second assumption is the notion of ICTs having a multipurpose character, making them suitable to fit into the needs of any economic activity, in such a way that the promotion of ICT and other technological programmes may be enough to achieve a horizontal policy affecting all sectors without any sector discrimination. This assumption ignores the complementarity between technological and non-technological innovation though, which has proved to be important (see, e.g., Bresnahan et al., 1999; Brynjolfsson and Hitt, 2003).

A third assumption often mentioned for having generic or horizontal rather than service-specific innovation policies is the huge heterogeneity among services. This would make it almost impossible to address different service innovation particularities without inferring too much into the markets. The idea then is that it is better to have a horizontal policy that follows the established and tested logic of supporting technological innovation without distorting competition, than to have a specific services innovation policy supporting service innovations which are seen as 'too close to the market' and might be in conflict with competition policies.[8] In our view this assumption underestimates the complexity of modern economies, the interrelationship between economic activities and the proximity to the markets existing in many goods innovations in manufacturing industries and, increasingly so, in the policies supporting these.

The following macroeconomic or contextual arguments can be given for including service innovation more explicitly in innovation policy-making (most of them already included in Rubalcaba, 2006):[9]

- Innovation has been proved to be an essential factor for economic growth (Schumpeter, 1939; Griliches, 1986; Fageberg, 1988; Freeman, 1994). Given the fact that services represent around 70 per cent of more advanced economies, service innovation will be a key factor for economic growth.

- A sluggish productivity growth in services – behind goods productivity rates – may be the consequence of a low performance of service innovation: structural changes require new innovative efforts to balance specialisation changes.
- Service innovation is a stimulant for innovation generally, and for investment in intangibles and knowledge, factors of endogenous growth and total factor productivity. There is empirical evidence proving the importance of service innovation in productivity and economic performance (e.g. Cainelli et al., 2006; Rubalcaba and Gago, 2006).
- There is relatively low productivity and performance in many service sectors (European Commission, 2003) and reduced use of ICT in some important services branches in Europe, as detected for ICT service users in O'Mahony and van Ark (2003).
- The relatively low participation of services companies – with some exceptions – in R&D programmes. This raises in an EU context the question whether the Lisbon Strategy and the aim to achieve 3 per cent of gross domestic product (GDP) in R&D investments in Europe can be attained without including services R&D and innovation more explicitly. Raising the R&D level in services would contribute considerably to accomplishing this goal.
- The lack of formalisation and organisation of service innovation, which requires the promotion of new instruments of business support. Beyond R&D, other inputs and drivers should be analyzed, monitored and, if appropriate, promoted.
- The recent deregulation and liberalisation in many service sectors, which means that businesses forsaking their protected market niches need to find new strategies to boost competitive levels. Innovation-driven growth in these once sheltered markets will need to rise.
- The current phenomenon of relocating services to lower-cost countries or countries with a higher specialisation demands that businesses in advanced countries should find new competitive strategies based on innovation.

26.3.2 Market failure rationale

Macroeconomic arguments for not paying attention to service innovation in innovation policies are mirrored in more microeconomic arguments focusing typically on the notion of market failure. As stated in a previous contribution (den Hertog et al., 2008), among innovation policy-makers there is a received wisdom that services perform less technological R&D.

Service firms are less focused on technological innovation compared to manufacturing firms. The result is that all too often it is concluded that no specific attention to R&D and innovation in services needs to be paid. This is at least remarkable as there are numerous service firms that do invest in technological R&D and innovation and these need at least to be treated in a similar vein as their peers in manufacturing.

Another argument used for not developing R&D and innovation policies that are more suited to service innovators is that innovation in services typically takes place close to the market. This would imply that the case for intervention aimed at facilitating innovation is less obvious. Put differently, service innovation is not sufficiently fundamental to be supported and policy-makers fear that intervention might distort competition. Further, it is believed that there is little scope for spillovers across firms as services innovation involves so much organisationally specific development. Put differently, externalities from investments in non-technological R&D and innovation are less obvious than externalities from investments in technological R&D and innovation.[10] Typically, these are the sort of arguments on the basis of which, for example, national ministries of finance judge the acceptability of R&D and innovation schemes.

Den Hertog (2010) continues to argue that it can be questioned whether the line of argument above is based on the right assumptions. Firstly, service firms and service industries are more active in technological R&D than is mostly anticipated. R&D and innovation in business-related services and in particular knowledge-intensive business services are performing substantial technological R&D, mostly considerably higher compared to the average for manufacturing firms (RENESER, 2006). These services are more likely to engage in R&D than other service industries and than most firms in manufacturing industries. Standard market failure argumentation[11] is as relevant to these industries as it is to manufacturing firms and industries.

Secondly, it seems to be forgotten that investing in technological R&D is just one of the ways through which firms become more innovative (and eventually more productive and competitive). There are simply innovations that do require relatively more investments in non-technological innovations such as new organisational or marketing concepts, new client interfaces, new types of delivery organisations or new smart combinations of service and product elements. These investments are more difficult to pinpoint and to assess, but they are as real as investments made in technological R&D. Further, as these intangible investments trigger innovations that eventually lead to economic growth, there is no fundamental reason to not facilitate these. However, a prerequisite is still that there should be externalities involved and these softer types of innovations do come at a cost and require serious investments, i.e. social benefits are higher than private benefits and the individual entrepreneur would have to invest substantially.

Thirdly, it can be seriously questioned whether technological R&D and innovation and non-technological R&D and innovation can be treated separately. These two are in economic reality increasingly difficult to disentangle and treat separately, as most innovations today are multidimensional, i.e. smart combinations of new or advanced technology in combination with new service elements or smart services enabled by innovative use of technology.[12]

Typically, the market failure argumentation used in standing R&D and innovation policy and its applicability towards services R&D and innovation needs to be reviewed. Do market failures inhibit new innovations in services R&D and innovation? We also need empirical research to see if these alleged market failures can be found in practice. Van Dijk (2002) studied three different types of market failures in service innovations: externalities, market power and asymmetric information. Cruysen and Hollanders (2008) used a different set of market failures in services innovation and linked these to possible policy intervention, namely: market power; externalities; nature of certain goods or the nature of their exchange; resource immobility; and market failures associated with property rights. Rubalcaba (2006) has, in a somewhat different vein, also reviewed the applicability of the market failure argument to services innovation and linked this more directly to policy options. He makes a similar differentiation between uncertainty and risk; externalities; scale economies; and market power.

Market failures can be analysed in the following way.

Uncertainty and asymmetric information Uncertainty is developed around innovation since this is produced in a context where expectations and information are distributed in a very asymmetric way (Dosi, 1988; Stiglitz, 1991). Since innovation requires different means and levels of interaction between seller and purchaser, the problems of asymmetric information could hamper innovation, as one party might be distrustful about unknown features of the other party (e.g. attitude, skills). The problem of uncertainty does not only justify public investment – to take risks derived from a potential failure – but also financing intermediation (soft credits, grants). The particular case of asymmetric information limits the demand for new services and requires public action affecting demand and market transparency and information; not only action directed towards the supply side is requested. Perceived risk and uncertainty of market players may justify public investment, some financing facilities, or attempts either to boost demand for new services or to increase the transparency of markets. Lack of information or overinformation in service markets may act in a similar way. An example of misinformation is the one given by the banking system when credits and loans are negotiated. The lack of tangible assets is used as an excuse for underfunding service innovation. Accurate information about intangible assets may solve part of the problem.

However, this market failure may lead to another type of failure in which lack of coordination between business accountancy, financial practices and the legal environment reveal a more systemic failure. A similar asymmetric information problem is revealed when public administrations are in a better position to assess the potential growth of strategic technologies or sector (Krugman and Obstfeld, 1994), or the reverse, when their position is worse due to their distance from the real world. This has important policy implications, for example, when trying to promote high dynamisms in service sectors like KIBS that may offer strategic growth areas for innovation. For this purpose statistics and analyses are needed to base decisions upon.

Externalities When societal returns to innovations exceed the private returns, firms may innovate too little, because innovations may 'leak' to competitors due to imitation or employees switching jobs. This is related to appropriability, but also to entry and exit conditions for firms and individuals. Externalities are derived from the public nature of knowledge and its spillovers, which generate problems of appropriability and use of innovation without the need to pay their market value (free-riding). This market failure justifies the intellectual property protection policies (intellectual property rights) on the one hand, and direct government intervention on the other hand, although the latter is only justifiable in the case that the intervention implies the maximisation of net social welfare. Market failure theories advocate policy actions to cope with externalities only when they clearly produce higher social benefits and cannot be appropriated by private enterprises (Heijs, 2001). Given the fact that information is a good hardly to be appropriated (Arrow, 1962), economic activities very intensive in knowledge face a greater problem. As far as services are concerned, appropriability problems seem to be even greater than in the case of goods, due to the limited use of patents, the insufficient protection offered by copyright systems and the intensive role of information in KIBS. Therefore, externalities arguments may support services innovation policies even more than goods innovation ones. From a policy perspective, a reinforcement of the appropriability system in services should consider the potential negative impacts that may have on potential innovation growth or market competition. This leads to consideration of service innovation policies beyond IPR issues.

Scale economies These are related to the indivisibility of technological activities requiring a minimum critical mass. The problem of indivisibility in the world of services is probably smaller than in the case of goods, where R&D processes are better structured and require a higher quantity of inputs. As innovation processes are more diffused in production processes, as in the case of services, it seems to be easier to reach a critical mass. On the contrary, when innovation effort concerning inputs is put into qualified human capital, an active policy is justified in the field of education and training. At the same time, services operate in small and medium-sized enterprise (SME) markets to a greater extent than in the case of goods. This reinforces the traditional justification of SME-oriented policies, where reaching a critical mass and sufficient human capital is more difficult. Most service companies are SMEs, in proportion even higher than in goods sectors, so the critical mass obstacle may be higher than in goods.

Market power This may be the result of high sunk costs, natural monopoly, low transparency or high switching costs. This is clearly related to industry market structure. Market power also has a particular interest for services, where the lack of competition can act as a disincentive for generating innovation. As many services operate in very segmented markets with a high monopolistic power, this market failure is particularly important.

Differences among service industries

All these four market failures affect services, but not all to the same degree in all services activities. Heterogeneity matters according to what was included in den Hertog et al. (2008). In that contribution, we surveyed several authors such as van Dijk (2002), who concluded that the strong heterogeneity in services makes it difficult to generalise as to the prevalence of these market failures. He therefore developed an alternative, more functional typology of service industries,[13] to show the large variety in service innovation and to be better able to identify possible market failures. He further concluded that in each type of service industry a different mix of market failures is likely. According to van Dijk, externalities are most visible in innovations in transport, post and telecommunications, financial services and some personal services. Market power can be observed in (again) post and telecommunications, banking and financial services and business and personal services. Asymmetrical innovation was found to be most prevalent in telecommunications, financial services and business services. He noticed that these market failures can be reduced through policies, but that may be harder as service innovation is difficult to define and in some cases the costs of these policies may be higher than the benefits.

Whether or not the intervention argument for technological innovation also applies to non-technological innovations[14] – i.e. market failure – depends, according to van Ark et al. (2003), on two questions:

- whether or not a (long and hence expensive) learning process precedes the non-technological innovation;
- whether or not other firms can easily use or copy the non-technological innovation; in other words, are there knowledge spillovers?

They argue that non-technological innovations may require extensive learning processes for individual firms (hence considerable investments), while future returns are uncertain. They indicate that non-technological innovations come about as a learning (by doing) process, and are not easily implemented and transferred. For example, a study by Kox (2002) showed that firms in the business service industry are quite often strongly dependent on inherent tacit skills of key employees in the firm. Such tacit skills take much time to acquire and cannot be easily transferred to other employees in the firm. As the lead time to obtain the returns from investment in skills can be quite long and the future returns themselves are quite uncertain, firms may be supported to make such investments nevertheless.

Van Ark et al. (2003) indicate that they find it harder to assess whether knowledge spillovers are present regarding non-technological innovation. This depends for instance on the type of non-technological innovations, whether these are internally focused or not, and how easily they can be observed from the outside. Put differently, how easily and at what costs can non-technological innovations be copied by other firms? Typically, new marketing concepts are directly visible and can easily be copied. New organisational types of innovation also easily transfer between firms as employees move between firms, and when firms cooperate in networks or within value chains. Van Ark et al. (2003) also observe that:

> However, non-technological innovations are also strongly correlated to each other. Hence when one type of non-technological innovation is easily imitated, it implies changes in other innovations as well. Still imitating firms can perhaps find a way to implement such additional changes at lower costs than the originally innovating firm, although adaptation to the specific firm-culture may still be costly. In any case the original innovating firm may refrain from engaging in non-technological innovations when the social returns of these innovations are much higher than the private returns. On this basis non-technological innovations may become the subject of government intervention, e.g. by assisting in the adjustment of the organisational structures when private incentives are lacking.

They conclude that if long and costly learning processes are, in practice, mostly combined with relatively easy imitation (at least the costs of making the 'stolen innovation' work should be considerably lower as there is also the lead time and possibly a reputation advantage for the innovator) it would most likely result in economy-wide underinvestment in service R&D and innovation. This would make it more logical to consider an R&D and innovation policy that more explicitly pays attention to services R&D and innovation.[15] At least we may conclude that – although more academic and policy research is needed on market failures in service industries – it cannot be taken for granted that market failure is perceived to apply exclusively to technological R&D and innovation and not to non-technological R&D and innovation. In practice, the two are more interlinked and hard to separate.

26.3.3 Systemic failure rationale

The technological bias in innovation policy is partly due to the persistency of the linear innovation model that sees technology as the key aspect between the use of R&D inputs and the final innovative products or processes. A different interactive innovation model was developed since the 1980s, for which the design, management, implementation and diffusion of results require continuous interactions and learning-by-doing among different actors. This has been used mostly to understand technological innovation in manufacturing better. However, services innovation has to a much lesser extent been conceptualised using this interactive and systemic context, let alone service innovation policies. In our view we need to look beyond the market failure argument and adopt a broader view on service innovation and the possible need or rationale for service innovation policies, i.e. possible systemic failures. There has been a growing body of thought that market failure argumentation is insufficient to deal with innovation dynamics and the rationale for R&D and innovation policies. In this section we look beyond market failure argumentation at another type of failure i.e. systemic failure. The argumentation used here is mostly based on evolutionary rather than neoclassical approaches to innovation (key references here are, e.g., Edquist, 1997, and linked to services, Metcalfe and Miles, 2000).

Observing the need to correct the systemic failures in wider innovation systems, the evolutionist theory suggests some innovation models or systems without simple one-way relationships between knowledge generation and absorption (O'Doherty and Arnold, 2003; Arnold and Kuhlman, 2001). Therefore, a systemic approach is necessary to understand the relationships between science, technology and innovation, as well as an evolutionist approach, indicating that there is a specific situation for each case according to the cumulative processes generating changes in the systems. It will be very difficult to cope with service innovation if innovation systems are not suited to facilitate and benefit from service innovation. In this subsection we will use the typology of four types of systemic failures as introduced by O'Doherty and Arnold (2003). For each of the four non-market types of failure,[16] we tentatively provide a number of examples to illustrate that there is some room to improve the way in which services, services R&D and service innovation is facilitated. This also implies making use of 'non-innovation policies', i.e. policies originally not designed to facilitate innovation which can be beneficial in furthering service innovation. Cruysen and Hollanders (2008, 8) define systemic failures as 'structural, institutional and regulatory deficiencies which lead to suboptimal investment in knowledge creation and other innovative activity'. They choose not to differentiate between different categories of systemic failure as they find many linkages between the different types of systemic failures. Based on a previous contribution (den Hertog et al., 2008) we present the four systemic failures as introduced by O'Doherty and Arnold (2003) separately, as we find that they are related, but nevertheless inherently different.

Capability failures These are defined by O'Doherty and Arnold (2003, 32) as 'inadequacies in potential innovators' ability to act in their own best interests'. If one translates this to the case of services (or rather service functions within innovation systems) one can think of:

- Service firms and their employees that might lack the right knowledge, skills, information and contacts to realise technological and non-technological innovations.
- Service firms that are not capable of identifying the actual needs of their clients.
- Service firms that are not capable of articulating their knowledge needs.

Some of these capability failures are more or less generic and may not be services-specific, but others are services-specific. The wider system should be designed in such a way as to reduce these typical capability failures.

Failures in institutions These are defined by O'Doherty and Arnold (2003, 32) as: 'failure(s) to (re)configure institutions so that they work effectively within the innovation system'. This is probably one of the systemic failures that does apply most clearly to the services case. Typical examples one can think of are:

- Schools that do not educate and train students (i.e. future service professionals) with the right set of capabilities for service firms.
- Innovation management courses that are biased towards manufacturing.
- Tax credit schemes that discriminate against service innovation.
- Financial and credit systems that do not always value the intangible assets of services companies. Such assets are quite often not registered in the businesses' balance sheets (Green et al., 2001). Despite the present efforts (Zambon, 2003) towards accountants acknowledging much better intangible assets, the current credit system penalises those activities of uncertain risk based on intangible assets, which are often considered as expenses and not as investments.
- Statistics that do not record services and service innovations properly.

We think that these examples illustrate how the institutional setup of an innovation system might lag considerably behind economic reality. Swift institutional adaptation can help considerably in laying the foundations for and facilitating much better services and hence services R&D and innovation.

Network failures This related category of failures related to the 'interactions among actors in the innovation system' (O'Doherty and Arnold, 2003, p. 32) similarly illustrate how the innovation system as a whole might not have adapted well enough to the increased role played by service firms in general and their (potential) role in R&D and innovation. Typical examples refer to problems such as:

- Public knowledge infrastructure that primarily caters for the needs of manufacturing firms. How come that in quite a few innovation systems, new intermediary centres of excellence have been created that are seldom about 'service technologies'?
- Government purchasing policies that do not challenge service firms (innovation is quite often not rewarded).
- The lack of an appropriate system for knowledge management and structural capital. These can be highly useful to cope with the non-technological innovations. Network infrastructures such as technological centres, scientific parks and other business services centres may continue to a large extent to deal with this type of network failure.
- An industry–science relations (ISR) debate that is strongly biased towards high-tech industries, but pays hardly any attention to the role of ISRs between the science base and services.

Framework failures These relate to the fact that 'effective innovation depends partly upon regulatory frameworks, health and safety rules, etc., as well as other background conditions, such as the sophistication of consumer demand, wider culture and social values' (O'Doherty and Arnold, 2003, 32). Under this label various failures are hidden that are real for quite a few innovative service firms and industries such as:

- All sorts of regulations that do not provide the right incentives for innovation in services (trade policies, spatial planning, environmental regulation, market regulation etc.).
- Particular failures and legal and financial obstacles hampering entrepreneurship and dynamism, mainly affecting SMEs.
- Consumers who are not prepared to pay for innovative services.
- Foresight and road mapping exercises aimed almost exclusively at high-tech and manufacturing industries.
- Governments that are not investing (enough) in innovative public services (which can act as 'role models').
- Mobility schemes that focus mostly on scientists and engineers.
- Innovation debates dominated by technological innovation.
- A lack of a services innovation culture.

This last category ultimately illustrates that there may be many failures which discriminate against services and services R&D and innovation. At the same time it demonstrates that the numerous contexts in which innovative service firms have to operate can be improved through various lines of policies ranging from typical R&D and innovation to educational, competition, government procurement and other lines of policy-making that are not primarily aimed at R&D and innovation.

Summarising the rationale for a service innovation policy

Most innovation systems – or at least how most innovation analysts and policy-makers typically think about their design and functioning – are lagging behind in adapting to an economy that is considerably service-driven.[17] However, bearing in mind the systemic perspective provides us at the same time with both analytical tools and a somewhat broader set of policy tools to make the innovation system as a whole better suited to support services R&D and innovation. Especially, some of the 'non-innovation policies' are identified as significant in providing the right framework for services R&D and innovation to take place.

In the above we have explained why in most countries R&D and innovation policies are still mostly biased towards technology and manufacturing R&D and innovation. A combination of factors can explain

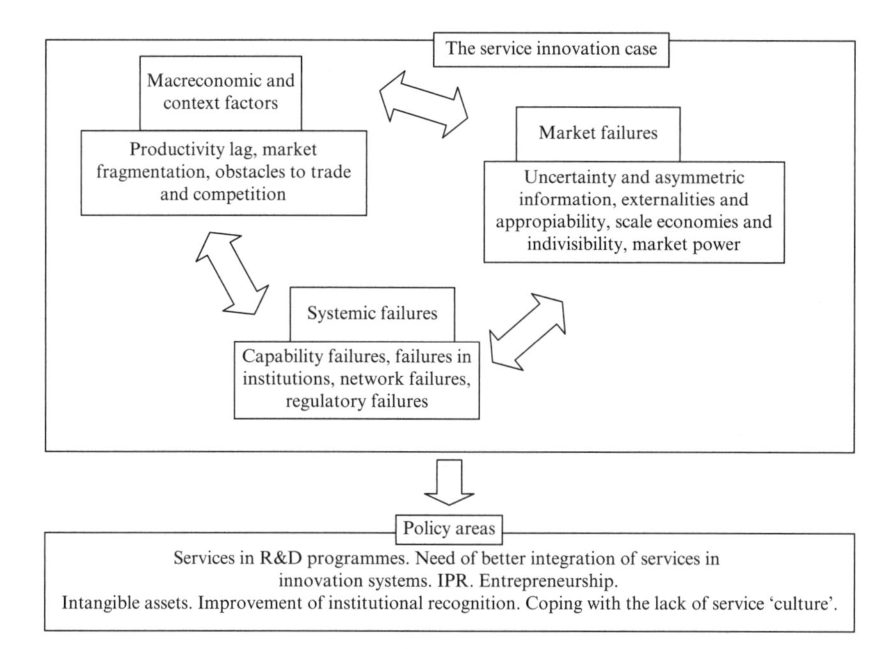

Figure 26.1 The rationale for a service innovation policy

this. Most dominant here are: the rather slow recognition of the pivotal role played by service functions and service innovation in the economic growth process; the dominance of market failure argumentation within government; and the hesitance to adopt the notion of systemic failures and apply these to the role of services in innovation systems. Figure 26.1 shows the most significant elements justifying a service innovation policy, not only from the neoclassical point of view of market failures, but also from the contextual facts that reveal challenges from macro and political changes and the systemic or evolutionist approaches. Obviously, the three types of arguments are interrelated and each one cannot be understood in isolation. For example, asymmetric information creates a natural barrier resulting in a competition deficit in many services markets with consequences in productivity and innovation; at an institutional level, these facts are not sufficiently recognised, and for this reason pro-innovative and pro-competitive actions are underdeveloped. Another example concerns the lack of recognition of intangible assets which is, at the same time, a market failure linked to the asymmetric information problem and a systemic problem linked to the functioning of financial and accounting systems. Between macroeconomic context-structural failures and the systemic failures there are obvious interrelations too. For example, regulatory framework obstacles affect different levels – micro, meso and macro – functions and conditions requested by service innovation development.[18]

26.4 Services firms in innovation programmes

In sections 26.2 and 26.3 we have argued that existing policy approaches and rationales for policy intervention are mostly less well developed and attuned to service industries and service innovation. The European Community Innovation Surveys (CISs) allow us to analyse the importance of public funding in innovative firms at sectoral level.[19] Generic policy programmes for supporting innovation would lead to similar rates of public funding in different economic sectors. However, recent evidence shows there is a significant bias in public funding against services in general and some services in particular. Both according to CIS3 and CIS4,[20] funding reaches a higher percentage of innovative companies in manufacturing compared to service industries.[21] CIS4 results show that overall funding of manufacturing firms is almost twice as high as for services firms (in terms of shares of innovative companies receiving public support). On average 28 per cent of innovative manufacturing companies versus 16 per cent of service companies receive public funding. This funding bias against services can be observed at all different levels at which innovation policies are being used, i.e. at the local, regional, national and EU level (Rubalcaba, 2007; Gallego and Rubalcaba, 2008).

It should be kept in mind, however, that this bias does not apply to the same extent to all cases, all countries and all subsectors. Regarding the type of innovation funding, a more balanced distribution between services and manufacturing can be found in the funding deriving from the European Union, as previously observed by RENESER (2006) for CIS3 and by van Cruysen and Hollanders (2008) for CIS4 data. This can be mainly explained by the active role of (mostly knowledge-intensive) business services in getting EU public funding. These business services in their turn represent a dynamic sector that receives important funds in all countries. Figure 26.2 shows that there is a significant bias in public funding against the whole set of services, but also that this bias is not true in the case of business services in a number of countries. Typical service industries with lower shares of public funding for innovation are distributive trade companies, transport and finance sectors.

The heterogeneity of services regarding public support is also illustrated in Table 26.1. It represents a distinctiveness coefficient, where different service branches are benchmarked with respect to the total of manufacturing industries. Figures are given for policy-related variables that are chosen among the set of possible CIS4 indicators. The percentage of

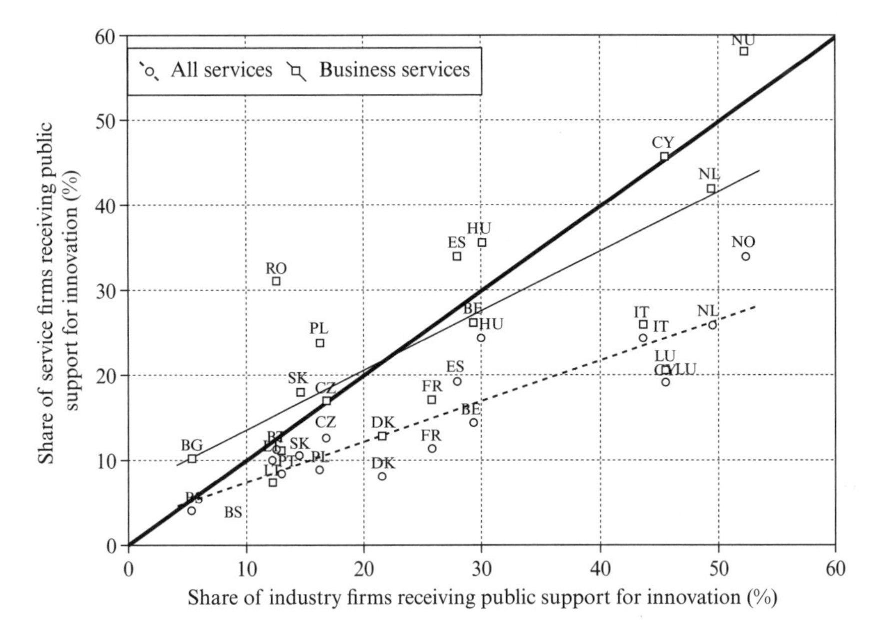

Source: CIS4 database, Eurostat.

Figure 26.2 Share of firms receiving public funding for innovation by economic sector

innovative enterprises is interesting in order to promote policies to spread and to disseminate innovative behaviour and attitudes among the business society. In this case, business services and financial services show high shares of innovative companies, unlike distributive trade and transport services. Concerning intramural and extramural ratios, business services use both R&D categories more than other services branches, except financial services which contract quite a lot of extramural R&D. In R&D only business and financial services offer similar or higher shares of companies than manufacturing, while other services show lower shares, in line with previous studies about the role of R&D in services (Miles, 2005; RENESER, 2006).

Concerning impacts, it is clear that innovation in service industries promotes less costs-related impacts compared to innovation in manufacturing industries, while the impacts related to the capacity to respond to customer needs is higher in service industries. In quality impacts the situation is more balanced among sectors. This suggests that the orientation of particular R&D and innovation funding should take into account the different expected impacts that such funding can result in. Possible

Table 26.1 Distinctiveness coefficient in some key policy-related indicators: services versus goods, Europe-16

	Total goods industries	Manufac-turing	Total services	Distributive trades	Transport and communi-cations	Financial services	Business services
% of innovative firms	1.00	1.004	0.773	0.699	0.625	1.204	1.070
Intramural R&D	1.00	1.060	0.791	0.601	0.627	0.815	1.213
Extramural R&D	1.00	1.017	0.964	0.932	0.873	1.142	1.112
Impacts on costs	1.00	1.005	0.677	0.656	0.841	0.888	0.576
Impacts on quality	1.00	1.010	1.033	0.907	1.063	1.118	1.170
Impacts on respond time	1.00	1.007	1.227	1.250	1.330	1.307	1.113
Patents	1.00	1.033	0.517	0.575	0.254	0.125	0.825
Copyright	1.00	1.014	1.598	1.065	0.531	0.764	3.632
Obstacles	1.00	1.005	0.901	0.878	0.799	1.004	0.989
Total public funding	1.00	1.005	0.574	0.470	0.463	0.239	0.944

Notes: Europe-16 refers to Belgium, Czech Republic, Denmark, Spain, France, Italy, Cyprus, Lithuania, Luxembourg, Hungary, Netherlands, Poland, Portugal, Romania, Slovakia and Norway.
In boxes: those coefficients for which service stand 20% below or above the total goods average.

Source: CIS4 database, Eurostat.

specific approaches may follow, even if overall problems for innovation are the same in goods and in services (a similar balance is obtained on the obstacles perceived for innovations). In other words, evidence on impact suggests that policies should not just take care of the volume of public spending addressed to services but the public programmes design as well. The latter may be in need of some customisation to the particular features of service innovation.

Differences are important in other policy-related variables such as the protection systems. As expected, patents are much more important in manufacturing than in services, while regarding the copyright system the opposite applies. Intellectual property rights (IPR) may be underdeveloped in services or/and services require another type of approach for protection of results. In any case, all this empirical evidence should be considered when the formulation of policy implications is envisaged. On the one hand, possible biases should be identified in order to assess explicit or implicit discrimination against services innovation in terms of total distribution of public funding and discuss rationale behind these biases. On the other hand, attention should be paid to the qualitative design of innovation policies; the problem is surely more qualitative than quantitative. The appropriate policy mix might be different for different industries, i.e. innovation policies – at least to some extent – need to be customised, even within service branches.

26.5 An emerging policy framework for service innovation[22]

Having discussed possible rationales for having services innovation policies we here provide a menu-approach or set of policy options available to policy-makers wanting to further service innovation. We offer a simple menu-approach taking the three approaches to services and service innovation – i.e. the assimilation, demarcation and systemic approaches – as outlined in section 26.2 as our point of departure.[23]

The assimilation approach – applied to policy-making and following den Hertog et al. (2008) – can be summarised as a 'no-regret' scenario as it does not imply a specific services R&D and innovation policy. It mainly is about increasing the service sensibility of existing generic R&D and innovation schemes. The demarcation approach as applied to furthering service innovation asks for services-specific R&D and innovation policies. The heterogeneity of the service sector at large, and the peculiarities of some specific service markets, in practice requires specific measures. In the synthesis or systemic approach services are perceived as a systemic dimension of economic activity and thus a potential focus of innovation practically anywhere. Policy initiatives originating from the systemic perspective have passed the services–manufacturing distinction, take a broad approach to innovation and see both R&D and innovation as well as 'non-innovation' policies[24] as vehicles for supporting R&D and innovation. Services are understood not only as an economic sector, but also as a horizontal dimension of the overall innovation system that may boost overall innovation and competitiveness.

A menu of some possible policy options in each of the three approaches is given in Table 26.2. We do differentiate here between R&D and already wider innovation policies, and policies not originally developed to spur R&D and innovation but having a considerable impact on the scope for service innovation (i.e. 'non-innovation' policies). The three approaches might suggest that policy-makers would have to make a choice between the three columns whereas in practice policy options from all three may be combined. The key issue is that there is not one single solution, as for every individual innovation system a (temporary) mix of instruments and policies will have to be developed to suit it. This mix not only depends on the economic structure, history and institutional setup of a particular innovation system, but also on the preparedness and willingness present among actors in such a system to experiment and invest in mutual learning. Further, in our view the development towards service innovation policies at the various policy levels will be evolutionary rather than revolutionary. We are already experiencing that there is no fixed recipe available as every individual innovation system (and for that matter sectoral innovation system) will need a customised (temporary) mix of instruments and policies that suits it.

26.6 Conclusions and some final remarks

In this chapter we have discussed policy frameworks and policy options for furthering service innovation. We have subsequently taken stock of the possible policy rationales for having service innovation policies in the first place, reviewed the participation of service firms in (EU) innovation programmes and some of the more recent (mostly EU) service innovation policy programmes and initiatives. Finally, we have offered some policy options along the lines of the assimilation, demarcation and systemic approach towards service innovation.

Our claim is that a systemic view on service R&D and innovation is needed where the link between an innovation environment or rather innovation system and service R&D and innovation is the point of departure. This linkage is two-sided. Firstly this perspective looks at how well the innovation system caters for or is adapted to the needs of service innovators. Does the overall innovation system provide the right incentives for triggering and diffusing service innovation or not. Secondly, the systemic perspective takes into consideration how services R&D and innovation

Table 26.2 Examples of assimilation, demarcation and systemic policies aimed at facilitating services R&D and innovation

	Assimilation policies	Demarcation policies	Systemic policies
R&D policies	• Increase accessibility of existing R&D support schemes • Inclusion of services in technological foresight & road mapping exercises • Include services firms in policies aimed at improving industry–sciences relationships	• Support public R&D in services • Introduce vertical R&D programmes aimed at service industries (logistics, trade, etc.) • Services IPR instruments • Create dedicated CoE in services R&D at RTOs and HEIs • Increase role of the humanities in service innovation	• Understand and support role of R&D services (KIBS) in innovation systems • Support for services R&D in and through hybrid firms • Integrated R&D programmes paying attention to technological and non-technological R&D and innovation
Innovation policies	• Increase accessibility of existing innovation support schemes • Innovation management training & practices more geared towards supporting all types of innovation in all industries • Mobility schemes no longer limited to qualified scientists and engineers	• Introduce courses on services innovation management • Awareness campaign on the importance of services innovation • Identify service innovation role models (including innovation in public sectors)	• Innovation & business support systems also supports services innovation • Availability and use of specialised services / KIBS • Increase transparency in KIBS markets • Insight into & international competitiveness of key service functions

Table 26.2 (continued)

	Assimilation policies	Demarcation policies	Systemic policies
			• Cluster and network type of policies that deliberately include services • Government procurement policies • Support role of users in innovation
Non-innovation policies	• Increase coverage of services in regular and R&D and innovation statistics	• Use deliberately policies such as trade, competition education & training policies for fostering R&D and innovation in services • Regulations that might trigger service innovations • Analyse offshoring in services	• Use regulation & standardisation to support innovation • Financial and credit systems that acknowledge intangible assets • Enhance high level service capabilities e.g. through education & training policies • Policies aimed at increasing entrepreneurship

Note: IPR= intellectual property rights; CoE= Centre of Excellence; RTOs= research and technology organisations; HEIs= higher education institutions.

Source: Hertog et al. (2008), copyright Inderscience and reproduced with permission.

contributes to the overall quality of the innovation system. We see the level and integration of service R&D and innovation as an intrinsic quality of innovation systems that might spur the overall quality in terms of innovativeness, competitiveness and economic growth and welfare created.

This systemic perspective implies that the rationale for having service innovation policies needs to be thought through much more thoroughly and go beyond the standard market failure argumentation and include wider macro-economic considerations and systemic failures as elements. In parallel the way service innovators benefit (or not) from existing – and more importantly future – innovation policies as well as other policy areas which jointly provide the framework conditions for spurring service innovation needs to be reconsidered as well. Service innovation policy frameworks and policy options need to be adapted accordingly. In this chapter we have provided a first analysis and some stylised conclusions are given below.

The three well-known approaches towards services innovation (the assimilation, demarcation and systemic approaches) are mirrored in existing and developing policy approaches towards service innovation.

Three categories of argumentation need to be taken into account when discussing service innovation policies, i.e. rationale based on market failure; contextual and macroeconomic rationale; and rationale based on systemic failures. The latter two categories are often ignored and market failure argumentation not always used properly.

All four categories of market failure (uncertainty and asymmetric information, externalities, scale economies and market power) may affect services, but not all to the same degree and in all services activities. Some service firms and service industries are more active in technological R&D than is mostly anticipated. Here market failure argumentation is as relevant as to manufacturing firms investing in technological R&D and innovation. It can further not be taken for granted that market failure is perceived to apply exclusively to technological R&D and innovation and not to non-technological R&D and innovation. In practice innovation is multidimensional, the two are interlinked and hard to separate. In cases where firms have to deal with long and expensive learning processes preceding non-technological innovation and knowledge spillovers (i.e. the risk of imitation of the non-technological innovation), market failure can be said to exist.

There are many macroeconomic reasons to consider service innovation policies. High-performing countries and industries around the world are increasingly based on services, high service performance, high levels of service innovation and the ability to successfully link complementary technological (associated with manufacturing, although not exclusively) and non-technological innovation (associated with services, although decreasingly so). Service innovation is a key factor for economic growth, and there is a clear need to improve the low productivity and performance in many service sectors, especially in once sheltered markets that are opening up to international competition.

It will be very difficult to cope with service innovation if innovation systems are not suited to facilitate and benefit from service innovation.

Therefore we consider the presence of systemic failures (capability failures, institutional failures, network failures and framework failures) as a relevant argumentation for considering (service) innovation policies. This also implies that the arsenal of potential policy options is widened, as next to R&D and innovation policies other types of policy actions – i.e. policies originally not designed to facilitate innovation – might prove beneficial in furthering service innovation.

Although 'fair shares' cannot be the starting point of any R&D and innovation scheme, participation in 'industry-neutral' standing schemes as measured in the EU through both the Community Innovation Surveys 3 (2000–2002) and 4 (2002–04) seem to point at discrimination against some service industries and, more widely, service innovation. However, the public funding bias against services does not apply to the same extent to all cases, all countries and all subsectors as diversity within the services sector is huge. However, this diversity within the service sector should not be used as a sufficient argument to deny its specificity and different needs (Rubalcaba, 2007).

Evidence on the role of services in R&D and innovation policies suggests that these should not just take care of the volume of public spending addressed to services, but the public programmes design as well. The design of individual schemes as well as the appropriate policy mix might be different for different industries, i.e. innovation policies – at least to some extent – need to be customised. Even where there are no significant differences observed in the obstacles experienced by innovating service firms as compared to manufacturing firms, this does not necessarily lead to the conclusion that there is no need for a services innovation policy, as not all firms experiencing similar problems are necessarily in need of the same solution.

In our view the development towards service innovation policies at the various levels will be evolutionary rather than revolutionary. A menu-approach – taking respectively the assimilation, demarcation and systemic approaches as the point of departure – was offered here. Eventually, in every individual innovation system, a (temporary) mix of instruments and policies will have to be developed that suits it best.

By way of conclusion, we reiterate that innovation systems need to be designed in such a way as to cater much better for the needs of service innovators. There are numerous examples of innovative services and service functions (importantly so in manufacturing industries and firms) that are key assets of dynamic and adaptive innovation systems. All businesses in all sectors need service innovation to improve their competitiveness and employment- and wealth-creating capacity. In our view the systemic approach to services R&D is especially most promising, as it recognises this key role of innovative service functions in developing competitive innovation systems. In our view the current debate on service innovation and service innovation policies is still too much dominated by the assimilation and demarcation perspectives and lacks a vision as to how services can be better embedded in innovation systems, and how innovative services may contribute to the overall innovativeness and competitiveness of these innovation systems. This missing vision is to an important extent due to the persistence of old and new myths regarding the services economy (low productivity, low innovativeness, low tradability; see Gallouj, 2002; Rubalcaba, 2007) and the still predominant 'technological view' on innovation (Gallouj, 2002). Developing the emerging systemic approach further is needed to counterbalance this myopic view, and for developing new and more effective services R&D and innovation policies.

Notes

1. This contribution is based on learning on the job when performing innovation and policy studies in service innovation. We would like to thank collaborators on numerous projects, and especially Bart van Ark, Faiz Gallouj, Jeremy Howells, Ian Miles, Jari Kuusisto, John Sundbo, Bruce Tether and Paul Windrum. Many of the insights presented here are the result of joint research with these and numerous other persons.
2. In den Hertog (forthcoming) it is observed that the communities of researchers and statisticians, policy-makers and practitioners do interact and learn from each other. For a long time the three communities have been trapped in the same dominant view or technologist paradigm or view on innovation and to a certain extent their interaction reinforced this dominant paradigm. Changing the dominant paradigm regarding service innovation has proved an uphill battle and nothing less than a paradigm switch.
3. This section is largely based on den Hertog et al. (2008) and Rubalcaba (2006).
4. Extensions of these approaches have been done by Drejer (2002), Nahlinder (2002), RENESER (2006), Rubalcaba (2006) and den Hertog et al. (2008). The latter is partly reproduced here.
5. However, the 'productive-sectors (i.e. goods industries)-are-what-really-matters-approach' still pops up every now and then. Typical arguments used are, e.g., goods-oriented R&D is led by the most advanced countries; intangibility of service innovation results prevents their evaluation; results of service innovation can hardly be appropriated; support for innovation could distort competition in the service sector, etcetera (based on Rubalcaba, 2007).
6. The less important role given to services stems from the classical economic tradition, for which services do not generate any 'value', so they contribute little to productivity and innovation. This approach is echoed in the well-known innovation taxonomy by Pavitt (1984) where services were mainly depicted as receivers of the innovations developed in other sectors, i.e. supplier-dominated. Although some services (computing or telecommunications) were later recognised in an innovative role, these cases were considered to be notable exceptions.
7. Research carried out by Preissl (2000), Gallouj and Weinstein (1997) and Sirilli and Evangelista (1998) points in this direction.
8. In OECD (2005) this line of argument is followed when it is observed that there are no particular obstacles affecting service innovation that cannot be found in technological innovation in manufacturing industries. So similar problems would lead to similar solutions. This assumption is based on the statistical indicators used by European surveys where obstacles are isolated from modes, drivers and effects. However, differences

between goods and services in terms of modes, drivers and effects of innovation are significant and may justify different policy actions. Similar obstacles according to CIS indicators cannot be a necessary condition for arguing policy horizontality. For example, two companies may have similar degrees of difficulty in protecting their innovation results, but the requested solutions may be different: one has difficulties in producing a patent while another is looking for a different instrument. Similar obstacles according to CIS indicators cannot be a necessary condition for arguing policy horizontality.

9. All these factors are based on a twofold view, macroeconomic and political, unlike market and systemic failures focusing more on micro and meso angles. Even if there are interrelations among the different types it is possible that the identification of challenges derived from structural change, e.g. leading to new evolutions in service productivity influenced by the use or non-use of service innovation.
10. The classical argument for supporting private and collaborative (technological) R&D in firms is that through these externalities, social returns to investments made in R&D are higher compared to private returns for the firms making these investments. The resulting underinvestment in technological R&D and innovation is a market failure which could be corrected through supporting private R&D efforts.
11. For an introduction into market failure argumentation and its applicability to service innovation see van Dijk (2002), van Ark et al. (2003), Cruysen and Hollanders (2008) and RENESER (2006).
12. At the level of firms and industries, the artificiality can be observed as well by manufacturing firms developing into hybrid firms realising a considerable part of their turnover in service activities, and service industries developing into firms with a sometimes impressive technological capability.
13. Van Dijk differentiates between the following (overlapping) categories: intermediary services; capital-sharing services; network services; knowledge- or skill-sharing services; trade and repair services; transport and storage; post and telecommunication; banking and insurances; staffing agencies; business services and personal services (see Van Dijk, 2002, 7–10).
14. The discussion on the applicability of this type of argumentation for supporting non-technological innovations included here is based on den Hertog et al. (2008).
15. See van Dijk (2002), Lipsey and Carlaw (2002) and Navarro (2003) for more on these issues.
16. O'Doherty and Arnold mention five systemic failures and include market failure here as just one of the five types.
17. This is particularly important in the case of some regions where service innovation capabilities are rather poor, partly explained by the high concentration of KIBS in more developed regions, and few endowments are located in less developed regions. In these regions market and asymmetric failures apply with a higher degree of intensity than in other regions.
18. It should be noted that many of the arguments discussed in this section also apply to other services-related policies, not just innovation policies. For example, employment and training policies or regulatory policies for services may be based on similar arguments (Rubalcaba, 2007).
19. Unfortunately official statistics hardly record service innovation within manufacturing firms, so this is a limitation of this analysis. Therefore we will have to stick for the moment to the sectoral categories in current statistics and focus on service industries.
20. CIS3 and CIS4 are not fully comparable since, for example, the definition of an innovative firm is different. In CIS3 the manufacturing- and technology-biased criterion is very strong while in CIS4 non-technological innovations such as marketing and organisational innovation are recognised. Anyway, the variables related to public funding are quite similar and allow for some comparative analysis.
21. This bias in public funding against services is more important than the differences in terms of shares of innovative firms. On the latter, differences between manufacturing

and services are markedly lower, as can be noticed from Table 26.1. This suggests that the possible discrimination against services is not due to the fact that services companies are less innovative.

22. It should be noted here that initiatives taken and studies funded by the European Commission (see for example European Commission, 2003, 2007; Expert Group on Innovation in Services, 2007; Howells and Tether, 2004; PREST et al., 2006; RENESER, 2006; Kuusisto, 2007; Arundel et al., 2007; Cunningham, 2007; Tekes, 2007), OECD (2005, 2006; Tamura et al., 2005; and earlier, Pilat, 2001) as well as in individual vanguard countries such as Finland, Germany, Norway and Denmark (see Tekes, 2007 for some examples) have played an important role in starting off the discussion on service innovation policies in the first place.
23. This subsection summarises the argumentation included in RENESER (2006) and den Hertog et al. (2008) and is also included in den Hertog (forthcoming).
24. The importance of other policy areas for innovation, and the need to develop more integrated policies, are argued at length in Lengrand et al. (2002).

References

Antonelli, C. (1999), *The Micro Dynamics of Technological Change*, London: Routledge.

Antonelli, C. (2000), 'New information technology and localized technological change in the knowledge based economy', in M. Boden and I. Miles (eds), *Services and the Knowledge-Based Economy*, London, UK and New York, USA: Continuum, pp. 170–191.

Ark, B. van, L. Broersma and P. den Hertog (2003), 'Service innovation, performance and policy: a review', Research Series No. 6, Ministry of Economic Affairs, The Hague.

Arnold, J. and S. Kuhlman (2001), 'RCN in the Norwegian Research and Innovation System', Background Report No. 12 in the Evaluation of the Research Council of Norway, Royal Norwegian Ministry for Education, Research and Church Affairs, Oslo.

Arrow, K.J. (1962), 'Economic welfare and the allocation of resources for innovation', in R.R. Nelson (ed.), *The Rate and Direction of Inventive Activity*, Princeton, NJ: University Press, pp. 609–625.

Arundel, A., M. Kanerva, A. van Cruysens and H. Hollanders (2007), 'Innovation statistics for the European Service Sector', INNO Metrics 2007 report, European Commission, DG Enterprise, Brussels.

Boden, M. and I. Miles (eds) (2000), *Service and the Knowledge-Based Economy*, London: Continuum.

Bresnahan, T.F., E. Brynjolfsson and L.M. Hitt (1999), 'Information technology, workplace organization and the demand for skilled labor: firm-level evidence', NBER Working Paper No. W 7136.

Brynjolfsson, E. and L.M. Hitt (2003), 'Computing productivity: firm-level evidence', MIT Sloan Working Paper No. 4210-01.

Cainelli, G., R. Evangelista and M. Savona (2006), 'Innovation and economic performance in services: a firm-level analysis', *Cambridge Journal of Economics*, **30**, 435–458.

Coombs, R. and I. Miles (2000), 'Innovation, measurement and services: the new problematique', in J.S. Metcalfe and I. Miles (eds), *Innovation Systems in the Service Economy*, Dordrecht: Kluwer, pp. 83–102.

Cruysen, A. van and H. Hollanders (2008), 'Are specific policies needed to stimulate innovation in services?', INNO Metrics 2007 report, European Commission, DG Enterprise, Brussels.

Cunningham, P. (2007), 'Innovation in services', INNO Policy Trendchart Thematic Report, http://www.proinnoeurope.eu/admin/uploaded_documents/Thematic_Report_Innovation_Services_Nov_2007.pdf.

Dijk, M. van (2002), 'Marktfalen bij innovaties in de dienstensector', CPB Memorandum No. 50, Netherlands Bureau of Economic Policy Analysis (CPB), The Hague.

Dosi, G. (1988), 'Sources, procedures, and microeconomic effects of innovation', *Journal of Economic Literature*, **26**, 1120–1171.

Drejer, I. (2002), 'Business services as a production function', *Economic Systems Research*, **14** (4) 389–405.
Edquist, C. (ed.) (1997), *Systems of Innovation: Technologies, Institutions and Organizations*, London: Pinter.
European Commission (2003), 'The competitiveness of business-related services and their contribution to the performance of European enterprises', Communication from the Commission to the Council and the European Parliament, COM(2003)747, European Commission, Brussels.
European Commission (2007), 'Towards a European strategy in support of innovation in services: challenges and key issues for future actions', Commission Staff Working Document SEC (2007) 1059.
European Commission, Brussels, available at: http://ec.europa.eu/enterprise/innovation/doc/com_2007_1059_en.pdf.
Expert Group on Innovation in Services (2007), 'Fostering innovation in services', http://www.europe-innova.org/
Fageberg, J. (1988), 'Why growth rates differ', in G. Dosi, C. Freeman, R. Nelson, G. Silveberg and L. Soete (eds), *Technological Change and Economic Theory*, London: Pinter, pp. 432–457.
Forfás (2006), 'Services innovation in Ireland: options for innovation policy', report commissioned by Forfás from CM international, Promoting Enterprise, Trade and Science, Technology and Innovation For Economic and Social Development, Dublin.
Freeman, C. (1994), 'Critical survey: the economics of technical change', *Cambridge Journal of Economics*, **18**, 463–512.
Gallego, J. and L. Rubalcaba (2008), 'Shaping R&D and services innovation in Europe,' *International Journal of Services Technology and Management*, **9** (3–4), 199–217.
Gallouj, F. (1994), *Economie de l'innovation dans les services*, Paries: L'Harmattan, Logiques Economiques.
Gallouj, F. (2002), *Innovation in the Service Economy: The New Wealth of Nations*, Cheltenham, UK and Northampton, MA, USA: Edward Elgar.
Gallouj, F. and O. Weinstein (1997), 'Innovation in services', *Research Policy*, **26**, 537–556.
Green, L., J. Howells and I. Miles (2001), 'Services and innovation: dynamics of service innovation in the European Union', Final report, PREST and CRIC, University of Manchester, Manchester.
Griliches, Z. (1986), 'Productivity, R&D and basic research at the firm level in the 1970s', *American Economic Review*, **76** (19), 141–154.
Heijs, J. (2001), 'Justificación de la política de innovación desde un enfoque teórico y metodológico', Documento de trabajo No. 25, IAIF: Instituto de Análisis Industrial y Financiero.
Hertog, P. den (2000), 'Knowledge intensive business services as co-producers of innovation', *International Journal of Innovation Management*, **4** (4), 491–528.
Hertog, P. den (2010), 'Managing the soft side of innovation: do practitioners, researchers and policy-makers interact and learn how to deal with service innovation?' in S. Kuhlmann, R. Smits and P.H. Shapira (eds), *The Theory and Practice of Innovation Policy: An International Research Handbook*, Cheltenham, UK and Northampton, MA, USA: Edward Elgar.
Hertog, P. den, L. Rubalcaba and J. Segers (2008), 'Is there a rationale for services R&D and innovation policies?' *International Journal of Services Technology and Management*, **9** (3–4), 334–354.
Hipp, C. and H. Grupp (2005), 'Innovation in the service sector: the demand for service-specific innovation measurement concepts and typologies', *Research Policy*, **34** (4), 517–535.
Howells, J. and B. Tether (2004), 'Innovation in services: issues at stake and trends', Inno Studies Programme (ENTR-C/2001), Commission of the European Communities, Brussels.
Kox, H. (2002), 'Growth challenges for the Dutch business-services industry: international

comparison and policy issues', Special Study No. 40, CPB Netherlands Bureau for Economic Policy Analysis, The Hague.
Krugman, P. and M. Obstfeld (1994), *International Economics: Theory and Policy*, 3rd edn, New York: HarperCollins.
Kuusisto, J. (2007), 'R&D in Services: review and case studies', paper submitted for the CREST R&D in Services Working Group, DG Research, Commission of the European Communities, Brussels.
Lengrand, Louis & Associes, PREST and ANRT (2002), *Innovation Tomorrow. Innovation Policy and the Regulatory Framework: Making Innovation an Integral Part of the Broader Structural Agenda*, Innovation papers No 28, Directorate–General for Enterprise, Innovation Directorate, EUR report no. 17052, European Community.
Lipsey, R. and K. Carlaw (2002), 'The conceptual basis of technology policy', Simon Fraser University Department of Economics Discussion Paper, No. 02-6.
Metcalfe, S. and I. Miles (eds) (2000), *Innovation Systems in the Service Economy: Measurement and Case Study Analysis*, London: Kluwer Academic.
Miles, I. (1999), 'Foresight and services: closing the gap?', *Service Industries Journal*, **19** (2), 1–27.
Miles, I. (2005), 'Knowledge intensive business services: prospects and policies', *Foresight: The Journal of Future Studies, Strategic Thinking and Policy*, **7** (6), 39–63.
Miles, I., N. Kastrinos, K. Flanagan, R. Bilderbeek, P. den Hertog, W. Huntink and M. Bouman (1995), 'Knowledge intensive business services: their roles as users, carriers and sources of innovation', Report to the EC DG-XIII Sprint EIMS Programme, PREST/TNO, Manchester/Apeldoorn.
Muller, E. (2001), 'Knowledge, innovation processes and regions', in K. Koschatzky, M. Kulicke and A. Zenker (eds), *Innovation Networks: Concepts and Challenges in the European Perspective*, Heidelberg: Physical-Verlag, pp. 37–51.
Nählinder, J. (2002), 'Innovation in KIBS: state of the art and conceptualisations', *Arbetsnotat*, No. 244.
Navarro, L. (2003), 'Industrial policy in the economic literature: recent theoretical developments and implications for EU policy', Enterprise Paper No. 12, Commission of the European Communities, DG-Enterprise, Brussels.
O'Doherty, D. and E. Arnold (2003), 'Understanding innovation: the need for systemic approach', *IPTS Report*, **71**, http://ipts.jrc.ec.europa.eu/home/report/english/articles/vol71/TEC3E716.htm.
O'Mahony, M. and B. van Ark (2003), 'EU productivity and competitiveness: An industry perspective. Can Europe resume the catching-up process?' European Commission, Brussels.
OECD (2005), *Enhancing the Performance of the Services Sector*, Paris: OECD Publishing.
OECD (2006), *Innovation and Knowledge-Intensive Service Activities*, Paris: OECD Publishing.
Pavitt, K. (1984), 'Patterns of technical change: towards a taxonomy and a theory', *Research Policy*, **13**, 343–373.
Pilat, D. (2001), 'Innovation and productivity in services: state of the art', in OECD (ed.), *Innovation and Productivity in Services*, Paris: OECD, pp.17–54.
Preissl, B. (2000), 'Service innovation: what makes it different? Empirical evidence from Germany', in J.S. Metcalfe and I. Miles (eds), *Innovation Systems in the Service Economy: Measurement and Case Study Analysis*, Boston, MA: Kluwer Academic Publishers, pp. 125–148.
PREST, TNO, SERVILAB, ARCS (2006), 'The nature of RRD in services: implications for EU research and innovation policy', European Commission, DG Research, Brussels.
RENESER (2006), 'Research and development needs of business related service firms', final report to European Commission, DG Internal Market and Services, Dialogic/Fraunhofer/PREST/Servilab, Utrecht/Stuttgart/Manchester/Madrid.
Rubalcaba, L. (1999), 'Business services in European industry: growth, employment and

competitiveness', Office for Official Publications of the European Community, European Commission, Luxembourg.
Rubalcaba, L. (2006), 'Which policy for innovation in services?', *Science and Public Policy*, **33** (10), 745–756.
Rubalcaba, L. (2007), *The New Services Economy: Challenges and Policy Implications for Europe*, Cheltenham, UK and Northampton, MA, USA: Edward Elgar.
Rubalcaba, L. and D. Gago (2006), 'Economic impact of service innovation: analytical framework and evidence in Europe', Conference on Services Innovation, 15–16 June, University of Manchester.
Rubalcaba, L. and H. Kox (eds) (2007), *Business Services in European Economic Growth*, London: Palgrave Macmillan.
Schumpeter, J.A. (1939), *Business Cycles: A Theoretical, Historical, and Statistical Analysis of the Capitalist Process*, New York, USA and London, UK: McGraw-Hill.
Sirilli, G. and R. Evangelista (1998), 'Technological innovation in services and manufacturing: results from Italian surveys', *Research Policy*, **27**, 882–899.
Stiglitz, J.E. (1991), 'The invisible hand and modern welfare economics', NBER Working Paper No. 3641.
Tamura, S., S. Sheehan, C. Martinez and S. Kergroach (2005), 'Promoting innovation in services', Working party on Innovation and Technology Policy, 14 October, OECD.
Tekes (2007), 'Innovation policy project in services', IPPS, October, Finland, available at http://akseli.tekes.fi/opencms/opencms/OhjelmaPortaali/ohjelmat/Serve/en/cooperation.html.
Tether, B. (2005), 'Do services innovate (differently)? Insights from the European Innobarometer Survey', *Industry and Innovation*, **12** (2), 153–184.
Wood, P.A. (2001), *Consultancy and Innovation: The Business Service Revolution in Europe*, London, UK and New York, USA: Routledge.
Zambon, S. (ed.) (2003), 'Study on the measurement of intangible assets and associated reporting practices', prepared for the Commission of European Communities, DG Enterprise, Brussels.
Zenker, A. (2001), 'Innovation, interaction and regional development: structural characteristics of regional innovation strategies', in K. Koschatzky, M. Kulicke and A. Zenker (eds), *Innovation Networks: Concepts and Challenges in the European Perspective*, Heidelberg: Physica-Verlag, pp. 207–222.

27 The innovation gap and the performance gap in the service economies: a problem for public policy

Faridah Djellal and Faïz Gallouj[1]

27.1 Introduction

Whether at micro- or macroeconomic level, the main purpose of innovation is to increase economic performance. In economies now largely dominated by services, an analysis of this relationship reveals a paradox that a British government agency (NESTA, 2006) describes as an 'innovation gap'. The innovation gap measures the difference between the reality of innovation produced in an economy and what traditional innovation indicators perceive. In the British case, this has been expressed in the 1990s and early 2000s by the observation that a relatively weak innovativeness (compared to other countries) causes relatively high economic performance. This gap is explained in particular by the fact that a significant part of innovation in services (in particular its non-technological forms) escapes measurement using traditional tools (for example research and development, patents). This argument is not new, as it has been at the heart of economic literature on innovation in services since the early 1990s (Gallouj, 1994, 2002; Sundbo, 1998; Metcalfe and Miles, 2000; Miles, 2002).

However, in a services economy, the problem is not only in the definition and measurement of innovation. It also lies in the definition and measurement of performance. One can thus identify a second 'gap', that we propose to call 'performance gap'. The performance gap measures the difference between the reality of performance in an economy and the performance assessed by traditional economic tools (mainly productivity and growth). The performance gap reflects the hidden performance, invisible to these tools (performance in terms of sustainable development, from the socio-economic and ecological viewpoint).

The existence of these two gaps means that post-industrial economies are more innovative and more successful (effective) than traditional assessments suggest (or at least that the sources of innovation and performance exploited are more numerous than generally imagined). But this double gap also helps to blur the relationship between innovation and performance, and to question the legitimacy of some public policies that support innovation.

This chapter is based on two research issues that we have explored separately, from a theoretical, empirical and methodological viewpoint: the issue of innovation in services (Gallouj and Weinstein, 1997; Gallouj, 1994, 2002) and that of productivity and performance in this same sector (Djellal and Gallouj, 2007, 2008). We would here like to put these studies in perspective, in order to discuss the question of the relationship between innovation and performance (theoretically at this stage).

The aim of this chapter, then, is firstly to explain the innovation gap in the light of studies devoted to innovation in services (section 27.2). We will then explain the performance gap, which tends to widen in economies which are dominated by intangible, relational and cognitive products, and which are concerned with a desire for sustainable development (section 27.3). Section 27.4 is devoted to an analysis of the innovation–performance relationship. We will see that the two gaps identified blur this relationship, and that they are the source of a certain number of paradoxes (in particular a new productivity paradox which expresses the idea that there is R&D and innovation everywhere, except in productivity statistics), which has to be accounted for. These gaps also lead us to question certain public policies that support innovation. Section 27.5 concludes.

27.2 Innovation gap: invisible innovation

The innovation gap is a characteristic that is common to all contemporary developed economies. It expresses the observation that innovation efforts have been underestimated. The explanation lies in the fact that post-industrial economies (of quality, of knowledge, of information), produce many more innovations than are accounted for by traditional definitions and measurement tools. The problem derives from the fact that in these (post-industrial) economies, innovation is perceived according to industrialist and technologist definitions, and measured using industrial indicators. These include indicators for R&D expenditure, and the number of patents. We can therefore formulate the hypothesis that the more economies are tertiarised (this is the case in Great Britain), the wider the innovation gap. This section is devoted to a discussion on the different characteristics of this gap. Indeed we note, firstly, that it is certainly closely linked to service activities (section 27.2.1), but that it transcends sectoral borders (section 27.2.2). We then note that its scope (or the perception of this) is sensitive to a certain number of variables (section 27.2.3): the 'services' profile of the economy under consideration; the actors concerned; and the manner of addressing the innovation effort (output or input approach).

27.2.1 The innovation gap and services

The question of innovation in services has long been almost exclusively associated with that of technological innovation. This idea of innovation, which prevails in manufacturing industry, links innovation to the production of material artefacts. This is the reason why we have described it as technologist or industrialist (Gallouj, 1994, 1998), which others have subsequently expressed by the term 'assimilation' (Coombs and Miles, 2000) (see Chapter 1 in this volume). In the services sector, the assimilationist perspective is coupled with a subordination perspective (Djellal and Gallouj, 1999, 2001). In fact, innovation is seen from the viewpoint of the adoption of technical systems and generally not from the viewpoint of their production. In other words, the services sector adopts technical systems that are produced in the really innovative and dynamic sector that is the manufacturing industry.

It is this dominant technologist concept of innovation in services that causes a significant share of the gap. Indeed, it does not take account of numerous efforts at innovation undertaken in services, which escape the traditional analytical tools.

The first research strategy implemented to fill the innovation gap opened by the assimilationist perspective, and to allow identification of the forgotten or hidden innovation forms, is the perspective that we have also characterised elsewhere as service-oriented (Gallouj, 1994, 1998), but that one could call a 'differentiation' or 'demarcation' perspective (see Chapter 1 in this volume). This research programme firstly emphasises the specificities of the nature of innovation.

This specificity of nature can be approached deductively. Indeed, the theoretical characteristics of services (in particular, their intangibility, their interactivity, etc.) are ideal-types, which allow one to formulate a certain number of hypotheses on the specificities of innovation in services. Thus, the vague and 'dynamic' nature of the output has several consequences. It leads to a blurring of the boundaries between the different common analytical categories (product, process, organisation); it leads to problems in counting the innovations, and problems in evaluating the economic impacts of innovation. It also facilitates imitation. Likewise the interactive (or co-produced) character of service has consequences for the nature of innovation and its modes of organisation and appropriation.

But, of course, this specificity is above all addressed in an inductive manner. Empirical studies are becoming more numerous; these seek to emphasise, in various services sectors, the particular forms of innovation which elude traditional (assimilation) perspectives (see Chapter 1 in this volume for a survey). A certain number of theoretical studies have also emerged which put forward local theories (that is, adapted to certain sectors of service) or which question existing sectoral taxonomies (see also Chapter 1 in this volume).

27.2.2 The innovation gap beyond services

This underestimation of innovation concerns not only services. In other sectors as well, certain innovation efforts escape traditional definitions and assessments, thereby helping to fuel the innovation gap. The general explanation for this phenomenon is the growing increase in power of the service dimension in all economic activities, and the trend towards blurring the borders between goods and services. New information and communication technologies (NICTs) (as a technical system shared between manufacturing and services) contribute to this 'blurring' (Broussolle, 2001).

In recent years, studies showing a certain natural convergence (integration) between goods and services have become more numerous. Indeed, service or information is the main component of many goods. A certain number of research activities have therefore been devoted to identifying and measuring the informational or service value of goods, whether industrial goods, for example automobiles (Lenfle and Midler, 2003), agricultural goods (Muller, 1991; Le Roy, 1997; Nahon and Nefussi, 2002) or construction (Carassus, 2002; Bröchner, 2008a, 2008b).

Moreover, numerous other studies have emphasised the transition from an economy of production and consumption of goods to an economy of production and consumption of hybrid solutions or packages. This means that goods and services are less and less sold and consumed independently, but are increasingly sold as solutions, systems, complexes or functions. This mode of the integrating perspective is favoured by the works of Bressand and Nicolaïdis (1998), which identify a shift from an economy of products to an economy of functions; the works of Furrer (1997), which emphasise 'services around products' (services provided as complements to tangible products); and the works of Barcet and Bonamy (1999), which analyse goods and services in terms of 'usage rights' or 'credence rights' (see also Chapters 28 to 31 in this volume).

Thus, in view of the evolution of the nature of the activity, numerous 'non-technological' innovations, which are implemented in manufacturing or in agriculture, also elude measurement. In the case of manufacturing industry, for example, there are many service innovations around product, whether in pre-sales, at point-of-sale, in after-sales services or independent of sales services[2] (Mathieu, 2001; Vandermerwe and Rada, 1988; Davies, 2004). In the case of agriculture, one can quote examples of the many innovative modes of rural tourism.

27.2.3 Some variables that influence the innovation gap

As we emphasised in the introduction, the innovation gap depends on the scope of 'problematic' activities. It is, therefore, correlated with the level of tertiarisation of the economy, tertiarisation that is defined not only as an evolution of the share of the tertiary sector but also, more generally, as an evolution in the service content of activities, beyond the tertiary sector. But qualitative variables also affect this gap (or the perception of it). Here we refer to three of these.

Firstly, insofar as certain service activities are more innovative than others, the sectoral distribution of services in an economy has an impact on the scope of the gap. The innovation gap is both vertical and horizontal, one could say. Thus we could assume that an economy characterised by a higher proportion of knowledge-intensive business services (KIBS) than another will be more affected by the gap. Indeed, amongst the services, KIBS are not only the most innovative, but they also contribute to innovation carried out by their customers. But in the two cases (own and induced innovation), the non-technological (invisible) forms occupy a key position (Djellal and Gallouj, 1999). It would thus be interesting, from the innovation viewpoint, to identify the different worlds of services, by being inspired by the work carried out by Gadrey (2005). Depending on the distribution of services in an economy, one would therefore be able to identify the more or less innovative profiles of service societies (in which the gap would be more or less pronounced).

Secondly, awareness of the scope of the gap varies according to the actors concerned (managers, public authorities, social science researchers). Indeed, firms' actors often (if not always) are very aware of the importance of their activities for adapting and changing, even if they do not always call them innovation. Academic works have a lower degree of awareness and are often subject to analytical inertia. The latter are, nevertheless, ahead of international statistical institutions and the public authorities. An analysis of bibliographical references would not find it difficult to measure the temporal gaps between suggestions made by the theoretical literature and taking these into account in national or international measuring institutions, and in public policies. The different revisions of Organisation for Economic Co-operation and Development (OECD) manuals mentioned in the following paragraph bear witness to this to some extent.

Thirdly, efforts have been made in recent years to reduce the innovation gap. But these efforts were more concerned with the reduction in the (innovation) output gap than with that of the (innovation) input gap.

Empirical studies carried out in the last few years (see section 27.2.1) have contributed towards making people aware of the extent of 'invisible' innovation (using traditional tools) in services. Thus, typological studies have been undertaken on different service activities (consultancy, financial services, hospitals, retailing, transport, etc.), aimed one way or another at exploiting, in the definition of innovation, the Schumpeterian opening tradition. Some of these concerns have been included in successive revisions of the OECD Oslo Manual (manual of innovation indicators).

Thus, in its 1992 edition, the *Oslo Manual* (OECD, 1992) only covers technological process and product innovation. Applying its guidelines in a certain number of surveys devoted to services encounters numerous difficulties, notably that of the distinction between product and process innovation. This leads to a systematic underestimation of innovation in services, particularly when it is not directly linked to a material technology. The *Manual* that was revised in 1997 (OECD, 1997) shows obvious concerns on integrating services into the surveys, but in the end it retains restrictive and technologist definitions of innovation in services. Taking account of services is finally expressed by three main amendments: the introduction of a paragraph which indicates that the term 'product' is used to refer to both goods and services; a warning against the difficulty, in certain cases, of distinguishing product innovations from process innovations; and the introduction of a box giving examples of innovation in services. Finally, in its latest version (OECD, 2005b), which still has very reduced empirical applications, as well as product and process innovations,[3] the *Manual* distinguishes marketing innovation and organisational innovation.

Thus, successive Community Innovation Surveys (CISs) have been opened up to an increasingly large number of services and to certain 'non-technological' forms of innovation. The output gap has tended to reduce, but progress still remains to be achieved, particularly in the area of social innovation (Dandurand, 2005; Chapter 9 in this volume) and in certain forms of frequent ad hoc and tailor-made innovation, in particular in the area of knowledge-intensive services (consultancy, some aspects of financial services). But this improved recognition of the diverse forms of innovation increases the difficulty of the problem of innovation appropriation regimes. The inadequacy of measuring innovation by patent is increased as the 'intangible' forms of innovation are integrated. The gap is then reduced on one side, but widens on the other.

On the other hand, regarding innovation inputs, the specificities of services are still insufficiently taken into account. Contrary to the *Oslo Manual*, which was able gradually to integrate the non-technological dimensions of innovation (in particular organisational), the latest version of the *Frascati Manual* (OECD, 2002) continues to be characterised by a technicist and scientist bias. But R&D activities in services often have a composite character, mixing aspects of sciences and technologies, social sciences, organisational engineering, etc. (Djellal et al., 2003). The last but one dimension mentioned is not sufficiently taken into account, and the last is not considered at all, which contributes towards widening the input gap.

27.3 Performance gap: hidden (or missing) performance

A country's performance is generally measured by the growth rate of its gross domestic product (GDP), which is considered to be closely linked to productivity gains. The National Endowment for Science, Technology and the Arts (NESTA) analysis (NESTA, 2006), mentioned in the introduction, is based on this definition of performance, thus considering that the second variable of the innovation–performance relationship does not pose (too much of a) problem. But this is far from being the case. Indeed, economic performance, as well, raises serious problems of definition and measurement, and one can also identify hidden performances here. These hidden performances are also linked to the service nature of activities. Contemporary developed economies are therefore not only faced with an 'innovation gap', but also with a 'performance gap'.

Critical analyses of concepts of productivity and growth are often dealt with in similar terms, insofar as, in the two cases, it is the nature of the product which is mainly at stake. The terms of this critical debate can be divided into two groups of arguments, one concerned with errors of measurement and the other, more fundamentally, with its conceptual invalidity (Gadrey, 1996; Djellal and Gallouj, 2008). The first set of arguments calls into question the results of studies and suggests corrections; the second challenges the concept itself and suggests it should be abandoned.

27.3.1 The measurement error and correction argument

These measurement errors have been spectacularly highlighted by some studies. In the USA, for example, the Boskin Commission Report (Boskin, 1996) confirmed that the consumer price index had been seriously overestimated and that productivity gains and growth had consequently been underestimated. It goes without saying that all the economic policies and scenarios developed on the basis of these erroneous data are problematic; if not actually doomed to failure, they are subject at the very least to considerable uncertainty.

The measurement errors can be explained by factors that are exogenous or endogenous to the indicators used; these factors may of course be combined. In a given socio-economic environment (in which the exogenous factors are stable), the endogenous factors are linked to the characteristics of the indicators used and to the difficulty of compiling (reliable) data, particularly on services. The numerous technical difficulties encountered in defining and measuring output, input, etc. and the difficulties of aggregating data (especially but not solely in services) give rise to measurement errors. These problems are the reason why there is such a diversity of techniques for measuring output volumes and productivity, particularly in national accounts. They also cast doubt on some international comparisons. Thus it would seem, for example, that the choices made by various countries in respect of the base year adopted, calculation of the price index and the adjustments required to take account of variations in quality, give rise to not insignificant differences in the values for national growth rates (Eurostat, 2001). These differences become problematic, for example, in the context of the 'Stability and Growth Pact' adopted by the European Council in July 1997, which requires member states to keep their public deficit below 3 per cent of GDP. These are fundamental problems in economic and monetary policy that spurred the European Commission to draw up a *Handbook on Price and Volume Measures in National Accounts* (Eurostat, 2001). The doubts raised here relate to the methods and conventions used in the calculations and not to the indicator itself. The raising of these doubts has led to the adaptation and harmonisation of the statistical tools used and to the correction of the measurement errors.

The exogenous factors, for their part, concern the fundamental changes affecting contemporary economies, which are causing chronic difficulties for the indicators used to measure productivity. To put it very simply, we are dealing here with the transition from a Fordist to a post-Fordist economy based on high-quality production and knowledge. The indicators in use are rapidly being rendered obsolete by the dynamic of contemporary economies (extremely rapid changes in quality; principle of permanent innovation).

These exogenous factors are important sources of measurement errors. The difficulties of constructing indicators are becoming real headaches in 'quality' and knowledge economies. Consequently, all the stops have to be pulled out in order to find technical solutions and to correct the habitual errors. However, these exogenous factors also sometimes cast doubt on the conceptual validity of the notion of productivity.

27.3.2 The (total or partial) conceptual invalidity of the notion of productivity and its abandonment

In some situations, the concept of productivity quite simply loses its validity. No amount of technical adjustments can resolve this problem. The only solution is to stop using this concept in order to evaluate the performance of an individual, a team or an organisation. At the microeconomic level, such a situation may arise in areas characterised by considerable informational asymmetries where moral hazard comes into play. This is the case, for example, with certain support functions, such as maintenance and IT development, and with intellectual planning and steering functions. Such a situation may also arise in areas characterised by strong service relations (particularly social and civic relations). In these various areas, service quality and productivity may become contradictory objectives.

The customer or user qualitative structure has effects on the nature of the service provided and on productivity.

In reality, several different cases can be identified. In the first, the concept of productivity has no meaning, since it is irrelevant to the main issues at stake, which lie elsewhere (creativity, quality of solution, etc.). This applies to the wind quintet concert suggested by Baumol, as well as to all forms of artistic creation.

In the second case, the concept of productivity does not necessarily lose all its validity, but no longer retains its position of supremacy. This might be described as a partial invalidation. This case reflects the difficulties that arise when the industrial concept of productivity comes up against what is known as the information or knowledge economy or society. The knowledge society is, after all, characterised by a sharp increase in the cognitive content of economic activities (knowledge being not only their input but their output as well) and by a proliferation of service relationships between providers and clients. The problem this raises is how to measure the productivity of social relations, on the one hand, and of knowledge, on the other. Now in such an economy (which Karpik, 1989 calls a 'quality economy'), the quantities or volumes of output and prices matter less than their long-term useful effects, otherwise known as outcomes. A lawyer's productivity is of no significance if it ends in judgments that are unfavourable to his or her clients; that of a doctor is of little importance compared with the results of the treatment provided; and a researcher's productivity means nothing unless it is compared with the quality of the results obtained. In all these cases, in which the outcome is subject to considerable uncertainty (where there is a high level of informational asymmetry), the mechanisms that produce trust are more important than any measurement of output or productivity. The (partial) conceptual invalidity argument now applies to many more economic activities than the total invalidity one. After all, the knowledge society seems to be a universal phenomenon. It manifests itself not only in services but also in manufacturing industry, where there has been an increase in service activities that has been described as an 'intensification of the symbolic activities and social interactions implied by the productive process' (Perret, 1995). Although this partial conceptual invalidity argument may apply to very diverse activities, it particularly affects knowledge-intensive service activities.

A third case is that in which the concept of productivity could possibly be meaningful if the environmental variables could be taken into account. In other words, the concept loses its validity when applied to interorganisational comparisons and benchmarking exercises. However, it could retain its validity if comparable organisations were to be compared or if environmental variables were taken into account (although in doing so we would be replacing measurement by productivity with a multi-criteria evaluation process).

27.3.3 The need for a multi-criteria evaluation

Nevertheless, nobody is suggesting that the criterion of productivity (or, at the macroeconomic level, the closely associated one of growth) should be abandoned completely. The usual recommendation is to abandon the absolute power (whether on the theoretical or operational level) of a single ratio (productivity or growth) and replace it with a pluralist and flexible evaluation system (in which simply abandoning the concept of productivity would, under certain circumstances, be a possible, albeit extreme option).

Abandoning the absolutism of productivity (and of growth) is justified by a number of arguments, outlined above, that cast doubt on the validity of the concept in certain situations. Regardless of the activity in question, indeed, productivity is always inaccurately estimated (although to varying degrees depending on the activity). It suffers from chronic mismeasurement. However, there are other arguments that also cast doubt on the absolutism of productivity (and of growth) and militate in favour of a pluralist approach.

Thus, firstly, in a given economic activity, performance is not an objective category but rather is considered in different, even contradictory terms depending on the actors concerned (individuals, firms, political authorities). The subjective nature of performance, which certainly applies to tangible goods, is particularly pronounced in the case of the 'goods' produced by the information and knowledge economy, which are based on intangible, abstract and socially constructed factors of production.

Secondly, account also has to be taken of the perverse effects of certain goals or targets. For example, at both the macro- and microeconomic level, the drive for growth and productivity generates negative externalities. It may give rise to certain social or environmental costs (stress and other health problems on the one hand, environmental degradation on the other) that are not taken into account in estimates of growth and productivity (Jex, 1998; Karasek and Theorell, 1990; Lowe, 2003). At the microeconomic level, the frequently criticised link between overly aggressive productivity strategies and a deterioration in quality is well known. In the administration of justice, attempts to rationalise processes (reduction in time taken to deal with cases) are acceptable only if they can be achieved without detriment to the rights of the accused. A productivist approach could sow the seeds for wrongful convictions, for example by generating excessive pressure to obtain confessions.

Thirdly and more generally, the level of production of goods and services is not the only indicator of a society's well-being. Nor is it necessarily the best one. Alternative macroeconomic indicators of development are now being developed, which could be adapted for use at the level of firms and organisations (for a survey, see Gadrey and Jany-Catrice, 2007). One of the best-known of these indicators is probably Osberg and Sharpe's index of economic well-being (see Sharpe, 2000) which is made up of variables associated with the following four components of economic well-being: consumption flows, capital accumulation, inequality and poverty and economic insecurity. Others include the Index of National Social Health developed by the Fordham Institute as an alternative to GDP, and various indicators of sustainable well-being (ISEW, Index of Sustainable Economic Welfare).

Fourthly, for other activities (particularly at the intra-organisational level), comparisons of productivity are unfair, counterproductive and discouraging for the units in question. This is because they carry out their activities in environments that may differ considerably from each other, making mechanical comparisons very difficult. This applies, for example, to comparisons of post offices or schools located in very different socio-economic environments.

In certain cases, fifthly and finally, the concept of productivity loses its validity or, without losing its relevance entirely, becomes insignificant compared with other aspects of performance. Thus the productivity (technical efficiency) of health and social services is a secondary issue compared with outcomes as essential as containing outbreaks of serious epidemics for examples.

27.3.4 A multi-criteria framework for evaluating output and performance

We hypothesise that the various purposes or 'outputs' of service activities can be linked to different 'worlds' (that is, sets of outputs or of concepts of outputs and criteria for evaluating those outputs). Drawing freely on the work of Boltanski and Thévenot (1991), it is suggested that services can be defined and evaluated on the basis of different sets of justificatory criteria, which equate to the following six worlds:

- the industrial and technical world, the outputs of which are described and estimated mainly in terms of volumes, flows and technical operations;
- the market and financial world, the 'output' of which is envisaged in terms of value and monetary and financial transactions;
- the relational or domestic world, which values interpersonal relations, empathy and relationships of trust built up over time and

Table 27.1 A multi-criteria framework for analysing service output and performance

	Industrial and technical world	Market and financial world	Relational or domestic world	Civic world	Innovation world	Reputational world
Direct output (short term)						
Performance relative to direct output						
Indirect output (long term)						
Performance relative to indirect output						

regards the quality of relationships as a key factor in estimation of the 'output';

- the civic world, which is characterised by social relations based on a concern for equal treatment, fairness and justice;
- the world of innovation (the world of creativity or inspiration);
- the world of reputation (the world of brand image).

Table 27.1 illustrates this framework, which does justice to the multiplicity of service 'products' or 'outputs' by combining space–time analysis with symbolic space.

This digression on product diversity is intended to highlight what is our primary concern here, namely the diversity of performance. After all, if the 'generic outputs' are different, and given that performance is defined as the improvement in the 'positions' or 'operating efficiency' relative to the various outputs, it is not difficult (at least in theory) to accept the existence of a plurality of (generic) performances associated with (generic) outputs considered in their two facets ('volume' and quality).

Just as with outputs, therefore, several types of performance can be identified, depending on the 'families' of criteria adopted for the purposes of definition and valuation: industrial and technical performance (in which the main criteria are volumes and flows), market and financial performance (in which the emphasis is on monetary and financial operations), relational performance (in terms of interpersonal ties), civic performance (in terms of equality, fairness and justice), innovation performance (in terms of the planning and implementation of innovative projects) and reputation performance (in terms of brand image).

The question of performance can also be considered in terms of the time-frame of the evaluation (short term, long term) or even from the point of view adopted in making the evaluation (the user's or the service provider's).

Contrary to certain prejudices, civic, relational, reputational and innovation performance are not immune to all forms of quantification. True, it may seem paradoxical, for example, to consider social or civic relations (which are generally associated with disinterested attitudes or the gift–counter-gift principle) in terms of performance (a notion with strong technical and commercial connotations). The intention is not of course to measure relationship intensities, particularly since sociologists have drawn attention to the composite nature of the service relationship, which is regarded as a locus for the verbal exchange of technical and market information and signs of civility and mutual esteem (Goffman, 1968). On the other hand, there is no reason why the length of time spent in the relationship or even, once the content has been examined, the quantity of relations of each type, cannot be measured. Thus improvements in (internal or external) customer satisfaction indicators and reductions in user turnover can be regarded as indicators of relational performance, while the evolution of the production and share of social quasi-benefits make it possible to some extent to monitor the evolution of civic performance. Similarly, the rate of (incremental) innovations introduced and the resolution rate for problems encountered during the test phase of an innovative project or even the share of solutions codified (routinised) and transferred to a generalised application can serve as indicators of innovation performance.

Table 27.1 depicts 12 different concepts of performance, which may mutually reinforce each other or, conversely, contradict each other, or at least may do so once a certain threshold has been crossed. For example, an increase in technical performance may give rise to an increase in market performance. It is likely, therefore, that an increase in the number of accounts opened per employee in a bank will be accompanied by an increase in net banking income per employee. Similarly, an improvement in relational performance (manifested, for example, in an increase in the customer retention rate) may have a positive influence on market performance. These different types of performance may also be negatively linked, since individual pairs of them may clash with each other. For example, a good civic performance (high rate of social quasi-benefits) may lead to a deterioration in a competitiveness or productivity (technical performance) indicator.

Overall, at the micro level as at the macroeconomic level, defining and measuring performance pose serious problems, and these result in a performance gap. These problems, as with those linked to the innovation gap, should be regarded with care when drawing up public policies.

27.4 The double gap, the innovation–performance relationship and public policies

In a service economy, the definition and measurement of innovation, as performance, raises numerous difficulties, as we have just observed. They are the cause, not only of an innovation gap, but also of a performance gap. We will now compare these two gaps and, at the theoretical level, examine their consequences on the fundamental relationship between innovation and performance; and at the operational level, their implications in terms of public policies.

The fundamental hypothesis of the analysis is that innovation efforts in a post-industrial economy are always underestimated. A consensus now seems to have been established on this point, as an increasing number of theoretical and empirical works bear witness but also, and particularly, the many revisions of the OECD official manuals (see section 27.2). The specificities of innovation in services are recognised, even if the inertia of our analytical tools and technical difficulties can prevent them from being taken into account, for example in surveys. On the other hand, a consensus on the nature, scope and challenges of the performance gap is far from being achieved. It is true that performance, considered from the viewpoint of productivity and growth, has always been at the heart of all economic theories, whatever they are, old or new, orthodox or heterodox. It is therefore subject to a major effect of cognitive irreversibility.

In view of these differences in the perceptions of gaps, it is necessary to consider several possible scenarios, to examine the consequences of these on the innovation–performance relationship, all other things being equal. The first case (the most frequent) is that in which one believes that performance is defined satisfactorily by productivity and growth. Public policies supporting innovation are based on this canonical scenario. The second case is where one assumes that the performance is badly defined (and underestimated), in other words that there is a performance gap. We will examine these two scenarios, as well as their consequences for public policy.

27.4.1 Performance is (considered to be) well defined

National and international policies supporting innovation (as, also, the economic theories that inspire them) are based on this hypothesis, according to which performance can be reduced to growth (and to productivity).

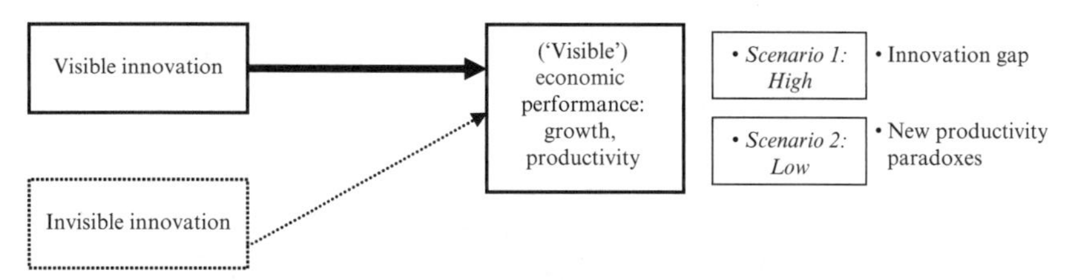

Figure 27.1 The innovation gap and the innovation–performance relationship

The discussion and possible theoretical or operational problems only therefore concern the innovation variable. For a given innovation effort, this hypothesis allows one to consider two interesting scenarios, which differ depending on the levels of performance achieved (Figure 27.1).

The first scenario is that which corresponds to a high economic performance for a given (visible) innovation effort. This scenario may wrongly lead to the impression being given of a high output of a country's visible innovation effort, while in fact part of the performance is explained by invisible innovation. In the case of Great Britain, for example, NESTA (2006) observes a high economic performance in the 1990s and early 2000s for a lower level of innovation than in other countries. For example, research and development (R&D) per capita expenditure in Great Britain is two times lower than in Sweden and in Finland. It is lower than in France or in Germany. The number of patents per inhabitant is much higher in Germany, Japan and the United States than in Great Britain. The explanation of this paradox lies in the British innovation gap. In fact, part of the performance can be explained by the invisible innovation effort carried out in particular (but not exclusively) in the services sectors.

The second scenario is that which corresponds to a weak economic performance (growth) for a given innovation effort. In fact the situation is then still more unfavourable than it appears (and it will be necessary to draw conclusions in terms of public innovation policies), since the level of real innovation is higher than the measures considered indicate. Invisible innovation efforts combined with visible efforts are not effective. Therefore, to paraphrase the Solow paradox, we can here formulate a new productivity paradox: there is innovation everywhere (including invisible innovation) except in performance statistics. The NESTA report (2006) does not take account of this second scenario, which does not correspond to the British situation during the period covered. On the other hand, it is possible that it illustrates the French situation. This new paradox of productivity can take a particular form if one considers innovation from the restrictive viewpoint of R&D input. Indeed, the concept of R&D as

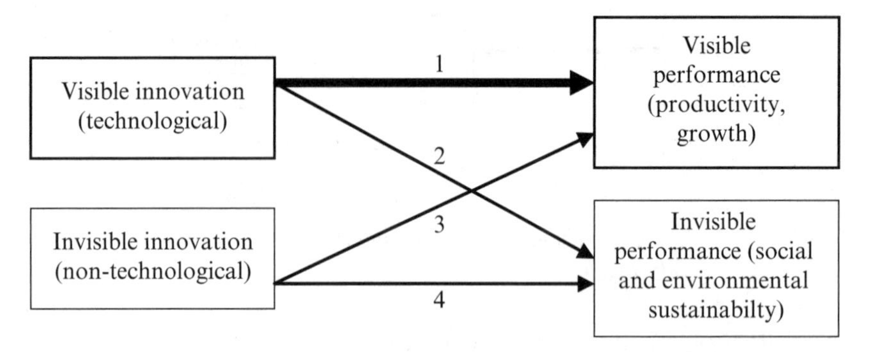

Figure 27.2 Innovation gap, performance gap and innovation–performance relationship

defined in the *Frascati Manual* is not adapted to services. If we accept an improved definition of this (such as that which we proposed in section 27.2.3), one can infer from this that R&D efforts in services are underestimated, and that therefore there is R&D everywhere (including in services) except in the statistics of productivity and growth.

27.4.2 The performance is badly defined

As we have previously emphasised, a certain number of recent studies question the dictatorship of productivity, of GDP and growth, by considering that they are neither the only, nor the best indicators of the economic performance of a country. Thus, just as there is invisible innovation, so there would be invisible performance. This invisible performance mainly concerns the field of socio-economic and ecological sustainability. It expresses concerns in terms of human development, social cohesion, equality, equity, environmental protection and outcomes rather than outputs. This second hypothesis is not taken into account by the NESTA (2006) analyses, whereas it can change the economic diagnosis and lead to important consequences for public policy (Figure 27.2)

This question of taking into account invisible performance is at the heart of a certain number of works focused on international comparisons, which are not concerned with the question of innovation and the innovation–performance link. This is the case, for example, for works devoted to international comparisons of the levels of well-being and development. These works identify the sometimes dramatic differences between GDP growth and the evolution of other indicators, in particular human development and social progress indicators. These kinds of comparisons often emphasise the superiority (from this viewpoint of alternative 'growth') of the Scandinavian socio-economic models, compared to the

Anglo-Saxon models. Here we are interested in the role of innovation in this performance, and notably in the (theoretical) consequences (on the analysis of the innovation–performance relationship) of taking this new gap into account.

Thus visible innovation certainly leads to visible performance (relationship 1), but it can also result in an invisible performance with regard to socio-civic and ecological sustainability (relationship 2). Technological innovation can indeed also be a source of social, civic and ecological benefits, and certain technological trajectories are more guided than others by the search for socio-economic or ecological sustainability. The informational paradigm, for example, is often considered not only as a source of economic growth, but also as a source of socio-economic and ecological sustainability. Insofar as the ICTs are considered to be weak MIPS[4] technologies, it may be thought that they favour sustainability and that, more generally, the information society is congruent with sustainable development. We can quote more precise examples of this relationship between ICT and sustainable performance: the substitution of the videoconference for business travel; the introduction of new ways of working (for example teleworking). The ICTs also operate in other dimensions of sustainability (in particular social). They thus allow the public authorities to be challenged and to mobilise citizens rapidly. Apart from the informational paradigm, technological innovation, whatever it is, can produce a more or less significant environmental or social benefit, beyond traditional growth. Within these material technologies, then, one can distinguish environmental technologies and social technologies. For example, technological innovations responding to the problems of the elderly (domestic robots, smart home, electronic surveillance) represent a powerful innovation trajectory in ageing service societies (Djellal and Gallouj, 2006).

Relationship 3, which links invisible innovation to visible performance, means that the non-technological forms of innovation are also a source of growth (visible performance). It is the reason (when invisible innovation efforts are significant) for the incorrect interpretation of the innovation–performance relationship (mentioned previously), which takes account of high growth for a relatively weak innovation effort.

Relationship 4, finally, which links invisible innovation to invisible performance, assumes a favoured relationship between non-technological innovation and invisible performance. There seems to be a strong correlation between the invisible component of innovation and the invisible component of performance. Indeed proximity services, for example (in particular when they are implemented by non-profit-making organisations and public services, in particular local services), are the setting for significant social innovation activity, which escapes traditional indicators, whereas their role in the resolution of social problems is fundamental. More generally, if one considers performance from the viewpoint of sustainability, one notes that although they are not dramatic, many non-technological, and particularly social, innovations play a significant role in this. There are many examples of non-technological (invisible) innovations which contribute towards sustainable performance and which cover all kinds of services. Amongst others, we can mention certain forms of sustainable tourism, the many innovative initiatives in the field of care for the elderly, childhood, social integration, and in the financial field, micro-credits to respond to the problem of banking exclusion, and government-subsidised loans to encourage firms to invest in environmentally friendly equipment. We can also mention local authorities developing 'one-stop shop' services for people in difficulty. The innovations produced by knowledge-intensive business services (KIBS) often come under 'environment-friendly' and sustainable trajectories, particularly when they concern ad hoc solutions provided by consultants to environmental or social problems, investment in new areas of expertise (for example, environmental law, social law, advice on sustainable development) or methodological innovations (for example, the MIPS indicator mentioned previously).

Because of the existence of hidden performance, innovation efforts can be more effective than the measures indicate. Thus, for given innovation efforts, an apparently weak (traditional) performance can be enhanced from the viewpoint of alternative performance. Conversely, an apparently high (traditional) performance can be put into perspective, insofar as growth and productivity gains are tarnished by ecological or social damage.

27.4.3 The double gap: a challenge for public policies

In view of the two gaps (innovation and performance) identified in this work, one can assume that public innovation policies are, to a certain extent, inadequate. Indeed, they rely on a partly inaccurate analysis, and consequently suggest solutions that could prove to be inappropriate.

In order to carry out their diagnosis, public policies generally favour relationship 1, which links visible innovation (mainly technological innovation: that which is based on R&D and which gives rise to patents) to visible performance (growth, productivity). One of the major indicators of the Lisbon Agenda (that is, the 3 per cent target for R&D/Gross Domestic Product) perfectly reflects such a relation. Figure 27.2 illustrates well all the errors in analysis and the paradoxes that can follow from such a hypothesis. We can thus identify (all other things being equal) a weak innovation effort at the same time as a high (growth) performance. This is the diagnosis achieved by NESTA (2006) in Great Britain for the 1990s and early 2000s. We can also identify an apparently higher innovation effort, which does not fulfil its promises on performance. This is the case for France in the same period. To establish a satisfactory analysis, it is necessary to take into account all the other relationships between innovation and performance (relationships 2, 3 and 4), which can contribute to different interpretations of innovation efforts and levels of performance achieved.

In view of the diagnosis established on the basis of relationship 1, the solutions recommended by the public authorities naturally consist of promoting technological innovation, that which is based on scientific and technical R&D activities and which can be appropriated by patents. These strategies mainly concern public research and the industrial sectors, in particular high technology. As regards training systems, policies will consist of favouring scientific and technological training. As the OECD (2005a) emphasises, the innovation policy of member countries was mainly considered to be an extension of R&D policies. Thus, in economies that are, however, largely dominated by services, these technologist and industrialist policies have also been transposed to services. In the same way as economic analysis (see section 27.2.1), public policies of support for innovation in services are dominated by an assimilationist perspective (Rubalcaba, 2006).

The main lesson to draw from the preceding analyses in terms of public policy is that, to take into account the double gap that has been identified, the public authorities should break with their technologist orientation and try to promote invisible innovations and performances.

It is thus necessary to emphasise innovation and R&D policies that are specific to services (perspective of differentiation); in other words, policies that are not content with supporting technological innovation and R&D, but which also favour non-technological forms of innovation and R&D. As we have emphasised in section 27.2.2, as far as the source of the gap is not confined to services, it is also necessary to support innovations in services within the manufacturing and agricultural sectors. If it happens, this recognition of invisible innovation in public policies should also redirect priorities on education policy. Indeed, one should also support the development of the necessary skills in non-technological forms of innovation, whether this is skills that produce or that absorb these innovations. These skills do not only concern an elite, they should be disseminated to all levels of the population. This is particularly obvious with regard to social innovations that can be produced and implemented in the informal and domestic sphere (voluntary work, community organisation) as in the formal sphere (or social entrepreneurship). All services of course are concerned by these innovation policies. But some sectors appear to be more concerned than others. This is the case with the KIBS, which contribute strongly to the innovation gap, both through their own internal non-technological innovation, but also by that which they produce for their customers. This is also the case for the numerous proximity services, where many social innovations are implemented.

If one considers performance in terms of sustainable development, one again notes that it is the technologist or assimilationist perspective which dominates. Most of the public policies of induction of sustainable innovation fall within such a perspective, which consists of supporting sustainable technological innovations in different ways: funding, taxation (for example, by granting tax credits for clean technologies or which save energy), public orders, and the dissemination of information. In order to favour invisible performance more, it is also necessary here to implement demarcation policies which emphasise the specificities of sustainable innovation in services and in particular social innovations, examples of which we have given previously (see section 27.4.2).

27.5 Conclusion

The relationship between innovation and performance (equated with growth) is a major economic relationship, which has been the subject of an extremely extensive literature. In post-industrial economies, the two terms of the relationship raise several problems, which have been the subject of a separate literature. In a highly tertiarised economy, service innovation partly escapes the tools of traditional economic analysis. One therefore observes an innovation gap. Performance continues to be defined in terms of growth and productivity, while other forms of assessing performance prove to be necessary. One therefore observes a performance gap.

Economic analysis and public policies favour the relationship between visible innovation (identified by traditional definitions, R&D and patents) and visible performance (equated with growth). They therefore emphasise technological innovation that is a source of growth. However, the double gap that has been identified reveals much more complex relationships, which can question the relevance of diagnoses and the validity of public policies supporting innovation. It therefore appears that these policies should adopt a demarcation perspective, which allows one to take account of and support the specific forms of innovation (in particular in services) and the most dynamic and most strategic sectors (for example, the KIBS), but also a certain number of economic sectors that are sources of social innovations (proximity services). These policies, whatever the form of innovation (technological or non-technological), should also favour less visible performance (sustainable performance). The analysis proposed in this chapter has tried to clarify the different relationships between innovation and performance on a strictly theoretical level. It should be continued by a quantified assessment of relationships.

Notes

1. This chapter is part of research supported by the EC within the context of the FP7 ServPPIN Project (The Contribution of Public and Private Services to European Growth and Welfare, and the Role of Public–Private Innovation Networks).
2. These are services that are independent of the product and the production process, for example child care, sporting and recreational services.
3. The term 'technological' has been dropped.
4. The MIPS indicator (Material Intensity Per Service Unit) measures the degree of utilisation of natural non-renewable resources to produce a good or a service.

References

Barcet, A. and J. Bonamy (1999), 'Eléments pour une théorie de l'intégration biens/services', *Economies et Sociétés, EGS Series*, **1** (5), 197–220.

Boltanski, L. and L. Thévenot (1991), *De la justification. Les économies de la grandeur*, Paris: Gallimard.

Boskin, M.J. (1996), 'Towards a more accurate measure of the cost of living', report for the Senate Committee by the Advisory Commission to study the Consumer Price Index, December, US Senate.

Bressand, A. and K. Nicolaïdis (1998), 'Les services au cœur de l'économie relationnelle', *Revue d'Economie Industrielle*, **43**, 141–163.

Bröchner, J. (2008a), 'Construction contractors integrating into facilities management', *Facilities*, **26** (1–2), 6–15.

Bröchner, J. (2008b), 'Client-oriented contractor innovation', in P. Brandon and S.-L. Lu (eds), *Clients Driving Innovation*, Chichester: Wiley-Blackwell, pp. 15–136.

Broussolle, D. (2001), *Les NTIC et l'innovation dans la production de biens et services: des frontières qui se déplacen*t, 11th Reser international conference, ESC-Grenoble, October.

Carassus, J. (2002), *Construction: la mutation: de l'ouvrage au service*, Paris: Presses des Ponts et Chaussées.

Coombs, R. and I. Miles (2000), 'Innovation measurement and services: the new problematique', in S. Metcalfe and I. Miles (eds), *Innovation Systems in the Service Economy: Measurement and Case Study Analysis*, Boston, MA: Kluwer, pp. 85–103.

Dandurand, L. (2005), 'Réflexion autour du concept d'innovation sociale, approche historique et comparative', *Revue française d'administration publique*, **3** (115), 377–382.

Davies, A. (2004), 'Moving base into high-value integrated solutions: a value stream approach', *Industrial and Corporate Change*, **13**, 727–756.

Djellal, F., D. Francoz, C. Gallouj, F. Gallouj and Y. Jacquin (2003), 'Revising the definition of research and development in the light of the specificities of services', *Science and Public Policy*, **30** (6), 415–430.

Djellal, F. and F. Gallouj (1999), 'Services and the search for relevant innovation indicators: a review of national and international surveys', *Science and Public Policy*, **26** (4), 218–232.

Djellal, F. and F. Gallouj (2001), 'Patterns of innovation organisation in service firms: postal survey results and theoretical models', *Science and Public Policy*, **28** (1), 57–67.

Djellal, F. and F. Gallouj (2006), 'Innovation in care services for the elderly', *Service Industries Journal*, **26** (3), 303–327.

Djellal, F. and F. Gallouj (2007), 'Les services publics à l'épreuve de la productivité et la productivité à l'épreuve des services publics', *Revue d'Economie Industrielle*, **120**, 25–54.

Djellal, F. and F. Gallouj (2008), *Measuring and Improving Productivity in Services: Issues, Strategies and Challenges*, Cheltenham, UK and Northampton, MA, USA: Edward Elgar.

Eurostat (2001), *Handbook of Price and Volume Measures in National Accounts*, Luxembourg: European Commission.
Furrer, O. (1997), 'Le rôle stratégique des "services autour des produits"', *Revue française de gestion*, March–May, 98–107.
Gadrey, J. (1996), *Services: la productivité en question*, Paris: Desclée de Brouwer.
Gadrey, J. (2005), 'Les quatre "mondes" des économies de services développées', *Economies et Sociétés, EGS Series*, **39** (11–12), 1925–1970.
Gadrey, J. and F. Jany-Catrice (2007), *Les nouveaux indicateurs de croissance*, Paris: Repère, La Découverte.
Gallouj, F. (1994), *Economie de l'innovation dans les services*, Paris: Editions L'Harmattan, Logiques économiques.
Gallouj, F. (1998), 'Innovating in reverse: services and the reverse product cycle', *European Journal of Innovation Management*, **1** (3), 123–138.
Gallouj, F. (2002), *Innovation in the Service Economy: The New Wealth of Nations*, Cheltenham, UK and Northampton, MA, USA: Edward Elgar.
Gallouj, F. and O. Weinstein (1997), 'Innovation in services', *Research Policy*, **26** (4–5), 537–556.
Goffman, I. (1968), *Asiles*, Paris: Edition de Minuit.
Jex, S.M. (1998), *Stress and Job Performance: Theory, Research and Implications for Managerial Practice*, Thousand Oaks, CA: Sage Publications.
Karasek, R. and T. Theorell (1990), *Healthy Work: Stress, Productivity and the Reconstruction of Working Life*, New York: Basic Books.
Karpik, L (1989), 'L'économie de la qualité', *Revue française de Sociologie*, **30** (2), 187–210.
Le Roy, A. (1997), *Les activités de service: une chance pour les économies rurales? vers de nouvelles logiques de développement rural*, Paris: L'Harmattan.
Lenfle, S. and C. Midler (2003), 'Innovation in automative telematic services: characteristics of the field and management principles', *International Journal of Automative Technology and Management*, **3** (1–2), 144–159.
Lowe, G.S. (2003), 'Milieux de travail sain et productivité: un document de travail', Division de l'analyse et de l'évaluation économiques, Santé Canada, April.
Mathieu, V. (2001), 'Service strategies within the manufacturing sector: benefits, costs and partnerships', *International Journal of Service Industry Management*, **12** (5), 451–475.
Metcalfe, J.S. and I. Miles (eds) (2000), *Innovation Systems in the Service Economy*, London: Kluwer Academic Publishers.
Miles, I. (2002), 'Services innovation: towards a tertiarization of innovation studies', in J. Gadrey and F. Gallouj (eds), *Productivity, Innovation and Knowledge in Services*, Cheltenham, UK and Northampton, MA, USA: Edward Elgar, pp. 164–196.
Muller, P. (1991), 'Quel avenir pour l'agriculture et le monde rural?' *Economie rurale*, 67–70, 202–3.
Nahon, D. and J. Nefussi (2002), 'Les services au cœur de l'innovation dans la production agricole: l'exemple de la pomme de terre', in F. Djellal and F. Gallouj (eds), *Nouvelle économie des services et innovation*, Paris: L'Harmattan, pp. 285–300.
NESTA (2006), 'The innovation gap: why policy needs to reflect the reality of innovation in the UK', National Endowment for Science, Technology and the Arts, Research Report, October.
OECD (1992), *Proposed Guidelines for Collecting and Interpreting Technological Innovation Data, Oslo Manual*, Paris: OECD.
OECD (1997), *Proposed Guidelines for Collecting and Interpreting Technological Innovation Data, Oslo Manual*, Paris: OECD.
OECD (2002), *Proposed Standard Practice for Surveys of Research and Experimental Development, Frascati Manual*, Paris: OECD.
OECD (2005a), *Governance of Innovation Systems*, Synthesis Report, Volume 1, Paris: OCED.
OECD (2005b), *Proposed Guidelines for Collecting and Interpreting Technological Innovation Data, Oslo Manual*, Paris: OECD.

Perret, B. (1995), 'L'industrialisation des services', in G. Blanc (ed.), *Le travail au XXIe siècle: mutations de l'économie et de la société à l'ère des autoroutes de l'information*, Paris: Dunod, pp. 37–38.
Rubalcaba, L. (2006), 'Which policy for innovation in services?' *Science and Public Policy*, **33** (10), 745–756.
Sharpe, A (2000), 'A survey of indicators of economic and social well-being', Canadian Policy Research Networks, research report, March, Ottawa, Canada.
Sundbo, J. (1998), *The Organisation of Innovation in Services*, Copenhagen: Roskilde University Press.
Vandermerwe S. and J. Rada (1988), 'Servitization of business: adding value by adding services', *European Management Journal*, **6** (4), 314–324.

PART VII

SERVICE INNOVATION: BEYOND SERVICE SECTORS

28 Service innovation and manufacturing innovation: bundling and blending services and products in hybrid production systems to produce hybrid products

John R. Bryson

28.1 Introduction

Since the 1960s there have been simultaneous quantitative and qualitative changes in the workings of the capitalist economy. Symptoms include alterations in the synergies that exist between manufacturing and service activities and in the ways in which manufactured goods are produced, sold and consumed. Service-type functions and service innovations have assumed a more critical role in processes of production as well as in the division of labour. There has been an increase in service-related occupations within the manufacturing sector and especially in the group of 'other professionals' that includes occupations such as business, finance and legal professionals (Pilat and Wölfl, 2005: 12). In some countries more than 50 per cent of manufacturing workers are engaged in service-related occupations (Pilat and Wölfl, 2005, 36). What has occurred and continues to occur is a shift in the relationship between service and manufacturing inputs in the production process. This is not a new shift, but it is one that has intensified since 1990. Part of this shift involves the creation of hybrid production systems and hybrid products that blend manufacturing and service processes and functions together to create value and to enhance competitiveness (Bryson, 2009).

It is important to emphasise that this is an old rather than a new process, but it is a process that is becoming more important for the competitiveness of firms. On 27 April 1846 Benjamin Smith and his son Josiah were granted a lease by the Earl of Stanhope to extract ironstone, coal and fireclay in the parishes of Dale and Stanton-by-Dale, Derbyshire, England. This eventually became the Stanton Ironworks Company, a company that operated blast furnaces and foundries and manufactured cast iron pipes and other objects made of iron. During the 1920s this company began to produce pre-stressed concrete pipes and concrete columns for street lighting. The Derbyshire plant closed on 24 May 2007.

The Stanton company produced goods that are traditionally associated with heavy industry rather than with services. Yet in a company history published in 1959 the company noted that 'behind all this production, whether it is iron or concrete, are two groups who might perhaps forgive the label of "backroom boys [sic]" (Lewis, 1959, 37), and these backroom employees were service workers. For these backroom workers, 'there is no end-product to their highly skilled work; but end products would not be possible without them' (Lewis, 1959, 37). This statement highlights the complex symbiotic relationships that exist between service and manufacturing innovations. For this company product innovation would have been impossible without the inputs from highly skilled service workers. It is impossible to understand this firm's business model without appreciating the complex ways in which service functions were an integral part of this firm's production process. It is worth exploring the two groups of backroom workers that played such an important role in the Stanton Ironworks Company.

First were the designers, draughtsmen, civil, mechanical, electrical and fuel engineers and related technical staff (Lewis, 1959, 38). These were the people who kept the plant working by servicing equipment, but they were also responsible for the design and installation of new plant. These service functions can be classified as producer service inputs into this company's production processes; 'producer services' provide intermediate inputs into the activities of client companies (Illeris, 1996). For Stanton, at this time, these services were provided by in-house employees, but they could also have been outsourced to specialist independent contractors. There is a developed literature on producer services, or on knowledge-intensive business services (KIBS), as this group of activities has grown rapidly over the last 20 years (Bryson et al., 2004; Rubalcaba and Kox, 2007).

Second were the activities of the research, chemical laboratory and metallurgical departments. These provided a range of technical services that included keeping up to date with technical developments and for innovations that involved improved methods of manufacture and new product development. These departments were also responsible for control over the raw materials that were used, over processing and with the quality of the finished products. It is worth noting that: 'these departments provide a service not only *to* Stanton but *by* Stanton, for a great deal of much appreciated work is carried out for customers' (Lewis, 1959, 38). This is an important distinction and is a distinction that has often been ignored in the service literature. This manufacturing company supplied goods to its customers, but it also supplied services and for many customers the manufactured product was worthless without the embedded services. The question is, what were these embedded services? Two types can be identified:

technical services; training and education. The technical services department worked with clients on special orders, on customised products and on special castings, for example pipework to fit tunnels. This department produced drawings and also designs; the function was one of technical interpretation to bridge the interface between a customer's precise requirements and the ability of the firm to manufacture the product. Training and education was provided by service engineers who visited sites to demonstrate pipe jointing techniques, provide advice on pipe-laying, testing and the efficient use of pipes. This group of service workers also prepared the technical data required for the production, sales, research and publicity departments and also represented the company on the appropriate British Standards committees.

There were many other service functions performed by this manufacturing company: sales offices, 74 agents representing the firm in 95 countries, transport planners and a publicity department that employed photographers, designers and editorial staff. The publicity department produced material in many languages, made technical films, designed and staffed trade exhibitions, operated a cinema and maintained an exhibition building. These service workers were responsible for creating and projecting the company's image and reputation. This was a manufacturing firm that employed many service workers and this was not an unusual firm. The distinction made by Lewis (1959) between 'services to Stanton' and 'services by Stanton' is extremely important and it is central to the analysis developed in this chapter. Stanton was directly providing a set of intermediate technical services that supported the manufacturing activities of this company. These can be conceptualised as a set of production-related services and these services can also be classified as producer services. They are termed 'production-related' as these functions are essential for value creation in production processes that involve the co-creation of goods and services. Essentially these are inward facing services as the client is Stanton rather than an external consumer. This is a fine distinction as many of these inward facing services could be outsourced to specialist suppliers. Stanton was also providing another set of outward facing services; services that were intended to enhance the quality of the relationship the firm had with its clients. These services provided clients with customised products and Stanton with the ability to sell service-enhanced manufactured products. These services can be called product-related services as they are customised services that are designed to meet the needs of a particular customer. The category of product-related services highlights the creation of hybrid products that blend service and manufacturing inputs together through a hybrid production system. This group of product-related services is extremely important as they create value for the supplier and the customer, provide opportunities for trust-based relationships to form between representatives of the producer and the consumer and can also partially ensure that a supplier maximises value creation from its existing customer base.

The example of the Stanton Ironworks Company highlights the ways in which service functions are an integral part of all production processes. This chapter explores these service functions by, first, reviewing the existing literature. There have been a number of important theoretical contributions especially in relation to the development of the concept of 'complex packages' or 'compacks' (Bressand, 1989) and also research that has explored product and service bundles. Second, the chapter explores the development of a new conceptual framework based around the concept of a 'service duality'. The duality refers to the identification of two service-informed or service value creation moments in production processes: the creation and consumption of production-related and product-related services. In section 28.4 product-related services are explored and a typology developed. Section 28.5 concludes the chapter.

28.2 'Compacks', bundling and encapsulation of service and manufacturing functions

One of the most difficult questions facing the social sciences concerns isolating and understanding the differences that exist between services and goods. This question keeps returning to challenge the ways in which academics and policy-makers conceptualise the ways in which wealth is created through the production and sale of goods and services (Hill, 1977; Gallouj, 2002; Bhagwati et al., 2004). We must accept that: 'although the objective of manufacturing industry is the production of commodities many people employed in manufacturing are not directly employed in the actual production process' (Crum and Gudgin, 1977, 3). In 1971, in the UK, every two production workers needed to be supported by rather more than one 'non-production' worker (Crum and Gudgin, 1977, 5); the term 'non-production' was used to identify people within a manufacturing firm who are not directly involved in the actual production process (managers, designers, sales team, research scientists).

Classical political economists like Adam Smith equated the category of 'service work' with 'unproductive labour' or labour that does not add 'to the value of the subject upon which it is bestowed' (Smith, 1977 [1776], 429–30). Productive labour was considered by Smith to produce value as: 'the manufacturer has his wages advanced to him by his master [sic], he, in reality, costs him no expense, the value of those wages being generally restored, together with a profit, in the improved value of the subject upon which his labour is bestowed' (Smith, 1977 [1776], 430).

Furthermore, Smith emphasises that time is important in differentiating between services and goods. Time spent creating a physical good is fixed and stored within the good and the value of the time can be realised some time later; however, 'the labour of the menial servant, on the contrary, does not fix or realize itself in any particular subject or vendible commodity' (Smith, 1977 [1776], 430). Smith was writing at a time during which the notion of 'service' was easily confused with a servant class. He failed to appreciate that many service functions and processes are essential for the production of physical goods. The simple bipolar classification of labour based upon the concepts of 'productive' and 'unproductive' labour is outdated and very much an eighteenth- and nineteenth-century conceptualisation of the economy. Nevertheless, the simple division of economic activity into processes that produce goods and those that produce services still underpins many conceptualisations of economic activity.

An academic literature does exist that explores the difficulties that come from conceptualising economic activity as segmented into manufacturing and service industries (Daniels and Bryson, 2002; Bryson, 2009). This literature is based on the premise that services cannot be delivered without the support of material goods, and vice versa. This dependency or interrelationship has led Barcet (1987) to suggest that a goods–services continuum exists consisting of five levels:

1. Pure goods that have a use value that is independent of any service. One example of this type of good might be food, but even this product is usually accessed via a service transaction (retailing). Food also increasingly comes with cooking instructions, suggested recipes and helplines. Thus, it would be possible to argue that many manufacturers of food are distorting the boundary that exists between the delivery of a 'pure good' and the provision of additional services.
2. Mixed goods with use values that are only realised through the application of service products. A good example would be a machine that can only be effectively used by reading an instruction manual; instruction manuals are the product of a set of service functions.
3. Goods–services complexes in which both elements are mutually dependent on one another. A mobile phone has no use value without access to mobile telephony services or a computer has limited value without software.
4. Services that can only be delivered with the support of goods.
5. Pure services, for example management consultancy or a range of technical services.

This approach highlights the complementarities that exist between production processes that are designed to produce goods and those developed to create services. The Barcet (1987) approach to understanding the relationship between goods and services was an important conceptual breakthrough. It was, however, a breakthrough that was largely ignored outside the French academic community. This was unfortunate as Barcet's conceptual contribution was extremely sophisticated and well ahead of its time.

Another approach to addressing the difficulties that come from separating service functions from manufacturing began by noting that the application of simple descriptive indicators to measure the service economy is 'an intellectual dead end' and that such an approach was only required when services were 'truly invisible not just to economists but even to many service companies themselves' (Bressand et al., 1989, 17). In their analysis Bressand et al. (1989) develop an integrated approach to understanding the wealth-creation process by challenging 'the traditional concept of "products" and to suggest instead that corporations were actually engaged in the production and sale of complex packages ("compacks") in which goods coexist with services' (Bressand et al., 1989, 17). This conceptual framework was largely ignored in the academic literature perhaps because the initial development of the ideas occurred in a report that was presented to the European Commission (Bressand, 1986). A good example of a 'compack' is found in the case of air transportation services. This industry provides 'packages that include the information management services now needed to make efficient use of the airline system. Such packages or "compacks" in turn open new strategic opportunities for service providers, notably through the use of bundling and unbundling' (Bressand, 1989, 53). 'Compacks' consisted of bundles of service and manufactured inputs. This conceptual development revolved around a reordering of the input–output relations in economies whereby '"soft" functions are increasingly provided by separate companies' (Bressand et al., 1989, 35). It also argued that successful companies would concentrate on 'service value-adding' activities that would revolve around three linked dimensions: information networks, customer relationships, and 'thoughtware'. Information networks were highlighted as being increasingly important in the relationship that a firm had with its suppliers and customers and that effective use of such networks provided firms with opportunities to 'maximise the value added by their "soft" assets, or thoughtware' (Bressand et al., 1989, 36). Customer relationships were identified as important in forming trust with consumers as trust would become a priceless asset as consumers would find it increasingly difficult to differentiate between competing products and especially intangible services. Thoughtware refers to the shift towards corporate competitiveness based on 'thought power' rather than muscle power.

The 'compack' approach is part of a broader debate on product and service bundling. The concept of 'bundling' can be applied to goods or services as well as to mixtures of goods and services. Bundling of products and services is an important marketing and sales strategy for producers and retailers. Two types of bundling strategy have been identified: pure or mixed (Adams and Yellen, 1976). Pure bundling describes a situation whereby a company offers for sale a bundle of elements (components, items, services), but consumers can only purchase the complete bundle rather than individual elements. Mixed bundling provides consumers with the option to purchase the bundle or individual components. Bundling is often used as a competitive strategy by companies trying to differentiate their products and services in the marketplace. An automotive company may sell its products with the option to purchase an accessories package or bundle. Much of the academic literature on product–service bundles explores the effect of product and service bundling on purchasing decisions (Yadav and Monroe, 1993; Hermann et al., 1997).

The literature on 'bundling' is complemented by the work on 'splintering' or the process by which part of the value added in a production process is subcontracted to other firms (Bhagwati, 1984). This is a process of fragmentation, or to confuse matters, of 'unbundling' elements of a production process (Baldwin, 2006). Splintering is a confusing concept as it 'occurs when part of the manufacturing value added, such as, say, painting a car, is done by contracting it out to a separate painting firm, and the painting value added then becomes part of the service sector' (Bhagwati et al., 2004, 95). This is a misleading example as it is debatable whether painting in this context is part of the service sector as a physical and tangible transformation to the car occurs. Within an automotive factory the physical process of painting would be just part of the production process and identifying it as a service element just confuses matters.

The 'splintering' concept has been developed and refined and these developments are perhaps best reflected in the recent work on the fragmentation of value chains (Baldwin, 2006; Bryson and Rusten, 2008). The process of unbundling (Baldwin, 2006) or fragmentation highlights the possibilities of separating service functions from production functions. The terminological confusion here is unfortunate in that the word 'bundle' is used in two different contexts. Fragmentation draws attention to the possibilities for the development of a new international division of service expertise that complements the international division of labour that occurs within manufacturing (Bryson and Rusten, 2006; Bryson 2007). The distinction between expertise and labour is important as Bryson and Rusten (2006) suggest that these represent two different formations of work activity. This distinction between expertise and labour and between service and manufacturing functions highlights that these are separate but interrelated functions.

The relationship between service and manufacturing industries has also been explored by Howells (2000, 15) when he identified two different methods by which manufactured products are not offered to consumers in their own right (see Bryson et al., 2004, 56–58), but rather as part of a package that includes service components:

1. Manufactured products provided with closely aligned services, for example finance, insurance maintenance warranties, repurchase clauses and service agreements.
2. The manufactured product is supplied to consumers as a vehicle for accessing services. In this case the product is not the end point of the transaction, but only the beginning of the relationship between the consumer and producer.

The second strategy is the more complex. In many respects, Howells' analysis resonates strongly with the work of Barcet (1987) and Bressand et al. (1989). Consumers are increasingly offered not the manufactured product but rather the outcomes that would come from owning such a product (Howells, 2002). A good example is car leasing where the consumer benefits from the use of the car, but does not own it and does not have to insure and service the car, or computer companies that provide processing time rather than equipment. In this case the computer company will take over the provision of in-house computing services whilst the client no longer has to be concerned with computer obsolescence or the problems of running a computer division.

These types of service product relationship represent forms of what is termed 'service encapsulation' (Howells, 2000) in which services are wrapped around or embedded in products and in which services can produce innovations in other sectors of the economy. New services can be developed to increase the profitability of physical products and increasingly manufactured goods are not offered to consumers in their own right; they can only be purchased with associated services. The increasing importance of service encapsulation highlights the ongoing transformation of the capitalist production process as it tries to develop new ways of enhancing profitability. In this new world, consumption is not a one-off transaction between the consumer and producer but is the beginning of a long-term relationship (Daniels and Bryson, 2002). It is through these types of long-term relationships that trust is developed and the producer may be able to lock consumers into their technology and their services. This is not a new process as machines have been sold or leased to

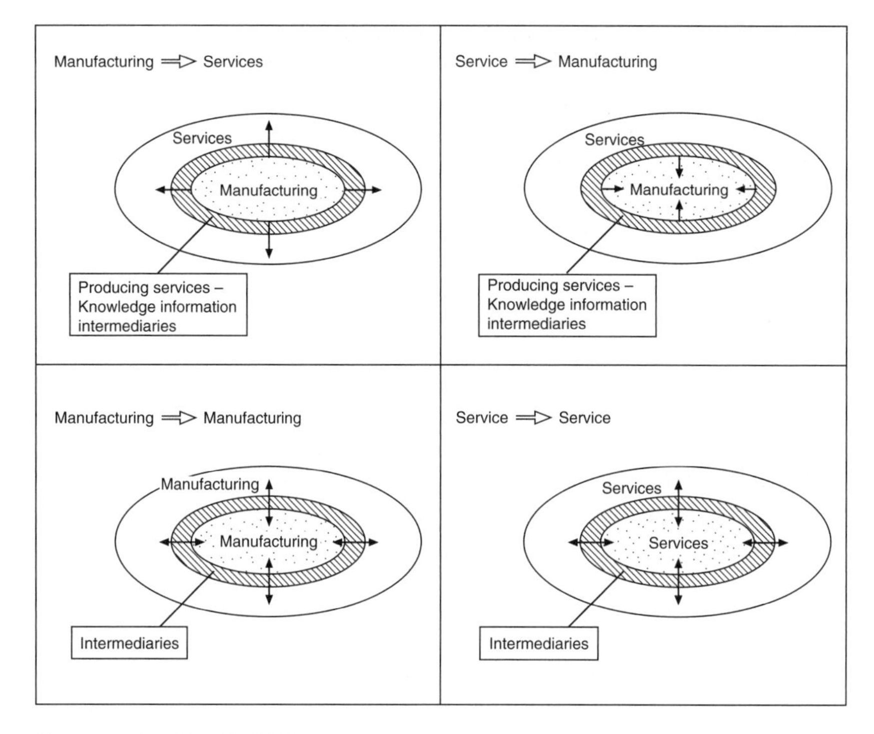

Source: after Howells (2002).

Figure 28.1 Different types of service and manufacturing encapsulation

consumers on service agreements for more than 100 years, for example telephones and reprographic machines, but maybe what is new is the opportunities that this offers to lock consumers into long-term relationships with providers of services that are linked to goods.

At least four types of encapsulation can be identified with each type representing a different way in which services are being positioned into the production process to create innovations (Figure 28.1). First, new products, existing products or alterations to existing products can create service innovations or possibilities for the delivery of new services. A new type of computer system can introduce a different way to deliver a service. Products have been designed, for example washing machines, with embedded software that can be upgraded to meet changes in washing technology, but this upgrade can only be undertaken by the manufacturer's technicians.

Second, the interaction between the service and manufacturing sectors of the economy leads to innovations in the manufacturing process and in the development of new physical products. In this case, a market research company may identify a product alteration that feeds into the production process. A good example of this process is the design of Jaguar sports cars. Market research identified that up to 25 per cent of the firm's sports cars would be purchased by women and that in the important American market this figure rose to 33 per cent. The design team for new cars was altered to include both male and female designers and the car's ergonomics altered to produce a comfortable drive for males and females. Thus, door clearances were designed so that a woman wearing a skirt or dress could get in and out without difficulty, and door pulls, release catches and switches designed so that they could be used without breaking long fingernails (Bryson, 1997, 380). Rather than engineers being fully responsible for the design of cars, designers, market researchers and marketing experts are instructing engineers in what they should design.

Third, the development of new physical products stimulates the modification of existing products or the creation of new physical products that may encourage the creation of new services; manufacturing firms are encapsulated by products provided by other manufacturing and service firms. Encapsulation is not just about service-informed innovations. Fourth, service providers are dependent upon a series of service inputs that can either be provided in-house or externally and the development of new service products also provides opportunities for the creation of new services.

All these forms of encapsulation include intermediaries, producer service firms, service providers and manufacturing companies, who operate as bridges to new forms of thinking and new knowledge and information. It is important to note that producer service firms have a major advantage over in-house staff. They work for many different clients and are able to identify client innovations that they are then able to transfer to other client companies (Bryson, 2000). They can be conceptualised as a form of knowledge transfer agent with clients gaining knowledge, but at the same time losing knowledge (Bryson, 2000).

As we have seen there have been a number of attempts to explore the interrelationships that exist between service and manufacturing functions. This literature contributes to ongoing debates regarding innovation and services as well as innovation and manufacturing. There is a developing theoretical literature into the relationship between services and innovation (Bryson and Monnoyer, 2004). Gallouj (2002, 1) classifies this literature into three basic types: technologist; service-orientated and integrative. Technologist approaches reduce or equate innovation with the introduction of new technologies into service firms and organisations. Service-orientated approaches try to identify differences between manufacturing innovation and service innovation; whilst the integrative approach explores the blurring of the boundaries between goods and services.

Gallouj argues that these three approaches can be considered to represent a 'natural life cycle of theoretical concerns'. The first technologist phase is the obvious starting point for research into the relationship between services and innovation, but this phase declines and is gradually replaced by attempts to identify the characteristics of service innovation. This phase may even overstate the differences. This second stage represents a mature approach to studying services and innovation, but this is gradually being replaced by the integrative approach; a theoretical perspective that is currently attempting to combine goods and service production innovative systems.

This life cycle approach, whilst being an extremely useful conceptual tool for understanding the evolution of the service innovation literature, fails to acknowledge that, in part, the integrative approach preceded both the technologist and service-orientated approaches. The empirical analysis provided by Crum and Gudgin (1977) and Barcet's (1987) conceptual contribution are especially important in this context. It is perhaps worth noting that Barras's ground-breaking paper on service innovation was published in 1986 and his work made an extremely important contribution to the technologist phase of the service innovation debate.

28.3 Hybrid production systems to produce hybrid products

The conceptual work that has been undertaken to explore service and manufacturing interactions provides a useful foundation for the development of a new approach to understanding the ways in which services and manufacturing functions are combined to create value. It is to this task that we now turn our attention. In this context it is important to remember that a production process consists of a set of processes and operations. A process refers to the collection of activities that combined together produce a product whilst an operation refers to an activity that is performed at a particular point in a production process (Blackstone et al., 1997, 602). The production of goods and services should be conceptualised as a process that consists of a complex and evolving blending of manufacturing and service operations. This can be conceptualised as a simple equation in which:

$$\text{Production } (P) = \text{Manufacturing Operation } (M) + \text{Service Operation } (S).$$

It is impossible to manufacture without services and services cannot be created or delivered without manufactured products (Bryson et al.,

2008). It is therefore important for academics and policy-makers to begin to identify and conceptualise the complex interrelationships that occur between different elements of production processes that together create value. This 'coming together' can occur within the same company or can be part of a coordinated value chain of independent companies that are managed by a company or even an individual to create a product (physical product or a service). The control of a process should take precedence over operations and all operations must be subordinate to the requirements of the process. Operations may be geographically distributed as a production process may be designed around the benefits that can accrue from an international division of labour (Bryson and Rusten, 2006, 2008). The production process must be controlled so that the constraints or limitations on the process can be identified and taken into consideration. This also means that individual operations may have to be controlled using different systems to take into consideration the nature of the activity, for example service or manufacturing inputs into the overall process, and the complexity that is associated with a geographically distributed production system.

To complicate matters it is possible to argue that production has evolved but our understanding of it has not or perhaps more correctly 'manufacturing has evolved, but our understanding of it has not' (Livesey, 2006, 1). It is important not to underestimate the sophistication and knowledge-intensity of manufacturing activities. Fingleton argues that: 'those who advocate post-industrialism overestimate the prospects for post-industrial services, but they greatly underestimate the prospects for manufacturing. A major problem with the argument of the post-industrialists is that they do not understand how sophisticated modern manufacturing truly is' (Fingleton, 1999, 3). The Department of Industry (UK) makes a very similar point, suggesting that:

> The modern manufacturing product cycle can now encompass functions from market research to distribution, branding and after-sales service, through to end-of-life disposal. The boundary between manufacturing and services is becoming increasingly blurred. For every factory producing machine tools, there is demand for collaboration with designers, software specialists, financial experts, caterers and other service providers. As new manufactured products are constantly introduced and upgraded, so new services are generated around them. (DTI, 2004, 11)

The point is that the manufacturing value chain does not solely consist of shop floor or fabrication activities, but now includes a complex blending of production and service operations with the latter including design, engineering, sales and marketing and after sales services (Figure 28.2).

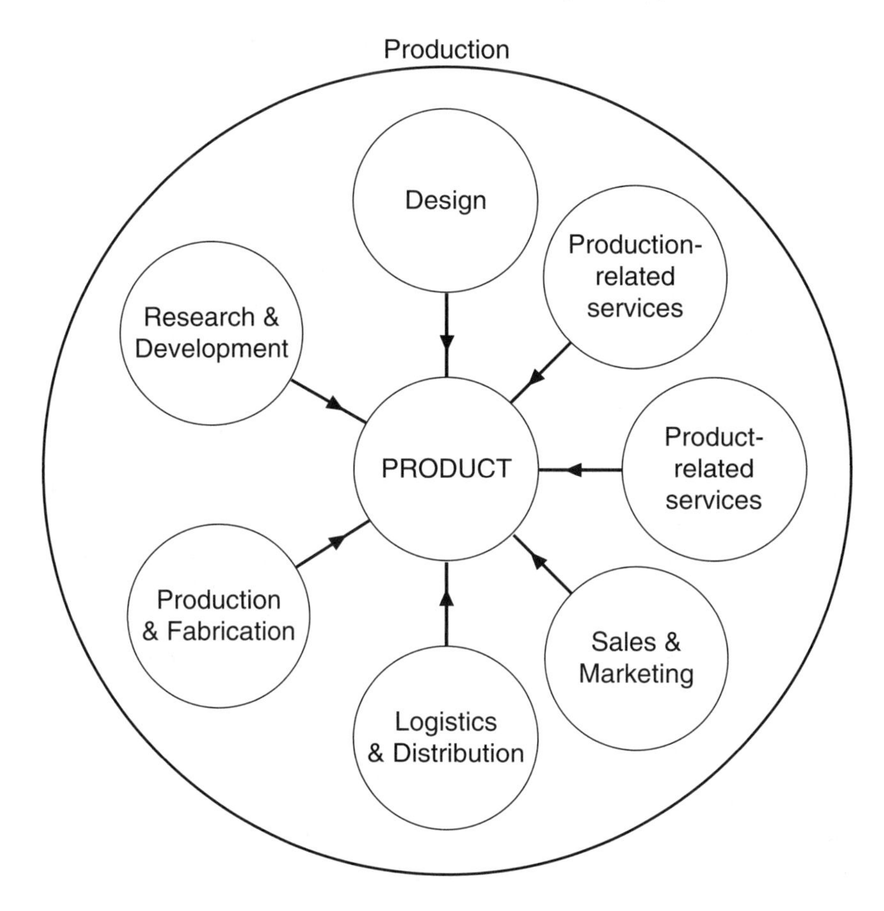

Figure 28.2 The elements of a manufacturing value chain

This conceptualisation of manufacturing has led to the development of an extended definition of manufacturing that includes services and production processes (BERR, 2008).

The extended definition of manufacturing includes service operations that contribute to the production of goods or services, but also service operations that contribute service functions to products (goods or services). This is similar to the compacks approach developed by Bressand et al. (1989) in which products are conceptualised as consisting of complex bundles of different services and products. Conceptual clarification is required here and this explains the importance of differentiating between production-related and product-related services. The former include all technical, business and professional services or producer services, or in other words, providers of intermediate inputs. The latter include service functions that directly and indirectly support consumers in their purchase and use of a product (Figure 28.3).

To prevent confusion, it is important to remember that the production process consists of a number of elements: manufacturing or fabrication, the provision of services that support fabrication, customer-targeted services and the provision of pure services. There is a danger that manufacturing is equated with production rather than conceptualised as one element of a much more complex production process; a production process that hybridises services and manufacturing operations to produce a complete product. It is worth noting, however, that all production systems are hybrid production systems, but that not all hybrid production systems produce hybrid products. A pure manufactured good can be purchased that comes with no embedded or attached services, but will still have been produced by blending service and manufacturing inputs together. The danger exists that too much attention is focused on service functions without exploring the contributions that they made to the overall production processes of which they are a part. The production of products and services should be conceptualised as a process that consists of a complex and evolving blending of manufacturing and service processes. Some of these service functions (production-related services) directly support the manufacturing or fabrication process whilst others support final consumption (product-related services). This implies that the simple production equation can be revised as follows:

Production (P) = Manufacturing Processes (M) + Production-Related Services (PRS) + Product-Related Services ($ProRS$)

This is a simple equation to describe a very complex process. Profit is created from the differential that must exist between the exchange value of the product (service/good/hybrid product) and the cost of production.

The identification of two service-informed or service value creation moments in the production process highlights the existence of a 'services duality' (Bryson and Daniels, 2009). In some respects the concept of a services duality is influenced by the 'innovation value chain' (IVC) approach that has been recently proposed by Hansen and Birkinshaw (2007) as a general framework within which firms' innovation activities can be considered. The IVC approach conceptualises innovation as a three-stage sequential process that involves knowledge investment, innovation process capability and value creation capability (Hansen and Birkinshaw, 2007, 122). This is a very structured and phased approach

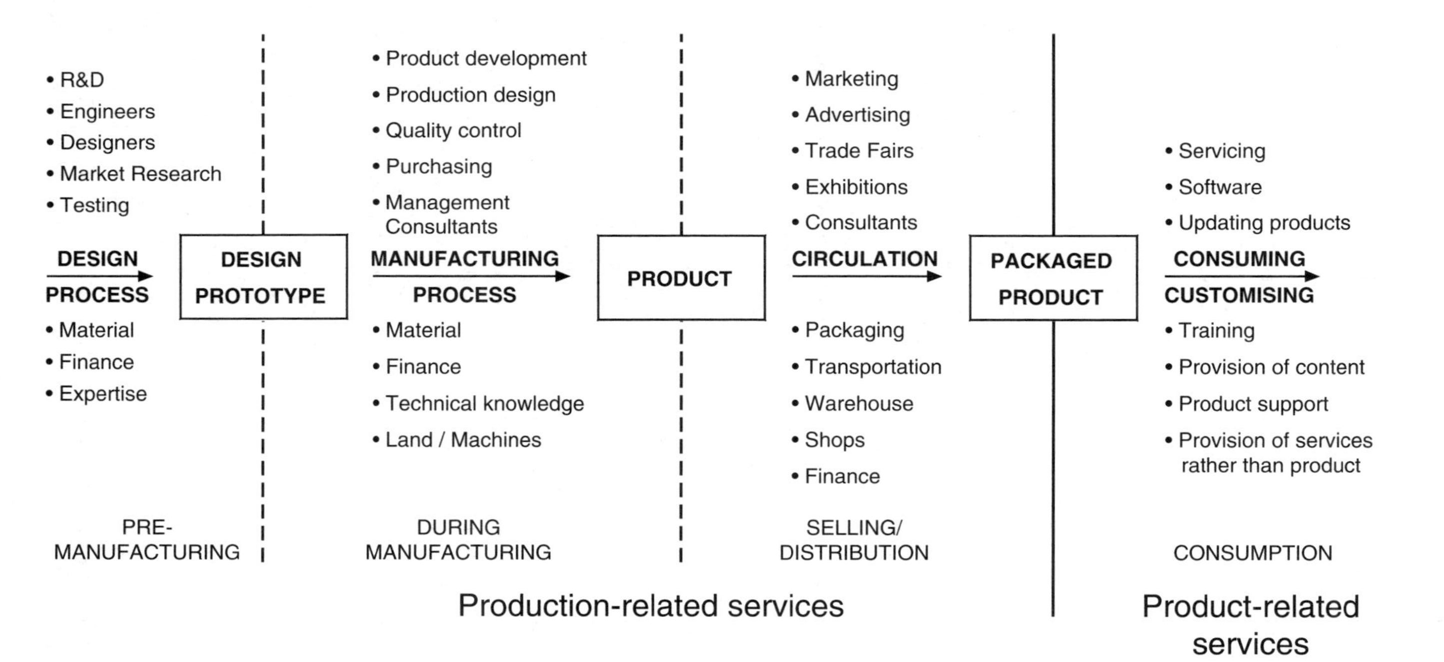

Figure 28.3 The services duality: production- and product-related services

towards understanding innovation within firms. The 'services duality' approach recognises that services are entwined within production processes, but at many different stages.

The concept of the services duality highlights the importance of both the activities of production-related services and services that are developed to support products (product-related services). This is to argue for a sequential analysis of the role that services play within the process of production as intermediate inputs and combined with products for final consumption. It is interesting to note that considerable academic research has been undertaken into exploring the economics and operational dynamics of production-related services (Rubalcaba-Bermejo, 1999; Rubalcaba and Kox, 2007) but that product-related services have been largely ignored. This approach therefore provides a conceptual tool for operationalising the extended definition of manufacturing. We now want to turn our attention to exploring product-related services and the ways in which they add value to production processes.

28.4 Product-related services and the development of hybrid products

This chapter began with an account of the Stanton Ironworks Company in which a distinction was made between 'services to Stanton' and 'services by Stanton' (Lewis, 1959, 38). This statement mirrors the services duality and the importance that has been given in this chapter to distinguishing between production-related (services to Stanton) and product-related services (services by Stanton). It is worth noting that the Stanton company history was published in 1959 and that, at this time, this company was hybridising service and manufacturing operations in production processes as a competitive strategy. This draws attention to the fact that academics have failed to appreciate the complex ways in which service operations have contributed to innovation and value creation in production processes for some considerable time. The service duality is not a new process but it is one that can perhaps be traced back to the early years of the twentieth century. This is a piece of economic history that has still to be researched and written.

Stanton was offering a range of services associated with the design, customisation and fabrication of goods made of steel. Its business model may have been ahead of its time, but there is no evidence to assume that this was the case. To develop this point, further research would be required into the activities of Stanton's competitors. What is interesting is that since 2000 steelmakers have begun to develop 'service offering' or product-related services. The evidence for this is found in the activities of many of the large steel companies (Siemens, Mittal Steel, etc.). Chiang Yao-Chung, President of China Steel, the Taiwanese steelmakers, noted recently that this steel company has 'to become better at service' and that 'steelmakers can no longer simply produce steel; they need to back up their production with services such as design, distribution and "tailoring" of products to meet their customers' more demanding requirements' (cited in Marsh, 2006, 10). Yao-Chung is highlighting the importance of a series of service operations that are targeted at customers and are intended to enhance competitiveness and value creation. China Steel is developing a business model based around the creation of hybrid products that blend service and manufacturing functions in a similar way to the Stanton Ironworks.

Product-related services are perhaps the most interesting and important part of the services duality. Detailed further research is required to explore the ways in which companies are weaving service and manufacturing functions together. Manufacturing firms can develop innovative training courses targeted at consumers as well as transform themselves into manufacturing firms that are also retailers. They can also transform themselves into project managers and installers. Rautaruukki, a Finnish steelmaker that manufactures roofs and gutter components, has begun to undertake the role of project manager in building contracts. The company supplies the roofing materials and is also responsible for the management of the installation. According to Sakari Tamminen, Chief Executive of Rautarukki: 'in some instances of work we do for customers in construction of steel, only 20 per cent (of the contract price) is accounted for by the material cost. The rest comes from what we are providing through design, intellectual property and management expertise' (Marsh, 2006, 10). It is possible to produce a typology of product-related services or hybrid products (Table 28.1). This identifies two extremes. First, are basic hybrid products that involve the customisation of goods and, second, pure hybrid products in which a good is transformed into a service. It is to these pure hybrid products that we now turn our attention.

Rolls-Royce, the provider of power systems for use on land, at sea and in the air, has developed a service-based business model. In the financial year 2006–07, 55 per cent of this company's revenues were derived from the delivery of services (Rolls-Royce, 2007, 15). In 1987 Rolls-Royce 'supported our engines in service by offering repair and overhaul arrangements which often failed to align our interests with those of our customers' (Rolls-Royce, 2007, 14). At this time, services were considered as a supporting set of functions rather than as an integral element within the firm's business model. Since 1987, Rolls-Royce has transformed itself into a provider of power rather than a provider of engines. This transformation has occurred in the firm's core market segments ranging from civil aviation to

Table 28.1 A typology of product-related services or hybrid products

	Product-related Services	Characteristics
Basic hybrid product	Customisation of manufactured product	Design and fabrication of a product to meet client requirements
	Development of additional product-related services	Servicing, service packages, provision of training, finance packages, updates
	Project management	Design, fabrication, installation and project management of manufactured goods supplied by a company
	Facility management	Provision of a good and also a contract to manage servicing and updating of the good
	Provision of content	Company supplies a product, but the primary source of value creation is through the delivery of service content, for example the Apple iPod and iTunes, or the manufacture of a computer printer and the subsequent supply of toner or ink
Pure hybrid product	Conversion of a good into a pure service	Company supplies a service that is provided by a product; the client does not own the product, but has access to a set of service outputs, for example the provision of heat, dehumidified air or power

defence aerospace and marine engines. The transformation has involved the company in developing:

> comprehensive through-life service arrangements in each of our business sectors. These align our interests with those of our customers and enable us to add value through the application of our skills and knowledge of the product. In 2007, underlying aftermarket service revenues grew by nine per cent and represented 55 per cent of Group sales. This growth has been achieved partly as a result of the introduction of new products, but also because our ownership of intellectual property enables us to turn data into information that adds value to our customer (Rolls-Royce, 2007, 14).

A good example of this shift is the mission-ready management solutions (MRMS ®) package provided by Rolls-Royce. MRMS provides the military with customised solutions that include total support packages and 'Power by the Hour'®. With the latter package, major airline and defence customers pay a fixed warranty and operation fee for the hours that an engine runs. Contract performance is measured against the performance of the fleet and in terms of ready for issue engine availability. In this case, Rolls-Royce no longer sells a good (an engine), but a service (power by the hour).

Rolls-Royce offers three types of service solution. First, TotalCare is based upon an agreed rate per engine flying hour and this enables customers to engage in accurate financial forecasting. This package is designed for airline fleets and it transfers the technical and financial aspects of fleet maintenance away from the customer to the service supplier. At the same time, it converts Rolls-Royce into a service provider or, more precisely, a provider of hybrid products. Second, CorporateCare is intended for corporate and business jet customers and is designed to ensure that the aircraft is available when required and also may result in increased residual value. Third, MRMS is targeted at defence customers and provides them with engine management and maintenance to ensure 24/7 operational capability. These types of hybrid products (good and service) have transformed Rolls-Royce from a company that designs and manufactures engines to a provider of turnkey engine power (Table 28.2). To maximise profitability, Rolls-Royce must now focus on the effective management of an extended manufacturing value chain or its hybrid production system. This includes the design and development of engines, installation, after-sales maintenance, repair and overall services and parts availability and management.

28.5 Conclusion

This chapter has explored an important trend that is altering the ways in which goods, products and services are produced and consumed. The development of hybrid production systems and hybrid products is a radical innovation in the production process, but not a completely new innovation. On the one hand, this represents a breakdown of the long-standing simple bipolar distinction that is made between services and manufacturing. This division is increasingly no longer helpful as manufactured products acquire many of the characteristics of services. On the other hand, this represents the intensification of a process that perhaps began during the early years of the twentieth century.

The identification of a service duality provides an interesting conceptual tool for theorising the ways in which services contribute value to production processes. In this context, it is important to differentiate between production and operations; production refers to the complete process, whilst operations refers to the manufacturing and service functions that

Table 28.2 The transformation of Rolls-Royce from a provider of a good to a provider of hybrid products

Good	⟶			Service
Delivery of engine				Delivery of power
Traditional Support	Enhanced Support	Advanced Support	Total Support	Extended support
Spare parts Repair & Overhaul	Data & forecasting services Technical & logistics support Customer training	Comprehensive package integrating elements of basic and enhanced support Spares inclusive Repair & Overhaul contracts	Complete, availability-based services. Can cover all off-aircraft, and some one-aircraft activity Configuration management and reliability enhancements covered	Partnered capability Turnkey service Non-propulsion related support solutions
Customer responsibility	⟶			Service provider responsibility

Source: after Rolls-Royce (2006).

are required to create products (goods and services). This distinction between production and operations highlights the importance of distinguishing between service operations that are directly integrated into the process of production, and service functions that are intended to add value during and after the moment of final consumption. This distinction reflects the service duality and the identification of a set of production- and product-related services. An interesting and important issue that needs to be explored concerns the identification of the differences and similarities that might exist between the innovation processes that operate within the services duality. Product-related services would appear to be especially interesting as the innovation process blends elements from manufacturing and services together, or perhaps the good just represents a service delivery mechanism.

A substantial body of research has already been undertaken into production-related services (Bryson et al., 2004), but product-related services have been neglected. It is time for academics to begin to explore the complex ways in which companies innovate and create value by blending service and manufacturing functions together in hybrid production systems to create hybrid products.

References

Adams, W.J. and J.L. Yellen (1976), 'Commodity bundling and the burden of monopoly', *Quarterly Journal of Economics*, **40**, 475–88.

Baldwin, R. (2006), 'Globalisation: the great unbundling(s)', Secretariat of the Economic Council, Finnish Prime Minister's Office: Helsinki.

Barcet, A. (1987), 'La montée des services: Vers une économie de la servuction', PhD thesis, Université Lyon-Lumière.

Barras, R. (1986), 'Towards a theory of innovation in services', *Research Policy*, **15**, 161–73.

BERR (2008), *Manufacturing: New Challenges, New Opportunities*, London: BERR.

Bhagwati, J. (1984), 'Splintering and disembodiment of services and developing nations', *World Economy*, **7** (2), 133–144.

Bhagwati, J., A. Panagariya and T.N. Srinivasan (2004), 'The muddles over outsourcing', *Journal of Economic Perspectives*, **18** (4), 93–114.

Blackstone, J.H., L.R. Gardiner and S.C. Gardiner (1997), 'A framework for the systemic control of organizations', *International Journal of Production Research*, **35** (3) 597–609.

Bressand, A. (1986), 'Europe in the new international division of labour in the field of services: the need for a new paradigm', Report No. 2 to the European Commission, PROMETHEE: Paris.

Bressand, A. (1989), 'Computer reservation systems: networks shaping markets', in A. Bressand and K. Nicolaïdis (eds), *Strategic Trends in Services: An Inquiry into the Global Service Economy*, New York: Harper & Row, pp. 51–64.

Bressand, A., C. Distler and K. Nicolaïdis (1989), 'Networks at the Heart of the Service Economy', in A. Bressand and K. Nicolaïdis (eds), *Strategic Trends in Services: An Inquiry into the Global Service Economy*, New York: Harper & Row, pp. 17–32.

Bryson, J.R. (1997), 'Business service firms, service space and the management of change', *Entrepreneurship and Regional Development*, **9**, 93–111.

Bryson, J.R. (2000), 'Spreading the message: management consultants and the shaping of economic geographies in time and space', in J.R. Bryson, P.W. Daniels, N. Henry and J. Pollard (eds), *Knowledge Space, Economy*, London: Routledge, pp. 157–175.

Bryson, J.R. (2007), 'A "second" global shift? The offshoring or global sourcing of corporate services and the rise of distanciated emotional labour', *Geografiska Annaler*, **89B** (S1), 31–43.

Bryson, J.R. (2009), *Hybrid Manufacturing Systems and Hybrid Products: Services, Production and Industrialisation*, Aachen: University of Aachen.

Bryson, J.R. and P.W. Daniels (2009), 'Dualidad de los serviciou y economía semiindustrial: la interactión entre serviciou e industria desde un análisis de producción, proyectos y tareas' ('Services duality' and the 'manuservice' economy: a production, projects and tasks approach to understanding interactions between services and manufacturing), *Papeles de Economia Española*, **120** (May), 186–199.

Bryson, J.R., P.W. Daniels and B. Warf (2004), *Service Worlds: People, Organizations, Technologies*, New York, USA and London, UK: Routledge.

Bryson, J.R. and M.C. Monnoyer (2004), 'Understanding the Relationship between Services and Innovation: the Reser review of the European service literature on innovation', *Service Industries Journal*, **24** (1), 205–222.

Bryson, J.R and G. Rusten (2006), 'Spatial divisions of expertise and transnational

"service" firms: aerospace and management consultancy' in J.W. Harrington and P.W. Daniels (eds), *Knowledge-based Services, Internationalisation and Regional Development*, Aldershot: Ashgate, pp. 79–100.
Bryson, J.R. and G. Rusten (2008), 'Transnational corporations and spatial divisions of "service" expertise as a competitive strategy: the example of 3M and Boeing', *Service Industries Journal*, **28** (3), 307–323.
Bryson, J.R., M. Taylor and P.W. Daniels (2008), 'Commercializing "creative" expertise: business and professional services and regional economic development in the West Midlands, UK', *Politics and Policy*, **36** (2), 306–328.
Crum, R.E. and G. Gudgin (1977), *Non-Production Activities in UK Manufacturing Industry*, Regional Policy Series 95, No. 3, Brussels: Commission of the European Communities.
Daniels, P.W. and J.R. Bryson (2002), 'Manufacturing services and servicing manufacturing: changing forms of production in advanced capitals economies', *Urban Studies*, **39** (5–6), 977–991.
Department of Trade and Industry (DTI) (2004), *Competing in the Global Economy: The Manufacturing Strategy Two Years On*, London: Department of Trade and Industry.
Fingleton, E. (1999), *In Praise of Hard Industries: Why Manufacturing, Not the New Economy, is the Key to Future Prosperity*, London: Orion.
Gallouj, F. (2002), *Innovation in the Service Economy: The New Wealth of Nations*, Cheltenham, UK and Northampton, MA, USA: Edward Elgar.
Hansen, M.T. and J. Birkinshaw (2007), 'The innovation value chain', *Harvard Business Review*, **85** (6), 121–130.
Hill, P. (1977), 'On goods and services', *Review of Income and Wealth*, **23** (4), December, 315–338.
Hermann, A., F. Huber and R.H. Coulter (1997), 'Product and service bundling decisions and their effects on purchase intention', *Pricing Strategy and Practice*, **5** (3), 99–107.
Howells, J. (2000), 'The nature of innovation in services', report presented to the OECD Innovation and Productivity Workshop, Sydney, Australia, October.
Howells, J. (2002), 'Innovation, consumption and services: encapsulation and the combinational role of services', paper presented at the Services and Innovation 12th International RESER Conference, 26–27 September, Manchester.
Illeris, S. (1996), *The Service Economy: A Geographical Approach*, Chichester: John Wiley.
Lewis, V. (1959), *The Iron Dale*, Nottingham: Stanton Ironworks Company.
Livesey, F. (2006), *Defining High Value Manufacturing*, Cambridge: Institute for Manufacturing.
Marsh, P. (2006), 'Dawn of steel's flexible future hybrid manufacturing', *Financial Times*, 14 November, p. 10.
Pilat, D. and A. Wölfl (2005), 'Measuring the interaction between manufacturing and services', Statistical Analysis of Science, Technology and Industry, STI Working Paper 2005/5, OECD, Paris.
Rolls-Royce (2007), *Annual Report 2007: A Global Business*, London: Rolls-Royce.
Rolls-Royce (2006), *MRMS® Mission Ready Management Solutions*, London: Rolls-Royce.
Rubalcaba, L. and H. Kox (2007), *Business Services in European Economic Growth*, Basingstoke: Palgrave.
Rubalcaba-Bermejo, L. (1999), *Business Services in European Industry: Growth, Employment and Competitiveness*, Luxembourg: Office for Official Publications of the European Communities.
Smith A. (1977 [1776]), *The Wealth of Nations*, Harmondsworth: Penguin Books.
Yadav., M.S. and K.B. Monroe (1993), 'How buyers perceive savings in a bundle price: an examination of a bundle's transaction value', *Journal of Marketing Research*, **30**, 350–358.

29 A customer relationship typology of product services strategies

Olivier Furrer

29.1 Introduction

With the advent of the service economy (Gadrey, 2005), product services (i.e. services offered as complements to tangible products) have taken on critical roles in the competitive arsenal of many manufacturing firms (Furrer, 1997, 1998; Gebauer et al., 2005; Malleret, 2006). For example, IBM has become a service provider more than a manufacturer of tangible products (*BusinessWeek*, 2005). Following Anderson and Narus (1995), this chapter considers product services to include much more than after-sales service, such as technical problem-solving, equipment installation, training or maintenance. Rather, product services also include programs that help customers design their products or reduce their costs, as well as rebates or bonuses that influence how customers conduct business with a supplier.

Despite their increasing managerial importance, academic research on the strategic role of product services remains embryonic (see Bowen et al., 1989; Dornier, 1990; Furrer, 1997; Horovitz, 1987; Mathe and Shapiro, 1993), and the concept still appears vague and ambiguous. Nor has existing research integrated product services into a coherent conceptual framework. Therefore, this chapter further refines the concept of product services and integrates it into a relationship marketing framework (Berry, 1995; Sheth and Parvatiyar, 1995), which suggests a consistent and managerially relevant typology of product services strategies.

The remainder of this chapter is organized as follows. First, in section 29.2, I define product services and discuss their strategic role, which depends on their position on the tangible product–service continuum. In section 29.3 I present a typology of four product service strategies – discount strategy, relational strategy, individual strategy and outsourcing strategy – illustrated by best practice examples that highlight the value-creation mechanisms. I then present in section 29.4 the required conditions for successful implementations of product service strategies. Section 29.5 concludes.

29.2 Product services: a definition

One of the first definitions of the concept of product services, proposed by Caussin (1955), highlights the measures that a supplier takes to facilitate the choice, purchase and use of a tangible product. Later, Horovitz (1987) defined product services further as all the benefits expected by a customer that go beyond the core product. Similarly, Davidow and Uttal (1989) refer to product services as the features, acts and information that increase customers' ability to leverage the value of a tangible or intangible core product. On the basis of an extensive literature review, Furrer (1997, 99) proposes the following comprehensive definition:

> Product services are services that are supplied complementary to a product to facilitate its choice and its purchase, to optimize its use and to increase its value for customers. For the firm providing them, they are a direct and indirect source of profit: direct because they are often more profitable than the product they surround and indirect because when expected by customers they induce demand for the product and are a source of differentiation on the firm's offering.

This definition highlights the strategic role of product services as direct or indirect sources of profit and competitive advantage, as well as stressing that product services should be valuable not only for the firms providing them but also for customers. This latter element is critical for integrating services in a relationship marketing framework, because relationship marketing aims to establish, develop and preserve long-term relationships that are profitable for both firms and their customers (e.g. Berry, 1995; Berry and Parasuraman, 1991; Sheth and Parvatiyar, 1995). Product services create value for firms by helping them attract new customers and reducing customer turnover, as well as by providing direct profits. For customers, product services create value because they offer benefits such as cost reductions, time saving, information, risk reduction, and ease of use (Furrer, 1997).

29.2.1 The strategic roles of product services

Product services' strategic roles rely on the assumption that the services are connected to the tangible products to which they are added. If services can be profitably detached from the products they surround, they become undifferentiated from regular 'pure' services and can be competitively provided by independent service firms. In such a case, firms that provide the core products may lose any competitive advantage.

However, for such firms, product services directly contribute to firm profits because the firm earns the fees charged to customers for the services they receive. Product services also indirectly contribute to a firm's profit because they induce and leverage sales of tangible products. In this sense, product services represent both a requisite condition for the sale of core products and a catalyst in the relationship between the firm and its customers (Mathe, 1990). Moreover, some services can minimize competitive

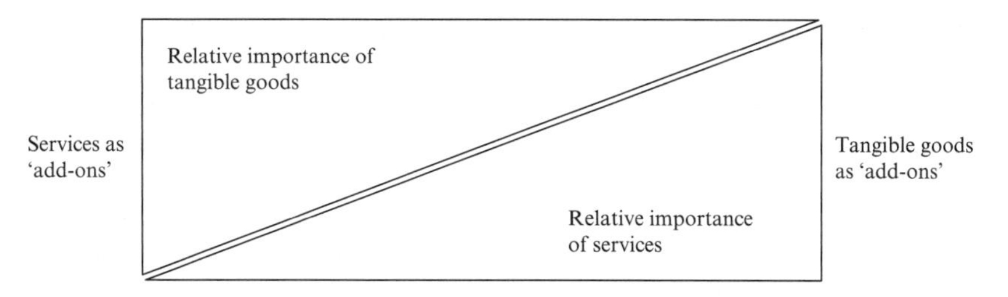

Source: adapted from Oliva and Kallenberg (2003).

Figure 29.1 The tangible products–services continuum

pricing pressures by differentiating a firm's products (Bircher, 1988; Levitt, 1980; Porter, 1985). Because the investments required to provide them are often high and risky, product services also erect entry barriers (Davidow and Uttal, 1989). Finally, product services help firms establish a dependence relationship with customers because of the switching costs they create for those who might wish to change suppliers (Vandermerwe and Rada, 1988).

29.2.2 Product–service combinations

Firms might combine their core products and services in different ways to achieve their strategic objectives. Product–service combinations thus can be arranged on a tangible product–service continuum (Gebauer et al., 2005; Gebauer and Friedli, 2005; Neu and Brown, 2005; Oliva and Kallenberg, 2003). One end of the continuum is dominated by the tangible elements of the combination, such that services are only 'add-ons' to the tangible products, whereas the other end consists mainly of services, and tangible products appear only as add-ons to these services (Figure 29.1).

When the provider of a tangible product also offers add-on services, it can use the services as a source of differentiation (Furrer, 1997; Levitt, 1980). With such a combination, the firm's benefits and revenues mainly come from the sales of the tangible products, whereas the financial contribution of the service elements remains relatively marginal. Service-oriented firms that offer products as add-ons to their services, in contrast, earn only a small part of their total value from tangible products and instead rely mostly on services.

Increasingly, manufacturers of tangible products, such as IBM and General Electric (GE), are moving along the continuum to become service providers by proposing tangible product–service combinations dominated by the service part (Gebauer and Friedli, 2005; Neu and Brown, 2005; Quinn et al., 1990). By 2000, GE generated approximately 75 per cent of its revenues from services (General Electric, 2000). Such shifts represent responses to two powerful forces (Dornier, 1990; Furrer, 1997; Mathe, 1990). On the one hand, producers of tangible products offer more services because services represent potential sources of sustainable competitive advantage. On the other hand, customers are asking for more services, which provide them with benefits in terms of cost reduction, time savings, increased knowledge and information, risk and uncertainty reduction, ease of use and reinforced image, social status or prestige.

Another traditional manufacturing firm, Caterpillar (CAT)[1] still offers a dominant tangible offering (farm and construction equipment, such as tractors and bulldozers) but also provides add-on services to improve its competitive position. For CAT, service does not end with the sale of the equipment. Rather, with the purchase of equipment, CAT offers experience and after-sales service, such that the sale represents only the beginning of a long-term relationship. To improve performance among its business customers, CAT offers solutions that minimize downtime, reduce costs and guarantee optimal operating use of equipment. Moreover, CAT dealers provide local services to customers to help them select the right equipment for their work and use the latest technology to monitor their equipment. In addition, CAT dealers have developed experience, knowledge and training and possess the tools necessary to handle all of their customers' maintenance and repair needs throughout the life of the equipment. By offering a wide range of solutions, services and products, CAT dealers help customers lower their costs, increase their productivity and manage their business more effectively.

On the other end of the continuum, Starbucks provides an example of a service company that sells tangible products in addition to serving coffee in an attempt to improve customers' total experience. The company recently began selling jazz and blues CDs, which in some cases include special compilations put together for Starbucks to use as store background music. The Starbucks CDs, some even specifically tied in with new blends of coffee the company promotes, represent an addition to the company's core products.[2]

29.2.3 A relationship marketing framework

To be successful, a product service strategy must change a firm's focus from a transactional type of customer interaction to a relational type (Gabauer et al., 2005). Building relationships with customers appears as one of, if not the, most important tasks in marketing. Some authors even herald a paradigm shift to indicate the importance of relationships within marketing theory and practice (e.g., Grönroos, 1994; Sheth and Parvatiyar, 1995; Vargo and Lusch, 2004; Webster, 1992). In turn,

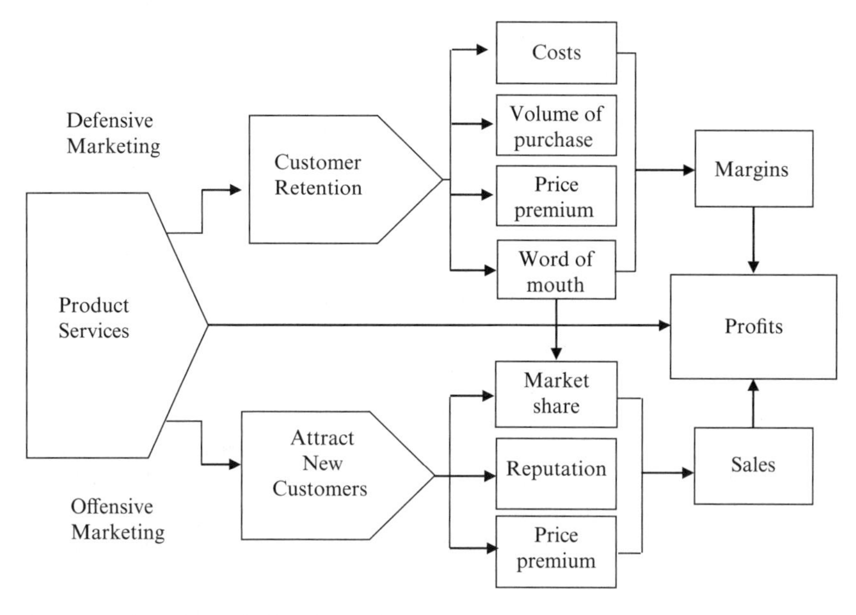

Source: adapted from Zeithaml et al. (2006)

Figure 29.2 Product services' strategic roles

modern marketing literature contains an enormous amount of research on relationship marketing that reveals how closer relationships result in higher profitability and lower costs (e.g. Kalwani and Narayandas, 1995). In 2004, the American Marketing Association even adopted a definition of marketing that explicitly includes customer relationships as a crucial element (AMA, 2004): 'Marketing is an organizational function and a set of processes for creating, communicating and delivering value to customers and for managing customer relationships in ways that benefit the organization and its stakeholders.'

Relationship marketing consists of establishing, developing, and preserving long-term relationships that are profitable for both firms and their customers (e.g. Berry, 1995; Berry and Parasuraman, 1991; Sheth and Parvatiyar, 1995). To integrate product service strategies into the relational marketing framework, we can organize product services strategic roles around two key customer relationship dimensions: attracting new customers and minimizing customer turnover (Fornell and Wernerfelt, 1987; Rust et al., 1995; Zeithaml et al., 2006) (see Figure 29.2). The first dimension consists of an offensive marketing strategy aimed to attract new customers and increase customers' purchase frequency; the second dimension represents a defensive marketing strategy concerned with minimizing customer turnover by retaining current customers and increasing their loyalty. Offensive marketing strives to attract competitors' dissatisfied customers, whereas defensive marketing is geared toward managing dissatisfaction among a firm's own customers (Fornell and Wernerfelt, 1987).

Attracting new customers Product services enable firms to attract new customers through an offensive marketing strategy (Furrer, 1997). Various types of firms, in addition to manufacturers, can provide services to support tangible products, including distributors and dealers, customers themselves or independent service firms (Kotler, 2000). However, for firms that also manufacture tangible products, services help leverage the products' characteristics and increase their attractiveness by better integrating tangible and service elements (Dornier, 1990; Furrer, 1997; Mathe and Shapiro, 1993). Firms that provide services along with their tangible products maintain tighter control over the key activities of the value chain (Porter, 1980; Quinn et al., 1990; Wise and Baumgartner, 1999), which helps them earn higher profits and sustain their competitive advantage. Such an offensive strategy not only increases the sales turnover these firms enjoy but also improves their market share and reputation, which enables them to charge premium prices (Zeithaml et al., 2006). The results of profit impact of market strategy (PIMS) studies empirically demonstrate the financial value of this competitive advantage (Buzzell et al., 1975; Buzzell and Wiersema, 1981) because they reveal that high-quality service companies manage to charge more, grow faster and make more profit on the strength of their superior service quality compared with less service-oriented competitors. Rust et al. (1995) also show that satisfied service customers contribute to the image and reputation of a firm through positive word of mouth, which attracts new customers, increases market share and enhances profitability.

Customer retention Firms also offer product services to develop bonds with their existing customers and differentiate their core products from those of competitors (Levitt, 1980; Porter, 1985). Several empirical studies indicate that customer profitability increases with the length of the customer relationship (e.g. Reichheld, 1993; Reichheld and Sasser, 1990). Therefore, the objective of a product service defensive strategy is to develop and maintain long-term customer relationships (Fornell and Wernerfelt, 1987), in line with the basic argument that the cost of attracting a new customer exceeds the cost of retaining an existing customer. A product service defensive strategy entails two components: improving customer satisfaction and erecting switching barriers. Greater customer satisfaction increases customer consumption and willingness to pay (Zeithaml et al., 1996), and switching barriers make it more costly for customers to switch to alternative suppliers (Colgate and Lang, 2001). Different types of costs (e.g. search, learning and emotional), cognitive effort and risk factors (e.g. financial, psychological and social) constitute switching barriers from the customer's point of view (Colgate and Lang, 2001). In this sense, the firm's profitability improves when it focuses on existing customers because satisfaction leads to lower costs, greater retention and higher revenues (Fornell and Wernerfelt, 1987).

Because of their intangibility, heterogeneity, inseparability and perishability (Zeithaml et al., 1985), as well as the absence of ownership (Judd, 1964; Lovelock and Gummesson, 2004; Rathmell, 1966, 1974), product services generally are more effective than tangible products for establishing long-term relationships and developing customer loyalty (Czepiel and Gilmore, 1987). Indeed, product services help firms stay in touch with their customers more regularly and transform transactional interactions into continuous relationships. By multiplying the occasions of contacts with customers, product services enable firms to remain better informed about the evolution of customer expectations, needs and preferences and to establish a better position from which to offer them other products or services (cross-selling). Cross-selling increases customer loyalty, because when a customer can acquire additional services or products from the same supplier, the number of points on which customer and supplier connect increases, which in turn increases switching costs (Kamakura et al., 1991). Furthermore, loyal and satisfied customers tend to share their experiences with family and friends more than do less loyal or less satisfied customers (Grönroos, 2000). Thus, positive word of mouth attracts new customers to the firm (Rust et al., 1995; Zeithaml et al., 1996).

A defensive marketing strategy can lower total marketing expenditures by substantially reducing the cost of the offensive marketing strategy (Fornell and Wernerfelt, 1987). Providing product services also creates a dependence relationship for customers, such that the comparison of complex and integrated product–service combinations is more difficult, which increases the costs of multiple transactions and therefore makes close relationships comparatively more attractive (Williamson, 1975).

Profitability direct improvement Finally, a direct relationship exists between product services and profitability, because services are often more profitable than the tangible products they support (Furrer, 1999; Gebauer et al., 2005). Tangible products can be too easily bypassed, reverse engineered, cloned or slightly surpassed to offer a source of real, sustainable competitive advantage (Quinn et al., 1990). Furthermore, whereas the direct competition of tangible product characteristics reduces margins, the intangibility of services makes them more difficult to compare and therefore less sensitive to competitive pressures (Bircher, 1988; Porter, 1985). When comparison is difficult, the likelihood of a price war between competitors declines, and margins could increase. In addition, the sale of tangible products with a long life cycle often involves only punctual transactions, repeated several months or years later. For the supplier, this situation represents a source of irregularity and uncertainty in sales turnover. Because the sale of the services that support installed products' base is continuous and more stable (e.g. maintenance contracts, after-sales services), product services allow firms to generate regular incomes, which facilitates their cash-flow management and provides them with the ability to face business cycle downturns (Dornier, 1990; Furrer, 1997; Gebauer et al., 2005).

29.3 Product services strategies

In the previous section, we focused on reasons that support the development of product services offerings and their advantages. In this section, we propose a typology of product service strategies that is based on the types of bonds (or links) that product services create between firms and their customers. Because product services allow firms to create relational bonds, they contribute to the development and maintenance of long-term relationships with customers. Berry and Parasuraman (1991) and Zeithaml et al. (2006) identify four types of relational bonds at four different levels that can distinguish four generic types of product service strategies.

That is, product service strategies may operate at four different relationship levels, each of which results in tighter customer bonds (Zeithaml et al., 2006): financial, social, customization and structural (Figure 29.3). Financial bonds offer financial benefits and advantages, such as price discounts or financing solutions, to customers (Levitt, 1980); social bonds provide relational benefits, including confidence benefits, social benefits and special treatment benefits (Gwinner et al., 1998; Hennig-Thurau et al., 2002); customization bonds seek to establish an individual relationship with each customer by personalizing the core products and services (Furrer, 1997, 1998); and structural bonds result from providing services that are designed into the value chain and service delivery system of each customer (Zeithaml et al., 2006).

Each type of bond provides the cornerstone of a different product service strategy: the discount strategy seeks to establish financial bonds with customers; the relational strategy uses product services to establish social bonds; the individual strategy exploits product services to establish

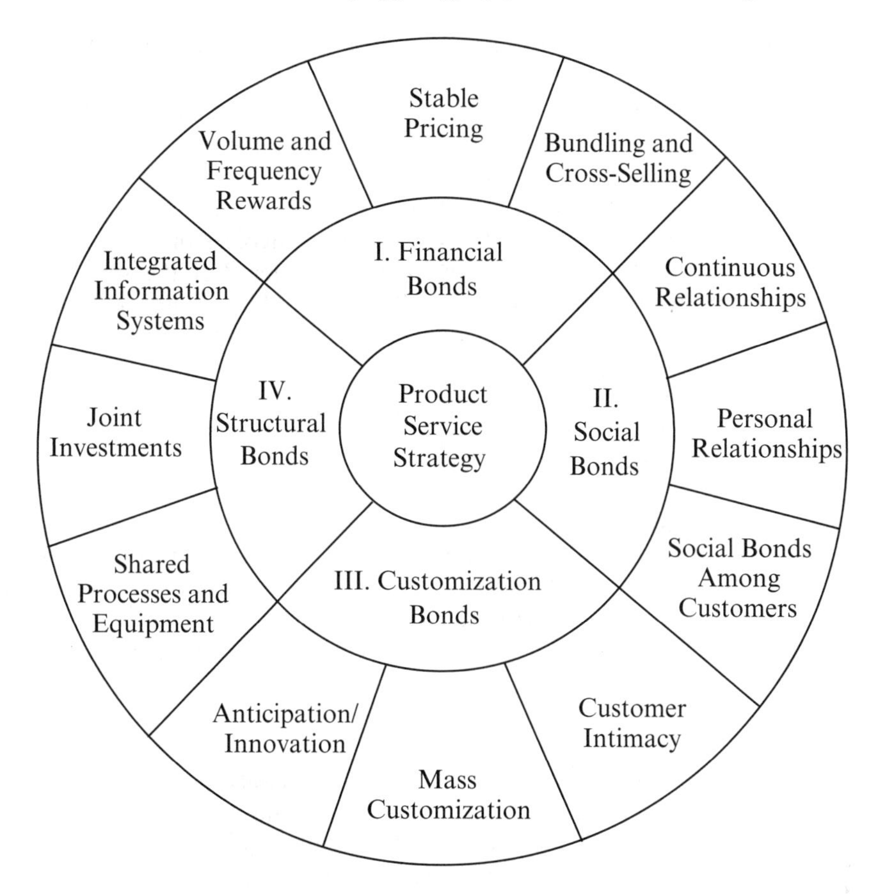

Source: adapted from Berry and Parasuraman (1991) and Zeithaml et al. (2006).

Figure 29.3 Different types of bonds

personalized relationships with each customer; and the outsourcing strategy attempts to establish structural bonds with customers. In the next subsections, we describe each of these strategies and illustrate them with best-practice examples.

29.3.1 The discount strategy

At the first relationship level lies the discount strategy. Firms using a discount strategy mainly seek to establish financial bonds with their customers, with the assumption that customers engage in relationships with providers for financial reasons and prefer financial benefits such as price reductions, promotions and free gifts. The main objective of this strategy therefore is to attract new customers by offering them financial advantages and retain them by creating switching barriers, such as termination costs to split from the current service provider or joining costs for a new service provider (Colgate and Lang, 2001; Fornell and Wernerfelt, 1987).

The discount strategy contains two variants, depending on the position of the firm's offering along the tangible products–services continuum. When tangible products dominate and services are only add-ons, firms offer free or discounted services. For example, retailers could offer a free guarantee extension for electronic equipment or an automobile dealer could buy back the customer's used car if that customer purchases a new model. Because firms' profits and revenues mainly come from the sales of tangible products, the role of product services is mainly to attract new customers and develop customer loyalty through the advantages associated with continuing the relationship, such as a loyalty card with extra benefits.

In 2007, Wal-Mart, the world's largest retailer, announced that it would offer a host of financial services to its customers through Wal-Mart MoneyCenters (Gogol, 2007). As part of its services, as of 2008 the company issued the Wal-Mart MoneyCard, a prepaid Visa card, at a cost of US$8.95. It can be used like a credit card in Wal-Mart stores and to shop online. By facilitating payments and credits, the Wal-Mart MoneyCard encourages customers to buy more from Wal-Mart and, at the same time, creates switching barriers. In Switzerland, Migros and Coop, the two largest national retailers, have proposed their own unbranded credit cards with the same rationales.

At the other end of the continuum, when firms mainly offer intangible core services and add tangible products or equipment to support the services, the discount strategy consists of offering tangible products for free or at a discount price to encourage the sales of the more profitable services. This strategy is most recognizable in the mobile telecommunication service industry, in which firms offer free mobile phones to customers who subscribe to services for one or two years. Some Internet access providers even offer free computers or laptops to service customers in a similar type of arrangement. In 2005, the French satellite television company Canal+ promised a free satellite dish and installation to new customers who signed up for a one-year subscription to 'Canal+ le Bouquet'. In this case, offering free equipment or tangible products simplifies customer access to the core services and works as an exit barrier, because customers are bound to the firm by their subscription or long-term contract.

The discount strategy thus is a popular product service strategy, because it is relatively easy to implement and has an immediate positive effect on sales. Unfortunately, customers' financial motivations, which represent the center of this strategy, do not generally develop into sustainable competitive advantages, unless firms combine this strategy with another. The discount strategy prevents firms from differentiating their offerings for a long period of time, because price is the easiest marketing mix component to imitate (Kotler, 2000). Moreover, customers who are most sensitive to the financial aspects of this strategy are also those most easily swayed by competitors that propose a new product–service combination with better financial advantages. Therefore, the discount strategy is particularly suited to attracting new customers but less efficient at keeping them loyal or reducing customer turnover. Moreover, offering products or services free of charge or at a discounted price has a negative direct effect on the firm's profitability. Some customers even may ask for more free services and support than the firm can afford (Lovelock, 1994). In such a situation, knowing when and how to say no can make the difference between a winning discount strategy and a losing one.

29.3.2 The relational strategy

At the second relationship level, relational strategies seek to establish social bonds between the firm and customers, so that customers relate to the firm not only through financial bonds but also through social and interpersonal bonds that characterize an emotional relationship and entail personal recognition of customers by employees, the customer's familiarity with employees, and the creation of friendships between customers and employees (Gwinner et al., 1998). With this strategy, customers are no longer considered as numbers; instead, they are viewed as individual human beings with desires and needs that firms must understand to satisfy. Social bonds allow firms to establish trusting relationships with their customers, who then become more committed to the firm, more loyal and more likely to spread more positive word of mouth (Berry, 1995; Hennig-Thurau et al., 2002). The focus of this strategy therefore is on keeping existing customers, reducing customer turnover, increasing customer commitment and loyalty, and encouraging customers to buy more; it is less concentrated on trying to attract new customers.

To develop and maintain social bonds, it requires a combination of human interaction and technology at both ends of the tangible product–service continuum. For example, Prada uses radio frequency identification (RFID) technology to enrich the shopping experience for customers in its New York City store by offering supplementary services with its products. The RFID smart labels identify merchandise and customers, and link shoppers with information about specific products. The devices also control video screens throughout the store, which demonstrate appropriate products on the runway and display collection photographs and designer sketches. The video screens provide more in-depth information about the color, cut, fabric and materials used to create Prada merchandise. In the Prada dressing rooms, RFID readers identify all merchandise a customer brings inside and display information about the garments on the interactive video touch-screen display. As soon as the garments get hung up in dressing rooms, the wireless readers capture information from their smart tags, which then appears on a plasma screen beside the closet. A customer can choose to view, for example, information about the designer or other colors available (Green, 2002; Hill, 2002; Schoenberger, 2002).

Firms that implement a relational strategy use product services to facilitate interactive communication with their customers, through which they receive information about their customers' preferences and the way they use the products. In turn, they can use this information to develop new products and services or increase customer satisfaction. To better understand its customers' needs when it comes to the style and characteristics of its Punto, Fiat developed a Web-based software that allows customers to evaluate several automobile concepts (Iansiti and MacCormack, 1997). Customers may fill in a survey to indicate their preferences in automobile design and characteristics, such as style, comfort, performance, price and security. The survey then asks them to describe what they hate most in a car and make suggestions for new features. Finally, with the software, customers can conceive of and visualize the car of their dreams by choosing from a variety of body styles, wheel designs and style options for the front and rear of the automobile, as well as different types of headlights, details and features. Thus, customers interactively experiment with different designs and see the results immediately on the screen. Fiat received 3000 survey responses in a three-month period, which provided the carmaker with important, detailed information that it could integrate into its development of the new generation of Punto (Iansiti and MacCormack, 1997).

Nespresso, a subsidiary of Nestlé selling premium coffee, offers an example of a firm with a successful relational strategy that relies on a combination of an integrated coffee machine and coffee capsule system, a selection of premier coffees and a club that maintains close relationships with customers. The Nespresso Club represents the cornerstone of the Nespresso system and offers a wide range of services, including coffee capsules and Nespresso accessories, available through the Club worldwide. Open 24 hours per day, seven days per week, the Nespresso Club guarantees delivery of mail, Internet, telephone and fax orders within 48 hours; quickly attends to comments and suggestions by its members; offers customized service; and provides a center of expertise regarding the varieties of coffee, the system, the use of the machine, and maintenance. The Club also provides a direct link with customers, which enables the firm to collect precious information from members – and anyone who buys a Nespresso machine automatically becomes a member of the Club. The Club's more than 1.2 million active members benefit from its services in the main European markets (Austria, Belgium, France, Germany, the United Kingdom, Italy, Norway, Portugal, the Netherlands, Spain and Switzerland), as well as in Australia, Israel, Japan, Russia, the United States and Canada.[3]

Although social bonds alone cannot permanently tie a customer to a firm, because of their interactive nature, especially when based on both technology and human relationships, they are much more difficult to imitate than are financial bonds (Berry and Parasuraman, 1991). If customers lack sufficient reasons to change suppliers, the social bonds created by product services may push them to maintain the relationship they have with their current supplier.

However, a weakness of the relational strategy pertains to customer heterogeneity, such that some customers may not want to be too closely involved in relationships out of fear that they will become too dependent on the supplier (Pillai and Sharma, 2003). Even when a product seems to fit customers' needs perfectly, after the purchase, it may not be possible for the firm to extend and deepen the relationship if the customer is unwilling to do so or unable to assume the costs (monetary or non-monetary) associated with the services over the long term. For example, medical doctors and consultants often see their patients and customers lose interest in their services once they realize what is involved in attaining the desired solution (i.e. long, painful and/or expensive treatments). These customers no longer perceive enough value to maintain the relationship with the service provider. Even if the firm senses a temptation to continue the relationship, it can suffer opportunity costs if it invests time and resources in a relationship with a customer who is willing to switch to another provider at any time.

29.3.3 The individual strategy

At a third relational level, in addition to financial and social bonds, firms seek to establish customization bonds and treat each customer individually (Peppers and Rogers, 1993). With this strategy, firms use their product services to individualize their tangible offerings according to the needs and wants of each individual customer (Dornier, 1990; Riddle, 1986). Developing customization bonds requires an intimate knowledge of each customer and the ability to develop solutions adapted to those individual needs. For example, in some of its stores, Levi Strauss offers customers the possibility of personalizing their jeans using a body scanner and an information processing system that links stores to factories. An employee measures the customer's dimensions with the scanner and asks the customer to try on a pair of jeans to find the ideal size. The information processing system sends these data to the factory, which makes the pair of jeans on demand (McKenna, 1995).

More recently, Adidas[4] unveiled its newest, interactive, high-tech retail store, the Adidas 'mi Innovation Center' (mIC), on the Champs Élysées. The Adidas mIC, a large-scale, futuristic computer, provides a focal point for innovation and customer interaction, including customized technology, style and design, using new technologies such as a configurator, lasers, infrared technologies, commands generated by gesture translation, virtual mirrors, a digital 3D universe and RFID. At one terminal, customers can customize their own 'mi adidas' through a high-tech process that enables each individual to get the same VIP treatment as elite-level Adidas athletes. After selecting the running, tennis, training or soccer options, customers customize their footwear, both aesthetically and according to their personal fit and performance needs. Specifically, consumers run toward the terminal's 'cube' on a computerized catwalk, followed by a virtual runner that records their running style. Sensors embedded in the track record the pressure of their footfalls and gauge running posture. These data, along with the customer's personal information, ensure that both Adidas shoes fit perfectly. Next, the consumer can customize the shoe's aesthetics to their exact specifications. A large flat-screen configurator allows each person to alter the finest details of the shoes with the point of a finger. Laser and infrared technology then translate the gestures into commands. The virtual mirror shows the customer the personalized shoe on their foot, using camera tracking and highly specialized software that merges a digital, 3D show model with its mirror image. Finally, if the consumer places an order for the individually designed shoes, in a few weeks, the shoes will be delivered to their doorstep.

In another terminal, a digital, 3D universe called 'Info Space' allows customers to enjoy Adidas innovations and films in real time by simply pointing to the desired objects. In addition to the design cube, the mIC also features a 'Consulting Zone' and a 'Scan Table' that displays information about the newest Adidas footwear styles. At the table, customers move a sliding carriage over the desired shoe to bring up specific product information on the screen, through the use of RFID technology. Throughout their mIC experience, customers are accompanied by specially trained Adidas experts who, like a personal trainer, give advice on nutrition, exercise and products. With a portable handheld PC, the Adidas experts record each customer's personal details and desires and thus create a user profile that the person can view at their convenience online. Personalized customer service thus represents a critical component of this 'shop of the future'.

The individual strategy comprises two variants, based on firms' positions on the tangible products–services continuum. When the offering is dominated by tangible products and services are only add-ons, firms develop mass customization systems to individualize their customers' products. Dell, for example, launched build-to-order and mass customization trends in the computer industry, which traditionally sold only standard models. Dell allows customers to adapt their computers to their personal needs, which has significantly contributed to the company's success. Mass customization of this sort represents a means to manufacture and sell products by decomposing the characteristics of products and offering them as choices to individual customers. Dell includes approximately 20 characteristics that customers may choose from to create an individualized computer (e.g. RAM, disk size, CPU frequency, modem, operating system) that is adapted to their individual needs.

At the other extremity of the continuum, the individual strategy consists of a service-dominant firm engaging in customer intimacy. Wiersema (1996) illustrates the advantages of customer intimacy with an example. Two types of tailors in Hong Kong are known for their talent and their speed. The best require three fitting and adjusting sessions to create a suit. The others require only one session. Few customers see a priori the need to visit a tailor three times to obtain one outfit, but if they feel confident about the tailor, they come to recognize that these sessions enable them to discover their own tastes and what they really expect. At the end, the customer is happy not only because they have obtained exactly what they wanted at the beginning, but also because their specific needs have benefited from the maximum amount of care.

Entering into customer intimacy thus consists of providing a service and a product that go beyond the explicit requests indicated by the customers. Because of its external and detached point of view of the customer's situation, the provider can help customers define their real needs, even if customers cannot express them. In addition, the interaction with customers enables the firm to anticipate customer needs better. The close and customized relationships created by this strategy also are difficult to replace among the customers who become more loyal and more profitable for the firm.

More and more firms are implementing such individual strategies. For example, Andersen Windows manufactures customized windows adapted to any house. Customers can get their names printed or embroidered on almost anything (e.g. T-shirts, sneakers). However, to implement an individual strategy efficiently and profitably, firms must possess really unique and distinctive capabilities and operational competences (Zipkin, 2001). To gain a sustainable competitive advantage, firms that follow this strategy must make the different elements of their strategy work together as well as individually. Of these elements, elicitation (which allows companies to interact with customers and obtain necessary personal information; Zipkin, 2001), process flexibility (e.g. technologies that allow firms to manufacture products on demand) and logistics (i.e. processes that firms use to deliver the right products to the right customers without error) are particularly crucial. The individual strategy also can be particularly risky because many customers will not pay the additional costs related to customization and may be sufficiently satisfied with cheaper, standardized products and services offered by competitors.

29.3.4 The outsourcing strategy

Finally, product services can enable firms to establish structural relationships with their customers, such that the tangible product becomes an add-on to the outsourcing service. This strategy provides services to the customer that are designed to fit right into its value chain and service delivery system (Zeithaml et al., 2006), and thus attempts to attract new customers and retain them by creating dependence and switching barriers. Furthermore, with this strategy, firms can make better use of some of their distinctive competences and develop economies of scale. The main risk related to this outsourcing strategy, however, lies in customer selection; it simply may not be profitable to establish structural relationships with just any customer.

This product service strategy also is the most difficult for competitors to imitate, because it entails structural bonds in addition to financial, social and customization bonds. Such structural bonds emerge when the firm offers services directly within customers' production systems across the tangible product–service continuum. For example, CAT Rental[5] offers its customers the opportunity to rent equipment rather than buying it. Renting out equipment allows CAT to maintain long-term relationships with customers and thus obtain better knowledge about their needs. Renting also has numerous advantages for customers compared with the purchase of material: equipment is always up to date and reliable, they suffer neither maintenance nor storage costs, and they need not make any costly financial investments. Renting also allows customers to try material prior to buying it and to benefit from the associated services.

One of the leading firms using such a strategy is IBM (*BusinessWeek*, 2005); Marathon Oil recently benefited greatly from IBM's services. In 2002, Marathon Oil's managers wanted to trim costs in the financial department and put dashboards in place to follow operations on a daily basis and thereby make quick adjustments. They called in consultants and researchers from IBM to talk about the issue. When IBM analyzed Marathon Oil's business processes, it offered suggestions to reduce the accounts payable and other processes from 18 days to eight, then built a dashboard on the managers' computers to help them monitor the business evolution. Although some customers prefer to keep control of their business processes and technologies, even those supplied by IBM, Marathon Oil decided to hand over a large part of its financial operations directly to IBM. Other customers even go further and outsource human resources management or customer services. IBM then develops new products and services that it can sell to other customers by using knowledge it acquired with its early customers, such as Marathon Oil.

29.4 When to use a product service strategy

In this chapter, we have defined the concept of product services and distinguished different strategies that firms offering product services can implement. However, most manufacturing firms find it difficult to implement such strategies successfully. In many cases, the transition to a product service strategy results in a larger service offer and higher costs but not necessarily higher profits (Gebauer et al., 2005; Gebauer and Friedli, 2005). Instead of tailoring their service offer to attract new customers or minimize customer turnover, many firms simply add layer upon layer of services to their offerings, without really knowing the value of these services to customers or the costs of providing them (Anderson and Narus, 1995). An important question thus emerges: when is it adequate to pursue a product service strategy? What conditions are best adapted to the development of such a strategy?

According to Dornier (1990), a strategy based on services added to tangible products offers an effective way to gain a sustainable competitive advantage in situations in which competition is strong and markets are saturated and rapidly changing. Furrer (1999) further argues that among the necessary conditions for a successful product service strategy, the competitive environment plays a crucial role; that is, not all competitive situations are suitable for implementing such a strategy. In some cases, a strategy based on low prices, higher-quality products or technological innovation is preferable. However, product service strategies are more effective when markets are saturated, because such markets allow firms to offer services that will increase their revenues and profitability. This benefit also emerges when the competitive environment is characterized by price wars. In such a case, product service strategies allow firms to reduce pressures on their margins (Bircher, 1988) and instead differentiate their tangible products from those of their competitors (Levitt, 1980). Yet these conditions are not insurmountable constraints, and corporations may be inclined to try to differentiate their products to gain a competitive advantage by innovating and developing new product service strategies.

The characteristics of tangible products also can determine the potential success of a product service strategy. Certain tangible products simply require customer and after-sales services. Products that are complex (requiring customer information and training), evolve quickly (requiring updates and upgrades), are radically innovative (requiring customers to be confident about their value), seem durable (requiring repairs and maintenance) or are commodities (requiring differentiation) induce greater demand for product services, which can reduce the uncertainty and risk that customers perceive at the time of purchase or during product use. In such situations, a product service strategy is no longer an option; it is a minimum requirement that the firm must meet to remain in business.

29.5 Conclusion

Product services represent powerful competitive tools for manufacturing firms. They can be combined in different proportions to develop four different types of strategies. Consistent with a marketing relationship framework, these four strategies are based on different types of customer bonds: financial, social, customization and structural. Furthermore, by using these different strategies, firms can attract new customers and reduce customer turnover, as well as increase their profitability. However, despite their power, these strategies may be difficult to implement, especially in particular conditions. Responding to pressures of two powerful forces, manufacturing firms today provide more services to their customers – namely, because such services are a source of sustainable competitive advantage, and because customers are asking for services from which they can benefit.

Notes

1. www.cat.com.
2. www.Starbucks.com.
3. www.nespresso.com.
4. www.adidas.com.
5. www.cat.com.

References

AMA (2004), 'Marketing', *Dictionary of Marketing Terms*, American Marketing Association, http://www.marketingpower.com/mg-dictionary.php.

Anderson, James C. and James A. Narus (1995), 'Capturing the value of supplementary services', *Harvard Business Review*, **73** (3), 75–83.

Berry, Leonard L. (1995), 'Relationship marketing of services: growing interest, emerging perspectives', *Journal of the Academy of Marketing Science*, **23** (4), 236–245.

Berry, Leonard L. and A. Parasuraman (1991), *Marketing Services: Competing Through Quality*, New York: The Free Press.

Bircher, Bruno (1988), 'Dienste um die Production. Informieren, Spezialisieren, Risiken tragen, Systeme verkaufen', in H. Afheldt (ed.), *Erfolge mit Dienstleistung–Initiativen für neue Märkte*, Stuttgart: Puller Verlag, pp. 55–70.

Bowen, David E., Caren Siehl and Benjamin Schneider (1989), 'A framework for analyzing customer service orientations in manufacturing', *Academy of Management Review*, **14** (1), 75–95.
BusinessWeek (2005), 'How IBM's business–services strategy works', 18 April.
Buzzell, Robert D., Bradley T. Gale and Ralph G.M. Sultan (1975), 'Market share: a key to profitability', *Harvard Business Review*, **53** (1), 97–106.
Buzzell, Robert D. and Frederick D. Wiersema (1981), 'Modeling changes in market share: a cross-sectorial analysis', *Strategic Management Journal*, **2** (1), 27–42.
Caussin, Robert (1955), *Service au client*, Brussels: CNBOS.
Colgate, Mark and Bodo Lang (2001), 'Switching barriers in consumer markets: an investigation of the financial service industry', *Journal of Consumer Marketing*, **18** (4), 332–347.
Czepiel, John A. and Robert Gilmore (1987), 'Exploring the concept of loyalty in services', in Carole A. Congram, John A. Czepiel and James Shanahan (eds), *The Service Challenge: Integrating for Competitive Advantage*, Chicago, IL: American Marketing Association, pp. 91–94.
Davidow, William H. and Bro Uttal (1989), 'Service companies: focus or falter', *Harvard Business Review*, **67** (July–August), 77–85.
Dornier, Philippe-Pierre (1990), 'Emergence d'un management de l'après-vente', *Revue Française de Gestion*, **79** (June–August), 12–18.
Fornell, Claes and Birger Wernerfelt (1987), 'Defensive marketing strategy by customer complaint management: a theoretical analysis', *Journal of Marketing Research*, **24** (4), 337–346.
Furrer, Olivier (1997), 'Le rôle stratégique des "services autour des produits"', *Revue française de gestion*, **113** (March–May), 98–108.
Furrer, Olivier (1998), 'Services autour des produits: l'offre des entreprises informatiques', *Revue française du marketing*, **166** (1), 91–105.
Furrer, Olivier (1999), *Services autour des produits: Enjeux et stratégies*, Paris: Economica.
Gadrey, Jean (2005), *L'économie des services*, Collection Repères, Paris: La Découverte.
Gebauer, Heiko, Elgar Fleisch and Thomas Friedli (2005), 'Overcoming the service paradox in manufacturing companies', *European Management Journal*, **23** (1), 14–26.
Gebauer, Heiko and Thomas Friedli (2005), 'Behavioral implications of the transition process from products to services', *Journal of Business and Industrial Marketing*, **20** (2), 70–78.
General Electric (2000), 'Company Data 2000', www.ge.com.
Gogol, Pallavi (2007), 'Why Wal-Mart will help finance customers', *BusinessWeek*, 20 June, www.businessweek.com/bwdaily/dnflash/content/jun2007/db20070620_604513.htm.
Green, Heather (2002), 'The end of the road for bar codes', *BusinessWeek*, 8 July, www.businessweek.com/magazine/content/02_27/b3790093.htm.
Grönroos, Christian (1994), 'From marketing mix to relationship marketing: towards a paradigm shift in marketing', *Asia-Australia Marketing Journal*, **2** (1), 9–29.
Grönroos, Christian (2000), 'Relationship marketing: the Nordic school perspective', in Jagdish Sheth and Atul Parvatiyar (eds), *Handbook of Relationship Marketing*, Thousand Oaks, CA: Sage, pp. 95–118.
Gwinner, Kevin P., Dwayne D. Gremler and Mary Jo Bitner (1998), 'Relational benefits in services industries: the customer's perspective', *Journal of the Academy of Marketing Science*, **26** (Spring), 101–114.
Hennig-Thurau, Thorsten, Kevin P. Gwinner and Dwayne D. Gremler (2002), 'Understanding relationship marketing outcomes: an integration of relational benefits and relationship quality', *Journal of Service Research*, **4** (3), 230–247.
Hill, Kimberly (2002), 'Prada uses smart tags to personalize shopping', 24 April, www.crmdaily.com/perl/story/17420.html.
Horovitz, Jacques (1987), *La qualité de service: A la conquête du client*, Paris: InterÉditions.
Iansiti, Marco and Aland MacCormack (1997), 'Developing products on Internet time', *Harvard Business Review*, **75** (September–October), 108–117.
Judd, Robert C. (1964), 'The case for redefining services', *Journal of Marketing*, **28** (January), 59.

Kalwani, Manohar U. and Narakesari Narayandas (1995), 'Long-term manufacturer supplier relationships: do they pay off for supplier firms?', *Journal of Marketing*, **59** (1), 1–16.
Kamakura, Wagner A., Sridhar N. Ramaswami and Rajenda K. Srivastava (1991), 'Applying latent trait analysis in the evaluation of prospects for cross-selling of financial services', *International Journal of Research in Marketing*, **8** (4), 329–349.
Kotler, Philip (2000), *Marketing Management*, 10th edn, Upper Saddle River, NJ: Prentice Hall.
Levitt, Theodore (1980), 'Marketing success through differentiation – of anything', *Harvard Business Review*, **58** (January–February), 83–91.
Lovelock, Christopher (1994), *Product Plus: How Product + Service = Competitive Advantage*, New York: McGraw-Hill.
Lovelock, Christopher and Evert Gummesson (2004), 'Whither services marketing? In search of a new paradigm and fresh perspectives', *Journal of Service Research*, **7** (1), 20–41.
Malleret, Véronique (2006), 'Value creation through service offers', *European Management Journal*, **24** (1), 106–116.
Mathe, Hervé (1990), 'Le service mix', *Revue Française de Gestion*, **78** (March–May), 25–40.
Mathe, Hervé and Roy D. Shapiro (1993), *Integrating Service Strategy in the Manufacturing Company*, London: Chapman & Hall.
McKenna, Regis (1995), 'Real-time marketing', *Harvard Business Review*, **73** (4), 87–95.
Neu, Wayne A. and Stephen W. Brown (2005), 'Forming successful business-to-business services in goods-dominant firms', *Journal of Service Research*, **8** (1), 3–17.
Oliva, Rogelio and Robert Kallenberg (2003), 'Managing the transition from products to services', *International Journal of Service Industry Management*, **14** (2), 160–172.
Peppers, Don and Martha Rogers (1993), *The One to One Future: Building Relationships One Customer at a Time*, New York: Currency Doubleday.
Pillai, Kishare Gopalakrishna and Arun Sharma (2003), 'Mature relationship: why does relational orientation turn into transaction orientation?' *Industrial Marketing Management*, **32** (8), 643–651.
Porter, Michael E. (1980), *Competitive Strategy*, New York: Free Press.
Porter, Michael E. (1985), *Competitive Advantage*, New York: Free Press.
Quinn, James Brian, Thomas L. Doorley and Penny C. Paquette (1990), 'Beyond products: service-based strategy', *Harvard Business Review*, **68** (March–April), 58–67.
Rathmell, John M. (1966), 'What is meant by services?', *Journal of Marketing*, **30** (October), 32–36.
Rathmell, John M. (1974), *Marketing in the Service Sector*, Cambridge, MA: Winthrop.
Reichheld, Frederick F. (1993), 'Loyalty-based management', *Harvard Business Review*, **71** (2), 64–73.
Reichheld, Frederick F. and W. Earl Sasser (1990), 'Zero defections: quality comes to services', *Harvard Business Review*, **68** (5), 105–111.
Riddle, Dorothy (1986), *Service-Led Growth: The Role of the Service Sector in World Development*, New York: Praeger.
Rust, Roland T., Anthony J. Zahorik and Timothy L. Keiningham (1995), 'Return on quality (ROQ): making service quality financially accountable', *Journal of Marketing*, **59** (29), 58–70.
Schoenberger, Chana R. (2002), 'The Internet of things', *Forbes*, 18 March, www.forbes.com/global/2002/0318/092.html.
Sheth, Jagdish N. and Atul Parvatiyar (1995), 'Relationship marketing in consumer markets: Antecedents and consequences', *Journal of the Academy of Marketing Science*, **23** (4), 255–271.
Vandermerwe, Sandra and Juan F. Rada (1988), 'Servitization of business: adding value by adding services', *European Management Journal*, **6** (4), 314–324.
Vargo, Stephen L. and Robert F. Lusch (2004), 'Evolving to a new dominant logic for marketing', *Journal of Marketing*, **68** (1), 1–17.
Webster, Frederick E., Jr (1992), 'The changing role of marketing into the corporation', *Journal of Marketing*, **56** (4), 1–17.

Wiersema, Fred (1996), *Customer Intimacy: Pick Your Partners, Shape Your Culture, Win Together*, San Monica, CA: Knowledge Exchange.
Williamson, E. Oliver (1975), *Market and Hierarchies: Analysis and Antitrust Implications*, New York: Free Press.
Wise, Richard and Peter Baumgartner (1999), 'Go downstream: new profit imperative in manufacturing', *Harvard Business Review*, **77** (September–October), 133–141.
Zeithaml, Valarie A., Leonard L. Berry and A. Parasuraman (1996), 'The behavioral consequences of service quality', *Journal of Marketing*, **60** (2), 31–46.
Zeithaml, Valarie A., Mary Jo Bitner and Dwayne D. Gremler (2006), *Services Marketing: Integrating Customer Focus Across the Firm*, 4th edn, Boston, MA: McGraw-Hill.
Zeithaml, Valarie A., A. Parasuraman, and Leonard L. Berry (1985), 'Problems and strategies in services marketing', *Journal of Marketing*, **49** (Spring), 33–46.
Zipkin, Paul (2001), 'The limit of mass customization', *MIT Sloan Management Review*, **42** (3), 81–87.

30 Innovation in product-related services: the contribution of design theory

Sylvain Lenfle and Christophe Midler

30.1 Introduction

Management researchers and practitioners are currently confronted with a paradox. While much research very clearly shows that our economies are increasingly dependent on tertiary activities (Gadrey, 2003), the substantial amount of literature that has developed around the question of the structure of the design process deals for the most part with development and innovation in the area of tangible goods (Brown and Einsenhardt, 1995). As Thomke (2003) points out, this focus on the question of product innovation probably explains the relative dearth of confirmed methods for the development of new services. The existing studies on innovation in services are in agreement on the following point: the innovation and design process is still largely unformulated in service companies, which partially explains the problems (e.g. the inability to meet deadlines, unsatisfactory quality, etc.) frequently observed during the development of new services (Gallouj and Gallouj, 1996; Jallat, 2000; Flipo, 2001).

This situation is problematic in a context in which competition through innovation is spreading to all sectors and in which the massive deployment of information and communication technologies (ICTs) has a profound impact on service activities (Barras, 1986, 1990; Bancel-Charensol, 1999). This is all the more true given that innovation in services is not limited to 'pure' service companies (e.g. banks, insurance, transportation, business-to-business services, etc.). Manufacturing companies are also associating more and more services with their products in order to differentiate their product ranges by offering their customers solutions better adapted to their needs (Furrer, 1997). The question of management methods adapted to innovation in services thus constitutes both a practical and a theoretical problem. It opens up a new field of investigation for researchers working on the question of the organization of the design and/or management of services.

To study these questions, in 2001 we made contact with one of the principal European car manufacturers, here identified as Telcar for reasons of confidentiality. After a presentation of our previous research on managing innovative projects (Lenfle and Midler, 2002), Telcar gave us permission to study the case of the Emergency Call service, whose development was then just beginning. Following the paradigm of grounded research (Miles and Huberman, 1994; Eisenhardt, 1989; Yin, 2003) our analysis is thus built on detailed field notes – interview notes, transcripts of project meetings, company documents – compiled into detailed case studies for each phase of the design process. This process was iterative as the cases were frequently updated after follow-up discussions with respondents.

The automotive industry is a good example of an industry offering a product–service combination (Eiglier and Langeard, 1987). Although historically in this sector the product–service pair has been dominated by the product, manufacturers are offering their customers a full range of services aimed at facilitating the purchase of the vehicle (i.e. credit), its upkeep (i.e. maintenance), its availability (i.e. assistance in the event of a breakdown) or a package of all three (i.e. a monthly lease including these different services). This tendency to develop services related to the vehicle has recently been strengthened:

1. By creating service packages that until recently were offered independently.[1] The goal here is to get the customer to accept a monthly lease payment for the vehicle.
2. By using ICTs to offer customers new types of 'telematics services' (address transmission, navigational aids, remote maintenance, emergency and breakdown calls, etc.). This opening up of potential areas of development as regards service constitutes an important innovation for manufacturers who are venturing into a field of which they do not yet have a full understanding.

It is this second trend that we have had the opportunity to study at Telcar by participating in the design process for new product-related services. In a preceding article (Lenfle and Midler, 2003), we have presented the challenges related to the development of telematics services and the organizational solution adopted by Telcar. At that time, we analysed the interest of deploying a cross-disciplinary team (the telematic platform) dedicated to the exploration of this field of innovation.

In this chapter, we would like to continue the analysis of the new service design process by focusing on the unfolding of the process and the means of managing it, rather than on the organization of the project. After briefly introducing the problem of innovative service design, we will analyse the contributions of the literature on services. This will lead us to propose a model of the design process for a service based on recent trends in design theory. Using the emergency call service proposed by Telcar as an example, we will then show the various possible applications of the proposed model and its interest for designers.

30.2 Designing innovative services: a review of the literature

Innovation in services encounters the generic problems characteristic of any innovation process, which can be roughly broken down into two categories (Van de Ven, 1986):

1. The generation of innovative ideas and new concepts, i.e. the invention stage, strictly speaking.
2. The management of the innovation project. In other words, once the invention stage has been 'completed', the challenge is to market a product/service using this invention. This implies both coordinating the design process, which refers back to the management of innovative projects (Lenfle, 2004, 2008), and establishing a favourable environment for innovation (Akrich et al., 1988a and 1988b).

The contributions of the literature on the development of new services on these two questions are very different.

30.2.1 Generating concepts for new services

The first point is infrequently dealt with in the literature on services. The studies on the question generally rely on methods developed for the generation of product concepts (for a review of these methods see Ulrich and Eppinger, 2004) by insisting, in view of the specific characteristics of the service, on the importance of observation of the customer and of the involvement of customer contact personnel. Edvardsson et al. (2000) thus show the interest of ethnographic methods for understanding the needs of users (in the broadest sense of 'customer' and 'customer contact personnel').

This question of the generation of concepts is relevant to the example of telematics services. This is, in fact, a typical example of a field conducive to innovative design, i.e. a field in which neither utilization values, nor the areas of competence required, are defined (Le Masson, 2001). The difficulty is therefore simultaneously to explore these two dimensions. Our involvement in the telematics project offered us the opportunity to observe two different approaches for exploring the field, conducted by multidisciplinary groups (i.e. marketing, information systems, advanced studies, etc.):

- The first is based on the technical possibilities offered by ICTs for the development of new services and/or the enrichment of existing services.

- The opposite approach involves investigating the requests of customers to define the services and technical features of onboard equipment.

In practice, the two approaches are complementary. The first defines the full range of options in light of the technical possibilities of the available equipment, while the second focuses on devising a package of services that will appeal to customers. The exercise is especially delicate, however, inasmuch as the market, by definition, does not yet exist. Customers cannot therefore be requesters of telematics services. Moreover, these approaches are always partial and limited by more or less pressing operational objectives. They are unlikely to provide a broad view of the field of telematics services or to suggest which areas of innovation should be explored. This lack of methods for exploring innovation in services is a problem for those involved. It is this vacuum that we are endeavouring to fill with this chapter.

30.2.2 Managing the innovation process

On the question of the process itself, the literature on innovation management in the service sector offers a number of contributions on the process itself, its content and the difficulties encountered. One research project has investigated the design process for services by drawing widely from studies of the development of new products (Lovelock, 1984). These studies, which are usually prescriptive in nature, offer a general view of the design process for a service by distinguishing the broad stages of the process (see, for instance, the frequently quoted overview provided by Scheuing and Johnson, 1989). The models vary depending on the number of stages involved, but draw freely on the major stages of any design process (i.e. generation of ideas, selection, development and testing of the concept, industrialization, marketing). This is both their strength and weakness. In fact, even though the framework supplied is relatively widely accepted, the models are generally rather poor concerning content, the exact operation of the process and the specific management problems of the various stages.[2] Furthermore, central questions such as the organizational structure required to manage the process or the relationships between the various stages (sequential or concurrent) are rarely dealt with.

Studies on service production systems, however, can shed some light on the content of the design process. In fact, they provide a description of the constituent elements of the service and, as a result, on those elements that need to be developed. The work of Eiglier and Langeard (1987) on 'Servuction' is particularly useful in this context. They distinguish six elements in a servuction system:

1. The customer.
2. The physical support structure (i.e. all of the equipment required for the delivery of the service by specifying the instruments required for the customer and/or the contact personnel, and the environment).
3. Contact personnel.
4. The service, defined as the 'result of the interaction between customers, the physical support structure and contact personnel'.
5. The internal system of organization responsible for the operation of the servuction system.
6. Other customers included in the system.

They then specify that the concept for the service, the target customer segment and the servuction system must be considered simultaneously during the process of service innovation. Their work makes it possible to clarify what is meant by service design, which in addition to the innovative concept, must deal with the question of the type of servuction system to be deployed. In the same spirit, Shostack (1981, 1984) has proposed a tool, with his 'blueprints', for representing the servuction system by distinguishing the front and back office. The objective is to diagram the operation of the service on paper prior to its implementation, in order to anticipate any problems and specify performance and quality criteria. This contribution is interesting because, as Bancel-Charensol and Jougleux (1997) have pointed out, the servuction model considers the back-office process to be a 'black box'. But, as we will see, its implementation raises some formidable design problems when it is based on innovative techniques such as those provided by ICTs.

Finally, a third set of studies is focused on the difficulties observed in the functioning of the design process. There are two distinct schools of thought in this regard.

The first focuses on a particular aspect of the service design process. The central question concerns the impact of the intangibility of the service and of its co-production by the customer. Under these conditions, how does one test the service and ensure that it is performed satisfactorily? The tricky question of the inclusion of the customer in the design process (Edvardsson et al., 2000; Le Masson and Magnusson, 2002) and the question of experimentation when no tangible goods are produced (Thomke, 2003; Abramovici and Bancel-Charensol, 2004) have thus been developed in a particularly interesting way.

A second school of thought uses case studies of innovation, dealing for the most part with 'pure' services (e.g. banks, insurance companies, hotels, business consulting services, etc.), to reveal the difficulties encountered by service companies during the design process (Eiglier and Langeard, 1987;

Gallouj and Gallouj, 1996; Jallat, 2000; Flipo, 2001). Their conclusions are remarkably convergent. They cite a lack of structure in the process (which underlines the normative nature of the models mentioned above), the infrequency of testing, the emphasis placed on technical problems to the detriment of customer involvement, the difficulty in mobilizing contact personnel and so on – all of which we observed on the Telcar project (Lenfle and Midler, 2003).

As we have seen, the literature on the development of new services sheds some light on the various dimensions of the new service design process. However, from the point of view of the designer, this heterogeneity is a problem. The studies presented are in fact either too general to be actually put into practice, or too focused on a particular dimension of the process, or limited to a particular department (generally marketing, or more rarely production[3]). The project team in charge of innovation therefore runs the risk of not having an integrating framework that would allow it to conceive of the process as a whole and include these contributions in the strategic management of the process.

30.3 Design theory and service innovation

To overcome these difficulties, it is necessary to develop a model of the design process for a new product/service that is general enough to provide an overview of the process, and which at the same time will allow the designer to identify potential areas for innovation. The C-K theory, in our opinion, can serve this very function. After presenting the theory, using automobile services as an example, we will show how it allows us to clarify the question of service design.

30.3.1 The C-K theory and the design tree concept

The question of how to represent the design reasoning process has been looked at in numerous studies since the pioneering work of Simon. We will here draw on the theory developed by Hatchuel (1996) and Hatchuel and Weil (2002, 2003). For these authors, at the start of any design process, those involved have at their disposal a knowledge base (K), composed of a body of heterogeneous fields of knowledge (objects, rules, facts, etc.). The design process begins from the moment a question appears that cannot be resolved using the knowledge that we have at the moment. Hatchuel and Weil use the word 'concept' to describe this trigger element for the design process. It is: '1) an object included in K (otherwise, no progress is possible), 2) that we wish to define in such a way that it has properties not present in K or properties also formulated as concepts (e.g. designing a "telephone for teenagers", a "flying boat", and so on)'. This 'semantic disjunction' between the universe of concepts and that of specific fields of

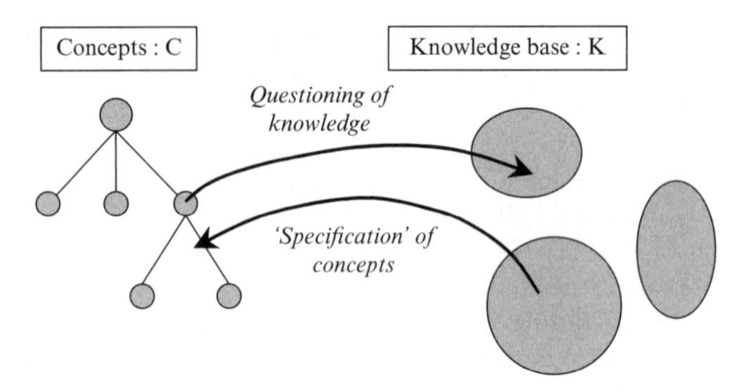

Figure 30.1 The two dimensions of the design process: C and K

knowledge represents the possibility of an action that is unrealizable given the knowledge that we have at the moment. The design process consists in the transition from this desired state to the concrete realization of this state. It takes place simultaneously in two different dimensions.

The knowledge bases (K) make it possible to explore gradually the initial concept and to state it. The authors then show that this exploration is achieved by dividing the original concept into subconcepts, which can be evaluated and, in turn, 'subdivided'. The flying boat thus requires wings, or propellers, or both. In this way, a 'design tree', which portrays the genealogy of the design process, is gradually constituted through the generation of alternatives. But, at the same time, the concepts search through the available knowledge bases. They reveal gaps in the knowledge of those exploring them and thus trigger the development of new knowledge.

The design process therefore involves a continuous interaction between the universe of concepts, which take shape gradually, and the universe of the knowledge bases that are being developed (see Figure 30.1 above). This involves a transition from a concept (the flying boat) to an object (the hydrofoil). The path is obviously not linear. Certain divisions lead to dead ends, requiring the designer to retrace his steps through a process of 'undivision', resulting in 'a more comprehensive concept allowing him to regenerate the divisions implicit in the concept used'. This dynamic design tree thus allows one to maintain one's bearings as regards the history of the design inasmuch as it makes it possible to trace the various solutions found back to the original 'disjunctive' concept.

30.3.2 Application to the case of automotive services

The theory thus invites us to reason simultaneously in two different dimensions (i.e. concepts and knowledge), which was not the case for earlier theories according to which the two dimensions were not explicitly related to each other.

The conceptual dimension The question of the concepts at the origin of the design process refers back to the types of services that the company offers or would like to offer its customers. In this regard, an evolution can be observed in the automobile manufacturing business from the production and sales of cars to the production, sales and leasing of cars and related services. A manufacturer such as Telcar thus offers its customers an increasingly complete range of product-related services that can be broken down into five categories:

- Financing: traditional loans, lease with option to buy, etc.
- Maintenance: service as needed or in the form of package deals, etc.
- Emergency assistance: round-the-clock coverage with the loan of a vehicle when necessary or accommodation if the customer is stranded far from home.
- Insurance.
- Fleet management services.

It should be noted that these categories are not mutually exclusive. Service contracts increasingly combine financing, maintenance and assistance in an all-inclusive package, allowing customers better to manage their automobile budgets. We can thus see an evolution in the concept of the automobile, with the manufacturer increasingly selling a mobility service, for which the support (in the meaning of Eiglier and Langeard, 1987) is an automobile.

The knowledge dimension: defining the design variables for a service Although much has been written on the question of concepts in service innovation (Eiglier and Langeard, 1987), much less has been written, or at least not explicitly, on the knowledge that needs to be developed to market the concepts in question. This implies that the variables involved in the design of a service must be determined or, in other words, those elements that must be developed by the project team to fill out the concept. We have isolated six such elements that summarize and complement the contributions of the studies presented previously, and in particular those concerning service production systems (Eiglier and Langeard, 1987; Bancel-Charensol and Jougleux, 1997). We will examine them now.

First, the target users and uses. The customer obviously constitutes the first variable in the design of a service. Eiglier and Langeard (1987) thus insist on the importance of precisely defining the customer segment targeted to ensure a good understanding of the uses and thus the success of the innovation (in the case of the automobile, is the service targeted at fleets or individuals? What types of individuals? etc.). But, in more general terms, the literature on service innovation shows both the absolute necessity of including customers in the process, since they are co-producers of the service, and the difficulty of achieving this in actual practice (Thomke, 2003; Abramovici and Bancel-Charensol, 2004). This explains the low level of attention accorded to the customer in the examples observed (Jallat, 2000; Eiglier and Langeard, 1987). Once a customer segment has been selected, the challenge is then to understand the needs of customers and the uses that they will make of the proposed service concept, and this takes us back to the question of experimentation conducted upstream of the design process (Thomke, 2003; Le Masson and Magnusson, 2002).

Second, a support product. In their 1987 study, Eiglier and Langeard point out the important role played by the physical support structure in the servuction system. They designate by this term all of the equipment required for the delivery of the service, as well as the service environment, and they emphasize the increasingly important role played by technology in servuction, due to the automation of front-office processes, among other things. The deployment of ICTs has reinforced this trend and, increasingly, the delivery of the service depends on the existence of a product that makes the service possible (e.g. a two-way navigational system in the case of telematics services, a GPRS (General Packet Radio Service) or UMTS (Universal Mobile Telecommunication System)-compatible mobile phone for the new services proposed by telecom operators, etc.).

Innovation in services therefore increasingly requires the development of innovative support equipment. The problem is then to coordinate two separate design processes that do not necessarily have the same timeframes, and which may not necessarily be managed by the same teams.

Third, a contract. The legal dimension of the service is not dealt with in service innovation studies, even though certain authors underline the specific issues pertaining to services in this area. Barcet and Bonamy (1999, 200) show that services, as opposed to physical goods, 'do not involve the transfer of ownership rights, but the creation of a lien, implying a mutual commitment between the producer of the service and the beneficiary'. The problem is then to define the conditions under which this lien may be exercised. Our experience at Telcar shows that this question is not trivial and raises complex legal problems when the proposed concepts engage the company's liability (emergency services, for instance). It is even conceivable that the arrival of ICTs on the scene raises new legal questions, as has been shown by numerous articles on the difficulties of online payment or the impact of online music sales on artists' royalties, to cite two recent examples. To overlook this dimension can totally block or slow down the design process.

Fourth, a front-office process. Much has been written about the nature of the service relationship and the problems raised by co-production (Gadrey, 2003). For the developer, the difficulty is to design the method(s) of interacting with the customer, which is what we generally understand by the front-office process. This is a complex problem since the front-office concept refers back to a set of very different realities, ranging from the traditional bank teller's window to various processes involving complex interactions via multiple channels (shops, the Internet, call centres, etc.). Thus, in the case of automotive services, it is useful to differentiate between two different processes: (1) the process through which the product is marketed; and (2) the process that ensures the co-production of the service (call centres, for instance). They can be provided by the same entities, but this is not necessarily the case.

The task of designing the front office must therefore involve the formulation and testing of the means through which this interaction will take place (will it take place through the salesperson in the dealership or through a call centre? etc.).[4] This is where the importance of the internal marketing process comes into play, which serves to integrate contact personnel into the process (Flipo, 2001).

Fifth, a back-office process. The back office is glaringly absent from the servuction model. But, as Bancel-Charensol and Jougleux (1997) or Balin and Giard (2006) have pointed out, it plays a crucial role in the performance of the service in production. The design of the back office, on which the delivery of the service depends and which has rarely been studied in the literature, is therefore of fundamental importance. This involves the information systems required (for contract management, invoicing, risk analysis, etc.) and the deployment of internal services and/or partners responsible for producing the service in support of the front office. When the information and communication technologies are involved, through the deployment of a telecommunications infrastructure, for instance, the task of the designers is considerably complicated.

Sixth, an economic model for the financing of the service. The question of the economic model for financing the service appears to be the second element that is glaringly absent from the servuction model, even though it is obviously a central question. It is included, however, in the innovation models. Scheuing and Johnson use the term 'business analysis' to describe this task, which involves estimating the costs related to the development of the service, the market potential and the revenue sources. We consider this to be an especially important point since the innovation may originate with the method of financing the service (e.g. it could be free when financing is provided by third parties).

These six variables serve to represent the design work that must be accomplished by the team in charge of the development of a new service. It goes without saying that they are interdependent. The process of designing a service involves working on these various dimensions simultaneously (some of which may be given at the start of the project), otherwise important aspects of the service may be neglected. Furthermore, this simultaneity is made necessary by the co-determination of the different variables. The choices made in one field will, in fact, influence the decisions on the other variables (the choice of a type of interaction with the customer, for instance, will determine the back-office tools required and vice versa).

Thus the team's ability to integrate these various dimensions will not only determine the smooth functioning of the design process, but also the quality of the final service. The work of Zeithaml et al. (1990) clearly shows that the quality perceived by the customer depends on the consistency of the choices made on the different design variables (understanding of usages, the quality of the support product, the suitability of the contract, the performance of the front- and back-office process and the economic model).

30.4 Applications of the model

We now have a model representing the service design process. In this last section, we intend to show how this model can be applied for the management of the service innovation process. We have thus distinguished four potential uses for the model.

30.4.1 Characterizing the innovation

Enriching existing typologies First of all, this model of the service design process can be used to characterize the type of innovation with which the company is confronted. The model thus shows that the innovation can originate with the concept, from one or several design variables or in all probability, from both dimensions at the same time. As a result, it offers a rather detailed characterization of the new features of a service.

Let us consider the example of the Emergency Call service developed by Telcar. The concept consists, as its name implies, in offering customers a service that allows them to call a number that triggers assistance or an emergency breakdown service in the event of an accident or breakdown.[5] The call makes it possible to determine the exact location of the vehicle. At first glance, the service may seem to be an extension of the assistance services offered for some time by automobile manufacturers. However, when the service is broken down into the six design variables, it becomes clear that there is a clear break concerning several dimensions simultaneously. In fact, the project requires:

1. The development of a support product to provide communication with the vehicle and determine its location, even in the event of an accident.
2. The resolution of the legal problems inherent in providing an emergency service (who is responsible if there is a problem in the assistance chain?)
3. The setting up of the front office with the understanding that the entity offering the service (the dealership network) is not the same as the entity responsible for its operation (a partner company specializing in emergency assistance).
4. The design and implementation of a complex back-office system composed of:
   a. telecommunications infrastructure that can locate the vehicle in extremely short time periods and with a high reliability rate;
   b. Information systems capable of recording service contracts, managing invoices, and then processing the data internally to manage the customer relationship.
5. The development of an economic model for the financing of a service that is acknowledged by all concerned to be 'difficult to market'.[6]

The proposed model defines precisely in what ways the proposed service is innovative. In that regard, it is complementary to existing typologies. For instance, the typology proposed by Gallouj and Weinstein (1997) does not include legal and economic questions, just as our model makes it possible to break down their notion of 'internal technical characteristics' into several components (support product, front office, back office), which facilitates its exploitation.

Identifying design situations Continuing with our reasoning, the design variables that we have identified also allow us to distinguish between the various situations that can occur in service design. For this purpose, we will refer to the complex constituted by the back and front office as the 'service infrastructure', and we will also take the 'support product' variable into account. At the start of the project, each of these elements may either already exist or need to be developed. This gives us a typology of the four possible configurations that may be encountered in the design of new services (Table 30.1).

1. Case A: the 'ideal' situation in which the existing equipment and infrastructure (adapted where needed) act as supports for the new service.

Table 30.1 A typology of design situations

Support product → **↓ Service Infrastructure**	**Existing**	**New**
Existing	A: The ideal situation, little new investment required, marketing of existing installations	B: Development/ adaptation of the support product
New	C: Deployment in a new country or process innovation	D: The most complex situation

The investment required will therefore be limited. This situation most likely corresponds to a strategy of gradual enrichment of a range of services by the addition of new features.

2. Case B: this time the new service relies on the existing infrastructure but requires the development of a new support product. Thus the importance of product/service coordination.
3. Case C is particularly interesting because it corresponds to two types of situations frequently encountered in service design:
   a. The deployment of the service in a new country, where one cannot rely on the initial infrastructure (the initial suppliers do not operate in the country, there are different information systems, etc.).
   b. Process innovation with the objective of streamlining the operation of the infrastructure without modifying the service, either to improve its operation or to reduce costs.
4. Finally, Case D corresponds either to the situation where the launch of a new service requires both the development of new equipment and the implementation of a new infrastructure, or to the situation where the existing service is redeployed using new equipment and a new infrastructure. This is obviously the most complex situation to manage.

This typology allows us to identify various design situations and evaluate the difficulty of the work required of the project team.

30.4.2 Providing a structure for the exploration of fields of innovation

In addition to facilitating the characterization of innovation, the model makes it possible for us to structure the thought process on the future services in two ways.

First of all, it shows that the exploration must combine ongoing reflection on the relevance of the concepts with consideration of the means of acquiring the necessary knowledge. Each new service concept implies the development of areas of competence in the six defined fields while, in return, the development of new knowledge bases facilitates the exploration of the concepts. Any approach failing to combine these two dimensions would be overly simplistic.

On this basis, the model provides a structure for the reflection process on future services. Let us take the example of telematics services in the automotive sector. We now have a framework that allows us to characterize the avenues for development opened up by the introduction of telematics. We will use the development tree concept for this purpose (Hatchuel and Weil, 2002, 2003 and the left side of Figure 30.2), which allows us to represent the various possible options when confronted with a question or design problem. We have used it both for the search for innovative concepts and for the previously identified design variables. We will now show how this model can facilitate the management of the innovation process.

The conceptual dimension What are the concepts that will guide the development of the future services? The use of ICTs has led to an explosion in the number of potential new services. Just about everything becomes conceivable, from the improvement of existing services (location-based roadside assistance, for instance) to ground-breaking innovations (mobile offices, remote navigation, real-time traffic information, etc.). The difficulty is then in organizing this profusion of potential new services and in giving it meaning.

The concept (in the meaning of C-K theory) that we consider to be central is the 'mobility service' concept. But expressed in this way, it is still rather vague. It is therefore necessary to continue the reflection process and break it down into sub-concepts that represent potential areas for exploration. This leads to the following design tree (see the left side of Figure 30.2). Each branch of the tree is aimed at offering the customer a certain kind of value: optimization of the customer's automobile budget in the first case, and of transportation time in the third case,[7] and so on. But it is still possible to refine the concept and continue the construction of the tree.

Let us consider the 'available mobility service' branch, for example, which constitutes a concept within the meaning of C-K theory. It can be 'divided up' even further. The 'available mobility service' concept is not, in fact, defined *ex ante*. It opens up a wide field of possibilities. The translations of this concept in terms of services can therefore be quite varied. The

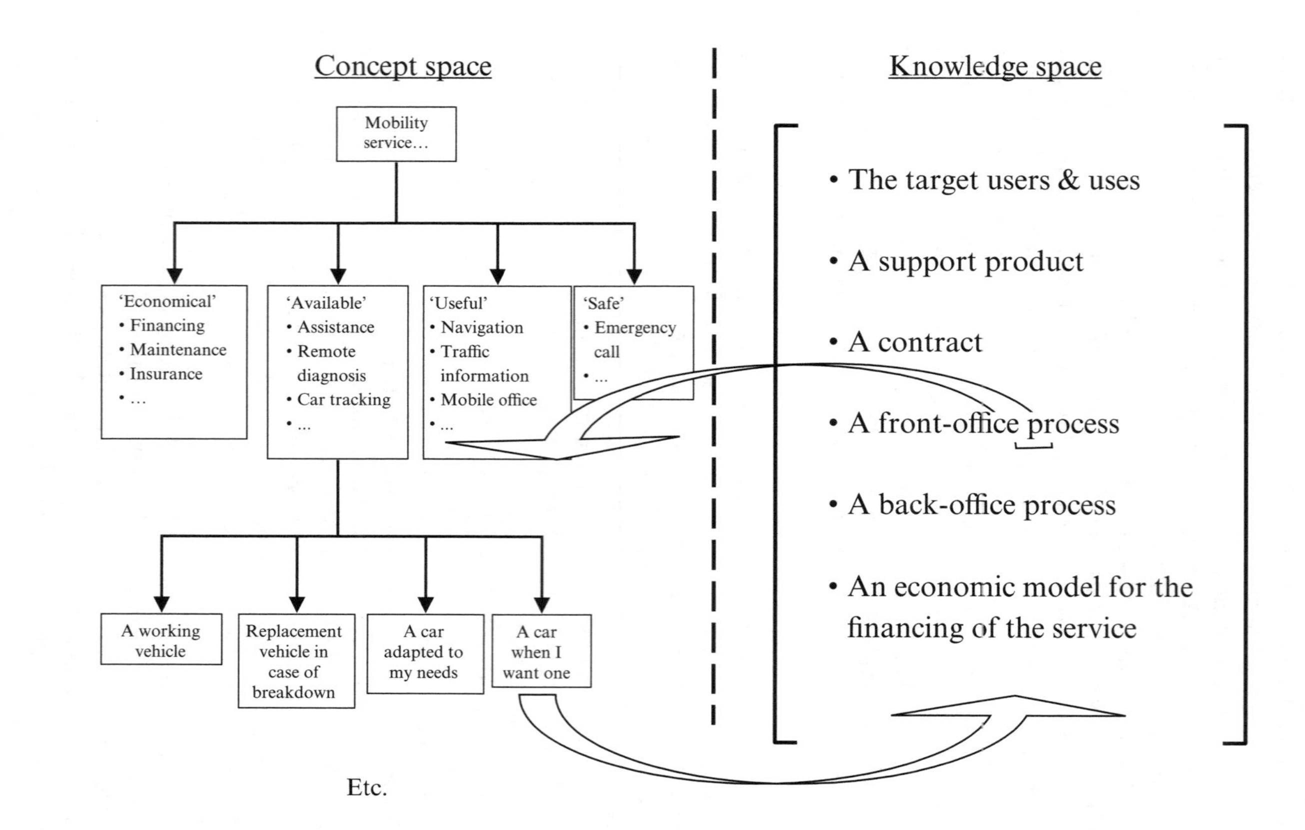

Figure 30.2 Organizing the exploration of a field of innovation: the example of automobile telematics services

first two cases refer to traditional solutions already implemented by Telcar (i.e. maintenance and assistance). The two branches on the right, however, constitute entirely new services:

- 'A car adapted to my needs', which allows customers to change cars depending on their objective: a compact vehicle during the week and a minivan when on holiday.[8]
- 'A car when I want one' is based on the idea of offering customers the car of their choice, wherever and whenever they so choose.

This demonstrates the interest of the design tree concept for representing potential areas for exploration and for developing new services. It should be noted that the various branches are not mutually exclusive.

The knowledge dimension Innovation is not limited to concepts. One must, in fact, determine simultaneously whether or not these potential areas for exploration are likely to be fruitful, which takes us back to the dimension of available knowledge or knowledge that must be acquired if we are to assess the relative fruitfulness of one of these areas of exploration.[9] An exploration method based on the C-K theory thus allows us to establish a link between the desired target (the concept) and the learning process that it implies (the knowledge bases). The relevance of a concept will therefore be evaluated not only for its intrinsic originality but also for the learning process on which it depends (complex or simple, limited to one variable or highly multidisciplinary, protracted or brief, etc.). But the link is not unilateral, and the development of new knowledge (the new information and communication technologies, for example) can give rise to new concepts.

To illustrate our remarks, let us return to the example of automotive services and consider the front-office process. Thus far, the preferred point of interaction with the customer has been the dealership network. But there too, one can conceive of other solutions that would offer more flexibility (Figure 30.3). The establishment of a direct relationship between the customer and the brand, via the Internet or a call centre, has already been implemented by Telcar. This type of interaction should become increasingly important in the context of Telcar's CRM strategy and in light of current automobile distribution trends in Europe.

But one could also conceive of more innovative solutions: the implementation of a sales network dedicated to the marketing of services, for instance, or the deployment of interactive terminals (in service stations, for example) capable of uploading data, updating software, and so on (this last is already used by Toyota in Japan with its Gbook service). This is a

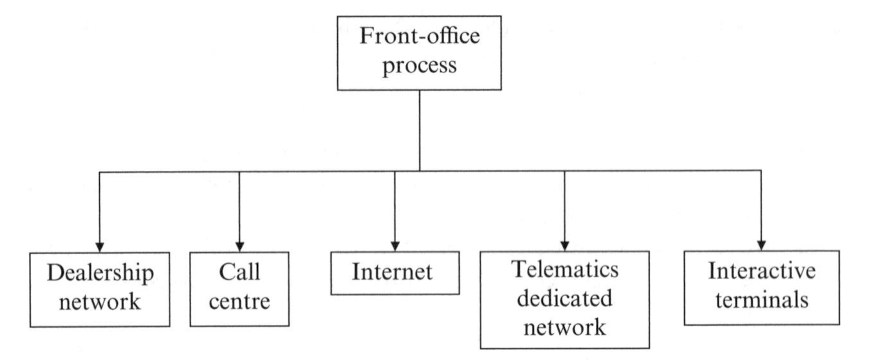

Figure 30.3 Potential areas for front-office innovation

crucial venture for Telcar and a real challenge since this type of customer relationship is completely new.

The same type of reasoning could be used for the various design variables. One can thus see how this model could:

- Provide a structure for the exploration of the field of innovation opened up by telematics, for instance.
- Or how it could be used to clarify the company's strategy. The design tree offers a means of recording the design choices made for future reference, as well as the reasons that led to those choices.[10] By keeping a record of the decisions made, it reminds us that the path on which the project is engaged (interaction via a call centre, for instance) is only one solution among all the other possible options (the Internet, a dedicated network, etc.). Moving further up the tree can thus open up new potential areas for innovation, with regard to both concepts and knowledge bases.

30.4.3 Analysing and managing the innovation process

Finally, the model presented can serve as a support for the construction of management tools for the new services development process. The preceding model already constitutes a management tool for the exploration of the field of innovation. But the model also has other potential uses.

First of all, it is useful for specifying the content of the various stages of the process. Let us take the example of the 'concept development' stage that is found in all service innovation models. In their 1989 article, Scheuing and Johnson defined the concept as 'a description of a potential new service. A typical concept statement would include a description of a problem that a prospect might experience, the reasons why the new service is to be offered, an outline of its features and benefits, and the rationale behind its purchase' (p. 31). Our model allows us to specify the content of the stage which, ideally, should not only make it possible to clarify the concept, but also to explore its impact on the various design variables. The goal of this upstream stage is therefore to map the possible solutions for implementing the concept without making a definite decision between the various options at this stage in the process.

This reasoning can then be extended to the entire process. The design of a new product or service is typically a process of knowledge creation (Clark and Fujimoto, 1991; Nonaka, 1994). In this perspective, our model supplies a diagram of the various fields of knowledge that need to be acquired during the process. The various stages must contribute gradually to:

- clarifying the concept;
- developing the required knowledge, which will require that studies be performed, that prototypes and testing be carried out under conditions that will little by little approach those of the 'final' servuction system.

On this basis, a process management tool can be designed that cross-references the stages on the one hand, with the concept and knowledge bases on the other hand. A management chart of this type can then assist the designers with the task of overseeing the project:

- What stage are we at concerning the various dimensions of the service (theoretical studies, internal testing, customer testing using a prototype or testing on the future servuction system, etc.)?
- Are we sure that we have not neglected a dimension that may cause a problem later on?
- Have we involved participants with the necessary skills? The drafting of the contract, for instance, would require the participation of legal specialists, and the development of the front-office system would require the participation of the sales department and contact personnel, etc.
- Etc.

We can then back up traditional project management tools with the support of an ad hoc service design model. For instance this model allows us to analyse the unfolding of the launch of the Emergency Call service (Lenfle and Midler, 2009). Using this model we were able to understand: (1) the low subscription rate for the service; (2) the gap between the performance of the 'technical' and 'sales' dimension of the service. The late and incomplete involvement of the sales department appears to be an important factor explaining this situation. In the terminology of our model an important dimension of the service, designing the sales process (which involves front and back office processes, but also contract design), has been neglected.

30.5 Conclusion

In view of the fact that service design has become a central question in the current economic context, this chapter has attempted to show both the contributions of the literature and the interest of recent developments in design theory. The work of Hatchuel and Weil, by distinguishing the conceptual dimension from the knowledge dimension, offers us a framework for a comprehensive view of the design process for both products and services. In the latter case, the literature on services and our research at Telcar have led us to identify the various design variables for a service. We then demonstrated how this model could be used to provide a structure for the exploration and generation of new concepts, and to manage the design process.

These proposals must now be further clarified. The application of this model to another example and/or in another context would allow its relevance and reliability to be tested. Indeed, this model has been developed in the context of product-related services. Its relevance for 'pure' services remains a research question. However, in a context where the frontier between physical goods and services tends to be more and more fuzzy (Balin and Giard, 2006), we believe that its application could cover a wide spectrum of situations. Finally, another potential application, that has been explored with Telcar, involves using the model to develop management tools, the rationale for which we have briefly touched upon. This brings up the central question of the need to clarify the means of acquiring knowledge about the various service design variables.

Notes

1. Plans combining assistance, financing and maintenance included in a monthly lease payment are becoming increasingly common (Tooty on the A-Class, Smart Box, Smart Moove, etc.).
2. The 'concept development' stage, present in almost all models, is not analysed in detail: what is a concept in the case of a service? How do we generate them? Who has the required skills? What aspects of the service should be explored at this time? Etc.
3. The sidestepping of this question by Flipo (2001, p. 148) is symptomatic of this 'marketing bias' of many studies on innovation in services.
4. Lovelock (1984, 61) has rightly pointed out that a tool such as the 'blueprints' proposed by Shostack must also be able to describe 'the service encounter from the customer's perspective'. As a result, there are two 'blueprints', and not one, for the same service.
5. Assistance is triggered automatically in the event of an accident.
6. The reader has undoubtedly noticed that the 'customer' design variable is missing.

On this point, in fact, the service is not very innovative; like the existing services, it is addressed to all Telcar customers provided that they have the onboard equipment.

7. By reducing it, through the use of navigational and traffic information systems, for example, or by making the best use of it (through mobile office type solutions or games for passengers, etc.).
8. This idea was part of the SMART project as it was originally conceived by N. Hayek.
9. Does the concept of 'a car adapted to my needs' appeal to customers? How much are they willing to pay? How do we know this? Is it profitable?
10. The model can then serve as a knowledge management tool: what have we learned about the back office? about the reactions of various types of customers? about the front office? etc.

References

Abramovici, M. and L. Bancel-Charensol (2004), 'How to take consumer into consideration in service innovation projects?', *Service Industries Journal*, **24** (1), 56–78.

Akrich, M., M. Callon and B. Latour (1988a), 'A quoi tient le succès des innovations? Deuxième épisode: l'art de choisir les bons porte-parole', *Gérer & Comprendre*, **12**, 14–29.

Akrich, M., M. Callon and B. Latour (1988b), 'A quoi tient le succès des innovations? Premier épisode: l'art de l'intéressement', *Gérer & Comprendre*, **12**, 4–17.

Balin, S. and V. Giard (2006), 'A process oriented approach to the service concepts', IEEE SSSM06 Conference (Service Systems and Service Management), Troyes.

Bancel-Charensol, L. (1999), 'NTIC et systèmes de production dans les services', *Economie et Sociétés, série Economie et Gestion des Services*, **5** (1), 97–116.

Bancel-Charensol, L. and M. Jougleux (1997), 'Un modèle d'analyse des systèmes de production dans les services', *Revue Française de Gestion*, **113**, 71–81.

Barcet, A. and J. Bonamy (1999), 'Eléments pour une théorie de l'intégration biens / services', *Economie et Sociétés, série Economie et Gestion des Services*, **5** (1), 197–200.

Barras, R. (1986), 'Toward a theory of innovation in services', *Research Policy*, **15**, 161–173.

Barras, R. (1990), 'Interactive innovation in financial and business services: the vanguard of the service revolution', *Research Policy*, **19**, 215–237.

Brown, S.L. and K.M. Eisenhardt (1995), 'Product development: past research, present findings and future directions', *Academy of Management Review*, **20** (2), 343–378.

Clark, K. and T. Fujimoto (1991), *Product Development Performance: Strategy, Organization and Management in the World Auto Industry*, Boston, MA: Harvard Business School Press.

Edvardsson, B., A. Gustafsson, M.D. Johnson and B. Sanden (2000), *New Service Development and Innovation in the New Economy*, Lund, Sweden: Studenlitteratur.

Eiglier, P. and E. Langeard (1987), *Servuction. Le marketing des services*, Paris: Edisciences International.

Eisenhardt, K.M. (1989), 'Building theories from case study research', *Academy of Management Review*, **14** (4), 532–550.

Flipo, J.P. (2001), *L'innovation dans les activités de services*, Paris: Editions d'Organisation.

Furrer, O. (1997), 'Le rôle stratégique des services autour du produit', *Revue Française de Gestion*, **113**, 98–107.

Gadrey, J. (2003), *Socio-économie des services*, Paris: La Découverte.

Gallouj, C. and F. Gallouj (1996), *L'innovation dans les services*, Paris: Economica.

Gallouj, F. and O. Weinstein (1997), 'Innovation in services', *Research Policy*, **26**, 537–556.

Hatchuel, A. (1996), 'Théories de la conception: trois approches', Ecole Nationale Supérieure des Mines de Paris.

Hatchuel, A. and B. Weil (2002), 'La théorie C-K: fondements et usages d'une théorie unifiée de la conception', seminar on Sciences de la conception, Lyons.

Hatchuel A. and B. Weil (2003), 'A new approach to innovative design: an introduction to C/K theory', *International Conference on Engineering Design (ICED)*, Stockholm.

Jallat, F. (2000), 'Le management de l'innovation dans les entreprises de services: spécificité des processus et facteurs de performances', in A. Bloch and D. Manceau (eds), *De l'idée au marché*, Paris: Vuibert.
Le Masson, P. (2001), 'De la R&D à la RID modélisation des fonctions de conception et nouvelles organisations de la R&D', École Nationale Supérieure des Mines de Paris: Paris.
Le Masson, P. and P. Magnusson (2002), 'Towards an understanding of user contribution to the design of mobile telecommunication services', 9th International Product Development Conference, Nice, France.
Lenfle, S. (2004), 'Peut-on gérer l'innovation par projet?', in G. Garel, V. Giard, and C. Midler (eds), *Faire de la recherche en management de projet*, Paris: Vuibert, pp. 35–54.
Lenfle, S. (2008), 'Exploration and project management', *International Journal of Project Management*, **26** (5), 469–478.
Lenfle, S and C. Midler (2002), 'Stratégies d'innovation et organisation de la concepdin dans les enterprises amont Enseignements d'une recherche cheg Usinor', *Revue française de Gestion*, **140**, 89–106.
Lenfle, S. and C. Midler (2003), 'Innovation in automotive telematic services: characteristics of the field and management principles', *International Journal of Automotive Technology and Management*, **3** (1–2), 144–159.
Lenfle, S. and C. Midler (2009), 'The launch of innovative product-related services: lessons from automotive telematics', *Research Policy*, **38** (1), 156–169.
Lovelock, C. (1984), 'Developing and implementing new services', in W. George and C. Marshall (eds), *Developing New Services*, Chicago, IL: American Marketing Association.
Miles, M. and M. Huberman (1994), *Qualitative Data Analysis*, 2nd edn, Thousand Oaks, CA: Sage Publications.
Nonaka, I. (1994), 'A dynamic theory of organizational knowledge creation', *Organization Science*, **5** (1), 14–37.
Scheuing, E. and E. Johnson (1989), 'A proposed model for new service development', *Journal of Service Marketing*, **3** (2), 25–34.
Shostack, G.L. (1981), 'How to design a service?' in W. George and J. Donnelly (eds), *Marketing of Services*, Chicago, IL: American Marketing Association.
Shostack, G.L. (1984), 'Designing service that deliver', *Harvard Business Review*, **62** (1), 133–139.
Thomke, S. (2003), 'R&D comes to service: Bank of America's pathbreaking experiments', *Harvard Business Review*, **81** (4), 71–79.
Ulrich, K. and S. Eppinger (2004), *Product Design and Development*, 3rd edn, New York: Irwin-McGraw Hill.
Van de Ven, A. (1986), 'Central problems in the management of innovation', *Management Science*, **32** (5), 590–607.
Yin, R. (2003), *Case Study Research: Desing and Methods*, 3rd edn, Thousand Oaks, CA: Sage Publications.
Zeithaml, V., A. Parasuraman and L. Berry (1990), *Delivering Quality Services*, New York: Free Press.

31 Innovation in construction

Jan Bröchner

31.1 Introduction

Studies of innovation in the construction industry have usually treated construction as akin to manufacturing but nevertheless have produced findings more consistent with those of the rapidly growing body of knowledge related to innovation in services. At present there are a few monographs exclusively devoted to construction innovation: Gann (2000) takes a broad view of innovation in the sector and offers numerous examples, Miozzo and Dewick (2004) are good on strategic issues also for the largest contractors, while Barrett et al. (2008) concentrate on the smaller firms found within the sector. There is a wealth of articles, often with an empirical base in case studies and often published in construction-oriented journals, little read by mainstream service innovation researchers. This chapter reviews many of these contributions, emphasizing those authors who should be of interest from a services perspective when they analyse patterns of innovation and their causes in construction firms. Here, the starting point is the construction firm, basically the construction contractor, although many studies have a broader scope and include providers of knowledge-intensive business services (KIBS) that are related to construction, such as engineering and design firms.

In official statistics, the construction industry is usually classified apart from the service sector. However, there is no obvious reason why construction innovation should be fundamentally different from patterns found in the service sector. Rethinking industry classification in a demand perspective would place construction in a different context (Dalziel, 2007). There is considerable diversity in and between service industries, and the same can be said of construction itself. Already the size range of construction enterprises and the structure of the industry can be expected to lead to varying patterns and mechanisms of innovation. Taking the EU only, there are more than 2.7 million construction enterprises (Eurostat, 2004), and 92 per cent of these have fewer than ten employees. At the same time, there are a number of large international construction groups with more than 50,000 employees.

The range of activities inside the construction industry and often within one and the same contractor enterprise stretches from mass-produced speculative house-building with little interaction with the individual customer, minor maintenance and repair work and, at the other extreme of the scale, major infrastructure projects such as bridges and dams. It is likely that this variety of processes and products will differ in their patterns of innovation.

This chapter relies on an initial analysis of why construction fits or does not fit current definitions of a service industry and how comparative statistical surveys allow the identification of specific features of construction. These features are related to characteristics of built facilities. Furthermore, the project nature of construction is explored and the typical nexus of various services, including KIBS, that make up construction projects is identified. Innovation in project-dominated firms raises issues of decentralization, which have been addressed either as typical of complex products and systems (CoPS) or as of loosely coupled systems. After looking at broader studies of how contractors pursue internal strategies for innovation, the role of information and communication technology (ICT) progress for construction innovation is highlighted. Technology diffusion in mass customized house-building is discussed. The chapter moves on to find that clients are important collaborators when construction firms innovate; patent activity is low in construction, and the reasons for the openness of innovation in this industry are reviewed. As an alternative to collaboration with other firms, construction contractors tend to diversify their activities, sometimes to create and deliver new bundles of services. A Swedish survey of innovation among construction contractors is reported. Furthermore, this chapter explores how construction innovation is subject to public intervention, in addition to what the client role of government implies. The final section underlines the necessity to distinguish between types of innovations, to understand their interaction, and the consequences for innovation in a stable nexus of services.

31.2 Construction as a service industry

Within the service sector as a whole, as defined in the current statistical classifications, it is not always obvious which features the subsectors have in common. Hill (1999) and Gadrey (2000) have proposed generic definitions that can be applied to construction as a test case to see whether it should be considered as part of the service sector. This highlights the specific features that might explain why it remains different in a deeper sense. In brief, it appears that both issues of ownership and of land (as opposed to goods) complicate the assignment of construction enterprises to the one or the other category. Although most buildings are produced to order, Hill (1999) sees them as goods that can be traded as desired once they have been produced; however, the producer of a building is seldom the owner, and the dilemma is worse when he emphasizes that goods can be consumed or used long after they are produced, at locations which are remote from their place of production. But if we follow Hill in thinking that it matters who owns the goods involved, who controls the timing and location of production, who bears the capital costs and who assumes the risks, it is difficult to escape the conclusion that most construction contractors are service providers. The distinction between goods (such as building materials and components), typically owned by the construction enterprise during the production process, and land, typically owned by the client, also complicates the definition proposed by Gadrey (2000). Thus the construction contractor intervenes on land owned by the customer, rather than on goods owned by the customer.

But construction is not unique in creating borderline problems for service definitions. Restaurants traditionally sell combinations of goods and services, and there are computer packages of hardware and software (Illeris, 2007, 27). At one end of the goods–services continuum, building maintenance and refurbishment display a full range of service characteristics. The long turn-round time, the uncertainty, complexity and large number of participants in housing refurbishment have been identified as essential features of housing refurbishment as a service (Holm, 2000). When it comes to mass-produced housing, the service character is less prominent.

However, all such attempts to catch the essence of services are bound to be compromises related to various purposes for which the resulting statistics will be used. Whoever wishes to develop policy recommendations for stimulating service innovation will perhaps prefer another set of characteristics than those who are studying employment patterns in services.

31.3 Comparative surveys

Clues to what is lacking in the definitions and needed for demarcating the service sector from an innovation viewpoint are offered by European innovation surveys where construction is included. It is only in recent years that comparative data have been collected and that non-technological innovation has been included in the survey definitions.

31.3.1 The Innobarometer surveys

The European Innobarometer 2002 survey (European Commission, 2002) surveyed companies with at least 20 employees in the at that time 15 EU member countries, distinguishing between four sectors of activity: construction, industry (manufacturing), services and trade. Where does construction stand out, in the sense that the average figure for construction firms lies outside the range defined by the three other activity sectors? See Table 31.1, which immediately reveals how new products are insignificant

Table 31.1 Construction innovation aspects compared to industry (manufacturing), services and trade

Aspect	Construction lowest of all four sectors	Construction highest of all four sectors
Importance of innovation	Percentage of turnover from new or renewed products or services introduced during the last two years	
Share of investment	Percentage of investment dedicated to innovation in products, processes or organization	
Focus	Percentage for new products	Innovation efforts more concentrated on new organizational changes
Access	Importance of in-house R&D	Acquisition of machinery or equipment
Needs	Feeling access to innovative markets as a need	Feeling unsatisfied needs for finding and mobilizing financial resources Need to improve basic skills of the workforce
Training		Popularity of semi-public institutions for training
New approaches next 2 years	Developing new products or service characteristics	Developing the relationship to suppliers or users
Advice		Importance of public and semi-public advisory centres for advice on new methods or approaches to management
Cooperation with other firms	Will become more useful in subsequent years	To launch new products or services or to introduce new processes
Effects of European single market	Improved regulations, cheaper or better available supplies, better access to new technologies	Better access to new markets, better cooperation to innovate

Source: European Commission (2002).

for construction companies. On the other hand, organizational innovation and the relation to other organizations appear to be important. The main impression left by construction respondents was however that their companies behaved like service companies.

Two years later, the European 2004 Innobarometer survey (European Commission, 2004) collected responses from companies with 20 to 499 employees in the 25 member states, again covering the four sectors as in 2002, but this time omitting larger companies. The questions also differed. Within this size range, construction firms were less likely to have conducted market research studies for introducing new products or services in the previous two years, or to have introduced new or significantly improved processes for delivery systems, production or logistics. They were less likely to have applied for patents or international trademarks, or carried out in-house research. More than in other sectors, they sold their products or services in the region where the companies are located. They were less likely to have obtained outside advice for their innovation activities. Construction firms were more prone to attribute a competitive disadvantage for their new products or services to safety regulations and regulations that determine product characteristics. For construction firms that had introduced new or improved processes, environmental and safety regulations were felt to have placed the firm at a disadvantage in relation to competitors. Government agencies were important buyers for products or services delivered by construction firms, in particular for public transport and traffic, public education and public water, sewage and pollution management.

31.3.2 The UK Innovation Surveys

The Community Innovation Surveys (CIS) do not cover construction in more than a few member countries, but the UK is one of them. UK Innovation Surveys from 1997, 2001 and 2005 provide data for firms with at least ten employees. The 1997 survey, included in the CIS2, placed construction at the lower part of the services technology activity table, with 1.35 per cent of turnover spent on technological innovation in construction (Miles, 2002). Similar figures were obtained for real estate and retail.

Reichstein et al. (2005) have analysed the construction sector responses from the 2001 UK Innovation Survey, comparing with low-tech as well as high-tech manufacturing. Their definition of the construction sector includes not only the construction industry, but also KIBS activities such as architecture, engineering consultancy and urban planning. They distinguish between firms with 10 to 50 employees and those with 50 or more. The analysis covers innovative performance (for product innovation and process innovation), factors hampering innovation (economic, internal, other factors) and sources of knowledge for innovation (internal, market, institutional, other, specialized). It was found that construction, in particular the smaller firms, resembled traditional service industries. Local market orientation is seen as a major explanation for low levels of innovation activity in the UK construction industry.

More precisely which of the service industries does construction look like? The 2005 UK Innovation Survey (DTI, 2006), breaking down enterprises into 20 divisions, of which construction is one, indicates answers to this question. For the whole range of survey questions, it is possible to identify which other divisions have a similar indicator response profile. It is easy to see that the divisions that most closely resemble construction in their innovation profiles are retail trade (excluding 'cars and bikes') and 'hotels and restaurants'. There are also similarities to post and courier activities due to generally low cooperation, whereas there is a high level of local cooperation. Also, there are similarities to real estate activities, and furthermore, but to a lesser extent, to renting of machinery, equipment, personal and household goods. Turning outside the service sector, it is remarkable that there are more similarities to mining and quarrying than to any division within manufacturing; like mining and quarrying, construction is characterized by values that are low for cooperation, low for increase in range of goods, low for acquisition of external knowledge, as well as low for marketing expenditure. Moreover, reduced environmental consequences are important for mining and quarrying as well as for the construction industry. It is hazardous to assume that firms within a more broadly defined construction sector share a specific pattern, since the survey reveals almost no similarities between construction and 'architectural and engineering activities and related technical consultancy', except for the impression that they share a context with more important relations to government research institutes and little cooperation with competitors.

31.3.3 Other comparative surveys

A 1998 survey of construction innovation in Germany showed that clients were the most important cooperation partners of inventors, whereas German manufacturing industry was more dependent on cooperating with universities and research institutes (Cleff and Rudolph-Cleff, 2001).

A Dutch investigation (de Jong and Marsili, 2006) has reconsidered Pavitt's taxonomy in a study of Dutch firms with less than 100 employees. In this context, the distribution over the four clusters for firms classified under 'construction' resembles that of those service industry firms that are classified as 'transport', while service firms that belong to 'engineering and architecture' are more similar to those who provide 'economic services' (accountancy, consultancy) if we look at distribution over four clusters:

supplier-dominated, specialized suppliers, science-based and resource-intensive. In all four clusters, the percentages for construction fall outside the ranges spanned by any type of manufacturing but remains within the range for services firms. Nevertheless, this leaves the question open whether large construction firms are more like manufacturers.

31.4 Characteristics of built facilities

One way of explaining why construction firms deviate from other firms is to consider the characteristics of constructed facilities: immobility, complexity, durability, costliness and high risk of failure (Nam and Tatum, 1988). The intensely localized character of the production process for buildings and built infrastructure, similar to that of mining, goes a long way towards offering explanations to the odd features of construction in the innovation perspective according to the surveys just discussed.

The process of building and the operation of built facilities have environmental and social consequences that often lead to government intervention. On the other hand, local climate and weather changes during a construction contract period create particular risks.

The durability of built facilities can be expected to discourage technological innovation if there is a difficulty to assess the long-term quality consequences of new products and processes; additionally, there is the potential for hidden faults that manifest themselves after many years of building use.

In most countries, there is government intervention with both intended and unintended effects on construction innovation. Government itself is a major customer, partly because built facilities such as bridges constitute technical monopolies.

Understood as outbursts of activity in a diversity of locations, construction takes place in temporary organizations (projects). This is believed by many to be crucial for understanding how innovation occurs.

31.5 Projects

With the possible exception of mass-produced housing, the activities of construction firms have always been dominated by projects. This dependence on temporary organizations has consequences for innovation.

31.5.1 The project nexus of service contracts

Construction projects form a dynamic nexus of services: architects, structural engineers and other KIBS providers are linked to construction clients and contractors, who typically rely on a range of services provided by specialized subcontractors and equipment rental services; once the built facility is finished, facilities management services are bought; ultimately, there might be demolition contractors in operation. The successive contracts follow a project and facility life cycle (Anderson, 2005). Some of the contractual links between firms can be more durable and survive from project to project, which led Eccles (1981) to identify the 'quasifirm' as a stable organizational unit after having studied relations between general contractors and specialized trade contractors. However, for innovative technologies, negative consequences of separating design services from construction itself were a major theme already for Bowley (1966) in her pioneering work on construction innovation. Often the stability of the nexus lies in institutionalized and at least partly standardized interfaces between categories of construction-related service providers. Such interfaces include standard forms of contracts and model specifications, to take only two examples.

If the temporariness of construction projects thus might create barriers to technological innovation, the rate of organizational innovation in the nexus arrangements is also very low compared to other industries. Recognizable elements of the nexus were in place already in ancient Roman times (Anderson, 1997). Particular interest has been generated by the slow changes in professional roles, not least that of architects (Powell, 1996) or the emergence of a UK profession such as quantity surveyors.

Over the centuries, government has been the source of an organizational trend towards stronger coordination functions among private contractors. Already the phenomenon of general contractors, who have a single contract with their client and who coordinate a number of subcontractors, appears to have been encouraged by government action, in the more advanced countries during the eighteenth century. In recent times, there is a development in the direction of letting the general contractor bundle design and construction in design–build contracts; one policy reason for this organizational innovation was precisely to encourage technological innovation. The 1990s introduction, or reintroduction, of concession projects where design, construction and operation of the built facility are bundled is a further step in this direction.

The project nature of construction makes it misleading to treat it as an industry like what is found in manufacturing (Groák, 1994). Seen in isolation, the 'construction industry' is bound to lag behind in innovation statistics and cannot be the object of meaningful comparison with manufacturing (Winch, 2003).

However, it is common that innovation research treats firms that make up the nexus of services under names such as the 'construction cluster' (Dahl and Dahlum, 2001; den Hertog and Brouwer, 2001) or the 'construction sector system' (Carassus, 2003), but given that KIBS are tackled elsewhere in the *Handbook*, the focus remains here on construction enterprises in a narrow sense.

31.5.2 Projects and decentralization

Construction firms are not unique in depending heavily on projects. We could expect such firms to be decentralized and that their degree of centralization affects the type and intensity of innovations that occur.

With inspiration from studies of the management of complex projects, the concept of complex products and systems (CoPS) is at least partly applicable to construction, insofar as construction (civil engineering) with novel technologies and skills that reveal a high engineering content can be classified as examples of CoPS (Hobday, 1998). What is analysed is the project as a temporary coalition of organizations and its output, together with the links between them. Hobday's list of product dimensions of CoPS includes features often associated with services, notably high degree of customization and intensity of user involvement. Users are likely to provide the incentives to conduct R&D, given that there is not an anonymous arm's-length market relationship but rather bidding and negotiations involved in CoPS.

That the CoPS innovation model, distinguished by: (1) many interconnected and customized elements hierarchically organized; (2) non-linear and continuously emerging properties where small changes to one element can lead to large changes elsewhere in the system; and (3) a high degree of user involvement in the innovation process, is relevant for construction has been argued by Winch (1998). In the same vein, Gann and Salter (2000) studied more than 30 organizations (design, engineering and construction firms) in the UK and carried out in-depth case studies with two of these organizations. They explored the mechanisms by which technical support was mobilized to projects from central resources within firms. In general, it must be kept in mind that CoPS is applicable primarily to large firms and large projects.

With more emphasis on decentralized innovation, and as is often the case in service sector firms (Sundbo and Gallouj, 2000), construction firms can be thought of as loosely coupled systems. If each construction site appears to be an experimental workshop (Dubois and Gadde, 2002), the theory should be applicable. An important tenet is that the loosely coupled system may contain novel solutions to new problems, while the structure that allows these mutations to flourish may also prevent their diffusion. This conceptualization has been applied to both technological and non-technological innovation in construction.

Using the distinctions between various types of intra-project and inter-project couplings outlined by Dubois and Gadde, Dorée and Holmen

(2004) studied three bridge projects built by a large Dutch contractor which applied its offshore assembly technology to large bridges. They found that some couplings can be tightened without loosening others.

The transfer of non-technological knowledge from the project level to the firm level in construction-related activities has been investigated by Drejer and Vinding (2006). They measured whether Danish firms used post-project reviews to anchor construction process knowledge and project knowledge; they also measured variables related to partnering: common goal-setting among partner firms, using open calculations and accounting, using self-governing site teams, attention to end-user needs. There was also a question whether firms had increased their use of partnering during the preceding three years. It was found that firms with 20 or fewer employees relied less on formal methods for anchoring or partnering, but for architects and consulting engineering firms, these methods are typical. Next, firms had been asked whether they had engaged in (any type of) innovative activities and this was found to be linked to the use of anchoring and partnering.

The loosely coupled nature of construction organizations, with their project-based work and devolved control patterns, can explain how organizational receptivity to innovation is related to long-term effects. The introduction of a new project performance evaluation tool, the 'dashboard', in a construction firm has been analysed by Bresnen et al. (2004), who found that the diffusion and embedding of new management knowledge in project-based organizations is influenced by a complex interplay between structural conditions within the organization and existing project management practices. Decentralization, short-time emphasis on project performance and distributed work practices are important. This is confirmed by Gray and Davies (2007), who investigated the Demonstration Project process, involving a wide section of the UK construction industry. Here company representatives perceived that current performance was satisfactory as to continuous improvement, teamwork and developing teams; on the other hand, managers considered target-setting and innovation performance measurement to be important, but this is where they felt that performance was less satisfying.

Against this background of the effects of decentralization, there is good reason to turn to how construction contractors develop explicit strategies for innovation.

31.6 Innovation strategies in firms

One stream of literature takes as its starting point the business strategies pursued by construction firms. It is reasonable that the innovation process within construction firms is influenced by the size of the firm, the type of innovation and the breadth of application of the innovation (Laborde and Sanvido, 1994).

The size of construction firms appears to determine their strategies. Miozzo and Dewick (2004) interviewed the largest three or four contractors in five European countries, analysing how corporate governance leads to long-term strategies of innovation, or to a concern with short-term interests of shareholders. At the other end of the scale, Barrett et al. (2008) have investigated innovation in small construction firms.

The approach advocated by Seaden et al. (2003) is that the business environment and business strategy are assumed to lead to innovative practices, in technology and in business, which lead to business outcomes measured as profitability and competitive advantage, all this depending on size of firm and sector of activity. This is the thinking that underlies the Canadian 1999 survey of innovation in construction (Anderson and Schaan, 2001), where firms were asked to indicate planned or current use of 18 advanced technologies and 12 business practices. However, drawing on single-year data for the pre-tax operating margin of firms, there were only weak relations between this measure of profitability and the technology and business innovative practices in use; if anything, the presence of innovative business practices showed a negative correlation with profitability in small firms and a slightly lower but positive correlation for large firms, although these findings are uncertain. There were higher levels of technology innovation associated with medium-sized firms whose pre-tax operating margins were above the median. Thus, firm size appears to be important for the financial mechanisms related to innovation. When firms were asked about which technological or business practice change had made the biggest impact on their business during the preceding three years, information technology was far more important than other business practices and construction technology, regardless of respondent firm size. Similar surveys with basically the same underlying theory have been carried out in Australia (Manley, 2005).

The introduction of a new role in a construction organization, such as a regional technology manager, has been analysed by Bresnen et al. (2003), who invoke social patterns, practices and processes, leading to the adoption of a community-based approach to managing knowledge. With less emphasis on conscious strategies, other researchers have attempted to identify drivers of innovation in construction firms. Bossink (2004) interviewed 66 construction experts in the Netherlands and identified four types of innovation drivers: environmental pressure, technological capability, knowledge exchange and boundary-spanning. Still after 2000 years, this is recognizably close to what Lancaster (2005, 166–181) sees as the ancient Roman drivers for construction innovation in context:

accumulated knowledge, evident need, economic ability and social, cultural and political acceptability.

31.7 Innovative ICT applications

While the construction industry did not emerge as a particularly strong user of ICT and e-business in a 2006 survey (European Commission, 2006), it is nevertheless clear that progress in ICT and the successive introduction of new applications is important for more than one type of construction innovation. It also appears to create stronger links between technological and non-technological innovation in construction. One example of the ICT factor is how the coordination needed for an increased reliance on prefabricated components is supported by better logistics systems. Another is embedded ICT in construction equipment. For infrastructure projects such as tunnels, a new construction method is often followed by new construction equipment, and a new principle for organizing work can be triggered by the implementation of new or improved equipment (Hartmann and Girmscheid, 2004). There appears to be a strong potential for successive linkages between building design, logistics, control of equipment during the construction phase and into the operation of facilities, although progress is slow. Advances in information technology have multiplied the capability for simulating design outcomes in construction (Dodgson et al., 2007), although this is arguably of greater importance for other firms in the typical construction project nexus. It could also be claimed that innovative ways of organizing customer contacts, such as open-book accounts as an element in partnering between contractors and clients, would be difficult in the absence of IT developments.

Studying what drives the diffusion of new information technologies in construction firms, Mitropoulos and Tatum (2000) pointed to competitive advantage, process problems, technological opportunity and institutional requirements. The technologies studied in their case studies were 3D CAD (three-dimensional computer-aided design), EDM (electronic document management) and EDI (electronic data interchange) systems. Moreover, interviews with seven small construction sector companies in the UK led to the insight that successful innovations tended to be ICT applications or clearly influenced by the development of ICT (Sexton and Barrett, 2003; Sexton et al., 2006). None of the successful innovations identified by these small companies belonged to construction technology proper. Instead, the innovations introduced with success were partnering (a management practice), computerized accounts (twice) and wages, mobile phones (twice), cordless power tools, and computer-aided design. The conclusion reached was that successful innovations for small construction companies tend to be generic in nature, explicit in technology composition and enabling in character. On the other hand, unsuccessful innovations were found to be construction specific, to depend on tacit knowledge and to be of a critical or strategic character rather than enabling. Taylor and Levitt (2007) have compared diffusion rates in the US and Finland for innovations in 3D CAD. They found that the traditional allocation of activities in the project nexus of services was important: in Finland, design was detailed by architects, whereas this was usually split between firms in the US, where the rate of diffusion was lower.

31.8 Supplier-dominated house-building

For construction contractors, the sources of innovation vary in strength with the category of facilities that are produced. Pavitt (1984) claimed that supplier-dominated firms, one of his three major categories of firms, could be found for example in house building. Supplier-dominated firms were seen as generally small and having weak in-house research and development (R&D) and engineering capabilities: 'They appropriate less on the basis of a technological advantage, than of professional skills, aesthetic design, trademarks and advertising.' Pavitt concluded that the technological trajectories of supplier-dominated firms were defined in terms of cutting costs. However, his empirical data which covered a large part of UK manufacturing, but not even all of manufacturing, did not include the construction sector.

The reality seems to be more complicated: Slaughter (1993) found that user-builders rather than component manufacturers were innovators, studying 34 innovations related to stressed-skin panels in US residential construction. User innovation was analysed in a learning-by-doing perspective. Nevertheless, there remains a clear difference for technological innovations when we compare construction contractors with their suppliers of materials and components (Bougrain, 2003). Pries and Dorée (2005) in their study of a century of Dutch construction innovations, using earlier literature and trade publications as sources, found that suppliers, often in the chemical industry, were responsible for almost 80 per cent of all product innovations but only for half of all process innovations in construction. Smaller enterprises tended to be more involved in process innovation while the larger enterprises had a higher proportion of product innovation. Improving productivity was much more important, throughout the period, than responding to specific market demands, although this had been increasing more recently.

In addition to bespoke construction projects, many residential construction firms engage in speculative building where the element of customer co-production is reduced and the builder may be the owner of the land. Here, the firm that engages in mass production of housing can be expected to be more similar to a manufacturer. A study of whether and how US home-builders used ten technical innovations showed that larger firms were more likely to use more innovations, and also those who built over a larger geographical area; however, those who built higher-priced houses were less likely to use innovations (Blackley and Shepard, 1996). Greater reliance on subcontracting had no significant effect on the use of innovations, contrary to what the authors expected.

Another investigation of the adoption of technological innovations by small and medium-sized home-building firms was based on interviews with more than 100 home-builders across the US (Toole, 1998). Those builders who are more apt to adopt non-diffused technological innovations tap into more sources of information about new products from portions of their organizational environments than do non-adopters. Reducing uncertainty plays a key role for adoption of technological innovations in residential construction.

Even in the speculative house-building industry, there is scope for innovation along the lines typical of mass customized services for individual customers (Ozaki, 2003). Studying UK housebuilders and their customers, Ozaki identified three key aspects: good service, customized house design on top of certain quality standards, and good information flows. The duration of the production process allows for dialogue and interaction.

31.9 Collaboration with clients

Construction firms differ little from most service industries in their customer orientation. Anderson and Schaan (2001) in the Canadian survey of construction and related industries found that more than two-thirds of responding businesses rated three factors to be of high importance for business success: (1) building and enhancing relationships with existing clients; (2) attracting new clients; and (3) hiring experienced employees. There is no reason to believe that the central position of clients is typical just of the Canadian context.

Early studies of product innovation in construction stressed the customer's role as initiator (Nam and Tatum, 1989) and as champion (Nam and Tatum, 1997). A 1998 survey of construction innovation in Germany showed that clients were the most important cooperation partners of inventors, whereas German manufacturing industry was more dependent on cooperating with universities and research institutes (Cleff and Rudolph-Cleff, 2001).

For certain types of construction, notably roads and similar types of infrastructure, it is tempting to claim that their technological innovation is customer-dominated, seeing that there is a long history of strong and competent public clients. While the emergence of strong public clients in Classical Greece may be due to labour market conditions (Davies, 2007, 353), the accumulation of skills among government officials leads to the pattern analysed by Lancaster (2005) for vaulted construction as an ancient Roman innovation. Public clients would also be the non-technological innovators, as when the curator of the aqueduct system of Rome proudly announces that he has introduced day-to-day scheduling of activities (Frontinus, 2004, Chapter 117). Still today, there are public sector clients who preserve or develop internal capabilities to innovate. A study of the non-residential building and civil sectors in three Australian states illustrates this: following immediately after trade contractors, clients were responsible for about one-fifth of all 'new to industry' technological innovations (Manley, 2006). As to adoption of advanced practices, clients were leading, just as in R&D efforts.

It seems intuitively appealing that highly competent customers would be more willing to embrace innovation than those with lower internal capabilities. However, customers who possess strong internal capabilities might emphasize the future operation of facilities rather than having a primary focus on the construction phase. A study of electrical power plant projects in North America found that 'technological solutions embedded in a power plant project which are widely perceived as novel by the industry', quoting the innovation definition used by the authors, were less frequent when owners had internal design capability (Lampel et al., 1996). Three UK case studies showed that risks perceived by clients were important explanations why there is less client-led innovation than expected (Ivory, 2005).

A typical 1990s innovation is reorganizing the relation between customers and contractors in construction. The literature on construction partnering has grown rapidly; there is a great variety of customers (clients) and as a consequence, of actual practices for collaboration. The improved communications for resolving problems, developing interorganizational and interpersonal trust as well as promoting a culture of innovation, are all stressed by Barlow (2000) in his case study of a complex offshore oilfield construction project. Improvements in IT were also found to have been important for the project: a new interactive system for designing structural steelwork was brought in.

31.10 Openness of construction innovation

A high degree of customer co-production is likely to complicate the protection of intellectual property, especially if the producer belongs to an industry that does not rely much on patents, trademarks and similar legal safeguards. There are at least two mechanisms which lead to openness of construction innovation and have to do with the client relation.

First, there is the behaviour of clients, typically but not only in the public sector, who call for competitive tenders. Here it seems that tenderers perceive a risk that the customer will leak unprotected or weakly protected intellectual property to competitors in the course of tender negotiations or a renewed call for tenders (Hartmann and Girmscheid, 2004). Client behaviour of this kind would obviously weaken incentives for contractor innovation. The second mechanism, which is less related to the degree of competition in tendering, is that the openness of construction innovation can be used by contractors to signal their hidden qualities to potential clients. Results from interviews with Singapore contractors, clients and national institutions (Lim and Ofori, 2007) can be interpreted as that innovations which support contracter branding, with intangible benefits due to higher credibility, have tended to be overlooked. Another openness mechanism, which does not involve the customer, is one of horizontal spread of innovative knowledge: the transparency of process innovation which follows from the visibility of activities on construction sites may lead to leakage between construction contractors.

In the historical perspective, it is wrong to see construction as lagging behind other industries because of its low intensity of patenting. A major technological breakthrough, the introduction of reinforced concrete, took place under a regime of patents and licences, as outlined by Addis (2007). However, this was a matter of decades, and in the UK case, the era of concrete patents was over in 1915. A few years later (1921) is also when organized government R&D into construction technologies was initiated, which later was organized as the Building Research Establishment (Sebestyen, 1998, 65). Once government had decided to conduct research in the field, a pattern that was soon copied in other industrialized countries, there would be a policy emphasis on disseminating findings and innovations freely to the construction industry.

Nevertheless, there is a flow of construction-related patents, and there are a few studies of the phenomenon. If we distinguish between technological regimes according to conditions of opportunity, appropriability, cumulativeness and properties of the knowledge base, and use patent data from six countries for 49 technological classes, of which 'Civil engineering, Infrastructure' is Class 30, this class appears as clearly following the Schumpeter Mark I type of innovation (Malerba and Orsenigo, 1996). The concentration ratio of the top four innovators is low, and the distribution of innovative activities is asymmetric among firms; the stability in the hierarchy of innovators is low, and the entry of new innovators is relatively high. Class 30 inventions tended to show relatively low spatial concentration but a high level of spatial cumulativeness, suggesting that civil engineering inventions exhibit both a collective and a localized nature. This is similar to the pattern for mechanical engineering, according to a study of four European countries by Breschi (2000).

31.11 Innovating by diversifying

Apparently in contrast to the traditional reliance of construction firms on a project nexus of contracts, they also innovate by entering new activities and creating new businesses. In three years, 11 per cent of all construction firms in the 1999 Canadian survey had set up a new business or division (Anderson and Schaan, 2001). Furthermore, a study of 11 Turkish contractors (Dikmen et al., 2005) revealed that market innovations, either entering a new international market or diversifying into a new related or unrelated market, were the innovations most frequently implemented, followed by process innovations. How can this be explained?

31.11.1 Theories of diversification

While the general tendency to vertical integration seems to be low in the construction industry, there are many examples of diversification, according to a study of 35 large UK contracting firms (Casson, 1987). When these firms did integrate vertically in the direction of their supplies, they went into activities that would be of particular importance for the overall costing of the projects on which they worked; in other words, integration could be seen as a symptom of risk management. Economies of scale might also explain backward integration into activities such as concrete fabrication. Casson was unable to identify generic explanations for the diversification strategies pursued by the contractors he surveyed; sometimes it was just that activities acquired as a result of taking over another company had remained within their business. Ultimately, it emerged that the type of knowledge possessed by the firm would determine the scope of its activities across the construction industry.

Later studies of construction integration and diversification have followed in Casson's footsteps and sought a dynamic combination of transaction cost analysis with a capabilities-based view of the firm: Bridge and Tisdell (2004) with a seven-level model of the capability and competence spectrum; as well as Cacciatori and Jacobides (2005), who concluded that firms in the UK construction industry were reintegrating to protect their positions, to enter new and related markets or find new ways to leverage their capabilities. The recent positioning of construction contractors as facilities managers and all-in-one providers was highlighted as a typical case of reintegration. Why a number of Swedish construction contractors chose to do so appears to be related also to a general pattern of diversification and competence (Bröchner, 2008a). It remains to be seen whether the global forces of specialization will prevail.

31.11.2 Concession projects

The basic principle of bundling a construction contract with a long period of facilities management is very old, but in the particular form it has taken since the 1980s with BOT (build–operate–transfer) infrastructure projects and even more since 1992, when the UK government began granting concessions under the Private Finance Initiative, it is possible to speak of an organizational innovation. Bundling should be efficient when it reduces suboptimization and exploits economies of scope if there are technological opportunities for choosing initially more costly construction alternatives that are expected to give lower long-term costs for operating the facility (Martimort and Pouyet, 2008).

Perhaps the private sector mechanisms for controlling technology risks might reduce the will to profit from total optimization of construction and operations costs over the concession period (Leiringer, 2006); if this is true, it is unclear how the four main arguments for public–private partnerships – design freedom, collaborative working, risk transfer and long-term commitment – affect technology innovation. Four UK case studies by Eaton et al. (2006) showed that several factors impeded technological innovation: the coalition nature of the industry, strict processes and procedures, the format of project contract, rigid project demands, the segmentation of project disciplines, unrealistic expectations for productivity, and financial constraints. Summing up, it has to be said that there is no consensus yet on the links between this organizational innovation and technological innovation.

31.12 Horizontal collaboration and knowledge brokers

Instead of or in addition to collaborating with clients or diversifying, contractors may engage in collaborative innovation efforts together with other contractors. Sometimes there is a history of government policies encouraging joint research of this type, although stable networks have arisen in some cases because of industry initiatives, such as in 1983 for the Development Fund of the Swedish Construction Industry (SBUF) (Bröchner and Grandinson, 1992; Widén and Hansson, 2007). The Dutch long-term data on construction innovations (Pries and Dorée, 2005) also indicate that cooperation between firms gained in importance from the 1980s, whereas collective, public supported programmes remained unimportant. Four stages of co-innovation strategy development in the Dutch house-building sector were identified by Bossink (2002): autonomous strategy-making, cooperative strategy management, founding an organization for co-innovation, and realization of innovations.

A closely related phenomenon is that of brokers, understood as agents handling knowledge flows through the construction sector, being neither the original sources of innovations, nor their implementers (Winch, 1998). Organized broker functions in many countries for construction innovation, ranging from government bodies to industry funds like SBUF, have been mapped and analysed by Winch and Courtney (2007).

31.13 A construction-as-services innovation survey

Findings from a Swedish 2006 survey of construction contractors can be used to illustrate relations between types and levels of innovation. Carassus (2003) has suggested the applicability of the framework outlined by Gallouj (2002) to construction, and inspired also by how Djellal and Gallouj (2005) have applied it to hospitals, it was chosen for mapping the innovation dynamics of Swedish construction contractors (Bröchner, 2008b). A questionnaire was sent early in 2006 to the 50 largest Swedish construction contractors, as measured by turnover in 2004. Six of these firms had more than 1000 employees. Responses were received from 44 contractors; the study was retrospective for research, development and innovation during a three-year period. Recent Organisation for Economic Co-operation and Development (OECD) Frascati and Oslo definitions of R&D and of innovation were applied. A distinction was made between innovation new to the firm and innovation that is perceived to be unique nationally.

While knowledge/routines was found to be the most frequent medium or trajectory for R&D in these firms, material/technology grows in importance as innovations 'new to the firm' and is the dominant mode for innovations 'unique nationally'. Thus the definitions of R&D and of levels of innovation are crucial. There is considerable variety of principal activities among firms classified as construction contractors. On average, each firm had 4.6 principal activities. The five firms active in quarrying and manufacturing exhibit a stronger orientation toward R&D and innovation, together with a greater reliance on employees with a research degree and on linkage with universities. The two most frequent external linkages for R&D and innovation were clients and IT suppliers for the largest firms, those with 1000 and more employees, while clients and consultants were the most frequent for the rest of the firms in the survey. But the priority list of collaboration partners also varies with the principal activities. Some contractors had added a constituent service by diversifying into facilities management, and their characteristics differ from those who had not (Bröchner, 2008a). Most frequently mentioned service characteristics related to customer utility functions were 'better ability to communicate with client' and 'higher technical quality'. Among the perceived internal effects of R&D and innovation, 'more efficient production process' and 'better image of the company' received top scores. Once again, the ranking was quite sensitive to mode of R&D and innovation. From a policy viewpoint, it is noted that university linkage is strongly related to materials and technologies R&D and innovation. European research collaboration was only found among the very largest contractors.

31.14 Government as regulator and innovation support provider

It has been mentioned already that local and central government customers are important for construction contractors, being responsible for the majority of contracts for several types of construction, and that they can be sources of both technological and non-technological innovation. Government is also an important regulator because of the environmental, health and safety aspects of construction. Furthermore, there is a tradition dating back to the 1920s in some leading countries that government supports construction innovation. There are differences between various construction activities because there are differing rationales for government intervention: social policies have dictated the need for policies that are intended to lead to cheaper and better housing construction.

31.14.1 Regulations

The 2004 Innobarometer survey (European Commission, 2004) showed that contractors who had introduced new or improved processes felt affected by environmental and safety regulations. A recurrent theme in studies of construction innovation has been the effects, predicted or observed, of public authorities shifting from a detailed prescriptive style to a performance style when formulating their regulations or requirements more in general (Gann et al., 1998; Sexton and Barrett, 2005). Moreover, the effect of regulations should be seen in a context of government-provided information and subsidies, as Miozzo and Dewick (2004) have underlined in their analysis of decisions to adopt new sustainable technologies where they have considered thermal insulation and active solar heating systems. Finally, the Dutch long-term data on construction innovations (Pries and Dorée, 2005) confirm that the role of building regulations or codes is a very important one – over 30 per cent of all innovations were found to be the result of new regulations, a figure rising over the years and reaching a stable high level from the 1980s, when safety and environmental impacts came to dominate the scene.

31.14.2 Public policies for construction innovation

Early policies for government support of construction innovation tended to be based on the linear, technology-push model of investigator-led research (Seaden and Manseau, 2001). Government institutes and universities pursuing R&D were expected to be the primary sources of innovation. However, the expansion of government construction research institutes was over by the mid-1970s (Sebestyen, 1998, 66). Over time, more collaborative arrangements emerged, often with part funding from the industry. A study of Singapore contractors (Na et al., 2007) indicates a strong link between a surge beginning in 2003 in national institutions' and the simultaneous rise in contractors' expenditure on construction R&D, but at the same time, a weaker relation between R&D expenditure and the full-time equivalents of R&D personnel among local construction contractors. The role of public procurement, especially the methods of specification as just mentioned, other award criteria than lowest price for contracts and the treatment of intellectual property, grows along with an increasing political reliance on market mechanisms.

There has also been a shift of focus from products to process-related issues, a trend that is supported by the potential for new ICT applications. These developments are common to many national policy contexts, although there is diversity as shown in the overview of 15 countries provided by Manseau and Seaden (2001).

31.15 Concluding remarks

'What other industry could cope with the variation in workload and constant changes – delivering virtually a hand-made product?' (construction SME representative, in Davey et al., 2004). There should now be little doubt that the answer lies in the service sector rather than in manufacturing, and that there are consequences for innovation. Which are the lessons that services researchers can learn from studies of innovation in construction? The importance of interaction between technological and non-technological types of innovation is one of these; another lesson that needs more exploration is how innovations arise and spread in projects that take place in a largely fixed nexus of specialized services.

Much of the research into construction innovation has dealt with phenomena in small and medium-sized enterprises (SMEs). It is not well understood how large contractors, who often form part of even larger groups with a range of activities also outside the construction sector, innovate. As there remain administrative barriers to the spread of new construction technologies and business practices, both within the European Union and within North America, there remains a commercial potential as well as questions for government policies intended to support innovation.

References

Addis, B. (2007), *Building: 3000 Years of Design Engineering and Construction*, London: Phaidon.

Anderson, F. (2005), 'Measuring innovation in construction', in A. Manseau and R. Shields (eds), *Building Tomorrow: Innovation in Construction and Engineering*, Aldershot: Ashgate, pp. 57–80.
Anderson, F. and S. Schaan (2001), 'Innovation, advanced technologies and practices in the construction and related industries: national estimates', Statistics Canada Working Paper, February.
Anderson, J.C. (1997), *Roman Architecture and Society*, Baltimore, MD: Johns Hopkins University Press.
Barlow, J. (2000), 'Innovation and learning in complex offshore construction projects', *Research Policy*, **29** (7–8), 973–989.
Barrett, P., M. Sexton and A. Lee (2008), *Innovation in Small Construction Firms*, London: Taylor & Francis.
Blackley, D.M. and E.M. Shepard (1996), 'The diffusion of innovation in home building', *Journal of Housing Economics*, **5** (4), 303–322.
Bossink, B.A.G. (2002), 'The development of co-innovation strategies: stages and interaction patterns in interfirm innovation', *R&D Management*, **32** (4), 311–320.
Bossink, B.A.G. (2004), 'Managing drivers of innovation in construction networks', *Journal of Construction Engineering and Management*, **130** (3), 337–345.
Bougrain, F. (2003), 'Innovations in the French building and construction industry: the case of material suppliers, manufacturers of building components and equipments and contractors', *Wirtschaftspolitische Blätter*, **50** (3), 366–371.
Bowley, M. (1966), *The British Building Industry: Four Studies in Response and Resistance to Change*, Cambridge: Cambridge University Press.
Breschi, S. (2000), 'The geography of innovation: a cross-sector analysis', *Regional Studies*, **34** (3), 213–229.
Bresnen, M., L. Edelman, S. Newell, H. Scarbrough and J. Swan (2003), 'Social practices and the management of knowledge in project environments', *International Journal of Project Management*, **21** (3), 157–166.
Bresnen, M., A. Goussevskaia and J. Swan (2004), 'Embedding new management knowledge in project-based organizations', *Organization Studies*, **25** (9), 1535–1555.
Bridge, A.J. and C. Tisdell (2004), 'The determinants of the vertical boundaries of the construction firm', *Construction Management and Economics*, **22** (8), 807–825.
Bröchner, J. (2008a), 'Construction contractors integrating into facilities management', *Facilities*, **26** (1/2), 6–15.
Bröchner, J. (2008b), 'Client-oriented contractor innovation', in P. Brandon and S.-L. Lu (eds), *Clients Driving Innovation*, Oxford: Wiley-Blackwell, pp. 125–136.
Bröchner, J. and B. Grandinson (1992), 'R&D cooperation by Swedish contractors', *Journal of Construction Engineering and Management*, **118** (1), 3–16.
Cacciatori, E. and M.G. Jacobides (2005), 'The dynamic limits of specialization: vertical integration reconsidered', *Organization Studies*, **26** (12), 1851–1883.
Carassus, J. (2003), 'Construction sector system and innovation in stock management', in B.O. Uwakweh and I.A. Minkarah (eds), *Proc. 10th Symposium Construction Innovation and Global Competitiveness*, Boca Raton: CRC Press, pp. 44–56.
Casson, M. (1987), *The Firm and the Market: Studies on Multinational Enterprise and the Scope of the Firm*, Oxford: Basil Blackwell.
Cleff, T. and A. Rudolph-Cleff (2001), 'Innovation and innovation policy in the German construction sector', in A. Manseau and G. Seaden (eds), *Innovation in Construction: An International Review of Public Policies*, London: Spon, pp. 201–234.
Dahl, M.S. and B. Dahlum (2001), 'The construction cluster in Denmark', in P. den Hertog, E.M. Bergman and D. Charles (eds), *Innovative Clusters: Drivers of National Innovation Systems*, Paris: OECD, pp. 179–202.
Dalziel, M. (2007), 'A systems-based approach to industry classification', *Research Policy*, **36** (10), 1559–1574.
Davey, C.L., J.A. Powell, I. Cooper and J.E. Powell (2004), 'Innovation, construction SMEs and action learning', *Engineering, Construction and Architectural Management*, **11** (4), 230–237.

Davies, J.K. (2007), 'Classical Greece: Production', in W. Scheidel, I. Morris and R. Saller (eds), *The Cambridge Economic History of the Greco-Roman World*, Cambridge: Cambridge University Press, pp. 333–361.
de Jong, J.P.J. and O. Marsili (2006), 'The fruit flies of innovations: a taxonomy of innovative small firms', *Research Policy*, **35** (2), 213–229.
den Hertog, P. and E. Brouwer (2001), 'Innovation in the Dutch construction cluster', in P. den Hertog, E.M. Bergman and D. Charles (eds), *Innovative Clusters: Drivers of National Innovation Systems*, Paris: OECD, pp. 203–228.
Department of Trade and Industry (DTI) (2006), 'Innovations in the UK: indicators and insights', DTI Occasional Papers No. 6, July.
Dikmen, I., M.T. Birgonul and S.U. Artuk (2005), 'Integrated framework to investigate value innovations', *Journal of Management in Engineering*, **21** (2), 81–90.
Djellal, F. and F. Gallouj (2005), 'Mapping innovation dynamics in hospitals', *Research Policy*, **34** (6), 817–835.
Dodgson, M., D.M. Gann and A. Salter (2007), 'The impact of modelling and simulation technology on engineering problem solving', *Technology Analysis and Strategic Management*, **19** (4), 471–489.
Dorée, A.G. and E. Holmen (2004), 'Achieving the unlikely: innovating in the loosely coupled construction system', *Construction Management and Economics*, **22** (8), 827–838.
Drejer, I. and A.L. Vinding (2006), 'Organisation, "anchoring" of knowledge, and innovative activity in construction', *Construction Management and Economics*, **24** (9), 921–931.
Dubois, A. and L.-E. Gadde (2002), 'The construction industry as a loosely coupled system: implications for productivity and innovation', *Construction Management and Economics*, **20** (7), 621–631.
Eaton, D., R. Akbiyikli and M. Dickinson (2006), 'An evaluation of the stimulants and impediments to innovation within PFI/PPP projects', *Construction Innovation*, **6** (2), 63–77.
Eccles, R.G. (1981), 'The quasifirm in the construction industry', *Journal of Economic Behavior and Organization*, **2** (4), 335–357.
European Commission (2002), *Innobarometer 2002: Flash Eurobarometer 129*, EOS Gallup Europe, October.
European Commission (2004), *Innobarometer 2004: Flash Eurobarometer 164*, EOS Gallup Europe, November.
European Commission (2006), 'ICT and e-business in the construction industry: ICT adoption and e-business activity in 2006', e-Business W@tch Sector Report No. 7/2006.
Eurostat (2004), 'Innovation in Europe: Results for the EU, Iceland and Norway', Office for Official Publications of the European Communities, Luxembourg.
Frontinus (2004), *De Aquaeductu Urbis Romae*, ed. R.H. Rodgers, Cambridge: Cambridge University Press.
Gadrey, J. (2000), 'The characterization of goods and services: an alternative approach', *Review of Income and Wealth*, **46** (3), 369–387.
Gallouj, F. (2002), *Innovation in the Service Economy: The New Wealth of Nations*, Cheltenham, UK and Northampton, MA, USA: Edward Elgar.
Gann, D.M. (2000), *Building Innovation: Complex Constructs in a Changing World*, London: Thomas Telford.
Gann, D.M. and A.J. Salter (2000), 'Innovation in project-based, service-enhanced firms: the construction of complex products and systems', *Research Policy*, **29** (7–8), 955–972.
Gann, D.M., Y. Wang and R. Hawkins (1998), 'Do regulations encourage innovation? The case of energy efficiency in housing', *Building Research and Information*, **26** (5), 280–296.
Gray, C. and R.J. Davies (2007), 'Perspectives on experiences of innovation: the development of an assessment methodology appropriate to construction project organizations', *Construction Management and Economics*, **25** (12), 1251–1268.
Groák, S. (1994), 'Is construction an industry? Notes towards a greater analytic emphasis on external linkages', *Construction Management and Economics*, **12** (4), 287–293.
Hartmann, A. and G. Girmscheid (2004), 'The innovation potential of integrated services

and its utilisation through co-operation', *Engineering, Construction and Architectural Management*, **11** (5), 335–341.
Hill, P. (1999), 'Tangibles, intangibles and services: a new taxonomy for the classification of output', *Canadian Journal of Economics*, **32** (2), 426–446.
Hobday, M. (1998), 'Product complexity, innovation and industrial organisation', *Research Policy*, **26** (6), 689–710.
Holm, M.G. (2000), 'Service management in housing refurbishment: a theoretical approach', *Construction Management and Economics*, **18** (5), 525–533.
Illeris, S. (2007), 'The nature of services', in J.R. Bryson and P.W. Daniels (eds), *The Handbook of Service Industries*, Cheltenham, UK and Northampton, MA, USA: Edward Elgar, pp. 19–33.
Ivory, C. (2005), 'The cult of customer responsiveness: is design innovation the price of a client-focused construction industry?', *Construction Management and Economics*, **23** (8), 861–870.
Laborde, M. and V. Sanvido (1994), 'Introducing new process technologies into construction companies', *Journal of Construction Engineering and Management*, **120** (3), 488–508.
Lampel, J., R. Miller and S. Floricel (1996), 'Information asymmetries and technological innovation in large engineering construction projects', *R&D Management*, **26** (4), 357–369.
Lancaster, L.C. (2005), *Concrete Vaulted Construction in Imperial Rome: Innovations in Context*, Cambridge: Cambridge University Press.
Leiringer, R. (2006), 'Technological innovation in PPPs: incentives, opportunities and actions', *Construction Management and Economics*, **24** (3), 301–308.
Lim, J.N. and G. Ofori (2007), 'Classification of innovation for strategic decision making in construction businesses', *Construction Management and Economics*, **25** (9), 963–978.
Malerba, F. and L. Orsenigo (1996), 'Schumpeterian patterns of innovation are technology-specific', *Research Policy*, **25** (3), 451–478.
Manley, K. (2005), 'BRITE Innovation Survey', Cooperative Research Centre (CRC) for Construction Innovation (Australia).
Manley, K. (2006), 'The innovation competence of repeat public sector clients in the Australian construction industry', *Construction Management and Economics*, **24** (12), 1295–1304.
Manseau, A. and G. Seaden (eds) (2001), *Innovation in Construction: An International Review of Public Policies*, London: Spon.
Martimort, D. and J. Pouyet (2008), 'To build or not to build: normative and positive theories of public–private partnerships', *International Journal of Industrial Organization*, **26** (2), 393–411.
Miles, I. (2002), 'Services innovation: towards a tertiarization of innovation studies', in J. Gadrey and F. Gallouj (eds), *Productivity, Innovation and Knowledge in Services: New Economic and Socio-Economic Approaches*, Cheltenham, UK and Northampton, MA, USA: Edward Elgar, pp. 164–196.
Miozzo, M. and P. Dewick (2004), *Innovation in Construction: A European Analysis*, Cheltenham, UK and Northampton, MA, USA: Edward Elgar.
Mitropoulos, P. and C.B. Tatum (2000), 'Forces driving adoption of new information technologies', *Journal of Construction Engineering and Management*, **126** (5), 340–348.
Na, L.J., G. Ofori, F.Y.Y. Ling and G.B. Hua (2007), 'Role of national institutions in promoting innovation by contractors in Singapore', *Construction Management and Economics*, **25** (10), 1021–1039.
Nam, C.H. and C.B. Tatum (1988), 'Major characteristics of constructed products and resulting limitations of construction technology', *Construction Management and Economics*, **6** (2), 133–148.
Nam, C.H. and C.B. Tatum (1989), 'Toward understanding of product innovation process in construction', *Journal of Construction Engineering and Management*, **115** (4), 517–534.
Nam, C.H. and C.B. Tatum (1997), 'Leaders and champions for construction innovation', *Construction Management and Economics*, **15** (3), 259–270.

Ozaki, R. (2003), 'Customer-focused approaches to innovation in housebuilding', *Construction Management and Economics*, **21** (6), 557–564.
Pavitt, K. (1984), 'Sectoral patterns of technical change: towards a taxonomy and a theory', *Research Policy*, **13** (6), 343–373.
Powell, C. (1996), 'Divide and rule? Division of labour and the position of designers in the UK construction industry', *Industrial and Corporate Change*, **5** (3), 863–882.
Pries, F. and A. Dorée (2005), 'A century of innovation in the Dutch construction industry', *Construction Management and Economics*, **23** (6), 561–564.
Reichstein, T., A.J. Salter and D.M. Gann (2005), 'Last among equals: a comparison of innovation in construction, services and manufacturing in the UK', *Construction Management and Economics*, **23** (6), 631–644.
Seaden, G., M. Guolla, J. Doutriaux and J. Nash (2003), 'Strategic decisions and innovation in construction firms', *Construction Management and Economics*, **21** (6), 603–612.
Seaden, G. and A. Manseau (2001), 'Public policy and construction innovation', *Building Research and Information*, **29** (3), 182–196.
Sebestyen, G. (1998), *Construction: Craft to Industry*, London: E&FN Spon.
Sexton, M. and P. Barrett (2003), 'Appropriate innovation in small construction firms', *Construction Management and Economics*, **21** (6), 623–633.
Sexton, M. and P. Barrett (2005), 'Performance-based building and innovation: balancing client and industry needs', *Building Research and Information*, **33** (2), 142–148.
Sexton, M., P. Barrett and G. Aouad (2006), 'Motivating small construction companies to adopt new technology', *Building Research and Information*, **34** (1), 11–22.
Slaughter, E.S. (1993), 'Innovation and learning during implementation: a comparison of user and manufacturer innovations', *Research Policy*, **22** (1), 81–95.
Sundbo, J. and F. Gallouj (2000), 'Innovation as a loosely coupled system in services', *International Journal of Services Technology and Management*, **1** (1), 15–36.
Taylor, J.E. and R. Levitt (2007), 'Innovation alignment and project network dynamics: An integrative model for change', *Project Management Journal*, **38** (3), 22–35.
Toole, T.M. (1998), 'Uncertainty and homebuilders' adoption of technological innovations', *Journal of Construction Engineering and Management*, **124** (4), 323–332.
Widén, K. and B. Hansson (2007), 'Diffusion characteristics of private sector financed innovation in Sweden', *Construction Management and Economics*, **25** (5), 467–475.
Winch, G. (1998), 'Zephyrs of creative destruction: understanding the management of innovation in construction', *Building Research and Information*, **26** (4), 268–279.
Winch, G. (2003), 'How innovative is construction? Comparing aggregated data on construction innovation and other sectors – a case of apples and pears', *Construction Management and Economics*, **21** (6), 651–654.
Winch, G.M. and R. Courtney (2007), 'The organization of innovation brokers: an international review', *Technology Analysis and Strategic Management*, **19** (6), 747–763.

Index